T0181774

Agricultural Development Policy

Agricultural Development Policy
Concepts and Experiences

Roger D. Norton

John Wiley & Sons, Ltd

Copyright © 2004   Food and Agriculture Organization of the United Nations

Published 2004   John Wiley & Sons Ltd, The Atrium, Southern Gate, Chichester,
                     West Sussex PO19 8SQ, England
                     Telephone (+44) 1243 779777

Email (for orders and customer service enquiries): cs-books@wiley.co.uk
Visit our Home Page on www.wileyeurope.com or www.wiley.com
For further information on FAO's publishing programme visit our Home Page on www.fao.org

The designations employed and the presentation of material in this publication do not imply the expression of any opinion
whatsoever on the part of the Food and Agriculture Organization of the United Nations concerning the legal status of any
country, territory, city or area or of its authorities, or concerning the delimitation of its frontiers or boundaries.
The designations 'developed' and 'developing' economies are intended for statistical convenience and do not necessarily
express a judgement about the stage reached by a particular country, territory or area in the development process.
The views expressed herein are those of the authors and do not necessarily represent those of the Food and Agriculture
Organization of the United Nations.

All rights reserved. Reproduction and dissemination of material in this information product for educational or other non-
commercial purposes are authorized without any prior written permission from the copyright holders provided the source is
fully acknowledged. Reproduction of material in this information product for resale or other commercial purposes is
prohibited without written permission of the copyright holder. Applications for such permission should be addressed to the
Chief, Publishing Management Service, Information Division, FAO, Viale della Terme di Caracalla, 00100 Rome, Italy or by
e-mail to copyright@fao.org
© 2003. FAO

Other Wiley Editorial Offices

John Wiley & Sons Inc., 111 River Street, Hoboken, NJ 07030, USA

Jossey-Bass, 989 Market Street, San Francisco, CA 94103-1741, USA

Wiley-VCH Verlag GmbH, Boschstr. 12, D-69469 Weinheim, Germany

John Wiley & Sons Australia Ltd, 33 Park Road, Milton, Queensland 4064, Australia

John Wiley & Sons (Asia) Pte Ltd, 2 Clementi Loop #02-01, Jin Xing Distripark, Singapore 129809

John Wiley & Sons Canada Ltd, 22 Worcester Road, Etobicoke, Ontario, Canada M9W 1L1

Wiley also publishes its books in a variety of electronic formats. Some content that appears in print may not be available in
electronic books.

Library of Congress Cataloging-in-Publication Data

Norton, Roger D., 1942–
    Agricultural development policy : concepts and experiences / Roger D. Norton.
        p. cm.
    "Food and Agriculture Organization of the United Nations".
    Includes bibliographical references and index.
    ISBN 0-470-85778-1 (ppc : alk. paper) – ISBN 0-470-85779-X (pbk. : alk. paper) – ISBN (FAO)
    1. Agriculture and state. 2. Sustainable agriculture. I. Food and Agriculture Organization of the United Nations.
    II. Title.
    HD1415.N88 2003
    338.1'8 – dc22   2003049480

British Library Cataloguing in Publication Data

A catalogue record for this book is available from the British Library

ISBN 0-470-85778-1 (HB)
    0-470-85779-X (PB)
    FAO Edition: 92-5-104875-4

Typeset in 10/12 Times by SNP Best-set Typesetter Ltd., Hong Kong

This book is printed on acid-free paper responsibly manufactured from sustainable forestry in which at least two trees are
planted for each one used for paper production.

Contents

Foreword   ix
Acknowledgements   xi

Introduction   1

1 Agriculture and Economic Development: Basic Considerations   3
1.1 Beginnings   3
1.2 The Agricultural Sector and Economic Growth   4
1.3 Agricultural Growth and Poverty Reduction   9
Discussion Points for Chapter 1   12

2 Strategies and the Agricultural Policy Framework   15
2.1 Strategies and Policies   15
2.2 The Nature of Agricultural Policy Instruments   19
2.3 Objectives of Agricultural Policy   22
2.4 The Role of Government   25
2.5 Implementation of Strategies and Policies   30
Discussion Points for Chapter 2   32

3 Broad Issues of Agricultural Policy   35
3.1 Agricultural Policy and the Macroeconomic Framework   35
3.2 Fiscal Expenditures and Subsidies   37
3.3 Improving the Incomes of the Rural Poor   44
3.4 Gender and Agricultural Development   46
3.5 Selected Issues in Privatization   48
3.6 Principal Aspects of the Legal Framework   51
Discussion Points for Chapter 3   53

4 Policies that Influence Producer Incentives   55
4.1 The Setting   55
4.2 Agricultural Prices and Their Determinants   56
4.3 Trade Policy   58
4.4 Exchange Rate Policy   74
4.5 Fiscal Policy and Agricultural Prices   80
4.6 Macroeconomic Policy Options for Agriculture   82
4.7 Sectoral Policies that Influence Agricultural Prices   86
4.8 Food Security, Agricultural Prices and the Rural Poor   98
4.9 Observations on Price Stabilization and Agricultural Development   103
Discussion Points for Chapter 4   104

5 Land Tenure Policies   109
  5.1 Introduction   109
  5.2 The Importance of Land Tenure   110
  5.3 Objectives of Land Tenure Policies   111
  5.4 Overview of Issues and Trends concerning Land Tenure   116
  5.5 The Nature of Land Rights   121
  5.6 Communal, Collective and Individual Rights to Land   130
  5.7 Experiences with Land Reform   142
  5.8 Policies for Land Markets   149
  5.9 Improving Access to Land for the Poor and for Women   176
  Discussion Points for Chapter 5   191

6 Water Management Policies in Agriculture   197
  6.1 Introduction   197
  6.2 Policy Objectives for the Irrigation Sector   203
  6.3 Strategic Planning for Irrigation as Part of Water Resource Management   207
  6.4 Strategic Issues in Irrigation Development   211
  6.5 Principal Policy Issues in the Irrigation Sector   223
  6.6 Institutional and Process Issues in Water Management   248
  6.7 Irrigation as a Tool of Rural Development   269
  Discussion Points for Chapter 6   271

7 Policies for Agricultural and Rural Finance   277
  7.1 The Role of Finance in Agricultural Development   277
  7.2 Policy Objectives for Rural Finance   287
  7.3 Keys to the Sustainability and Efficiency of Financial Intermediation   295
  7.4 The Regulatory Framework for Rural Finance   300
  7.5 Structural Considerations for Rural Financial Institutions   317
  7.6 Approaches to Managing Rural Financial Institutions   330
  7.7 Macroeconomic Policy to Support Rural Financial Intermediation   340
  7.8 Elements of a Strategy for Developing Rural Finance   346
  Discussion Points for Chapter 7   351

8 Policies for Agricultural Technology   357
  8.1 Introduction: The Role and Context of Agricultural Technology   357
  8.2 Issues in Agricultural Research   363
  8.3 Issues in Agricultural Extension   375
  8.4 New Directions in Agricultural Research   390
  8.5 New Approaches to Agricultural Extension   404
  Discussion Points for Chapter 8   420

9 Agricultural Development Strategies: Process and Structure   425
  9.1 The Roles of an Agricultural Strategy   425
  9.2 Participatory Processes for Developing Strategies   428
  9.3 Structure and Consistency in a Strategy   442
  9.4 Substantive Orientations of an Agricultural Strategy   450
  9.5 Rural Development and Poverty Alleviation   460
  9.6 Implementation of a Strategy   474

9.7 Concluding Observations 475
Discussion Points for Chapter 9 476

Annex: National Economic Policies and Irrigation in Yemen 483
1 Water in Yemen's Development Process 483
2 Demand Management vs. Water Supply Options 484
3 The Origins of the Water Crisis 485
4 Principal Policy Instruments Available for Water Demand Management 486
5 Bearing the Costs of Change 487
6 Schematic Presentation of Policy Options 488
7 Fiscal and Strategic Issues 488
8 Summary 490

Index 493

Foreword

Hunger and malnutrition continue to persist in a world of wealth, and this fact is repeatedly brought forward by both individual countries and the international community. The eradication of extreme poverty and hunger is the first of the eight United Nations Millennium Development Goals, global targets set by the world's leaders at the Millennium Summit in New York in September 2000. The ambitious agenda of this first goal is to cut the number of poor and hungry people in half by the year 2015, a target initially set at the World Food Summit organized by the Food and Agriculture Organization of the United Nations (FAO) in Rome in 1996.

FAO leads international efforts to defeat hunger. Its mission is to help build a food-secure world for present and future generations. Achieving food security for all is at the heart of FAO's efforts – to make sure that people have regular access to enough high-quality food to lead active, healthy lives. Because the vast majority of the hungry and undernourished people live in rural areas, FAO pursues the goals of reducing the numbers of chronically undernourished people and of ensuring that agriculture and rural areas become economically viable. This in turn will contribute to economic and social progress and to the well-being of all. In carrying out its work towards these goals, FAO strives to ensure that natural resources are used sustainably.

Serving both developed and developing countries, FAO acts as a neutral forum where all nations meet as equals to negotiate agreements and debate policy. FAO is also a source of knowledge and information. It helps developing countries and countries in transition to modernize and improve agriculture, forestry and fisheries prac-tices and to ensure good nutrition for all. Since the founding of the Organisation in 1945, it has focused special attention on developing rural areas, home to 70 percent of the world's poor and hungry people.

One part of FAO's work is to encourage experience sharing in the fields of agriculture and rural development worldwide, and to contribute to decision-making capacities for sustainable agricultural development in developing countries. Economic adjustments and liberalization policies in many countries have brought about renewed interest in and lent an increased sense of urgency to the task of formulating policies and strategies for the agricultural and natural resource sectors, including forestry and fisheries.

Agricultural Development Policy – Concepts and Experiences is part of FAO's work to ensure an enabling policy environment for agriculture at both a global level, in the context of international regulatory frameworks and commitments, and at a country level through appropriate strategies and policies. This book offers lessons of international experience and research, showing how agricultural policies need to be incorporated fully into a broader framework of economic policies, linking them with macroeconomic as well as sub-sectoral policies.

The book highlights that agricultural growth is crucial for economic development. History has shown that the development of an agricultural sector is a prerequisite for the subsequent progress of a country. Furthermore, as this sector is deeply interconnected with every other aspect of the economy, agricultural development is a major determinant for the growth of rural economies, including the rural non-farm sector. Agriculture is thus essential for rural and urban poverty reduction, and it remains a key sector for the economies of many developing countries. Even if its share in economic growth declines with development, it will remain a crucial sector for food security.

It is our hope that *Agricultural Development Policy – Concepts and Experiences* will widely contribute to improving the policy environment for agriculture at national level, and to policy and regulatory frameworks at the international level, including investment support and a trading environment favorable to the agriculture sector of developing countries. By drawing attention to the closely linked effect of agriculture on all other aspects of the economy, and thus of agriculture's central place in economic policies, the book proposes a vision that goes beyond the institutional confines of Ministries of Agriculture to encompass a multiplicity of stakeholders, interests and aspirations.

Santiago Funes
Director
Policy Assistance Division
Food and Agriculture Organization of the
United Nations

Acknowledgements

I am very indebted to the Food and Agriculture Organization (FAO) of the United Nations, whose staff conceived the idea of a book like this and invited me to draft it and provided information and many helpful comments throughout the process. Maurizio Garzia of the Agricultural Policy Support Service launched the project and assisted it in its early stages, and Maria Grazia Quieti gave support and wise counsel that were crucial to bringing it to conclusion.

Many others in FAO contributed in their technical areas. The following FAO staff deserve special mention for the comments and guidance they provided for the development of the book:

Isabel Alvarez-Fernández, Chief, Research and Technology Development Service

Lorenzo Bellu, Agricultural Policy Support Officer, Agricultural Policy Support Service

Jelle Bruinsma, Chief, Global Perspective Studies Unit

Maximiliano Cox, Director, Rural Development Division

Sissel Ekaas, Director, Gender and Population Division

Adriana Herrera, Agrarian Analysis Officer, Land Tenure Service

Materne Maetz, Senior Agricultural Policy Support Officer, Agricultural Policy Support Service

David Palmer, Land Registration and Cadastre Officer, Land Tenure Service

Fernando Pizarro, Land and Water Development Senior Engineer, Investment Centre

Kalim Qamar, Senior Officer for Agricultural Training and Extension, Extension, Education and Communication Service

Ramesh Sharma, Senior Economist, Commodity Policy and Projections Service

Anthon Slangen, Senior Officer, Agricultural Management, Marketing and Finance Service

Aysen Tanyeri-Abur, Economist, Agriculture and Economic Development Division

Marcela Villarreal, Chief, Population and Development Service

Esther Zulberti, Chief, Extension, Education and Communication Service

I also wish to thank the referees who reviewed the manuscript for the publisher and provided valuable suggestions for improving it.

Many other persons have contributed to the making of this book. In good measure it is the result of a very large number of illuminating conversations and collaborative processes over the years, in all regions of the world. I am very grateful to the numerous colleagues in developing and transitional countries who shared their concerns openly on many occasions and contributed unstintingly of their time and insights but are far too numerous to list individually. I have learned a great deal from all of them and hope this volume can repay a small part of their generosity.

This book is dedicated to my wife Fabiola, without whose sustained support and encouragement it could not have come to fruition.

Introduction

Throughout the developing world and regions in transition, interest in formulating and carrying out new agricultural policies and strategies has not been diminished by the emphasis in recent years on economic liberalization. On the contrary, in many instances the implementation of programs of economic adjustment has lent increased urgency to the task of finding policies to invigorate the agricultural and natural resource sectors. The motivation arises in part from the need to redefine the role of agriculture in a manner consistent with new visions of the economy, and at the same time to ensure that the needs of rural populations are addressed in that context. Rural poverty is a persistent and pervasive issue, and agricultural growth is the most effective way to address it.

For these reasons, and because of the inherent importance of agriculture in economic development, agricultural policy is experiencing a period of ferment and evolution throughout the world, from Latin America to Africa to the Newly Independent States and the Middle East and South and East Asia. As this book shows, many new approaches to agricultural policy are being developed, refined and implemented. Lessons are being learned and adapted. Yet in the available literature there is little systematic guidance for agricultural policy makers in the form of distillations of the findings of international experience and research. Often Ministries of Agriculture and other public agencies seek new solutions, attempt-

ing to move away from traditional forms of government intervention in the sector, such as the use of support prices, State marketing arrangements, State ownership of assets, subsidized credit, import controls, and centralized provision of agricultural services, but concrete guidelines for possible new policy orientations are not always readily available.

The aim of this book is to provide a systematic exposition of several important classes of policy issues and strategic considerations for agriculture, plus the emerging international consensus on viable approaches to those issues. For each area the conceptual foundations are developed, key contributions to the literature are cited, and illustrations are presented of policies that have worked, and some that have not, with explanations of why or why not. The book may serve as training material for policy staff of national and international institutions, as a reference for agricultural policy makers on issues in the sector and alternative approaches to them, and also as a source book for teaching about agricultural development.

The topics covered in the early chapters of the book include agriculture's role in economic development, the objectives of agricultural policy and strategies and the nature of policy instruments used to fulfill them, and recurring issues such as fiscal policy for agriculture, debates over the role of subsidies, policies for improving the conditions of rural poor, gender issues, privatisation, and the

Agricultural Development Policy Concepts and Experiences. R. D. Norton
© 2004 Food and Agriculture Organization of the United Nations
ISBNs: 0-470-85778-1 (HB)   0-470-85779-X (PB)   FAO Edition: 92-5-104875-4

role of legislative frameworks. The succeeding chapters cover in detail the linkages between macroeconomic and agricultural policy, land tenure policies, water management policies for agriculture, policies for the agricultural financial system, and policies for agricultural technology development and dissemination.

In most experiences in policy formulation, a number of strategic issues must be confronted and resolved while seeing the process through to fruition. They range from the technical to the institutional and the social and political. The more these issues can be anticipated, and options analysed, the greater is the likelihood of a successful outcome of the process. For this reason, the basic orientations of agricultural policy are often developed in the context of a long-term national agricultural strategy. The final chapter of the book reviews issues related to the formulation of strategies and requirements for making them successful, with respect to both their content and the processes by which they are developed. Participatory approaches are discussed, and in this strategic context frameworks for rural development are reviewed as well.

A central message of this book is that in the final analysis successful strategies and policies for agricultural growth and poverty alleviation must arise out of each country's own context. In spite of globalization, conditionality of international agencies, the exigencies of world trade agreements, and the role of transnational corporations, considerable space for creative policies exists in each country. Examples from elsewhere can stimulate the policy process and suggest new avenues of thinking, but in the end the solutions have to be tailored to each country's circumstances.

As a training and teaching tool, this book is intended to serve as a basic reference for a frequently encountered policy topics and principal considerations that can assist the process of developing policy proposals. For practitioners, it is hoped that the experiences summarized in it will provide stimulus in their continuing search for appropriate answers to issues in agricultural policy, and reinforce their conviction that broadly based and sustainable agricultural development is achievable.

1

Agriculture and Economic Development: Basic Considerations

1.1 Beginnings   3
1.2 The Agricultural Sector and Economic Growth   4
1.3 Agricultural Growth and Poverty Reduction   9
Discussion Points for Chapter 1   12

1.1 BEGINNINGS

The importance of sound agricultural policy has been recognized since earliest times in all cultures. In China in the sixth century BC, Lao Tze wrote:

> There is nothing more important than agriculture in governing people and serving the Heaven.

In addition, he admonished rulers who neglected the agricultural sector:

> The imperial palaces are magnificent, while the farmland is allowed to lie in waste and the granaries are empty. The governors are dressed elegantly, wearing sharp swords and eating luxurious food. The properties they own are more than enough; they show off like robbers. How far away from the Tao![1]

In other traditions, hallowed texts also remind us of our intimate, ineluctable links with the earth. In the Judaeo–Christian tradition, the Lord said, 'dust you are, to dust you shall return'.

In Shakespeare, death is clothed in dust: 'And all our yesterdays have lighted fools the way to dusty death'.

The prospects for life itself are also tied to the earth in our earliest writings. In Genesis (13:16), it is said: 'if anyone could count the dust upon the ground, then he could count your descendants'. To put it in a slightly different way, it has been the addition of water to dust – yielding mud – that has sustained the increase of human life on the planet.

Mankind's ability to nurture mud has made possible the creation of what we know as society, and the economy:

> Mud, the ubiquitous mud of the alluvial plains of southern Mesopotamia, was the material from which the world's first civilization was built. Mud, formed into uniform rectangular blocks, was used in the construction of houses, temples, and city walls. Mud, rolled flat into tablets, was the medium on which citizens recorded their commercial transactions, their laws, and their religious rituals. Mud, formed and fired, produced cooking and storage utensils. Mud, molded into human and animal figurines, represented the early sculptors' view of the world. But above all, mud provided the fertile topsoil that nourished the crops on which cities depended.

---

1. Lao Tze, *Taode Jing*, Chapters 53 and 59.

Agricultural Development Policy Concepts and Experiences. R. D. Norton
© 2004 Food and Agriculture Organization of the United Nations
ISBNs: 0-470-85778-1 (HB)   0-470-85779-X (PB)   FAO Edition: 92-5-104875-4

If the stands of wheat and barley failed, so did the city. And not just through lack of food. An agricultural surplus freed farmers from the field, allowing them to become artisans or traders; the organization of essential irrigation projects provided a hierarchy of rulers and administrators; the export of grain paid for the import of luxury goods; and the subsequent rise in wealth attracted immigrants and merchants from the surrounding countryside. Jobs, government, things to buy, and people to meet – the hallmarks of any modern city – all, ultimately, depended on mud.[2]

This basic condition of human existence was not lost on early economic theorists. As noted in a recent lecture by D. Gale Johnson, Adam Smith perceived 'a significant relationship between productivity improvement in agriculture and the wealth of nations'.[3] He quoted the following observation by Smith:

... when by the improvement and cultivation of land the labor of one family can provide food for two, the labor of half the society becomes sufficient to provide food for the whole. The other half, therefore, or at least the greater part of them, can be employed in providing other things, or in satisfying the other wants and fancies of mankind.[4]

In fact, the performance of agriculture over the centuries has made a fundamental contribution to present standards of living:

Generally speaking, *labor productivity growth in agriculture has been greater than in other sectors of the economies in the industrial countries*.... From 1967–68 to 1983–84, for 17 of 18 industrial countries for which there were [adequate] data.... the unweighted average

annual growth rate for agriculture was 4.3 percent compared to 2.6 percent for other sectors.... *The growth of total factor productivity in the OECD countries in agriculture has been greater than in manufacturing during the past quarter century or more.* The difference has not been small: one study indicates that total factor productivity growth was approximately 2.7 percent in agriculture compared to 1.5 percent in manufacturing for the period from 1960 to 1990....[5]

Thus, increases in agricultural productivity have constituted a principal source of improvements in overall economic well-being in modern societies. The sector's productivity has increased more rapidly than that of manufacturing both in terms of output per unit of labor and in terms of output per unit of all factors. This has not only put more food on tables in cities as well as countrysides but, as will be shown, it has contributed to more rapid economic growth and employment creation overall.

1.2 THE AGRICULTURAL SECTOR AND ECONOMIC GROWTH

Because higher productivity in farming can release labor for other sectors, for several decades of the twentieth century this relationship between agriculture and overall economic growth became distorted into a doctrine of pursuing industrialization even at the expense of agricultural development, with the result of undercutting agriculture's possibilities of contributing to overall development. The sector was viewed as playing a supporting role to industrial development, which was considered the most essential aspect of a growth strategy. In fact, industry was thought to be so important to a nation's long-run economic prospects that it was common practice

2. Editors of Time–Life Books, *The Rise of Cities*, Time–Life Books, Alexandria, VA, USA, 1990, p. 7.
3. D. Gale Johnson, 'Agriculture and the Wealth of Nations', Richard T. Ely Lecture, *American Economic Association Papers and Proceedings, American Economic Review*, **87**(2), May 1997, p. 2.
4. Adam Smith, *An Inquiry into the Nature and Causes of the Wealth of Nations*, Modern Library Edition, New York, 1937, p. 37 (cited in D. G. Johnson, 1997, p. 2).
5. D. G. Johnson, 1997, pp. 9–10 [emphasis added].

to subsidize it, at the expense of taxpayers and other sectors.

This was the doctrine of the first generation of economic development strategies. The practice of favoring and subsidizing industrial development was especially pronounced in Latin America and some countries of Asia. Perhaps the best-known early Latin American exponent of that tradition was Celso Furtado. In words that have an odd ring today, Furtado said, regarding sectoral priorities in Brazilian development:

Government action as a source of ample subsidies for industrial investments through foreign exchange and credit policy permitted the expansion, acceleration and broadening of the industrialization process. Without the governmental establishment of basic industries (steel, petroleum) and without the foreign exchange subsidies and the negative interest rates of government loans, industrialization would not have achieved the speed and the breadth that evolved during that quarter of a century.[6]

In this approach to development, the role of agriculture was seen as that of a provider of 'surpluses' (of labor, foreign exchange and domestic savings) to fuel industrial development. It was not looked to as a source of income growth in its own right. On the contrary, the implementation of subsidies for industry meant levying a tax, implicit or explicit, on agriculture which was likely to depress its growth prospects. Furtado commented in another context that in Mexico:

... since 1940, agricultural policy has systematically pursued the objective of increasing the agricultural surplus extracted for urban consumption or export.[7]

This vision of the limited role of agriculture in economic development was not confined to Latin American economists. It was a central part of the 'dual-economy model' of John Fei and Gustav Ranis.[8]

Anne Krueger summed up the early thinking in development economics as containing:

several prevalent and dominant strands ... : (1) the desire and drive for 'modernization'; (2) the interpretation of industrialization as *the* route to modernization; (3) the belief in 'import substitution' as a necessary policy to provide protection for new 'infant' industries; (4) the distrust of the private sector and the market and the belief that government, as a paternalistic benevolent guardian, should take the leading role in development; (5) related to (4), a distrust of the international economy and pessimism that exports from developing countries could grow.[9]

While not calling for subsidies to industry, Hollis Chenery and Moises Syrquin emphasized that agriculture should be expected to provide transfers of capital as well as labor to urban areas to promote the general development of the economy.[10] Even agricultural economists have subscribed to this thesis in the past:

---

6. Celso Furtado, *Obstacles to Development in Latin America*, Anchor Books, Doubleday and Company, New York, 1970, p. 144.
7. Celso Furtado, *Economic Development in Latin America*, 2nd Edn, Cambridge University Press, Cambridge, UK, 1976, p. 259.
8. John C. H. Fei and Gustav Ranis, *Development of the Labor Surplus Economy: Theory and Policy*, Irwin Publishing Company, Homewood, IL, USA, 1964.
9. Anne O. Krueger, 'Policy Lessons from Development Experience since the Second World War', in J. Behrman and T. N. Srinivasan (Eds), *Handbook of Development Economics*, Vol. IIIB, North-Holland Publishing Company, Amsterdam, 1995, p. 2501.
10. Hollis Chenery and Moises Syrquin, *Patterns of Development, 1950–1970*, published for the World Bank by Oxford University Press, London, 1975.

... agriculture must provide major increases in agricultural production, but it must also make significant net contributions to the capital needs of other sectors of the economy. . . .[11]

and:

The contribution of the rural sector to capital formation may be marshaled. . . . through the medium of taxation. . . . [and through] a relative decline in agricultural prices. . . . Taxes are most easily levied on export crops[12]

Today, agricultural policy makers often struggle to arrest declines in real agricultural prices and profitability. In addition, it is now recognized that commodity-specific taxes reduce the sector's growth prospects, not only by causing reductions in the profitability of investment and production, but also by distorting the allocation of resources among commodities.

Bruce Johnston and John Mellor developed a more complete vision of the agricultural development process and advocated policies that favored development of the smallholder sector. Theirs was the first agricultural development strategy to emphasize the importance of increasing productivity, even among smallholders. They described a long-term growth process in which the type of technological innovation varies by phase of the process. Nevertheless, their view of agriculture's role was to support the development of the other sectors in the economy, mainly by supplying goods and factors of production to the rest of the economy. That role includes supplying labor, foreign exchange, savings and food to the economy, as well as providing a market for domestically produced industrial goods.[13]

Therefore, far from advocating support for agriculture, much of the thinking in the past 50 years about agriculture's role in development advocated taxing the sector, directly or through pricing policies, to provide resources for the development of the rest of the economy, and in some cases using the resultant resources to subsidize industry. Among other concerns that would be raised today about this approach, a basic question is how far can agricultural incomes be reduced by pricing and taxation mechanisms before rural poverty reaches unacceptable levels and agricultural production stagnates for lack of profitability?

Until recently, the success of East Asian economies reinforced, for many observers, the conviction that industrialization was the route to the creation of national wealth, and it contradicted the earlier pessimism about the possibilities of export expansion in developing countries. Over the years a discussion has been carried on regarding the extent and success of government interventions to promote the industrial sector in the East Asian economies, with divergent conclusions. An exhaustive analysis of that experience by a World Bank team concluded that industrial credit subsidies sometimes (but not always) contributed dynamic benefits to the industrialization process in those countries, and that subsidies for exports were more successful in that regard:

Whether these interventions contributed to the rapid growth made possible by good [policy] fundamentals or detracted from it is the most difficult question we have tried to answer . . .

The experience of both the northern high-performing Asian economies . . . and the Southeast Asian newly industrializing economies . . . suggests that economies that are in the process of trade liberalization would benefit from providing specific incentives to manufactured exports. Modest subsidies to exports could be linked, for example, to the

11. John W. Mellor, *The Economics of Agricultural Development*, Cornell University Press (paperback edition), Ithaca, New York, 1966, p. 5.
12. *Op. cit.*, pp. 84 and 92.
13. Bruce F. Johnston and John W. Mellor, 'The Role of Agriculture in Economic Development', *American Economic Review*, **51**, 1961, pp. 566–593.

bias against exports in the domestic economy and bound by strict time limits. . . .[14]

A lesson drawn from the East Asian experience is that export growth is a key to economic development and, in these carefully circumscribed cases, the dynamic benefits from export subsidies have offset the static welfare losses, but that other types of interventions by governments did not succeed in compensating for those losses. These conclusions should be clearly distinguished from the prescriptions of Furtado, who favored strong protection for import-substitution industries and State ownership of heavy industry.

This qualified conclusion in favor of export subsidies, and the consensus that protection of import substitution industries does not work, were both reached primarily on *empirical grounds, through review of experience*. In addition to the rapid rise of the East Asian economies on the back of a vigorous expansion of exports, another experience that provoked this re-thinking was the decades-long stagnation of the Argentine economy under a policy regime that favored industrial import substitution, and a shorter experience of the same kind in Brazil.

It is now accepted as obvious that industries shielded from international competition have little incentive to improve their efficiency, and hence their productivity growth is likely to be very low, whereas exporting industries, by definition, must maintain competitiveness in world markets in order to survive. Accordingly, a related policy prescription would be that subsidies for export promotion should not be too large or long-lasting, or otherwise exporting industries will come to depend on continuing largesse from the national treasury instead of on improving their own economic efficiency. In fact, in East Asia, 'reliance shifted from export subsidies and tax credits to use of the exchange rate itself to provide export incentives'.[15] However, independently of the mode of encouraging exports, the potentials of a dynamic agro-export sector have not played an important role to date in thinking about development paradigms.

The East Asian experience also puts a different light on the supporting role of agriculture in development:

As in other economies, agricultural sectors in the high-performing Asian economies were a source of capital and labor for the manufacturing sector. But in East Asia these resources were generally *pulled into manufacturing* by rising wages and returns, rather than *squeezed out of agriculture* by high taxes and stagnant or declining relative incomes. As a result, urban–rural income differentials were smaller in the high-performing Asian economies than in most other developing economies.[16]

In short, in East Asian economies, policy did not attempt to force the transfer of resources out of agriculture, but rather such transfers as occurred were a natural part of the development process, a process in which agriculture played an important role even though those economies are best known now for their success in industrializing. These experiences conformed to what Vernon Ruttan has called the 'urban-industrial impact model' of agricultural development.[17]

The conception of agriculture as a handmaiden to the development of the rest of the economy, as a store of labor and capital to be exploited, is increasingly being supplanted by the view that agricultural development should be pursued in its own right, that at times it can be a leading sector in the economy, especially in periods of economic adjustment. The World Bank's *World Development Report, 1990*, highlighted a number of cases

14. The World Bank, *The East Asian Miracle: Economic Growth and Public Policy*, published for the World Bank by Oxford University Press, New York, 1993, pp. 354 and 360.
15. A. O. Krueger, 1995, p. 2517.
16. The World Bank, 1993, p. 352 [emphasis added].
17. Vernon W. Ruttan, 'Models of Agricultural Development', in Carl K. Eicher and John M. Staatz (Eds), *International Agricultural Development*, 3rd Edn, The Johns Hopkins University Press, Baltimore, MD, USA, 1998, pp. 155–162.

of adjustment programs in which agriculture responded quicker than other sectors to the new policy regime and grew faster than other sectors for a period of four to five years, leading the economies out of recession. In Chile and Brazil, agriculture grew faster than manufacturing through most of the 1990s. In Chile, agriculture was the main source of new scientific, technical, professional, managerial and administrative jobs during that decade.[18]

When agro-processing industries, agricultural input sectors and marketing activities are taken into account, agriculture's contribution to total GDP typically lies in the 35 to 45% range for low- to middle-income developing countries, much greater than the share of primary agricultural output alone would indicate, and almost always much greater than that of manufacturing alone. The bulk of poverty is often found in rural areas and therefore for poverty alleviation purposes, as well as to avoid a swelling of urban shanty towns, agricultural development can claim a place in national priorities.

One of the most important lessons to emerge from the re-thinking of agriculture's role in economic development is that, although agriculture historically has generated surpluses that permit the flourishing of the rest of the economy, this relation does not imply that policy should tax agriculture more heavily, or try to reduce its prices relative to other prices in the economy, in order to extract even greater surpluses from it. However, until the mid- or late 1980s it was common to repress agricultural prices through a variety of policy measures, and the practice continues even today in many countries. Research on this topic found that:

Discrimination against agriculture was significantly greater than had been earlier realized, and was the consequence not only of policies which were oriented toward agriculture, but also was the outcome of overall trade, payments, and macroeconomic policies. A lesson that is valid for all sectoral policies, and not only those pertaining to agriculture, is that the overall thrust of macro economic policy significantly affects the incentives and responses of each segment of economic activity. . . .[19]

Such policies are self-defeating, in that they reduce agricultural growth and its surpluses and create a greater poverty problem for society. A reduction in agricultural growth means a reduction in overall economic growth. A comparative international study by Anne Krueger, Maurice Schiff, Alberto Valdés and others showed that *there is a strongly negative relationship between a policy of taxing agriculture (through both explicit and implicit means) and an economy's overall growth rate.*[20]

One of the earliest quantitative, country-specific analyses to investigate this relationship was carried out for the case of Argentina, and it came to similar conclusions. As a point of departure, the study reported that the combination of a low real exchange rate and high industrial protection constituted a substantial implicit tax on agriculture:

The result of the implicit tax was to extract, on average for the period 1940–73, 50 percent of agricultural output. . . .[21]

---

18. Roger D. Norton, 'Critical Issues Facing Agriculture on the Eve of the Twenty-first Century', in IICA, *Towards the Formation of an Inter-American Strategy for Agriculture*, San José, Costa Rica, 2000, p. 260.
19. A. O. Krueger, 1995, p. 2527.
20. See Anne O. Krueger, Maurice Schiff and Alberto Valdés, 'Agricultural Incentives in Developing Countries: Measuring the Effect of Sectoral and Economy-Wide Policies', *The World Bank Economic Review*, 2(3), September 1988. The negative relationship between agricultural taxation and economic growth is summarized in Maurice Schiff and Alberto Valdés, *The Plundering of Agriculture in Developing Countries*, The World Bank, Washington, DC, USA, 1992, pp. 10–11.
21. Domingo Cavallo and Yair Mundlak, *Agriculture and Economic Growth in an Open Economy: The Case of Argentina*, Research Report No. 36, International Food Policy Research Institute, Washington, DC, USA, December 1982, p. 14.

Then the study used an econometric model to construct an alternative scenario of how the economy would have evolved under different policies that included a devaluation of the exchange rate as well as trade liberalization. The new scenario was associated with very substantial increases, of 30 to 40%, in both agricultural and non-agricultural output, as compared to the actual course of the economy. Food prices also increased more rapidly than urban wages under the new scenario, and so one of the conclusions of the study was that a combination of urban food subsidies and a higher real exchange rate should be explored. The policy actually followed of taxing agriculture through trade and exchange rate policies gave highly negative results for all sectors of the economy.

There now is growing agreement that *agricultural growth is a key to expansion of an entire economy*. In support of this consensus, Mellor has written that:

Typically high growth rates are achieved when agriculture grows rapidly. That is because the resources used for agricultural growth are only marginally competitive with other sectors and so fast agricultural growth tends to be additive to growth in other sectors, as well as being a stimulant of growth in the labor surplus non-tradable sector. . . . Block and Timmer's model of the Kenyan economy[22] shows the multipliers from agricultural growth to be three times as large as those for non-agricultural growth.

The explosion in trade and global income mean that agriculture can grow at 4 to 6 percent rate (50 percent higher than what was conceivable three decades ago) even while domestic incomes are too low to make large markets for high value commodities.[23]

Other reasons for the strong effect of agricultural growth on growth in the entire economy arise from the structure of income and consumption in rural areas: (1) since rural populations are on average poorer than urban groups, their propensity to spend additional income rather than save it is higher than in urban areas, and (2) the composition of their spending is proportionately more weighted toward domestic goods rather than imported goods, as compared to the behavior of urban consumers. These fundamental facts underlie the high income multiplier effects that have been detected in many countries for increases in agricultural and rural incomes.

Part of the beneficial stimulus of agricultural growth takes the form of creating markets for non-agricultural rural goods and services, diversifying the economic base of rural areas. As economies grow, non-agricultural economic activities acquire increasing importance in rural areas. Their development, however, depends in part on agricultural growth. The two are complements, not substitutes, for rural development.

1.3 AGRICULTURAL GROWTH AND POVERTY REDUCTION

In the past decade, solid empirical evidence has emerged that *agricultural growth is not only effective in alleviating rural poverty, but it is more effective than industrial growth in reducing urban poverty*. Researchers have begun to assemble and study richer data sets on rural and urban income distributions than were previously available. Martin Ravallion and Gaurav Datt analyzed data from 33 household surveys in India over the period from 1951 to 1991, and they came to the following unambiguous conclusions:

Both the urban and rural poor gained from rural sector growth. By contrast, urban growth had adverse distributional effects within urban areas, which militated against the gains to the

22. Steven Block and C. Peter Timmer, 'Agriculture and Economic Growth: Conceptual Issues and the Kenyan Experience', mimeo, Harvard Institute for Economic Development, Cambridge, MA, USA, 1994.

23. John Mellor, 'Faster More Equitable Growth: The Relation between Growth in Agriculture and Poverty Reduction', CAER II Discussion Paper No. 70, Harvard Institute for International Development, Cambridge, MA, USA, May 2000, pp. 10, 13 and 29.

Thomas Vollrath summarized empirical evidence indicating that agricultural growth contributes more to the economy than other sectors' growth does:

Upon examining the contemporary record ... Houck (1986) ascertained that agricultural growth had a more pronounced impact on increases in developing-country income than did growth in the nonagricultural sector. He found that a 10 percent rise in agricultural productivity was associated with a 9.0–10.2 percent increase in per capita GDP. By contrast, a 10 percent rise in manufacturing productivity was associated with only a 1.5–2.6 percent increase in per capita GDP across countries. ... Hwa's (1988) cross-country empirical analysis. ... found that agricultural growth contributed more to economic growth than did export growth. ... Bautista (1990) empirically examined agricultural growth linkages with the rest of the economy among 34 food-deficit developing countries. He found the growth–linkage elasticity to be greater than unity. ... [at] 1.3 for the 1961–84 period and 1.4 for 1973–84.

Reprinted from Food Policy, *19(5), Thomas L. Vollrath, 'The role of agriculture and its prerequisites in economic development: a vision for foreign development assistance', p. 473, Copyright (1994), with permission from Elsevier. His references are: R. M. Bautista, 'Agricultural growth and food imports in developing countries: a reexamination', in Seiji Naya (Ed.),* Economic Development in East and Southeast Asia, *East–West Center, Honolulu, HI, USA, 1990; J. P. Houck,* Foreign Agricultural Assistance: Ally or Adversary, *Staff Paper P86–50, Department of Agricultural and Applied Economics, University of Minnesota, Minneapolis–St. Paul, MN, USA, 1986; E. C. Hwa, 'The contribution of agriculture to economic growth: some empirical evidence',* World Development, *16(11), 1988, pp. 1329–1339.*

urban poor. And urban growth had no discernible effect on rural poverty. ...

Both primary and tertiary sector growth reduced poverty nationally and within urban and rural areas. By contrast, secondary sector growth had no discernible positive effect on the poor in either urban or rural areas. ...

Our investigation points clearly to the quantitative importance of the sectoral composition of economic growth to poverty reduction in India. Despite the rising urbanization of Indian poverty, it is likely to remain true for many years to come that – from the point of view of India's poor – it is the dog (the rural economy) that wags the tail (the urban sector), not the other way around. Fostering the conditions for growth in the rural economy – in both primary and tertiary sectors – must thus be considered central to an effective strategy for poverty reduction in India.[24]

Klaus Deininger and Lyn Squire at the World Bank assembled a multi-country time series of household income data which permit analysis of the relationship between agricultural growth and poverty reduction for more countries.[25] Peter Timmer combined the observations of highest quality from these data with time series on real per capita incomes (for entire countries) adjusted for purchasing power equivalents, to analyze the agriculture–poverty reduction linkage for a sample of 27 countries. His analysis explored the relationship between income per agricultural worker and poverty levels over time, whereas Ravallion and Datt had looked at income per unit of agricultural land.

Timmer uses a model that tries to capture long-term relationships between economic growth and income of the poor, as opposed to measures of how the poor are affected by short-term economic fluctuations. He found different results for countries that had relatively even income distributions and those that had very skewed income

24. Martin Ravallion and Gaurav Datt, 'How important to India's poor is the sectoral composition of economic growth?', *The World Bank Economic Review*, **10**(1), January 1996, p. 19.

25. Klaus Deininger and Lyn Squire, 'A New Data Set Measuring Income Inequality', *The World Bank Economic Review*, **10**(3), September 1996, pp. 565–591.

distributions. In the former group of countries, *increases in agricultural income per worker lead to increased incomes at the economy-wide level in all income strata* (where rural and urban households are aggregated together), with the greatest effect seen in the lowest strata. Thus, in these economies improvements in labor productivity in agriculture generate growth throughout the economy, and more so for the poor, so they make the income distribution even more equitable over time. However, in economies that start out with highly skewed income distributions, the richer strata benefit substantially from improvements in agricultural productivity, but the poor benefit much less from productivity growth in both agriculture or in non-agriculture, so the income gap continues to widen regardless of the sectoral composition of growth.[26]

For a larger sample of 35 countries, with all types of income distributions, Timmer found that:

A one percent growth in agricultural GDP per capita leads to a 1.61 percent increase in per capita incomes of the bottom quintile of the population in 35 developing countries. A similar one percent increase in industrial GDP increases the incomes of the poor by 1.16 percent.[27]

He noted that the statistical significance of these results is poor because of 'noise' in the data, but concluded 'they do suggest that, on average, the sectoral composition of growth affects the strength of the linkage between economic growth and poverty'.

These results may not appear very different numerically, but extrapolated over a decade or decades they potentially represent a large difference in the incomes of the poor. The importance of agricultural growth may be seen by comparing

the effects on the incomes of the bottom quintile of 4% annual growth in per capita income, sustained over 20 years, in industry versus agriculture. As a result of industrial growth at a rate of 4% per capita, in 20 years the incomes of the poorest would have increased by 2.5 times, according to Timmer's results. In contrast, with agricultural growth of 4% per capita, in 20 years those incomes would have increased by 3.5 times.

In a subsequent piece, Timmer argues on the basis of observed experience that 'the East and South-East Asian approach of "growth with redistribution", relying heavily on stimulation of the rural economy, in combination with a policy to stabilize domestic food prices, is the fastest approach to managing this escape [from hunger and famine]'.[28] He concludes by amplifying a theme of Johnston and Mellor in their early work, regarding the role of small farms, and addresses the issue of relative prices:

It is clearly difficult to find a way to structure the growth process so that the poor gain in relation to the rich. Historically, the only way to do that has been a rural-oriented development strategy that raises productivity and incomes of the broad population of small farmers and other rural workers. . . .

Such a strategy, however, requires significant price incentives to create the rural purchasing power that, in turn, stimulates the rural growth needed to make the strategy consistent with overall macroeconomic performance. . . .

This 'food price dilemma', in which poor consumers have their food intake threatened in the short run in order to fuel a long-run growth process that removes them from poverty, has been emphasized before. . . . But experience in East and Southeast Asia since the 1970s shows that such a strategy, when implemented in the context of large-scale investments in rural

26. C. Peter Timmer, 'How Well Do the Poor Connect to the Growth Process?', CAER II Discussion Paper No. 17, Harvard Institute for International Development, Cambridge, MA, USA, December 1997, in particular pp. 16–22.

27. *Op. cit.*, p. 3.

28. Reprinted from *Food Policy*, **25**(3), C. Peter Timmer, 'The macro dimensions of food security: economic growth, equitable distribution, and food price stability', p. 291, Copyright (2000), with permission from Elsevier.

infrastructure, human capital, and agricultural research, can lead to economic growth and an increase in average incomes per capita of 5 percent per year or more, with the rate of growth in the bottom two quintiles faster than that in the top. . . .[29]

Mellor has synthesized the survey data analysis of Timmer, Ravallion and Datt, and other studies, arriving at the following broad conclusions:

It is now clear that high rates of economic growth may rapidly reduce the proportion of the population in absolute poverty. In low-income countries, rapid overall growth is likely to be accompanied by rapid growth of the agricultural sector, because virtually all low-income countries have large agricultural sectors encompassing the majority of the population. There has been a tendency to generalize that *economic growth* reduces poverty, when in fact it is *the direct and indirect effects of the agricultural growth that account for virtually all of the poverty decline*. . . .

it is notable that agricultural growth reduces inequality among the poor as well as lifting the poor above the poverty line. . . .

emphasizing agriculture in order to improve income distribution does not result in slow growth. The sectors are more complementary than competitive. . . .[30]

A corollary to these lessons about the importance of agriculture, both in the overall economy and for poverty reduction, is that appropriate agricultural policies are crucial for the entire development process. In the words of Gale Johnson:

It is hard to understand how the role of policies can be ignored, given the enormous differences in the economic performances of the planned and market economies between 1950 and 1990 and the sharp change in the rate of growth in China following the reforms of the late 1970s.

. . . much of the concern over future world food supplies is based on the assumption that land is the limiting resource. This is putting the emphasis in the wrong place. The major factors that may limit the growth of food production in developing countries are knowledge and research, the availability of nonfarm inputs at reasonable prices, and the governmental policies that affect incentives. If policies provide for the first two and do not discriminate against agriculture in trade and macroeconomic policies, farmers will do the rest.[31]

In today's international economic environment, many observers have expressed concerns about the effects of globalization and free markets on agriculture, and hence on levels of rural poverty, in developing countries. For this reason, sometimes a recommendation is made to return to pervasive State controls over developing agriculture. It is a central contention of this book that the negative effects of globalization and distortions in international markets can be largely corrected without incurring the economic costs of centralized controls, by using indirect instruments of national economic policy, and that appropriate policies can create a favorable economic environment for agricultural growth and for the reduction of rural poverty. By now, there are many examples of pro-agriculture and pro-poor policies in the developing world, and one of the aims of this book is to present them and discuss them in a systematic way.

DISCUSSION POINTS FOR CHAPTER 1

• Although historically agriculture has generated surpluses of labor, savings and foreign exchange that have facilitated the growth of other sectors, policies that attempt to tax agriculture in order to extract more surpluses, implicitly and explic-

29. *Op. cit.*, p. 293.
30. J. Mellor, 2000, pp. 1, 8 and 10 [additional emphasis included].
31. D. G. Johnson, 1997, pp. 10 and 11.

itly, have proven counterproductive for growth and employment generation.

- Agricultural growth contributes proportionately more to economic development than industrial growth does, because the multiplier effects of agricultural growth on the domestic economy are greater.

- Studies have shown that agriculture is the most effective sector for reducing both rural and urban poverty, although the poor benefit less from agricultural growth in economies whose existing income distribution is very unequal.

- Sustaining adequate real price levels at the farmgate is one of the keys to agricultural growth and hence to poverty reduction. Attempting to pursue a policy of cheap food worsens the poverty problem in the long run.

2

Strategies and the Agricultural Policy Framework

2.1 Strategies and Policies   15
   2.1.1 Sectoral Strategies   15
   2.1.2 Why Agricultural Policy?   17
2.2 The Nature of Agricultural Policy
   Instruments   19
   2.2.1 Types of Agricultural Policies   19
   2.2.2 A Taxonomy of Agricultural Policies   21
   2.2.3 Policies, Programs and Projects   21
2.3 Objectives of Agricultural Policy   22
   2.3.1 National and Sectoral Objectives   22
   2.3.2 Principles of Agricultural Policy   23
2.4 The Role of Government   25
2.5 Implementation of Strategies and Policies   30
Discussion Points for Chapter 2   32

2.1 STRATEGIES AND POLICIES

2.1.1 Sectoral Strategies

Policy reforms are often made individually, addressing a single issue at a time. However, because each issue has implications in other areas, reforms are sometimes more effective if they are designed and implemented jointly as part of an integral package, or strategy, for the sector. A *strategy* constitutes both a vision of what the sector should look like in the future and a 'road map' showing how to fulfill this vision. Its point of departure is the current situation of the sector and the issues that it faces. It should be firmly grounded in both history and an assessment of the sector's future potentials. The development of an agricultural strategy may be motivated by an economic crisis in the sector or by other problems that catalyze a decision to make fundamental changes. In some cases, it has been conceived as the principal supply side of a structural adjustment program, to stimulate production response to offset the otherwise deflationary effects of macroeconomic structural adjustment in the short term. Whatever the motive for developing a strategy, it usually needs the support of the principal actors in the sector, i.e. the farmers, if it is to be successful.

There can be as many strategic visions as there are observers of the sector's performance, so one of the defining characteristics of a strategy is *by whom* and *for whom* it is developed. A strategy should represent a commitment by the sector's authorities to carry out specified reforms, and hence participation of the government is a *sine qua non* in its elaboration. On the other hand, a strategy that is developed in part by the producers of the sector, and is responsive to their principal concerns, is more likely to be made truly operational than one which is developed solely by government officials or academic experts. Equally, a government's fiscal goals and other national development concerns need to be reflected accurately in a sectoral strategy to ensure its viability, and so the most successful strategies generally are those that are a collaborative effort among different institutions and groups in society.

A strategy needs to be realistic, but its vision of the future should be based on *the sector's*

Agricultural Development Policy Concepts and Experiences. R. D. Norton
© 2004 Food and Agriculture Organization of the United Nations
ISBNs: 0-470-85778-1 (HB)   0-470-85779-X (PB)   FAO Edition: 92-5-104875-4

strengths and opportunities. It also needs to clearly identify the constraints to be overcome in order to realize the opportunities. A strategy that does not offer a vision of a better future, backed up by concrete policies for realizing such a vision, cannot motivate the rural population to participate in its implementation. At the same time, the more realistic it is, and the better its analytic underpinnings, then the greater are the chances of attaining its objectives.

In some agricultural strategies the vision for the future is quantified, in terms of cultivated area to be irrigated, area planted in principal crops, and so forth. Whether or not numerical projections are made, the vision should define the anticipated *directions of change* in the future, the new emphases that will characterize future growth. One of the best guides to the directions of change is the sector's *comparative advantage*, that is, the lines of production in which it can most effectively compete on world markets in the long run. However, it should be borne in mind that market conditions can change rapidly, and in the end producers are always in a better position than governments to choose the specific product mix.

Identification of a sector's comparative advantage is a basic step along the pathway to defining explicitly the national *objectives* for the sector. In one form or another, at the most general level they must include *increasing productivity* – the main prerequisite for increasing incomes – and *reducing poverty*. These are the classic economic goals of efficiency and equity. Other broad objectives may be included as well, such as reducing gender inequality. The productivity objective embraces both technological improvements (yield increases) and shifts in production patterns toward higher value crops or other products.

The next sequence of steps in the exercise of developing a strategy normally consists of the following: (i) identifying specific constraints to be overcome (problems to be solved) in each area, (ii) specifying operational sub-objectives in each area for overcoming the constraints and achieving the overall objectives, and, on the basis of the above (iii) drafting a set of policies to address the constraints and achieve the sub-objectives. It is vitally important that a *technical justification for*

the policy recommendations be developed, otherwise a strategy may run the risk of being seen as only another set of opinions.

While it is important to visualize scenarios about how the structure of the sector may evolve, unless the policy framework and concrete policy instruments are also specified, and the farmers' role in making such decisions is taken into account, it may be difficult to implement a strategy. *Policies and public sector investments* constitute the means for implementing a vision. These are the primary links between national objectives, on the one hand, and the decentralized workings of a market economy, on the other hand.

All over the world, governments increasingly strive to use *indirect instruments of policy*, rather than direct controls. Accordingly, the type of vision expressed in strategies is shifting away from projections of specific future cropping patterns or production levels. Their place is being taken by the use of indirect policy instruments which improve the sector's markets, for purposes of promoting both equity and efficiency. Principal areas of policy instruments include agricultural incentives policies (operating primarily through trade policy and macroeconomic policies) and marketing systems, the land tenure framework, irrigation policies, the rural financial system and the system for development and transfer of farming technology. Concepts and experiences in each of these areas are described in Chapters 4–8 of this volume.

Increasingly, the issues raised and analyzed in a strategy have to do with *institutions, laws, markets and resource endowments.* The latter include not only land, irrigation supplies and field labor, but also *farmers' managerial and administrative capacities.* It is increasingly acknowledged that agriculture is but one part of the rural economy, and that a strategy should contain a program for rural development in a more complete sense, including the creation of economic opportunities for the rural landless. Chapter 9 is devoted to a discussion of these broader concerns.

In its operational form, an agricultural strategy is an integrated package of policies for the sector, complemented by an investment program. Some of its policies may be designed to take effect immediately or in the short term, but most of them typically represent deep reforms whose effects will be felt increasingly throughout the sector over a period of many years. The chief advantages of developing policy reforms in the context of a comprehensive strategy are that (a) the policies are derived from, and support, specified national objectives and a clear vision of the future, (b) they are designed to be mutually consistent across all facets of the sector and with macroeconomic policy, (c) no important areas of policy reform are overlooked, and (d) the effort of developing a strategy represents an opportunity to work out a consensus among principal interest groups in the sector.[1]

Additional characteristics of agricultural strategies, and the tasks involved in developing them, are reviewed in more detail in Chapter 9. That chapter also addresses issues of civic participation in strategy formulation.

2.1.2 Why Agricultural Policy?

Sometimes, macroeconomists and government officials question the *raison d'etre* of agricultural policy. It is argued that the basic requirements for a successful economic transition or development experience are correct macroeconomic policy, privatization of government assets to the maximum extent, and the elimination of regulatory barriers and other counterproductive government interventions. According to this viewpoint, there is no need for sectoral policy *per se*, once markets are freed up and macroeconomic stabilization is assured. What makes agriculture different, it is asked, from the textile industry, the cement industry or the restaurant industry, in economic policy terms?

In response, sometimes it is asserted that agriculture's priority for policy makers derives from the fact that it produces food, second in importance only to fresh water for human survival. This argument is valid to a certain degree. It holds mainly for the poorest rural areas of the lower-income countries, which tend to produce mainly for their own consumption, but it becomes less applicable as the world economy becomes increasingly integrated and food imports and exports expand in almost all countries. Increasingly, it is recognized that the nutrition levels of most poor families depend more on their income levels and health conditions than on whether they produce basic foodstuffs. An example is found in the hillside farmers of Central America, who can raise their family incomes significantly by switching from producing corn and beans to cash crops.

Agriculture, of course, is the main source of income and employment in rural areas; indeed, in poorer countries it is often the principal employer in the entire economy. As pointed out in Chapter 1, agricultural growth is also the main way to reduce poverty, in *both* rural and urban areas. Poverty alleviation is universally recognized as a proper concern, indeed a responsibility, of policy.

There are other basic reasons for treating agriculture as a central topic of economic policy. No other sector is so deeply interconnected with the rest of the economy. Agriculture uses – sometimes abuses – vital resources which are limited and depletable, e.g. water, soils, forests and fish stocks. These are precisely the natural resources whose use has proven to be the most difficult to subject to the rule of markets. A completely hands-off approach, or *laissez faire* policy, for such resources has not proven to be tenable in any country, for it invariably leads to their over-exploitation.

In addition to being a major employer of the labor force, the sector plays a similarly large role in the balance of payments in many countries, and it is the largest user of a country's fertile land. Questions about the societal role and legal status of land impinge heavily on agriculture. Likewise, environmental pollution of land and water arising from farming and ranching activities can severely

---

1. In some contexts, the usage of the terms 'strategy' and 'policy' is reversed, so that the broader concept is called an 'agricultural policy', and it consists of 'operational strategies' in each area. The important point is that an overall strategy or policy document should contain both *a vision for the sector* and the *concrete instruments of governmental action* that are needed to implement it.

affect urban communities if appropriate policies for ameliorating the damage are not in place.

These observations indicate that agriculture's performance has a significant effect on the rest of the economy, and vice versa. However, there is yet another, more fundamental, reason why agriculture is different from other sectors in the economy. Labor and capital can be moved from one industry to another, or one service activity to another, with varying degrees of ease or difficulty, and back again if circumstances should so dictate. However, once labor is moved out of agriculture, it is costly and extremely difficult to shift it back again. Many countries have discovered this lesson through their own experiences, as did Nigeria in the 1970s and 1980s, Mexico in the 1970s, and China during the Cultural Revolution. The intersectoral movement of labor, between agriculture and the rest of the economy, is practically an *irreversible flow of resources*. The reasons for this irreversibility may be as much social and cultural as economic, but nonetheless they are powerful.

A related concern is that such an intersectoral movement of labor gives rise to large societal costs per migrant for investing in additional infrastructure in urban areas: new housing, water and sewage facilities, and transportation networks, among other items. In addition, rural–urban migration in excess of cities' capacities to create gainful employment generates severe social problems. A study for El Salvador concluded that rural–urban migration costs the country between $159 million and $189 million annually, in terms of new infrastructure (roads, housing, potable water, sewage systems and electricity generation), and that the cost would be much higher if other types of infrastructure and social problems were taken into account. The authors pointed out that it would cost less than a tenth of this conservative estimate to improve the infrastructure, including housing and roads, to acceptable levels in rural areas if the migrants chose not to relocate.[2]

Therefore, a policy of discouraging agriculture prematurely and promoting primarily the growth of urban sectors can entail irreversible effects on the economy and society, and it can be costly as well. In light of these considerations, it is important for national economic policy to consider carefully the role of agriculture in the nation's development prospects and to design appropriate policies. This is not to argue that agriculture should be subsidized at the expense of other sectors' growth prospects, but an appropriate balance should be sought, consistent with an accurate assessment of agriculture's growth prospects.

Another reason for developing a set of agricultural policies – or a strategy for the sector – is that in most countries the economic institutions are generally less developed, and the economic rules of the game less clearly articulated, in rural areas than in urban–industrial areas. In effect, the economic environment in rural areas may be less well adapted to the requirements of economic growth. The geographical dispersion of farmers and poorly developed road networks and lack of other infrastructure may mean that farmers' access to markets is uncertain and expensive, and banks may not possess much expertise in evaluating agricultural projects or knowledge of their clients, to mention only two ways in which the entrepreneurial environment is weaker in the countryside than in cities.

In some cases, the economic environment in rural areas may be conditioned in part by the historical legacy of a different economic era, as in cases of pervasive State ownership of agricultural land or antiquated rural land registry systems. In contrast, land ownership or long-term leases are almost universally available in urban areas. Whatever the reasons for the differences between rural and urban economic environments, they exist. Reforming rural economic institutions so that they are more conducive to business activity, and at the same time facilitate a reduction in poverty, is normally a long-term undertaking but no less essential for that reason.

Thus, there are many persuasive reasons for assigning a high priority to the development of

2. Roger D. Norton and Amy L. Angel, *La agricultura salvadoreña: políticas económicas para un macro sector*, FUSADES, San Salvador, El Salvador, 1999.

appropriate agricultural policies. Such policies are strongly interdependent among themselves. For example, policies oriented toward improving marketing channels are usually linked to international trade policies, to competition policy (as applied to agro-industry) and to rural financial policy. Strengthening institutions dedicated to rural finance in turn frequently depends in part on providing greater security of land tenure, and so forth. Hence, a policy reform program in agriculture often has to be fairly comprehensive, dealing with policies in several important areas. This underscores the importance of developing an agricultural strategy.

2.2 THE NATURE OF AGRICULTURAL POLICY INSTRUMENTS

2.2.1 Types of Agricultural Policies

After the *why* of agricultural policy, the next major question is *what* does it consist of? The content of macroeconomic policy is unambiguous: the fiscal deficit, the money supply, instruments required to move those variables to their target levels, such as government spending and revenue collection, bond issuance, monetary targets, interest rates, reserve requirements, banking regulations and the like, and, in many economies, the exchange rate. In spite of the great antiquity of agriculture, and of government interventions in agriculture, there is not a similar consensus about the substance of agricultural policy.

The conception of what constitutes agricultural policy is undergoing change in all parts of the world. Historically, one of the major instruments of agricultural policy has been *government expenditure*. In all countries, fiscal outlays have been made for a variety of purposes in the sector. A few of the more prominent forms of expenditure have been investment in infrastructure for purposes such as irrigation, crop storage, transportation and marketing, direct provision of credit to producers and subsidization of private credit, the funding of research, extension and seed production, financing the deficits incurred by programs of purchasing grains from farmers at high prices and selling them to consumers at lower prices, and direct payments under land set-aside programs or other support programs.

The second major class of policy interventions often has consisted of *controls*, primarily on prices and trade, but sometimes also on access to land and irrigation water and on production levels themselves. The use of support prices, and administered prices for both consumers and producers, has been widespread in all regions of the world, but while it remains common practice in Europe and East and South Asia, it is being phased out in most of Latin America and Africa and reduced in the Middle East. The third main class of policy instrument in many countries has been *direct management of production and marketing* through State-owned enterprises which have spanned a wide range, from production collectives, sawmills and fisheries corporations to banks and marketing enterprises. The tendency in most parts of the world has been to reduce State ownership of assets in the sector, although the pace of those changes varies from region to region.

Now that there is a growing international consensus that direct government interventions in the economy generally should be reduced, along with fiscal expenditures, the question of what is agricultural policy (and what is the role of a Ministry of Agriculture) comes more sharply into focus. It is put into even higher relief by the fact that price levels and trade volumes respond essentially to macroeconomic policy and international market conditions. What remains in the purview of a Minister of Agriculture except perhaps to run a research and extension service and administer phytosanitary controls? More to the point, how can a constellation of agricultural policies be defined that will boost the growth rate of the sector, or raise income levels of the rural poor? Are there any constructive degrees of freedom left at the sector level, between macroeconomic policy and field-level programs? In a liberal economy, is *agricultural policy* an oxymoron?

In summary form, the answer is that there is a great deal for agricultural policy to do in all countries, but for the most part it requires different approaches than those of the past. A principal task of agricultural policy is *to improve the func-*

tioning of product and factor markets in rural areas, with special attention to poor families' access to these markets and the conditions of their participation in them. In some cases, this requires infrastructure investment, but it almost always requires good policies as well. Factor markets include those for land, financial capital, labor and, in some cases, irrigation water and environmental characteristics. Rural families' participation in labor markets, for example, is enhanced by programs of training and agricultural extension.

Today more than ever agricultural policy must be co-ordinated with other areas of policy and the agencies in charge of them: for example, with the Central Bank and Ministry of Finance in the case of rural financial policy, with the Ministry of Economy or Trade for agricultural trade policy, with the Ministry of Finance for expenditure programs on irrigation and research, and with a Ministry of the Environment or Natural Resources for irrigation policy. In the sphere of government, agricultural development policy is increasingly a topic that concerns many Ministries and agencies.

Equally, good design and implementation of agricultural policy require participation from local governments, producers' associations, water users' associations, NGOs, regional offices of the Ministry of Agriculture and other decentralized organizations. The *policy coordination role* of the Ministry of Agriculture increasingly occupies center stage.

One of the main tasks of modern agricultural policy is to promote the *development of adequate institutions* to fulfill the requirements of a growing rural economy, from marketing to provision of farm services to supplying production finance. Even though many such activities may logically belong to the private sector in the long run, the public sector has a large responsibility in fostering the development of the needed capacities, and in overseeing the launching of those capacities and ensuring their functioning in an initial period. In economies in which the government has been managing trade in food products, for example, it often is found that the private sector is not necessarily prepared to step in and take over that responsibility at short notice. It may lack the financial capacity and the technical

commercial expertise, and it may hesitate to enter the field because it is not convinced that the government will not intervene again.

Developing institutional capacity includes *developing and refining the rules of the game for a market economy* and fostering respect for those rules. In societies in which the rule of law is shaky in rural areas or judicial systems are weak, and there are not sufficient means for adequate enforcement of contracts, this vital task can be very difficult. It can be an undertaking of many years if not decades. However, that is all the more reason to emphasize it.

Another broad task of policy is to ensure that the *legislative framework* is appropriate for the sector's development, that it encourages economic activity rather than inhibiting it, and at the same time provides the right measure of protection to the interests of producers, consumers and the environment. This task can involve an extensive review of existing legislation, ranging from the labor code, the commercial code and consumer protection laws to land tenure laws, resource management decrees and many other kinds of legislation.

A few concrete examples of agricultural policy instruments in a market economy include import tariffs, rebates on tariffs on inputs for exporting industries, farm support prices, certificates of deposit to finance grain storage by small farmers, food quality regulations, regulations mandating public auction of forestry concessions, fishery licenses, statutes for land funds, regulations for land registry systems, legislation concerning land tenure, policies on the structure and functioning of agricultural extension services, policies for devolving ownership and management responsibilities for irrigation systems to users' associations, policies for privatization of other State assets, laws for the protection of the rural environment, policies for providing food assistance to the poor, and legislation governing rural financial systems. The list can be extended very considerably, and it will necessarily differ from country to country, since in the final analysis each country's approach to agricultural policy has to be consistent with its own history, traditions and overall economic policy.

2.2.2 A Taxonomy of Agricultural Policies

Given the diversity of agricultural policies, it can be helpful to review policy needs from the viewpoint of the requirements of a producer. In order to operate successfully, a producer needs three basic things, i.e. adequate *incentives* to produce, a secure *resource base* (farmland and water) and *access* to markets for outputs and inputs, including technology. Accordingly, agricultural policy comprises the following three broad components:

* *Pricing policy*, which in a market economy is determined in large part, but not entirely, by macroeconomic policies.
* *Resource policy*, including *land tenure policy* and *policies for management of resources* (land, water, forests and fisheries).
* *Access policy*, including access to agricultural inputs, output markets and technology. Rural financial policy is an important part of access policy, since finance in many cases is a prerequisite for obtaining inputs and marketing products.

The divisions among these three broad classes of policies are not cut and dried. For example, policy measures designed to improve marketing channels (improve *access*) are likely also to raise farmgate prices (and hence are part of *pricing policy*). A broader conception of resource policy includes the basic resource of *human capital*, for which rural education and training programs are vital. The role of land tenure policy is to provide security of access to land resources, which can be as important as physical access to land.

Most classes of policies are relevant to the entire sector, or most of it, and generally are not specific to particular crops. In this sense, there is not a cassava policy or a corn policy or a groundnut policy, or a millet policy or a plantain policy. Good policies facilitate the work of the farmer, part of which is the selection of the crop and livestock mix. Differentiating policies by product runs the risk of creating greater incentives for some than for others, and governments usually are not in the best position to designate which crops have the best future possibilities. The combination of markets and farmers' judgments can make such choices more reliably.

However, it is common to establish government-supported *programs* for important products: a program for the renovation of coffee plantations, a dairy development program and a rice-improvement program. Such programs represent one of the means of implementing policies.

2.2.3 Policies, Programs and Projects

Programs are limited in time and resources. They require the active participation of the government (even if implementation is contracted out to the private sector), and when the funding terminates, the program ends. In contrast, *policies* are permanent, at least until a new policy regime is designed and put into effect. They do not always require government expenditures. A law eliminating import restrictions, for example, does not require expenditures or a staff to implement it, and it is permanent unless in the future new limitations to free trade are legislated. Although policies may not represent a cost to government, they often imply costs for users of public services, producers in general, marketing agents, consumers and other groups in the economy. Part of the art of policy making is balancing these costs with the benefits conferred by new policies.

Programs consist of directly managed activities (usually with sizeable staffs), involving face-to-face interaction with farmers, financial institutions and other private economic agents. Many policies, by contrast, exercise their influence indirectly and consist of definitions of economic rules of the game, through laws, decrees and regulations; in principle, many of them can be put into effect with the aid of a handful of specialists on a ministerial staff.[3]

Projects, like programs, are limited in time and are staff-intensive. They usually involve a significant *investment* component. They utilize the government's capital account budget, whereas

---

3. It should be noted that the word *program* is sometimes used in another sense, that of a *policy program* or a set of interrelated policies.

programs utilize the current account budget. However, some programs also contain investment expenditures, and so the distinction between programs and projects is not always clear-cut. This is particularly true in the area of training (human capital formation), where current outlays are used to create capital.

Frequently, both programs and projects are required to implement the policies in a strategy. If they are not derived directly from a sector-wide or sub-sectoral strategy, they should be made consistent with it. In the hierarchy of governmental decisions, programs and projects normally are subordinate to and derived from policies, and the latter, in turn, often are developed within the framework of a strategy. In the real world of decision making with multiple and conflicting interests and actors, things do not work out so neatly, but the attempt to co-ordinate policies, programs and projects can make them all more effective. An irrigation investment is more productive if it is accompanied by legislation that facilitates the formation and operation of water users' associations (Chapter 6). A livestock investment gives greater returns to farmers if rural financial systems are strengthened so that future funding is available on a continuous basis for adequate herd management (Chapter 7). A community development program can be more effective if a policy decision has been made to decentralize agricultural research and make research and extension more participatory (Chapter 8).

2.3 OBJECTIVES OF AGRICULTURAL POLICY

2.3.1 National and Sectoral Objectives

The foregoing paragraphs discuss the *why* and *what* of agricultural policy. The question of *for what* must also be addressed. To what end is agricultural policy designed?

Agriculture is not an island in the economy. Its ultimate objective should be to support national development. In agriculture, as in other areas, economic policy responds to national imperatives and to a social and political vision. It is designed to promote the achievement of societal aims that are not exclusively economic in character. Therefore, the basis of a strategy, or set of policies, should be the enunciation of broad social, or societal, goals for agricultural and the rural sector. Fundamentally, they should be related to the promotion of *human development*. Specific objectives for the agricultural sector can be derived from this overarching goal.

In most economies, the ways in which agriculture can most effectively support human development are (a) ensuring that *nutrition and other basic material needs* are met in rural areas, and (b) contributing indirectly to the satisfaction of those needs in urban areas. In some transition economies, nutrition levels are high enough that they no longer are a general concern, but meeting other material needs is very much an issue, given the prevalence of poverty in rural areas. In many developing countries, nutrition levels are still deficient among a significant part of the rural population, although it is important to recognize that, for the world as a whole, the share of the population in poverty has dropped markedly over the past three decades.

What *sub-objectives*, if achieved, will best enable agriculture to meet these overall objectives? In many parts of the world, it has long been the practice to define the aim of agricultural development strategy as increasing production levels. Frequently, that objective has been stated in narrower terms, as increments in the production of staple food crops, usually grains and sometimes principal root crops. However, while producing more staple foods can be important, a physical target of that nature is not sufficient for promoting the goal of human development, or even the objective of raising levels of material well-being. Production alone is not necessarily the best indicator of the economic status of rural households. *Income* is a better indicator, for it takes into account the prices farmers receive and their costs of production. Even more relevant is *real income*, which adjusts net income levels for the rate of inflation, in order to measure the *purchasing power of rural households*.

Therefore, agriculture can make its most effective contribution to nutrition and other basic material needs by generating more real income for rural households. This contribution, in turn, depends on three factors, namely **production**, **real farmgate prices**,[4] and **non-farm employment in rural areas**. Real prices are almost always beyond the control of farmers themselves but can be influenced by policies. Production is a function of the land area cultivated (including that in pastures) and productivity, or unit yields. As limits on the availability of cultivated land are being reached, and sometimes exceeded, in many places in the world, production increases in the future will increasingly depend on technology to deliver improvements in productivity.

Figure 2.1 illustrates this hierarchy of objectives and sub-objectives for the agricultural sector, including indirect contributions that the sector makes to development in urban areas, by providing foreign exchange earnings and creating demand for processed foods and other manufactured products.[5]

It should not be overlooked that the level of nutrition of a rural family can also depend on the degree of control over production exercised by women in the household. The discussions in Chapters 5–8 point to various ways in which rural women can be empowered.

2.3.2 Principles of Agricultural Policy

In addition to the identification of objectives and means of policy, a strategic framework should also be based on **principles that guide policy actions**. In other words, the objectives of policy will not be pursued at any cost. The principles represent conditions or limits on the kinds of actions – means – that will be employed in attempting the fulfill the strategic objectives.

There are five basic principles for making an agricultural strategy sustainable over the long run, as follows:[6]

- **Economic sustainability.** The strategy must find ways to deliver real economic benefits to the rural sector. Although fiscal discipline is important, this means, among other things, not simply subjecting the sector to the fiscal retrenchment of a structural adjustment program. It is worth recalling the discussion in Chapter 1 about the importance of agricultural development for the growth of the entire economy.
- **Social sustainability.** The strategy also must improve the economic well-being of lower income groups and other disadvantaged groups, including women. Otherwise, it loses social viability.
- **Fiscal sustainability.** Policies, programs and projects whose complete sources of financing are not identified should not be undertaken. In an era of increasing budgetary stringency in all governments, application of this principle encourages a search for new sources of fiscal revenue and ways in which beneficiaries of the policies, programs and projects can contribute to their financing, i.e. ways to foster cost recovery.
- **Institutional sustainability.** Institutions created or supported by policy should be robust and capable of eventually standing on their own. For example, financial institutions which are just credit channels to farmers and ranchers, and which do not have deposit-raising capabilities of their own, are not likely to survive over the longer term. Equally, research and extension services that are supported mainly by international loans and grants are not sustainable in the long run.
- **Environmental sustainability.** Policies should be developed to bring about sustainable management of forests and fisheries stocks and reduce to manageable levels agricultural pollution of

---

4. *Real* agricultural prices, and real incomes, are agricultural prices and incomes deflated by an index of economy-wide prices. Thus, real farmgate prices are such prices *relative to* others in the economy.
5. This figure is taken from R. D. Norton, 'Integration of Food and Agricultural Policy with Macroeconomic Policy: Methodological Considerations in a Latin American Perspective', FAO Economic and Social Development Paper No. 111, Food and Agriculture Organization of the United Nations, Rome, 1992.
6. These principles were applied, for example, in developing Guyana's *National Development Strategy* and Estonia's *National Strategy for Sustainable Agricultural Development*.

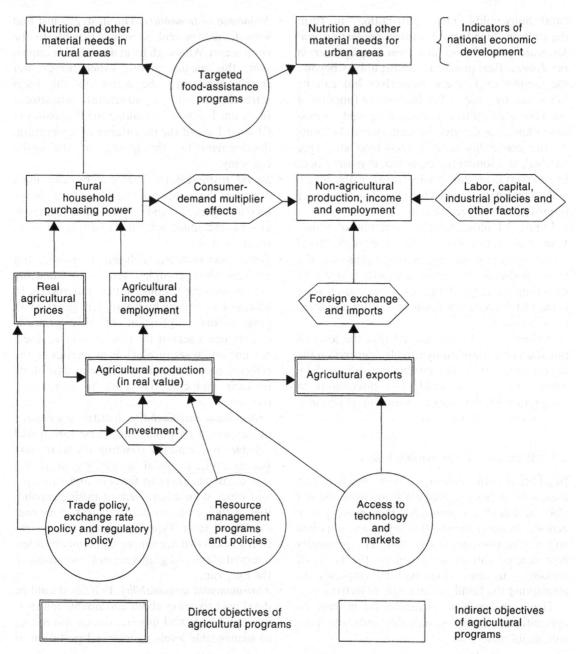

Figure 2.1 The role of agricultural programs in national economic development.

water sources and degradation of soils. A major challenge for agricultural policy in some countries is to slow or stop the expansion of the 'agricultural frontier', the zone in which cultivation is possible only by felling trees.

Some observers may prefer to call these principles objectives, and the decision is partly a matter of individual taste. According to the circumstances of each country, it may be desired to add additional principles to the above list, in order to guide the formulation of strategies and policies for the sector.

2.4 THE ROLE OF GOVERNMENT

The consensus on the appropriate role of government in the agricultural economy has shifted over time toward less direct management of economic activities and fewer controls on prices and quantities of factors and outputs. Although the concept of 'market failure' in the private sector has long been recognized in economics, the awareness of 'government failure' is now much greater than before. Much of government failure can be attributed to inappropriate institutional incentives and weak fiscal policies, but nevertheless it is a reality. Rapid turnover of Ministers of Agriculture and principal staff members can greatly hinder the process of developing and implementing consistent policies. On the other hand, government needs to play a leading role in mobilizing consensuses about appropriate development policies and translating them into concrete actions. It also needs to correct with indirect measures the most egregious kinds of market failure.

Uma Lele, Robert Emerson and Richard Beilock, drawing on contributions by Joseph Stiglitz, have aptly summarized the state of the debate on the role of government in agricultural development:

The new institutional economics stresses that the nature of contractual arrangements, and the income and wealth distribution, matter because they affect the incentives and multiplier effects from agriculture (Stiglitz, 1993)[7]. . . . Therefore Stiglitz argues that the advice to adopt market systems is too simplistic when problems of unequal land distribution, imperfect information and incomplete risk markets are serious. . . . Notwithstanding these risks, agriculture under centrally planned systems of management has worked less well than under market conditions, for many reasons. First, under the public sector, organization and management of agriculture, and institutional surrogates for markets, tend to be poorly organized to adapt to information and incentives. Also, information tends to be poorly processed owing, in part, to the hierarchical relations which vest decision-making authority away from the scene of economic activity. Shirking becomes a central problem, individual initiative tends to be lacking, soft budget constraints replace hard budget constraints, and job and salary security inhibit quick responses to new and critical information. Thus the reduction of risks faced by farmers tends to be achieved at huge costs under the public system of management.

Then what should be the role of government, if not in production or distribution? The non-controversial roles of government have been clear enough: *protection of property rights, enforcing contractual obligations to foster competition, and the provision of public goods such as agricultural research, technology, information and infrastructure*. The more controversial roles involve *redistributing assets through forced mea-*

7. Joseph E. Stiglitz, 'Incentives, Organizational Structures and Contractual Choice in the Reform of Socialist Agriculture', in A. Braverman, K. M. Brooks and C. Csaki (Eds), *The Agricultural Transition in Central and Eastern Europe and the Former USSR*, The World Bank, Washington, DC, USA, 1993.

sures, stabilizing prices, absorbing risks and providing credit. If the government goes where private markets fear to tread, it needs to do so cautiously and with considerable safeguards.[8]

Gale Johnson has provided a clear definition of six traditional areas in which government action is needed. One of them is correcting market failures, but he is cautious in his prescriptions for entry into this area, emphasizing mainly the provision of education. His remarks, excerpted, are as follows:

Provision of public goods

... there are some goods and services that a competitive market would not supply at all or would provide in less than optimal amounts. These include public goods where consumption is not exclusive, such as maintenance of law and order, protection of civil rights, national defense, public parks, agricultural research, and some forms of communication. . . .

Competitive markets may not be optimal

There . . . are goods and services that societies believe will either not be provided in adequate amounts by competitive markets owing to economies of scale (public utilities), or not be utilized in socially optimal amounts by certain segments of the population. The latter is the argument for providing free primary and secondary education. . . .

Because of the path-breaking work of . . . Theodore W. Schultz (1964)[9], the important role of investment in human capital in economic growth has received increasing consideration. The evidence is now very strong that investment in human capital through universal access to primary and secondary education contributes to economic growth, while at the same time limiting or preventing increases in inequality as economic growth occurs. Taiwan and the Republic of Korea provide important lessons for developing countries in this regard. Each gave quite early emphasis to making primary and secondary education universally available. They not only enjoyed rapid growth but each achieved growth without a significant increase in inequality.

Development of infrastructure

Rural infrastructure has not received the same attention and emphasis as it has in urban areas. This was especially the case in socialist countries where the neglect of roads left a high percentage of villages and farm households inaccessible . . . throughout the year, or for a large part of the year, owing to the poor quality of existing roads. . . .

Support of research

The recent unparalleled rates of growth of food production in developing countries owe much to the provision of a public good, agricultural research. There is overwhelming evidence of the high rates of return to investments in agricultural research (Evenson *et al.*, 1979)[10]. . . .

Information

Market information is essential for the efficient functioning of markets. In developing economies and those in transition, such information is unlikely to be adequately supplied by market institutions and, unfortunately, little is

8. Uma Lele, Robert Emerson and Richard Beilock, 'Revisiting Structural Transformation: Ethics, Politics and Economics of Underdevelopment', in G. H. Peters and D. D. Hedley (Eds), in *Agricultural Competitiveness: Market Forces and Policy Choice*, Proceedings of the 22nd International Conference of Agricultural Economists, International Association of Agricultural Economists, Dartmouth Publishing Company, Aldershot, UK, 1995, pp. 183–184 [emphases added].
9. Theodore W. Schultz, *Transforming Traditional Agriculture,* Indiana University Press, Bloomington, IN, USA, 1964.
10. R. Evenson, P. E. Waggoner and V. W. Ruttan, 'Economic Benefits from Research', *Science*, **205**, 1979, pp. 1101–1107.

being done, especially in economies in transition, to provide it systematically. . . .

Institutions

Governments also must actively limit the role of monopolies, either through a positive competition policy (including liberal policies towards international trade) or by regulation where competition is not a viable alternative. . . . public monopolies can be far more powerful and damaging than even the strongest private monopolies. . . .

The clear definition and enforcement of property rights has a major role to play in creating the incentives for an efficient agriculture. . . . The assigned rights must be enforceable against arbitrary actions of governments as well as of private entities.[11]

The definition of property rights can be expanded to include legislation and regulations that guarantee rights agreed to under contracts, such as regulations for bonded storage warehouses, rules for weights and measures in markets, legislation supporting land rental contracts, and bank supervision legislation for microfinance institutions. These and many other measures of this type are aimed at *improving the functioning of rural markets*.

In a modern economy, the private sector and government play mutually supporting roles, and the boundary between them is not always clearly defined. As Stiglitz has said:

The new agenda. . . . sees government and markets as complements rather than substitutes. It takes as dogma neither that markets by themselves will ensure desirable outcomes nor that the absence of a market, or some related

market failure, requires government to assume responsibility for the activity. It does not even ask whether a particular activity should be in the public or the private sector. Rather, in some circumstances the new agenda sees government as helping to create markets – as many of the East Asian governments did in key aspects of the financial market. In other circumstances (such as education) it sees the government and private sector working together as partners, each with its own responsibilities. And in still others (such as banking) it sees government as providing the essential regulation without which markets cannot function. . . .

Government can help promote equality and alleviate poverty, policies that in East Asia contributed to growth. . . . the exact role of government will change over time.[12]

From a different perspective, Lawrence Smith has also stressed that the dividing line between the roles of the public and private sectors is not hard and fast:

Parastatal reform is often considered in terms of a choice between public and private sector provision of goods and services. However . . . this is a false and sterile dichotomy because the *provision* of any good or service can be disaggregated into four components . . . namely:

* *Funding the provision*
* *The actual production process*
* *Regulating the provision*
* *Consumption*

The important point to appreciate is that, in theory, there is no necessity for all of these functions to be carried out by the same organization or, indeed, by the same sector of the economy. . . . As [the World Bank] states:

11. D. Gale Johnson, 'The Limited but Essential Role of Government in Agriculture and Rural Life', Elmhirst Memorial Lecture, in G. H. Peters and D. D. Hedley (Eds), *Agricultural Competitiveness: Market Forces and Policy Choice,* Proceedings of the 22nd International Conference of Agricultural Economists, International Association of Agricultural Economists, Dartmouth Publishing Company, Aldershot, UK, 1995, pp. 16–18.

12. Joseph E. Stiglitz, 'An Agenda for Development in the Twenty-First Century', Keynote Address in Boris Pleskovic and Joseph E. Stiglitz (Eds), *Annual World Bank Conference on Development Economics, 1997,* The World Bank, Washington, DC, USA, 1998, pp. 22–23.

Public production or marketing of seed may well be a necessary first step to show the existence of a market and to make seed available, pending the development of private seed companies.[13]

... [There would be] a possible need for the government to continue to *fund*, wholly or partially, some activities even though these may not be carried out exclusively by the parastatal. For example, if the seed industry is mainly concerned with the production of high yielding varieties of *self-pollinated* seed ... farmers can save seed for several generations without a major decline in yields. This makes it difficult for seed companies to recover the full cost of improved seed from seed sales. ... Without a subsidy the wider benefits to society stemming from improved seeds would be lost. While this funding could be injected through private sector companies, the private sector may still be unwilling to invest sufficient resources in these activities and this might require a permanent parastatal involvement.[14]

Bardhan has underscored the role of government in stimulating agricultural development in the modern context, and has described issues in the decentralization of governmental functions:

Even with all its limitations (of administrative capacity as well as vulnerability to wasteful rent-seeking processes) the State can play at least as a catalyst in the initial stages in pump-priming agricultural finance and underwriting risks (while being careful to avoid the associated moral hazard of encouraging dependency). It can take the initiative in establishing commodity exchanges, generating and disseminating

information, allowing for contingent contracts, and arbitration in contract disputes. ...

In the post-reform period in many developing countries public investment in agriculture has been on the decline. Given the undoubted complementarities between public and private investment in this field, it is not surprising that private investment has been slow to make up for this deficiency. In particular, falling public investment in agricultural research and development in many countries is slowing the rate of technological progress in agriculture, and the decline of investments in maintenance and repair of irrigation and drainage systems, rural roads and in prevention of soil erosion have curtailed the effectiveness of earlier investments in agriculture. (Recent IFPRI projections for China suggest that each yuan invested in the coming decades in research and irrigation could yield returns of between 3.6 and 4.8 yuan.) The issue of public investment will be increasingly important also in the case of biotechnology research to develop technologies in plant and livestock breeding and in native crops suited to local conditions (sorghum in Africa, millet in India), a need that is likely to be neglected by the patent-protected multi-national biotechnology companies ...

In public investments in agriculture the emphasis is shifting from massive State investment in large dams ... to better local management of existing irrigation systems and minor irrigation projects under some form of community control to improve the effectiveness of investments. In a comparison of the mode of operation of canal irrigation bureaucracy in [South] Korea and India, Wade (1997)[15] finds the former to be more sensitive to the needs of the local farmers and thus more effective. The

13. Operations Evaluation Department, *The Seed Industry in South Asia*, The World Bank, Washington, DC, USA, 1996.
14. Lawrence D. Smith, 'Agricultural Parastatal Reform," paper prepared for the Policy Assistance Division, Food and Agriculture Organization of the United Nations, August 31, 1998, pp. 7–8 [emphasis in original].
15. R. Wade, 'How infrastructure agencies motivate staff: canal irrigation in India and the Republic of Korea', in A. Mody (Ed.), *Infrastructure Strategies in East Asia*, Economic Development Institute, The World Bank, Washington, DC, USA, 1997.

Indian canal systems are large, centralized hierarchies in charge of all functions. . . . Their ways of operation . . . and source of finance . . . are totally insensitive to the need for developing and drawing upon local social capital.[16] In contrast, in Korea there are functionally separate organizations in the canal systems: the implementation and routine maintenance tasks . . . are delegated to the Farmland Improvement Associations, one per catchment area, which are staffed by local part-time farmers . . . knowledgeable about changing local conditions, dependent for their salary and operational budget largely on the user fees paid by the farmers, and continually drawing upon local trust relationships.

The same problems of low accountability to the local population affect the volume and particularly quality of provision of local public goods and services in many developing countries. . . . Fisman and Gatti (1999)[17] document a significant negative correlation in cross-country data between the subnational share of total government spending and various measures of corruption, controlling for other factors, suggesting that decentralization can mitigate corruption. Of course, the adverse effects of lack of local accountability on the quality of public goods and services show up in even less tangible forms of leakages and targeting failures than what the measures of corruption indicate.

Going beyond the impact of local accountability on the quality of service in publicly supplied facilities, it is important to note that a local community organization, if it has a stable membership and well-developed structures for transmitting private information and norms among the members, may have the potential for better management of common property resources. . . .

A major problem that hinders most schemes of decentralized governance is related to distributive conflicts. In areas of high social and economic inequality, the problem of 'capture' of the local governing agencies by the local elite can be severe, and the poor and the weaker sections of the population may be left grievously exposed to their mercies and their malfeasance. . . . Similarly, collusion among the elite groups may be easier at the local level than at the national level. . . .[18]

At any level, an effective government for agriculture has an active commitment to the sector's development and yet is aware of the limitations of government action and the danger of creating serious distortions in the economy if the policies are not well conceived. It is a government that supports development intelligently, by diagnosing accurately and continuously problems that emerge and playing a facilitating or enabling role for the sector's growth. It is a government whose institutions are strong and have developed out of a country's own historical experience and social context. While most governments provide infrastructure, agricultural research services and other public goods as indicated by Gale Johnson, the legitimate role of government in promoting development may be framed in a broader way. Dani Rodrik has provided an illuminating perspective on the role of government and the nature of its institutions and the policy reform process:

[There are] five functions that public institutions must serve for markets to work adequately: pro-

16. Author's note: see, however, the experience of Andhra Pradesh in the decentralization of management of large numbers of irrigation systems, described in Chapter 6, where the topic of local management of irrigation is treated in some depth.

17. R. Fisman and R. Gatti, *Decentralization and corruption: evidence across countries,* World Bank Working Paper, The World Bank, Washington, DC, USA, 1999.

18. Pranab Bardhan, 'Institutions, reforms and agricultural performance', in K. G. Stamoulis (Ed.), *Food, Agriculture and Rural Development: Current and Emerging Issues for Economic Analysis and Policy Research*, Economic and Social Department, Food and Agriculture Organization of the United Nations, Rome, 2001, pp. 155–158.

tection of property rights, market regulation, macroeconomic stabilization, social insurance, and conflict management. In principle, a large variety of institutional setups could serve those functions. We need to be skeptical of the notion that a specific institution observed in a country (like the United States, say) is the type that is most compatible with a well-functioning market economy.

Partial and gradual reforms have often worked better because reform programs that are sensitive to institutional preconditions are more likely to be successful that those that assume that new institutions can be erected wholesale overnight. Learning and imitation from abroad are important elements of a successful development strategy. But imported blueprints need to be filtered through local experience and deliberation. . . .

The lesson of the 20th century is that successful development requires markets underpinned by solid public institutions. Today's advanced industrial countries – Japan, the United States, Western European nations – owe their success to having evolved their own workable models of a mixed economy. While these societies are alike in the emphasis they place on private property, sound money, and the rule of law, they are dissimilar in many other areas: their practices in labor market relations, social insurance, corporate governance, product market regulation, and taxation differ substantially.

All these models are in constant evolution, and none is without its problems. . . . What is true of today's advanced economies is also true of developing countries. *Economic development ultimately derives from a homegrown strategy, not from the world market.* Policymakers in developing countries should avoid fads, put globalization in perspective, and focus on domestic institution building. They should have more confidence in themselves and in

domestic institution building and place less faith in the global economy and the blueprints that emanate from it.[19]

2.5 IMPLEMENTATION OF STRATEGIES AND POLICIES

As the foregoing discussion has indicated, there are a large number of types of decision-makers in developing agriculture. Apart from the basic decision unit of the rural household, decision-makers include many kinds of village and producer associations (often referred to as 'social capital'), local and foreign enterprises, NGOs, local governments, provincial governments, decentralized agencies, central governments and international development agencies. It is increasingly accepted that decisions should be devolved to the lowest hierarchical level possible, that is, the level closest to those with a stake in the decision. This tends to ensure better accountability in decision-making. However, exceptions can occur when the weight and presumed neutrality of the central government may be needed to offset pressures by local elites placed on local governments and other organizations. In such cases, as in local government tax collection, the central government may work alongside local government institutions.

Implementation of a strategy requires the assent and active participation of many kinds of decision-makers. The central government usually has the responsibility of taking the lead in implementation, but it needs to act in co-ordination with local decision-makers. In the framework of a central or provincial government, implementation of policy decisions occurs only in the following five ways:

- New *legislation*
- *Administrative decisions and decrees* of the executive branch that alter the rules governing the economic environment for agriculture and change institutional structures

19. Dani Rodrik, 'Development Strategies for the 21st Century', in B. Pleskovic and N. Stern (Eds), *World Bank Annual Conference on Development Economics, 2000*, The World Bank, Washington, DC, USA, 2001, pp. 87 and 105–106 [emphasis added].

- Allocations of public investment, or *capital account funding*, some of which may come from external partners in development
- Allocations of the *current account budget* of the government
- *Voluntary participation* in implementation by the private sector and civil society

Implementation may occur through more than one of these channels simultaneously, as in the case of programs that require both investment and current account spending and are supported by new administrative decrees. Policy implementation is a major undertaking, a challenge that is sometimes underestimated. To be successful, it requires *conviction, consensus and co-ordination*. Conviction is required on the part of those who are promoting reforms, for in most cases they will face many obstacles along the road to implementation. Consensus is required among government agencies and above all with the private sector and local decision making entities, for without their active support it is difficult to implement policy reforms fully. They also must be co-ordinated with international agencies that are supporting the development effort. In addition, co-ordination is required among the reforms themselves and among the implementation efforts, to ensure consistency all along the way and to maintain the pace of reforms on schedule.

Roger Douglas, who was New Zealand's Finance Minister from 1984 to 1988, when that country carried out what was widely hailed as a very successful structural reform program, commented thus on the need for quick action in a reform program: 'Speed is essential. It is impossible to move too fast. The total program will take some years to implement even at maximum speed. . . . Move too slowly and the consensus that supports reform can collapse before the results are evident. . . . It is uncertainty, not speed, that endangers structural reform programs' (The Wall Street Journal, January 17, 1990).

To achieve the required consensus and co-ordination, often it is necessary to undertake extensive negotiations with interest groups, both within the sector and outside the sector, and with international agencies as well. Through this process, a strategy will be molded to political realities, but the clearer the initial vision and the more cogent the technical justifications, then the better the chances that the main thrusts of the strategy – or at least some of them – will emerge intact from such a process. In the end, a nation's political authorities, including both executive and legislative branches, through negotiation with the private sector and civil society and international agencies, will determine the main lines of the country's development strategy. However, the attempt to develop a coordinated and internally consistent strategy, involving the participation of many of those actors, can elevate the level of the national debate over policy options and lead to improvements in policies. Participatory approaches to strategy formulation are reviewed in Chapter 9.

A common procedure for implementing a strategy at the central government level is through the development of *annual implementation plans*. Such plans co-ordinate both the current account and investment budgets of the government, as well as the implementation activities of all of the agencies involved and the participating bodies of the private sector.

Even the simplest of reforms usually require 10 or more, sometimes as many as 30, significant steps, and so implementation plans must assign responsibilities and target dates and provide for close monitoring of the progress. The complexity of the implementation process is one of several reasons why it is important to move the process forward as rapidly as possible, while still maintaining co-ordination with all interested parties.

The dividing line between policy design, dissemination (attaining consensus) and implementation is not a sharp one. Effective dissemination is essential for putting new policies into effect, and during the dissemination phase feedback is often received which results in changes in the design of the policies. In general, the following six stages can be identified in the phase of dissemination-cum-implementation, that is, after an initial design for policy reform has been proposed:

- Promotion of the new policies among all concerned policy makers (including in Ministries other than Agriculture).
- Dissemination of the new policies and their benefits among producers and the public at large and to international development agencies.
- Drafting of the implementation plan or program for implementation through the five channels indicated above. (The more concrete the policy proposals are in the strategy, then the easier is this task.)
- Implementation of the policies.
- Monitoring the implementation.
- Evaluation of the results and modification of the implementation plan.

In any of the first five of these stages, the policy design itself is subject to modification as additional elements of reality are brought to bear on the proposals.

In spite of the flexibility that such design changes require, it is important to maintain conceptual consistency over time in regard to the main lines of the proposed reforms. Uncertainty about the basic directions of a development strategy discourages consensus and weakens the prospects for implementation, plus it may discourage investment and production in the sector as well.

Policy implementation is a demanding task, and for that reason it merits careful attention and a systematic approach. Implementation capacity often is a very scarce resource in the public sector. Without successful implementation, even the best designed policies are of no value.

DISCUSSION POINTS FOR CHAPTER 2

- Strategies can provide useful frameworks for policy reforms. A strategy constitutes both a vision of the sector's future and a 'road map' for getting there. Public sector investments are one of the means for implementing strategies,

but increasingly strategies concentrate on issues related to institutions, legal frameworks, the functioning of markets, and resource endowments. Both equity and efficiency objectives should be taken into account in a strategy.
- Sometimes, it is asked why an agricultural policy is needed – what sets agriculture apart from other sectors of the economy? A traditional justification for an agricultural strategy is the size of the sector. In most developing countries, it is the largest employer of labor and the largest single earner of foreign exchange, and it is usually larger than the manufacturing sector alone. It also places major demands on a country's natural resource base, especially water and soils. A more fundamental justification for an agricultural policy is that transfers of rural labor out of agriculture represent essentially irreversible flows of resources in the economy, and hence neglect of agriculture can give rise to permanent and not necessarily desirable structural changes in the economy. Another consideration is that economic institutions are generally less well developed in rural areas than in cities, and therefore special policy attention is required to strengthen those institutions.
- The instruments of agricultural policy increasingly are indirect and multi-sectoral in nature. Governments are now less directly involved in production and marketing activities and more concerned to ensure the adequate functioning of factor and product markets in rural areas, and the adequate development of economic institutions. Hence, the regulatory and legislative role of policy has acquired greater importance.
- The basic areas of policy for agricultural growth, including the supporting policies for the macroeconomy and for other sectors, include (1) policies that influence relative prices, (2) policies for natural and human resources, and (3) policies to promote access to inputs, output markets and agricultural technology.
- In order that public sector programs and investments may have their greatest benefits, they should be designed in the context of a strategy or a policy framework. They are means of implementing policies.

- The most fundamental objective of agricultural policy is not increasing production but rather human development, helping to meet basic human needs. Increasing rural household incomes is a key to promoting fulfillment of this objective, and this, in turn, requires emphasis on productivity increases and adequate real farmgate prices.
- The principles that underpin sound agricultural policy include economic sustainability (delivering real benefits), social sustainability (reducing poverty), fiscal sustainability, institutional sustainability and environmental sustainability.
- The role of government has evolved considerably in recent decades. It is now a more indirect role than before but it is no less powerful. There is a consensus that a government's responsibilities include provision of public goods, definition and protection of property rights (of many kinds), promotion of competition, improving the functioning of markets in other ways through regulation and institutional development, provision of social insurance, and economic stabilization. Sometimes, the improvement of functioning of markets requires policies that facilitate better risk management. Conflict resolution also is accepted as an important role for government. A more controversial role, but still accepted in some circumstances, is the redistribution of economic assets among economic strata. In many cases, governments are best placed to promote and fund activities, while leaving their implementation and management in the hands of the private sector or NGOs.
- Decision makers in developing agriculture are diverse and include many kinds of local organizations, businesses, NGOs and local governments and decentralized governmental agencies as well as central governments. International development agencies also play decision-making roles in many circumstances.
- At the central government level, policy implementation occurs through five channels: new legislation, administrative decisions and decrees, investment projects (capital account expenditures), programs (current account expenditures) and voluntary implementation by the private sector and civil society. Implementation is a demanding task that requires conviction, consensus and coordination. Implementation plans and monitoring are essential components of successful implementation.

3

Broad Issues of Agricultural Policy

3.1 Agricultural Policy and the Macroeconomic
Framework   35
   3.1.1 The Price Linkage   35
   3.1.2 A Spurious Argument on Prices   36
3.2 Fiscal Expenditures and Subsidies   37
   3.2.1 Fiscal Expenditures in Agriculture   37
   3.2.2 The Pros and Cons of Subsidies   38
   3.2.3 The Issue of Support for Agriculture   42
3.3 Improving the Incomes of the Rural Poor   44
3.4 Gender and Agricultural Development   46
3.5 Selected Issues in Privatization   48
3.6 Principal Aspects of the Legal Framework   51
Discussion Points for Chapter 3   53

3.1 AGRICULTURAL POLICY AND THE MACROECONOMIC FRAMEWORK

3.1.1 The Price Linkage

The relationship between agriculture and the macroeconomy gives rise to a number of important questions for policy making and strategic planning. Can a successful agricultural policy be designed and implemented independently of the macroeconomic framework? Conversely, are there circumstances in which macroeconomic policy must be modified if agricultural policy is to achieve its aims? If so, what are the tradeoffs at the economy-wide level? Would it be in the broader interest of society to accede to those modifications in the macroeconomic framework? Are there other sectors, such as industry, that also would benefit from an adjustment of macroeco-

nomic policies in a direction favorable to agricultural growth, or do some macroeconomic choices pit agriculture against the rest of the economy?

In the long run, all sectors of the economy benefit from a macroeconomic environment characterized by relatively stable prices and which is conducive to savings and investment and opportunities for foreign trade. However, to date many developing and transitional economies have not yet reached an optimal macroeconomic configuration in this sense, and the paths taken in attempting to reach that optimum differ across countries. Hence, it is important to ask whether different macroeconomic strategies have different implications for agriculture.

Classically, the tradeoff between macroeconomic goals and agricultural development was thought to be a choice between stability and growth. Greater fiscal expenditures for agriculture were presumed to stimulate expansion of the sector, even at the risk of creating or deepening a fiscal deficit and therefore feeding inflation. In reality, this tradeoff was more imagined than real, for two reasons. First, experience worldwide in the last decade has confirmed that economic stability, through its reduction of economic uncertainty and its stimulus to saving and investment, is a powerful influence for growth in its own right.

Secondly, it is apparent that many fiscal expenditures in the sector have been inefficient in achieving their aims. They often have not been

Agricultural Development Policy Concepts and Experiences. R. D. Norton
© 2004 Food and Agriculture Organization of the United Nations
ISBNs: 0-470-85778-1 (HB)   0-470-85779-X (PB)   FAO Edition: 92-5-104875-4

targeted on the poorest groups in rural areas, and their stimulus to production has been weak in relation to the volume of expenditure. A common example is subsidized credit, which frequently is characterized by a poor repayment record, thus creating the need for further subsidies to sustain the same amount of lending, and which also is diverted in part to non-agricultural end-uses. Another typical example is expenditure on State-owned grain storage facilities, which tend to be operated at low rates of capacity utilization and therefore generate a low or negative return on the funds invested.

In reality, the tradeoffs for the sector are different than stability versus growth. The type of macroeconomic policies adopted can have a strong influence on the development prospects of the agricultural sector in the following ways: (a) they can affect the intersectoral terms of trade, or intersectoral *relative prices*, and hence both the incentives for production and the real incomes of agricultural households, (b) they can create greater or lesser *incentives for agricultural exports*, and (c) they can also influence levels of capital formation in agriculture by creating an economic environment which is more or less *conducive to rural financial activity and investment*. In addition, sound fiscal policies can generate funding for vital infrastructure investments in rural areas.

A stable macroeconomic environment favors investment in the sector provided that the rates of return are sufficiently high. These returns depend more than anything else on the trends in real agricultural prices, that is, agricultural prices relative to non-agricultural prices. Usually, macroeconomic policy has a decisive effect on real agricultural prices. The linkage between macro and sectoral levels via relative prices is powerful and is often the dominant one. In other words, the main policy tradeoff for the sector is not agricultural growth versus the rate of inflation, but rather relative prices which are favorable or unfavorable for agriculture. Normally, macroeconomic policies that favor agriculture in this sense also favor manufacturing, including agro-industry, at the expense of the service sector.

The fact that both agriculture and agro-industry can see their real prices improved by some types of macroeconomic policies bears emphasis, for normally their preferences are mutually opposed, because agro-industry wishes to see lower prices for its raw materials and agriculture, and higher prices for its output. This tension will always be present, but classes of policies exist which will improve the profitability of both sectors.

From the viewpoint of the agricultural sector, the principal instruments of macroeconomic policy are the following: (i) exchange rate policy, (ii) trade policy (the degree of openness of the economy to international trading possibilities), (iii) tariff policy, (iv) taxation policy, (v) fiscal expenditure policy, (vi) interest rate policy (or monetary policy, which influences interest rates), and (vii) the regulatory framework for finance and contractual relations in general. All of these instruments can affect the real returns to agricultural production, but the first four are especially important in determining real prices of agricultural outputs. The relationships between macroeconomic policies and the sector's performance are explored further in Chapter 4, but first it is worth devoting some attention to a common assertion about agricultural prices.

3.1.2 A Spurious Argument on Prices

On occasion, it has been said that a reduction in real price incentives for agriculture is a healthy effect, since it is presumed to stimulate improvements in the sector's productivity. It is true that policy should encourage productivity increases, because in the long run the standard of living of farming families depends strongly on agricultural productivity. Nevertheless, in countries in which yields and total economic productivity are low in agriculture, often it is found that agricultural profit margins are already very low, and thus producers would not have the capacity to make the investments required to raise productivity – and banks will not lend to them for that purpose either when rates of return are low.

The argument that the economic screws have to be tightened on agriculture is flawed in two other respects. First, if it is beneficial to reduce real prices and returns, why is not the same policy applied to other sectors, including banking services, insurance, advertising, legal services and so forth? Secondly, if reducing real prices is an effective medicine for the sector, where is the dividing line between a healthful dose and an overdose which may kill the patient? Is it a reduction of real prices by 25%, by 50%, by 80%, or by another magnitude? Answers to these questions have not been provided in any country – a fact which indicates that the argument lacks a foundation.

This contention that a policy to lower real prices for the sector necessarily stimulates productivity improvements rests on confusion about the effects of competition. When a new firm with lower costs enters a market, it brings with it an improved technology (whether in management, marketing or production itself) which provides the basis for reducing output prices. With or without the new technology, existing firms will be forced to reduce prices as well, and those that survive are likely to adopt either the new technology or some other improved way of doing things.

In the words of Christopher Adam, 'a contestable market is one in which any firm is constantly exposed to actual or threatened competition from more efficient producers who can enter the market easily, undercut the incumbent's price and acquire market share. The threat of this profit-reducing competition is thus the spur to efficient operations by all firms in the market'.[1] This is the mechanism by which competition reduces prices, and it should be noted that in agriculture the entry of new producers is particularly easy, as compared with some other sectors.

However, *a real price reduction imposed on the sector from the outside (by policy), instead of through competition, does not bring with it the technological basis for producing at lower cost*. Hence, it does not function in the way that a competitive market does to reduce costs. In fact, the price reduction makes more difficult the adoption of technological improvements, since it reduces the capacity of producers to finance such improvements. Granted, some firms may manage

to stay afloat under an externally driven price reduction, but innovations may arise equally, or more so, out of the normal workings of competition in the sector, without an imposed price reduction. In addition, under the externally generated fall in real prices, some firms that could have survived may collapse economically. In a context in which most of the 'firms' are in fact low-income rural households, the equivalent of collapse is deeper impoverishment.

The studies mentioned in Chapter 1 have shown conclusive evidence that a policy of reducing real agricultural prices leads to slower growth not only in the sector but also in the entire economy.

3.2 FISCAL EXPENDITURES AND SUBSIDIES

3.2.1 Fiscal Expenditures in Agriculture

Government expenditures have long been regarded as a principal instrument of agricultural policy, both capital and current-account expenditures. They become *subsidies* in cases in which their costs are not recovered from the beneficiaries of the policies. Fiscal subsidies are *explicit subsidies*. Agricultural policy also confers many kinds of *implicit subsidy*. A very common example is the legislation of an import tariff, which shields domestic producers to a degree from international competition and therefore implicitly subsidizes higher costs of production. Price controls are implicit subsidies – or the reverse (implicit taxes) – because frequently they are established at levels that effectively subsidize consumers and tax producers.

Implicit subsidies are also found in the field of natural resource management, when a user of a natural resource does not pay the full damage to it. For example, a widespread practice that constitutes an implicit subsidy is the issuance of licenses for fishing boats that are not priced sufficiently high to restrain over-fishing of the stocks, or simply are given out free, or do not

1. Christopher Adam, 'Privatization and Structural Adjustment in Africa', in *Negotiating Structural Adjustment in Africa*, Heinemann, Portsmouth, NH, USA, 1994, Chapter 9, p. 139.

exist. The lack of such licenses also constitutes a subsidy, for it leads to an understatement of the true, long-run costs of fishing in a sustainable manner.

The benefits that implicit policy subsidies confer on producers are sometimes called *economic rents*. To economists, rents are flows of income that derive from the mere fact of ownership of an asset, above and beyond the normal returns to the labor and management that would accrue to operating the asset in a competitive market. A common example is a license for a radio or television station, while another is a license for distribution of alcoholic beverages. Both kinds of license are restricted in supply relative to the potential demand for them, and so the owners can reap monopoly or oligopoly profits.

In the same sense, access to special regulatory niches can generate economic rents. Access to a support price for grains, if it is set above the market-clearing level, creates rents for producers who enjoy such access.

Some of the more common examples in agriculture of explicit subsidies, that is, fiscal expenditures that are not fully recovered from the beneficiaries, are the following:

- Construction of irrigation systems, whose full cost is rarely recovered from the users.
- Subsidized sales or grants of State-owned farmland, with or without provision of full title. Often, this transfer of land takes the form of the State's acquiescence in squatting on the land. If a full market price or rental rate is not paid for the land, then its transfer is subsidized.
- Production credit from State-owned banks, or credit from commercial banks that is supported by government rediscount lines at lower-than-market interest rates. The subsidy may be embodied in the interest rate or in a failure to exert normal efforts for loan recovery.
- Provision of research and extension services free or at less than cost. (It often is argued that the externalities arising from agricultural research justify its subsidy.)
- Purchase of harvests at above-market prices and their subsequent sale at below-market prices.

- Construction of feeder roads in rural areas.
- Provision of seeds, seedlings, veterinary services and other inputs at below-market prices.

Historically, pressures to make these kinds of outlays from the government budget have arisen in part from a desire to compensate the sector for unfavorable trends in real prices. In some cases, domestic price trends have been a consequence of downward trends in world market prices, and in other cases the price decline has been accentuated by macroeconomic policy. In yet other cases, government price controls for some commodities have held down their prices to producers, for the purpose of subsidizing consumers. It is also argued frequently in international forums that subsidies for production and exports in richer countries have depressed world prices for agricultural commodities, requiring some kind of compensatory action by governments of poorer countries. Whatever the reasons, there is little doubt that such fiscal expenditures have come to be regarded as a legitimate, and perhaps the most important, instrument of agricultural policy in many countries.

3.2.2 The Pros and Cons of Subsidies

Given the central role of fiscal expenditures in agriculture, and the fact that many of them constitute subsidies to the sector, it is useful to develop guidelines to help decide when they are justified. One of the most universal justifications for a subsidy is poverty of the beneficiaries. However, before accepting this argument as a basis for programs and policies in the sector, it is necessary to ask how well the subsidies are targeted on poorer households. Normally, considerable efforts are expended to target direct food assistance at poor households, but agricultural programs themselves tend to be surprisingly regressive in their incidence over income groups, as illustrated in the box about Honduras.

The reasons for a disproportionate distribution of benefits to larger farms are fairly obvious. In the case of price supports, for example, a larger farmer is likely to have a truck for sending freshly harvested grain to the government's collection point, whereas most small farmers do not. By the same token, large farmers are likely to know the officials in charge of the price support program, and probably the Minister himself, and so it is a simple matter for them to pick up the phone and arrange for their shipments to be received promptly. In contrast, small farmers often tell tales of waiting days at the collection point for their grain to be received, and of sometimes having to leave without consummating a sale. In the same vein, large farmers can offer lunch and other benefits to agricultural extension agents who visit their farms in a timely manner and who spend as much time as necessary providing technical advice.

The lesson of these kinds of experiences is that in practice it is very difficult to target generalized agricultural programs on poorer farmers unless special measures are taken.

In reviewing the arguments for and against subsidies, it is useful to start by recalling a basic result of economic theory: that interventions which affect market prices (for outputs or inputs) invariably lead to a loss of economic welfare. While producers and consumers each may gain, the loss to society is greater than the sum of the gains. This is called the 'static welfare loss'. While it is an abstract principle, its effect in practice is to lead to reduced economic growth rates, because resources are no longer allocated to their most efficient uses. There are more compelling arguments of a practical nature, both for and against subsidies, and so the theoretical argument will not be pursued in this context, but it is well to bear it in mind, for it too has empirical relevance.

Major ***arguments against the use of subsidized public expenditures*** include the following:

• ***Subsidies tend to be allocated to the least competitive industries***, for they are usually the ones that press the government hardest for favors. Subsidies rarely are allocated to industries and products that have a comparative advantage.

A survey in Honduras funded by the European Commission and the French Government provided quantitative information about the incidence of benefits (implicit subsidies) in three major agricultural programs. Among several farm size strata, it was found that only 0.2% of the smallest farms (<2.5 ha) sold their grain harvests at the official support price, whereas 13.1% of the largest farms (>50 ha) sold at that price. In other words, the largest farms had a sixty-five times greater (13.1/0.2) chance of access to the support price. Regarding credit, the survey asked if the respondent had been denied credit from the State agricultural bank for lack of loan guarantees. Of the smallest farms, 75.8% said 'yes', while among the largest only 12.0% said 'yes'. Regarding extension services, it was asked if the service was timely and whether it was good, fair or poor in quality. For the Ministry of Agriculture's service, 39% of the smallest farms (this time, <10 ha) said it was both timely and good. Of the largest farms, 72.7% so responded. For the Agrarian Reform Institute's service, the dispersion in such replies was even greater: 20.2% of the smallest farms and 81.7% of the largest farms. (Source: G. Gálvez et al., Honduras: Caracterización de los Productores de Granos Básicos, Secretaría de Recursos Naturales, Honduras, November 1990.)

Accordingly, over time subsidies tend to shift the economy's allocation of productive resources toward those industries which are inherently least competitive, thus prejudicing a country's long-run growth prospects.

• Once established, *a subsidy is hard to eliminate*. Economic and political interests arise in its defense, and so the cost to the government may continue for many years and even increase.[2]

2. 'The study of the history of agricultural policies shows that many existing distortionary agricultural policies in OECD countries have been implemented initially as "temporary measures" to overcome a specific (time-limited) problem. If we have learned anything, it is that agricultural programs tend to create their own constituency and tend to persist long afterwards, because for political economy reasons they are very hard to remove once they have been implemented'. Reprinted from *Food Policy*, **24**(1), Johan F. M. Swinnen and Hamish R. Gow, 'Agricultural credit problems and policies during the transition to a market economy in Central and Eastern Europe', pp. 44–45, Copyright (1999), with permission from Elsevier.

- The *fiscal cost of subsidies* implies a higher tax burden or a reduction in government expenditures elsewhere. In an era of increasing fiscal stringency all over the world, this is a primary consideration.
- The *presence of subsidies tends to keep high-cost producers in business*, disguising the need for improvements in productivity (reductions in costs), thus contributing to making the economy less competitive internationally and its products more costly to domestic producers and consumers.
- A government policy which relies significantly on subsidies tends to encourage producers to invest time and resources in soliciting further favors from the government (*rent-seeking behavior*), rather than emphasizing efforts to increase the productivity of their own operations.
- In practice, the *benefits of subsidies often tend to be regressive*, that is, they accrue in a disproportionate manner to upper-income groups rather than to the poor in society.[3]
- The existence of subsidized modes of operation may *reduce the possibilities of developing institutions that do not rely on subsidies* and hence could be more viable in the long run. For example, the availability of subsidized credit through government banks may make it difficult for private banks or microfinance institutions to develop lending operations in the same areas.
- Sometimes, legislated subsidies are not backed up with sufficient government funding, and producers may hold back decisions on productive investments in the expectation of eventually obtaining the subsidy, and yet in some instances it may never materialize, and so the promise of the subsidy may have the perverse effect of delaying investments in the sector.[4]
- A more subtle but pervasive drawback of subsidies is that they tend to *encourage an anti-economic mentality* among the beneficiaries, which has the effect of hampering the development of efficient institutions and ways of operating. Again, an example is found in subsidized credit, which sometimes fosters a lax attitude about repayment obligations on the part of farmers, and this syndrome in turn makes it more difficult for them to become clients of commercial banks.

In light of these strong arguments against the use of subsidies, their justification has to be very solid indeed if they are to be employed as an instrument of policy. Nevertheless, there do exist cases in which the arguments in favor of subsidies are also compelling. The principal arguments *in support of subsidies* are the following:

- The role of subsidies in *poverty alleviation* finds virtually universal acceptance. The important questions in this regard are (a) how to target such programs effectively at the poor, and (b) how to find ways to assist the poor that will increase their own capacity for future economic improvement, and not just alleviate the most pressing current symptoms of poverty. The latter approach leads to a continuing need for the assistance, and promotes an attitude of dependence on the assistance on the part of the

---

3. This effect has been confirmed recently in the case of Egypt: 'Subsidized bread is available for virtually every household in Egypt at a fixed price in unlimited quantities. . . . Cooking oil and sugar are made available to consumers in monthly quotas through ration cards, covering close to 70 percent of the population. . . . In practice, there is no strong correlation between household income and access to subsidies through the ration card system'. Reprinted from *Food Policy*, **26**(1), Hans Löfgren and Moataz El-Said, 'Food subsidies in Egypt: reform options, distribution and welfare', p. 67, Copyright (2001), with permission from Elsevier.
4. This effect was observed in Colombia in 2002, where the unrealized hope of obtaining access to the program called 'Incentives for Rural Competitiveness', which could potentially have represented a 40% reduction in investment costs, caused deferral of a decision to invest in a major irrigation project whose funding had been guaranteed through the financial instruments of the National Agricultural Commodity Exchange (Bolsa Nacional Agropecuaria).

recipients. The former approach permits an eventual phasing-out of the assistance. This principle is widely recognized but sometimes difficult to put into practice.

- Cases of *environmental and economic externalities* provide another justification for subsidies, on both theoretical and practical grounds. Farmers who plant trees or build terraces to control soil erosion provide benefits not only to themselves, but to others in the same watershed, and so there exists an argument for society's sharing of the costs of such investments. The same argument holds for larger-scale tree planting for the purpose of sequestration of greenhouse gases.
- Temporary subsidies may be necessary at times to *facilitate a transition to a less-subsidized regime of policies*. A recent example on a large scale has been the use of subsidized privatization vouchers for the public in Eastern Europe, in order to bring to an end the continuing drain on the fiscal budget that was caused by State ownership and management of productive enterprises.
- In *emergencies*, such as those caused by natural disasters, benefits are usually provided to those who are affected without expecting repayment. However, care must be taken in using this instrument. For example, in many legal systems, declaration of an emergency in cases of, say, drought may exempt farmers from the requirement to repay their production loans, even if the crop loss is not total, and this, in turn, may cause difficulties for banks and discourage them from expanding their future operations in agriculture. A debate over precisely this issue developed in Nicaragua during the drought that was apparently caused by the El Niño phenomenon in the fall of 1997.
- Subsidies can be used to compensate for cases of *imperfect information*. A common example

concerns the case of farmers in remote areas who do not have reliable information about market prices, and therefore the government underwrites the cost of regular radio broadcasts of price data. For radio broadcasts to this group of clients, it may be difficult to find advertisers willing to cover the costs.
- Other kinds of *market failure* may require intervention with subsidies. However, care must be exercised. Often, regulation is a more appropriate response, and not all instances of market failure require government action. The phenomenon of 'asymmetric information' between lenders and borrowers (regarding a borrower's ability and willingness to pay) has received much comment in the literature,[5] and a common policy response is to develop a financial regulation framework that favors the growth of microfinance. However, in some cases the response has taken the form of subsidizing rural branches of banks, to encourage them to move closer to their potential rural client base.[6] Education and training are classic examples where government financial support is justified because an educational provider in the market cannot capture all of the broader benefits (externalities) to society, as mentioned in the quotation from D. Gale Johnson in Chapter 2.

These lists of the *pro* and *contra* arguments concerning subsidies are not necessarily exhaustive of the subject, but additional arguments should be scrutinized carefully before being added to the lists. In most practical decision situations, consulting the above lists should help clarify the advantages and disadvantages of a proposed subsidy.

The lists can be interpreted to develop guidelines applicable in specific circumstances. For example, subsidies for particular crops or agro-industries definitely should be avoided. They

5. See, for example, Karla Hoff and Joseph Stiglitz, 'Introduction: Imperfect Information and Rural Credit Markets – Puzzles and Policy Perspectives', *The World Bank Economic Review*, **4**(3), September 1990, pp. 235–250.
6. Mark Wenner observes that 'Temporary subsidies to defray the costs of establishing branch networks may be provided', in M. Wenner, *Rural Finance Strategy*, Sector Strategy and Policy Paper Series, Sustainable Development Department, Inter-American Development Bank, Washington, DC, USA, December 2001, p. 14.

create the problems indicated in the 1st, 2nd, 4th and 5th arguments against subsidies. The only exception that may be justified is a transitional subsidy to facilitate the privatization of an agro-industry (according to the 3rd argument in favor of subsidies), especially when large numbers of farmers become shareholders in the privatized entity.

On the other hand, subsidies for poverty alleviation do not have to be unrestricted. They can take the form of cost-sharing for productivity-enhancing services for small farmers. An example that increasingly finds favor in various countries is the provision of vouchers or other forms of subsidies to poor farmers for the purchase of private extension services, in a context in which better-off farmers pay the full cost of such services.

3.2.3 The Issue of Support for Agriculture

The arguments in the two foregoing lists usually would be considered in the context of proposals for individual policies or programs. There is another class of reasoning that applies to the sector as a whole. If the discussion about the uniqueness of the agricultural sector in Chapter 2 is correct, especially in the sense that allowing agriculture to falter would create irreversible labor–market effects and impose high costs on all of society as a result of excessive rural–urban migration, then there are grounds for considering a policy of generalized support for the sector. Indeed, almost all of the more industrialized economies subsidize their agricultural sectors, many of them extensively. The irony is that the less-developed economies, which have a greater problem of rural poverty, often implicitly tax their agriculture rather than supporting it. Part of the historical reasoning for this strategic approach was reviewed in Chapter 1, but today it appears inappropriate in most instances.

The World Trade Organization (WTO) accords make provision for domestic support for agriculture, as long as it does not distort prices and markets. The issue is the extent to which developing countries can and will avail themselves of this opportunity. The case for generalized support to agriculture has been advanced not only by politicians but also in the economic literature. Peter Timmer has written:

> Because the [international] prices ... are depressed by the dumping of surplus commodities generated from subsidies in rich countries, the undervaluation of the agricultural sector in poor countries is even more severe than would be apparent in a world of free trade. ...
>
> It has long been clear that political discrimination pushes the domestic value of agriculture in developing countries below its value in markets at the border. ... however ... border prices *themselves* undervalue the contribution that agriculture can make to growth in the early stages of development. If agriculture is critically important to stimulating and sustaining rapid economic growth, those countries that fail to correct this discrimination exact a heavy toll in economic performance. Furthermore, the poorest countries will suffer the most. ...
>
> There is widespread agreement that agricultural protection in rich countries depresses world prices for many commodities. ... World prices for staple grains do not reflect the importance to countries of maintaining food security. ... The special role of the agricultural sector in alleviating poverty is ignored in the market value of agriculture ...[7]

If these arguments for support for the sector in general are accepted, then the operational questions become the following: (a) *how much support is appropriate?*, and (b) *what are the most efficient ways to extend the support?*

In asking how much support is appropriate, it must always be borne in mind that part of the population must pay for the support. It is the urban population and, in terms of sectors, mostly the service sectors who would pay, either through

---

7. Reprinted from *Food Policy*, **20**(5), C. Peter Timmer, 'Getting agriculture moving: do markets provide the right signals?', pp. 456 and 459–561, Copyright (1995), with permission from Elsevier.

higher food prices or through higher taxes, or through both mechanisms. This consideration alone will tend to limit the amount of support for the sector, through the normal workings of the political process. In reality, net support to agriculture has tended to be very low and even negative in many developing countries, especially when the effects of exchange rate and tariff policies are taken into account.

If agricultural support is proposed in order to offset the consequences for commodity prices of international subsidies, then calculations of the quantitative effects of those subsidies can be made, on the basis of published estimates of their effects on world prices, combined with data on the quantities of production of the affected commodities in the country concerned. For example, an OECD study concluded that eliminating agricultural subsidies in all countries would raise the price of wheat by 30%, of coarse grains by 19%, of sugar by 59%, of tea by 17.5%, of dairy products by 53%, and of cotton by 16%. Some other commodities' prices would rise by less, for example, rice by 6%, and some would decline (mainly coffee and cocoa).[8] More recent estimates show distortion levels that are lower but still significant. As of 2000, elimination of all subsidies in wheat would have raised its international price by 18%, in rice by 10%, in other grains by 15%, in oilseeds by 11%, in sugar by 16%, and in meat and milk products by 22%.[9] Accordingly, for a wheat-producing country, multiplication of its average quantity of wheat produced by 18% could constitute one estimate of the amount of support for agriculture. Of course, other commodities should be included in such calculations as well – and the estimates of the international market effect of subsidies would have to be updated.

If poverty alleviation is to be accomplished by programs and policies that stimulate agricultural growth, rather than transitory forms of assistance, then a possible indicator of the fiscal magnitude of the programs would be the 'poverty deficit', namely the difference between poor households' income levels and the poverty line, or minimum acceptable income level, summed over rural households. For establishing the magnitude of support for specifically agricultural programs, the relevant sum should be the *difference* between the poverty deficits in rural and urban areas.

These are illustrations of possible ways to make calculations of budgetary magnitudes that would correspond to the concept of generalized agricultural support, as justified by the arguments in this present chapter and Chapter 1. They are minimum estimates as they do not take into account the economic externalities that arise from slowing the rate of rural–urban migration, also mentioned previously. Those externalities constitute additional grounds for supporting the sector's development.

Efficiency considerations should be foremost in designing programs of agricultural support. They are increasingly taken into account in all agricultural strategies and policies. For example, they underlie the recent moves of the European Union to reduce the kinds of support that affect commodity prices and to increase direct income support to farmers. Direct support to factors of production does not distort price relationships from their market levels, and therefore it does not induce farmers to invest in products which may have unpromising prospects for being competitive in the long run. For this reason, it does not interfere with the efficiency of the market in allocating productive resources.

The arguments against subsidies listed above are relevant to efficiency concerns. In other words, the fiscal support provided to the sector should not create the types of problems indicated by these arguments. The risk is that the sector could

8. I. Goldin, O. Knudsen and D. van der Mensbrugghe, *Trade Liberalization: Global Economic Implications*, OECD, Paris, 1993. The US Department of Agriculture and other institutions provide regular updates of these estimates.

9. Mary E. Burfisher, ed., *The Road Ahead: Agricultural Policy Reform in the WTO–Summary Report*, Market and Trade Economics Division, Economic Research Service, U.S. Department of Agriculture, Agricultural Economic Report No. 797, January 2001, p. 8.

be made less competitive and accordingly its growth prospects reduced.

In conclusion, in developing and transitional countries there are valid reasons for making fiscal outlays to support agricultural development, more so than for other sectors, but great care must be exercised in the design of the policies and programs which would convey support to the sector. The topic of appropriate design of policies is the main subject of this volume.

Equally, attention should be directed to possibilities for raising more revenues to support infrastructure development, agricultural research and other programs in the sector. Commodity taxes are not advisable because of their distortive effects on incentives. Often efforts are made to improve the administration of income taxes, but with the lack of reliable accounts on most farms in developing countries, that route to revenue collection always will be difficult in rural areas. Among the more viable options are rural land taxes (on a per hectare basis), which are discussed extensively in Chapter 5, and partial user fees for services. In effect, the decentralization to users of the operation and maintenance of irrigation services is a form of levying higher user fees. Similarly, under privatization of extension services, measures can be taken to require that medium- and large-scale farms pay at least part of the cost of the services. Other revenue measures, including the participation of farmers in the cost of agricultural research, are discussed throughout this volume. Thus, fiscal policy for agriculture should not be viewed only from the expenditure side, but nevertheless the case for at least a degree of net support to the sector is a strong one.

3.3 IMPROVING THE INCOMES OF THE RURAL POOR

A central question for poverty alleviation programs is how to devise a set of policies that will help place the rural poor on a self-sustaining growth path, rather than simply continue to provide handouts to meet their immediate needs, urgent though these may be. Programs of food assistance and health care are required for the poorest groups in the population and must be continued, but they do little to assist them to develop capabilities to eventually satisfy their needs through their own efforts. In addition to programs that respond to the *symptoms* of poverty (such as malnutrition and high incidences of disease), policies are required that diminish the *causes* of poverty, i.e. that will enhance the earning capacity of low-income households. Better education is often cited as fundamental in this regard. In the agricultural sector, other possibilities arise, including, but not limited to, improved access to cultivable land, improved access to technology and better farmer training, and improved access to production credit.

This question has broader ramifications, for it is part of the concern of **how to design policies that not only promote growth in general but which improve the lot of the poor**, or at least prevent a deterioration in their condition while other groups prosper. Traditionally, stabilization and growth have been viewed as falling in the purview of *policy*, but the alleviation of poverty has been relegated to the realm of *programs* and *projects*. The question is, can *policies* be developed that simultaneously promote income growth in general and growth for the poor?[10]

Part of the answer is found in the research reported in Chapter 1 which concludes that agricultural growth is more effective than industrial growth in reducing poverty as well as contributing to the growth of the entire economy. However, given the magnitude of the rural poverty problem in most developing countries, that answer alone is not sufficient. Hence agricultural policies need to incorporate a special focus on poverty alleviation.

Another part of the answer to this strategic question can be found in seeking to target more effectively the fiscal subsidies that are present in

---

10. An example of the pervasiveness of this question is found in Chile. Valdés comments: 'Perhaps the most complex unfinished business in Chilean agriculture after 15 years of reform is how to address the needs of small farmers, who are geographically scattered, usually located in disadvantaged areas, and remain separate from the sector's newly found dynamism' (Alberto Valdés, 'Mix and sequencing of economy-wide and agricultural reforms: Chile and New Zealand', *Agricultural Economics*, **8**(4), June 1993, p. 302).

all economies. One of the keys is identifying where these subsidies lie and what their incidence is (over income groups), and another key is finding ways to reorient that incidence. In addition to fiscal outlays, many other examples of policies can be found that, in effect, discriminate against the poorer rural families. For example, in Honduras until 1992 it was illegal to give full title to farms with less than five hectares, in spite of the fact that the vast majority of the farms in the country were that size or smaller. This situation, of course, made it much more difficult for small farmers to obtain production loans than those with larger farms who had full ownership. A similar situation still prevails in Estonia, where the small household plots, a heritage of the era of collective farms, are both the most numerous and the most productive forms of agriculture in the country, but under the existing legislative framework there is no mechanism for titling them.

In Honduras, El Salvador, the Dominican Republic and other countries, the majority of beneficiaries of the agrarian reform have been obliged to receive farmland in a collective form, without being consulted as to their preferences. In the case of El Salvador, a survey showed that the productivity of collective units was markedly inferior to that of farms distributed to individuals.[11] Additionally, in Honduras before 1992 it was very difficult for rural women to obtain access to land through the agrarian reform, which was oriented almost exclusively toward men. These issues and others concerning access to land are explored in Chapter 5.

In effect, these kinds of policies, which were embedded in legislation, made the rural poor into second-class citizens, without the full property rights that other citizens enjoyed. Changing such policies can help bring poor rural families into the economic ambit which prevails for most of the rest of the citizenry, thus giving them improved opportunities to better their economic condition through their own efforts.

Other examples of *agricultural growth policies aimed at the poor* include the following:

- Systems of certificates of deposit (crop liens) which small farmers can participate in. Normally, the existing banking legislation enables farmers with large harvests to obtain loans for crop storage but often it is not adapted to the situation of small farmers (it may require, for example, separating stored grains by owner, which is impractical when dealing with many small lots).
- Creation of a second-storey land fund, which can finance purchases of small farms, with a subsidy in the price to the buyer. Because of the transaction costs, and uncertainty about legal backing for foreclosure on small units in the event of default, many banks are reluctant to issue land mortgages for small farms, and thus there is scope for government to complement the workings of the market in this area.
- Market-assisted land reform, which is a broadening of the concept of a land fund, so that communities participate in identifying beneficiaries and, in some cases, potential beneficiaries are required to put forth a farm development project.
- Improvement of land rental markets, since rentals are a principal avenue for poor rural families to gain more access to land.
- Privatization of grain silos and other facilities in a way that enables small farmers, and large farmers too, to become shareholders in the enterprise.
- Provision of subsidies via vouchers to enable poorer farmers to purchase private extension services.
- Decentralizing agricultural research, and involving farmers more directly in agricultural research and extension through participatory approaches.
- Development of a viable rural financial system oriented toward the needs of small-scale producers.
- Investments in small-scale irrigation and formation of water users' associations for participatory management of irrigation projects of all sizes.

11. R. D. Norton and M. Llort, *Una Estrategia para la Reactivación del Sector Agropecuario en El Salvador*, Fundación Salvadoreña para el Desarrollo Económico y Social (FUSADES), San Salvador, October, 1989.

- Institutional reforms to improve respect for contracts, since poor farmers frequently are victims of breaches of contract, for example, by export brokers. In some cases, it may be necessary to develop a system of rural or agrarian tribunals that can act swiftly, and at low cost to plaintiffs, in reinforcing the integrity of contracts.
- Education and training programs oriented toward poor rural families, especially toward rural women.

These are examples of policies that reduce the economic distance between small farmers and the institutions that serve the sector, providing them with more nearly equal economic opportunities, as compared to those enjoyed by large farmers and their urban counterparts. They and other policies for poverty alleviation are discussed throughout this volume.

3.4 GENDER AND AGRICULTURAL DEVELOPMENT

Gender issues receive considerable attention throughout this text, for two reasons, namely, (a) discrimination on the basis of gender is very widespread in developing agriculture, and (b) in addition to questions of justice and fairness, the evidence is now clear that gender biases against women hinder agricultural development and reduce the nutritional status of rural households.

Gender bias is manifested in many different ways, including diminished access to land and credit, little attention to women's needs as producers by agricultural research and extension services, exclusion from most decision-making regarding irrigation systems, and less access than men have to agricultural inputs.[12] Bias sometimes is embodied in legal codes that, for example, may recognize only the head of household for many purposes, or which give women unequal inheritance or divorce rights. Equally, bias is present in traditional, unwritten codes of conduct and conflict resolution. It is frequently found in the design and implementation of agricultural services and projects. Agricultural extension services, for example, typically deal almost exclusively with male farmers, and extension agents do not schedule their visits at a time which is convenient for women in light of the many household duties that women shoulder in addition to work in the fields. Many examples of this bias are cited in succeeding chapters (especially in Chapters 5, 7 and 8). Typical illustrations, characteristic of many countries, are found in Uganda and India:

> ... civil law in Uganda provides for equal rights in divorce – but customary law prevails in the division of conjugal property, and divorced women are unable to retain access to land.[13]

> ... most daughters in all Indian States do not inherit land, though legally eligible. ... in Bihar, India, some Ho women remain unmarried to keep this access.[14]

Studies have shown that rural women's time is exceedingly scarce and therefore is valuable. As a consequence, agricultural research that is directed toward ways of reducing the time requirements of household duties results in higher agricultural growth rates because more women's time is liberated for agricultural labor (Chapter 8). Country-level studies have shown that:

> by hindering the accumulation of human capital in the home and the labor market, and by systematically excluding women or men from access to resources, public services, or productive activities, gender discrimination diminishes an economy's capacity to grow and

---

12. See, for example, FAO, *SEAGA Macro Handbook: Gender Analysis in Macroeconomic and Agricultural Sector Policies*, Food and Agriculture Organization of the United Nations, Rome, draft, March 2002, pp. 39–40.

13. The World Bank, *Engendering Development – Through Gender Equality in Rights, Resources and Voice*, Policy Research Report, The World Bank, Washington, DC, USA, 2002, p. 16.

14. IFAD, *Rural Poverty Report 2001: The Challenge of Ending Rural Poverty*, Oxford University Press, Oxford, UK, 2001, p. 86.

raise living standards. . . . In households in Burkina Faso, Cameroon, and Kenya more equal control of inputs and farm income by women and men could raise farm yields by as much as a fifth of current output.[15]

Women's education is one of the key factors in reducing undernourishment and increasing economic growth:

. . . a recent study by IFPRI, which examines the relationship between a variety of factors and reductions in the number of underweight children in 63 developing countries between 1970 and 1995 . . . indicates that the statistical explanation of lower numbers of underweight children centers on [among other factors] level of women's education (43 percent) . . . [and] women's status in society (12 percent).[16]

Low investment in female education also reduces a country's overall output. One study estimates that if the countries in South Asia, Sub-Saharan Africa, and the Middle East and North Africa had started with the gender gap in average years of schooling that East Asia had in 1960 and had closed that gender gap at the rate achieved in East Asia from 1960 to 1992, their income per capita could have grown by 0.5–0.9 percentage points higher per year.[17]

In the long run, some observers feel economic development itself helps correct gender inequalities:

Rising income and falling poverty levels tend to reduce gender disparities in education, health and nutrition. Higher productivity and new job opportunities often reduce gender inequalities in employment. And investments in basic water, energy, and transportation infra-

structure help reduce gender disparities in workloads.[18]

However, to reduce gender bias in the short and medium term, and also reduce its drag on economic growth, fundamental reforms are required in institutions and legislation, in ways of designing and carrying out programs and projects in rural areas, and in monitoring and evaluation of those activities and policy reforms. Large-scale training and capacity building efforts are required in order to effect these changes, accompanied by a strong political commitment Isolated projects for gender improvement may not be useful because other barriers to women's participation remain in place. Therefore the only viable approach is *gender mainstreaming*, starting with comprehensive gender analyses of the sector.[19]

The importance of capacity building cannot be overemphasized:

Although most governments and their partners have explicit commitments to the integration of gender into agricultural strategies, there has been little capacity building in gender analysis for the agricultural sector as a whole. Many of the gender inputs are oriented to micro level issues without linkage to overall agricultural priorities and processes. There is still a need to strengthen sector-wide gender capacity in most Ministries of Agriculture and among the policy formulation and management units of donor institutions. A recent review by the World Bank[20] shows that gender analysis capacity is generally weak in Ministries of Agriculture. A gender-sensitive institutional analysis of Ministries of Agriculture should assess their capacity to integrate gender into the agricultural policy process (research and

---

15. The World Bank, 2002, p. 11.
16. FAO, *The State of Food Insecurity in the World 2001*, Food and Agriculture Organization of the United Nations, Rome, 2001, p. 7.
17. The World Bank, 2002, p. 11.
18. *Op. cit.*, p. 2.
19. FAO, 2002, pp. 41–43.
20. The World Bank, *Gender, Growth and Poverty Reduction*, Washington, DC, USA, 1999.

strategy setting, policy formulation and implementation).[21]

One of the most valuable steps that international development organizations can take in the gender area is to ensure that the design of all of their projects commences with a gender analysis of constraints and issues in the domain of operation of the project. The concluding section of Chapter 5 presents a partial list of the kinds of questions that a proper gender analysis would confront in regard to land tenure issues. Through measures such as gender analysis, a much greater awareness of the seriousness of gender constraints can be developed, and building awareness is the first step to solving the problem.

3.5 SELECTED ISSUES IN PRIVATIZATION

Privatization is carried out for many reasons, some of the most important of which are decreasing the burden on the government's budget, bringing into the firms more skilled management and improved technologies, and attracting private investment. Improvements in the performance of an enterprise after its privatization do not necessarily arise from the nature of ownership *per se*, but rather because:

the switch from public to private ownership results in more precise, and more measurable, objectives on the part of the owners which then create the environment and incentives to monitor and control management more effectively. An important additional aspect of this argument is that, under private ownership, firms will only remain in existence as long as they are viable. If they are not their resources will be re-allocated (through the market mechanism) to more efficient uses. In contrast, un-economic State-owned enterprises (SOEs) are frequently long-lived, maintained as a result of access to soft credit combined with political and other non-economic pressures which, in additional to constituting a drain on

Private ownership itself makes a difference. Some State-owned enterprises have been efficient and well managed for some periods, but government ownership seldom permits sustained good performance over more than a few years. The higher probability of efficient performance in private enterprise needs to be considered in choosing whether to invest public funds in SOEs or in health, education and other social programs. (S. Kikeri, J. Nellis, and M. Shirley, Privatization: The Lessons of Experience, *The World Bank, Washington, DC, USA, 1992, p. 1.)*

the financial resources of governments, also constrain efficient allocation of scarce financial, and especially human, resources.[22]

Apart from ideology, one of the common motives for *not* privatizing has been fear of potentially monopolistic behavior on the part of firms that are transferred to private ownership. In short, the fundamental reasons *for* privatization are the fiscal cost of maintaining SOEs – although it should be remembered that improving the *net* flow of revenues to the Treasury, including taxes, requires improvement in the profitability of the privatized entity – and the need to improve the enterprises' productivity and economic growth performance. The basic counterargument is the concern about creating a non-competitive market structure. In agro-industry, such a concern is, concretely, that monopolistic or oligopolistic processing firms will offer lower prices to producers and charge higher prices to consumers.

Many SOEs around the world have performed well for considerable periods of time, but it is difficult to sustain that performance indefinitely. One of the reasons is that government committees or Ministers may control investment decisions, rather than the enterprise itself. Another is that selection of managers is not always carried out on grounds of technical competence, while a

---

21. FAO, 2002, p. 45.
22. C. Adam, 1994, p. 138.

third reason is that rates of staff remuneration are often not linked to performance.

On the other hand, privatization carried out improperly can lead not only to the ills of market concentration but also to windfall gains for the new owners or managers, fueling social discontent with a policy of opening toward the market. At stake here is the nature of the distribution of wealth in society, and how privatization policy may affect it. Kikeri *et al.* consider that satisfaction of two fundamental conditions can make privatization work well: existence of a reasonably competitive market, and a capacity in the government to regulate industry. Only if the market is truly competitive can the regulatory capacity factor be ignored.[23]

These guidelines may be applicable in many parts of the world, but in countries where State ownership has been extensive, as in Eastern Europe and the Former Soviet Union countries, experience suggests that perhaps the drawbacks of non-competitive market forms are not as damaging to the economy as the inefficiencies of continued State ownership:

> Comparative analysis of the Czech and Slovak, Polish, Russian and Lithuanian mass privatization models indicate that during the initial steps of privatization it is better to run the risks of imperfect competition and markets, and to accelerate the process, than it is to delay and possibly derail privatization.[24]

The same authors point out, however, that in such cases privatization must be followed by structural reform: 'Even where mass privatization occurs and the sell-off process is accelerated . . . there is clear recognition that this is only the first phase of structural reform. Where privatization occurs with little else in the way of structural reform, as it has in Russia, the process is vulnerable to charges that it has failed. . . . post-privatization structural change and adjustment

issues will require careful thought and prudent planning' (*op. cit.*, p. 47).

Finally, Lieberman *et al.* emphasize that 'privatization needs to be viewed as part of a more comprehensive reform program' to 'create the base for a market economy' (*ibid*). This is perhaps the most fundamental objective of privatization programs.

In practice, the operational question often is not so much **whether** to privatize as **how** to privatize. For agricultural storage and processing facilities, it can be a choice between a strategy to attract the investor with the deepest pockets versus one oriented at encouraging a broad base of ownership through creating widely dispersed shareholdings. The former is often the aim of privatization conducted via public auctions, and the latter is accomplished through special legislation which defines the types of shares and their rules for distribution and sale.

If there are concerns about potential monopoly power, sale to a single bidder is likely to exacerbate them. In contrast, sale of shares to, say, a large number of farmers would tend to prevent their being exploited by the newly-privatized agro-processing enterprise. On the other hand, in principle sale at public auction is the only way to guarantee that the asset being sold commands a market price, that is, a price which reflects its true economic value.

If it is anticipated that privatizing State grain silos will lead to reinforcement of an existing oligopoly in domestic grain trade, is it preferable to sacrifice the theoretical advantages of selling the facilities at public auction, in favor of a direct sale to groups of farmers? If so, are subsidized sales terms justified as a way to end the continuing annual subsidy in the State's operation of the silos? The conceptual literature offers no firm guidance on such questions. In the blunt words of Stanley Fischer, 'Given the magnitude of the task [of privatization], it would be a mistake to discourage any potentially viable form of privatiza-

23. S. Kikeri, J. Nellis and M. Shirley, *Privatization: The Lessons of Experience*, The World Bank, Washington, DC, USA, 1992, p. 5.

24. I. W. Lieberman, A. Ewing, M. Mejstrik, J. Mukherjee and P. Fidler (Eds), *Mass Privatization in Central and Eastern Europe and the Former Soviet Union, A Comparative Analysis,* Studies of Economies in Transformation, No. 16, The World Bank, Washington, DC, USA, 1995, pp. 47–48.

tion that is not theft'.[25] In Honduras in the early 1990s, policy makers opted for sale of silos to large numbers of producers, at a subsidized price, and the experience was considered a success. In that case, it was clear that sale of the facilities to the highest bidder would indeed have resulted in reinforcing an existing oligopoly in grain trade.

One of the keys to success of privatization is building adequate regulatory institutions and capacity. In the words of Pranab Bardhan:

> While the process of deregulation should continue [in agricultural reform programs], in some respects the regulatory powers of the State have to be enhanced, for example, in ensuring the implementation of the avowed purpose of reforms to increase competition. Otherwise privatization often involves replacement of a public monopoly by a private monopoly.[26]

Desirable as privatization may be in many cases, implementing it is not necessarily a simple task. For assets which are divested at public auction, it is essential to take pains to ensure that the process is transparent, and that the winning bid is honored.

An issue which sometimes arises is that the domestic private sector may not have the managerial capacity or financial depth to take over ownership of a significant number of facilities. This problem has arisen in acute form in Malawi,[27] Mozambique, Guyana and other countries where the State has had a preponderant role in managing the economy until recent years. Francesco Goletti and Philippe Chabot have commented on this issue for the privatization of agricultural marketing, in the Central Asian context:

> If market reforms are undertaken, a thriving and efficient private sector will not necessarily develop and engage in those functions previously performed by the public sector. In the presence of market failures and infrastructure bottlenecks, the effects of market reforms on agricultural marketing may be adverse. Sometimes, governments give exclusive rights to one major private or domestic firm, limiting the access of technology to farmers, and institutionalizing barriers to entry. In other cases, imports are limited to particular brands, resulting in a restricted access by farmers to the broader range of choices available in the international markets. The private sector may not have an incentive to participate in the marketing of agricultural inputs because of thin markets or lack of credit. In the Central Asian case, the major impediments seem to be regulatory and physical (*i.e.*, the presence of large grain elevators and cotton ginneries inherited from the Soviet era).[28]

The alternative of allowing foreign capital to acquire most of the privatized facilities is not always considered acceptable. One solution is to provide generous terms on which new shareholders may acquire their participation in facilities, but it cannot be a complete solution since one of the aims of privatization is to bring in significant amounts of new capital. An optional solution is to combine foreign and domestic shareholding, by auctioning a given portion of the enterprise's assets and using a formula for dis-

---

25. Stanley Fischer, 'Privatization in Eastern European Transformation', Working Paper IPR6, Institute for Policy Reform, Washington, DC, USA, March, 1991.

26. Pranab Bardhan, 'Institutions, reforms and agricultural performance', in Kostas G. Stamoulis (Ed.), *Food, Agriculture and Rural Development: Current and Emerging Issues for Economic Analysis and Policy Research*, Economic and Social Department, Food and Agriculture Organization of the United Nations, Rome, 2001, p. 155.

27. For comment on this issue in the case of Malawi, see C. Adam, 1994, pp. 150–151.

28. Reprinted from *Food Policy*, **25**(6), Francesco Goletti and Philippe Chabot, 'Food policy research for improving the reform of agricultural input and output markets in Central Asia', pp. 675–676, Copyright (2000), with permission from Elsevier.

tributing another portion of them among the domestic public. An Eastern European solution has been to foster the creation of holding companies, by providing the general public with tradeable privatization vouchers, and allowing them to be used for investing in the holding companies, or funds, instead of directly in the privatized facilities.

Privatization issues are pervasive in the agricultural sectors of many countries. Creating security of land tenure, of one form or another, for private farmers is still a challenge throughout the world. Farmer participation in irrigation systems also can involve considerations of privatization. It now is widely accepted that management of such systems should be devolved to the local level, generally to groups of irrigation users. They often are required to finance part or all of the maintenance costs through fees levied on themselves for that purpose. The two central ideas are straightforward, namely (a) that farmers will be more willing to pay maintenance fees if they can manage the corresponding expenditures themselves, and thus be reassured that the fees go to the intended purpose in an efficient manner, and (b) maintenance is likely to be more effective if it is carried out by those who have the most direct interest in the system's long-term viability.

An issue that is not recognized as often is who should be the owner(s) of the system – national government, local government, farmers or someone else? It can be argued that as long as farmers are not full owners of the system, their interest in its maintenance will not be a strong as it could be. If farmers were shareholders in the system (main canals, pumps, etc. – all components excepting the canals within each property), then they could sell their shares along with their land, in the event of a decision to leave farming or leave the area. Equally, their children could inherit their shares. Hence they would have an interest in the potential *capital gains* of the system as well as in its year-to-year functioning for irrigating their fields. This additional interest would be expected to increase their commitment to the system's proper maintenance and management.

Nevertheless, the most typical policy is to leave system ownership in government hands and require users to shoulder the burden of system maintenance. Since this kind of arrangement leaves incomplete the incentives for maintenance, there would appear to be grounds for considering alternative approaches. This question is discussed in Chapter 6 where examples of farmer ownership of irrigation systems are given.

Privatization considerations can be extended to the area of farm services as well, and they are very relevant to questions of the structure of the agricultural financial sector. These issues are also considered in subsequent parts of this volume.

3.6 PRINCIPAL ASPECTS OF THE LEGAL FRAMEWORK

Sound legislation and a well-functioning judicial system are essential underpinnings to all economic activity. Judge Richard Posner has made the following observations about legal systems and economic growth:

If it is not possible to demonstrate as a matter of theory that a reasonably well-functioning legal system is a necessary condition of a nation's prosperity, there is empirical evidence showing that the rule of law does contribute to a nation's wealth and its rate of economic growth. . . . It is plausible, at least, that when law is weak or nonexistent, the enforcement of property and contract rights frequently depends on the threat and sometimes the actuality of violence . . . , on family alliances that may be dysfunctional in the conditions of a modern economy, and on cumbersome methods of self-protection. These are costly substitutes for legally enforceable rights, as are the discredited 'command-and-control' methods used in communist economies. The hidden costs of these substitutes are a bias against new firms, which have no established reputation to persuade clients that they are reliable, and a bias in favor of simple, simultaneous exchanges over more complex transactions because it is unlikely that legal remedies can be invoked against nonperformance. . . . The cumulative costs of doing without law in a modern economy may be enormous.[29]

In a market economy, the basic concepts of *ownership* and *contractual obligations* need to have clear legal support, and the judicial system needs to offer appropriate opportunities for rapid redress in cases of transgression. *Forms of economic association* also require a solid legal framework. Oddly, in many countries the concept of joint stock companies, or limited liability companies, is not fully spelled out in the corpus of legislation for the agricultural sector. Preference is sometimes given to co-operative forms of association, partly because of tradition and partly out of concern that one individual or a few could come to dominate a joint stock company by purchasing the shares of others.

Co-operatives have been successful for generations of farmers in some countries and subsectors, as in the case of grain marketing in Western Canada and dairy marketing in Denmark and elsewhere. Marketing and input purchase have been the strongest areas for co-operatives, while production co-operatives have had a much more mixed record, although some of them have been successful as well. On the other hand, co-operatives have two distinct disadvantages, i.e. (a) the rules for distribution of the gains among members are not always clear, nor are they always linked to the intensity and effectiveness of each member's contributions, and (b) they are not attractive to lenders, because a co-operative may be able to escape its repayment obligations simply by dissolving and reconstituting itself under a different name. For this last reason, co-operatives often find it more difficult to raise funds than joint stock companies.

There would appear to be room for more creativity in crafting the legislation governing forms of association in agriculture. For co-operatives, the rules governing the distribution of net earnings could be made more precise, and also responsibility for financial obligations could be placed on each member. For joint-stock companies, rules preventing the concentration of shares in few hands could be drawn up, and current members of the enterprise could be given the first option to bid for the shares of a member who has decided to leave, thus minimizing the chances that persons outside the entity would come to control it. Such legislation has, in fact, been put in place in some countries.

The objective in this area should be to improve the *entrepreneurial character* of the association.

Major types of legislation that can be relevant to agriculture in this regard and others include *the commercial code, the labor code, gender aspects of legislation (especially, but not only, family legislation and land tenure legislation) and tax legislation.* Another set of laws that is pertinent to agriculture's development prospects is *consumer protection legislation.* It is increasingly important all over the world to legislate and enforce adequate safeguards regarding food quality and safety, not only out of domestic concerns but in order to be able to penetrate export markets. However, there is another realm in which caution needs to be exercised, and that is the putative protection offered to consumers by anti-hoarding laws. Such laws are intended to prevent 'speculators' from causing undue price increases during the season of scarcity of staple crops. However, they can have the perverse effect of discouraging investment in storage and marketing facilities, and therefore increasing the amplitude of the seasonal fluctuations in prices. Storage and marketing are important economic activities that relocate products in time and space. Accordingly, they have a cost, and that cost needs to be remunerated if those activities are to be encouraged. Attempts to legislate limits on private trading in food products is one of the main causes of underdeveloped marketing systems in many lower-income countries.

Legislation is crucial in *the financial field.* To facilitate agricultural lending, the concept of *collateral* needs adequate definition. Often its legal

---

29. Richard A. Posner, 'Creating a legal framework for economic development', *The World Bank Research Observer*, **13**(1), February 1998, p. 3.

definition excludes crops and livestock, thus severely limiting a farmer's chances of borrowing to cover production costs. More generally, **bank supervision and prudential norms** is another critical area for legislation. In agriculture, where local savings and loan co-operatives may provide an important source of production finance, those norms need to be defined in a flexible, but nonetheless solid, way in order to avoid stifling the growth of such co-operatives or associations. Among the financial laws, one of the most important is the **bankruptcy law**. Without clear, firm rules governing the disposition of insolvent enterprises, the development of finance will remain hampered.

Enforcement of laws is often weak in developing countries, and poor families are the most disadvantaged in access to legal resources. Judicial systems urgently need strengthening, and it can be important to establish rural tribunals in order to make justice swifter and more accessible to all.

These and other legislative issues are explored extensively in the succeeding chapters in the context of specific areas of agricultural policy. Chapter 5, for example, reviews many aspects of the legal framework for land tenure policy, while Chapter 7 comments at length on the legislation for bank regulation in the rural sector.

DISCUSSION POINTS FOR CHAPTER 3

- Macroeconomic policy strongly influences both the incentives for agricultural production and the real incomes of rural households through its effects on real or relative prices, or the inter-sectoral terms of trade in an economy.
- Macroeconomic policy also influences the incentives for agricultural exports and can establish a framework that is conducive to rural finance and investment.
- From the viewpoint of agricultural development policy, the most important instruments of macroeconomic policy are exchange rate policy, trade policy, tariff policy, taxation policy, fiscal expenditure policy, monetary policy and the regulatory framework.

- Falling real agricultural prices are injurious to production and rural standards of living. In the long run (as pointed out in Chapter 1), a policy of lowering domestic food prices worsens the problem of rural poverty and reduces agriculture's ability to contribute to the economy's overall growth rate.
- Both explicit and implicit subsidies have played important roles in agricultural policy in the past, sometimes for the purpose of offsetting biases against agriculture in other areas of macroeconomic policy. However, often such subsidies are regressive in their effects over income classes.
- Other arguments against subsidies include the following: they tend to be allocated to the least competitive industries, subsidies are hard to eliminate in the future, their fiscal cost to the nation can be high, they encourage rent-seeking behavior, they reduce the possibilities of developing institutions that are viable in the long run without subsidies, the expectation of obtaining subsidies can lead to the deferral of worthwhile projects and, especially in the case of subsidized credit, they can foster counterproductive economic attitudes among the beneficiaries of the subsidies.
- Arguments in support of subsidies include the following: they can be crucial for reducing poverty, they are justified when they compensate for environmental externalities, when used in a transitory way they can help facilitate a transition to an economic regime that is less dependent on subsidies in the long run, they are necessary in cases of natural disasters, and they can compensate for instances of imperfect information and market failures.
- Given agriculture's unique contributions to growth and poverty alleviation, and the social costs of rural–urban migration, there are solid arguments in favor of generalized support for the sector.
- Basic questions for policy makers in each country include the amount of such sectoral support and the channels through which it is extended, bearing in mind efficiency considerations and the associated costs to other sectors.

- Historically, attempts to reduce poverty have been carried out via *programs* and *investment projects* but there has been little emphasis on designing *growth policies for poverty reduction*. In agriculture, there are many examples of policies whose primary beneficiaries would be the poor.
- Gender bias is pervasive in legislation, programs and projects in developing countries. Such biases are not only unfair but they also reduce a country's capacity for economic development.
- Studies have shown that greater attention to the education of women increases a country's nutrition levels and economic growth rate, and that providing rural women with greater control over farm inputs and decisions increases farm yields.
- Many projects funded by international agencies also suffer from gender bias. A starting point for reducing gender bias is to carry out a gender analysis at the beginning of the design stage of each project and program.
- There are many State-owned enterprises in developing countries, but often they represent a drain on fiscal resources and a diversion of the country's energies away from more productive kinds of investment. Provided that concerns about the potential creation of private monopolies and oligopolies can be surmounted, privatization often improves a country's growth prospects.
- The manner in which privatization is carried out is crucial. A strong regulatory framework is a prerequisite for successful privatization. In addition, it needs to be borne in mind that in many developing and transitional economies the private sector may not have, in the short and medium run, the financial or managerial capacity to assume control over many State-owned enterprises. In some cases, transitional subsidies can play a role in the privatization process, particularly if large numbers of farmers or citizens are expected to become shareholders in the newly privatized facilities.
- Inadequate legal frameworks and inconsistent enforcement of laws is a major impediment to economic development. The basic concepts of ownership and contractual obligations require strong legal support in a market economy.
- The legislative framework also is fundamental in regard to forms of economic association, gender, land tenure, commercial relationships, finance and collateral, bankruptcy provisions, consumer protection and other areas.

4

Policies that Influence Producer Incentives

4.1 The Setting   55
4.2 Agricultural Prices and Their Determinants   56
  4.2.1 Basic Price Concepts   56
  4.2.2 Structural Determinants of
     Agricultural Prices   57
4.3 Trade Policy   58
  4.3.1 Basic Issues   58
  4.3.2 Tariffs in Developing Countries   62
  4.3.3 Export Incentives   67
  4.3.4 Trade Restrictions   68
  4.3.5 Policies for Food Aid   72
4.4 Exchange Rate Policy   74
  4.4.1 The Role of the Exchange Rate   74
  4.4.2 Exchange Rate Policy for Agricultural
     Development   77
4.5 Fiscal Policy and Economic Prices   80
4.6 Macroeconomic Policy Options for
  Agriculture   82
4.7 Sectoral Policies that Influence Agricultural
  Prices   86
  4.7.1 Price Controls   86
  4.7.2 Farm Support Prices   87
  4.7.3 Strategic Reserves and Grain Market
     Liberalization   89
  4.7.4 Instruments of Grain Storage Policy   92
  4.7.5 Market Development   93
  4.7.6 Farm Prices and Agro-Industries   94
  4.7.7 Input Markets   96
  4.7.8 Additional Considerations about
     Prices and Markets   97
4.8 Food Security, Agricultural Prices and the
  Rural Poor   98
4.9 Observations on Price Stabilization and
  Economic Development   103
Discussion Points for Chapter 4   104

4.1 THE SETTING

Food prices have been a source of concern for policy makers in all parts of the world and in all historical periods. They affect the welfare of farmers, consumers, middlemen, agro-industries, exporters and importers and, through indirect or multiplier effects, incomes and employment in many other sectors. More than two thousand years ago, the Governments of China began to buy grains from farmers at pre-established prices, and Roman emperors were always concerned to have the silos of Rome filled with wheat.

In poorer countries, expenditures on food constitute half or more of the household budget for a significant share of the population, and so food prices acquire a social and political importance that is difficult to appreciate from the perspective of richer countries. Increases in the price of bread or corn or rice or meat have caused riots in Egypt, Tunisia, Jordan, Nigeria, Zambia, Poland, the Dominican Republic, Venezuela, Indonesia and other countries in the last decade or so.

The price of bread played its part in the French Revolution:

> The four-pound loaf [of bread] that formed the staple of three-quarters of all French men and women and which, in normal times, consumed half their income, rose in price from eight sous in the summer of 1787 to twelve by October 1788 and fifteen by the first week of February. ... The doubling of bread prices ... spelled

Agricultural Development Policy Concepts and Experiences. R. D. Norton
© 2004 Food and Agriculture Organization of the United Nations
ISBNs: 0-470-85778-1 (HB)   0-470-85779-X (PB)   FAO Edition: 92-5-104875-4

destitution. . . . It was the connection of anger with hunger that made the Revolution possible.[1]

Every government in the world has acted to influence food prices in one way or another. In the market economy of the United States the price of milk is as much determined by decisions in Washington as by dairy farmers. The price of sugar is far above what a free market would establish. The European Community has so manipulated food prices that at times it has been embarrassed by mountains of surplus grains and dairy products and lakes of surplus wine. Japanese consumers pay for rice several times what they would were their government to allow unrestricted imports of that grain.

As these casual examples suggest, there is a fundamental irony in the world's patterns of governmental intervention in food markets: the richer countries, whose farmers constitute a very small portion of the population, tend to subsidize farmers and penalize consumers through artificially high food prices, and the poorer countries in which a much larger share of the population earns its livelihood from agriculture, often try to hold food prices below their international equivalents, further impoverishing their farmers in order to deliver benefits to their urban consumers. As stated by Jo Swinnen and Frans van der Zee, 'There exists a general tendency to discriminate against farming in poor countries and to subsidize farmers in rich countries'.[2]

In today's world economic environment, direct governmental interventions in food prices as well as other economic matters are being cut back. Nevertheless, political and economic concerns about prices of foods, and prices of other agricultural products and agricultural inputs, have not diminished, and more indirect approaches now are being utilized to respond to those concerns.

4.2 AGRICULTURAL PRICES AND THEIR DETERMINANTS

4.2.1 Basic Price Concepts

Agricultural prices can be viewed from many perspectives: at the farmgate, at the rural and urban wholesale levels, and at the consumer level; at harvest time and in the season of relative scarcity; at the border in the case of imports and exports, or at inland locations; by quality of the product, and so forth. Some classes of policies are aimed at reducing the difference between producer and consumer prices, through improvements in the efficiency of the marketing chain. Others try to reduce the seasonal fluctuations in prices, through better access to storage facilities and improved mechanisms for timely arrival of imports when needed. Yet others try to induce farmers and rural traders to improve the quality of the delivered product, thus obtaining a better average price.

These kinds of policies have an important place in the pantheon of sectoral policies, but there are policies that can influence the entire constellation of agricultural prices, moving them all upward or downward together. For this purpose, agricultural prices are best viewed from another perspective, that of *relative, or real, prices*. As mentioned earlier in this volume, real agricultural prices are calculated by dividing the nominal, or raw, agricultural prices by other prices, i.e. by those of other sectors, or by those of the economy as a whole. Real agricultural prices can be calculated for any level in the marketing chain, but for analyzing producer incentives, they usually are calculated on the basis of farmgate prices.

Which other prices should be used as the denominator in the calculations? Indexes are required here, since the concepts refer to weighted averages of many prices. The most commonly used price index is the consumer price index. Thus, the real price of rice may be expressed as the price of rice divided by the consumer price index.

1. Simon Schama, *Citizens: A Chronicle of the French Revolution*, Knopf, New York, 1989, pp. 306–308.
2. Jo Swinnen and Frans A. van der Zee, 'The political economy of agricultural policies: a survey', *European Review of Agricultural Economics*, **20**(3), 1993, pp. 261–262.

This concept measures the purchasing power of a unit of rice harvest, in terms of all goods and services in the economy. Since it is a ratio and an index, its value at any moment in time is not meaningful, but its year-to-year changes show changes in the purchasing power of farmers.

For policy analysis and decision-making, it is useful to construct sector-wide and sub-sectoral *indexes of agricultural prices*, for both nominal and real prices, rather than leaving the analysis at the level of prices of individual products. In this way, measures of the movements in prices at the level of the entire sector can be created and monitored. An index of all real agricultural prices can be calculated as the index of nominal agricultural prices divided by the consumer price index. This concept shows what is happening to producer incentives in regard to their purchasing power. The data necessary for calculating a real agricultural price index are available in every country (average farmgate prices and total quantities produced for each product). Often, they are issued in published form, but in a surprising number of cases such an index is not calculated or maintained up to date. In order to keep decision makers informed about the most fundamental trends in the sector, it should be calculated annually.

In these calculations, the consumer price index may be replaced with other indexes: a producer price index, a GDP deflator, an index of agricultural input prices, an index of non-food consumer prices, and so forth. Each definition of real agricultural prices measures a somewhat different concept, but all of them constitute numerical expressions of trends in the purchasing power of agricultural output.[3] Deflating an agricultural price index by an index of prices of agricultural inputs would give a real index which indicates trends in profitability of the sector's production, abstracting from productivity changes. On the other hand, deflating it by the consumer price index yields an index of the purchasing power of farm households as consumers.

Under any of the alternative definitions, these indexes provide an empirical dimension to the discussion of the effects of policies on agricultural incentives, and a basis for monitoring the sector's performance over time in the price dimension.

4.2.2 Structural Determinants of Agricultural Prices

Real agricultural price trends are powerfully influenced by structural factors, and these factors in turn place limits on the extent to which policies can influence the sector's prices. The balance of supply and demand is the most obvious of these factors. For products oriented toward the domestic market, a poor harvest almost invariably leads to a increase in real agricultural prices. Abstracting from such short-run fluctuations, which tend to even out over time, the longer-run price trend is influenced by trends in supply growth relative to real demand growth. The responsiveness of food demand to income growth (income elasticity of food demand), for all households and food products taken together, tends to be consistent across countries at values in the range of 0.6 to 0.7. This implies that, in a highly simplified economy with no foreign trade in food products, aggregate real income growth in the economy of 5% would generate demand for growth in food production of 3% to 3.5%. Faster growth in food production would tend to depress real agricultural prices and slower growth would tend to push them up.[4]

This relationship is altered when imports can compensate for shortfalls in production and exports can provide an outlet for excess supply,

3. Different definitions of price indexes and their interpretations are reviewed more fully in the author's *Policy Analysis for Food and Agricultural Development: Basic Data Series and Their Uses*, Training Materials for Agricultural Planning No. 14, Food and Agriculture Organization of the United Nations, Rome, 1988.

4. This illustration is also simplified in other ways. Among other considerations, the demand for food will grow as a function of increases in *per capita* incomes and population, expressed as separate variables, and the relevant concept of income for this kind of projection is not GDP but rather something closer to national income.

although not all staple foods are readily importable or exportable. Exports allow agriculture to grow significantly faster than the limits imposed by growth of domestic demand alone. When trade possibilities exist, international transport and handling costs will establish a gap between the fob export price and the cif import price of any commodity in a given country – a range within which domestic supply and demand and policy measures may influence prices.

The existence of international trading options subjects domestic prices to another structural factor: the influence of trends in world market prices. In most periods in this century, owing to consistent agricultural productivity growth on a world scale, international agricultural prices have not kept up with prices of industrial goods. They have declined in real terms. Binswanger *et al.* found that real international agricultural prices declined by 0.5 to 0.7% per year from 1900 to 1984.[5] Independently of domestic policies, these trends have tended to depress real agricultural prices within each country. In addition to this factor, agricultural subsidies in developed economies have lowered the prices of their agricultural exports to poorer countries, thereby affecting world market prices considerably, as discussed in Chapter 3.

In summary, three important structural factors influence the trends in domestic real agricultural prices in all economies: the trends in domestic supply and demand, the secular, or long-run, trends in international prices, and the presence of subsidized exports in world markets. Notwithstanding the influence of these factors, domestic economic policy also has an influence on real agricultural prices, through both macroeconomic and sectoral instruments. The most important macroeconomic policy influences on real prices at the sectoral level are tariff and trade policy, exchange rate policy and fiscal policy. Each is discussed below in turn.

4.3 TRADE POLICY[6]

4.3.1 Basic Issues

Trade policy can confer powerful incentives or disincentives for production, through its influence on prices and quantities of competing products that are imported into the country and through its effects on domestic prices received for exports. Policies that make imports more expensive in the domestic market are said to provide economic protection. The main instruments of trade policy are tariffs and quotas on the side of imports and various kinds of incentives on the export side. In some cases, a combination of quotas and tariffs is used ('tariff-rate quotas' (TRQs)), under which tariffs rise when imports exceed a specified quantity.

Trade policy has been the subject of intensive international negotiation for decades. Ever since the disastrous tariff wars of the 1930s, the aim of the negotiations has been the progressive dismantling of barriers to international trade. There is an international consensus that high rates of protection not only invite retaliatory protection measures by trading partners but also lead to inefficiencies in a country's own productive structure, by removing pressure for productivity increases and for reallocating a country's productive resources to its more competitive product lines.

Many developing countries have benefitted substantially from increased international trade in recent decades. For example, Eugenio Díaz-Bonilla and Lucio Reca have pointed out that

5. H. Binswanger, Y. Mundlak, M.-C. Yang and A. Bowers, 'On the determinants of cross-country agricultural supply', *Journal of Econometrics*, **36**, 1987, pp. 111–131 (cited in Yair Mundlak, 'The Dynamics of Agriculture', The Elmhirst Lecture, The XIII International Conference of Agricultural Economists, Sacramento, CA, USA, August 10–16, 1997).

6. Some of the next few paragraphs have been adapted from the author's 'Critical Issues Facing Agriculture on the Eve of the Twenty-First Century', in IICA, *Towards the Formation of an Inter-American Strategy for Agriculture*, San José, Costa Rica, 2000. This section is written from a viewpoint of the economics of trade policy for development. Where WTO rules are relevant to the discussion, they are mentioned.

Latin America and the Caribbean have run a positive net agricultural trade balance for decades, and by 1996 it had increased to about US$20.2 billion.[7]

The benefits that derive from greater volumes of international trade give developing countries an interest in promoting it and in ensuring fairness in international trading rules. However, since the completion of the Uruguay Round (UR), developed countries have increased their exports more than developing countries have, and concerns have been raised about the continuance of agricultural protection measures in developed countries. Issues of this nature are figuring prominently in the latest round of trade negotiations.

Agricultural protection has been reduced much more slowly than industrial protection in the last decade. Timothy Josling has analyzed the prevailing patchwork of tariffs in the world and has written:

> . . . manufacturing tariffs are now at modest levels in most industrial countries and in an increasing number of middle- and low-income countries. Many of these tariffs are around 5 to 10 percent. By contrast, agricultural tariffs average above 40 percent, with tariff peaks (megatariffs) of over 300 percent . . . which effectively block trade. . . .
>
> Canadian dairy imports are a well-known example of such megatariffs: the tariff on butter is 351 percent and on cheese is 289 percent. Even by the year 2000, these will still be at 299 percent and 246 percent, respectively. . . . poultry tariffs are also above 200 percent in Canada. The United States has megatariffs for sugar and dairy products, as does Japan for grains, sugar and dairy products.[8]

In a detailed analysis of tariff developments after the Uruguay Round's Agreement on Agriculture, N. Hag Elamin explains in what ways the application of that Agreement has been unfavorable to developing countries:

> Having in place the new set of rules on market access is a remarkable achievement of the Agreement on Agriculture (AoA) and a significant contribution to the predictability and security of trade. However, what is important for actual trade to take place . . . is the level of tariffs and other access conditions which are country-specific. . . . High tariffs on temperate-zone food products and low rates on tropical products [is] a typical pattern of the post-UR tariff profile of many developed countries. The reason is simple – imports of temperate-zone products compete with domestic production while others do not. This difference is also reflected in reduction rates from the base to final levels – while tariffs on tropical products as a whole were cut by an average of 43 percent, the reduction rates on other product groups were lower, e.g. 26 percent on dairy products. The AoA rule required a 36 percent overall simple average reduction for agricultural products – the developed countries surpassed this target by one percentage point.
>
> There is a catch here worth noting – tariffs on many tropical products were already very low (e.g. 5–10 percent) and so it was feasible to reduce these even more sharply (e.g. by 50 percent) without disrupting domestic markets. Some commentators have expressed doubts whether these relatively sharp reductions on tropical products added anything substantial in terms of access. In many cases, these levels have fallen to what is called 'nuisance tariffs' of 1–2 percent. On the other hand, the base period tariffs on temperate-zone products were very high – in many cases over 100 percent and so a further reduction of 20–25 percent from that level still leaves considerable border protection.

7. E. Díaz-Bonilla and L. Reca, 'Getting Ready for the Millennium Round Trade Negotiations, Latin American Perspective', Focus 1, Brief 2 of 9, *2020 Vision*, International Food Policy Research Institute, Washington, DC, USA, April 1999.

8. Timothy Josling, *Agricultural Trade Policy: Completing the Reform*, Policy Analyses in International Economics No. 53, Institute for International Economics, Washington, DC, USA, April 1998, pp. 6–8.

There were a number of reasons why bound tariffs on these products turned out to be very high:

- The choice of 1986–88 as the based period was an important factor. At that time, world market prices were very low and as a result the computed tariff equivalents, or the gaps between domestic and world prices, were very high. These high tariff equivalents were used to set base period tariffs. A 15–20 percent or even 36 percent reduction from these high base levels would still result in high bound rates by the year 2000.
- There is some evidence that several [developed] countries set their base tariffs for some products at much higher rates than justified by computed tariff equivalents, particularly on temperate-zone products such as cereals, dairy, meat and sugar.
- The simple average formula used in the UR allowed countries to make smaller (the 15 percent minimum required) cuts on some commodities (e.g. the so-called 'sensitive' products) combined with larger cuts on others (e.g. tropical products) in order to arrive at the simple average [reduction] of 36 percent. . . .

A recent OECD study . . . showed that actual border protection to agriculture was higher in 1996 compared to 1993 in eight of the ten OECD countries (EU as one) covered by the study, the two exceptions being Australia and New Zealand.[9]

Another major concern of poorer countries is the prevalence of agricultural export subsidies in the richer nations. They have the effect of reducing prices for farmers in poor countries, and therefore they tend to aggravate the problem of rural poverty. In Josling's words:

The use of export subsidies in agricultural markets poses serious problems for countries trying to develop competitive agricultural sectors.[10]

This effect arises not only from explicit export subsidies. Agricultural production subsidies of various types in richer nations contribute to over-supply of agricultural outputs, exacerbating the downward trend in real agricultural prices on world markets.

Developing countries can respond to these circumstances with a two-pronged approach: by strengthening their joint participation in international trade negotiations (as they are now doing), and by implementing domestic policy measures to limit the economic damage caused by policy distortions in other countries. The greatest source of resistance by farmers to trade liberalization is fear that low-priced imports will drive down domestic prices to unprofitable levels or, for subsistence farmers, to levels that mean privation for their families. These concerns are legitimate, especially for developing countries with already serious problems of rural poverty. The challenge for policy is to respond to these concerns without falling into the self-defeating trap of protectionism.

Staple crops that are the main source of food and income for poor families have a claim to special treatment in trade policy. The benefits of liberalized trade arise from the ability of workers to learn more efficient ways to produce and to learn new occupations. Poor rural families are generally the least well educated and would have the greatest difficulty in learning new occupations. In addition, well-known deficiencies in rural credit and land markets make it difficult for them to invest in improved agricultural technology. These families have no social safety net to assist them in making adjustments, as they would in more developed countries. Partly for these reasons, Mexico negotiated a 100% tariff on corn for 15 years in the context of NAFTA, and in the

---

9. N. Hag Elamin, 'Market Access I: Tariffs and Other Access Terms', *Multilateral Trade Negotiations on Agriculture, A Resource Manual*, Vol. II, Agreement on Agriculture, Module 4, Food and Agriculture Organization of the United Nations, Rome, 2000, pp. 58–60.
10. T. Josling, 1998, p. 120.

period 1999–2001 El Salvador, Nicaragua and Panama raised tariffs on imported grains after several years of moving tariffs only in a downward direction.

Central Asian agriculture also has experienced difficulties during a process of market liberalization. With movement toward a market system in regard to marketing, trade and prices, sector performance has been poor and productivity low, many farms have become financially unviable, irrigation systems have fallen into unusable condition, and rural poverty has increased.[11]

Perhaps with these kinds of issues in mind, Dani Rodrik has articulated a thoughtful and pragmatic approach to trade policy for developing countries in which the stage of a country's development should be a factor determining the rate of liberalization:

... the nature of the relationship between trade policy and economic growth remains very much an open question. The issue is far from having been settled on empirical grounds. There are in fact reasons to be skeptical that there is a general, unambiguous relationship between trade openness and growth waiting to be discovered. The relationship is likely to be a contingent one, dependent on a host of country and external characteristics. The fact that practically all of today's advanced countries embarked on their growth behind tariff barriers, and reduced protection only subsequently, surely offers a clue of sorts. Moreover, the modern theory of endogenous growth yields an ambiguous answer to the question of whether trade liberalization promotes growth. The answer varies depending on whether the forces of comparative advantage push the economy's resources in the direction of activities that generate long-run growth (through externalities in research and development, expansion of product variety, upgrading of product quality, and so on) or divert them from such activities.

Indeed, the complementarity between market incentives and public institutions that I have repeatedly emphasized has been no less important in trade performance. In East Asia the role of government in getting exports out during the early stages of growth has been studied and documented extensively.... Even in Chile, the exemplar of free market orientation, post-1985 export success has been dependent on a wide range of government policies, including subsidies, tax exemptions, duty drawback schemes, publicly provided market research, and public initiatives fostering scientific expertise. After listing some of the pre- and post-1973 public policies promoting the fruit, fishery and forestry sectors in Chile, Maloney[12] ... concludes that 'it is fair to wonder if these, three of the most dynamic export sectors, could have responded to the play of market forces in the manner they have without the earlier and concurrent government support'.

The appropriate conclusion to draw from all this is not that trade protection should be preferred to trade liberalization as a rule. There is no evidence from the past 50 years that trade protection is systematically associated with faster growth. The point is simply that the benefits of trade openness should not be oversold. When other worthwhile policy objectives compete for scarce administrative resources and political capital, deep trade liberalization often does not deserve the high priority it typically receives in development strategies. This is a lesson of particular importance to countries in the early stages of reform, such as those in Africa.[13]

11. See Francesco Goletti and Philippe Chabot, 'Food policy research for improving the reform of agricultural input and output markets in Central Asia', *Food Policy*, **25**(6), December 2000, p. 662.

12. Reproduced from William F. Maloney, 'Chile', in L. Randall (Ed.), *The Political Economy of Latin America in the Postwar Period*, University of Texas Press, Austin, TX, USA, 1997, pp. 59–60, with permission of University of Texas Press.

13. Dani Rodrik, 'Development Strategies for the 21st Century', in B. Pleskovic and N. Stern (Eds), *Annual World Bank Conference on Development Economics, 2000*, The World Bank, Washington, DC, USA, 2001, pp. 102–103.

Among the 'other worthwhile policy objectives' alluded to by Rodrik, reduction of rural poverty would figure high on the list. The following sections discuss specific issues in trade policy and review policy options that would help secure the benefits of liberalized trade while at the same time contributing to poverty alleviation.

4.3.2 Tariffs in Developing Countries

Although a tariff system confers economic protection, it can be a mixed blessing for domestic producers. In the first place, *exporters generally suffer from a tariff regime*, since it raises the costs of their inputs, directly and indirectly, but it does not allow their export price to rise correspondingly. This was well documented for the case of Colombia in a classic study by Jorge García García.[14] Secondly, *even an import-competing sub-sector can be hurt by a tariff regime*, if tariff rates are not uniform and they are higher on its inputs than on the products that compete with its outputs. In analytic terms, this effect is captured by measuring *effective protection rates* as opposed to *simple (nominal) protection rates*, and effective rates can be negative if protection is higher on inputs than on outputs. Thirdly, it is well known that *tariffs at significant levels can undermine the competitiveness of domestic industries and sectors*, as the additional economic profits delivered by the tariff protection tend to weaken the resolve to increase productivity.

For this last reason, it now is an accepted principle that *tariff rates should not be high in general, and if they are, a program should be put in place to scale them downward over time*. Usually, free trade agreements incorporate clauses to put these objectives into effect. As noted, in the case of NAFTA, up to 15 years have been allowed to eliminate some agricultural tariffs, but an agreement for their eventual elimination has been sealed by a treaty.

In addition to avoiding high tariff levels, *the second basic principle for tariff systems is that their rates be relatively uniform over sectors and products*. For the sake of fostering economic efficiency, this principle is exceedingly important. Three exceptions to it are discussed below, but otherwise the more uniform the tariffs are, then the better for the sake of promoting economic growth. Uneven tariff protection favors some industries or sub-sectors over others, and often those which are so favored turn out to be the least competitive over the longer run. They may be the industries which feel the pinch of competition the most and therefore have applied the greatest political pressure for protection. This is an example of the rent-seeking behavior induced by subsidies (implicit subsidies in this case) that was mentioned in Chapter 3. Empirical documentation of the strongly inverse relationship between the competitiveness of products and their rate of tariff protection has been developed for Honduras where, for example, coffee, one of the most competitive products, has received negative economic protection and sugar, one of the least competitive, has received highly positive protection.[15]

In essence, a system of tariffs that are unequal across products places the government in the position of 'picking winners', and experience has amply demonstrated that governments are much less able to do that successfully than the market is. A common variant of a non-uniform tariff system is a set of graduated tariffs, with the lowest rate reserved for primary products, a higher rate applied to industrial intermediate goods, and the highest rate applied to manufactured final consumption goods. Such a system was applied, for example, in the Republic of Korea in the early decades of its economic take-off, and more recently in Guatemala. This system emerged from the early school of development thought which emphasized industrialization as the road to economic improvement (Chapter 1). It discriminates

---

14. J. García García, *The Effects of Exchange Rates and Commercial Policy on Agricultural Incentives in Colombia*, Research Report No. 24, International Food Policy Research Institute, Washington, DC, USA, 1981.

15. See Roger D. Norton and Magdalena García U., *Tasas de Protección Efectiva de los Principales Productos Agrícolas*, Serie Estudios de Economía Agrícola No. 4, Proyecto APAH, Abt Associates Inc., Tegucigalpa, DC, Honduras, May 1992, p. 22 and tables.

against agriculture. If a country has a comparative advantage in some agricultural products, it may be asked, why should the tariff system be designed to penalize those products and implicitly subsidize industry?

Provided that tariffs are in the low-to-moderate range, it can be said that it is at least as important, from an economic development perspective, to make them as uniform as possible, as to continue reducing them. *In order for a country to exploit its comparative advantage to the fullest, and hence maximize its growth prospects, it is very important to align domestic relative prices with international relative prices as much as possible.* A uniform tariff policy marks a major step in this direction.

Another common practice is to deviate from uniformity by setting tariffs on basic food products at zero, either by granting tariff exemptions to a State importing agency or by simply legislating the rates at zero. The purpose of such a policy is to make foods such as dairy products and cereals more affordable for the poor, but it can be argued that tariffs are not the most appropriate instrument for achieving this aim. In the first place, subsidizing foods through tariff exemptions means that farmers bear the entire burden of the subsidy, and thus domestic production is likely to decrease relative to imported sources of supply. In the words of Valdés, 'A predictable and well-documented result of cheap food policies is that self-sufficiency in the commodities in question, that is, those that are implicitly subsidized, decreases rapidly'.[16]

Secondly, this kind of subsidy is completely nontargeted, so that all families receive it in proportion to their levels of food consumption, regardless of their incomes. In that sense, it usually turns out to be a regressive subsidy. The

The beneficial effect for the rural poor, including the landless, of rising food prices was found in Honduras by Dean Schreiner and Magdalena Garcia. The devaluations carried out in stages between 1988 and 1990 had a marked positive effect on real agricultural prices. The production response was immediate and sustained. All rural income strata benefitted, and the poorest strata benefitted the most. (See the study by D. Schreiner and M. García, Principales Resultados de los Programas de Ajuste Estructural en Honduras, *Serie Estudios de Economía Agrícola No. 5, Proyecto APAH, Tegucigalpa, DC, Honduras, June 1993.)*

value of the subsidy is greatest for the most well-off families. For example, in Kenya it was found that 'subsidies on sifted maize meal in urban Kenya were disproportionately captured by the more well-off strata of the urban population'.[17]

Thirdly, *tariff exemptions on food usually exacerbate the problem of rural poverty*. This effect occurs because reducing production incentives affects producers of all farm size classes and accordingly also reduces employment for rural landless laborers. In most lower-income countries, the bulk of the poverty is found in rural areas, so the problem is a serious one from a national perspective. Exceptions to this conclusion may occur when the proportion of rural landless to rural landed is very high, as in Bangladesh.[18]

A recommended alternative to eliminating tariffs on food imports is a program of targeted

16. Alberto Valdés, 'Explicit versus Implicit Food Subsidies: Distribution of Costs', in Per Pinstrup-Andersen (Ed.), *Food Subsidies in Developing Countries: Costs, Benefits and Policy Options*, The Johns Hopkins University Press, Baltimore, MD, USA, 1988, Chapter 5.
17. Reprinted from *Food Policy*, **22**(5), T. S. Jayne and G. Argwings-Kodhek, 'Consumer response to maize market liberalization in urban Kenya', p. 456, Copyright (1997), with permission from Elsevier.
18. See Raisuddin Ahmed, *Foodgrain Supply, Distribution and Consumption Policies within a Dual Pricing Mechanism: A Case Study in Bangladesh*, Research Report No. 8, International Food Policy Research Institute, Washington, DC, USA, 1979.

food assistance, funded through fiscal channels.[19] In this way, the beneficiaries are only, or almost only, the poor, while those who bear the burden of financing the programs are taxpayers in general. If the taxation system is even modestly progressive, both the benefits and cost side of such programs will have a progressive incidence on the well-being of the population.

The political calculus for eliminating tariff exemptions on foods can be complicated by the interests of agro-industries which rely on imported inputs, such as the poultry and feed concentrates industries, which often use imported yellow corn and sorghum, and industries which use powdered milk as a raw material. Therefore, it is important to point out clearly to decision-makers that such tariff exemptions constitute non-targeted and regressive subsidies, perhaps more so if they represent responses to pressures from agro-industries. (Some countries have negotiated tariff-quota agreements between agro-industry and farmers associations, under which tariffs fall drastically, sometimes to zero, after all the domestic crop is sold.)

A complicating factor is that the agreements for international food assistance programs such as PL 480 generally prohibit the imposition of border taxes on the products they supply. However, in Honduras it was found to be possible to impose a tax on the PL 480 products *after* the government imported them and when they were sold to domestic agro-industries, thus compensating for the tariff exemption.[20]

In short, there are several valid reasons for *making tariffs as uniform as possible and for not providing tariff exemptions on food*. In addition to the considerations cited above, moving toward a policy of more uniform tariffs usually is beneficial for agriculture, since the biases in the non-uniform systems generally favor industry. In the case of Brazil, for example, for the period 1966–1983 it was found that tariffs and subsidies for imports and exports of non-agricultural goods had the result that crops were discriminated against in economic terms.[21] In the case of Honduras in the mid-1980s, Julio Berlinsky found that the average effective protection rate for industry was 99%, while the above-referenced study of Norton and García found that effective protection for agriculture was nil and even negative for many crops.[22]

The three justifiable exceptions to a uniform tariff policy are as follows:

(1) *The case of international subsidies that reduce world market prices.* These subsidies are determined by decision-makers in a few, better-off countries, and the international prices for those commodities reflect those decisions. Given the essentially irreversible nature of the flow of labor

---

19. For a guide to such programs, see Margaret Grosh, *Administering Targeted Social Programs in Latin America: From Platitudes to Practice*, The World Bank, Regional and Sectoral Studies, Washington, DC, USA, 1994.

20. According to Section 401(b) of the PL 480 legislation, called the 'Bellmon Amendment', there must be assurance that the delivery of commodities under the program does not create disincentives for production in the receiving country. Nevertheless, it is logically impossible to satisfy both this requirement and the provisions of Sections 103(c) and 103(n) which refer to the 'uniform marketing requirements'. The latter require that PL 480 shipments be *in addition to* what the country normally would have imported; however, if this is the case, then clearly such shipments will drive the domestic price below what it would have been otherwise. This point is brought out in R. D. Norton and C. A. Benito, 'An Evaluation of the PL 480 Title I Programs in Honduras', a report prepared for the USAID Mission to Honduras, August, 1987. In El Salvador in the early 1990s, the Government terminated the PL 480 program rather than accept the provision of tariff exemptions.

21. Antonio Salazar P. Brandâo and José L. Carvalho, 'Brazil', in A. O. Krueger, M. Schiff and A. Valdés (Eds), *The Political Economy of Agricultural Pricing Policy:* Volume I, *Latin America*, The Johns Hopkins University Press, Baltimore, MD, USA, 1991, Chapter 3, p. 67.

22. Julio Berlinsky, *Honduras: Estructura de Protección de la Industria Manufacturera*, UNDP, Buenos Aires, Argentina, July 1986.

from agriculture to urban occupations (Chapter 2), it is difficult to argue that a developing country should accept relative prices determined in other, more developed countries when those prices may drive its labor permanently out of agriculture into the army of unemployed and underemployed. If the more developed countries eventually reduce their subsidies substantially (which seems likely), then a less developed country would face a daunting challenge in trying to create conditions for a recovery of its agriculture. It would be a very difficult task once a significant part of the rural labor force has left for the cities. In addition to these considerations, international subsidies can worsen rural poverty in low-income countries, and the social and economic costs of excessively rapid rural–urban migration are high, as noted previously, and so it may be asked, why should a poor country acquiesce in policy decisions made elsewhere, when the consequences would include those costs? This is the essence of Peter Timmer's argument, mentioned earlier in Chapter 3.

Sometimes, the question of compensating domestic producers for the 'international subsidies' is viewed through the lens of anti-dumping provisions, in both the WTO agreement and domestic legislation. That legal route is an option. Pursuing it has the aim of gaining internationally recognized legal authorization to apply compensatory tariffs, or 'countervailing duties' as they are called in that context. However, to prove that dumping has occurred is time-consuming and costly, and the process itself gives rise to frictions among trading partners. The procedure is basically designed for cases of *firms* selling at below cost, rather than to subsidies provided through national policies. A simpler expedient for a developing nation would be to simply legislate a tariff surcharge (preferably on the base of an otherwise uniform tariff rate) which is equivalent to the international price distortion caused by subsidies to producers in other countries, as mentioned in Chapter 3. The WTO accords stipulate a tariff ceiling ('binding'), in most cases high enough to permit all or most of the international price distortion to be compensated by a surcharge.

If such a policy were implemented, it would be important to base the magnitude of the surcharges on estimates of price distortions that have been made or sponsored by international agencies, and also to legislate a provision which mandates their review at intervals of say, five years. The reviews could gather new evidence from the most recent calculations of the international price effects of subsidies. The surcharge should disappear when the price effects of international subsidies become negligible. In any event, such a policy would be applicable to only a few products, chiefly or entirely agricultural, which were affected by those subsidies.

This option would not benefit a developing country if it did not have a comparative advantage in the products affected by international subsidies, in the absence of those subsidies. In those cases, the developing country would distort its own resource allocation and affect negatively its agro-industries by the higher tariff.

(2) *The case of price bands for smoothing price fluctuations.* Properly designed, price bands are neutral on average with respect to economic protection. However, recently a panel of the WTO ruled that they are in violation of WTO accords in Chile, where they were first applied. The full reason for the ruling is not clear, although the panel may have wanted to ensure consistency with an earlier ruling outlawing the European system of variable levies – even though price bands are different. Given the valuable role that price bands have played in some developing countries, it is worth a brief review of them in the event that their potential use is revived in future trade negotiations, and because they are still in operation in places where they have not been challenged formally. First of all, unlike variable levies, price bands have no link to a domestic support price. They compensate for excessively high peaks in international commodity prices (for the benefit of consumers) and, in equal measure, for excessively low troughs (for the benefit of producers). They accomplish this by varying the tariff rate on, say, a bi-weekly basis, according to an automatic formula which, in turn, is based on the historical series of the relevant international price. When the international price rises above its historical

trend line by more than a pre-established percentage or amount (often set as one standard deviation), then the corresponding tariff begins to rachet downward. When the price fluctuates downward by more than the same amount, then the tariff is raised. All movements in the tariffs caused by the price bands are temporary, subject to modification in the following month or two-week period.[23]

Under the price band system, there is not a fixed floor price or ceiling price, but rather there are threshold prices which trigger tariff changes, and these threshold prices change each month as the most recent month is added to the moving time series of historical prices and an earlier one is dropped from the series.

In the case of the Salvadorean and Honduran policy reforms, implementation of a price band system proved to be a vital element in convincing producers to accept a free trade regime. They had been fearful of the economic damage that could be done by pronounced drops in international prices, even if they were temporary. The weakness of price bands in some cases has been a lack of fully transparent administration. If they were to be revived under international accords, provision should be made for a small panel of international experts to supervise the installation of the system and to monitor its functioning from time to time. A good explanation of how price bands are designed and a summary of Central American experience in operating them, as well as an explanation of some of common misconceptions about the bands, has been provided in two publications by Julio Paz Cafferata.[24]

(3) *When a crop is the main source of food and income for the rural poor.* This case is especially relevant to the conditions of recent years, when many agricultural prices have declined continuously and substantially. The reasons why the rural poor may not be able to readily adapt to new occupations or improve their agricultural yields, in the face of declining real agricultural prices, have been mentioned. If tariff surcharges have not been applied to a basic product grown by the poor for reason (1) above, and if a price band system is not functioning, it can be important to set tariffs on the products at or near the levels of the WTO bindings to help alleviate poverty in rural areas. A less distortive measure would be to provide the rural poor with direct income support, but given that they are usually numerous and geographically dispersed, and titles to their farms often are not registered, it is more difficult to target assistance on them than it is for the urban poor.

Only one or two products would merit higher tariffs for this reason, and the higher level of tariffs should have a time limit, as in Mexico. One of the costs of such a policy is that not all of the protection afforded by the tariff would go to the poor. It would be constructive to augment the policy by depositing at least some of the proceeds from the tariff in a special fund for improvement of technology on small farms, and to use those funds for farmer training, yield-enhancing investments, and investments in alternative crops or livestock products. It would also be important to target such assistance primarily on the women in poor families, to the extent feasible. The tariff could be visualized as an adjustment tariff if the revenues from it were effectively used for helping poor farm families make economic adaptations.

---

23. Price bands should not be confused with the variable levy system initiated by the European Union. A variable levy is designed to continuously ensure that border prices are equivalent to domestic support prices. Under a price band, there need not be a support price, as the movements in the tariffs are not linked to any domestic price, but rather to the historical pattern of international prices. Usually, a 60-month moving average of international prices is used to establish the trend line and to update it continuously.

24. (a) Julio Paz Cafferata, 'El Sistema de Bandas de Precio: Una Alternativa de Política de Precios para los Granos en Honduras', RUTA II, San José, Costa Rica, June, 1990. (b) Julio Paz Cafferata, 'La Experiencia de Banda de Precios para la Regulación de las Importaciones de Granos en Centroamérica', in *Mercados y Granos Básicos en Nicaragua: Hacia una Nueva Visión sobre Producción y Comercialización*, H. Clemens, D. Greene and M. Spoor (Eds), Escuela de Economía Agrícola y Programa Agrícola CONAGRO/IDB/ UNDP, Managua, Nicaragua, 1994, Chapter 5.

Each of these exceptions to a uniform tariff policy has a clear rationale and they should not be used to justify a policy of protectionism. At most, they would apply to three or four agricultural products, usually one or two. One of the mechanisms, the price bands, does not constitute protection at all over the longer run.

In addition to uniformity over products, subject to these three classes of exception, *a third basic principle is that tariff systems should be relatively stable over time*, except for downward adjustments in phases that have been programmed years in advance. In practice, this is one of the most difficult principles to gain adherence to, for political leaders sometimes succumb to the temptation to tinker with tariffs in response to perceived crises in industries or special interests. Varying tariffs at frequent intervals is very damaging to economic growth prospects, because it creates a large degree of uncertainty about future economic policy and therefore it discourages productive investments.

In light of these principles, it can be observed that the WTO rules have important weaknesses from a viewpoint of promoting economic development. First, they allow a fairly high ceiling on tariffs during a long transitional period,[25] and consequently they allow considerable variation among tariffs by product, with some items located at or near the lower end of the permissible range (which usually is zero) and others at the upper end. While this room for maneuver may be convenient from a viewpoint of imposing compensatory surcharges for a few agricultural products affected by international subsidies, in general it is quite damaging to allocative efficiency and therefore to economic growth prospects. Secondly, the rules do not require stability over time in national tariff systems.

Thirdly, they discriminate against exports, in favor of import-competing sectors, by allowing tariffs to be considerably higher than export subsidies, which, in fact, are supposed to be eliminated. For the poorest countries, a policy which

was neutral between the two classes of goods would permit uniform export subsidies in exactly the same percentage as a uniform tariff on imports. All countries will benefit in the long run from a worldwide reduction of tariffs and export subsidies, but during the lengthy period in which they are being phased out, allowing tariffs to be significantly higher than export subsidies introduces a bias that works against economic development for many countries. In light of these observations, although the Uruguay Round and other WTO negotiations have represented important advances toward liberalizing trade, with concomitant benefits for all countries, it can be seen that the WTO rules do not fully take into account the perspective of development policy.

Domestic policy can counteract these weaknesses to a considerable extent, by making tariffs uniform except for the three cases noted above, by forging a national agreement to maintain stability in the system over time, and in some cases by implementing a program of export subsidies up to the 10% limit if tariffs are at that level or higher. Since the aggregate value of imports almost always exceeds that of exports in developing economies (because developing countries are net capital importers), such subsidies could be financed out of tariff revenues, with room to spare.

4.3.3 Export Incentives

Export subsidies, as well as tariffs, are generally discouraged under WTO rules. They have been a major issue of contention among industrial trading nations, and the intent of international negotiations has been to reduce them over time. Nevertheless, in view of the above-mentioned bias against exports in the WTO regime, and their importance for developing countries, it is worth considering measures that would encourage them in the context of sound economic policy. Many countries have adopted a form of export incentives, through tax exemptions and drawbacks of duties paid on imported inputs.

---

25. Referring to the process of converting quantitative trade restrictions into tariffs, the FAO has noted that 'the reform process often resulted in levels of protection (at least potentially) above even the high levels that had been in effect in the mid-1980s' (FAO, *The State of Food and Agriculture 1995*, Food and Agriculture Organization of the United Nations, Rome, 1995, p. 248).

In developing-country agriculture, export products frequently are more labor-intensive than import substitutes, and therefore the bias against exports is especially prejudicial to prospects for raising rural employment and reducing rural poverty. If tariffs and export incentives were to be approximately equalized, normally the revenues from the tariffs would be more than sufficient to cover the support to exports for the reason mentioned above.

A principal problem with existing measures of export support is that they don't reach the majority of farmers. Most poor farmers don't use significant quantities of imported inputs and therefore cannot perceive benefits from duty drawbacks. Equally, most farmers do not earn (or report) enough income to pay taxes and therefore tax exemptions are not relevant for them. Typically, export incentives are applied at the level of the exporter, and if they reach farmers at all they are considerably diluted. Thus, implementation deficiencies have become a principal obstacle to an effective program of export incentives. For these reasons, in the design of policy it is important to give consideration to the administrative mechanics of an export subsidy, especially for products that are processed before being exported.

Some of the relevant factors can be illustrated by an example. Through a producers' association, coffee growers could be given coupons with X monetary units per kilo of coffee, coupons that would have no value unless they were sold to an exporter. For an exporter, the same coupon would be worth more, say $X(1 + s)$. The value of s might range from, say, 0.20 to 0.50, or more if the exporter also were to process the coffee. The exporter would buy coupons only for the quantity that could be sold on international markets. Then the exporter, in turn, could exchange the coupons for cash in a bank upon presentation of the export receipt documentation. The exporter would receive the higher value, retaining a net incentive of sX units per kilo after paying X per kilo to the growers.

This kind of mechanism would be self-enforcing because even if coffee growers were given an excessive amount of coupons, exporters would purchase only the number of coupons that corresponded to the volume of their exports. Each coupon or voucher would have three segments, one of which would remain with the farmer, one of which would be retained by the exporter, and one of which would be given to the bank at the time of redemption of the coupon. The banks would be required to remit their coupon stubs and export documentation to a governmental authority for monitoring purposes.

This simplified illustration would apply if the exporter were the coffee roaster. The scheme could be modified to allow for intermediaries who did the roasting but did not export themselves. Then, in gross value, the coupon could be worth X to farmers, $X(1 + s)$ to roasters, and $X(1 + s)(1 + t)$ to exporters. One of the necessary conditions for the successful operation of this kind of scheme would be the existence of a strong producers' association. It would be expected to help design, implement and monitor the program.

The point of this brief illustration is to focus attention on the need for adequate mechanisms for providing incentives to agricultural exports from developing countries, which seem to be the orphan of international trade negotiations. Eventually, it may be possible to reduce both tariffs and export subsidies to insignificant levels in all countries, but that prospect is for the very long run. In the meantime, attention to the development of viable mechanisms for export incentives is required.

4.3.4 Trade Restrictions

The term 'national trade policy' often includes tariff regimes, but when it is treated separately it refers to, apart from trade agreements, the policy regarding the degree of openness of international trade, i.e. the degree to which controls on imports and exports have been eliminated. These restrictions take various forms, including import quotas, licensing requirements for imports and exports, export bans in some cases, restrictions on the availability of foreign exchange, and sometimes phytosanitary requirements. More subtle forms of controls on imports include requirements for making deposits of foreign exchange in the banking system far in advance of the actual importation, and increasing the amount of advance deposit required.

There is an international consensus that trade restrictions have markedly negative effects on development prospects over the long run.[26] In terms of their price effects, import controls (referred to generically as 'non-tariff barriers') are equivalent to tariffs which are highly variable over time, all the more so if they are imposed arbitrarily and with little advance notice. Depending on the degree of scarcity of a product in the domestic market, and the amount of imports allowed in, the domestic price can rise very high in response to an import control measure. Conversely, limiting or banning exports leads to repression of the domestic price and thus usually is resisted strongly by producers.

Nevertheless, there is increasing awareness that opening an economy rapidly and fully (eliminating trade controls and reducing tariffs drastically) can worsen the problem of rural poverty in the short and medium run. Care must be exercised, as indicated by the foregoing comments of Rodrik. The transition from a system of trade controls to a liberalized system is not easy, as indicated by the Central Asian experience cited in Chapter 3 (Section 3.4). When there is a shortfall in a harvest, political points can be scored in the eyes of consumers by promoting a ban on the export of the product, as has happened several times in the case of beans in Central American countries. However, such policies hinder agricultural development and thus worsen the poverty problem.

In periods of surplus production, sometimes the corresponding policy response has been an import ban. 'In Kenya, all imports of maize and wheat, except those for humanitarian purposes, were suspended in mid-1994 for six months. . . . In early 1994, the Government of Nigeria imposed a ban on imports of maize, barley and rice. . . . At the end of 1994 the authorities of the copper belt in Zambia banned the export of maize out of the area . . . '.[27] Mali has recently banned the export of hides and skins, in order to promote an intensification of the domestic processing of these products. Guyana maintained a controversial ban on the export of raw hardwood logs for many years, while Latvia implemented one for its coniferous logs.

The case of the ban on log exports illustrates well the potential economic repercussions of such measures. The objective of an export ban is to promote further processing of raw wood by domestic industry, thus increasing the value added resulting from each unit of forest extraction. The objective is laudable, but this particular means of achieving it can have counterproductive results, for by eliminating the export market for logs it makes them artificially cheap in the domestic market, thus fostering the creation of a domestic wood processing industry that is conditioned to a cheap supply of raw material. Such industries generally are not competitive under world market conditions, so their growth prospects are limited to the domestic market. Furthermore, if the export ban is eventually lifted, some of them may collapse because of the higher price of logs. In the words of Jan Laarman, 'reluctance to open markets to external competition results in the inefficient use of forest raw materials. Employment is provided, but at a high social cost per job'.[28]

26. 'High tariffs, quantitative restrictions and other non-tariff barriers encourage inefficiency', in FAO, *Agro-Industrial Policy Reviews, Methodological Guidelines*, Agricultural Policy Support Service, Policy Analysis Division, Food and Agriculture Organization of the United Nations, Rome, 1997, p. 35.
27. FAO, *The State of Food and Agriculture 1995*, Food and Agriculture Organization of the United Nations, Rome, 1995, pp. 82–83.
28. Jan G. Laarman, *Government Policies Affecting Forests in Latin America: An Agenda for Discussion*, Inter-American Development Bank, Social Programs and Sustainable Development Department, Washington, DC, USA, March 1997, p. 38. For further discussion of this issue, see Robert D. Kirmse, Luis F. Constantino and George M. Guess, *Prospects for Improved Management of Natural Forests in Latin America*, LATEN Dissemination Note No. 9, Latin America Technical Department, The World Bank, Washington, DC, USA, December 1993, p. 24.

In Guyana, the effects of the ban were aggravated by the uncertainty as to whether the log exports would be taxed or banned, and by the delays in reaching decisions which sometimes led to the logs rotting before they could be shipped. Administratively created uncertainty is as deleterious for economic development as uncertainty about fundamental policy orientations.

For a country which has had a log export ban in place, lifting the ban can cause disruption to the existing industry. In Ecuador, where the ban was lifted, 'domestic sawmills that use eucalyptus logs now face higher prices as increased competition reduces their supply'.[29] The development-oriented answer, nevertheless, would not be to maintain the ban but rather to provide transitional assistance to the industries affected by its removal.

Chile's forestry sector eschewed a ban on such exports, and it found that such exports did not rise to exceptionally high levels, but rather they stabilized at about 10% of total exports of wood and wood products.

In the case of the temporary export bans on beans and other products, they have had the effect of depriving producers of the potential profits associated with the higher prices, so leaving them with greater losses in years of surplus production. *The net effect of export bans is to discourage production of that crop, thus exacerbating shortages in the future* depending on their competitiveness and the state of world markets. They also tend to create corruption as producers attempt to evade them.

In regard to the effects of market liberalization on the rural poor, a simulation study of a potential relaxation of rice export controls in Viet Nam led to the conclusion that an export-oriented strategy can be consistent with food security and with smallholder production.[30] However, these authors point out that Viet Nam's rural sector is characterized by a relatively equal distribution of land (with very few landless families) and good marketing infrastructure.

On the import side, it is sometimes argued that controls are needed to protect investments in domestic industries, as in oilseed processing plants, grain mills and sugar mills. Imports of processed products may be viewed as a threat to these industries. From an economic growth viewpoint, it may be preferable to eventually reallocate the labor and capital involved in these industries to other activities which enjoy brighter long-run prospects. To not do so is to condemn the labor force to a life in an industry in which prospects are dim for output expansion and for productivity increases, and therefore for real wage increases. This is the true economic cost of protection, in whatever form it is given. The investments already made in the industries in question are sunk costs and should not have a bearing on forward planning. The appropriate operational question in these cases is not whether the shift should be made, but how to manage the transition. This is another instance in which the case for *transitional subsidies* is strong; they can ease short-term economic pain, and therefore facilitate the required political decisions, and they also can assist in the nurturing of sub-sectors which may have better growth prospects and thus will help give the economy a stronger foundation in the future.

Phytosanitary controls on imports are sometimes abused and turned into disguised import controls including, it is alleged, by industrialized countries. In cases in which shipments of products arrive that are apparently inferior in quality but not unsafe to the consumer, the preferred remedy is the development of an accurate national product grading and labeling system, with a corresponding informational campaign for consumers.

Since import controls of any kind create unpredictable and sometimes large effects on domestic prices, policy becomes more transparent if they

29. Laarman, *op. cit.*

30. Nicholas Minot and Francesco Goletti, 'Rice Market Liberalization and Poverty in Viet Nam', Research Report No. 114, IFPRI, Washington, DC, USA, December 2000, Abstract, p. 2.

are replaced with tariffs, a process known as 'tariffication'. Tariffs have a clear, stable effect on prices, and so expectations of future profits can be more readily formed by investors and producers under a tariff regime as opposed to a system of trade controls.

A resurgence of trade controls, disguised or open, may be a sign that the process of tariff reduction is proceeding faster than the economy can support, or that the transitional funding and other transitional measures are insufficient. *It may be preferable to move slowly in reducing tariffs, provided that steps are also undertaken toward making them more uniform, and avoid the emergence of trade controls, than it is to move rapidly for some products, while tariffs remain high for others and controls remain or re-emerge.* Generally, tariff reforms have been accompanied by a reduction of the dispersion in tariff rates, but exceptions have occurred, and they are a signal that the process may need rethinking. Moving too fast in tariff reduction may provoke instability in the tariff regime, thus undermining one of the pillars of efficient resource allocation, and may make it more difficult in the future to achieve a consensus on the need for stable and modest tariffs. This is a very real concern in many countries. 'Trade reforms are . . . not only difficult to implement, they are also difficult to sustain, and there are strong pressures for a return to greater protection of the import-competing sectors'.[31]

The recent experience of the countries of Central and Eastern Europe underscores the potential for instability in trade regimes and other agricultural policies when market conditions become adverse:

In the face of falling world market prices and rapidly increasing imports, several countries in the region responded by raising import tariffs, export subsidies, minimum prices, intervention purchases and direct payments. For example,

several [countries of the region] increased subsidies on agricultural exports in 1998, in particular the Czech Republic, Lithuania, Hungary, Slovakia and Slovenia.[32]

It has been mentioned that the greatest source of resistance by farmers to trade liberalization, even if they are raising their productivity, is fear that low-priced imports will drive down domestic prices to unprofitable levels or, for subsistence farmers, to levels that mean extreme poverty for their families. These fears can be addressed through the three exceptions to a uniform tariff policy that were mentioned above: the surcharges to compensate for the price effects of international subsidies, the price bands to counteract the effects of fluctuations in international prices, and higher tariffs for long transitional periods on basic crops produced by poor farmers. Conversely, attempting to respond to the concerns through trade controls introduces distortions in the agricultural economy which are damaging to future growth prospects in the entire economy.

Another prevalent form of non-tariff barrier has been the granting to the State of a monopoly on the import or export of specified commodities. In effect, this is a form of trade control, for a State agency makes the decisions on how much of each commodity is to be imported or exported each year, as well as determining the timing of the shipments. This kind of control often goes hand-in-hand with tariff exemptions for the import of agricultural products. For the same reasons that apply to import and export licenses and other types of trade restrictions, such State monopolies have had deleterious consequences on economic efficiency and therefore on the sector's growth prospects. In addition, State agencies generally do not function with agility in decision-making, and the grain trade is notoriously demanding of experience and quickness. As a result, imports through those agencies sometimes arrive at inopportune

31. Alain de Janvry, Nigel Key and Elisabeth Sadoulet, 'Agricultural and Rural Development Policy in Latin America: New Directions and New Challenges', Working Paper No. 815, Department of Agricultural and Resource Economics, University of California, Berkeley, CA, USA, 1997, p. 11.
32. OECD, *Agricultural Policies in Emerging and Transition Economies, 1999*, Vol. I, Organization for Economic Co-operation and Development, Paris, 1999, pp. 149–50.

The trend worldwide is to eliminate State monopolies on trade in agricultural products, and frequently to eliminate State participation in such trade altogether. In Colombia in the 1990s, 'import restrictions were eliminated, including the government monopoly on imports of most grains and oilseeds' (FAO, The State of Food and Agriculture 1996, Food and Agriculture Organization of the United Nations, Rome, 1996, p. 178). In Jordan, 'a trade liberalization program was launched to remove the public monopoly on trade, including marketing and distribution of agricultural commodities. The Government has removed the Agricultural Marketing and Processing Company's ... import monopoly of potatoes, apples, onions and garlic. It has also removed import and export licence requirements on many fresh and processed agricultural products (FAO, The State of Food and Agriculture 1995, Food and Agriculture Organization of the United Nations, Rome, 1995, pp. 161–162). State agricultural trade monopolies were removed in El Salvador in 1990, in Honduras in 1991 and not long afterward in Peru.

moments, exacerbating, for example, a downward seasonal trend in producer prices. Domestic State trading monopolies have equally deleterious effects on the marketing system. It was pointed out for the case of Zambia that the main effect of such a monopoly has been to inhibit the development of markets in rural areas.[33]

4.3.5 Policies for Food Aid

A sub-set of trade policy concerns food aid. Such aid has been criticized on grounds that it weakens incentives for domestic producers. In the long run, a recipient country may benefit more from financial aid than from food aid, using the funds to enable poor families to purchase their food requirements on the domestic market. Shlomo Reutlinger has presented forceful arguments for the advantages of financial aid over food aid:

> Putting it bluntly, food aid is the product of an era in which governments, in both the industrialized and the developing countries, were expected to intervene in the production and marketing of food in a big way. ... The most sweeping challenge to the notion that food, more than other aid is needed, arises from the now widely accepted view that poverty, and not the food supply or under-performing food markets, is at the root of hunger and malnutrition. ... If the cause of hunger is not lack of food in the market, the choice between food and financial aid can be based solely on the basis of efficiency calculations. ...
>
> People could have obtained more, and better suited food to their needs, if given cash to purchase food in local markets. Not only this, it is increasingly recognized that malnutrition can not be prevented by food alone. Aptly, food aid is wastefully used when people have to convert food into cash (because in some emergency situations, the only aid provided is food aid).[34]

However, as long as food aid is present, the concessional component of the aid always represents a net economic gain for the recipient country, and so the principal challenge is how to use the aid to fortify domestic production and marketing systems. This concern, and a constructive approach to addressing it, have been described for the case of Mozambique:

> ... large amounts of [imported] low-priced yellow maize can restrict the urban market for domestically produced white maize and have adverse effects on Mozambican producers. ... instability in yellow maize prices, caused by

33. Shubh K. Kumar, 'Design, Income Distribution, and Consumption Effects of Maize Pricing Policies in Zambia', in *Food Subsidies in Developing Countries: Costs, Benefits, and Policy Options*, P. Pinstrup-Andersen (Ed.), The Johns Hopkins University Press, Baltimore, MA, USA, 1988, Chapter 21, p. 295.
34. Reprinted from *Food Policy*, **24**(1), S. Reutlinger, 'Viewpoint: from "food aid" to "aid for food": into the 21st century', pp. 7 and 12–13, Copyright (1999), with permission from Elsevier.

irregular food aid arrivals, has been transmitted to the white maize market. . . . the management of commercial and emergency food aid created large rents for consignees prior to the southern Africa drought, and significant losses for at least a year after that time. These conditions are clearly not conducive to the development of an efficient and effective maize production and marketing system in the country.

On a more positive note, the food aid program's emphasis beginning in 1992 on creating competitive marketing conditions for food aid facilitated the growth of the informal marketing system and the small-scale milling industry. Both now play key roles in linking Mozambican producers and consumers and providing consumers with affordable maize products. This is an important example of how food aid can be used to help develop markets, and represents a major accomplishment of the food aid program in Mozambique.[35]

In addition to directing the proceeds of food aid to the most constructive uses in this sense, it is important to avoid conferring tariff exemptions on the products imported through the food aid program, as pointed out above, in order to minimize the disruptions of incentives for domestic producers.

The advantages and disadvantages of food aid have been assessed by the Overseas Development Institute (ODI), which has attempted to summarize the growing consensus on the topic. First, the ODI makes a basic distinction:

Food assistance describes any intervention designed to address hunger, in response to chronic problems or short-term crises. Food assistance may involve the direct provision of food, for example in supplementary feeding or food for work projects. Equally, it may involve financial interventions, for example to support food subsidies or price stabilization schemes. Food assistance may be funded largely internally, as in India, or be supported by internationally sourced food and financial aid, as in Bangladesh or Ethiopia.

Food aid is commodity aid that is used either to support food assistance action or to fund development more generally, by providing balance-of-payments support in substituting for commercial imports, or budgetary support through the counterpart funds generated from sales revenue. Food aid transfers are required to meet the Development Assistance Committee (DAC) criteria for official development assistance (ODA) – grants or loans with at least 25% concessionality.[36]

Then, the ODI synthesizes research findings on food aid that point to a need to re-think its role and plan its use in a broader context:

Relief food aid plays a clear and crucial role in saving lives and limiting nutritional stress in acute crises caused by conflict or natural disaster. However, there is frequently a lack of robust evidence quantifying its impact, much evidence of ineffectiveness, and some evidence of late-arriving, inflexible relief hampering the recovery of local economies affected by natural disaster.

Developmental food aid has proved relatively ineffective in the 1990s as an instrument for combating poverty and improving the nutritional and health status of vulnerable people. *Program food aid*, which is provided to governments for sale, is a particularly blunt instrument for these purposes. Robust evidence on impacts of *project food aid*, which provides food directly, is lacking because of inadequate performance monitoring, in particular of the effectiveness of targeting and impacts on human resource development.

---

35. Reprinted from *Food Policy*, **21**(2), D. Tschirley, C. Donovan and M. T. Weber, 'Food aid and food markets: lessons from Mozambique', pp. 205–206, Copyright (1996), with permission from Elsevier.
36. Reproduced from Overseas Development Institute, *Reforming Food Aid: Time To Grasp the Nettle?*, ODI Briefing Paper, Overseas Development Institute, London, January 2000, pp. 1 and 2, with permission of the Overseas Development Institute.

Financial food aid is a more efficient way in most circumstances of funding activities such as school meals or food-for-work, or providing balance-of-payment or budgetary support for general development of food security. Hence the massive fall in program food aid and the slow decline in WFP [World Food Program] development activities.

Success in mitigating the effects of natural disasters and conflicts indicates that food aid has a continuing role in emergency relief and post-crisis rehabilitation, though with considerable scope for improved performance. It can also be useful as targeted assistance to highly food-insecure people in situations of poorly functioning fragile markets and serious institutional weakness. However, it has not proved an effective or efficient instrument for supporting poverty reduction strategies more generally.

The implications are clear. Hunger remains an important problem, and one that needs a comprehensive package of food assistance measures, devised and implemented nationally, and with international support. Food aid has a positive but limited role to play in this task, especially in emergencies. It needs to be planned and managed in the wider, food assistance context. Unfortunately, current rules and institutional arrangements continue to treat food aid as a special case.[37]

4.4 EXCHANGE RATE POLICY

4.4.1 The Role of the Exchange Rate

At one time or another, exchange rate policy has been a central topic of discussion in a very large number of countries. In the course of the sometimes convoluted analyses and debates, it is easy to lose sight of the fact that the exchange rate is simply another price. It is the price of *foreign exchange*. It is more accurate to speak of a family of exchange rates, one for the currency of each trading partner, but analyses and recommendations usually are simplified to deal with the concept of *an* exchange rate.[38]

The exchange rate of a country reflects its supply of and demand for foreign exchange, where the supply arises primarily from exports and capital inflows, and the demand from the need to import goods and services. Speculation about the country's future balance of payments often enters into the determination of the rate at any moment in time. In most circumstances, ***the exchange rate also responds to domestic inflation rates***, for the following reason: an increase in domestic prices, above that of price increases in trading partners, will make a country's exports less competitive and its imports more attractive. Therefore, other factors being equal, it will diminish the future supply of foreign exchange relative to the demand for it, and hence will tend to cause the exchange rate to ***depreciate*** (more units of domestic currency required per unit of foreign exchange). 'In most countries [with flexible exchange rate regimes], the official exchange rate is changed frequently in accordance with . . . the difference between the domestic and foreign rates of inflation'.[39]

In this very simple sense, abstracting from capital flows, an uncontrolled exchange rate will tend to move over time in consonance with the differential between domestic and external inflation, thus maintaining 'purchasing power parity' between the country and its trading partners. That is a long-term trend, and there may be considerable short-term variations around the trend, especially in response to fluctuations in capital flows.

Since a depreciation in an exchange rate makes imports more expensive, the movements in the

37. *Ibid*, reproduced with permission of the Overseas Development Institute.
38. Quantitative analyses employ the device of a weighted average of all of the country's exchange rates, where the weights are the respective trade volumes.
39. Peter J. Montiel and Jonathan D. Ostry, 'Targeting the Real Exchange Rate in Developing Countries', *Finance and Development,* The World Bank and the International Monetary Fund, Washington, DC, USA, March 1993, p. 38.

exchange rate further feed domestic inflation. However, the exchange-rate-driven increments in inflation tend to be less than the depreciation itself in percentage terms (usually 50 to 70% of the depreciation). Therefore, if inflation is brought under control by sound fiscal and monetary policy, the movements in both the exchange rate and the inflation rate will diminish and eventually cease, with the result of price stability.

It is important to note the basic economic *chain of causality*: fiscal and monetary policy determine the inflation rate, and the latter in turn plays a large role in determining the exchange rate (ignoring capital flows as well as international price movements for the moment). However, given that movements in the exchange rate generate temporary effects in the opposite direction, adding to inflation rates, governments seeking stability are sometimes tempted to fix the value of the exchange rate, or restrain its movements, in order to bring down the inflation rate in the short run. In many developing countries, this is a popular policy with the urban middle class, the main purchaser of imported consumer durables. However, it is a policy which runs against the direction of causality, and consequently it is hard to sustain. It undermines the competitiveness of a country's exports and makes it difficult for producers to compete against imports in the domestic market. Under a fixed exchange rate, increases in domestic costs and prices are transmitted in the same proportions to the price of exports expressed in foreign currencies, and hence the exports become less competitive in foreign markets. As a result, an exchange rate which is fixed in spite of domestic inflation, or does not move enough to maintain purchasing power parity, can become unsustainable, and the policy can explode, as happened in December of 1994 with the Mexican peso.

When an exchange rate depreciates by less than the differential between domestic and external inflation rates, it is said that the *real exchange rate appreciates*. This can occur for natural economic reasons, as when a country has a large, continuing source of foreign exchange inflows that is more or less immune to domestic inflation rates. The classic example of this occurs when impor-

tant deposits of gas or petroleum are discovered. The ensuing inflows of foreign exchange 'support' the exchange rate in the sense of keeping it from depreciating, regardless of domestic inflation rates. However, the combination of domestic inflation (which raises costs to producers) and a stable exchange rate (which holds export prices stable in domestic currency), i.e. an appreciating real exchange rate, usually undermines the competitiveness of domestic agriculture and the more price-sensitive domestic industries. This economic phenomenon first was noticed with Holland's discovery of natural gas deposits in the 1950s, and hence it has come to be known as the 'Dutch disease', in reference to its debilitating effects on other sectors of the economy.

These windfall gains of foreign exchange can take other forms, such as workers' remittances in the case of El Salvador, or subsidized sugar and rice exports to the European Union under the provisions of the ACP Protocol and the Lomé Convention, as in the case of Guyana. Whenever this phenomenon occurs, it raises special difficulties for policy makers, as they are obliged to look for ways to compensate for the damage caused to other sectors by the boom in the major export sector. After Nigeria began exporting oil on a significant scale, it went from large net exports of agricultural goods to large net imports, with a reversal of the agricultural balance of payments by more than $1.5 billion between 1970 and 1980.

The most effective response to the Dutch disease syndrome is what the government of Kazakhstan has done, to purchase foreign exchange and invest it abroad in a long-term trust fund for social and economic development, thus putting the opposite presssure on the exchange rate in the short and medium run.

When policy tries to hold the exchange rate at an artificially overvalued level, preventing it from depreciating to its purchasing-power-parity level by, for example, raising interest rates to attract short-term foreign capital or by rationing access to foreign exchange, then the same consequence occurs for agriculture: it becomes less competitive on export markets and *vis-à-vis* imported products.[40] This kind of distortion became very common in the 1980s and 1990s and it constituted the main source of policy-induced bias against agriculture in developing nations. Combined with other sources of bias, such as tariffs that were higher for industrial goods than for agricultural goods and taxes on agricultural exports, this exchange rate distortion was found by The World Bank to contribute to a net 'tax' (through reduced real prices) on import-competing agriculture of 7% and a net tax of 35 to 40% on export agriculture, for a sample of 18 low-income countries.[41] The same study found that the countries that discriminated most against agriculture via policy were those which experienced the slowest overall growth.

Appreciating real exchange rates persisted into the 1990s and, in some cases, into this century. For a sample of eight countries, Valdés found that exchange rate policy was the principal cause of declines in real producer prices.[42] In essence, this is the nature of the damage to agriculture brought about by *appreciating real exchange rates: they cause real producer prices to drop*. That decline in prices has been measured at about 50% in Estonia since the exchange rate there was fixed in relation to the deutschmark, in June of 1992, to the end of 1996, and at 40% in El Salvador during the 1980s.

Thus, *the exchange rate is the most powerful policy influence on relative prices within an economy*, and its effect on real agricultural prices normally far outweighs the effects of other kinds of price intervention. This occurs because agriculture is typically the sector which is most

The maintenance of overvalued exchange rates is of particular significance as they impose a tax on exports and subsidize imports. This tool has been used at high cost to stabilize and hold down domestic food prices for urban consumers at the expense of the domestic producers of import-competing and exportable agricultural products, often in the face of severe domestic inflation which has been poorly controlled or exacerbated by economic policy measures. In the long term, therefore, the effects are damaging for food security as structural changes in the tastes and preferences of urban consumers that do not take account of real international prices, as well as increasing urban incomes, exert pressure to maintain and increase food imports; the ability to pay for those imports has been reduced by depressing the expansion of agricultural and food exports which, for many low-income countries, are the main source of export earnings; . . . and greater exchange rate overvaluation is related to lower GDP growth. (*FAO*, The State of Food and Agriculture 1996, *Food and Agriculture Organization of the United Nations, Rome, 1996, p. 294.*)

exposed to the influence of foreign trade: almost all of its products are either exported or importable, or they are close substitutes in production or consumption for products which are exportable or importable. Hence, agriculture's prices are largely determined by those of international markets and by the filter through which the

---

40. In addition, high real interest rates do disproportionate economic damage to agriculture, given its relatively high capital intensity and its unique dependence on production finance, which is a result of the long time lag between the investment in land preparation and the harvest. Yair Mundlak (1997, p. 19) has pointed out that agriculture is more sensitive to the interest rate than non-agriculture is.

41. Anne O. Krueger, Maurice Schiff and Alberto Valdés, 'Agricultural incentives in developing countries: measuring the effect of sectoral and economy-wide policies', *The World Bank Economic Review*, **2**(3), September, 1988, p. 266. See also the summary of those findings, and further development of some of the themes, contained in Maurice Schiff and Alberto Valdés, *The Plundering of Agriculture in Developing Countries*, The World Bank, Washington, DC, USA, 1992.

42. Alberto Valdés, *Surveillance of Agricultural Price and Trade Policy in Latin America during Major Policy Reforms*, World Bank Discussion Paper No. 349, The World Bank, Washington, DC, USA, 1996.

latter are transmitted to the domestic economy, which is the exchange rate. In contrast, the infrastructure and service sectors largely produce outputs which are neither imported nor exported, and so their domestic prices can rise with inflation, while those of agriculture are held down by the external influences and an appreciating exchange rate. It is for this reason that the exchange rate strongly influences *relative prices* in the economy.[43]

4.4.2 Exchange Rate Policy for Agricultural Development

In regard to the aforementioned difficulties in sustaining an open trade policy, it should be noted that one of the most effective ways to forestall pressures by farmers for the imposition of import restrictions is to ensure that the exchange rate remains in equilibrium in the sense of purchasing-power parity, because farmers are very aware of the prices of competing imports and an overvalued exchange rate cheapens them in domestic currency units. A policy of overvaluing the exchange rate makes it difficult to implement a policy of agricultural trade liberalization, and of industrial trade liberalization as well. In this sense, *it is very important to co-ordinate trade and exchange rate policy and to sequence the measures appropriately*.[44] This lesson was drawn out clearly from a careful study of the experiences in economic reform in New Zealand and Chile:

The most fundamental issue arising from reform in both Chile and New Zealand is that agriculture, composed largely of tradable goods, is most sensitive to shifts in trade and macroeconomic policies. The main elements are sound fiscal policy and exchange rate management. The level and stability of the real exchange rate in both countries were strategic. A real appreciation of the currency is not conducive to stimulating agricultural output and can create considerable resistance by farm lobbies against trade liberalization, and create strong pressures for special treatment after the major reforms have been implemented.[45]

The case of the CFA devaluation of 1994 illustrates the effects of the exchange rate on agriculture and also on economy-wide growth. The FAO found that:

in spite of the social and economic management problems caused by the initial price shock ... the devaluation contributed to a strong expansion of export earnings and a reduction of external deficits. ... Agricultural exporters benefitted significantly from the devaluation. ... For instance, Côte d'Ivoire. which had been suffering economic recession since the mid-1980s, increased its economic activity by close to 2 percent in 1994 and is expected to expand it by a further 5 percent in 1995, largely because of booming export performances. ... Economic growth is also expected to accelerate in Senegal, from 1.8 percent in 1994 to 3.5 percent in 1995, a significant contributing factor being groundnut exports. ...[46]

The most important factor in determining these economies' responses to the devaluation was the

---

43. Further discussion and examples concerning the exchange rate and relative prices are found in Roger D. Norton, *Integration of Food and Agricultural Policy with Macroeconomic Policy: Methodological Considerations in a Latin American Perspective*, Economic and Social Development Paper No. 111, Food and Agriculture Organization of the United Nations, Rome, 1992.

44. This is one of the principal messages to emerge from the study by Demetrios Papageorgiou, Armeane M. Choksi and Michael Michaely, *Liberalizing Foreign Trade in Developing Countries: The Lessons of Experience*, The World Bank, Washington, DC, USA, 1990.

45. Alberto Valdés, 'Mix and sequencing of economywide and agricultural reforms: Chile and New Zealand', *Agricultural Economics*, **8**(4), June 1993, p. 307.

46. FAO, *The State of Food and Agriculture 1995*, Food and Agriculture Organization of the United Nations, Rome, 1995, p. 79.

degree to which the revised export prices were passed on to producers. In some cases, the government controlled the price and passed on only part of the increase:

> in the Peanut Basin. . . . the government raised producer prices for peanuts twice – a combined increase of 71% over pre-devaluation levels, but less than the 100% increase in the CFA value of the world price. Passing on only a portion of the increase in the export price was a common strategy of Sahel governments after the devaluation for sub-sectors with strong government intervention. . . . in Mali, there was a rise in the profitability of Malian irrigated rice, whose output price was allowed to rise faster than input costs. . . . In the cotton zone of Mali, the net real return to farm output (cotton, maize, sorghum) increased by 14% for the south Sudanian zone and 20% for the north Guinean zone in the first 2 years after the devaluation. . . . This resulted from higher maize prices and from the government's decision to 'pass through' to producers the higher cotton prices resulting from devaluation gains.[47]

As a consequence of these policies, since 1994 Mali's agricultural sector has been one of the fastest-growing in Africa. Other countries in the region benefitted as well from the devaluation. According to the *New York Times Service* in March of 1996, in the Ivory Coast the 'devaluation of the [CFA] has given the economy a new competitive edge'. The same *New York Times* dispatch quoted an Ivorian agro-industrialist, Laurent Basque, on the subject: 'Now we are in a full-blown expansion and attacking new markets. The devaluation has allowed us to grow, and the same is true for most exporters'. In Mexico, after the exchange rate adjustment of 1994, a national federation of farmers issued a public statement vowing that they would never permit the exchange rate to become overvalued again.[48]

A lesson from the CFA experience is that, since a real exchange rate devaluation has almost entirely positive effects on agriculture, it is important not to dilute or nullify those effects by controlling prices to producers, restraining their response to the devaluation.

Since exchange rate policy can be a contentious subject, it is important to point out that a country cannot 'devalue its way to growth' in the longer run. Trading partners will respond to devaluations that are made artificially for the purpose of enhancing the country's international competitiveness, with the net effect of no gain for anyone and higher inflation all around. The conclusion that is being drawn here is that artificially controlling the exchange rate in the reverse direction, of overvaluing it in an unsustainable manner, has serious negative consequences for agricultural development. If such deviations from an equilibrium exchange rate are corrected sooner rather than later, the magnitude of the required adjustment, and hence the shock to consumers, is less, and the benefits for agricultural and national development greater.

The negative economic consequences of allowing the exchange rate to become overvalued are in fact more pervasive. In the words of T. L. Vollrath, 'exchange-rate overvaluation . . . lowers returns to agriculture, penalizes exports, induces capital flight, depresses foreign exchange earnings, dis-

---

47. Reprinted from *Food Policy*, **22**(4), T. Reardon, V. Kelly, E. Crawford, B. Diagana, J. Dioné, K. Savadogo and D. Boughton, 'Promoting sustainable intensification and productivity growth in Sahel agriculture', pp. 320–321, Copyright (1997), with permission from Elsevier. It should be noted that Mali tried to control many prices initially after the devaluation but soon abandoned that policy.
48. Examples of the effect of the exchange rate on agriculture are abundant. In the case of Ecuador, a model-based study concluded that 'macroeconomic policy, via the exchange rate, was responsible for a large part of the reduction in the share of agriculture in the economy between 1971 and 1981'. (From Grant M. Scobie and Veronica Jardine, 'Macroeconomic Policy, the Real Exchange Rate and Agricultural Growth: The Case of Ecuador', paper presented to the Annual Conference of the Australian Agricultural Economics Society (New Zealand Branch), Blenheim, New Zealand, July 1988, p. 12.)

courages domestic savings, crowds out productive investment and creates import restrictions that allocate resources to unproductive rent-seeking activity'.[49] In brief, *it is hard to overestimate the importance of an appropriate exchange rate policy for economic development, and not only for the agricultural sector*. India's strong economic growth in the 1990s was in good measure a result of a substantial real depreciation of the rupee achieved through nominal devaluations.[50]

A special case of exchange rate policies is found in *currency boards*, which not only fix the exchange rate with respect to an international reserve currency but also tie monetary policy by fixing the rules for determining the money supply.[51] Bulgaria, Estonia, Hong Kong, Djibouti and, until early 2002, Argentina are the main examples of the few countries that adopted this approach to macroeconomic policy in the past decade. Under a currency board, in effect, monetary policy loses all discretionality and disappears for all practical purposes.[52] The noteworthy advantage of this system is that it undoubtedly delivers price stability, albeit after a lapse of time which may last several years (in the case of Estonia). The guarantee of eventual price stability that is inherent in the system helps it overcome the usual tendency of overvalued exchange rates to encourage capital flight, and in fact currency boards have proven able to stimulate capital inflows, at least in the short run.

Their disadvantage is their rigidity: economic adjustments are forced to occur in the realms of interest rates and the labor market, and hence the consequences can include very high rates of unemployment. The system requires exceptional bank solidity, and therefore there is an increased risk of bank failures in the early years of its implementation (e.g. Estonia) and in crisis situations. There is a real danger that the system's rigidity can convert a recession into a depression, as happened in Argentina. This country saw its rates of unemployment rate rise to 18% in the years immediately after the introduction of the currency board, and the rate rose to explosive levels by the beginning of this decade, with the result of abandonment of the system. Lack of fiscal discipline at the national and provincial levels was a major reason for the collapse of the Argentine currency board, but that outcome only reinforces the point that such systems become straitjackets on policy in the face of economic shocks and stresses.

The most important operational concern about currency boards from an agricultural viewpoint is that the exchange rate should not be pegged at too strong a level (too appreciated a value). The reason is that there almost always is a considerable amount of residual inflation that needs to work its way out of the system before price stability is achieved. In the face of a fixed exchange rate, several years of continuing inflation, even if its annual rate is declining, can result in a considerable appreciation of the real exchange rate. This effect, of course, drives down real agricultural prices. The same prescription applies to economies that 'dollarize', or accept another major currency as their own: care should be taken with the rate of conversion of currencies, to avoid handicapping domestic producers in both agriculture and industry.[53]

---

49. T. L. Vollrath, 'The role of agriculture and its prerequisites in economic development: a vision for foreign development assistance', *Food Policy*, **19**(5), October 1994, p. 476.

50. D. Rodrik, 2001, p. 94.

51. A thorough explanation of currency boards from an advocacy standpoint is found in Steve H. Hanke and Kurt Schuler, *Currency Boards for Developing Countries: A Handbook*, Sector Study No. 9, International Center for Economic Growth, San Francisco, CA, USA, 1994.

52. The Argentinian currency board allowed exceptions to this rule which, for example, permitted the Central Bank to increase the liquidity in the banking system after the devaluation of the Mexican peso in 1994.

53. The decision to convert East German marks to West German marks at 1:1 made many industries in Eastern Germany uncompetitive and cost West German taxpayers very large sums over 10 years to compensate the East in the form of unemployment insurance and other subsidies.

> *The importance for productive sectors of pegging the exchange rate appropriately has been underscored by Sebastian Edwards: 'The adoption of an exchange rate-based stabilization program, where the nominal exchange rate is either pegged, or its rate of change is predetermined at a rate below ongoing inflation, carries a serious danger of provoking a major overvaluation. This can even happen if the fiscal deficit is fully under control. . . . This suggests then that in instances when exchange rate pegging indeed becomes part of the anti-inflation program, the initial starting point should be one of undervaluation. . . .' (Sebastian Edwards, 'Stabilization and Liberalization Policies in Eastern Europe: Lessons from Latin America', Working Paper IPR4, Institute for Policy Reform, Washington, DC, USA, 1991, p. 42.)*

In the Estonian case, the Bank of Estonia estimated that, between the date of fixing the exchange rate, June of 1992, and June of 1996, cumulative inflation was over 450%[54] and the real exchange rate appreciated by 268%. This was the main reason why real agricultural prices declined by about 50% during that period, *after* having declined substantially in the preceding period as a consequence of the elimination of Soviet-era price controls. The consequences included widespread bankruptcies of agricultural enterprises, abandonment of agricultural land and intensification of rural poverty.[55]

This experience highlights another lesson: that currency boards should not be implemented when inflation still is at very high levels. In Estonia, inflation for the entire year 1992 was about 1000%. The boards may be effective in bringing down inflation from moderate levels, and in keeping it down, but implementing one in the midst of high inflation practically ensures a strong subsequent appreciation of the real exchange rate, with very negative consequences for real agricultural prices.[56]

4.5 FISCAL POLICY AND AGRICULTURAL PRICES[57]

Fiscal expenditure policy can affect producer prices in agriculture through the provision of key elements of infrastructure: port facilities that allow better access to export markets, collection centers and storage depots that increase access to domestic markets, rural roads that reduce the cost of transport to market, irrigation that enables production to take place in the dry season or in arid zones, provision of new crop varieties that increase yields, and so forth. Such investments influence prices by increasing marketed supplies. Sometimes, they increase producer prices by providing farmers with access to new markets, and by reducing marketing costs. Especially for products which are not readily importable or exportable, they tend to reduce prices over the longer run by increasing supplies in domestic markets.

The effects on prices are direct and immediate in the case of particular kinds of revenue policy: commodity taxation and the regulation of prices in parastatal monopolies for purposes of revenue collection. It has been a common practice to tax export crops, partly for convenience of revenue collection in countries where income tax systems are not well developed, and partly out of conviction that such taxation would not affect production levels very much. In past decades, Argentina

---

54. Another factor explaining the continuing Estonian inflation was the rapid accumulation of foreign exchange reserves which, under the currency board system, automatically translated into increases in the money supply.

55. Ministry of Agriculture, Estonia, *National Strategy for Sustainable Agricultural Development*, Tallinn, Estonia, 1997, Chapter 2.

56. To the extent that industry is not protected by trade and tariff policies, its real prices will suffer in approximately the same degree as agriculture's in such a case.

57. This section is adapted from Roger D. Norton, *Integration of food and agricultural policy with macroeconomic policy: methodological considerations in a Latin American perspective*, FAO Economic and Social Development Paper no. 111, Rome, 1992.

was well known for a policy of taxation of its beef sector, which reduced prices and incomes to ranchers.[58] The Dominican Republic taxed its sugar sector and other plantation crops, and many other countries implemented similar taxation policies.

However, *selective commodity taxes distort incentives in the same way as unequal tariffs*: they give rise to economic protection rates that differ over products, thus breaking the linkage between internal relative prices (among commodities) and external relative prices. Implicit taxation through the pricing of parastatals has the same effect and has been widespread. In Haiti in the 1980s, parastatal agencies bought from producers at artificially low prices and sold to consumers at inflated prices, in order to generate additional fiscal revenues.[59] In the Dominican Republic, the State marketing and processing agency implicitly taxed vegetable oils, through a low price to producers, in order to finance subsidies on imported grains, thus depressing producer prices for grains as well.[60]

When capacity for tax collection is weak, commodity taxes appear to be an option that is administratively convenient, particularly for traditional exports since they are processed and sold in relatively few locations. In addition to their ease of revenue collection, three other arguments sometimes are cited in defense of such taxation practices:

(1) The short-run responsiveness of production of traditional export crops (their supply elasticity) is thought to be very low.

(2) For products facing international quotas, it is claimed that producers earn rents (profits above the normal rate) which can be transferred to the public sector via taxes, without reducing the supply of those products.

(3) Agriculture often does not contribute to fiscal revenues to the extent that the urban sectors do, and so commodity taxes offer a way to compensate for that paucity of tax payments.

Each of these statements is reviewed in turn. The first argument overlooks the fact that the long-run supply responsiveness of agricultural products, including plantation crops, is far from negligible. In fact, most of the long-run supply elasticities are much closer to 1.0 than to zero, thus suggesting that production will rise or fall approximately in proportion to price changes.[61] A second deficiency of the argument is that *relative prices have consequences for the intersectoral distribution of economic welfare*. In other words, even if there were no supply responsiveness to price, taxing agricultural exports would reduce agricultural incomes relative to those of other sectors. Given that rural households are poorer than their urban counterparts in most countries, this kind of tax is regressive.

The second argument, i.e. that the supply of products covered by international quota agreements is fixed, may hold in special cases, and within a limited range of price variation, but it is

58. Adolfo Sturzenegger (with Wylian Otrera), *Trade, Exchange Rate, and Agricultural Pricing Policies in Argentina*, World Bank Comparative Studies, The World Bank, Washington, DC, USA, 1990.

59. Roger D. Norton, 'Haitian Agriculture: Domestic Resource Costs and Pricing and Fiscal Structures', prepared for the Latin American and Caribbean Regional Office of The World Bank, Washington, DC, USA, 1984.

60. Consejo Nacional de Agricultura, 'Alternativas para una Nueva Política de Intervención de los Precios: El Caso de INESPRE', in *Compendio de Estudios sobre Políticas Agropecuarias en República Dominicana, 1985–1988*, Tomo I, Santo Domingo, Dominican Republic, 1990, Chapter 1.

61. Summaries of statistical evidence of supply elasticities are found in the following: (a) H. Askari and J. T. Cummings, *Agricultural Supply Response: A Survey of the Econometric Evidence*, Praeger Publishers, New York, 1976; (b) Shida Rastegari Henneberry, *A Review of Agricultural Supply Responses for International Policy Models*, prepared for the USAID Policy Analysis Project, Department of Agricultural Economics, Oklahoma State University, Stillwater, OK, USA, 1986; (c) Isabelle Tsakok, *Agricultural Price Policy, A Practitioner's Guide to Partial Equilibrium Analysis*, Cornell University Press, Ithaca, NY, USA, 1990, Appendix D.

not generally true. In El Salvador, the negative economic protection on coffee, which reached an extreme value of –67% in 1987,[62] sharply reduced coffee output. In 1975, Honduras and El Salvador exported approximately the same quantities of coffee (of the same type), and by 1988 El Salvador's coffee exports had fallen to about one-fourth of those of Honduras. (This trend was not the result of the war, as coffee-growing areas were not generally targeted by the contending armies.) As a consequence, Honduran coffee quotas were renegotiated upward and Salvadorean quotas downward. Thus, quotas do not necessarily determine amounts supplied, nor are they immutable.

With respect to the transfer of economic rents to the public sector, many producers of plantation crops are smallholders – coffee growers in Honduras, El Salvador and southern Mexico, harvesters of wild oil palm in Nigeria, banana growers in the Eastern Caribbean, and plantain growers and contract growers of bananas in Honduras, to mention but a few examples – so transfers of presumed rents out of the sector often come at the expense of relatively poor households. Following a sustained drop in the prices of sugar (another international quota crop), many sugar mills were closed in the Dominican Republic, Peru, Panama and other countries. If the rent element had been significant in the price, then price declines would not have caused those closures.

The third argument, i.e. that agriculture does not pay its proper share of taxes, may be correct in many cases, but it should be evaluated in the context that pricing mechanisms, both direct and indirect, frequently bring about implicit taxation of agriculture. If collecting more fiscal revenue from agriculture is a national goal, then it is important to bear in mind that *a tax on primary factors of production (land, labor and capital), or on income, which is the joint return to all factors,* *is not distortive of resource allocation decisions among products*. Therefore, from a viewpoint of economic efficiency, it is preferable to tax primary factors rather than commodities.

Thus, it can be seen that the typical justifications for commodity taxes in agriculture are lacking in foundations. Apart from the basic concern about fiscal burdens, such taxes are harmful to the sector's economic efficiency and therefore to its growth prospects, and they tend to make more unequal the rural–urban income distribution as well. As discussed in Chapter 5 of this volume, in many circumstances the most appropriate form of taxation in agriculture is a land tax.

4.6 MACROECONOMIC POLICY OPTIONS FOR AGRICULTURE

For the past five decades most economic policy in the developing world has displayed a pronounced bias against agriculture. T. L. Vollrath, citing the work of R. Bautista and A. Valdés, reported that 'the trade, macroeconomic, and sector-specific pricing policies adopted in the developing countries since the early 1950s have given rise to the following incentive biases: against the production of tradable goods and in favor of non-tradables; within the tradable goods sector, against exports as compared with import-competing goods; within the export sector, against agricultural products compared with manufactured goods; and within agriculture, against export crops compared with food crops'.[63] Furthermore, the 'empirical record shows that agricultural growth had a more pronounced impact on increases in developing-country income than did growth in the nonagricultural sector. The reason for the differential impact is that developing countries focusing on agricultural development experienced more broad-based income growth and increases in the demand for domestically-produced goods . . . '.[64]

62. Calculation by Mauricio González Orellana of the Salvadorean Foundation for Economic and Social Development (FUSADES).
63. Thomas L. Vollrath, 1994, p. 474.
64. *Op. cit.*, p. 469.

India presented a classic case of these anti-agricultural biases in policy, and as a consequence after the initiation of policy reforms in 1991 these biases had become the subject of extensive debate within India. The situation, which is illustrative of the macroeconomic framework for many countries today, was summarized as follows before the real devaluations of the rupee:

• The inward-looking, import substitution development strategy, which was aimed at rapid industrialization, shifted resources from tradable agriculture to industry by turning the terms of trade [relative prices] against agriculture.
• The overvaluation of the exchange rate subsidized imports and adversely affected all exports, especially agricultural exports.
• Most sector-specific policies at all stages of production, consumption and marketing of agricultural produce, worked against agriculture. For example, the price policy was in practice designed primarily to help the consumers. Farmers were generally given low administered prices in the name of helping the urban poor even when they had to pay higher prices for domestically produced inputs because of the protection given to local industry. In addition, a major proportion of the costs of the inefficient functioning of parastatal organizations, such as the Food Corporation of India, were borne by farmers.

In addition, 'large subsidies given on agricultural inputs. . . . failed to compensate the farmers for the negative impact of lower administered output prices, discrimination against agriculture because of overvalued currency and higher input prices caused by the excessive protection given to industry'.[65]

The case of Madagascar illustrates very well how agriculture can be caught in the vise of contradictory policies. A liberalizing reform process which started in the 1980s came to a halt in 1991 and 1992 amid social unrest and the reimposition of foreign exchange restrictions and import surcharges. Subsequently, 'attempts to combat inflation – itself fueled in considerable part by preferential (and non-performing) loans extended to public agricultural enterprises – led to continuing taxes on agricultural exports, an overvalued exchange rate, and attempts to keep agricultural raw material prices low for domestic agroprocessors'.[66]

Sometimes, the sequencing of well-intended policy reforms has deleterious effects on agriculture, as illustrated by the liberalization of markets for wheat and inputs in Kazakhstan:

Kazakhstan eliminated grain procurement in 1995. However, prices for farm inputs such as fuel and fertilizers were liberalized starting in 1993. Because of this sequencing gap, the grain sector experienced a severe terms of trade shock over a 2-year period. . . . Because producers had to pay market prices for inputs while receiving below-market prices for outputs, they incurred significant debts as they sought to meet State procurement orders for wheat. Indeed, many producers in Kazakhstan remain mired in debt. This example illustrates that while the end effects of liberalization policies can be positive, it is crucial to consider the sequencing.[67]

Anti-agricultural biases in policy may be starting to diminish with increased consciousness of these issues, but the movement is not always forward, and there still is a tendency in many

65. FAO, *The State of Food and Agriculture 1995*, Food and Agriculture Organization of the United Nations, Rome, 1995, pp. 120–121.
66. Bruce L. Gardner, 'Policy Reform in Agriculture: An Assessment of the Results in Eight Countries', Department of Agriculture and Resource Economics, University of Maryland, Baltimore, MD, USA, 1996, p. 8.
67. F. Goletti and P. Chabot, 2000, p. 670.

countries to allow the exchange rate to become overvalued and to decontrol agricultural input and output markets only partially.

In light of these circumstances, and the importance of the sector for economic development, the design of macroeconomic policy packages should take into account explicitly the effects on agriculture of policy changes. Principal macroeconomic options for agriculture, and their basic components, are outlined in the following paragraphs. Each of the five options provides at least some measure of incentives for agricultural development, or other kinds of measures to promote it, and each one utilizes a different class of policy instruments or a different combination of instruments.

(1) *Macroeconomic policy for promoting growth of the productive sectors – principal policy instrument: the exchange rate.* This option consists of eliminating, or avoiding, an overvalued exchange rate. If the currency is presently overvalued, returning it to an equilibrium level will have a short-run cost of additional inflation, but inflation can be eliminated over time if fiscal and monetary policies remain sound. This option is consistent with the objective of a stable exchange rate, but it would be achieved gradually through fiscal and monetary policy, rather than by trying to fix the exchange rate or influence its level through indirect means of rationing foreign exchange or artificially high interest rates to induce greater net short-term capital inflows. This option can also be called the approach of *maintaining the country's international competitiveness*. It has the advantage of providing incentives to agriculture without requiring additional fiscal expenditures, beyond those normally budgeted. It also is an option associated with lower real interest rates, *vis-à-vis* the scenario of an overvalued exchange rate, and it is favorable to industrial development as well as agricultural growth. If decisions on macroeconomic policy exclude this option, then one of the other four options listed below could be pursued in order to stimulate agricultural development.

(2) *A policy of favoring agricultural import substitutes – principal policy instrument: the tariff system.* This option can consist of up to four elements, or any subset of them: (a) making tariffs as uniform as possible over products, including eliminating tariff exemptions; (b) keeping the tariffs stable over time, except for planned gradual reductions in them; (c) applying tariff surcharges to products that face subsidized prices in the international marketplace and for a basic crop of the rural poor, as discussed previously; (d) applying a price band system to a few, relatively homogeneous products, such as cereals and powdered milk. In most cases, putting these policies into effect would improve effective protection rates for import substitutes in agriculture. The principal danger to be avoided is protectionism, through tariffs that are too high, which would make it difficult for the sector to eventually attain international competitiveness and would penalize consumers significantly without delivering growth benefits.

This policy clearly would have favorable effects on fiscal balances. Its principal disadvantage is that it discriminates against export products, and in agriculture these products typically generate more employment and income per hectare than the products oriented toward the domestic market. If this option were adopted, it should be made clear that it would be temporary, lasting only as long as the exchange rate were overvalued and international markets were distorted by subsidies in exporting nations. There is another danger here, of course: once in place, the tariffs may be difficult to remove or reduce significantly. In this regard, it can be important to tie part of the tariff revenues to a program of investments in productivity improvement for the protected products, and limit explicitly the duration of the program, say, to ten or fifteen years.

(3) *A policy of promoting both exports and import substitutes in agriculture – principal policy instruments: tariffs and fiscal expenditures.* This option is similar to the previous one, except that export subsidies would be instituted at the same level as import tariffs, to the extent permitted by WTO agreements. Since imports are almost invariably greater than exports in developing economies, the tariff revenues could finance the export subsidies. This option would eliminate a principal weakness

of the previous one, of an anti-export bias, and it could provide growth incentives to all products, provided that the protection levels were modest. High protection and subsidy levels would encourage inefficiencies. Like the previous option, this one should be viewed as a temporary compensation for aberrations in exchange rate policy and international markets. On the other hand, the presence of the Dutch disease syndrome, as in the case of Nigeria, could call for a correction by implementing this kind of policy on a long-term basis.

(4) *Fiscal compensation for the costs of the macroeconomic policy* – *principal policy instrument: fiscal outlays.* This option compensates for negative exchange rate effects on agricultural incentives through the mechanism of direct payments to producers, as a function of the amount of land tilled. Since the payments are made to a primary factor of production, they do not distort relative prices, in contrast to the two preceding options. This approach has been adopted in the European Union, as the McSharry Plan, and it was also put into effect in Mexico, where it is known as PROCAMPO, on the eve of Mexico's joining NAFTA. It has been adopted in Estonia and has been discussed in other countries. In Mexico, the payments were scaled at $100 per hectare, for land that had been cultivated in specified crops. However, the producers were free to change the crop composition after becoming eligible for the program. In the Mexican case, it was found that the distributional effects of the direct payments program were more equitable than both the agrarian reform and the guaranteed price program.[68]

Apart from its fiscal cost, the principal reservation about this type of program is that it is demanding in an administrative sense, and strict management and monitoring procedures would be required to avoid abuses. The optimal form of the program is to extend participation to all producers regardless of their cropping patterns, in order not to bias the selection of crops through the design of the program. Possible variants could include reducing the amount of payment per hectare for larger farms, and tying participation to observance of specified environmental guidelines. However, the latter variant would introduce a new dimension of administrative complexity, and so it would have to be planned with care.[69]

Implementation of the program would require an answer to the question of how to finance it. If the program replaces a price-support regime, then revenues formerly devoted to the latter could be redirected to the former. If this is not the case but the exchange rate has become overvalued, then in principle a case could be made for levying a special tax on the service sectors, since they would be the beneficiaries of such an exchange rate policy.

(5) *Institutional and structural reforms in agriculture* – *principal policy instruments: laws, regulations, Executive decrees and fiscal expenditures.* This is the structural reform option within agriculture, and it should be viewed as a complement to whichever of the previous four options is selected, and not as a substitute for them. It includes investments in the sector's infrastructure and also improvements in the land tenure arrangements, the rural financial system, the system of technology generation and transfer, and other structural aspects of the sector. While the previous options are aimed at creating *incentives* for growth in agriculture, this option attempts to promote the sector's growth by directly improving the *efficiency* of production and marketing

---

68. These findings were reported in Ramón Valdivia Alcalá, Jaime A. Matus Gardea, Miguel A. Martínez Damián and María de J. Santiago Cruz, 'Análisis comparativo de la distribución de la tierra y apoyos directos al productor: estudio de casos', *Políticas Agrícolas*, **IV**(3), 2000, pp. 93–127.

69. A skeptical view of the administrative capacity for such programs in developing countries is found in the review by John Baffes and Jacob Meerman, 'From Prices to Incomes: Agricultural Subsidization without Protection?', Policy Research Working Paper No. 1776, The World Bank, Washington, DC, USA, June 1997. While most of the paper's points are well taken, its negative view of the effects of direct support on the sector's growth prospects can be questioned.

and support services. In part, the programs under this policy approach seek to reduce what may be called the economic distance between producers and the markets and institutions on which they depend. In the language of economics, this is the approach of reducing transaction costs in product and factor markets. It is worth pointing out that such costs are almost always higher for the poorer farmers, and sometimes especially so for women farmers. A policy aimed at placing these farmers on the road to self-sustaining growth through their own efforts would emphasize precisely the measures needed to reduce the economic distance between them and rural markets and institutions.

These five options may help the clarify the broad panorama of macroeconomic policies as they are related to agricultural development. Under any option or combination of options chosen, it would be important to accompany them with an appropriate set of sector-specific policies – in effect, a full expression of option no. 5, of improving the productivity of farmers and the effectiveness of institutions. The remainder of this present chapter and Chapters 5 through 8 of this volume are devoted to issues concerning sectoral policies.

In closing this section, it should be emphasized that macroeconomic reforms are not necessarily adverse from the viewpoint of the agricultural sector. In fact, structural adjustment programs, which often remove any existing overvaluation of the exchange rate rather than exacerbate it, normally are favorable for the growth prospects of the sector. The analytics of agriculture's responses to adjustment packages are reviewed in a previous study by this author, and one of the conclusions was that 'agriculture is likely to benefit from in important ways from a typical structural adjustment program',[70] mainly because of the induced shift in relative prices in favor of

the sector. Other authors have reached the same conclusion, for example, Patrick Guillaumont for the case of sub-Saharan Africa. While noting that inconsistencies in implementation can dilute the benefits flowing to the sector, he observed that structural adjustment policies in principle favor agricultural development, because they aim at improving price incentives and productivity.[71]

4.7 SECTORAL POLICIES THAT INFLUENCE AGRICULTURAL PRICES

4.7.1 Price Controls

While the predominant instruments of agricultural pricing policy are those at the macroeconomic level, policies exist at the sector level which can also have an influence on real producer prices. The most obvious pricing instrument is ***administered prices***, or direct price controls. Although this instrument has been utilized in many countries, more for food prices than for non-food prices, it is increasingly being left by the wayside as policy reforms are implemented.

A fundamental problem with schemes of administered prices is that it is impossible for a centralized agency to continuously gauge with accuracy the balance of supply and demand, and therefore at most points in time there will be either excess supply or excess demand at the administered price. Excess supply is likely to mean that in reality producers will receive less than the producer equivalent of the administered price, or the government will accumulate costly stocks of the product. Through informal channels consumers may pay less, too. Excess demand translates into queuing, the classic symptom of shortages. It should be recalled that a basic role of prices is to equilibrate supply and demand, and prices need to have flexibility to be able to play that role.

When prices are fixed, the only element in the market process which can vary is quantity, and so

70. R. D. Norton, *Agricultural Issues in Structural Adjustment Programs*, FAO Economic and Social Development Paper No. 66, Food and Agriculture Organization of the United Nations, Rome, 1987, p. 42.

71. Patrick Guillaumont, 'Adjustment Policy and Agricultural Development', in G. H. Peters and B. F. Stanton (Eds), *Sustainable Agricultural Development: The Role of International Cooperation*, Proceedings of the XXI International Conference of Agricultural Economists, International Association of Agricultural Economists, Dartmouth Publishing Company, Aldershot, UK, 1992, p. 221.

shocks to the system (variations in harvests, for example, or declines in aggregate demand in the economy) are all converted into fluctuations in quantities, thus exacerbating the occasional problems of surplus and shortages.

Another problem with this policy is that, in order to sustain fixed prices, trade needs to be controlled. One kind of intervention into the market breeds another kind. For this reason, imports often cannot alleviate temporary shortages, for governments usually do not react swiftly enough to modify the trade controls, taking into account the lead times necessary to bring in imports. Furthermore, attempts to get around the surpluses and shortages created by price controls may lead to black markets and corruption.

It should be noted that administered prices are almost always used to favor consumers rather than producers, especially for basic foods such as cereals, dairy products and cooking oils.

Apart from these essentially operational problems, there is a more fundamental difficulty with price controls: they are almost certain to lead to a misallocation of resources, in the sense of over- or under-investment in the production of the commodities subject to the controls. A market price not only balances supply and demand, but if foreign trade is unrestricted also, then such a price encourages supply to be at the level indicated by the country's comparative advantage. The counter-argument is sometimes advanced that the presence of oligopolies and monopolies in food processing requires the imposition of price controls, but there is no assurance that administered prices will better approximate the outcome of a competitive market, especially as they are subject to political influences. Other policy responses are available for the problem of monopolies and oligopolies, as commented upon below.

For all of these reasons, price controls often worsen problems of shortages of supply, thus defeating in the long run the attempt to control inflation by use of this instrument. A study of the transition from planned to market economies in Eastern Europe concluded that price liberalization actually leads to lower inflation than a policy of price controls.[72]

A special case of administered prices is pan-territorial pricing which has been put into practice at times in Zambia, Peru and some other countries. This kind of policy tries to make a commodity's price equal throughout all regions of the country, by administrative decree. The fallacy in such pricing is that it tries to suppress the reality of transport and marketing costs, to move products from regions of surplus supplies to those of deficits. Since these costs are not allowed to be reflected in the product's price variations by location, they have to be paid by someone else, i.e. the government, producers or consumers. If the government pays, then usually it takes control of the marketing process, something that it is not well qualified to do efficiently. If the cost is assigned to producers or consumers, through the level of the price which is set, usually producers are the ones who pay, and that implies a negative incentive for production. On the other hand, if the price is set artificially high in major producing regions, then a surplus can result. Often, such a surplus can be exported only at a cost to the government's budget, since the producer price may have been higher than the equivalent international price, taking into account margins for transport and marketing.

One of the most important challenges in agricultural development is fostering the growth of an efficient private marketing system, and denying that this activity has a legitimate cost marks a step backward in terms of meeting that challenge.

4.7.2 Farm Support Prices

Support prices, or **guaranteed prices**, also have been widely implemented. They are cousins of controlled prices; they attempt to control prices in the downward direction only, while allowing them to rise without restriction. They suffer from the same conceptual and practical limitations that administered prices do. In addition, support

---

72. Martha de Melo, Cevdet Denizer and Alan Gelb, 'Patterns of transition from plan to market', *The World Bank Economic Review*, **10**(3), September 1996, p. 397.

prices are costly to the government budget, for usually they are designed to raise the price to producers above the market-clearing level while keeping prices for consumers at or below the market level. The government then pays the difference, or the inverse marketing margin.

Another source of budgetary expense arises from the network of collection points and storage facilities that governments feel obliged to construct and operate in order to manage a policy of farm support prices. Often these facilities are not managed effectively, which further increases the cost of the policy to the government. In China, for example, after a government policy review in 1998:

> The grain stations were charged with: i) being too slow to change management mechanisms; ii) having surplus staff; iii) being poorly managed; and iv) having diverted funds designated for grain purchases into other uses.[73]

In Honduras, it was found that State-owned silos were operated on average at only 10% of capacity over a period of many years.[74] In El Salvador, the government silos were effectively empty for a number of years before they were closed down.

In spite of these budgetary outlays, frequently there are not enough funds available to fulfill the commitment to purchase the crop at the guaranteed price, especially in years of abundant harvests. As a result, just when the guaranteed prices are most needed from the viewpoint of farmers, the government often cannot deliver them. In Russia, 'by 31 December 1995, only 1 million of the 8.6 million tonnes of grain due to be procured by the federal government had been bought. By 1 February 1996, only 1.6 million of the planned [amount] had been purchased.... A lack of budget funding ... was blamed ....'.[75]

Another common defect of these policies is that access to the guaranteed prices is biased in favor of the better-off farmers, who can truck their crop to the collection points or otherwise ensure favorable treatment. Statistical documentation of a strong bias in this sense, for the case of Honduras, is mentioned in Chapter 3.

A basic conceptual problem with support prices concerns the determination of their level. Even if it is accepted that they are not intended to be market-clearing prices, what should their level be? Often, the operational procedure is to try to set a support price at a level which covers the estimated cost of production of the crop concerned, and to raise the price from year to year as costs increase. However, this is equivalent to rewarding inefficiency.

In addition to this serious problem with the scheme, it may be asked: whose costs of production? After all, in reality a supply curve is composed of many thousands of points, each representing a different farmer and/or technology or region of production. Should the support price equal the average cost of production, the marginal cost, or some other cost? Setting it at the marginal cost would seem to favor inefficiency also, and provide rents to all producers save the least efficient one. Plus, in the heterogeneous real world of farms and their resource endowments, an average cost may be difficult to calculate with any degree of accuracy.

For all of these reasons, guaranteed prices are falling out of favor in an increasing number of countries, from Latin America to the Middle East to Sub-Saharan Africa. They have been more durable in South and East Asia, perhaps made more effective in their implementation by these countries' long traditions of public administration, and in Eastern and Western Europe. Nevertheless, the conceptual objections to the schemes hold in the Asian and European contexts as well, and so they are undergoing re-examination there also. Prices to farmers can be raised more effectively, and at a lower cost to the government, by eliminating tariff exemptions on food imports,

---

73. OECD, 1999, p. 134.

74. Secretaría de Recursos Naturales, Programa Nacional de Reactivación Agrícola: Desarrollo Compartido en el Agro, *La Nueva Política de Comercialización de Granos Básicos*, Tegucigalpa, Honduras, May 1991.

75. OECD, *Agricultural Policies, Markets and Trade in Transition Economies: Monitoring and Evaluation, 1996*, Organization for Economic Co-operation and Development, Paris, 1996.

ensuring the exchange rate is at an equilibrium level. For this reason, a lemma of an effective agricultural policy reform can be: ***moving from a controlled but penalized agricultural economy to a free but protected one***. In Honduras in the early 1990s, effective protection rates went up for farmers after reforms to the tariff and exchange rate systems, even while trade was being freed and guaranteed prices were being dismantled.

The foregoing are powerful arguments against the use of farm support prices. However, in some cases policy makers may either remain unconvinced or feel that a commitment to such policies cannot be altered, at least in the near term. If that is the case, then two important guidelines are suggested:

- The number of instruments should not be greater than the number of objectives. This basic rule was first enunciated by Jan Tinbergen in the 1950s[76] and still applies to economic policy. The cost of not heeding it is usually considerable economic inefficiency resulting from clashing policy instruments. In this case, the number of objectives is one – maintaining a floor price for the specified product. Hence, there should be one instrument – the support price. Adding the instrument of trade controls to tariffs can be counterproductive.
- The government purchasing agency should have adequate infrastructure and funding to carry out purchases at the support price as needed, in all parts of the country. Its management should be as professional as possible and endowed with clear operating guidelines.

The importance of observing these two guidelines is brought out by a comparison of the contrasting experiences of Sri Lanka and Indonesia, in a study by Frank Ellis, Piyadasa Senanayake and Marisol Smith:

This paper describes the food market intervention system in Sri Lanka, and poses the question whether the parastatal organizations that trade in rice and wheat play any useful role in contributing to food security and food price stability. Analysis of time-series price data for the rice market suggests that the private rice marketing system is competitive and efficient. . . . There is no reason to deduce that the domestic distribution of wheat flour could not similarly be performed efficiently by private sector agencies. . . . It is interesting to compare the Sri Lankan case to that of Indonesia. . . . In the case of Indonesia, the annual setting of a floor price in advance of planting decisions; its effective defense at peak harvest by an active purchasing agency, Bulog; the stability of annual procurement levels by Bulog at around 4–6% of the harvest; and the historical reluctance by the government to import rice, meant that the floor price approach appeared the most effective and low-cost means of achieving domestic price stability. In Sri Lanka, by contrast, the erratic and intermittent setting of the [support price]; the widely varying procurement levels of PMB [the State procurement agency]; the deteriorating infrastructure of PMB; and the flexible use of rice or wheat imports to keep staple food markets in balance, means that the [paddy rice support] price is a redundant policy instrument.[77]

4.7.3 Strategic Reserves and Grain Market Liberalization

Strategic reserves, almost always of grains, represent a response to the kind of concern felt by the Roman emperors mentioned at the beginning of this chapter: that the population might be left without adequate supplies of a basic food, if severe droughts or other unexpected events were to occur. Clearly, no government is willing to incur that risk, hence the tendency to store up strategic reserves. The pressure to accumulate strategic reserves can be strong for products with thin international markets, such as white corn, cassava and some varieties of beans.

---

76. Jan Tinbergen, *On the Theory of Economic Policy*, North-Holland, Amsterdam, 1952.
77. Reprinted from *Food Policy*, **22**(1), F. Ellis, P. Senanayake and M. Smith, 'Food price policy in Sri Lanka', p. 95, Copyright (1997), with permission from Elsevier.

Attempts to strengthen the private sector's participation in food marketing can founder on strong biases against the private sector in bureaucratic circles. As Joseph Ntangsi has pointed out for the case of Cameroon:

. . . there has been a long-standing prejudice against the private sector, especially by government officials, who consider private traders as unorganized and inefficient (since they operate in numerous small units) or as downright unscrupulous and exploitative of ordinary citizens. (Joseph Ntangsi, 'Agricultural Policy and Structural Adjustment in Cameroon', in G. H. Peters and B. F. Stanton (Eds), Sustainable Agricultural Development: The Role of International Cooperation, *Proceedings of the XXI International Conference of Agricultural Economists, International Association of Agricultural Economists, Dartmouth Publishing Company, Aldershot, UK, 1992, p. 267.*)

At the same time:

The inefficiency of State marketing is also evident in the marketing of foodstuffs and inputs; in both cases marketing costs have been higher than for private traders, and in the case of inputs farmers have had to face the additional problem of late delivery (op. cit., p. 272).

Strategic reserves are costly, however. In addition to the storage costs over time, government outlays must cover the difference between (a) the purchase price plus handling costs, and (b) the sales price, since the grains often are sold at less than full cost. Administration of the program and storage losses of the grains represent other costs. For beans in particular, storage for extended periods of time makes them harden so they become difficult to use. Finally, there is an eco-

nomic efficiency cost associated with government interventions in the marketplace, when purchases are made for the reserve and also when releases from the stockpile are deemed necessary. This cost includes discouraging the development of modern marketing systems in the private sector. Therefore, the question arises of whether the stated aim could not be achieved in other, less expensive ways.

Constitution of a strategic reserve represents a lack of faith that the market will prevent extreme shortages from occurring. However, experience has shown that markets usually are better at ensuring that the populace's needs are met than governments are. A study by John Mellor and Sarah Gavian showed that worldwide, apart from wars, famines have been caused more by misdirected policies than by natural disasters.[78] In principle, under a free trade regime, the private sector should respond to signals of impending shortages, and the quantities available on the world market have always been sufficient in the post-World War II period. That food has not always been available in the needed quantities and in a timely manner is attributable in part to weak distribution systems, and continuing government actions in markets for food products tends to inhibit the development of such systems.[79]

While the validity of these points is widely recognized, in low-income countries the private sector is not always prepared, as noted earlier, to take over all marketing and storage functions immediately. Capital and the requisite expertise may be in short supply. Above all, *entrepreneurs may be skeptical that the government will stay out of the field in the future*. For the case of Nicaragua, for example, it has been observed that private traders did not fill the gap left by the State when it ceased agricultural marketing activities.[80] A similar phenomenon was found in the course of an economic reform program in Malawi: 'Traders have not replaced [the parastatal marketing corporation] in the markets which were abandoned as a result of the rationalization of

78. J. Mellor and S. Gavian, 'Famine: causes, prevention and relief', *Science*, **235**(4788), 1987, pp. 539–545.
79. Chronic malnutrition, as opposed to temporary food shortages, is more a function of low income levels than of physical scarcities of food.
80. M. Spoor, 'Liberalization of grain markets in Nicaragua', *Food Policy*, **20**(2), April 1995, p. 99.

economic activities. . . . This has had adverse effects in most remote areas in terms of access to inputs and markets'.[81]

However, the private sector is not always slow to respond. In Somalia, 'Prior to reform in the mid-1980s, producers were obliged to deliver all surplus production to the Agricultural Development Corporation . . . at prices well below parallel market prices. . . . there appears to have been a significant production response to the removal of these controls and the legalization of private trade and storage. This was accompanied by a strong private sector trading response'.[82]

Furthermore, the same authors point out that 'A large number of studies of private traders' responses to food market liberalization have been carried out in several countries in [Eastern and Southern Africa] over the last five years. The general finding of most of this research is that private traders have been able to respond to increased market opportunities resulting from the relaxation of legal restrictions on their activities and the reduced role of the parastatal marketing agencies with greater success than pessimists might have anticipated' (*op. cit.*, p. 404). Nevertheless, the private sector's response was not always fully adequate in those countries: 'In most countries of the region, private sector trading activities have received little official support even after liberalization. This has inhibited the private sector's capacity to respond to the opportunities of a more liberalized environment. There is therefore an urgent need to find effective ways of assisting the development of the sector and the promotion of competition within it' (*op. cit.*, p. 406).

In circumstances in which the private sector is not quick to fill the gap left by elimination of governmental marketing programs, policy makers can be faced with a difficult choice: whether to allow shortages to develop far enough, with markedly rising food prices or input costs, so that the private sector will respond by importing food in the necessary quantities, or whether to continue to maintain a strategic reserve against such a contingency, and possibly to import foods on public sector account once prices begin to rise. The latter policy surely will delay the promotion of a private sector role in food marketing, but the former policy may entail real hardship in the short run for the population, at least for the urban poor.

This dilemma is basically a transition problem, a question of tactics rather than long-run strategy, but it is no less difficult for that reason. A solution has to be developed within the economic and historical context of each country, but perhaps adherence to two principles can assist in the development of an appropriate approach: (a) ***proper sequencing of policy reforms***, and (b) absolute ***clarity and transparency regarding the rules governing public sector interventions*** during the transition period.[83] An example of the proper sequencing of reforms is that a targeted food assistance program for the urban poor must be in place before all responsibility for managing the food marketing chain is turned over to the private sector. Another example was mentioned above for Kazakhstan, regarding the need to liberalize agricultural output markets before or simultaneous with a liberalization of input markets. Likewise, such a fundamental change should be preceded by extensive sensitization of private businessmen to the nature of the new policies and the role expected of them, through workshops, focus groups, publications in the news media and other techniques.

---

81. Reprinted from *Food Policy*, **18**(4), K. M. Mtawali, 'Trade, price and market reform in Malawi: current status, proposals and constraints', p. 306, Copyright (1993), with permission from Elsevier.
82. Reprinted from *Food Policy*, **17**(6), J. Beynon, S. Jones and S. Yao, 'Market reform and private trade in Eastern and Southern Africa', p. 401 (and pp. 404 and 406), Copyright (1992), with permission from Elsevier.
83. Based on the experience of Tanzania, it was found that successful implementation of grain market liberalization requires clarity and consistency in government policy and also the reform of financial services and the sale of unused storage capacity to the private sector (J. Coulter and P. Golob, 'Cereal marketing liberalization in Tanzania', *Food Policy*, **17**(6), December 1992, p. 420).

For ensuring clarity and transparency of the rules for government interventions during a transition period to a market economy, it is important that releases of grains from a strategic reserve be made *only* when, for example, the real prices of the grains rise from one month to the next by more than a specified percentage. Such a rule should be disseminated widely. On the side of purchases for the reserve, in order to avoid favoritism and to ensure that the market price is not disrupted more than necessary, the required purchases should be made in small lots in open auctions among producers and other sellers, in different locations of the country. Finally, a companion measure should be to gradually reduce the size of the strategic reserve, in a programmed way, until it is completely eliminated.

Alternatively, decision makers' concerns about food prices can be allayed by constitution of a *financial food security reserve* instead of a physical grain reserve. The financial reserve would be used to import grains under specified conditions regarding rises in real prices, again respecting the principle of clarity and transparency. An advantage for a government is that a financial reserve is cheaper than a physical reserve: there is no wastage, there are no management costs, and it would earn interest. A financial reserve also should be viewed as a transitional measure, maintained only until it is clear that the private sector can manage the grain trade adequately, but it may be the least costly of the alternatives.

Finally, when the problem of building up an adequate private sector marketing system is viewed in its fullest expression, there is an argument for transitional subsidies for marketing activities, in order to assist with the shift from a State-dominated system to a private one. Such subsidies can be implemented through privatization of State storage facilities on concessional terms (provided that private monopolies are not created) and perhaps through use of temporary credit rediscount lines to support investments in private capital goods for storage and marketing. This kind of support should be accompanied by training activities for private marketing agents.

4.7.4 Instruments of Grain Storage Policy

Traditionally, grain storage policy has been conceived as a program of building (State-owned) silos and collection points. Today, storage policy is more likely to emphasize adequate management of the facilities and the financing of storage costs incurred by farmers. Accordingly, its four main components are the following:

- Privatization of storage facilities, optionally with at least some capital participation by farmers, as discussed in Chapter 3.
- Development of a program of certificates of grain deposit, or crop liens, which is available to all classes of farmers.
- Establishment of an effective system of quality standards for grains.
- Transformation of grain handling to a bulk basis, instead of in bags.

A network of private grain storage facilities is a key component of a marketing system, and it goes a long way toward reducing the chances of extreme seasonal fluctuations in prices.

Regardless of whether the silos are owned by the State or by the private sector, certificates of grain deposit are essential in order that farmers of all income strata have access to the storage option. Storage not only carries its own direct costs, but it also represents a deferral of the income that results from sale of the crop. Many smaller farmers are obliged economically to sell their crops at harvest time, when prices reach their lowest point during the annual cycle. Some even sell the crop when it is still standing in the fields. In many circumstances, they could gain by postponing the sale for two or three months, until prices begin to recover, but cannot afford to wait that long to receive their major source of income for the year. Bankable certificates of grain deposit, along with a scheme of registered storage locations, can facilitate for them a payment that represents up to 80% of the value of the crop, through the banking system.

Certificates of grain deposit, in turn, cannot function without a system of grading standards for the products. A storage facility cannot be managed on an economically viable basis by depositing bags of grain with farmers' names on them. Grains have to be stored collectively, by grade. To put a grading system into effect requires a substantial effort in educating farmers. Among its benefits is the differentiation of market prices by grade, and this will provide incentives for quality improvement and eventually will assist domestic farmers to better meet the challenge of imported products, which tend to be uniform in grade.

An important implication of the move to storage by grade is that products can be transported in bulk, instead of in bags. Similarly, in the case of larger farms and co-operatives, modern harvesting equipment can be used that handles the product in bulk. Changing a country's grain handling and storage system to a bulk basis, in the links for which it is feasible, can reduce harvest and post-harvest costs and make the country's grain producers more competitive internationally. Argentina, Uruguay, Ecuador, Venezuela and most of Brazil have carried out such a transition to bulk handling of their rice production. It can be worthwhile to investigate the feasibility of making such a conversion.

Both certificates of grain deposit and the development of product grading systems to facilitate storage are examples of policies that improve farmers' *access to markets*.

Public policy also can assist in the establishment of adequate on-farm storage. The value of this contribution has been illustrated by V. Seshamani for the case of Zambia:

> ... lack of on-farm storage facilities and compulsory expenditures such as the payment of school fees and purchase of blankets due to the onset of winter at the very time of the ripening of the crops impel low-income farmers to sell their crops at very low prices to exploitative marketing agents. This results in further low incomes for the farmers rendering them unable

to find the wherewithal for the purchase of seed, fertilizer, etc., leading to lower crop output in the following season and further low incomes.

> Inadequate availability of marketing agents, inadequate finance and credit facilities, inadequate on-farm storage facilities, the poor state of the roads, inadequate transport facilities in the rural areas and slow information flows all contributed to the poor overall supply response to the policy changes. . . .[84]

4.7.5 Market Development

Opening new markets can be as beneficial to farmers as increases in prices in existing markets, and more so if the new markets are associated with higher-value crops. Achieving access to new markets can be difficult for individual producers, especially in the case of export markets, and therefore it may require specialized commercial expertise or effective farmer organizations that can acquire such expertise. Globalization is providing such opportunities in greater numbers and at the same time making the requirements for entry more demanding. Markets are increasingly stringent in regard to product quality, and therefore *greater market penetration usually requires close attention to quality issues*.

In the longer run, product quality will have a great bearing on prices. In both industrialized and developing countries, consumers are ever more conscious of considerations of food safety and the environmental conditions under which food is produced. Already certain products from some developing countries have been excluded from developed markets due to problems with food safety. New and stricter food safety standards for imports will come into effect in the European Union in 2006. Controlling levels of pesticide applications on farms is one of the challenges in meeting food safety standards.

---

84. Reprinted from *Food Policy*, **23**(6), V. Seshamani, 'The impact of market liberalization on food security in Zambia', pp. 549–550, Copyright (1998), with permission from Elsevier.

There are three major dimensions to product quality: phytosanitary conditions, or freedom from plant diseases and pests, food safety conditions, which refer mainly to freedom from chemical residues, and consumer preferences as reflected in product taste, size, shape, color, uniformity and suitability for preparation in kitchens.

Meeting phytosanitary and food safety conditions requires appropriate institutional development within the exporting country, so that laboratories and national certification processes can gain international recognition. The historical priority of phytosanitary regimes was to prevent the importation of plant and animal diseases and pests. That priority remains important, but now an additional priority is to adopt an outward-looking orientation and ensure compliance with international sanitary regimes in order to facilitate exports. Given that most export crops are more labor-intensive than import-substitution crops, this concern is relevant to poverty alleviation as well as to the balance of payments.

Fruit and vegetable exports are the most sensitive to phytosanitary and food safety regimes, and in many countries they are largely cultivated on small farms. The same is true of organic production, for which markets are growing and also ever more demanding in terms of product quality. For small farmers, once again, organization is a key for penetrating these markets. When organic certification is obtained through farmer organizations, its cost per producer is considerably reduced.[85] While there is a substantial price premium for organic products, market analysts have forecast that it will decline because of the rapidly increasing supply of organic products. However, at the same time, national markets for organic foods will be increasing within developing countries, as consumers become more aware of health issues in the diet.[86]

There appears to be a trend toward concentrating the export supply of particular products in a few countries, in effect rewarding those countries that made an early effort in this area. For example, as of 2002 Costa Rica exported around US$100 million annually in pineapples, Peru about US$15 million just in fresh mangoes, Argentina US$83 million in lemons and limes, and Mexico US$124 in mangoes and more than US$50 million in avocados.[87] The more a country delays in mounting the required efforts for quality certification in non-traditional products, then the more difficult it is to penetrate international markets.

Satisfying consumer tastes and delivery requirements depends on strong linkages with the marketing chain and the technical assistance that can be provided through it. Establishing relations of confidence in international markets – brand reliability – is difficult but can bring many economic benefits to producers when it is achieved.

4.7.6 Farm Prices and Agro-Industries

For many processed agricultural products the domestic market is characterized by a concentration of buying power – monopsony or oligopsony – either nationally or in regions within the country. Invariably, these are products that are milled or otherwise processed before reaching the consumer. A few of the most common examples include cotton, sugar, milk, rice, wheat, coffee, rubber, palm oil, tea and jute. The high unit cost of transportation, relative to the value of the raw product, makes it impractical for farmers to deliver their harvests to processing facilities located in distant parts of the country – much less to destinations abroad. Therefore, the local facilities frequently enjoy a quasi-monopsonistic position which gives them considerable power over prices of raw products.

This kind of issue is worthy of the attention of policy makers, given the importance of agricultural incentives for development, but it represents a difficult challenge. A free trade policy can help

85. See Octavio Damiani, 'Small Farmers and Organic Agriculture: Lessons Learned from Latin America', Office of Evaluation and Studies, International Fund for Agricultural Development (IFAD), Rome, 2002.

86. *Ibid.*

87. Luz Amparo Fonseca, 'Los mercados de frutas y hortalizas: Colombia frente al ALCA', mimeo, Bogotá, Colombia, 2002.

place an effective cap on the prices which processors charge to consumers, but it has no effect on the prices that they pay their suppliers. There are three approaches to this problem which can improve the situation to a degree, although none is a panacea:

(1) ***Producers' ownership of selected processing facilities.*** Facilities owned by the primary producers can offer a better price for the raw material, subject to the limits imposed by profitability considerations, and they can set an example in pricing policy for other facilities. Ownership can take the form of co-operatives, corporations or other types of proprietorship. Dairy co-operatives have had an illustrious history in countries such as the Netherlands, Canada and the United States, and they also have been formed in many lower-income countries. Producer ownership of some rice mills has been found to be effective also, in Peru, Guyana and other countries. In El Salvador, State-owned sugar mills were privatized in the early 1990s under the requirement that 55% of the shares be transferred to cane growers.

(2) ***Anti-monopoly legislation.*** Legal remedies for cases of abuse of the power of monopolies and oligopolies, and monopsonies and oligopsonies, can be made available, and should be. Nevertheless, proving abuse is likely to be a cumbersome, uncertain process, and the difficulties of doing so should not be underestimated, especially in circumstances where judiciaries are relatively weak.

(3) ***Pricing agreements endorsed by the government.*** In particularly difficult cases, it may be necessary for the government to apply 'moral suasion' and broker price agreements between associations of farmers and of co-operatives. There are political costs associated with risking the government's prestige in this way, and there also is a risk of introducing additional distortions in the market, and so it is not a step to be undertaken lightly, but if there appear to be clear-cut cases of abuse and no other solution is in sight, then such an approach may be useful.

In sub-sectors with multiple processing facilities, it would be justified to provide credit, through rediscount lines, to assist producers to buy one of them or construct a new facility of their own, to help change the market structure to one in which pricing more closely follows competitive patterns.

For the approach of tripartite price negotiations between producers, processors and the government, it is useful to have objective reference points concerning prices, such as international data on the relationships between prices of raw materials and processed products. Obviously, such relationships depend on many factors in each country, including the technical and economic efficiency of processing industries. Nevertheless, international norms may provide helpful points of reference. For example, a survey of all major sugar producing countries in the world conducted in the 1980s found an average ratio of 53.1% between the price of the sugar content of cane and the price of processed sugar.[88] It may be argued that the most relevant ratio is that which applies to the more efficient producers, i.e. a given country's policy should not emulate that of the more costly producers. The above mentioned ratio was recalculated as 54.8% for the 26 most efficient producing countries in the sample.

To illustrate how such information might be used, *if* a country's field costs were less than the international average (which was $193/mt in 1982, according to the cited study), and *if* the ratio between field costs and total costs were less than 54.8%, then an agreement might be reached to raise that percentage by a full or half percentage point per year, until the target level of 54.8% were

---

88. William McNally, Wilfred David and David Flood, *Sugar Study*, prepared for the Agriculture and Rural Development Department of The World Bank, Washington, DC, USA, July 1984. In calculating this average from their study, a sample of 53 countries was used, leaving out three extreme outliers that may have been the result of numerical errors.

reached. Such increases would cease if the field cost of sugar rose to the international average. This is only an example; more up-to-date data should be utilized. International organizations such as the FAO could assist in disseminating international information on this kind of price relationship.

There is a third approach which indirectly may assist producers to receive a better price from processors, and that is the previously mentioned policy of implementing product grading standards and a system of price differentials by grade. At first, some producers may receive lower or unchanged prices under such a system, but improvements in the quality of their product would eventually bring them better returns.

While producers and agro-processors may be opposed on price, co-operation among them can bring benefits in the form of lower transaction costs, higher quality of raw produce, and consensus on areas in the production and marketing chain that most need improvement and policy assistance. The approach of 'agricultural product chains' (*cadenas agropecuarias*) has yielded these kinds of benefits in Colombia.

4.7.7 Input Markets

In all countries, input markets are as important to farmers as output markets, and their liberalization can play a vital role in stimulating the sector's growth. A reform process faces many of the same issues that characterize liberalization of output markets. For example, is the private sector prepared to take on the responsibility? Is government committed to staying out of the markets? Nevertheless, governments are withdrawing from markets for all classes of purchased inputs, including improved seeds, where government's main continuing role is in the certification process.

Markets for agrochemicals present special issues because of the economies of scale in their operations. Yet the benefits of liberalization in this area are strong also, provided that appropri-ate complimentary policies are undertaken. A study of liberalization of the fertilizer market in Kenya shows some of the benefits and issues to be dealt with:

The results [of the study] confirm that liberalization of Kenya's fertilizer market has induced a vigorous response from the private sector in both high- and low-potential [agricultural] areas, implying that broad efficiency gains have accrued to the farming sector. But the results also suggest that there may be additional benefits to be reaped through policy initiatives. . . . two demand-side factors (agroecological conditions and food prices) and one supply-side factor (access to credit) dampen trade in [low-potential] areas. . . . continued support for research on soil fertility management problems in low-potential areas is vital. . . . public investment in improved rural infrastructure is likely to be important to unleashing potential on both the supply side and the demand side. . . .

As in many parts of the world, fertilizer trading in Kenya exhibits large economies of scale that require access to credit if they are to be realized. . . . A key policy challenge is to design mechanisms through which these economies – currently captured by large-scale traders and farmers – can be passed on to smaller traders and farmers. In theory, this was one of the roles to be played by the now moribund Kenya Farmers Association. . . .[89]

This last challenge is important for bringing down the price of chemical fertilizers to small- and medium-scale farmers. Small, poor countries import relatively small quantities of fertilizer by world standards, and therefore their farmers typically pay a much higher price for this input than farmers do in industrialized countries. Farmers' associations can attempt to play a commercial role in fertilizer markets, but in the end the fertilizer price will always be higher than in larger countries, and so it also is important to adapt technologies for soil management and production to reduce the degree of dependence on

---

89. Reprinted from *Food Policy*, **26**(1), S. W. Omamo and L. O. Mose, 'Fertilizer trade under market liberalization: preliminary evidence from Kenya', p. 7, Copyright (2001), with permission from Elsevier.

chemical inputs. In this regard, research and extension efforts must adapt to market signals as much as farmers do.

4.7.8 Additional Considerations about Prices and Markets

Investments in infrastructure have not been mentioned in more than a passing way in this section because the main topic of this volume is policies, but it should be stressed that investments in roads are one of the most effective ways of bringing better prices to the farmgate. Not only are transport costs reduced, but in some cases an improved road is necessary to provide any kind of access during the rainy season. If vehicular access to a locality is limited to only half the year, then clearly dairying, for example, would not be an option there. The role of such investments is discussed further in Chapter 9.

Systems of price supports may constitute policy responses to price variability as well as concerns over the secular trend or level in prices. If that is the case, then they can be considered inefficient responses. The set of more efficient policies may include commodity hedging on futures markets. A case for considering these kinds of policies has been made in a recent article about the wheat market in Pakistan.[90] More generally, the point needs to be made that a certain degree of price variability is normal and even necessary in grain markets. Without at least some increases in grain prices in the months following the harvest period, there would be no incentive for the private sector to invest in needed storage facilities. Policy concerns should be directed to the possibilities of extreme price fluctuations, and in the first instance try to identify misplaced policies which may in fact be contributing to the fluctuations. Reflexive calls for agricultural policy to adopt the goal of 'price stability' need to be

qualified (see Chapter 9 for a further discussion of this issue).

Several of the principal policy options explored in this section, and some additional ones, have been brought out in a paper by Lowell Hill and Karen Bender.[91] Pointing out that 'efficiency, equity and growth will be dramatically improved if the government provides the appropriate economic and regulatory environment', they list the following 'requirements for development of private commodity markets', along with provision of infrastructure and others:

Warehouse regulations [i.e. a system of certificates of grain deposit]. . . .

Regulated entry. Restricted entry may appear to be counter to the principle of competitive markets, but some restrictions are essential. . . . Licensing of grain dealers and marketing firms are needed to restrict entry by firms or individuals that are either financially or managerially unqualified to buy and sell grain. . . .[92]

Grades and standards. . . .

Information. Although private firms can provide much of the market information required for transaction and investment decisions, there are several types of data that require a central agency with authority to assure continuity and objectivity. Detailed benchmark data (such as a census of agriculture), production data on crops and livestock, and price and consumption data, along with analysis of these data, are essential information for decisions by private firms. . . .

Limiting market power. . . . effective anti-trust legislation, and regulations to prevent private firms from creating barriers to entry. . . .[93]

The interdependence between domestic policies and trade policies should be underscored. It has

---

90. R. Faruqee, J. R. Coleman and T. Scott, 'Managing price risk in the Pakistan wheat market', *The World Bank Economic Review*, **11**(2), May 1997, pp. 263–292.

91. Lowell D. Hill and Karen L. Bender, 'Developing the Regulatory Environment for Competitive Commodity Markets', paper prepared for the FAO/World Bank Seminar on Agricultural Price Stabilization, Santiago, Chile, October 18–20, 1993.

92. The same point applies to export brokers for non-grain products, as commented upon in Chapter 3.

93. Hill and Bender, 1993, pp. 3–4.

been noted that an overvalued exchange rate increases resistance to trade reforms; equally, a free trade regime is not consistent with domestic price controls or a major State role in marketing. Tobias Takavarasha noted for the case of Zimbabwe that trade liberalization by itself cannot substitute for sound domestic economic policies, which must be put in place before gains from trade can be realized.[94]

When the trade option is opened up, futures markets can be used to assist in reducing price uncertainty, as noted. The National Agricultural Commodity Exchange in Colombia, for example, actively supports futures contracts. The development of private commodity exchanges itself marks an important step toward reducing price uncertainty and disseminating grading standards among producers. They also help spread respect for the notion of contracts, which is fundamental to the development of agriculture. Their clients on the supply side are the more commercial farmers, but small farmer co-operatives also can participate.

Another dimension of the international sector concerns foreign investments. In most countries, they still don't play a large role in agriculture, but they are on the rise. International marketing companies provide technical advice and credit to small-scale growers of cantaloupe and other fruits in Central America and the Dominican Republic, and for potatoes in Colombia. In many countries, foreign investments are significant in the food processing sector. Creating an economic environment that is favorable to foreign direct investment is increasingly important for agricultural development.

Finally, the importance of consistency in policies over time cannot be stressed too much. It is essential for giving proper signals to producers and for attracting investment. To ensure consistency over time, a broad constituency in support of policies needs to be put together. In addition,

as mentioned earlier in the context of trade policy, it is preferable to take smaller steps in reforms and consolidate them, rather than take big steps and then backtrack. An extreme experience of policy volatility has been observed in Zambia:

Although liberalization had been announced in 1992, there have been numerous instances of vacillating policy stance. For instance, although the announcement of maize price liberalization had been made, until much later floor and ceiling prices continued to be announced. These were abolished only in 1995 when the government's control over maize prices was truly removed. However, when maize and maize meal prices began to rise in the wake of the full liberalization, anxieties were created and made the Republican President publicly admonish allegedly rapacious traders, who were told to reduce their prices. In the aftermath of the presidential dictate, sellers were compelled to bring down the prices the very next day by 10%. Again, while agricultural trade was liberalized, maize exports have been subsequently banned, unbanned and banned again. Another instance one can cite is in respect of fertilizer subsidy. Although it has been officially removed, a tacit subsidy continues to exist since most of the [fertilizer] has not been paid for by recipients. . . . All these instances of policy inconsistency create an uncertain policy environment which tells of the credibility of the policy regime which may tend to dampen private sector response.[95]

4.8 FOOD SECURITY, AGRICULTURAL PRICES AND THE RURAL POOR[96]

Food security has become a dominant aim of agricultural policy in most countries. In interna-

---

94. T. Takavarasha, 'Trade, price and market reform in Zimbabwe', *Food Policy*, **18**(4), August 1993, p. 290.
95. V. Seshamani, 1998, p. 548.
96. The first paragraphs of this section have been adapted in part from Roger D. Norton, *Integration of food and agricultural policy with macroeconomic policy: methodological considerations in a Latin American perspective*, FAO Economic and Social Development Paper no. 111, Rome, 1992.

tional and domestic forums alike it often is accepted reflexively as a principal goal, and yet the great variety of definitions it has been given tends to make its precise meaning elusive.[97] The principal questions are: what does it mean in operational terms, and what are appropriate policies for pursuing it?

In the 1970s, definitions of food security emphasized a nation's aggregate food production but since then they have stressed the ability of poor households to gain access to food in the necessary amounts. In lower-income countries, the context is one in which, for many families, food expenditures occupy the lion's share of the budget. In addition, food prices and availabilities tend to be more volatile than in industrialized countries, because marketing and trade mechanisms are less well developed. Temporary shortages, such as the unavailability of milk or sugar or other basic products in the markets for a week or more, are common experiences in developing countries although they have disappeared from the collective memories of industrialized societies. The fact that such shortages may occur as the inadvertent consequence of some classes of policy decisions does not lessen their impact on the population. Even without temporary scarcities, chronic undernourishment still is an unfortunate fact of life for large segments of the population in many lower-income countries.[98]

For purposes of designing programs to alleviate undernourishment, the FAO has pointed out that:

Meaningful action to end hunger requires knowledge of not just the number of hungry people around the world but also of the depth of their hunger.

Knowing the number of kilocalories missing from the diets of undernourished people helps round out the picture of food deprivation in a country. Where the undernourished lack 400 kilocalories a day, the situation is more dire than in a country where the average shortage is 100 kilocalories. The greater the deficit, then the greater the susceptibility to nutrition-related health risks. . . .

The diets of most of the 800 million chronically hungry people lack 100–400 kilocalories a day. Most of these people are not dying of starvation. Often, they are thin but not emaciated. The presence of chronic hunger is not always apparent because the body compensates for an inadequate diet by slowing down physical activity and, in the case of children, growth. In addition to increasing susceptibility to disease, chronic hunger means that children may be listless and unable to concentrate in school, mothers may give birth to underweight babies and adults may lack the energy to fulfil their potential.

In terms of sheer numbers, there are more chronically hungry people in Asia and the Pacific, but the depth of hunger is clearly the greatest in sub-Saharan Africa. There, in 46 percent of the countries the undernourished have an average deficit of more than 300 kilocalories per person per day. By contrast, in only 16 percent of the countries in Asia and the Pacific do the undernourished suffer from average food deficits this high.[99]

The FAO's latest estimates indicate that, in 1998–2000, there were 840 million undernourished people in the world, with 799 million of

97. 'At the last count, there were close to two hundred different definitions of the term' (S. Maxwell, 'Food security: a post-modern perspective', *Food Policy*, **21**(2), May 1996, p. 155).

98. The average daily per capita consumption of calories declined significantly from 1971–1972 to 1991–1992 in Angola, Malawi, Mozambique, Zambia and Zimbabwe. See P. Pinstrup-Andersen, R. Pandya-Lorch and S. Babu, 'A 2020 Vision for Food, Agriculture, and the Environment in Southern Africa', in *Achieving Food Security in Southern Africa: New Challenges, New Opportunities*, International Food Policy Research Institute, Washington, DC, USA, 1997, Chapter 2, p. 18.

99. FAO, *The State of Food Insecurity in the World, 2000*, Food and Agriculture Organization of the United Nations, Rome, 2000, pp. 1 and 3.

them in developing countries, 30 million in transition countries and 11 million in the industrialized countries. There has been a slowdown in the 1990s in the reduction of undernourished people in the world. The overall decline in the number of undernourished in poorer countries hides contrasting trends: only 52 of the 99 developing countries reviewed registered a decrease in their numbers of undernourished between 1990–1992 and 1998–2000.

In regard to strategies for reducing undernourishment, the FAO emphasizes productivity improvements and the strengthening of local institutions, themes that are developed throughout this book. It confirms that these approaches constitute a way forward for betterment of the conditions of the poor in both rural and urban areas.[100] The FAO also indicates that agricultural growth is a path for reduction of undernourishment, and in cases of natural and man-made disasters, utilizing food aid:

> Countries that perform well [in reducing undernourishment] may do so by following one or more routes. They may have devoted more resources to increasing agricultural production – the best option for the purposes of increasing economic growth and, if small farms and poor consumers are able to participate and benefit, for creating a more equitable society. Alternatively, they may have imported large amounts of food, either purchased on the international commodity markets or received as food aid. Countries afflicted by long-standing civil wars or recent short-term shocks may achieve better than anticipated performances by the latter route.[101]

In response to the kind of concern voiced in the ODI consensus statement about the effectiveness of food aid (cited above):

New country programs [for food aid] are being designed with more accurate geographic, sectoral and beneficiary targeting. . . . [They have] the following objectives:

- concentrate resources on areas with the highest incidence of food-insecure people
- focus on the most insecure populations (the hungry 'poor') within these areas
- carry out activities that address community needs and the root causes of food insecurity[102]

In framing objectives for long-run strategies, the two main expressions of the concern over food insecurity can be understood as sufficiency and access:

(1) that *sufficient supplies of staple foods always be available to the population*;
(2) that *poor families have access to sufficient food to avoid malnutrition*.

The autarkic approach to achieving the first goal – of attempting to achieve national self-sufficiency in producing basic foods – is now recognized as a costly approach for most countries: 'it has been impossible since the early 1980s to speak credibly of food security as being a problem of food supply, without at least making reference to the importance of access and entitlement' (Maxwell, 1996, p. 157). Rather than pursue complete national food self-sufficiency, generally it is much more economical for a country to produce, and also export, the classes of goods in which it has an international comparative advantage, and import some of its food needs. If a hectare of land generates twice the income in export crops as in basic food crops, both farmers and the nation's balance of pay-

---

100. FAO, 2000, p. 25.
101. FAO, *The State of Food Insecurity in the World*, Food and Agriculture Organization of the United Nations, Rome, 2001, p. 4.
102. FAO, 2001, p. 30.

ments will be better served by producing the former and importing the latter, at least at the margin.

The danger of the autarkic approach is that:

quests for self-sufficiency may turn to extracting rural produce cheaply to feed cities, creating perverse incentives, harming food output and employment and worsening undernutrition.[103]

The discussion in the foregoing sections has underscored the fact that achieving food security in the sense of the total availability of food depends on pursuing appropriate trade policies and on measures to develop marketing systems, sometimes more than on food production itself. In addition, given the fluctuating nature of all harvests, attempting to satisfy all food needs purely through domestic production would mean the population would be forced to endure occasional shortages and the producers, surpluses.

The concern about nutrition levels is not a concern about total national food self-sufficiency, but rather about the *food security of poor families*. In the short run, malnutrition can be alleviated with food assistance programs, while in the long run raising income and educational levels of poor households is the most sure way to eliminate it. Using international evidence, it has been shown conclusively that the three principal determinants of nutritional intake are income per capita, educational levels, and health, in that order.[104]

Hence food production by lower-income rural families contributes to their nutritional levels primarily by generating income (purchasing power) for them. If such families have opportunities to raise their incomes by switching to higher-value, non-food crops, then the nutritional levels of their families can be expected to increase. Such crops are often more labor-intensive than grains and thus are well suited to the high labor endowments per hectare of labor that characterize low-income farming families. However, in cases where cash crops are not a viable option, as in many parts of the Sahel, whether because of agronomic conditions, lack of full market access or other reasons, clearly the main route to improvements in household nutrition levels is found in increasing yields of food crops.

A survey of rural areas in Ethiopia found a lack of correlation between the nutritional status of children and the areas their households devoted to crop production, and neither the altitude of the farm nor the size of the household provided a correlation. The conclusion of the study was that health care and practices were more important to nutrition than family food production itself:

... the results suggest that household food security is positively associated with child nutritional status in some regions, negatively associated in some regions, and shows no consistent association in other regions. These results are found even when variation in household size and altitude is taken into account in the analysis ... Under this broader view of malnutrition, household food security is a necessary but not sufficient condition for maintaining adequate nutritional status. Health related conditions and child care and feeding are also necessary conditions. ...[105]

The fundamental lesson here is that *to achieve improved levels of nutrition in rural households, cropping patterns should be allowed to follow com-*

103. FAO, *The State of Food and Agriculture, 2000*, Food and Agriculture Organization of the United Nations, Rome, 2000, p. 237.

104. O. Knudsen and P. Scandizzo, 'The demand for calories in developing countries', *American Journal of Agricultural Economics*, **64**, February 1982, pp. 80–86.

105. Reprinted from *Food Policy*, **20**(4), D. L. Pelletier, K. Deneke, Y. Kidane, B. Haile and F. Negussie, 'The food-first bias and nutrition policy: lessons from Ethiopia', pp. 293–294, Copyright (1995), with permission from Elsevier.

parative advantage, and farmers should not be given artificial incentives to grow basic foods. The surest way to enhance family food security, which is a concern of the poor, is through generating increases in their incomes. The production of food may have little relation to the food security of the poor, again with the exception of the more remote rural areas with few cropping alternatives.

Some of these conclusions have been summarized well by Simon Maxwell, in the form of proposing a 'consensus strategy' for food security in Africa:

A primary focus on supplying vulnerable people and households with secure access to food: individual and household needs take precedence over issues of national food self-sufficiency or self-reliance. . . . The importance of poverty-reducing economic growth: poor rural and urban people need secure and sustainable livelihoods, with adequate incomes. . . . Within agriculture, growth strategies are needed which lay particular emphasis on generating jobs and incomes for the poorest groups, including those in resource-poor and environmentally degraded areas. Agriculture and rural development strategies should usually favor labor-intensity. . . . A balance between food crops and cash crops is likely to be the best route to food security, following the principle of long-term comparative advantage rather than of self-sufficiency for its own sake. . . . Efficient food marketing is needed, to store and distribute food, at reasonable prices, to all parts of the country in all seasons and in all years. . . . More effective and efficient safety nets need to be established, by strengthening community institutions, introducing new targeted food and nutrition interventions. . . . Finally, food security planning should follow a

'process' rather than a 'blue-print' approach, with large-scale decentralization. . . .[106]

In addition to these considerations, a basic factor in determining family nutrition is the education levels of women, as pointed out earlier in Chapter 3.

It may be asked, if aggregate quantities of domestic food production are not at issue from a food security viewpoint, then what about their prices? Clearly, higher food prices mean a more restricted ability to purchase food for many poor families. Here it is important to distinguish between rural and urban families. For the latter, food assistance programs often are essential in the short and medium term, while ways are sought to improve their income-generating power. In addition, such programs are all the more needed when food prices increase.

For the rural poor, who in developing countries usually constitute most of the low-income families, and also most of the extremely poor families, food prices may have a different significance. Magdalena García *et al.* have shown for Honduras that even families with as little as one hectare of land are net beneficiaries of price increases of basic food products, and for those with two hectares and more the benefits are significant even with small price increases.[107] For most countries, about one hectare of land is the amount needed for supplying a family's annual needs in a basic cereal, and additional land beyond that amount contributes increasingly to a farmer's participation in the market.

In addition, food price increases create incentives for more production and therefore create more employment, thus benefitting the landless as well. For this reason, unless the rural landless are very numerous in relation to those who have at least some land, increases in real agricultural prices are likely to benefit the rural poor as well

106. Reprinted from *Food Policy*, **21**(2), S. Maxwell, 'Food security: a post-modern perspective', p. 164, Copyright (1996), with permission from Elsevier.

107. Magdalena García, Roger Norton, Mario Ponce Cámbar and Roberta van Haeften, *Agricultural Development Policies in Honduras: A Consumption Perspective*, special research publication of the Office of International Cooperation and Development, US Department of Agriculture, Washington, DC, USA, 1988.

as the rural non-poor. As noted earlier, Schreiner and García found that the lowest-income strata in rural areas were proportionately the greatest beneficiaries of the real agricultural price increases in Honduras in 1989–1991 that were brought about by a real devaluation of the currency.[108]

Thus, the common reaction that higher food prices necessarily worsen the poverty situation in a country needs to be re-examined. Looking at the issue from an analytical viewpoint, Benjamin Senauer reached a similar conclusion:

The implication for nutrition is that higher food prices may actually lead to an improvement in the nutrition of farm household members, because of the effects of profits on income. Even if the consumption of the commodity for which the price has increased declines, the increased profits and income can be used to buy increased amounts of other foodstuffs, the result being an improvement in nutrient intake. . . . higher prices may improve the welfare and nutrition of agricultural households, and possibly even of rural non-farm families whose income depends on agriculture, such as agricultural laborers.[109]

In other words, as real agricultural prices increase, lower-income farm families may actually retain *less* of the crop for home consumption, and still they would be better off, in terms of both income and nutrition.

This conclusion supports a central thesis of this volume, for which empirical evidence was presented earlier: adequate real price incentives in agriculture are important not only for economic growth but also for alleviating rural poverty.

4.9 OBSERVATIONS ON PRICE STABILIZATION AND ECONOMIC DEVELOPMENT

In the last 10 to 15 years, macroeconomic policy throughout the developing world has placed strong emphasis on policies for price stabilization, often giving them precedence over growth policies in the short- and medium term. While the benefits of stabilization are undeniable, and almost everyone would prefer less inflation to more, care should be taken to ensure that the means of achieving stabilization do not prejudice growth prospects, as indicated in the discussion earlier in this chapter about exchange rate policies. When inflation is very high, clearly priority has to be given to bringing it down. When it is moderate, however, at times it may be more appropriate to have inflation control share priority with the creation of income and employment. This matter has been put into perspective by Joseph Stiglitz.[110] Some of his work on this and related subjects was summarized by Simon Maxwell and Robin Heber Percy in the following terms:

- high inflation (over 40 percent per annum) is seriously damaging, but lower inflation is not, and controlling inflation should not be a priority for many developing countries;
- there is too much concern with controlling budget and current account deficits. Deficits are to be sustainable;
- macro-economic stability is less important than stabilizing output or unemployment

108. Dean F. Schreiner and Magdalena García U., *Principales Resultados de los Programas de Ajuste Estructural en Honduras*, Serie Estudios de Economía Agrícola No. 5, Proyecto APAH, Abt Associates, Inc., Tegucigalpa, Honduras, June, 1993, p. 19.

109. B. Senauer, 'Household behavior and nutrition in developing countries', *Food Policy*, **15**(5), October 1990, p. 63; quoted in FAO, *State of Food and Agriculture 1995*, Food and Agriculture Organization of the United Nations, Rome, 1995, p. 65.

110. The relevant papers of Stiglitz are: (a) 'More instruments and broader goals: moving towards the post-Washington consensus', Annual Research Lecture, World Institute for Development Economics, Helsinki, January 7, 1998; (b) 'Towards a new paradigm for development strategies, policies and processes', 1998 Prebisch Lecture at UNCTAD, Geneva, October 19, 1998.

– which often requires microeconomic intervention. . . .

• rather than focusing simply on liberalizing trade, governments must intervene to create competitive export sectors;

• privatization needs to be supplemented by provision of institutional infrastructure, including regulatory bodies, and there are critical issues about both sequencing and scope. . . .[111]

Many countries such as Mexico through the latter half of the 1990s have managed to grow at a satisfactory rate while inflation was at levels above 10%. In this recent experience in Mexico, policy worked on gradually reducing inflation while maintaining stimuli to growth. The key to a successful stabilization policy is sustainable reductions in fiscal deficits, instead of artifices such as overvalued exchange rates, abrupt tariff reductions and monetary sterilization of liquid balances in the economy. It takes time to engineer an enduring reduction in inflation in the proper way, but many countries have done it, and the benefit can be more rapid growth of output and employment in the meantime. This is especially true of growth in the agricultural sector.

It may be time to shift the balance in macro-economic policy back toward promoting economic growth, which in the end is the most effective way to alleviate poverty, both rural and urban.

DISCUSSION POINTS FOR CHAPTER 4

• Agricultural price policy is concerned with *real* prices, that is, agricultural prices relative to other prices in the economy. To obtain an index of real agricultural prices, a nominal index of agricultural prices may be deflated by an economy-wide price index, such as producer prices, consumer prices or the GDP deflator, or by an index of input prices. Each deflator gives rise to a different interpretation of the real index, but in some way each version measures the purchasing power of agricultural output.

• Three *structural* factors strongly influence real agricultural prices: trends in domestic supply and demand, real trends in international agricultural prices, and the presence of subsidies to agriculture in exporting nations.

• Trade in agricultural products has brought important benefits to many developing countries. However, since the completion of the Uruguay Round of trade negotiations, developed countries have increased their agricultural exports more than developing countries have.

• Worldwide, agricultural tariffs have been reduced much more slowly than non-agricultural tariffs have, and megatariffs of 200–300% still exist in developed countries.

• The extent of benefits to trade liberalization depend in part on how effective policies and other economic forces are in pushing an economy to reallocate resources in the direction of its comparative advantage. When such reallocations are slow, trade liberalization may have to proceed more slowly than other kinds of structural economic reforms that can improve the resource allocation process.

• High tariffs hurt a country's own competitiveness in international markets, by raising costs to all producers, including those who produce for export. Therefore, from a development viewpoint tariff levels should not be high, and if they are, a program should be put in place to scale them downward over time.

• It is equally important that tariffs be relatively uniform over sectors and products. This helps align domestic *relative* prices with international relative prices and is a way to encourage resources to be allocated to the lines of production that are most competitive, i.e. that have the best growth prospects.

---

111. Simon Maxwell and Robin Heber Percy, 'New trends in development thinking and implications for agriculture', in Kostas G. Stamoulis, 2001, p. 71.

- The practice of granting tariff exemptions to food imports is an obstacle to making tariffs uniform over products. Such exemptions constitute implicit subsidies that are not targeted on the poor, and in fact they usually turn out to be regressive subsidies. In addition, through their effects on production incentives they tend to exacerbate the problem of rural poverty.

- The three justifiable exceptions to a uniform tariff policy occur for (1) products whose international prices are distorted by subsidies in large exporting countries, (2) products whose international price fluctuations may be smoothed by price bands before they are transmitted fully to the domestic economy, and (3) one or two products that are the basic sources of food and income for the rural poor, since in the short and medium term their opportunities for finding alternative sources of income are very limited.

- Compensating for international price distortions is not advisable when the product in question does not have a comparative advantage in the developing country.

- Stability of tariffs over time is often elusive but is important for providing appropriate price signals for investment and production. It may be more important to reduce tariffs in small but sure steps instead of attempting drastic reductions only to reverse the reforms later, as has occurred in some countries.

- Export incentives are discouraged under the WTO regime, with a modest exception for the poorest countries, but since the WTO allows sometimes very high tariffs, the net effect of the world trade regime is to discourage developing country exports, relative to import substitutes. This creates a bias against the products that generally create the most employment and income per hectare cultivated.

- In practice, existing export incentive regimes rarely provide benefits to small- and medium-scale producers of export goods. Thus, the management of such schemes is a bottleneck and warrants review with the aim of developing more equitable systems of incentives.

- Trade restrictions, on both imports and exports, are damaging to growth prospects, in part because of the price uncertainty that they create.

- Except for cases in which emergency food aid is needed, financial food aid is generally more effective than commodity food aid, because it is now acknowledged that poverty is the principal cause of hunger and malnutrition.

- The exchange rate strongly influences *relative* prices in an economy and is the most powerful policy instrument for determining real agricultural prices.

- An appreciating real exchange rate reduces real agricultural prices. An appreciating real exchange rate can be caused by purposive policy and also by large inflows of foreign exchange that are not associated with the main employment-generating sectors, for example, oil and gas earnings and remittances from workers abroad. In economies characterized by such phenomena, the tasks of agricultural development and poverty alleviation are more difficult than otherwise because of the resultant price disincentives for the sector.

- A useful measure to ameliorate the Dutch disease effects on the exchange rate is for the Government to purchase foreign exchange and invest it abroad in a long-term development fund.

- Fiscal policy can influence agricultural prices through (1) investments in roads, ports and marketing facilities, (2) other basic investments that support increases in production, (3) the purchasing and selling policies of agricultural parastatal companies, and (4) commodity taxes.

- Selective commodity taxes on agricultural products distort incentives in the same way that unequal tariffs do and they reduce agricultural incomes relative to non-agricultural incomes. This text reviews arguments sometimes put forth in favor of agricultural commodity taxes and finds them generally lacking in solid foundations.

- Macroeconomic policy can support agricultural development through alternative combinations of the policy instruments mentioned. Maintaining a competitive exchange rate is a powerful policy which is neutral with respect to resource allocation in productive sectors. If this is not possible, then exchange rate distortions can be compensated through trade and/or fiscal policy. Relying on tariffs alone creates other distortions by biasing the sector's prices and production toward import substitutes, and so a policy of tariffs and export incentives would be preferable to tariffs alone. Direct fiscal outlays to support production, as per the experiences in the European Union, Mexico and Estonia, constitute a non-distortionary policy but are demanding in administrative terms. Finally, any macroeconomic policy for agricultural development should be accompanied by institutional and structural reforms in agriculture, in areas such as land tenure, water management, financial systems and agricultural technology. Reforms in these fundamental areas can improve the sector's efficiency and help reduce poverty.

- The sectoral policy instrument of price controls raises a number of issues, including the virtual impossibility of setting prices that continuously equilibrate supply and demand, the fact that maintaining price controls may create the need to impose trade controls as well, the consequence of inefficient resource allocations as a result of distorted relative prices, and the discouragement of the development of adequate private marketing mechanisms.

- Guaranteed crop prices, or farm support prices, encounter many of the same difficulties in addition to causing fiscal outlays. In most countries that have tried to support farm prices, State marketing agencies have proven to be inefficient and the benefits of the support prices tend to flow disproportionately to larger farms.

- When a government decides to reduce or eliminate its agricultural marketing interventions, the private sector may not always be fully prepared to step in and replace the government. It can be inhibited by limitations in both financial and managerial terms, especially in countries with a long tradition of State intervention in agricultural markets. To ease this problem, it is helpful to implement reforms in the proper sequence – which means having a targeted urban food assistance program in place and lib-eralizing agricultural output markets before input markets – and maintaining clarity and transparency regarding the management of food reserves and revised rules governing public sector interventions in markets. In general, financial food security reserves are more efficient than physical ones.

- Grain storage policy is an element of pricing policy because of its influence on the efficiency of marketing channels. Components of such a policy may include privatization of State-owned grain storage facilities, with the option of providing for capital participation by farmers, development of a program of certificates of grain deposit that are available to all classes of farmers, establishment of quality standards for grains, and conversion of harvest and post-harvest systems to bulk handling of grains where feasible. All of these measures effectively improve farmers' *access* to markets.

- Opening of new export markets can be an important avenue of agricultural development, and an outlet for labor-intensive production from small farms, but it requires close attention to issues of product quality. These issues include phytosanitary conditions, food safety requirements, and satisfaction of consumer preferences in taste and product presentation. Government institutional capabilities are needed in this area, and for smallholders an essential step for penetrating non-traditional markets is sound farmer organization.

- Agro-processing industries may enjoy quasi-monopolistic positions with respect to primary producers, within regions of a country. Extreme exploitation of this position is rare because producers and processors need a long-term working relationship, but in the short term issues can arise concerning the price offered by processors. This is a difficult problem for policy to deal with, but measures that may help reduce its magnitude can include support for producers' joint purchase or construction of processing facilities, anti-monopoly legislation, and government mediation of price negotiations, on the basis of international guidelines regarding relative prices of primary and processed products. In other respects, such as improving product quality and identifying needed areas of agricultural research and other public sector support, producers and processors can work together fruitfully through the product chain approach.

- Liberalization of input markets can be as important as liberalization of output markets for improving agricultural productivity. However, many farmers in developing countries will always pay higher prices for agrochemical inputs than their counterparts in richer countries do, because of the small scale of production and/or shipments in the former countries. Farmer co-operation in input purchase can reduce this price penalty.

- Other measures that can assist in improving the prices received by farmers include investments in transport infrastructure, market information systems, encouragement of agricultural commodity exchanges and futures markets, licensing of grain dealers, programs to improve product quality, and ensuring consistency in pricing and trade policy over time.

- Food security is most usefully interpreted in terms of access to food on the part of low-income households, rather than national self-sufficiency in producing basic foods. Chronic undernourishment is still very widespread in developing countries, affecting approximately 800 million people.

- Sound agricultural policy as a whole, including measures that promote productivity improvements and institutional strengthening, is important in the effort to reduce undernourishment. Equally, in food assistance programs it is important to improve efforts to target the most needy populations.

- At the household level, the most important determinants of a family's nutrition levels have been found to be income, education and health status, in that order. The education of the woman in the family is especially important for improving the entire family's nutrition.

- Food security of poor households is not necessarily correlated with their production of basic foods. When agronomic and market conditions are appropriate, production of cash crops as well as food crops can lead to improvements in household welfare. Allowing cropping patterns to follow regional and national comparative advantage is the surest path to reducing undernourishment.

- Higher farmgate prices generally reduce rural poverty. They increase the economic well-being of families with even small plots of land, and their stimulus to higher production creates more employment for the landless.

- While price stabilization is a worthwhile goal of economic policy, care should be exercised in regard to how it is achieved. Bringing down inflation rapidly through distortions of the exchange rate and excessively rigid monetary policy may have the consequence of reducing *real* agricultural prices, slowing the overall economic growth rate and worsening poverty. In many circumstances, it may be preferable to reduce inflation in a lasting, structural manner, through reductions of the fiscal deficit, even though the process takes longer, because in the meantime the poorest households may fare much better. This central issue of balance in the goals of economic policy deserves fuller consideration than it is often given in prescriptions for developing countries.

5

Land Tenure Policies[1]

5.1 Introduction	109	
5.2 The Importance of Land Tenure	110	
5.3 Objectives of Land Tenure Policies	111	
5.4 Overview of Issues and Trends Concerning Land Tenure	116	
5.4.1 Key Issues for Land Policies	116	
5.4.2 Historical Trends Regarding Rights to Use Land	117	
5.4.3 The Era of Landed Estates	120	
5.5 The Nature of Land Rights	121	
5.5.1 Forms of Property Rights	121	
5.5.2 Tenure Security and Land Rights	122	
5.5.3 Land Rights, Farm Size and Agricultural Productivity	125	
5.5.4 Transferability of Traditional Land Rights	127	
5.5.5 Transitional Issues in Moving from Customary to Formal Systems	128	
5.6 Communal, Collective and Individual Rights to Land	130	
5.6.1 Communal Lands	130	
5.6.2 Collective versus Individual Farms	134	
5.6.3 The State as Landowner	139	
5.7 Experiences with Land Reform	142	
5.7.1 Return to Origins	142	
5.7.2 Rationales, Failures and New Formulations	144	
5.8 Policies for Land Markets	149	
5.8.1 The Degree of Formalization of Land Rights	150	
5.8.2 Forms of Customary Title	156	
5.8.3 Conditions on Titling	157	
5.8.4 Land Leases and Rental	159	
5.8.5 Share Tenancy Contracts	163	
5.8.6 Policies Regarding the Sale of Agricultural Land	164	
5.8.7 Transforming Collective Farms: Corporations, Private Co-operatives and Individual Farms	167	
5.8.8 Transition Issues in the CIS and Eastern Europe	171	
5.8.9 Policies to Promote Co-operative Forms in Agriculture	174	
5.9 Improving Access to Land for the Poor and for Women	176	
5.9.1 Land Market Mechanisms and the Rural Poor	176	
5.9.2 Improving Access to Land for Women	185	
Discussion Points for Chapter 5	191	

5.1 INTRODUCTION

Few topics in economic policy have inspired as many controversies and conflicts over the centuries as agricultural land tenure. Perhaps this should not be surprising, since in most countries land is the principal form of wealth in rural areas, sometimes in the entire economy. It also can represent a source of social status and political influence. Land tenure policies, therefore, powerfully affect household income levels and the distribution of wealth, and even social and political struc-

---

1. A special note of thanks is due to Adriana Herrera and David Palmer for their very thoughtful and detailed comments on an earlier version of this chapter.

Agricultural Development Policy Concepts and Experiences. R. D. Norton
© 2004 Food and Agriculture Organization of the United Nations
ISBNs: 0-470-85778-1 (HB)   0-470-85779-X (PB)   FAO Edition: 92-5-104875-4

tures. In addition, since land is a principal productive resource in any country, national decision-makers frequently have felt a special responsibility for defining its policy framework, even for ensuring that it is efficiently used and for deciding on the uses to which it should be put. Customary forms of holding and managing land also strongly influence policy considerations. Probably in no other area of economic policy do traditional ways of doing things have such deep historical roots. Hence, ideologies and ideas about the appropriate direction of economic policy are combined with long traditions and the vital personal interests of a large part of the population in the potent mix that is land tenure policies.

Another aspect of the French Revolution illustrates well the difficulty of establishing land tenure policies:

> During the French Revolution . . . the members of the Convention hesitated to design and implement agrarian reforms 'because they were simply frightened of the immense complexity of the concrete rural issues'.[2]

In the circumstances, a set of international policy guidelines cannot expect to provide prescriptions that are valid universally. More than in any other field of agricultural policy, solutions have to be developed within the historical, social and political context of each country. What the guidelines can do is attempt to clarify the objectives that are at stake, identify a number of the main issues, review and assess alternative approaches, and offer examples of viable policies and reforms. An objective discussion of problems and illustrative frameworks in this sense may be useful to policy makers in devising appropriate variants for their own countries.

For dealing with land tenure issues, it is well to bear in mind at all times the cautionary words of Abhijit Banerjee, in his recent article about alternative approaches to land reform:

> . . . we need to know more. Making policies that can change the lives of large numbers of people is always daunting, but it is all the more so when it is based so heavily on speculation.[3]

5.2 THE IMPORTANCE OF LAND TENURE

Other than labor, land is the most important factor of agricultural production. Without clearly defined rights of access to land, or land tenure, production is more difficult to carry out and incentives are weakened for long-term investments in land to raise its productivity. Land tenure also is one of the organizing pillars of rural economies and societies that helps define economic and contractual relationships, forms of cooperation, and social relationships.

Paul Munro-Faure, Paolo Groppo, Adriana Herrera and David Palmer have written persuasively about the importance of land tenure. Their perspective was that of designing and implementing rural development projects, but their observations are relevant for agricultural and food policy in general:[4]

> In many cases, responses to concerns of environmental sustainability, social conflicts, and food security of the vulnerable are affected by land tenure and have an impact on land tenure. Failure to consider land tenure implications at

2. From A. Moulin, *Les Paysans dans la Société Française de la Révolution à Nos Jours*, Editions du Seuil, Paris, 1988 (cited in J.-P. Platteau, *Land Reform and Structural Adjustment in Sub-Saharan Africa: Controversies and Guidelines*, Economic and Social Development Paper No. 107, Food and Agriculture Organization of the United Nations, Rome, 1992, p. 291).
3. Abhijit V. Banerjee, 'Prospects and Strategies for Land Reform', in *Annual World Bank Conference on Development Economics, 1999*, B. Pleskovic and J. E. Stiglitz (Eds), The World Bank, Washington, DC, USA, 2000, p. 272.
4. Paul Munro-Faure, Paolo Groppo, Adriana Herrera and David Palmer, *Land Tenure and Rural Development Projects*, FAO Land Tenure Studies, Food and Agriculture Organization of the United Nations, Rome, 2002, pp. 2–3 [emphasis added].

the beginning of a project is likely to result in unanticipated outcomes. This failure may lead to the project not generating an improvement. In some cases, it may even worsen the situation, for example, by inadvertently dispossessing people of their rights to land. . . .

Eradicating hunger requires increasing the entitlements to food of a person or family. The extent to which individuals and families are able to increase their entitlements depends in large part on the opportunities they have to increase access to assets. An emphasis on building people's endowments of assets brings land tenure into the picture. People who have rights to land are more able to enjoy a sustainable livelihood than those who have only partial rights of access; those who have partial rights are, in turn, better off than those who are landless.

Property rights that provide access to land, together with labor, form the most common endowments used to produce food for home consumption as well as cash crops that allow the family or individual to pay for other needs (e.g. health, education, etc.). *Property rights to land are thus one of the most powerful resources available to people to increase and extend their collection of assets beyond land and labor to the full portfolio necessary for sustainable livelihoods* (i.e. natural resources, social, human, and financial capital as well as physical assets).

5.3 OBJECTIVES OF LAND TENURE POLICIES

The broad objectives of land tenure policies are not essentially different to those which guide the formulation of policies for any resource or sector, but historically the discussions of the subject have placed particular emphasis on the two overriding goals of *economic efficiency* and *equity and poverty alleviation*. In addition, *environmental* and *institutional sustainability* are concerns of increasing importance in developing such policies. In the context of land use, efficiency has short-run, or static, and long-run, or dynamic, dimensions. While it refers to encouraging allocations of land to uses which currently yield the highest economic

productivity, it also means stimulating appropriate management of the land resource and investments for sustaining and improving its productivity over time.

Achieving efficiency in the use of one of the sector's basic resources is an essential condition for agricultural growth to occur. A simple but complete equation of agricultural growth states that the rate of output increase is the sum of the rate of expansion of cultivated land and the rate of increase of productivity per unit of land. The last factor includes both changes in cropping patterns and increases in crop yields.

Given the increasing scarcity of suitable agricultural land in all parts of the world, and the concerns about deforestation and land degradation, the productivity factor will have to account for an increasing share of the sector's growth in the future, hence the crucial need to use land more efficiently. In addition, through multiplier effects agricultural expansion generates benefits throughout the economy, as commented upon earlier in this book, and so while sustained agricultural growth will not necessarily solve the poverty problem, it can make a uniquely valuable contribution in this regard in both rural and urban areas.

It should be noted that in determining national policies regarding land distribution, adopting the equity objective does not necessarily mean striving for an equal distribution of land. In operational terms, the emphasis usually is placed on providing mechanisms of access to land for as many rural families as possible, access to plots of at least a size that can sustain what is considered a minimum acceptable standard of living, along with reasonable conditions of access to some additional amounts of land for those who can work it.

Access to land is relevant to questions of poverty alleviation in part because poverty is usually more widespread in rural than in urban areas. It also is relevant because the distribution of land holdings, or of access to land, is one of the principal factors determining the degree of rural poverty. By extension, it also is a determinant of urban poverty when poor economic conditions in the countryside force large numbers of families to migrate to cities without secure prospects of employment.

In the words of Vijay Vyas:

The countries, and regions within large countries, where least impact has been made on the eradication of poverty often suffer from an inequitable agrarian structure, or marginalized and depleted resources, or both. . . . The rigors of small ownership holdings are eased if there is an active lease market, or when non-farm employment opportunities are present and expanding. In the absence of either of those two conditions a strong correlation exists between the inequality in ownership of land and the extent of poverty, as in South Asia, Southern Africa and large parts of Latin America.[5]

Issues concerning the rental and leasing of agricultural land, communal land rights and whether secure access to land is an adequate substitute for its ownership, are discussed subsequently in this chapter, but the correlation with poverty indicated by Vyas also holds with respect to the distribution of effective rights to use land, independently of whether they refer to ownership.

Unequal conditions of access to land, with consequent impoverishment of much of the population, are not exclusively a phenomenon of contemporary societies. In the case of pre-Columbian Mexico, for example, it has been observed that:

The land was clearly allocated . . . [into] diverse categories of possession and rights of usufruct; but in the real order of things it was concentrated in a few hands; it was the basis for the social preeminence, the wealth and the political influence of a select group. The king, the nobles and the warriors were the great landowners of the era. . . . The conquests, trade and political relations between different peoples, and the population increase itself gave rise in the cities and towns to an agglomeration of many persons who did not have any land at all and who were prohibited from acquiring it. In this way great masses of the disinherited were formed. How did they live? Orozco y Berra says it with utter clarity: '. . . as the contemptible masses, from the grains they harvested, of which three measures gave [them] one, one of each three they cultivated; their work was for the despot of Mexico; they were slaves of the land, and when they ate eggs, it seemed to them the king granted them a great favor, and they were so oppressed that almost all what they ate was rationed and the rest was for the king'.[6]

There were a great number of wage workers whose condition was as bad as today's farm laborers, perhaps worse, because the latter have the legal possibility of becoming owners of land. . . . The masses recognized and respected the unequal distribution of land, because they recognized and respected the social inequalities. The legal system maintained property rights in a drastic way, since changing the fences or boundary markers was punished . . . with the death penalty.[7]

It is not only an unequal distribution of access to farmland that makes it difficult to reduce rural poverty. Inappropriate land policies, and poor land management practices and regulations, can undermine the efforts of rural poor to improve their situation, as well as causing inefficiencies in land use for farmers of all income strata. *Both the initial distribution of land and the nature of land tenure regimes and policies have a direct bearing on the extent of rural poverty.*

5. Vijay S. Vyas, 'Agrarian Structure, Environmental Concerns and Rural Poverty', Elmhirst Memorial Lecture, in G. H. Peters and B. F. Stanton (Eds), *Sustainable Agricultural Development: The Role of International Co-operation*, Proceedings of the 21st International Conference of Agricultural Economists, Dartmouth Publishing Company, Aldershot, UK, 1992, p. 11.

6. Orozco y Berra, *Historia Antigua y de la Conquista de México*, Mexico City, Mexico. 1880, Volume I, p. 371.

7. Lucio Mendieta y Núñez, *El Problema Agrario de México y la Ley Federal de Reforma Agraria*, 19th updated edition, Editorial Porrúa, SA, México, 1983 [1st Edition, 1923], pp. 28–29 [author's translation].

It is sometimes asked why policy makers should be concerned with providing land to poor families, rather than providing them with support and subsidies of other kinds. As Banerjee has put it:

On the question of whether we should redistribute land rather than money, the instinctive answer among economists is that redistributing money must be better, all else being equal, since beneficiaries could always use the money to purchase land. . . . if the only reason the rural poor do not buy land is that they are too poor to do so, all poor rural residents would use a cash distribution to buy land and the productivity gains from land reforms would be realized. The case for redistributing land could thus be based on the belief that all beneficiaries want land and that redistributing land directly would eliminate some transactions costs. In all other cases, one could argue, it would be better to distribute money.

Redistributing money may not always be the best option, however, for several reasons. One is that land reform may help keep people in rural areas instead of moving to cities. . . . A more compelling argument is that land can be a permanent source of income for poor families. Heads of families may not always act in the collective interest of their families.[8] If there are conflicts of interest within the family or between current and future generations, the goal of redistribution may be better served by giving the family an asset other than money. Doing so might, for example, prevent a husband from decamping with financial assets. . . . Moreover, land may be a particularly good

asset to inherit, because fewer skills are needed to make use of that asset than other fixed assets, such as factories or shops. . . .

These arguments are obviously highly speculative. In the absence of better empirical support, they make what is at best a very tentative case for land redistribution as a way of benefitting the rural poor.[9]

In addition to Banerjee's arguments, it now is widely recognized that fiscal transfers alone cannot be expected to solve poverty problems, in good measure because tax administration in most developing and transitional economies is not strong enough to support that burden.[10] Furthermore, when the definition of the issue is broadened to encompass *access to land*, and not only ownership, there is scope for a wider variety of policies, and not just land redistribution, that can create benefits in terms of poverty alleviation and at the same time promote sectoral growth (improve economic efficiency).

It is widely acknowledged that *access to land is the most fundamental determinant of income-earning potential in rural areas in developing and transitional countries*, and so the equity dimension cannot be ignored in formulating land policies without the risk of generating serious social and political tensions as well as exacerbating poverty. An implicit and universal recognition of this imperative is provided by the customary practice of many African and Asian village societies of providing land use rights to all families belonging to the village,[11] and by the approach of 'land to the tiller' which has characterized land reform movements, from Taiwan and the Republic of

---

8. See Chapter 7 for evidence that in general women in rural households handle finances more responsibly than men do.

9. A. V. Banerjee, 2000, pp. 261–262.

10. Albert Fishlow also points out that 'The scope for progressive taxation is necessarily limited when simultaneously advancing private savings and investment', and observes 'Of the possibilities for redistribution of wealth, land stands out as the most potent variable used in the past' (in 'Inequality, Poverty and Growth: Where Do We Stand?', *Annual World Bank Conference on Development Economics, 1995*, M. Bruno and B. Pleskovich (Eds), The World Bank, Washington, DC, USA, 1996, pp. 32 and 29).

11. In almost all tribal systems in both Asia and Africa, 'Only by being formally expelled from the group itself can a member lose his right of cultivation of land' (Ester Boserup, *The Conditions of Economic Growth: The Economics of Agrarian Change under Population Pressure*, George Allen & Unwin Ltd, London, 1965, p. 79).

Korea to Peru and Central America and South Africa. It also underlay the 19th-century Homestead Act in the United States. How to promote wider access to land in an effective way is a major issue in land policies.

Security of land rights is as important as access. Without security, farmers are reluctant to invest in productivity-enhancing improvements, they may not be able to obtain financing for the improvements and for annual production inputs, and widows and children may be denied rights to the same land. As will be discussed subsequently, security of tenure can be provided in diverse ways, not only by outright land ownership with registered titles.

Therefore, the broad policy objectives of equity and efficiency translate into the following three operational objectives for the case of agricultural land policy: *equitable access to land, secure land rights, and smoothly functioning land markets and other allocative mechanisms*. The latter condition allows land to be flexibly and quickly reassigned to those who are able to extract higher returns

from it, thus enhancing *allocative efficiency* over time.

Environmental sustainability is an objective of increasing concern for land policies all over the world. Soil traditionally has been regarded as a fully *renewable resource*: leave it in a fallow state, apply organic fertilizers, terrace it to trap nutrients, etc., and it will regenerate. This process can work within a range of soil conditions, but if soils are abused to an extreme, their capacity to regenerate themselves is lost essentially forever. The damage may become *irreversible* and soil may become a non-renewable resource, at least for a span of several human lifetimes. A look at the denuded, rocky hillsides of Haiti, where there once was soil and vegetation, confirms the fragility of soils. Even if the capacity of land is not pushed to such an extreme, topsoils can be lost in sufficient quantity that their recovery could take several decades. In areas of South-East Asia, the Amazon Basin, Central America and other parts of the world, rates of soil loss are very high. It has been estimated that the rate of loss in some areas of El Salvador is equivalent to a total depletion of the topsoil in 19 or 20 years.[12] In today's land use patterns, a major threat to soils arises from the sacrifice of forest cover for annual crops in tropical regions. These trends also bring with them hazards for water supplies, biodiversity, and local and perhaps global climates. A serious threat also arises from inadequate maintenance and over-exploitation of existing agricultural lands, especially through overgrazing.

The sustainability and poverty issues are interrelated. Proper management of land, which usually involves investments (via use of capital and/or labor) or decisions to set land aside for a while, whether in tree crops or in fallow rotations, implies a sacrifice of some immediate benefits for the sake of benefits in the longer run. Thus, sustainable land management often requires the *ability to save*, or to forego some of the present consumption possibilities. Many poor families are not able to save and are driven to extract the

> *Strictly speaking, the agricultural soil must be gradually 'constructed' . . . as is evident from the history of agricultural development in Europe, Asia and pre-Columbian Latin America. In these continents, vast amounts of family and village labor have been used to build fences, pick stones, remove stumps, construct flood embankments, level and terrace land, drain water, and so forth. . . . Equally important is the fact that investment of labor is also required on a recurrent basis in order to maintain the land infrastructures once they have been built. Indeed, if the soil can be 'constructed', it can also be destroyed and the process of soil destruction is especially rapid in countries – such as those of Africa – where problems such as leaching, wind or water erosion and flooding are permanent threats (From J.-P. Platteau, 1992, p. 111).*

---

12. Roger Norton, Ricardo Arias and Vilma Calderón, *Una Estrategia de Desarrollo Agrícola para El Salvador, 1994–2000*, FUSADES, San Salvador, El Salvador, 1994, p. 9.

maximum current benefit from land, to the detriment of its future capacity. Swidden agriculture is the classic example of this syndrome. A vicious circle is created: poverty may aggravate the problem of degradation of land, and in turn the degraded land exacerbates poverty in the future.

Environmental sustainability is not only relevant to poor societies. In fact, it will become more important for land policy in developing and transition economies as they grow, as illustrated by the experience of Western Europe:

> At least in EU countries, (rural) land policy is increasingly becoming *environmental policy* with 'multifunctionality' of agriculture as its buzzword and justifying ongoing subsidies for rural areas. Some of the newly emerging functions of land are closely related to the provision of environmental goods and services: clean air and water, less intoxicated soils, carbon dioxide fixation, etc. ... based not only on national policy priorities but as well demanded by legally binding international Conventions and Regimes, such as on Climate, Biodiversity and Desertification.[13]

Institutional effectiveness and sustainability may be regarded as an operational objective also, i.e. a means for attaining the broader objectives. It is gaining increasing importance as land tenure regimes gradually make a transition from informal, customary systems to formalized systems involving recording of rights and bureaucratic procedures. *For developing countries whose national public institutions often are still weak, the challenge of implementing and maintaining, for example, modern land titling and registry systems can be very large indeed.* David Atwood has made the point forcefully:

The adjudication, sophisticated recording systems, precise boundary delineation, and mapping requirements of land registration or titling are quite costly in the use of legal, technical and managerial skills. These skills tend to be needed in a number of other high-priority areas in many African countries. Given existing constraints on the human and financial resources available to most African countries in the next few years, a decision to allocate skilled people to land registration is implicitly a decision to deny those skills to another important sector or activity.[14]

Klaus Deininger and Hans Binswanger have underscored the operational importance of this issue:

> The titling process requires a clear legal basis and a streamlined institutional infrastructure that is capable of administering the process efficiently. Numerous World Bank projects have either underestimated the complexity of the technical issues involved in titling or assumed that titling could be initiated even if agreement over complex policy issues had not been reached. Many countries have a plethora of institutions, programs, and projects – often with overlapping competencies and responsibilities, contradictory approaches, and high resource requirements – that make it impossible to administer a titling program effectively or to instill confidence in the validity of the titles issued.[15]

The large initial costs of titling may be borne in part by international development agencies, but a land titling system that is not maintained up to date loses most of its value. This is particularly an issue in Africa, where many customary

---

13. Michael Kirk, 'Review of Land Policy and Administration: Lessons Learnt and New Challenges for the Bank's Development Agenda', peer review provided at the electronic forum on Land Policy Issues and Sustainable Development, co-ordinated by the World Bank, Washington, DC, USA, March 5 to April 1, 2001, p. 5.

14. D. A. Atwood, 'Land registration in Africa: the impact on agricultural production', *World Development*, **18**(5), 1990, p. 666.

15. K. Deininger and H. Binswanger, 'The evolution of the World Bank's land policy: principles, experience and future challenges', *The World Bank Research Observer*, **14**(2), August 1999, pp. 260–261.

systems have prevailed up to present times, but it is also a concern in Latin America and many parts of Asia. Other kinds of land regulations, including taxes and restrictions on land use, also impose potentially difficult administrative requirements. Today, increasing recognition is given to the need to ensure that land policies do not impose administrative burdens that are beyond institutional capabilities, and that appropriate attention is paid to the needs for institutional strengthening in this area.

Institutional effectiveness is essential for ensuring that allocative efficiency is achieved, whether by land markets or other mechanisms. To achieve effectiveness in this sense, an appropriate legal and regulatory framework is required, as well as the institutional capacity to administer it. The regulations governing land rentals and sales, for example, are critical to allocative efficiency and are discussed later in this chapter. Institutional capacity is vital for ensuing an even-handed application of the rules for access to land, to families of all income strata and all political affiliations.

Another institutional dimension of the land problem comprises the entities and procedures for resolution of land disputes. Often those institutions are weak, or practically non-existent from the viewpoint of most farmers, since they may take the form of courts located in distant areas, and access to them would be prohibitively time-consuming and expensive for most rural persons who, in any case, may feel intimidated by their procedures. Judicial systems that guarantee swift, impartial and inexpensive resolution of land disputes are an important component of security of land tenure, as well as representing a bulwark against abuse, and therefore they deserve priority as an objective of policies and programs for land tenure.

5.4 OVERVIEW OF ISSUES AND TRENDS CONCERNING LAND TENURE

As illustrated by the foregoing discussion, issues concerning rural land tenure are numerous as well as complex. They are also very largely institutional in nature. They differ across regions of the world and from country to country. While they exhibit some common threads, the solutions have to be very specific to the context. A summary of some of the more frequently occurring issues include the following, drawing in part on a recent paper by Liz Alden Wily:[16]

5.4.1 Key Issues for Land Policies

(i) *Issues Concerning the Nature of Land Rights*

- How should unregistered, customary land-holding and common rights be dealt with in the law, and how much emphasis should be placed on converting informal land rights to formal rights?
- What is the most appropriate form or mix of land use rights among communal, co-operative and individual forms?
- How can the rights of women to land be promoted, especially given their major role in agriculture?
- What is the proper role of the State as an owner or steward of agricultural lands?
- How should major transitions in land tenure regimes be designed and guided: transitions out of collectivized forms of landholding in Eastern Europe and parts of Latin America, transitions away from concentrations of land in a few hands in some countries of Latin America, Africa and Asia, and transitions out of informal tenure systems into more formalized systems in Africa and elsewhere?

(ii) *Issues Concerning Land Markets and Their Regulation*

- What is the role of land markets and should there be restrictions on the transfer of land? This question applies to land rental and leasing as well as land sales.
- Is there a role for agrarian reform, and are there alternatives to it?

---

16. Liz Alden Wily, 'Land Tenure Reform and the Balance of Power in Eastern and Southern Africa', *ODI Natural Resource Perspectives*, No. 58, June 2000, in particular, p. 2.

- How can improved access to land for the poor be promoted; and how can financing for access to land be made more widely available?
- How can all rights to land be brought into a simple, efficient and accessible system of documentation and evidence, and how can subversion of such systems for personal interests be minimized?
- At what level in society and with what degree of autonomy from the Executive should property regulations be regulated and administered?

(iii) *Issues Concerning Environmental Sustainability*

- In light of the increasing scarcity of arable land in all continents, what policy measures can be designed to foster intensification of the use of existing cultivable lands?
- How can land tenure regimes influence the maintenance of forest cover and soils in the face of growing pressure on the land?

Most of these issues, and others, are taken up in the remainder of the chapter. To set the stage, a highly compressed review of pertinent historical background is provided.

5.4.2 Historical Trends Regarding Rights to Use Land

Rights to land take many forms, and unrestricted private ownership is only one of them. Almost all customary systems of land management designate both communal areas where, for example, all families in a village may graze their livestock, and areas in which a 'parcel' is reserved for independent cultivation by each family. In both cases, **rights of usufruct** are defined but not **ownership rights** that would permit the owners to transfer the land. The ownership of the land, such as it is, belongs to the collectivity (usually the village).

Pre-Columbian Aztec society again provides a reminder of the antiquity of these practices:

The *calpulli* [settlement] was the unrestricted owner of its [designated] lands; but the right to use the land belonged to the families who held it in perfectly well delimited plots. . . . the usufruct right was transferrable from fathers to sons, without limitation or time limit; but it was subject to two essential conditions; the first was to cultivate the land without interruption; if the family ceased to cultivate it for two consecutive years, the chief of the settlement summoned the family [to discuss it], and if in the following year [the situation] was not remedied, the family lost its usufruct right irredeemably.

The second condition was to remain in the settlement to which the land belonged. . . . [furthermore] only those who were descendants of members of the settlement were entitled to make use of the communal property.

When some land of a *calpulli* remained unused for whatever reason, the chief . . . with the agreement of the elders, divided it among newly formed families.[17]

All over the world, the rights to individual parcels have been found to undergo an evolution as population pressure on the land increases (i.e. as the value of land increases relative to the value of labor). When the land is in abundance, the rights are *general* in that a family does not necessarily have access to the same plot year after year, but rather it is only guaranteed access to land somewhere in the village's domain. In a typical pattern, a family may cultivate a plot for two consecutive years, after which it is left fallow – becoming communal land – and all families may put their animals to graze on it. After the fallow period, the plot may be assigned to another family, and the first family will in turn receive rights to a different plot.

As population pressure increases, land becomes relatively more scarce and families become motivated to hold onto the land they have been working, out of uncertainty about the prospects of obtaining another piece of land which is equivalent in quantity and quality. *Hence, over*

---

17. L. Mendieta y Núñez, 1983, p. 17 [author's translation].

time, rights to use the land tend to become specific to particular plots. Where the legal system of land tenure permits it, there is a corresponding tendency toward claiming private ownership of the land a family has been working.

Accordingly, Daniel Cotlear found the following trends in land tenure in indigenous communities of the highlands of Peru:

The original form of land ownership in the Andean communities was communal; the families of the community members had the right to cultivate and graze their livestock on the community's lands and all other families were excluded from those rights. Each family had a general right to be periodically assigned "fresh" land for cultivation and it retained exclusive rights to specific plots only as long as the cropping cycle lasted; these rights were lost when the land entered into fallow once again. . . .

The basis of the original system was the abundance of land. Increasing pressure on the land was what led to change. When the plots of land began to become scarce, the community members wanted to re-cultivate a given parcel before the normal period of fallow would have ended. . . . Each year the cropped area covered a larger portion of the land. . . . in these conditions, it became increasingly difficult to find plots of better land which were not already taken by another family. . . . In this way, under the increasing pressure on the land, the fallow period was reduced and the community members became more conscious of the necessity of developing special rights for particular pieces of land. From then on, the development of informal property rights followed rapidly upon increases in the cropping intensity.[18]

A similar pattern of evolution of customary land rights was observed in Zambia:

Most Zambians conduct their activities in accordance with and subject to customary law. . . . Two contending views are held on customary land tenure in customary law. One view suggests that land and land rights are not individual but commonly shared. The other, increasingly held view recognizes individualism in land relations and tenure. . . . Both views are valid because they arise from the dynamism of customary tenure, which has evolved from commonly shared land rights to individualization of croplands with continued commonly shared rights to grazing land, forests and fisheries. Individualization of croplands is a result of agricultural intensification, increase in population pressure and commercialization of agriculture.[19]

Boserup, in her seminal work, saw this trend as a component of an almost universal pattern of evolution of land tenure:

Both the physiocrats and the classical economists in Britain based their ideas of the effects of population growth in agriculture upon the assumption that private property in land emerges when agricultural land becomes scarce under the pressure of growing numbers of people. . . . that a class of private landowners would appear as soon as good agricultural land had become scarce. . . . The gradual disappearance of the general rights to clear new plots and to graze the animals freely in fallow and commons and the replacement of these rights by the permanent right of each cultivator family over particular pieces of land, is only one link in the chain of events which gradually changes the agrarian structure in such a way that private property in land becomes a dominating feature.[20]

18. Daniel Cotlear, *Desarrollo Campesino en los Andes: Cambio Tecnológico y Transformación Social en las Comunidades de la Sierra del Perú*, Instituto de Estudios Peruanos, Lima, Peru, 1989, pp. 48–49 [author's translation].

19. Vernon R. N. Chinene, Fabian Maimbo, Diana J. Banda and Stemon C. Msune, 'A comparison of customary and leasehold tenure: agriculture and development in Zambia', *Land Reform*, Food and Agriculture Organization of the United Nations, Rome, 1998/2, p. 91.

20. E. Boserup, 1965, pp. 78 and 86.

Land rights under Islamic law are similar in some respects, but not in all, to other traditional systems:

As with other indigenous systems, land belonged to the 'person who vivified it' . . . the act of cultivation, or boring and enclosing underlying streams, gave the person doing so a right of ownership. But Islamic law differed from other indigenous rules in two respects: First, once land had been appropriated, nonuse did not mean a loss of ownership; that could happen only through conquest or sale. Second, Islamic law provided for defined rules of inheritance for both men and women. . . . (G. Feder and R. Noronha, 'Land rights systems and agricultural development in sub-Saharan Africa', The World Bank Research Observer, 2(2), July 1987, p. 147.)

While not agreeing that such an evolutionary process must necessarily lead to private property, Platteau observed, 'population growth and commercialization of agriculture have always resulted in a process of individualization of land tenure. This means, broadly speaking, that the rights of individuals or nuclear (as opposed to extended) families have been gradually increased at the expense of the larger group's prerogatives'.[21] In his extensive survey of sub-Saharan Africa, he cites studies whose findings support this trend in pre-Independence societies in Ghana, Niger, Nigeria, Rwanda, Burundi, Madagascar, Tanzania and Senegal (although the trend proceeded at different speeds in different parts of Senegal).

Today, in Africa individual rights are more prevalent than communal rights on croplands in most cases. Atwood, citing several studies, comments that 'African farming is most often based on individual household or family units, not on larger communal farm operations. Land-use rights, most often to a specific plot of land, are held by individuals or households'.[22]

Nevertheless, traditional land rights can display great complexity and imagination in providing resource access to satisfy basic needs. Traditional systems have shown considerable flexibility in defining bundles of land rights, as documented by Tidiane Ngaido and Michael Kirk, drawing upon many observations in African and Asian societies:

. . . our broad definition of rangelands suggests the parallel existence of different property rights regimes. For example, communities may hold common property rights over their local pastures, use-rights over the routes and grazing corridors, access rights that are generally based on reciprocal arrangements with neighboring communities, and private rights on the fields that they cultivate on high potential areas of their pastures. . . . Ngaido (1998)[23] therefore, proposes the classification of property rights in rangeland as:

- *Private property rights*, which can be enjoyed by individuals or families, can be either on animals, water resources or high-potential tracts of rangelands (e.g. "wadis" or "bas fonds"). Private property assures full control over the resource including the right to sell or lend.[24]
- *Secured access rights*, which are dominant rights over pastoral resources, are use-rights granted to community members by traditional leaders on their common pastures and water resources for their production activities. Under such tenure regimes, individuals have only priority use-rights, which can be maintained for long time periods but do not entitle individuals to private property.

---

21. J.-P. Platteau, 1992, pp. 133–134.
22. D. A. Atwood, 1990, p. 661.
23. Tidiane Ngaido, 'Can pastoral institutions perform without access options?', paper presented at the International Symposium on Property Rights, Risk and Livestock Development, Feldafing, Germany, September 27–30, 1998.
24. Here, Ngaido might have said, 'full control to the extent permitted by law, formal or informal'.

However, the flexibility of this tenure system allows leaders to recapture unused land or redistribute land to needy members.

• *Access options* are bundles of options available to individuals and communities for securing their livelihood and production systems in response to the constraints they face. These options could be based on formalized or informal institutional arrangements such as reciprocity or on market relations like purchase of feed or rental of harvested fields. In eastern Sudan, written inter-ethnic contracts on reciprocal range utilization in times of crises co-exist with informal oral arrangements since the 19th century. They are increasingly complemented by market transactions on feed resources, such as agricultural by-products and water, as well as by manifold tenancy arrangements for feed resources based on fixed rent or sharecropping. . . .[25]

5.4.3 The Era of Landed Estates

Sometimes, customary land rights have not been transformed directly into individual holdings, but rather there has been an intermediate stage of creation of large estates in the hands of the ascendant social classes. This stage has often been followed by land reform. Boserup and Binswanger, Deininger and Feder have documented this trend well, and the associated interventions in labor and product markets, which mainly had the effect of creating bonded labor of various forms. According to Binswanger *et al.*, 'as population density increases private rights to land emerge in a slow and gradual process which exhibits great regularity'.[26] They present a useful schematic vision of the process, in which the first

stage normally was a *manorial system*, in which the peasant cultivators, who had their own rights of usufruct on portions of the manor, were required to pay tribute and/or provide labor services to the overlord.

Given the supervisory costs entailed in farming large expanses of land, coercive measures were required to ensure that the larger exploitations could be competitive against small owner-operated farms that used only family labor. Such coercion sometimes went as far as indentured labor contracts, serfdom and slavery, but it also could take the form of interventions in product markets (for example, limiting the right to market some products, only to the overlords).

In the course of economic development, non-agricultural employment opportunities opened up for the cultivators, and so it became increasingly expensive for an overlord to use their services to cultivate his own lands (the *demesne* of the estate). Hence, many manors, or *haciendas*, evolved into fully rented-out *landlord estates*. This occurred in Western Europe, Ethiopia, Eastern India, Iran, China, Korea and Japan. In contrast, the *hacienda system*, which continued to impose restrictions on peasant farmers and in which the overlord directly cultivated a portion of the estate, became the prevailing model in Latin America, some parts of Eastern and Northern Africa, and in parts of Eastern Europe.[27] A variant of this model, termed the *junker estate*[28] by Binswanger *et al.*, relied more on hiring workers from off the estate, mainly to avoid the threat of land reform which tended to arise out of a situation in which the peasants held well-defined rights of usufruct on the estate. Such estates first emerged in Prussia but also developed in many other parts of the world.

25. Tidiane Ngaido and Michael Kirk, 'Collective Action, Property Rights, and Devolution of Rangeland Management: Selected Examples from Africa and Asia', in Land Policy Network (www.worldbank.org/landpolicy), no date, ca. 2000.

26. Reprinted from J. Behrman and T. N. Srinivasan (Eds), *Handbook in Development Economics*, Vol. 3B, H. P. Binswanger, K. Deininger and G. Feder, 'Power, distortions, revolt and reform in agricultural land relations', p. 2666, Copyright (1995), with permission from Elsevier.

27. *Op. cit.*, p. 2678.

28. The term 'junker estate' was coined by the Russian rural sociologist A. V. Chayanov in the 19th century.

In a later era, junker estates often became specialized in industrial crops and tended to make the transition directly to large-scale mechanized farms and, in Eastern Europe and Nicaragua under the Sandinistas, to State farms. In contrast, haciendas and landlord estates almost always were transformed by processes of land reform, frequently into collective farms in Latin America, and into small owner-operated farms in East Asia. Now, the collective and State farms themselves are undergoing a transformation, often yielding to individualized rights to the land in Latin America, sometimes to private cooperatives, but the debate over their future is still underway in Eastern Europe, China and some other places.

5.5 THE NATURE OF LAND RIGHTS

5.5.1 Forms of Property Rights

The great variety of forms in which land is held in the world today can be simplified into the following six basic types:[29]

* *Open access lands.* In this kind of tenure no one can claim ownership to the land or resource and no one can be excluded from access to it. It sometimes applies to forest lands or range lands. Marine resources usually are of the open-access type.
* *Communal lands.* These refer to the jointly held community lands in *customary land regimes*. They are open to all members of the community, but there are community restrictions on their use and on access to them. Frequently, they are grazing lands.
* *Collective lands* are used for joint production by a group of farm families and are defined by a decision of central authorities through land reform. Their use is not controlled by traditional authorities but rather through newly created management structures. They can include individual plots and jointly worked plots. In most cases, the members of a collective farm have had no voice in deciding the form in which the land is to be held and exploited but rather the decision was made centrally.[30]
* *Individual land rights under associative tenure.* These rights embrace individual plots in customary and collective tenure regimes. In the former Soviet Union, collective farms usually contained 'household plots' of around 1–5 hectares each, on which families were entitled to grow food crops of their choice.
* *Private land rights.* These rights include *ownership* (with varying degrees of restrictions), and other usufructuary rights in a market context such as rental, leasing and sharecropping. They may also be subordinated, temporarily and partially, to group decisions through *voluntary* cooperation in selected farming tasks or in the procurement of farm services. Ownership brings with it the right to dispose of the land according to the owner's wishes: in sale, leasing, rental, inheritance, and to encumber it with contingent claims such as a mortgage.
* *State lands.* In this case property rights are assigned to an authority in the public sector, local or national.

---

29. This list is an amplification of the discussion of the topic found in Land Tenure Service, FAO, 'Land tenure, natural resource management and sustainable livelihoods', report prepared for the World Food Program, Food and Agriculture Organization of the United Nations, Rome, 2001, p. 6.
30. *State farms* are distinguished from collective farms by the State ownership of the land. In these cases, cultivators are employees of the State. However, in practice the cultivators of collective farms usually do not enjoy the typical rights of ownership, and the State tightly regulates the operations, and so the operational distinctions between the two forms are relatively narrow. Unless otherwise stated, the term 'collective farms' will be understood to apply to both forms. In Latin America, collective farms often have the juridical status of *production co-operatives*, as defined by agrarian reform legislation, but the members of the co-operatives are not vested with full property rights in the farm's assets. Therefore, this class of co-operative must be distinguished from private co-operatives (mostly for services) that are constituted under a country's commercial code.

Forms of tenure may overlap. For example, State lands may be given out to farm families in long-term, tradeable leases, which are a form of private land rights.

Ownership of land is rarely absolute. Property rights in land have developed that usually are not as absolute as rights in other commodities. Laws that allocate property rights can be complicated and involve a number of jurisdictional authorities. Land ownership may be viewed as a bundle of rights, rather than absolute control over the land.[31]

> ... when one party holds a property right in a resource, the party does not necessarily hold all the rights. The existence of a property right that does not contain some restrictions, which imply that some other entity also holds rights in the resource, seems rare. While restrictions are often placed on certain behavior by the State, restrictions also arise from other quarters (family, kinship group, religion). ... even the concept of 'ownership' in western societies, where the full rights in a resource belong to an individual after certain governmental reservations are taken into account ... recognizes that at least two parties have rights in the resource.[32]

The fact that rental and ownership rights may exist on the same piece of land illustrates the pervasiveness of multiple kinds of rights or 'sticks in the bundle of rights'. Herders, for example, may have rights to drive cattle across certain private lands; and the State may reserve the right to build roads or place power lines in an *easement* on private property. A community may reserve the right to define the kinds of activities that may take place on private lands, expressed through *zoning restrictions* or other regulations.

Banks may hold contingent rights to claim ownership of a plot in the event of default. In some traditional societies, persons may purchase or rent trees on a piece of land without having rights to the land itself. In Haiti, for example, it still is common for a farm family to rent a breadfruit tree for a season, thereby obtaining rights to its harvest.

Property rights display great diversity. The fundamental point is that **property rights are rarely absolute**. Therefore, the conditions imposed on their use may be more important than who or what institution has ownership rights.

5.5.2 Tenure Security and Land Rights

Pursuit of the objective of economic efficiency requires security of land tenure, security at least in the basic sense that (a) society recognizes an individual's claim to the land, and (b) legal or other institutional mechanisms exist to defend that claim without normally incurring undue

---

Examples of Facets of Property Rights

- *A right to use the property and prevent others from using it.*
- *A right to control how property is used (a community may exercise zoning control).*
- *A right to derive income from a property.*
- *Immunity from expropriation.*
- *A right to transmit a property to heirs.*
- *A right to transfer all or part of a property to others.*
- *A residuary right that takes effect when other rights lapse (i.e. owners may reclaim use of property when leases end).*
- *Rights in perpetuity or rights delimited in time.*
- *Contingent rights to ownership, such as those held by a creditor.*

(Adapted from Land Tenure Service, FAO, 2001, p. 7.)

---

31. Gerald A. Carlson, David Zilberman and John A. Miranowski, *Agricultural and Environmental Resource Economics*, Oxford University Press, New York and Oxford, 1993, p. 406.
32. Reprinted from *Journal of Development Economics*, **33**, B. A. Larson and D. W. Bromley, 'Property rights, externalities, and resource degradation', pp. 237–238, Copyright (1990), with permission from Elsevier.

costs. *Tenure security in turn requires clarity in defining a user's right to the land, and stability of that right over time.* Inadequately defined property rights have been blamed for many of the problems encountered by developing countries and transition economies.[33] With reference to Africa, Platteau has said 'efficiency and equity costs may arise equally [whenever] land laws are vague, use confusing or non-operational concepts, are badly applied or frequently modified, leave too much room for arbitrary decisions or involve too heavy procedures'.[34] Conversely:

> ... efficiency requires individual land rights to be recognized in a way that provides sufficient security (either in the form of long-term leases or land titles). That stage may not have been reached yet in parts of sub-Saharan Africa. But in other parts (sometimes only a region within a country), the justification for a change in land arrangements already exists.[35]

Feder and Noronha draw a distinction between security of the right to use a given plot of land and security in the stronger sense of also being able to transact the land. This distinction further illustrates the fact that land rights are rarely absolute:

> The term 'security' is often misunderstood in the literature. When it refers to the ability to use land for a certain period and for a defined purpose without disturbance, security of possession is usually ensured under indigenous systems. It is clear that in most sub-Saharan African societies, land under cultivation by an allottee cannot be taken away. Eckert (1980)[36] notes that in Lesotho the average period of

landholding is eighteen years which, adds Doggert, is 'more than that prevailing in the United States'.[37]

The situation, however, is entirely different when security is defined as the ability of an occupant to undertake land transactions that would best suit his interests – for example, to offer land as collateral for a loan.[38]

Different facets of tenure security have been described by Munro-Faure *et al.* in the following terms:

> Tenure security complements access to land by providing confidence that a person's claims to rights will be recognized by others in general and enforced in cases of specific challenges. Security implies certainty; the counterpoint of security is the risk that rights will be threatened by competing claims, and even lost as a result of eviction. ....
>
> Tenure security can be interpreted in several different ways. Security directly relates to protection from loss of rights. For example, a person may have a right to use a parcel of land for a 6 month growing season, and if that person is safe from eviction during the season, the tenure is secure.
>
> By extension, tenure security can related to the length of tenure, in the context of the time needed to recover the cost of investment. Thus the person with use rights for 6 months will not plant trees, or invest in erosion prevention measures and irrigation works as the time is too short for that person to benefit from the investment. The tenure is insecure for long-term investments even if it is secure for short-term ones.

33. A. Schleifer, 'Establishing Property Rights', in *Proceedings of The World Bank Annual Conference on Development Economics*, World Bank, Washington, DC, 1994, p. 93.
34. J.-P. Platteau, 1992, p. 163.
35. G. Feder and R. Noronha, 1987, p. 163.
36. J. Eckert, *Lesotho's Land Tenure: An Analysis and Annotated Bibliography*, Lesotho Agricultural Sector Analysis Project Special Bibliography 2, Lesotho Ministry of Agriculture and Department of Economics of Colorado State University, Maseru, Lesotho and Fort Collins, CO, USA, 1980.
37. Clinton L. Doggett, Jr, *Land Tenure and Agricultural Development in Lesotho and Swaziland: A Comparative Analysis*, Bureau for Africa, US Agency for International Development, Washington, DC, USA, 1980.
38. G. Feder and R. Noronha, 1987, pp. 158–159.

The importance of long-term security has led some to argue that full security can arise only when there is full private ownership (e.g. freehold) as under such tenure, the time for which the rights can be held is not limited to a fixed period. It is argued that only an owner enjoys secure rights, and holders of lesser rights, such as tenants, have insecure tenure because they are dependent on the will of the owner. It is then implied that security of tenure comes only with holding transfer rights, i.e. rights to buy, mortgage, etc. Equating security with transfer rights is true for some parts of the world but it is not true in many others. People in parts of the world where there are strong community-based tenure regimes may enjoy tenure security without wishing to sell their land, or without having the right to do so, or having strictly limited rights to sell, for example, sales may be restricted to members of the community.[39]

Tenure security in the sense of being certain to be able to use the land for a given period, and to an extent being able to transfer it, can be provided by customary land tenure systems:

The individual use rights characteristic of most African land tenure systems are most often secure. . . . They tend to be heritable. . . . Many African customary land tenure arrangements permit land transfers, and this tendency has increased with greater integration of rural areas into the market economy. Borrowing, pledging,[40] renting and selling are widespread and often have been for some time. . . .[41]

The tenure security characteristic of customary rights has been summarized by Platteau:

There is another important aspect of customary rights in sub-Saharan Africa that has been largely misunderstood by those holding the conventional view about the present inadequacy of these rights. Indeed as it is now admitted by many authors . . . security of tenure was usually quite satisfactory under traditional African systems of land rights. . . . except in extreme circumstances (as in the case of open conflict with the customary authorities or other extraordinary conditions justifying the exclusion of an individual from his social group), the allottee's right to use a given piece of land is safeguarded as long as he keeps cultivating it. In Lesotho, the chiefs and their subjects are even reported to be confident today that 'land would be reallocated only after several years' neglect, *however keen population pressure and the demands or rival claimants might be*' (Robertson, 1987).[42] Moreover . . . cases were not rare where the heirs are given the lands that were (more or less intensively) cultivated at the time of the death of allottees, even though their rights do not generally extend to lands which had been cultivated by the same but were under fallow when they died.[43]

Customary land tenure systems display considerable variety, and in some instances the rights extend only to plots of a size to cover subsistence needs and little more. Customary rights normally cannot serve as collateral to raise bank financing for investments because sale outside the community is not permitted. In this regard, customary access to land may be strictly tied to membership in a rural community. Such land cannot represent an economic asset whose value could be transferred to other sectors, as for an investment in starting up a small business in an

39. P. Munro-Faure, P. Groppo, A. Herrera and D. Palmer, 2002, pp. 18–19.
40. 'Pledging' is transferring use rights of the land for a temporary, but often extended, period in exchange for a loan.
41. D. A. Atwood, 1990, pp. 661–662.
42. A. F. Robertson, *The Dynamics of Productive Relationships – African Share Contracts in Comparative Perspective*, Cambridge University Press, Cambridge, UK, 1987 [emphasis added].
43. J.-P. Platteau, 1992, pp. 123–124.

urban area, nor can it generate much wage labor in rural areas:

> In customary tenure, farming is not a business but a way of life for the people. Therefore customary tenure only ensures a greater utilization of available labor. The strength of privatized tenure is greater employment creation for wage labor.[44]

Nevertheless, the point to be recognized is that customary systems may provide a degree of tenure security which is considered satisfactory for most production purposes. Chinene *et al.* summarize well the differences between customary and modern tenure, for the case of Zambia, but in words that are applicable elsewhere:

> Customary tenure has by and large been more successful than [long-term] leasehold tenure in meeting the needs of the people. The administrative procedures are simple and easily implemented. Land issues are dealt with efficiently and decisively. The problem, however, is that the land rights are never registered, although their recognition guaranteed. . . .

> Leasehold by its nature facilitates adoption of business approaches to farming. Because of the length of the lease, long-term investments are encouraged. The income-generating potential is therefore higher than under customary tenure. Greater access to markets facilitates income generation on [leasehold] land.

> A major advantage of leasehold tenure over customary tenure is that titles facilitate land sales which both generate income and move land resources to efficient producers.[45]

5.5.3 Land Rights, Farm Size and Agricultural Productivity

Research on the relation between security of land tenure and agricultural productivity has not been extensive; more results are available on the relation between farm size and productivity. The empirical evidence available regarding the role of tenure all has confirmed that the relationship between tenure security and measures of productivity is positive. Path-breaking work in this regard has been carried out by Gershon Feder and associates in Thailand. Feder and Tongroj Onchan analyzed data for more than 500 farms in three provinces of Thailand, and they concluded that land titling gave rise to greater on-farm investment in two of the provinces and to the acquisition of more farmland in the third.[46] Feder and Noronha cite research in Costa Rica and Brazil which showed a positive relation between ownership security and the amount of investment per unit of land.[47] Intensity of land use also has been shown to respond positively to tenure security for the case of Jamaica.[48] Atwood has cited research demonstrating a link between tenure security and farm income in the case of Kenya (D. A. Atwood, 1990, p. 659). Binswanger *et al.* (1995) also cite the results of a less formal study in Ecuador which supports the relationship between titling and farm income levels.

Feder and colleagues have found a clear, positive relationship between land titling and the availability of credit to the farmer. They present evidence from India, Thailand and South Korea which shows that both formal and informal institutions are more willing to lend to farmers when land is used as collateral.[49] In addition, they analyze data from Lop-Buri Province, Thailand,

---

44. V. R. N. Chinene, *et al.*, 1998, p. 93.
45. *Op. cit.*, pp. 92 and 94.
46. Gershon Feder and Tongroj Onchan, 'Land Ownership Security and Capital Formation in Rural Thailand', Discussion Paper ARU 50 (revised version), Research Unit, Agriculture and Rural Development Department, The World Bank, Washington, DC, USA, February 1986.
47. G. Feder and R. Noronha, 1987, p. 160.
48. *Op. cit.*, p. 161.
49. Reprinted from *Agricultural Economics*, **2**, G. Feder, T. Onchan and Tejaswi Raparla, 'Collateral, guarantees and rural credit in developing countries: evidence from Asia', pp. 234–236, Copyright (1988), with permission from Elsevier.

under two different econometric models, concluding that (p. 243) 'in the area studied, the pledging of land collateral increases the amount of institutional credit offered by 43%... or 55% ... as compared to a loan without a security'.

Feder and Noronha mention that a study by Seligson in Costa Rica[50] showed that titling increased the share of farms obtaining credit from 18% to 31.7%, and a study in Jamaica[51] showed that titling enabled almost half of the recipients to increase their borrowing.[52]

In an analysis of land titling experiences in Nicaragua, Deininger and Juan Sebastian Chamorro found that a program of conferring registered (legally binding) titles on small farms increased land values by an average of 30% and increased the propensity to invest in the land by 8–9%.[53]

For the relationship between farm size and productivity, perhaps the most thorough empirical investigation was carried out by Rasmus Heltberg for Pakistan. With farm-level panel data, he analyzed the effect of farm size on value added and land returns per hectare, controlling for differences in soil quality, market imperfections in labor and credit, and other variables. He obtains 'highly significant' statistical results that indicate a U-shaped relationship between farm size and each of the two measures of productivity, i.e. value added and land returns per hectare. However, in a range of farm sizes that account for 90% of all farms and 65% of farmland, his results show an inverse relationship. He hypothesizes (as other authors have done) that the main explanation for the inverse relationship is that hired labor is an imperfect substitute for family labor. In his own words, his main conclusions are as follows:

Three lines of criticism have been raised in the literature against the IR hypothesis [inverse relationship between farm size and productivity]: (a) that the empirical evidence is flawed due to omitted variable bias, (b) that the relationship may no longer hold after the Green Revolution, and (c) that a consistent explanation for the inverse relationship is missing.

With respect to the first criticism, the article has presented strong evidence to support the presence of an inverse size–output relationship even when soil and other heterogeneity is controlled for. . . . With regard to the second point of critique, it was found that small farms are still significantly more productive than big farms, also in irrigated and relatively developed areas of Pakistan. With respect to the third point . . . a set of reasonable hypotheses about labor, land, credit and insurance market failures was set up to account for systematic size–output and size–profit relationships. . . . the market imperfections framework conforms well with the data.[54]

Heltberg finds imperfections in the market for credit as well as farm labor. There also is a U-shaped relationship between credit used *per hectare* and farm size, with the credit variable declining over a considerable range of farm sizes. When credit per hectare begins to increase with farm size, he conjectures that it may contribute to the greater productivity of very large farms over medium-sized farms.

Another careful empirical study of the farm size–productivity relationship was conducted by Fidele Byiringiro and Thomas Reardon for Rwanda. Their principal results were consistent

---

50. M. A. Seligson, 'Agrarian reform in Costa Rica: the impact of the titles security program', *Inter-American Economic Affairs*, **35**(4), Spring 1982.

51. Inter-American Development Bank, 'Jamaica Land Titling Project Feasibility Report', Inter-American Development Bank, Washington, DC, USA, 1986.

52. Feder and Noronha, 1987, pp. 144–145.

53. Klaus Deininger and Juan Sebastian Chamorro, 'Investment and income effects of land regularization: The case of Nicaragua', The World Bank and the University of Wisconsin at Madison, Washington, DC and Madison, WI, USA, mimeo, January 2002.

54. R. Heltberg, 'Rural market imperfections and the farm size–productivity relationship: evidence from Pakistan', *World Development*, **26**(10), 1998, pp. 1823–1824.

with those of Heltberg, including the finding of a
U-shaped relationship over a wider range of farm
sizes. Their conclusions are as follows:

> We explored: (1) whether the smaller farms
> have greater average and marginal land pro-
> ductivity than the larger farms, and whether
> the smaller farms are less allocatively efficient;
> and (2) whether . . . soil erosion reduces, and
> soil conservation investments increase, land
> productivity. Both queries were answered with
> a strong affirmative. Moreover, the inverse rela-
> tionship is *not* mitigated by the smaller farms'
> being more eroded, despite their farming more
> intensively (with less fallow). In fact, smaller
> farms are *not* more eroded than larger farms.
> Moreover, the inverse relationship is *not* miti-
> gated by larger farms' using more nonlabor
> inputs or by their putting more of their land
> under cash perennials. In fact, larger farms do
> *not* do more of either compared to smaller
> farms.
>   . . . we find the marginal value product of
> land on smaller farms to be well above the
> rental price of land, implying factor use ineffi-
> ciency and constraints to land access. By con-
> trast, the marginal value product of labor on
> smaller farms was well below the market wage.
> This implies a 'bottling up' of labor on smaller
> farms, and constraints to access to labor
> market opportunities. . . .[55]

In addition to these econometric studies, tabu-
lations of data from agricultural censuses and
surveys in a number of countries show *markedly
higher productivity per hectare on small farms than
on large farms*. Binswanger, Deininger and Feder
cited early findings along these lines by Albert
Berry and William Cline which show that the
value of output per hectare is 5.6 times larger on
small farms than on the largest ones in Northeast
Brazil, 2.7 larger in Pakistan's Punjab, and 1.5
times larger in the Muda irrigation scheme of
Malaysia.[56] Similarly, for Honduras census data
for the mid-1970s showed that farming income
per hectare fell from 584 lempiras for farms of
0–2 hectares to 215 lempiras for farms of 10–20
hectares.[57] Comparable findings have been made
for a number of other countries. The inverse relationship between farm size and productivity per
hectare in developing countries is now widely
accepted as generally correct, although exceptions always may be found.

5.5.4 Transferability of Traditional Land Rights

In the earlier stages of traditional land systems,
transfers of land rights sometimes occur but
within the community and especially among close
kin. Transfers to outsiders are either not permitted or approval by the whole community is
required. However, tribal societies often have
found ways to circumvent formal bans of land
sales.[58] Empirical evidence for this has been
found. In Niger, for example, sales of land are on
the increase in spite of the fact that customary
regulations prohibit them.[59] In the Volta region of

55. Reprinted from *Agricultural Economics*, **15**(2) F. Byiringiro and T. Reardon, 'Farm productivity in Rwanda: effects of farm size, erosion and soil conservation investments', p. 135, Copyright (1996), with permission from Elsevier [emphasis in original].
56. H. Binswanger *et al.*, 1995, p. 2703 (citing R. A. Berry and W. R. Cline), *Agrarian Structure and Productivity in Developing Countries*, International Labor Organization of the United Nations, Geneva, Switzeland, 1979.
57. Magdalena García, Roger Norton, Mario Ponce and Roberta van Haeften, *Agricultural Development Poli- cies in Honduras: A Consumption Perspective*, Office of International Cooperation and Development, US Department of Agriculture, Washington, DC, USA, 1988, p. 33.
58. R. Noronha, 'A review of the literature on land tenure systems in sub-Saharan Africa', Discussion Paper No. 43, Research Unit of the Agricultural and Rural Development Department, The World Bank, Washington, DC, USA, 1985 (cited in H. P. Binswanger *et al.*, 1995, p. 2669).
59. This finding was reported in Amoul Kinni, *Etude sur la commercialisation du betail et de al viande dans le departement de Zinder, Niger*, University of Arizona, Arid Lands Natural Resources Committee, Tucson, AZ, USA, 1979.

Ghana, Nkunya[60] says that outright purchases are becoming more and more frequent.[61]

The evolution of land rights in response to increasing scarcity of land relative to labor has not been confined to Africa and pre-Colombian America. Gershon Feder and David Feeny summarize the historical process of formalizing land rights in Thailand in the past century:

> ... in land-abundant, labor-scarce, early-nineteenth-century Thailand, slaves rather than land served as collateral in financial markets. There was a well-developed legal system to govern transactions in labor commitments. In contrast, the system of usufruct land rights was not as extensively developed. ... As land became more valuable and frontier areas were brought under cultivation, land disputes became endemic. The Thai government responded with a series of procedural and administrative changes. A major new law on land rights was enacted in 1892. ... the lack of adequate surveys and record-keeping continued to inhibit the precise documentation of rights; land disputes continued. In 1896 the government responded by initiating a cadastral survey in an area in which important government officials were also landowners, and in 1901 created a formal system of land titling.[62]

However, it must be recognized that under most customary systems of land tenure, selling land still is difficult if not impossible. The *historical trend toward progressive individualization of land rights* not only raises the question of what the nature of land rights should be in a market economy, but to what degree and how they should be regulated. Regulation of land use does exist in traditional societies, particularly with respect to rights of transfer of land and fallow periods and other measures for maintenance of soil fertility; such regulation often is exercised by village councils instead of being promulgated through written laws and the actions of impersonal bureaucracies.

5.5.5 Transitional Issues in Moving from Customary to Formal Systems

Whether through formal titling or other means, the transition out of customary systems to more formal land tenure systems is occurring all over the world. *The process of transition itself, from one type of land tenure regime to another, can create uncertainty over rights to land, unless the transition is guided very carefully.* Such uncertainty undermines the security of land tenure.

A concern that arises during transitions of this nature is how to ensure that new institutions and procedures for land management are not manipulated by the better-off segments in society, in order to deprive lower-income families of land rights. For example, 'experience in the former Soviet Union shows that even after legal reform is completed, bureaucrats are difficult to dislodge from their posts and misinform landowners in efforts to retain control, creating confusion and insecure land rights'.[63] In the words of another observer:

> In Eastern Europe ... a major reason for inefficiency in the allocation of resources in that politicians and bureaucrats have excessive control rights over much of the economy. ... [in Russia] land reform has been ... stymied by government agencies with effective control over all land transactions.[64]

---

60. G. K. Nkunya, 'Land Tenure and Agricultural Development in the Angola Area of the Volta Region', Land Tenure Center Paper No. 120, University of Wisconsin, Madison, WI, USA, 1974.
61. G. Feder and R. Noronha, 1987, p. 155.
62. G. Feder and D. Feeny, 'Land tenure and property rights: theory and implications for development policy', *The World Bank Economic Review*, **5**(1), January 1991, pp. 137–138.
63. Stevan Dobrilovic, 'Land Policy and Administration', peer review provided at the electronic forum on Land Policy Issues and Sustainable Development, coordinated by the World Bank, Washington, DC, USA, March 5 to April 1, 2001, p. 5.
64. Andrei Schleifer, 1994, pp. 93 and 114.

The problem also emerges in the context of transformations of customary land tenure regimes into modern systems with formal title, as in the case of Zambia:

> There is a well-founded fear that proposed land reforms will make it easier for outsiders to get title deeds to land on reserves and trust land at the expense of the local people. When titling is introduced, wealthier and better-informed individuals may use their information advantages to claim land over which other, less informed individuals have customary rights.[65]

Platteau has commented:

> ... the superimposition of a new system of land rights backed by the State authority tends to create serious uncertainties about the application of indigenous rules while, conversely, the persistence of customary land arrangements has the effect of creating uncertainty about the actual validity of formal land legislation. A variety of efficiency costs, static and dynamic, result from this ambivalent situation of land rights dualism.[66]

S. Berry has called attention to the problem of 'politicized accumulation' of land in Africa.[67] Platteau notes that often urban sector groups, especially public officials, are much more knowledgeable about land titling procedures than the average farmer, and thus can use the corresponding procedures to their advantage:

> Real estate appears to be the privileged sector of accumulation for the political class and its allies. ... In Côte d'Ivoire, for example, the *Code foncier* [Real estate code] of 1971 has trig-

> *... the process of land titling or registration can, as in Kenya, permit some individuals to appropriate for themselves exclusive ownership rights in commonlands previously open to any member of the community, or commonly owned family lands. ... land titling or registration, when improperly administered or when customary rights have not served as the basis for registration, has allowed powerful people to take land from poor rural farmers in some African situations (D. A. Atwood, 1990, p. 661).*

> gered off a process of land appropriation by the politico-administrative élite. ... [titling] requests, which were first confined to the periphery of big cities, were gradually extended to rural lands and, as a consequence, customary lands were increasingly transferred from village communities to the urban élite. ... In Senegal ... under the [National Land Law] residents were allowed to establish title and request registration within six months from the date of passage of the law. This measure clearly worked to the advantage of the wealthier and the better informed people who used the transitory period allowed by the law to enlarge their land estates by resorting to every known trick: traditional pledging and borrowing of lands which, once brought under cultivation, would be considered as their property, when the [land law] comes into force; erection of makeshift structures on uncultivated lands; constitution of so-called agricultural associations or groupings purported to exploit the coveted land. ... By contrast, most rural people were generally unaware of the new law provisions and they were not notified to present claims.[68]

---

65. V. R. N. Chinene, *et al.*, 1998, p. 95.
66. J.-P. Platteau, 1992, p. 163.
67. S. Berry, 'The food crisis and agrarian change in Africa: a review essay', *African Studies Review*, **27**(2), 1984 (cited in J.-P. Platteau, 1992, p. 177).
68. J.-P. Platteau, 1992, pp. 177 and 180. It should be noted that this kind of problem can occur even in the absence of titling projects as, for example, when influential interests start to fence off parts of communal lands for their own use.

Binswanger, Deininger and Feder summarized this problem in general terms by pointing out that titling can lead to greater concentration of landholdings and to the dispossession of those who held land rights under a previous customary system. With the introduction of titling, wealthier and better connected individuals take advantage of their better access to information to lay claim to land over which other, less informed, individuals have customary rights.[69]

An important policy conclusion is that *the uncertainty associated with the change from one land tenure regime to another is lessened when the transition simply takes the form of legal recognition of customary rights*.

Another problem that sometimes emerges in the transition process is that when a decision is made to title customary lands in favor of the State, the farmers' rights to those lands can be weakened. For example, they (or the community) may lose the right to decide who inherits each plot, and the option of selling or renting out land may become prohibited. Alden Wily has summarized this problem in the context of Eastern and Southern Africa:

> Strategies for transforming customary land rights . . . have included . . . the subordination of customary rights [to] government land, as in the homelands of South Africa, the communal lands of Zimbabwe, Namibia and Malawi, the trust lands of Kenya, and until recently the 'public lands' of Uganda. These are customarily held lands vested in presidents and States, turning their inhabitants into landless 'occupants' on their ancestral lands.[70]

Even though the hazards associated with implementing a titling system are very real, *not* moving toward a titling system can create risks of

its own. One such risk has been indicated by Alden Wily in the paragraph above. Another risk of not titling the land (whether to groups or individuals) is that farmers' decisions to improve the land may be inhibited because of uncertainty about the value of improvements to the land, and local power struggles over land may be encouraged. The problems associated with lack of formal title tend to worsen as land values rise. Disputes over land and land grabbing by more influential persons have been observed to grow more frequent as land values rise.[71]

5.6 COMMUNAL, COLLECTIVE AND INDIVIDUAL RIGHTS TO LAND

5.6.1 Communal Lands

The area covered by communal lands (common property resources) has tended to shrink in recent decades, as the formalization of customary rights progresses, but these lands still exist in all parts of the world and in some countries are the dominant form of rural landholding. Normally, both their ownership and management are in the hands of the community, but Nadia Forni has commented that:

> . . . resources can be considered as common property irrespective of whether ownership is legally bestowed on the common property resource users, the State or another public body, provided the resources are actually managed according to common property norms. Village ponds, forests, rivers and rivulets, for instance, often fall under formal legal ownership of the State but their *de facto* management rests with the community. . . . Common property resources are management systems where resources are accessible to a

---

69. H. P. Binswanger, *et al.*, 1995, p. 2721.

70. L. Alden Wily, 2000, p. 3.

71. G. Feder and R. Noronha, 1987, p. 144. On the other hand, during the reforms to land laws initiated in Viet Nam in 1988, the central government appears to have effectively blocked attempts at land grabbing by local elites and to have ensured that poor households obtained their fair share of land allocations (see Martin Ravallion and Dominique van de Walle, 'Breaking up the Collective Farm: Welfare Outcomes of Vietnam's Massive Land Privatization', The World Bank, Washington, DC, USA, mimeo, November 12, 2001).

group of rights holders who have the power to alienate the product of the resource but not the resource itself.[72]

Ciriacy-Wantrup and Bishop, Platteau[73] and other authors have made a clear distinction between communal lands and open-access lands, arguing that the former are not necessarily the latter. *The 'tragedy of the commons' occurs on open-access lands, where no one has an incentive to conserve the resource and many may be motivated to use it as much as possible, to its detriment, before others do the same.* In contrast, access to communal lands, and other common-property natural resources, can be regulated by the community, at least in principle:

Common property, with the institutional regulation it implies, is capable of satisfactory performance in the management of natural resources, such as grazing and forest land, in a market economy.[74]

Forni stresses the potential benefits of common property regimes in some circumstances, especially in cases of fluctuating returns to the land:

. . . Common property resources are not evolutionary relics; they exist because they produce certain advantages. They are to be preferred to open access or private property regimes in cases where the resource could be split into individually controlled units but the cost of controlling sole ownership would be prohibitive. . . . Jodha (1992)[75] states that, in a sample of Indian villages, as much as 14 to 23 percent of income was derived from common property resource utilization, and the figures rise to 84 to 100 percent in the case of the poor. [It is difficult to apply] efficiency or optimality principles where there are significant yearly fluctuations of production, as is often the case in common property resources, and where short-term efficiency may undermine long-term sustainability.[76]

Larson and Bromley attribute the tragedy of the commons more to rural poverty than to the communal nature of the property, owing to the lack of effective incentives for sustainable use.[77] While it may be difficult to sort out with precision the causes of land degradation, and communal land management can prevent or arrest degradation in some cases, *in practice the community management of communal lands in developing countries often tends to be weak, and communal lands usually tend to become degraded faster than private lands.* In Zimbabwe, for example, a physical and chemical study of soil erosion patterns found that:

• *Commercial [private] grazing lands* with low densities of livestock and relatively complete vegetation cover had a low rate of erosion.
• *Commercial arable lands* demonstrated similar moderate rates of erosion. This can be attributed to the maintenance of physical conservation measures. . . .
• *Communal arable lands*, though adopting similar physical conservation measures, had considerably higher rates of erosion compared to commercial areas. This was largely due to their poor state of maintenance, open access to grazing. . . .

---

72. Nadia Forni, 'Common property regimes: origins and implications of the theoretical debate', *Land Reform*, 2000/2, Food and Agriculture Organization of the United Nations, Rome (pp. 2 and 3 in on-line version).
73. (a) S. V. Ciriacy-Wantrup and R. C. Bishop, ' "Common property" as a concept in natural resources policy', *Natural Resources Journal*, **15**, 1975; (b) J.-P. Platteau, 1992, pp. 120–121.
74. S. V. Ciriacy-Wantrup and R. C. Bishop, 1975, p. 721.
75. N. S. Jodha, *Common Property Resources: A Missing Dimension of Development Strategies*, World Bank Discussion Paper No. 169, The World Bank, Washington, DC, USA, 1992.
76. N. Forni, 2000, p. 3.
77. B. A. Larson and D. W. Bromley, 1990, p. 256.

• *Communal grazing lands* were the most eroded of all. Livestock numbers exceeded the carrying capacity of the land.[78]

Some observers feel that the degradation of communal lands in Zimbabwe is primarily attributable to the increased population density on the land, rather than to weaknesses in the community management of communal lands. It is likely that such management becomes less adequate as population pressures increase, and hence land values increase.

Ramón López found a clear tendency toward over-exploitation of common agricultural lands in Côte d'Ivoire. Some of his principal findings are the following:

The estimated contribution of biomass to gross agricultural revenues for Côte d'Ivoire is very similar to the values estimated for Ghana (López 1997).[79] . . . Moreover, several agronomic studies in tropical countries have shown that the fallow period makes a large contribution to agricultural productivity. . . . the agronomic evidence is consistent with the estimates here for the effect of biomass on farm revenues. . . .

According to the econometric estimates, the agricultural income of an average village could be increased by 14 percent in the long run if the total cost of biomass were internalized by individual cultivators. This represents a large loss, many times larger than the losses usually estimated for price or trade distortions. The main source of this income loss is the fact that land is overcultivated by about 23 percent.

Rural communities have apparently failed to maintain a system of incentives and controls over individual cultivators that would induce a socially optimal allocation of land among forest, fallow, and cultivation and thus avoid the 'tragedy of the commons.' These results give support to authors who have questioned the effectiveness of indigenous forms of property in achieving a socially efficient allocation of natural resources. . . .

In general, it appears that efficiency of the commons tends to be present in communities with low population density where the transaction and monitoring costs are low. . . . The paradox is that it is precisely in cases of high and rapidly increasing population density that collective action is most needed to achieve an efficient use of common resources.[80]

For Zambia, Chinene *et al.* have observed that communal grazing lands are poorly managed with overgrazing and erosion as a consequence.[81] Although not all empirical studies have reached the same conclusion as the studies of López and Chinene and colleagues, Platteau also acknowledges the failure of common property regimes to protect land resources as the prevailing reality: 'it cannot be denied that nowadays "tragedies of the commons" multiply to such an extent that they have become an important cause for concern in many African countries. . . . *customary property regimes tend to be gradually transformed into open-access regimes*'.[82]

In spite of these problems, communal lands still are important to many groups throughout the

---

78. David Norse and Reshma Saigal, 'National Economic Cost of Soil Erosion in Zimbabwe', in Mohan Munasinghe (Ed.), *Environmental Economics and Natural Resource Management in Developing Countries*, Committee of International Development Institutions on the Environment (CIDIE), distributed by the World Bank, Washington, DC, USA, 1993, Chapter 8, pp. 233–235.

79. R. López, 'Environmental externalities in traditional agriculture and the impact of trade liberalization: the case of Ghana', *Journal of Development Economics*, **78**(1), 1997.

80. R. López, 'The tragedy of the commons in Côte d'Ivoire agriculture: empirical evidence and implications for evaluating trade policies', *The World Bank Economic Review*, **12**(1), January, 1998, pp. 121, 123, 106 and 125.

81. V. R. N. Chinene, *et al.*, 1998, pp. 94–95.

82. J.-P. Platteau, 1992, p. 121 [emphasis added].

world,[83] including indigenous or tribal peoples in Asia and Latin America, as well as to groups in virtually every country in Africa. In the context of *a strategy for communal lands*, policies can be devised to assist these groups, first, by protecting them against loss of their lands and, secondly, by providing them with advice on options for management of communal lands and education of their membership in that regard. In the words of Binswanger *et al.*, 'a community title could be issued to ensure the community's security of ownership against . . . outside encroachment and prevent the poor from being excluded from communal property'.[84] This has been done in Mozambique.

Atwood is part of the consensus on this issue and is more specific on the nature of group forms of title:

If tenure insecurity does not affect transfer of land, but rather maintenance of existing land claims by local people, then other alternatives may be available. When insecurity is primarily with respect to people outside the community, reinforcing or creating group tenure with some legal legitimacy and enforcement may overcome the problem. *Land corporations, group titles*, and other forms of legally sanctioned common property may provide the needed security vis-a-vis the outside world. The advantage of such group tenure forms is that they are much less expensive, and therefore more sustainable, than land titling or registration. They rely on local, informal capacity for management, information, and dispute settlement, rather than on the mechanisms of the formal State legal system. Such group tenure systems need to *safeguard the rights of individuals*, so they require both clear criteria for individual rights and the settlement of disputes, and mechanisms to ensure the accountability of local leaders.[85]

Group titles, of course, have their limitations as a tenure instrument. *Group titles appear to be most appropriate for small communities*, in which achieving and implementing a consensus on rules for regulating use is relatively easy. Larger communities may find controlling communal use patterns to be more difficult, and also frictions are more likely to rise when new investment opportunities are created by the arrival of new production or post-harvest technologies.[86]

Ruth Meinzen-Dick provides a salutary reminder that the decision processes of groups that are given title to land may themselves be exclusionary in some respects:

While in general group registration would be preferable on equity grounds to leaving common property resources undefined, and therefore susceptible to either land grabs by outsiders or State expropriation, it is important to consider how groups are defined. Any time you strengthen the control rights (esp. management and exclusion rights) of one person or group, you reduce the use rights (access or withdrawal) of others. This can be good for environmental

---

83. Forni offers an alternative explanation of environmental degradation of communal lands: 'Overgrazing is frequently . . . a consequence of official restrictions on mobility. [In] Peru and Bolivia . . . indigenous practices were disrupted by obstacles to mobility [with] consequent overgrazing. In addition, the economic significance of herding, which included trading during transhumance and transporting . . . produce, was greatly reduced. Evidence from Morocco . . . seems to indicate that forage land is likely to be privatized when the potential value of the land is high enough to encourage investment for increases in productivity, whereas common property management is suitable in cases where investment is risky and the transaction costs of policing the area are more efficiently shared among larger user groups' (Forni, 2000, p. 7).

84. Reprinted from J. Behrman and T. N. Srinivasan (Eds), *Handbook of Development Economics*, Vol. 3B, H. P. Binswanger, K. Deininger and G. Feder, 'Power, distortions, revolt and reform in agricultural land relations', p. 2722, Copyright (1995), with permission from Elsevier.

85. D. A. Atwood, 1990, p. 667 [emphasis added].

86. Feder and Feeny (1991, p. 140) have discussed this point.

preservation, and may even strengthen the rights of the poor. However, the composition of the group is important. For example, if a young men's club is given the decision-making rights for community forestry, they may make decisions that protect the forest but decrease the ability of women to collect dry branches for firewood. Similarly, watershed management programs in India in which elite males dominate the decision-making have reduced access by women and pastoralists, with the benefits largely going to those with croplands that receive more water.[87]

An example of conferring communal legal protection to the lands of indigenous groups is found in the cited Honduran *Law for the Modernization and Development of the Agricultural Sector*. Article 65 of that law reads, in part:

Indigenous communities which can show occupation of the lands where they are settled for a period of no less than three years will receive titles to the property, completely free of charge, extended by the Agrarian Reform Institute [unofficial translation].

Policy options for communal lands are explored in more detail in Section 5.8. A major challenge for policy is to avoid undermining communal rights, and customary tenure systems in general, during the period in which more formal systems of rights are being put in place.

5.6.2 Collective versus Individual Farms

Collective and State farms, and their variant known as production co-operatives, almost always have been created by land reform movements. They have been based either on State ownership of the land or a very restricted form of group ownership by those who work the land. They did not evolve out of existing communal forms of production but in effect they have been imposed on the beneficiaries of the reforms by governments, with perhaps the only exception being the *kibbutzim* of Israel. Their formation generally has been motivated by a belief in the existence of economies of scale in farming, a pursuit of socio-economic equity within the rural population, and sometimes by an ideological bias against private property.

As economic enterprises, most collective farms were born with a number of handicaps. Usually, their members were denied the right to mortgage their assets or to sell or rent out part of them. They also were legally denied the option of obtaining production finance from private banks,[88] and all of their agricultural advisory services had to be supplied by the government. To offset these disadvantages, they usually have been heavily subsidized, especially through the provision of machinery and equipment, along with cheap credit from State institutions.

Whatever their benefits and constraining circumstances, their performance almost always has been weak in comparison with private farms. The experience of El Salvador with collective farms was evaluated after several years of operating experience and contrasted with the performance of small, individual farms adjudicated to former renters. The results of the evaluation are illustrative of the experiences of many countries with collective farms:

Among the cooperatives [i.e. collective farms] the land cultivated in collective form has steadily diminished and, to the contrary, the land cultivated in individual form, still prohibited by the Basic Law (Decree No. 153), has increased to almost 25 000 hectares, which represents almost 40% of the total of the lands

87. Ruth Meinzen-Dick, 'Peer Review of *Land Policy and Administration: Lessons Learned and New Challenges for the Bank's Development Agenda*', mimeo prepared for the World Bank, International Food Policy Research Institute, Washington, DC, USA, 2001.
88. The Honduran Agrarian Reform Law of 1975 was typical in this respect. It said, in Article 96, 'The lands adjudicated in conformity with this law cannot be used as collateral'. This clause and several others were eliminated or changed in the 1992 reforms.

cultivated in collective form. Likewise, the number of non-member campesinos who cultivate land rented [illegally] from the cooperatives continues increasing. In summary, the land cultivated in collective form fell from 91 361 ha in 1980/81 to 63 049 ha in 1986/87....

Since the first evaluation of the cooperatives ... carried out ... in 1983/84, the amount of cultivable land which is unused has increased from 16 000 to 20 500 ha and the land classified as natural pastures (which represent an under-utilization of the land resource) have risen from 34 000 ha to 51 000 ha ...

In spite of the fact that [those who received rights to small plots they had been working as renters] received lands markedly inferior in quality to those of [the cooperatives], they managed to increase their yields relative to those of the cooperatives ... in three of the four basic grains. ... The average size of the plots [received by the former renters] remained very small, at 1.05 ha, while the same figure is 5.46 ha [per beneficiary in the cooperatives]. Nevertheless, the net income per beneficiary is only 7% less [for the former renters] than for [the cooperative members].[89]

Michael Martin and Timothy Taylor conducted a survey of 28 collective farms of the agrarian reform in the grain-growing departments of Olancho and El Paraíso in Honduras. Among their findings were the following:

... most cooperatives have been financial failures. ... One of the main institutional problems confronting farmers in the agrarian reform sector has been the requirement to work in a collective manner. Farmers complain that the intensity of work suffers on collective parcels because some farmers feel no incentive to work as hard as they would on individual parcels. No group in the survey works completely collectively. Most have apportioned at

The author visited a collective farm in the Department of Puno, Peru, in the late 1980s. The members and their livestock were undernourished, the stables were in bad repair, and their one breeding bull was too old to perform his duties. They complained that although the farm had far more land than was necessary to support their existing herds, legally they could neither sell nor rent part of the land, which was what they most wanted to do in order to raise capital to improve the operation on the remaining land.

least some of the adjudicated lands to cultivate individually. Some have abandoned collective production altogether.[90]

In Peru, the collective farms of the agrarian reform in the highlands, which were formed out of expropriated haciendas, collapsed spontaneously during the latter part of the 1980s, after many years of severe economic difficulties. In a reversal of the normal historical trend, most of their lands were absorbed by the indigenous communities which had a customary system containing communal lands and individual plots operated under rights of usufruct. Accordingly, the agrarian reform law which had created the cooperatives was repealed and a new agrarian code was developed in its place. The new code is more market-oriented and at the same time respects customary rights.

In the case of China, the first stages of the agrarian reform on mainland China created individual owner-operated plots, and only subsequently were associative forms of farming developed and progressively forced into the mold of collective farms. That experience is summarized by Niu Ruofeng and Chen Jiyuan and is worth reviewing for the light it sheds on the relative efficiency of collective and individual farms:

89. Roger D. Norton and Mercedes Llort, *Una Estrategia para la Reactivación del Sector Agropecuario en El Salvador*, FUSADES, San Salvador, El Salvador, October 1989, pp. 13 and 14 [author's translation].

90. Michael Martin and Timothy G. Taylor, 'Synopsis of Human Capital Research in Olancho and El Paraíso, Honduras', mimeo, University of Florida, Gainesville, FL, USA, February 1990, pp. 1 and 13.

The land reform was completed in mainland China between 1950 and 1952. This reform abolished the feudal system, confiscated the redundant land of landlords, and distributed it among landless peasants, or to those peasants with insufficient land [free of charge]. . . . thus the control over this land shifted to the tillers. . . . about 300 million peasants in mainland China acquired 50 million ha and other means of production. In addition, the new owners were freed from all debts in terms of rent and usurious borrowings.

Small private farms . . . became almost the only management form after the agrarian reform. . . . Agriculture . . . destroyed during the protracted warfare prior to 1949, recovered and was quickly revitalized. Total output rose by 14 percent annually in the early 1950s.

The government encouraged small farmers to create mutual aid groups (MAGs) and elementary agricultural producers' cooperatives (EAPCs) to overcome some of the difficulties faced by individual farmholders, such as lack of farm implements and draught animals. . . . MAGs were based on private ownership. Farm households helped each other with manpower, farm tools and draught cattle in fieldwork, but they could decide what to produce on their own land. . . .

[Subsequently] the socialist transformation of agriculture was conducted and, by the end of 1956 . . . had on the whole, been realized. A total of 740 000 advanced agricultural producers' cooperatives (AAPCs) were set up nationwide, covering over 90 percent of all farm households. . . . AAPCs were characterized by a fully collective management system and two-level accounting (at the cooperative and team levels). The land, draught animals and large farm implements were collectively owned. The private ownership of the cooperative members over these means of production was abandoned. . . . Their income came only from the collectives according to their man-days of labor, and rights to income based on their share of the assets assigned to the cooperative were abolished.

The reorganization of the EAPCs into AAPCs reached a high tide in the summer of 1955 and the growth rate of agricultural production began to slow down. . . .

The people's communes were created country-wide in the summer and autumn of 1958 . . . There were no economic incentives to stimulate the enthusiasm of farmers. In general, agricultural production grew slowly during the 20 years of communes, resulting in stagnation of the rural economy, per capita output and consumption levels. . . . agricultural production dropped sharply during the first three years of the commune era. . . . Farmers suffered most severely in these years.[91]

Similar results were found through the estimation of a production function for Chinese agriculture by Justin Yifu Lin:

The findings indicate that the dominant source of output growth during 1978–1984 was the change from the production-team system to the Household Responsibility System. . . . many policymakers and scholars, not only in China but also in many other developing countries, consider collective farming an attractive method for land consolidation and productivity improvement. However, my findings suggest that the household farm has advantages of its own. Since the household farm leads to a more productive use of inputs, it may be a more appropriate institution for the growth of agriculture in developing countries, including China.[92]

91. Niu Ruofeng and Chen Jiyuan, ' "Small farmers in China and their development', in G. H. Peters and B. F. Stanton, 1992, pp. 621–624.
92. J. Y. Lin, 'Rural reforms and agricultural growth in china', *American Economic Review*, March 1992, **82**(1), pp. 47–48.

Table 5.1 Shares of land ownership, assets and value added (in %) by tenure type[94] in Hungary

Feature	Co-operative farms	State farms	Individual units
Land	76.2	14.3	9.5
Farm assets	64.6	21.2	14.2
Value added	40.1	8.9	44.4

While adverse pricing policies also influenced the agricultural sector's performance during the collectivized period in China, the consensus is that the land tenure regime was the principal obstacle. The World Bank stated that much of the success of Chinese agriculture since 1978 was due to the promotion of individual land rights through a system of explicit or implicit long-term leases.[93]

In Hungary, too, a direct comparison of the performances of collective and individual farms favors the latter. Ferenc Fekete, Tamas Fènyes and Jan Groenewald have presented striking data for the year 1989 which showed the shares of ownership of land and of farm assets, and of agricultural value added, by type of tenure (Table 5.1).[94] Collective farms failed also in Ethiopia in the period 1978–1985. According to research by Klaus Deininger, those collective farms ('co-operatives'):

> received modern inputs and credit at subsidized rates, paid less head-taxes than independent farmers, were favored recipients of State extension services and could impose labor (corvee) requirements on surrounding peasant communities. Despite these advantages, their productive performance remained dismal: Yields for the five main cereals were consistently lower

than yields obtained by smallholders. . . . Once government's difficulties forced it to loosen its grip on the cooperative sector in 1990, virtually all production cooperatives were rapidly disbanded. . . .

State farms were even more favored in terms of resource allocation. Although comprising only 4% of total area, they received 76.5% of chemical fertilizers, 95% of improved seeds, and 80% of the available credit. . . . In contrast to an average farm size of 0.7 ha in the family farm sector, the average area per worker in the State farm sector was 15 ha . . . The list of problems encountered by such farms is long, including slow centralized decision-making and high overhead costs, technical inefficiencies, use of inappropriate technologies . . . and severe motivational difficulties, attributed mainly to inappropriate bonus payment schemes and the inability of managers to dismiss workers. . . . Average annual losses incurred by the State farm sector in the 1978–1985 period amounted to $40 million . . . and it is agreed with the new government that abandonment, breakup, and liquidation of the remaining assets are the only economically feasible options.[95]

Deininger's paper reviews similarly disappointing performances of collective farms in Viet Nam, China, Cuba, Nicaragua, Peru and Israel. (In the case of the latter, the main economic problems with the *kibbutzim* have been the very large subsidies required to maintain them, and their high degree of indebtedness.) Poor performance also has characterized the production co-operatives in Cameroon, where it was observed that State management did not represent the interests of the

93. The World Bank, *From Plan to Market, World Development Report 1996*, The World Bank, Washington, DC, 1997, p. 58.

94. Ferenc Fekete, Tamas I. Fènyes and Jan A. Groenewald, 'Problems of Agricultural Restructuring in South Africa: Lessons from the Hungarian Experience', in Margot Bellamy and Bruce Greenshields (Eds), *Issues in Agricultural Development: Sustainability and Cooperation*, IAAE Occasional Paper No. 6, International Association of Agricultural Economists, Aldershot, UK, 1992, p. 232.

95. Klaus Deininger, 'Cooperatives and the Breakup of Large Mechanized Farms: Theoretical Perspectives and Empirical Evidence', World Bank Discussion Paper No. 218, The World Bank, Washington, DC, USA, 1993, pp. 33–34.

farmers.[96] In light of their unsatisfactory results, the tenure status of collective farms, and lands on which cultivators work as tenants of the State, also has been transformed radically in Mexico, Chile and El Salvador, giving more scope to individual land ownership.

In summary, the empirical evidence on the form of collectivized farms is strongly negative. In marked contrast, the agrarian reforms of Japan, the Republic of Korea and Taiwan, which gave land to the tillers and which established small owner-operated farms, were much more successful in raising productivity. According to Platteau, the experience of South Korea:

> . . . is a particularly instructive case to study, precisely because it has largely succeeded in achieving both equity and efficiency objectives not only in the short but also in the long term. . . . it created a genuinely egalitarian base in the agricultural system. Its main thrust was the replacement of tenancy by owner cultivation and not a radical redistribution of land. . . . the main impact of the reform was to affect production incentives positively as well as to widen the 'absorptive capacity' of the agricultural system for investment and technical change.[97]

Nonetheless, there have been exceptions, for example, in the case of some oil palm co-operatives in northern Honduras. Binswanger *et al.* point out that the conditions for successful collective farms, or production co-operatives, include economies of scale and a requirement for quick and well-co-ordinated movement of the product to port or market. Therefore, they suggest that oil palm and tea are especially amenable to plantation operations, and hence they are candidates for successful cultivation under production co-operatives or collective farms (Binswanger *et al.*, 1995, p. 2696).

The widespread failures of collective farms, under their several variants, may be attributed principally to the following factors:

(i) Lack of ownership of the property and hence the inability of the farm members to exercise the normal options of sale, rental and mortgaging; sometimes, sale or rental of only part of the property would be necessary to maintain the economic viability of the unit.

(ii) A pattern of State interference in the management of the units, so that decisions often were not made on grounds of sound management of a farm enterprise. In the reforms of the Mexican ejidos, 'the legal foundation for government involvement in ejidos was dismantled, thus bringing an end to the much criticized bureaucratic paternalism of the State'.[98] In the case of Cameroon, 'State control of [production] co-operatives has reached the point where they are run by government officials in an attempt to keep political power centralized and to use co-operatives as a source of patronage distribution'.[99]

(iii) The lack of an adequate internal structure of incentives to encourage farmers to work as diligently on collective plots as on their own. This is a fundamental deficiency, and consequently the only area in which the collective units have been able to be competitive with private units is in plantation agriculture, which in any case is operated on a basis of wage labor.

(iv) In some cases, incompetent management of the collective, either because the managers were political appointees without the appropriate qualifications or because of corruption.

96. Joseph Ntangsi, 'Agricultural Policy and Structural Adjustment in Cameroon', in G. H. Peters and B. F. Stanton, 1992, p. 272.

97. J.-P. Platteau, 1992, pp. 223–224.

98. *State of Food and Agriculture 1993*, Food and Agriculture Organization of the United Nations, Rome, 1993, p. 138.

99. Joseph Ntangsi, 1992, p. 269.

In addition, the paternalistic manner in which many agrarian reform production co-operatives have been managed by the State has given rise to other kinds of problems which cannot be solved simply by restructuring the enterprises. For example, the management of co-operatives often has given rise to an inappropriate attitude towards credit repayment on the part of farmers, as illustrated in the case of Zambia:

> . . . most cooperatives were communal production units but . . . were soon defunct. . . . While cooperative marketing is still practiced in Zambia, the idea of communal production units has largely been dropped. It proved to be an extremely expensive experiment because very few of the substantial loans for items such as tractors were ever repaid. The idea that loans do not need to be repaid became widely accepted amongst farmers and has continued to undermine efforts to develop viable credit programs.[100]

5.6.3 The State as Landowner

In almost all countries, the State is owner of significant amounts of rural lands, whether they be nature reserves, zones of commercial forestry, common grazing lands or agricultural lands. The worldwide trend, though, is for the State to reduce its holdings of agricultural lands. In most Asian countries, there has been a long tradition of private ownership of agricultural land. Latin America, except Cuba, emerged from an era of State ownership of land with a decisive shift in policies toward private ownership, although there are still residual State agricultural lands in some countries. Africa and the former Soviet Union are the principal regions in which State lands still play a reasonably prominent role.[101] As commented by Zvi Lerman:

> private ownership of land is not recognized in the Central Asian States (Kazakhstan, Kyrgyzstan, Tajikistan and Uzbekistan; in Turkmenistan private ownership of land is purely notional, without any rights of transfer), in Belarus (except for small household plots), and in a number of autonomous republics within the Russian Federation.[102]

Alden Wily has commented on the pervasiveness of State ownership of agricultural lands in Eastern and Southern Africa, and the problems that may be associated with that form of tenure:

> . . . all but one State (Uganda) has firmly entrenched primary ownership of land . . . in the hands of the State (or president), leaving to citizens only the right to own 'interests' in that land. From this proceeds power to designate, regulate, intervene and appropriate land at will – a facility used with unusual frequency, dubious purpose, and lack of accountability by many governments in the region in recent decades and one that, arguably, has promoted rent-seeking landlordism.[103]

In many cases, customary landholdings in Africa have passed into the hands of the State, including in Mauritania, Nigeria and Senegal, in

100. R. Watts, 'Zambia's Experience of Agricultural Restructuring', in Csaba Csaki, Theodor Dams, Diethelm Metzger and Johan van Zyl (Eds), *Agricultural Restructuring in Southern Africa*, International Association of Agricultural Economists in association with the Association of Agricultural Economists in Namibia, Windhoek, Namibia, 1992, pp. 373–374.
101. However, 'A high percentage of land in Sri Lanka is State land. A significant amount of this land is occupied by private persons through permitting, grant and leasing schemes. Unauthorized occupation of State land [is] a significant problem . . .' (Land Tenure Service, FAO, 2001, p. 49).
102. Zvi Lerman, 'Agriculture in ECE and CIS: From Common Heritage to Divergence', The Hebrew University, Rehovot, Israel, and The World Bank, Washington, DC, USA, January 1999, p. 3; available at [www.worldbank.org/landpolicy].
103. L. Alden Wily, 2000, p. 4.

addition to the countries in the Eastern and Southern part of the continent. This shift parallels historical experiences, especially in Europe and Latin America, noted in Section 5.4, in which customary lands became the property of large landlord estates or haciendas. In Senegal, in principle the National Land Law provided that allocations of land to cultivators, and termination of use rights, were to be managed by village councils, as they had been under the traditional systems.[104] In practice, the law proved to be vague in important respects, and its intended principles were not always followed.

In modern times, the reasons why the State has become an owner of agricultural lands are diverse, but for the most part they fall into five categories:

- historical reasons, e.g. that a colonial power's lands passed to the government after independence, and sometimes the colonial regime itself was statist in its economic policies;
- an *a priori* ideological or political conviction that basic resources such as water and farmland are too important to society to be controlled by individuals;[105]
- a concern that privatization of land may lead to concentration of landholdings in relatively few hands, thus exacerbating problems of rural poverty;
- a concern that private ownership will encourage speculative holding of land, with the consequence of leaving idle land that could be productive;
- a concern to assure that sustainable resource management practices are followed, as in the case of State lands with timber concessions.

In light of the frank appraisals of authors such as Platteau, Ntangsi and Alden Wily, it should be added that the possibilities of patronage may also have encouraged some governments to hold onto lands that they might otherwise have divested to the private sector. In these cases, lack of political will to privatize land is a major obstacle to change. On the other hand, in Estonia and other countries of the former Soviet Union, the State became a major landowner by default, since the process of privatizing farmland has been slower than anticipated and has not yet been completed.

The ideological position cannot be countered on technical economic grounds. After all, no country in the world permits individual citizens to own natural sources of water, but rather they may possess only rights to use it. Proponents of State ownership of land ask: is farmland fundamentally different in this regard? Each country has to develop its own answer to this question. In part the answer depends on the prevailing conception of the role of agricultural production in the economy and society. If it is regarded mainly as a source of income for producers, then policy is likely to treat land as any other productive resource, albeit always with some special restrictions, as noted previously. If, on the other hand, it is felt that agriculture must serve broader objectives such as socio-economic equity and food security, then sometimes private property will be assigned a more restricted role, with State ownership of land assuming a correspondingly more important role – although there exist alternative policy instruments for promoting those goals.

In reviewing issues associated with State lands, it is critical to bear in mind the distinction between ownership and usufruct rights. As in other activities involving productive resources, ***the State generally has not proven to be a capable manager of agricultural lands***. As noted in the cases of the collective farms, or production cooperatives, State land ownership usually means that those who are making decisions at the farm level are deprived of important economic options that private farmers have: renting in or out, buying and selling, and mortgaging. In land tenure regimes in which the State rents out its land to cultivators, a common problem is that the rental rates may bear little relation to (annualized) market values of land, as has been the case

---

104. J.-P. Platteau, 1992, pp. 153–159.
105. In some countries of Central Asia, there is a constitutional prohibition against private property.

in Guyana.[106] When the administratively determined rental rates are lower than market rates, queuing for access to State lands results, and illegal sub-letting of the land at market rates usually occurs as well (thus promoting 'rent-seeking landlordism', in Alden Wily's phrase). Those who are favored with initial access to the land perceive windfall gains, and the temptations for rent-seeking behavior leading to political manipulation of the process of land allocation are obvious.

When usufruct rights of individuals and communities on State lands are not clarified, confusion can result from the conflict with traditional forms of tenure rights, as has been the case in Mali:

> For land and natural resources, *State property* is the rule. Individual property is the exception. In this context, as legislation does not fit the norms and values of the local population, there are many cases of evasion, misinterpretation and misuse accompanied by another major problem, the one dealing with the superposition of multiple rights to land, woodlots, grazing lands and watersheds.[107]

Another problem is insufficient tenure security for those who farm State lands. Rental contracts often are not long enough to encourage farmers to invest in improvements in the land, or to protect its soil quality, and they cannot be used for collateral. For State lands in Namibia, it has been observed that 'The tenure security of settlers is weak. Settlers may have a right to use and occupy the land, but not to sell or bequeath residential and/or arable land, nor exclude others'.[108]

Because of these kinds of problems, the Government of Guyana is implementing the option of creating the legal formula of long-term tradable and renewable leases for plots on State lands.[109] The Government of Trinidad and Tobago initiated such a system in 1995. A variant of the lease option as promulgated in Guyana, and in Estonia as well, would allow leaseholds to be converted to freeholds (with or without a requirement of additional payments), after the cultivator has worked a plot for a given number of years. A similar approach was implemented in Zambia where, when all land was nationalized in 1975, most freehold titles were converted to 99-year leases.[110] Under the Ethiopian system of leases on State lands, landholders are permitted to sub-lease the land, sell or otherwise transfer the lease, and then the lease can serve as collateral.[111]

Thus, in light of international experience, the presumed benefits of State ownership and management of land are overshadowed by the disadvantages in terms of reduced levels of production and investment, and increased degradation of land. *If the government retains ownership of agricultural land, the only way to avoid these negative consequences is to remove the State from the realm of direct land management itself.* One option is to provide farmers with permanent, tradable rights of usufruct, through long-term, freely transferrable leases. The other principal option is to leave the management of State lands at the local level, through customary rights regimes, as suggested by Nadia Forni (2000) above.

In order to maintain transparency, i.e. to avoid bias in the initial allocation of land under such government leases, one approach is for the plots in newly available land areas be auctioned off locally to producers who live in the correspond-

---

106. Harold Lemel, 'Patterns of Tenure Insecurity in Guyana', Working Paper No. 43, Land Tenure Center, University of Wisconsin–Madison, Madison, WI, USA, April 2001, p. 7.

107. Land Tenure Service, Food and Agriculture Organization of the United Nations, 2001, p. 45.

108. Martin Adams and John Howell, 'Redistributive land reform in Southern Africa', *ODI Natural Resource Perspectives*, No. 64, Overseas Development Institute, London, January 2001, p. 5.

109. Ministry of Finance, Government of Guyana, *National Development Strategy: Shared Development through a Participatory Economy*, Government of Guyana, Georgetown, Guyana, draft, 1996, Chapter 29.

110. R. Watts, 1992, p. 374.

111. Land Tenure Service, Food and Agriculture Organization of the United Nations, 2001, p. 40.

ing areas, with a ceiling on the size of the plots to be auctioned. Another approach is to post lists of those eligible for leases with explanations of the criteria for selection. A third option is to select lessees by lottery among those who meet the eligibility requirements. For existing agricultural land, preference in issuing the leases normally is given to the families who have been tilling their plots for a given number of years without interruption (again subject to an upper limit on the size of plots leased out).

In all cases, it is vital that lease fees be set at market levels, to provide incentives for productive use of the land and to avoid creating incentives for 'landlordism' – rent-seeking behavior in the form of attempting to gain access to leases only for the purpose of sub-letting at a profit. As previously commented, this problem has been very widespread in Guyana. Another example of this phenomenon prevailed in Zambia, where lease fees on State lands have been very low: 'Leasehold land has been found to be considerably underutilized. Leaseholds are granted free of charge at a minimal ground rent, which is inadequate incentive to encourage optimal use of land' (Chinene *et al.*, 1998, p. 95).

In Taiwan, where the agrarian reform reinforced private ownership, State-owned lands were adjudicated in an equitable manner, in small plots (Platteau, 1992, p. 227). In the production conditions of Taiwanese agriculture, the small plots proved to be efficient as well as conducive to equity in the land distribution. This example illustrates the point that the manner of disposing of State lands is quite important.

Land reform in Honduras in the 1970s provided an example of State-owned lands that were not managed in a transparent way. About a third of all agricultural lands were owned by national and local governments, and after the passage of agrarian reform legislation they continued to be occupied by squatters without formal title, sometimes in plots that contained several hundred hectares of land. Under a different approach, the agrarian reform could have placed a ceiling on the size of these plots, insisted that their occupants pay for them (over time), and distributed the excess land above the ceilings to landless families.

Instead, the landless were tacitly encouraged to invade existing private lands, which then were expropriated after the fact, often on dubious legal grounds. Thus, the agrarian reform channeled the quest for access to land into the most conflictive channels, and created substantial uncertainty of tenure for private farmers, instead of using State lands to provide access to land for the rural poor.

As will be discussed in a subsequent section of this chapter, there are indirect policy instruments which can be deployed to discourage a concentration of landholdings and also to minimize the practice of leaving productive land idle. Once a strategy is defined for creating access to land for the rural poor, and adequate agricultural incentives, if such policies are implemented, then much of the economic justification for State land ownership *and* management weakens.

On the other hand, the role of the State is exceedingly important in defining land tenure policy, specifying its detailed regulations and instruments of implementation, and monitoring its functioning. A clear and effective land tenure regime is one of the most important prerequisites for sustainable agricultural development. Usually, it is more productive to orient the State's management capabilities in that direction rather than toward direct administration of land.

5.7 EXPERIENCES WITH LAND REFORM

5.7.1 Return to Origins

Land reform began as a liberal economic idea, based on willing sales of land to poor farmers, and only in the past century was it transformed into a concept based on coercion by the power of the State:

Modern concepts of agrarian and land reform probably have their most direct heritage in the agrarian transformation that began in Denmark in the late 1700s. Building on the ideas that were emerging, especially in Britain but also in France and Germany, reformers such as the Counts of Bernstorff and Reventlow initiated a program of consolidating their peasants' fields, introducing new technology and selling the land to their peasants. . . . the framers of this *reform* also recognized that the peasants-turned-landowners stood little chance of success without institutional protection. . . . the process also depended on the emerging cooperative movement to provide the economies of scale necessary for Danish smallholders to move from indentured servants of 1770 to prosperous farmers of 1870. . . . The Russian revolution of 1917 and a variety of national socialist as well as populist regimes between the First and the Second World Wars moved the ideology of agrarian reform in the Western world from a liberal economic process to a State-engineered way to redistribute land and achieve equity in rural areas. . . . Most of the land reforms in Latin America, Asia and the Near East were derived from this model.[112]

From a viewpoint of economic development, land reform has been viewed, by analysts as well as politicians, as a way to promote both equity and efficiency in rural areas, and coercion or government intervention of other types has been justified as necessary to make land reform work:

Most redistributive land reform is motivated by public concern about the rising tensions brought about by an unequal land distribution. The common pattern is concentration of landownership among relatively few large owners in an economy where labor is abundant and land is scarce. Thus the masses of landless laborers and tenants who derive their livelihoods from agriculture receive relatively less income because their only asset is labor. Redistributive land reform can also increase efficiency, by transferring land from less productive large units to more productive small, family based units. . . . land markets will not

typically effect such transformations of ownership patterns.[113]

Banerjee has summarized the argument for coercive land reform in similar terms:

The case for land reform rests on two distinct arguments: first, that a more equitable distribution of land is desirable and second, that achieving more equitable distribution is worthwhile even after a careful consideration of the costs associated with redistributing land and the alternative uses to which the resources [required to implement the reform] could have been put. . . . At the heart of the argument for more equitable land distribution is the observation that small farms in developing countries tend to be more productive than larger farms.[114]

In Africa the equity argument has a special resonance:

Existing property rights. . . . legalized more than a century of land grabbing by whites, an outcome strongly resented by Africans across the region.[115]

The results of coercive land reform generally have been disappointing, although a few, like those of East Asia in the 1950s and that of Zimbabwe's Phase I in the 1980s, have been considered successful on both equity and efficiency grounds. In most of Latin America, land reform has been in retreat in recent years. Herrera, Riddell and Toselli have stated that 'the type of agrarian reform that considers the redistribution of land from the rich to the poor either through confiscation or through pre-emptive buyouts belongs to the past'. The reasons include the fact that the political will and economic capacity to

112. Adriana Herrera, Jim Riddell and Paolo Toselli, 'Recent FAO experiences in land reform and land tenure', *Land Reform*, 1997/1, pp. 53–54.
113. H. Binswanger, K. Deininger and G. Feder, 1995, pp. 2730–2731.
114. A. V. Banerjee, 2000, pp. 253–254.
115. M. Adams and J. Howell, 2001, p. 1.

effect compensated expropriations does not exist – 'agrarian land policies can only take shape as part of a larger economic and political canvas' – and agrarian reform programs have been failures from a perspective of reducing poverty – 'rather than reduce poverty, they generally resulted in shared rural poverty. Subsidized services and inputs that were part of agrarian reform programs far too seldom benefitted agrarian reform beneficiaries. . . . and security of land tenure was not provided . . .'.[116]

Herrera *et al.* conclude that for agrarian reform to work it must be accompanied by targeted polies that assure 'development of the capacity among the reform beneficiaries for capital accumulation in terms of human capital (education, training) [and] social capital (civil society associations), as well as productive capital'.[117]

In regard to the waning interest in land reform in Latin America, Alain de Janvry, Nigel Key and Elisabeth Sadoulet have said:

> The general direction of reforms that codify access to land has been to end or greatly restrict the old systems of State-managed land confiscation and redistribution. Limits placed on the size of land ownership have been relaxed or removed.[118]

In light of these experiences, in recent years the approach to land reform has come full circle, returning to the eighteenth-century idea of negotiating with willing sellers of land in order to provide it to the poor, always with an element of government subsidy since the beneficiaries of a land reform usually cannot pay the full value of their land.[119] The new approach varies considerably from country to country in the few cases it has been tried. Its variants go under different names, including market-assisted land reform, negotiated land reform and (subsidized) land funds.

5.7.2 Rationales, Failures and New Formulations

Regardless of the approach taken, the rationale for land reform still exerts a compelling influence in some spheres. It is patently on the table in Eastern and Southern Africa, although Alden Wily points out that ' "political will" for reform often wavers, as has been the case so far in Uganda, Tanzania, Zimbabwe, South Africa, Malawi, Lesotho and Namibia'.[120] From a World Bank perspective, Deininger and Binswanger have stated that two of the 'four key principles' in the 'broad consensus underlying current thinking about land issues' are:

- The desirability of owner-operated family farms on both efficiency and equity grounds.
- The positive impact of an egalitarian asset distribution and the scope for redistributive land reform where nonmarket forces have led to a highly dualistic ownership and operational distribution of land, that is, a distribution characterized by very large and very small holdings.[121]

They develop further the rationale for land reform in the following terms:

> The practical difficulties associated with implementing land reform notwithstanding, the conceptual attractiveness of such a policy rests on three pillars: First, in situations where credit and product markets are incomplete, access to land can make a significant contribution to food security, households' nutritional well-

---

116. A. Herrera, J. Riddell and P. Toselli, 1997, pp. 53, 54 and 55.
117. *Op. cit.*, p. 55.
118. Alain de Janvry, Nigel Key and Elisabeth Sadoulet, 'Agricultural and Rural Development in Latin America: New Directions and New Challenges', Working Paper No. 815, Department of Agricultural and Resource Economics, University of California at Berkeley, Berkeley, CA, USA, March 1997, p. 18.
119. H. P. Binswanger, K. Deininger and G. Feder, 1995, p. 2731.
120. L. Alden Wily, 2000, p. 2.
121. K. Deininger and H. Binswanger, 1999, p. 248.

being, and their ability to withstand shocks. . . . Second, land ownership affects economic growth and poverty reduction through credit-financed investment. . . . And finally, several studies have argued that a more egalitarian distribution of assets (not necessarily land) would improved political stability.[122]

In its most condensed form, these authors present the argument for land reform in the following words:

Many of the impediments to a smooth functioning of land, labor and product markets date from the colonial era; because such long-standing barriers maintain a highly unequal distribution of land, large tracts of productive land lie idle, while peasants have to eke out a living on marginal and often environmentally fragile lands. In addition to reducing productivity, unequal land ownership is also linked to social unrest and violence.[123]

However, they add that most coercive land reforms undertaken during the last 20 to 30 years had political motivations and have fallen short of expectations.

Deininger and Binswanger attribute the lack of success of coercive land reforms to lack of government investment in complementary infrastructure, the tendency to group the reform beneficiaries into collective farms, the lack of entrepreneurial experience of beneficiaries, and their lack of start-up capital (lack of access to credit).

Herrera, Riddle and Toselli have drawn up the balance sheet on agrarian reform in the following words:

Even if we discount economic performance of the agrarian reform programs conceived during the 1960s, 1970s and 1980s as social welfare costs for the rural populations, they still have proved unsuccessful. Rather than reduce rural poverty, they generally resulted in shared rural poverty. Subsidized services and inputs that were part of agrarian reform programs far too seldom benefitted agrarian reform beneficiaries. Political support for land redistribution was seldom obtained. The economic costs of land distribution and land regularization were too high; and security of tenure was not provided owning to inadequate (or non-existent) land cadastre and land registration programs. . . .[124]

Land reforms often have been followed by a reconcentration of landholdings that were redistributed and substantial amounts of redistributed lands left idle, especially in production cooperatives. For example, 'a recent census of Brazilian land reform settlements reported that only about 60 percent of land reform beneficiaries were actually tilling their land'.[125]

Although the failure to put in place complementary policy initiatives and programs is part of the explanation for these results, the failure of coercive land reform has deeper explanations. To assess the phenomenon properly, the broader reverberations of a land reform in the society and polity need to be recognized. It is not simply one economic policy tool among many. *In normal circumstances, land reform cannot be conceived of as a policy instrument that governments would be able to select on purely technical grounds and implement rapidly on a significant scale.* Land reform is inherently a political process and usually acquires momentum during crises of a socio-political nature, precisely at a time a government's technical capacity to implement it may be undermined. In the words of Alden Wily:

Without exception, political change underwrites and directs land reform. This has been signaled in a wave of new independence or political regimes in the last decade (Eritrea,

---

122. *Op. cit.*, p. 256.
123. *Op. cit.*, p. 266.
124. A. Herrera, J. Riddell and P. Toselli, 1997, p. 55.
125. K. Deininger and H. Binswanger, 1999, p. 266.

Ethiopia, Rwanda, Mozambique, Namibia, South Africa, Zambia, Malawi and Uganda). Or it may arise through shifting socio-political relations within society itself, being realized through 'multi-party-ism', strengthening of devolutionary strategies, and heightening popular voice and demand. The thoroughly political nature of land distribution and security means that 'reform' readily becomes a focus in times of political uncertainty. As the current crisis in Zimbabwe illustrates, it may be all too readily used as a tool to prompt or control rising political opposition.[126]

On the other hand, Binswanger *et al.* are more inclined to emphasize the risk that land conflict, and in some cases civil war, may arise from 'the perils of incomplete land reform'.[127] Here, they are clearly treading in speculative territory, but whatever the cause-and-effect relationship, historical experience makes it clear that undertaking a coercive land reform is a path that often leads to, or is accompanied by, serious political unrest and even violence.

The implementation of land reform also faces formidable hurdles of a practical nature. In reviewing the experience of land reform in South Africa, Adams and Howell have pointed out that *technical, administrative and economic barriers tend to make the process a slow one, regardless of whether land reform is market-assisted land reform or expropriative*:

> Redistribution for the rural poor has been limited largely because of the technical and economic problems in subdividing large livestock-based farms in semi-arid areas. . . . Pastoral settlement schemes in Africa suggest that neither the subdivision of commercial ranches into family livestock farms, nor group or cooperative ranching are viable options. The costs of settling families with small herds and flocks on individual farms, with reasonable standards of social and economic infrastruc-

ture, are very high and both economic returns and environmental effects almost certainly negative. . . .

> Can land redistribution be achieved by encouraging landowners to offer farms for sale voluntarily, or should the government compulsorily acquire land for redistribution? Zimbabwe, Namibia and South Africa have adopted market-assisted land reform, although Zimbabwe now seems to have abandoned it. . . .

> In South Africa post-1994 the main aim was to contribute to the alleviation of poverty and injustices caused by previous apartheid policies. The redistributive content of the program was constrained by the government's grant conditions in the context of its willing-buyer policy. . . . On average only two thirds of a R15–16 000 grant was used for land purchase since it also had to cover capital investments necessary to make the land productive. Furthermore, since 1994 the Department of Land Affairs has consistently underspent its annual land reform capital allocation, largely because of inadequate administrative capacity. Even if the policy had been based on expropriation instead of market transactions, this would have been a binding constraint. *Land redistribution through due legal process is slow and administratively binding*. . . . its complexity in a constitutional democracy tends to be greatly underestimated by those who have not tried it. A numerous and widely deployed army of well-trained field staff is essential to inform people of their entitlements and to facilitate the many and complex legal, financial and administrative tasks involved. . . . Adequate post-settlement support must also be provided if new farmers are to succeed. . . . In both Namibia and South Africa there is an unbridgeable gap between the ambitious redistribution targets that have been announced and the financial and administrative resources for realizing them.[128]

126. L. Alden Wily, 2000, p. 1.
127. H. P. Binswanger, K. Deininger and G. Feder, 1995, p. 2693.
128. M. Adams and J. Howell, 2001, pp. 2–3 [emphasis added].

The experience of Zimbabwe in the 1980s – when 3.3 million hectares were redistributed to 52 000 families, with compensation of former owners – seems to contradict this pessimistic assessment for the scope of land reform. Both equity and efficiency improved in the areas affected. However, in the 1990s Zimbabwe also ran up against administrative and financial difficulties and consequently the pace of reform slowed drastically. A major legacy of recent attempts to accelerate the process is that 'the country's wider economic and social fabric has been severely damaged'.[129]

Heinz Klug has commented in the same vein on African experiences with land reform:

As in Latin America, it has been difficult to carry out [land] reform on a scale which fundamentally alters the structure of landholding. This has been due to a number of factors, including constitutional constraints, shortage of funds for land purchase, and shortage of funds, trained staff, etc. for resettlement.[130]

Similarly, in the Asian context Deininger, Pedro Olinto and Miet Maertens pointed out that financial restrictions have limited the pace of land reform in the Philippines. They commented that lack constitutes the single most important constraint to a faster and more wide-scale implementation of land reform.[131]

Implementation of the Nicaraguan land reform of the 1980s and 1990s reduced the government's institutional capacity to manage agricultural and rural development programs and generally weakened respect for law. A legacy of the expropriations conducted extra-legally was many thousands of unresolved conflicts over land rights and institutional overlap and ambiguity. As late as 2001, only 35% of the expropriated properties had been compensated and at least five kinds of land titles were being issued by different government agencies.[132] In that country's legal framework, the new occupants of redistributed land cannot be provided with fee-simple titles until the previous owners are compensated for the confiscation. A similar conundrum has slowed the pace of land titling in Honduras. *The possible consequence of negative effects on a country's institutional structure is not taken into account when*

The positive historical record of the land reforms which established owner-operated farms should not obscure the enormous difficulties of undertaking a land reform. Schuh and Junguito point out the 'widespread tendency to romanticize about the Mexican Revolution, which many people believe alleviated rural poverty. It fails to recognize the political difficulties and fiscal costs of bringing off a successful land reform. It also fails to recognize the extent to which the Mexican land reform has institutionalized rural poverty. . . . [and it] fails to recognize the extent to which the redistribution of land is a one-time gain, and the extent to which it fails to prepare the beneficiary for participation in a modern market economy. A key reason for the prevalence of poverty in Latin America is that governments in the region have grossly underinvested in the human capital of their rural populations'. (G. E. Schuh and R. Junguito, 'Trade and agricultural development in the 1980s and the challenges for the 1990s: Latin America', Agricultural Economics, 8(4), June, 1993, p. 398.)

---

129. *Op. cit.*, p. 4.
130. Reproduced from Heinz Klug, 'Bedevilling Agrarian Reform: the Impact of Past, Present and Future Legal Frameworks', in J. van Zyl, J. Kirsten and H. Binswanger (Eds), *Agricultural Land Reform in South Africa: Policies, Markets and Mechanisms*, Oxford University Press, Cape Town, South Africa, 1996, p. 197, with permission of the Oxford University Press, Southern Africa.
131. Klaus Deininger, Pedro Olinto and Miet Maartens, 'Redistribution, investment and human capital accumulation: The case of Agrarian Reform in the Philippines', mimeo, The World Bank, Washington, DC, USA, 2000.
132. Unpublished data compiled by Gustavo Sequeira in Nicaragua (2001).

land reforms are contemplated, but frequently it is a by-product of their implementation. Once a nation's institutional integrity is degraded, restoring it is an arduous, long-term task.

A conclusion that emerges clearly from these experiences is that *massive, coercive land* reform cannot be expected to be implemented within the confines of a stable legal framework and polity. It is inherently disruptive and is favored by chaotic political circumstances. The land reforms of Japan, South Korea and Taiwan, considered as the most successful examples, were carried out in wartime or during a process of political collapse and reconstruction. As remarked by Binswanger *et al.* (1995, p. 2683):

> Most large-scale land reforms were associated with revolts . . . or the demise of colonial rule. . . . Attempts at land reform without massive political upheaval have rarely succeeded in transferring much of a country's land.

The alternative of market-assisted land reform also may be forced to move at a gradual pace because of implementation constraints. Referring to both kinds of land reform, Adams and Howell concluded that:

> One lesson from attempts to transform land tenure in Africa over the last forty years is that wide departures from existing systems are rarely immediately feasible: evolutionary approaches are slow but, as Zimbabwe demonstrated in 2000, revolutionary approaches generate high social and economic costs.[133]

When land reform is carried out, some improvement of the economic status of the beneficiaries can be expected but it does not necessarily benefit the poorest rural families. The Philippine experience in land reform has elicited much review and commentary. Deininger, Olinto and Maertens analyzed the effects of land reform by carrying out a careful statistical comparison of the behavior of beneficiaries and non-beneficiaries for a small sample of farmers in Central Luzon province. Their principal findings included the following:

> . . . from a point of view of static productivity gains, the case for land reform in the Philippines is much weaker than would be expected. . . . we are unable to ascertain any significant relationship between tenure status and agricultural productivity. . . . [but] we find a stronger impact of asset ownership and land ownership status on income. . . . the educational advance of children affected by land reform was about 0.60 years higher than that of non-beneficiaries. . . . in 1985–1998, beneficiaries' income grew at a faster rate than that of non-beneficiaries. The difference in income growth between beneficiaries and non-beneficiaries is estimated at US$86, almost half of the original level of income.[134]

However, Deininger *et al.* (2000, pp. 15 and 24) also found that:

> land reform failed to benefit the landless but rather targeted share tenants and leaseholders. . . . government officials generally were not able to target the program to the poorest within the overall universe of agricultural cultivators. . . . To increase the supply of land, the government has prohibited share tenancy and imposed land ownership ceilings. These measures are not only costly to implement (and often circumvented by spurious subdivision) but also restrict access to land through the rental market and are likely to discourage land-related investment in labor intensive agro-export and plantation crops.

Yujiro Hayami found the Philippine reform to have serious flaws, including disincentives for existing producers and increasing rural inequality because of the lack of targeting of the poor-

---

133. M. Adams and J. Howell, 2001, p. 6.
134. K. Deininger, P. Olinto and M. Maertens, 2000, pp. 11, 12, 17 and 18.

est families that was documented by Deininger *et al.*:

> ... land reform has been successful in transferring much of the economic return to land from absentee landlords to former sharecroppers. However, the reform has created serious income inequality within village communities. . . . [and] by limiting program application mainly to tenanted land, the reforms created a strong incentive for landlords to evict their tenants and cultivate their land directly. . . . Therefore, the exemption of land under direct administration of landlords had the effect of reducing labor input per hectare below an optimum level, thereby reducing the income of the laborers. . . . Significant negative effects of land reform on agricultural production efficiency also occurred outside the rice and corn sector. The cash crop sector has not been covered by reform programs. . . . However, plantation owners fear that their land will eventually be expropriated. It is only natural that they have stopped investing in improvements in their land infrastructure, including planting and replanting trees.[135]

Although it is too early to assess the new, market-oriented approaches to land reform, they were designed to be more successful than coercive reforms in assisting beneficiaries to improve their economic well-being and in avoiding the creation of disincentives for the rest of the sector and other costs to society.[136] This expectation would be more realistic if they were complemented by appropriate policies on land rental and other tenure issues. In any case, the redistributive process appears destined to be a slow one unless governments place much greater emphasis on increasing administrative capacities for such programs at the local level. Colombia's recent experience with market-assisted land reform, in which a number of administrative complications brought the process to a halt, has not been encouraging.

Land reform is not the only way to promote better access to land by the rural poor. Tenure reforms themselves can be effective in this regard. In the context of South Asia, and concerning the policy framework for land tenancy in particular, Banerjee has pointed out that:

> Unlike land reforms, tenancy reforms do not attempt to change the pattern of ownership of land. They simply give tenants additional rights on the land. . . . Tenancy reforms work by making tenants more expensive. . . . There is not enough evidence to conclude that tenancy reforms are an effective substitute for land reforms. If, however, increasing the cost of tenant labor can bring about better incentives, then a range of interventions – what elsewhere are called empowerment strategies . . . will become relevant. . . . Were implementation not a constraint, traditional (coercive) land reform would have a number of clear advantages over alternative types of land reform. . . . Implementation is a constraint, however, and may indeed be the binding constraint in many cases. In such cases, market-assisted reforms or tenancy reforms may provide better outcomes.[137]

The topic of tenure reforms is explored further in the following section, as well as other avenues for increasing access to land on the part of poor families.

5.8 POLICIES FOR LAND MARKETS

As noted in Section 5.5, land rights almost always are restricted in the name of the broader public

---

135. Y. Hayami, 'Ecology, history and development: a perspective from rural Southeast Asia', *The World Bank Research Observer*, **16**(2), Fall 2001, pp. 191–192.

136. A detailed assessment of preliminary experience with market-assisted land reform in three countries is found in K. Deininger, 'Making negotiated land reform work: initial experience from Colombia, Brazil and South Africa', *World Development*, **27**(4), April 1999, pp. 651–672.

137. A. V. Banerjee, 2000, pp. 269–271.

interest, whether that interest be represented by the local community or the national government. Furthermore, governments frequently restrict land transactions as well.[138] Evaluating such restrictions requires confronting a series of questions, always in the context of the country concerned.

A fundamental policy question in all countries is, what should the nature of land rights be? Should customary land tenure regimes be retained? If so, how can they be protected against land grabbing? For land use systems that are emerging from customary regimes, in what form should land rights be formalized? For systems making the transition out of collective farming, what options should be provided to the members of the collectives? For countries with extensive State ownership of land, what kinds of customary and private land rights should be acknowledged or established, and how should allocations of State lands to users be made? For countries with traditions of private land ownership, are equity and efficiency objectives adequately served by the existing land tenure rules and regulations? In the interest of greater equity, what kinds of restrictions are appropriate for ownership of agricultural land, if any? What are the consequences for efficiency?

Platteau has argued strongly against promoting systems of unrestricted private property rights in sub-Saharan Africa:

> . . . a freely operating land market ought not to be allowed to emerge in sub-Saharan Africa today. The land market should instead be regulated, meaning that the rights of individuals to own, buy, sell and accumulate private land property should be constrained in the pursuance of objectives of social equity, economic growth and political stability.[139]

The opposing argument, in effect the general case for full property rights, has been stated by the World Bank as follows:

Property rights are at the heart of the incentive structure in market economies. They determine who bears risk and who gains or loses from transactions. In so doing they spur worthwhile investment, encourage careful monitoring and supervision, promote work effort, and create a constituency for enforceable contracts. In short, fully specified property rights reward effort and good judgment, thereby assisting economic growth and wealth creation. In addition, a wide distribution of property rights can counteract any concentration of power in the political system and contribute to social stability.[140]

In many developing countries, the choice between these two viewpoints is further complicated by the existence of customary tenure systems with long histories. Before discussing issues concerning the regulation of land markets, the question of whether or not to convert customary regimes into formal, registered property rights is reviewed. However, it should be borne in mind that the most fundamental criterion for choice of a land tenure regime should be what the users of the land want.

5.8.1 The Degree of Formalization of Land Rights

Apart from problems that may arise during the transition between customary and modern systems of land rights, a fundamental question is *whether full land titling is necessary in order to provide the degree of tenure security necessary to encourage agricultural development*. It is an issue that still generates passionate opinions. An example of the controversies generated over this topic has been provided for the case of Russia by Alexander Nikonov, as follows:

> The issue of private ownership of land with the right to buy, sell and mortgage arouses a fierce difference of opinion in Russian society. Grant-

---

138. H. Binswanger, K. Deininger and G. Feder, 1995, p. 2725.

139. J.-P. Platteau, 1992, p. 216.

140. The World Bank, 1997, pp. 48–49.

ing of land for lifelong possession with the right to inherit was resolved by the Supreme Soviet of the USSR and was accepted by peasants everywhere. But the issue of land buying and selling remains unresolved and arguments continue.[141]

To analyze the issue dispassionately, it should be recalled that the principal advantages of tenure security are four, as follows: guaranteeing the cultivator's continuing right to work the land (security of possession), encouraging improvements in and conservation of the land, facilitating access to credit, both for the improvements and for purchasing production inputs, and facilitating transfers of land among producers.

The existence of land transfers, i.e. a smoothly functioning land market, sometimes gives rise to concerns about the possibility of rural equity being reduced over time, through an increasing concentration of landholdings. Yet it contributes in an important way to economic efficiency, because it facilitates putting land in the hands of those who are able to make it yield the most (and who are not necessarily the largest-scale farmers), whether the mechanism is rental, pledging or sale–purchase. A land market is deemed efficient if the marginal productivity of different plots of the same soil quality are brought into approximate equality, in part through transactions in land. If their productivities are not equal, then there is scope for intensifying the production of the high-yielding plots, until at the margin their productivity begins to decline, and for increasing the productivity of less efficient plots, perhaps by placing them in the hands of other cultivators. In principle, this is one of the major reasons for promoting a land market, and usually it is considered that a prerequisite is that the rural properties be titled.

On the other hand, it has been observed that customary systems of land tenure can provide tenure security for the cultivator, unless the population pressure on the land becomes too great or dual systems of land rights emerge. It has been noted by Atwood that a good deal of technological improvement to farms has occurred under traditional tenure rights in Africa. Customary systems also can provide a degree of access to credit, through the informal credit market and the practice of 'pledging', although the evidence cited by Feder and others suggests that formal titling increases that access. Traditional systems also have fostered an extensive amount of land transfers, as noted previously.

Thus, in practical terms, a basic question for traditional tenure systems is not whether land transactions occur but rather is the following: 'Are the nature and scope of [land] transactions sufficient to permit more productive land to go to more productive farmers up to the point where marginal productivities are equalized?'.[142] Atwood provides a partial answer to his own question:

The work of Collier (1983)[143] in Kenya suggests that they are not, although it is unclear if this is an indictment of the indigenous, informal system of land transfers, the formal land registration system itself, or both. . . . A view of the problems which indigenous land tenure may pose for land sales is clearly presented by Johnson (1972).[144] The problem is not a prohibition of land sales, but certain social conventions which may discourage some efficiency-enhancing land transactions from taking place. . . . There is empirical evidence that Johnson's assertion may be valid in some African situations. Early colonial Uganda and early independent Côte d'Ivoire both enacted

---

141. A. A. Nikonov, 'Agricultural transition in Russia and the other former states of the USSR', *American Journal of Agricultural Economics*, **74**(5), December 1992, p. 1160.
142. D. A. Atwood, 1990, p. 662.
143. P. Collier, 'Malfunctioning of African rural factor markets: theory and a Kenyan example', *Oxford Bulletin of Economics and Statistics*, **45**(2), May, 1983.
144. O. E. G. Johnson, 'Economic analysis, the legal framework, and land tenure systems', *Journal of Law and Economics*, **15**, 1972.

laws which took substantial tracts of land out of the realm of traditional social authority and made it private property. It appears that in both cases the kinds of risks and transactions costs [arising from social conventions] Johnson discusses declined substantially and resulted in increased transfers, and eventually market sales, to more productive farmers, and a resulting boom in productive agriculture in both places. . . . In neither case, however, was a mechanism of land registration the moving force, at least initially. Rather it was the decrees to privatize land. . . .[145]

These pieces of empirical evidence would appear to suggest that formal titling systems are more likely to promote efficient use of the land than traditional systems are. However, the nature of traditional systems varies widely, and the process of introducing formal titling can be expensive and slow, in addition to introducing the risk that some small farmers may lose their usufruct rights, or herders may lose temporary grazing rights, during the transition process.

Atwood himself is skeptical about the efficiency benefits of widespread land titling *per se* in the African context:

It is likely that most African farmers are secure in their holding at present. That substantial investment in land improvement and new technology has taken place across a wide variety of African land tenure situations . . . supports this contention. . . . *There are many African situations where land titling or registration would not have the intended impact, would not be economically justifiable, or would even be counterproductive.*[146]

Traditional rights of usufruct can provide the degree of tenure security necessary to encourage production and even investment on the land, although they normally do not permit sales of the land nor do they usually guarantee inheritance rights. The main concern about such usufruct rights is their durability in the face of the socio-economic change which inevitably occurs in the development process. The debate over the benefits of traditional, or customary, systems of land rights, versus systems of formal land ownership titles, has been well summarized by William Kingsmill and Christian Rogg, in a kind of checklist that may be used by policy makers in deciding whether to strengthen or modify traditional tenure regimes:

Customary land tenure arrangements are often characterized as leading to a relatively equitable distribution of land but as being relatively inefficient. However, whether this is indeed the case in practice often depends upon the exact context and the precise nature of governance arrangements, and associated conditions of use:

- How far does population pressure restrict the amount of land available per household?
- How much discretion do traditional authorities have to allocate or withdraw use rights, and how accountable are they to all local people for the way that any discretion is exercised?
- In particular, what criteria are used when making a new allocation or reallocation of land?
- What constraints exist for customary rights to be inherited by the family of the deceased?
- Can women hold land in their own right?
- How far are individuals or households at liberty to delegate use rights through rental or sharecropping arrangements?

The answers to such questions will not only determine the opportunities for the poor to access land, and the security of such access. They will also affect matters such as the incentives for households to invest resources in improving the productivity of their land, and

---

145. D. A. Atwood, 1990, pp. 662–663.
146. *Op. cit.*, pp. 665 and 668 [emphasis added].

the incentives for taking action to avoid environmental damage.[147]

When there are doubts about the durability, efficiency or equity of customary regimes of land rights, the solution may lie in formalizing them and making any necessary modifications during the process of formalization. Alden Wily has noted the growing acceptance and formalization of customary land rights, as opposed to fee-simple titling, in Eastern and Southern Africa:

... the most dramatic transformation [in land tenure in the region] is being made in how unregistered, customary rights as a whole are being handled in State law. ... despite a century of purposeful penetration by non-customary tenure ideology and legal provision, unregistered, *customary tenure not only persists but is still by far the majority form of tenure in the region.*

Perhaps the most radical shift in tenure reform occurring in sub-Saharan Africa today is that for the first time in one hundred years, *States are being forced to recognize African tenure regimes as legal in their own right and equivalent in the eyes of national law to the freehold/leasehold culture.*

In Eritrea, majority customary rights have been reconstructed into lifetime usufructs, with guaranteed State protection (1994)[148]. ... In South Africa, how to clarify the millions of unrecorded rights in homelands and embed them in statute appears to be posing a much greater challenge than the 'simpler' mechanics of restitution and redistribution. However, the clearest lead of all is being given in the new tenure laws of Uganda, Tanzania and Mozambique. In different ways, these simply recognize customarily obtained land as fully legally tenured 'as is', in whichever form and with whatever characteristics they currently possess. In addition they may be made registerable entitlements if so desired. Zimbabwe and Malawi policy recommendations suggest these two States could follow suit (1999).

These developments undermine the very principles upon which property relations have been legally constructed over the last century. Previously, recording, registration and entitlement were all geared towards individual ownership; now, the link has been broken. Whilst certification remains indispensable as a founding route to land security, it is no longer necessarily tied to individualization. Accordingly, new tenure laws in South Africa, Mozambique, Uganda and Tanzania make provision for not just individuals but for two or more persons, groups, associations, and communities, to hold land in legal and registerable ways. *The certification process itself has to change: it may be verbal, and verbally endorsed (Mozambique). The community itself may conduct the adjudication, recording and titling processes (Tanzania).* As a matter of course, the local regimes through which these rights are created and sustained – customary land tenure systems – are being empowered in these ways. ...

... what was yesterday commonage, and with all the ills of open-access implied, is today legally registerable private (group) property. Commonhold itself is emerging as a new form of tenure.[149]

The international consensus is that the need for land titling is largely a function of the social and economic circumstances of the rural economy. It is a greater priority when doubts begin to arise about whether customary systems can regulate adequately pressures for land transactions, when

147. Reproduced from William Kingsmill and Christian Rogg, 'Making Markets Work Better for the Poor: A Framework Paper', Department for International Development (DFID), London, 2000, Annex 4, with permission of the Department for International Development.

148. However, in Eritrea the legislation has scarcely been implemented, an experience that illustrates the point that successful reform of land tenure requires much more than just passing a law.

149. L. Alden Wily, 2000, pp. 2–3 [emphasis added].

land values increase, and in situations in which more than one tenure regime is applicable to transactions. The circumstances or resources may not always exist for a shift to a full-fledged titling system, as indicated by Alden Wily. In addition, the option of introducing a formal land titling and registration system should be evaluated in terms of whether the efficiency gains offset the costs of implementing the new system.[150] According to one viewpoint, the formalization of land rights is essentially a pragmatic issue whose urgency depends in part on the scarcity of cultivable land. Both Platteau and the World Bank have expressed this view:

> Formalization of land rights through issuance of titles or other land-register documents is an urgent step to be taken in all the areas where competition for land has become so stiff as to impose high ex ante and ex post transactions costs on the agents.[151]

More recent analyses have tended to emphasize the advantages as well as weaknesses of traditional tenure systems. In the words of Deininger and Binswanger:

> When the community rather than the individual owns the land, whatever market exchanges (sale or rental) exist are normally limited to the community. Individuals have very secure and normally inheritable rights to land even after a period of absence, but they do not have permanent property rights to a specific plot, a limitation that may reduce investment incentives. In some cases, communal systems also permit periodic redistribution of land by the village chief to accommodate population growth.
>
> In the past, communal tenure arrangements were often considered economically inferior and equivalent to collective production. The establishment of freehold title and the subdivi-

sion of the commons were proposed to prevent the efficiency losses that were assumed to be associated with communal ownership. More intensive study of communal tenure systems in a broader framework and the recognition that these systems perform multiple functions has led to a reassessment of these recommendations, however.

> On the one hand, the efficiency losses associated with communal tenure systems may be more modest than generally assumed. . . . With arable land becoming increasingly scarce, many communal tenure systems either recognize a user's property rights if the land has been improved or compensate the user for improvements when the land is redistributed, thus attenuating tenure-related investment disincentives.
>
> *. . . in cases where there is no clear demand for demarcation of individual plots, communal titles that are administered internally in a transparent fashion could provide tenure security at a fraction of the cost of individual titles.*[152]

The new Rural Code of Niger is an example 'which recognizes and empowers customary land tenure practices and institutions. . . . It recognizes customary ownership rights and incorporates local land tenure and land management systems'.[153]

While it can be valuable to reinforce customary systems of land rights, it is important for policy makers to bear in mind that such rights are likely to undergo evolution, to monitor their adequacy, and to be prepared to adapt the tenure regime as needed. Local and community institutions that currently administer, or will administer, tenure regimes require strengthening. Pressure for more secure tenure rights, in whatever form they are provided, is likely to come above all from farmers. Deininger and Binswanger have underscored farmers' concerns in this regard:

---

150. G. Feder and R. Noronha, 1987, p. 163.
151. J.-P. Platteau, 1992, p. 292.
152. K. Deininger and H. Binswanger, 1999, pp. 257–258 [emphasis added].
153. A. Herrera, *et al.*, 1997, p. 62.

More secure land rights may be highly valued by cultivators even under conditions of relatively low population density. For example, in Zambia (with a population density of 12 people per square kilometer and where 75 percent of the land is suitable for farming), almost 50 percent of farmers feel their land tenure is insecure and would be willing to pay something (US$40 on average) for land titles. . . . Disputes, efficiency losses arising from limiting transfers and barring certain groups from land rights, investment disincentives, and land grabbing in anticipation of future appreciation are all indicators that existing land rights are inadequate. *Clarification and formalization of informal property rights in a process that increases the accountability of local leaders, establishes a transparent and implementable legal basis, and provides for adjudication of boundary disputes across communities must precede any effort to award formal titles.* Adopting a flexible institutional structure that gives communities freedom of choice in accomplishing these goals is therefore of great importance.[154]

A policy of fully titling customary and/or State lands carries with it demanding administrative requirements:

Agricultural modernization combined with population pressure will make land titling necessary. Traditional tenure systems need to be codified. . . . The transition to full titling will take time to achieve in most African countries and should be attempted only in response to demand by rural people. Nationally legislated rights are likely to conflict with customary rights. Judicial mechanisms for dealing with disputes between owners claiming traditional versus modern land rights are urgently required. As with other actions needed for agricultural growth, the critical element in any new land policy will be the administrative capability to manage it.[155]

In regions of the world where customary rights already have been weakened or superseded, and where the State is not the sole owner of agricultural land, the case for accelerated implementation of titling systems is strong. In such circumstances, they provide unambiguously improved security of tenure for the cultivator, facilitate eventual inheritance by his or her descendants, enhance the farmer's capacity to secure financing for production and investment, and convert the land into an economic asset which a farmer can utilize to more readily start life outside the agricultural sector in the event of such a decision. If a farmer does not have the right to sell land, he/she cannot derive benefit from the cumulative investments – use of own savings – made in the farm over time, in the event of a decision to leave the farm.

In practice, the shift to a modern land titling system is a very slow process, and it can be expected that both modern and customary systems will exist side by side for a long time, as illustrated by the experience of Zambia:

Customary tenure is the dominant system governing land administration on 94 percent of the land mass of Zambia. To eradicate rural poverty rural farmers must be transformed into business-oriented farmers. This shift will inevitably result in demands for changes in tenure systems. The land laws must be able to respond positively to these demands. . . .

Despite the strengths and weaknesses of the tenure systems in place, both are so entrenched that substitution of one for the other is not feasible or practical. The two systems are expected to coexist in the foreseeable future. A new land law should serve different interests in different parts of Zambia. It must recognize customary tenure, but the demarcation of the country into

---

154. K. Deininger and H. Binswanger, 1999, p. 259 [emphasis added].
155. The World Bank, *Sub-Saharan Africa: From Crisis to Sustainable Growth*, The World Bank, Washington, DC, USA, 1989.

State [leased] land and traditional land is no longer justified. . . .[156]

5.8.2 Forms of Customary Title

When traditional land rights are still in force, at least partly, a possible policy approach is to offer well-defined alternatives, including group titles as mentioned, with regard to systems for formalizing land rights. In addition to issuing group titles, *there are a number of policy options for titling communal lands*. In some cases, depending on the circumstance, groups with communal lands may wish to consider establishing community fees for use of such lands, the proceeds of which could be directed to land improvement and protection measures. When there is substantial concern about the degradation of common lands, it may be advisable to urge communities to consider participating in a program of titling land for individuals, with partial restrictions on sales of land in order to address concerns about losing community land to outsiders. One such option would be to require (by regulation) that before a sale of land to a person outside the community is consummated, members of the community have the right to try to match the terms of the sale ('first right of refusal', in the parlance of real estate). 'Where appropriate institutions for intragroup decision-making are available . . . permitting the community to limit sales and giving it the right to decide whether to eventually allow sales to outsiders may be an acceptable compromise between equity and efficiency concerns'.[157]

A first right of refusal was built into the legal framework governing land tenancy in Sri Lanka in its reforms of 1958, with positive and negative effects:

That reform allowed tenants to sell their right of tenancy, but only to local cultivation committees. The condition discouraged landowners

from trying to coerce the tenants to sell their land rights, but it also kept the committee from reallocating land rights to the best possible person. Setting effective criteria for how the land should be reallocated – perhaps by allowing bidding among potential tenants as long as they satisfy certain criteria – may make this system more effective.[158]

Yet another option would be to levy a significant tax, the proceeds of which would go to the community, on sales of 'privatized communal land' to parties outside the community. Such a provision was legislated for the agrarian reform sector as a whole in Honduras, in the *Law for Modernization and Development of the Agricultural Sector* (1992).

Communities also may wish to consider supporting legislation that would allow them to make a decision to lease out the land to outsiders or to enter into joint investment contracts of long duration, in which the outsider supplies capital for investing in the land and the village supplies the land and labor. This last option also was granted to the production co-operatives of the Honduran agrarian reform sector by the above-mentioned law and legalized in the reforms to the legislation governing the Mexican ejidos.[159]

In yet another approach, ownership of the communal lands can be transformed into a more entrepreneurial (corporate) form, with transferrable and inheritable shares held by community members (perhaps with a first right of refusal on sale of the shares to persons outside the community). The desire to maintain the value of the shares would constitute an incentive for improved land management. Clearly, any of these policy options, or variants of them, should be developed in close consultation with the communities involved and should be accompanied by an extensive educational campaign.

---

156. V. R. N. Chinene, *et al.*, 1998, p. 98.
157. H. P. Binswanger, *et al.*, 1995, p. 2726.
158. A. V. Banerjee, 2000, pp. 270–271.
159. L. Téllez, 1994, p. 257.

Forni emphasizes that the State sometimes can play a critical role in facilitating the functioning of common property resources:

> There is empirical evidence, particularly in the case of pastoral communal groups. . . . that transaction costs increase as a system increases in complexity, from self-contained common property resources to outward-looking ones, such as those based on movable assets, e.g. flocks grazing over large territories with float-ing borders. . . .
>
> It could thus be argued that State interven-tion is essential to coordinate and ensure the functioning of common property resources beyond the purely local level and. . . . where common property resource rights holders have to interact with other common property or private regimes over large territories. . . .
>
> In the case of pastoral common property regimes in the Islamic Republic of Iran, for instance, the schedule of transhumance movements along routes determined basically by natural conditions depended not only on the state of vegetation but also on the co-ordination of movements with other tribes. This was ensured by a complex intertribal organization.[160]

In summary, legislation and regulations for a land tenure regime could be designed to support any of the following kinds of alternative new rules for traditional, communal lands:

- Issuing either group titles or strictly individual parcels, or group titles for some lands and indi-vidual titles for the others.
- Giving legal backing to customary, verbal systems of conferring tenure rights.
- Within the option of group titles (or the lands titled jointly), titling corporate forms or com-munal forms of landholding.
- Giving rights holders permission to sell their plots to anyone, including outsiders or, alterna-tively, only to community members.

- Giving community members the first right of refusal in the event of a pending sale, and/or taxing sales made to persons or entities outside of the community.
- For title registration, the option of installing fixed boundary markers, following traditional descriptions of plot boundaries, versus the option of recorded maps.
- Developing institutions for resolution of land disputes at the local level or the district level.

When a package of land tenure options is offered to a community, the choice of later chang-ing the selected option, to another one in the package, also should be offered. However, not all directions of change should be allowed, i.e. moving from communal lands to joint lands held in corporate form, but not vice versa, and moving from a system which permits sales only to com-munity members to one which permits sales to anyone (possibly with first rights of refusal), but not vice versa.

5.8.3 Conditions on Titling

In all parts of the developing and transitional world, titling agricultural land has the potential to create undesirable side-effects. If such effects are anticipated, both administrative and judicial mechanisms can be put in place to avoid or minimize them. One hazard noted above is the appropriation of land rights by those who are well placed in the politico-administrative system. This hazard can be reduced by campaigns of rapid and widespread dissemination of informa-tion about a titling program and its implications, and the greater involvement of civil society (including NGOs) in those programs. Such cam-paigns can be supported by international agen-cies. Political will in the country concerned, expressed at the highest levels, also can do a great deal to reduce this danger.

Another potential side-effect is the creation of landlessness among families who have not had traditional rights of permanent possession of

---

160. N. Forni, 2000, pp. 10–11.

land but who have been renting or leasing land for long periods. The opposite danger also exists of dispossession of the traditionally recognized owners of land that is rented out. These are very real dangers and they can be exacerbated by prohibitions on rental in the new system or by ambiguity regarding the rights of renters.

In light of these concerns, *it is important that titling programs: (a) carry out local consultations, extensive if necessary, to confirm traditional rights of both ownership and rental (or sharecropping), and (b) contain procedures to record all existing secondary rights*. For example, titling systems should contain a mechanism for the registration of rental contracts and, in more traditional contexts, to record rights of third parties to graze animals on fields of stubble. These steps would help address the concerns of authors such as Stanfield[161] and Atwood that 'land titling or registration to replace traditional of customary tenure arrangements by its very nature will extinguish some secondary land rights'.[162] In this regard, a titling program can be supported by a decree that confirms all existing traditional land rights.

In addition, under agrarian reform programs that title land to the tiller, handling the transition is a delicate matter. It is essential that the legislation allow sufficient time, in each case, for the existing rental arrangement on a plot to terminate, and for the owner to resume direct cultivation if desired. Subsequently, failing a demonstration of interest in directly cultivating it on the part of the owner, then the plot could be titled to the tiller. Without such a provision, owners could be penalized for engaging in a practice which was legal under the prevailing tenure regime. The way in which the Honduran land reform was written imposed this kind of penalty on landowners; although the government had assured the public that the ownership of rented

properties would be respected, an agrarian reform law passed at midnight on December 31, 1974, declared expropriable immediately those lands that had been rented out. The reverse kind of problem occurred in El Salvador: 'About 45000 beneficiaries received land under [this phase of the agrarian reform], but many renters were forced off the land before [the government] could proceed, which increased the number of landless persons'.[163]

Financing the costs of titling programs is another issue that requires resolution. Often, the conventional approach is to require beneficiaries of titling to reimburse the full costs, on the same grounds that user fees are established for other public services: the user, who is the beneficiary, should pay rather than taxpayers at large, and the fees collected will permit widening the coverage of the program. Nevertheless, it should be borne in mind that titling costs can act as a deterrent for very low-income farmers. The same argument can apply to registration of land sales, in order to keep property registries up to date. It was commented by Atwood that in Kenya the level of fees discourages official registration of some sales.[164] The counter-argument to user fees for titling small farms is based on the poverty of the beneficiaries and also the fact that society as a whole benefits from a removal of the ambiguities in land ownership and from the establishment of the basis for a land market. On these grounds, *there is a justification for subsidizing part of the costs of title registration and also of registration of subsequent sales, for farms smaller than a given threshold in size*.

Perhaps the most important prerequisite for titling and implementing any land policy is, as has been emphasized throughout this chapter, adequate administrative capacity. This capacity necessarily must be developed at the local level, and

161. J. David Stanfield, 'Projects that Title Land in Central and South America and the Caribbean: Expectations and Problems', Land Tenure Center Paper, University of Wisconsin, Madison, WI, USA, 1984.

162. D. A. Atwood, 1990, p. 661.

163. Latin America and Caribbean Office of the World Bank and the Salvadorean Foundation for Economic and Social Development (FUSADES), *Estudio de Desarrollo Rural*, Vol. I, August 1997, San Salvador, El Salvador, p. 40 [author's translation].

164. D. A. Atwood, 1990, p. 663.

therefore the implementation of a land policy requires decentralization of administrative responsibilities. Local bodies increasingly are taking responsibility even for allocation of land (selection of beneficiaries) under programs of market-assisted land reform. Land registries need to be developed in the first instance at the local level, and their records needed to be co-ordinated nationally. Reforms in property registration systems are beyond the scope of this survey, but their importance for implementing land tenure policies warrants emphasis.

In spite of a long history of conflict over land rights, El Salvador now is one of the leaders in Latin America in titling and registering rural land. Decentralization, in an integrated manner, is one of the aims of its work in this area, even though the country is quite small in territorial extent:

> Rather than have three separate [national] entities [in charge of titling and registering land] . . . four integrated regional registry-cadastre offices will take their place. . . . These offices will, by means of computer networking, be interconnected, and will be able to share and access national Registry information rapidly. The purpose of the reform is to decentralize access to registry records, while at the same time unifying topographic and legal information.[165]

This kind of decentralization is needed in all land registration systems. Failing it, a significant number of transactions simply go unregistered.

5.8.4 Land Leases and Rental

Under systems of State ownership of land, when the customary land regimes no longer exist, tenure security can be provided by long-term, transferrable leases, as mentioned above. When this approach is adopted, the leases should be at least 25 years in duration, preferably longer, in order to encourage investment in the land, and they should be automatically renewable provided that the cultivator has met normal obligations in respect of payment of lease fees and taxes. The list of requirements can include putting into effect sound environmental practices for land management. In order that efficiency criteria be satisfied, *it is necessary that the leases be transferrable (saleable) without requiring government approval, and that the parties to the transaction be free to negotiate their own price.* The only procedural requirement would be that the new leaseholder be registered as such within a certain period after the consummation of the transaction. If, on the contrary, government approval for the transaction is required, then the leasehold would lose much of its value, and banks could not use it as collateral, out of uncertainty over the prospects of being able to take possession of the collateral in the event of default.

Leases can be stepping stones to a regime of private property, if a policy decision has been made to carry out such a transition. After the cultivator has shown interest in a plot by cultivating it for a given number of years, and keeping up to date on the lease fees, then the cumulative fee payments can be credited as a down payment on the land, and the lease contract can be converted into a mortgage. For reasons of sound financial management, it would be preferable that mechanisms be developed to have the mortgages assumed by private financial institutions, and the corresponding lease payments made to such institutions. The device of crediting cumulative lease fees toward payment for purchase of the land is a possible option for improving the *access to land ownership on the part of poor rural families.*

Retention of a State lease also could be made conditional on not leaving the land in fallow or otherwise idle for more than specified periods of time, although if the lease fee is related to the amount of land leased, it is unlikely that farmers would pay for more land than they could put under the plow.

---

165. Brian Trackman, William Fisher and Luis Salas, 'The Reform of Property Registration Systems in El Salvador: A Status Report', mimeo, Harvard Law School, Cambridge, MA, USA, June 11, 1999, pp. 13–14.

In summary, the instrument of a lease on State lands can be made to represent virtually the equivalent of private ownership, and over time it can be converted into private ownership if national policy foresees moving in that direction. To the contrary, if long-term, transferrable leases are not adopted, then it will not be possible to meet the normal requirements of tenure security for cultivators on State lands. In this regard, *under 'modern' (non-customary) land tenure systems, the choices are not many: there are only two basic options for providing tenure security and adequate incentives to improve the land: private ownership and long-term, freely transferrable leases*.

The rental of agricultural land, and share tenancy systems, have been prohibited in the agrarian codes of many countries. The apparent reasoning, more political than economic, is that rental and sharecropping arrangements are perceived to represent an exploitation of the tiller by the landowner. Under many of the original agrarian reform laws and land codes, from the South Korea and Taiwan to Senegal, El Salvador and Peru, land which was rented out in any form was subject to expropriation, and in those cases it usually was adjudicated to the person who was renting in the land.

In practice, land rental occurs in all countries. Atwood has pointed out (1990, p. 664) that 'While land rental is often not recognized or sanctioned in government land laws, it exists in many African situations'. James Riddell has written:

> In all agrarian societies land tenure rights are negotiated and transacted. Even in societies where sale, transfer, lease and so forth of land is said to be forbidden, people are observed to transact these rights for consideration.[166]

In Tajikistan, after the reform of the collective farms land leasing became common: 'Interfarm leasing is widespread: 50 000 collective and individual land leaseholders and 33 agricultural cooperatives control 93 000 hectares of land'.[167]

Since land rental practices appear to be unavoidable, declaring them illegal puts both parties to the transaction at risk and diminishes incentives for investments in improving the land. It also creates incentives for corruption.

Land rental and sharecropping commonly occur because they bring about four principal benefits:

- *The practice eases rural poverty by providing a mechanism through which low-income farming families can gain access to additional land and acquire farming experience*. Another part of the statement of Vijay Vyas quoted in the opening section of this chapter is: 'The rigors of small ownership holdings are eased if there is an active lease market . . .'. In the same lecture, he added 'There is a large body of empirical evidence, and plausible logic to explain it, which suggests that, even in circumstances of overcrowded agriculture, crop-sharing provides for the landless rural laborer and marginal farmers a better alternative to limited opportunities for wage-labor'.[168]

In the words of Neil Ravenscroft, 'Leasing offers a means for farming families with little or no land and capital to gain access to land. As such, leasing arrangements are an established part of the fabric of the agricultural sector, often to the extent that their significance has been overlooked, downplayed, or misunderstood'.[169]

166. James C. Riddell, 'Emerging trends in land tenure reform: Progress towards a unified theory', *SDdimensions*, Food and Agriculture Organization of the United Nations, Rome, June 2000, p. 5.
167. Reprinted from *Food Policy*, **25**(6), A. Tashmatov, F. Aknazarov, A. Jureav, R. Khusanov, K. D. Kadyrkulov, K. Kalchayev and B. Amirov, 'Food policy reforms for sustainable agricultural development in Uzbekistan, the Kyrgyz Republic and Tajikistan', p. 728, Copyright (2000), with permission from Elsevier.
168. V. Vyas, 1992, pp. 11 and 12.
169. Neil Ravenscroft, *Good Practice Guidelines for Agricultural Leasing Arrangements*, FAO Land Tenure Studies 2, Food and Agriculture Organization of the United Nations, Rome, 2001, p. 3.

The case of Java is illustrative: 'Land tenure in West Java is largely free of legal restriction. While approximately 70% of farm plots are owner-cultivated, the evidence shows that land leasing functions as an equalizer of farm size . . . The lease market has permitted an increase in access to land suitable for vegetable cultivation whereas, prior to vegetable growing, there was no leasing of land in these areas'.[170]

• Almost by definition, *rental or a share tenancy arrangement for a plot of land is likely to improve its productivity*, because it transfers the use of the plot from a party who is less interested in or capable of working it to one who is more interested or capable. This axiom is reflected in the empirical findings of Michael Lyne and colleagues in sub-Saharan Africa:

Renting will not only bring idle land into use but, where leasing out [land] is risky, it will also

> *Land rental is even found under customary tenure regimes:*
>
> '*In the Hausa villages of Nigeria, the lowland (fadama) area is almost entirely cultivated at present and irrigable land is in great demand. A land rental market has therefore emerged and in 1978 it was estimated that almost 40 percent of the plots were rented. . . . In Madagascar, the incidence of sharecropping or rent contracts is increasing, especially in traditional or modern irrigation schemes, in spite of legal prohibitions. . . . In the Senegal river basin, the land-lease market has developed on a significant scale during the last decades. . . .*' (*J.-P. Platteau, 1992, p. 142*).

tend to transfer land to farmers confident of their ability to cover the risk premiums charged by lessors. . . . The data show that renting transfers land to farmers who gross more income per hectare operated. . . . The data presented in this paper show that renting closes productivity gaps by transferring land to farmers who can use it more effectively, and that, far from damaging the interests of the poor, it sustains many households that would otherwise be destitute.[171]

• *The option of temporarily renting out land reduces income risk to landowners*, by providing a mechanism for obtaining at least minimal flows of income from the land in years when, for reasons of health, family finances or other factors, the landowner is incapable of cultivating the land. In this regard, it also provides an alternative to possibly having to sell the land, or losing it at foreclosure. This benefit accrues as much or more to small landowners as to large ones, because the latter in any case would be less likely to work the land directly.

• *For large landowners, it has been shown that rental or sharecropping contracts normally will be a more efficient way to have the land worked than hiring wage labor, mainly because of the reduced supervision costs.* Sharecropping is often an efficient way to deal with risk in yields and prices, and to compensate for imperfections in credit markets.

In light of these benefits, prohibiting land rental generally has perverse effects and often leads to underutilization of land. As pointed out by Lyne *et al.*:

---

170. *Op. cit.*, p. 5, citing S. Yokohama, 'Agricultural diversification and institutional change: a case study of tenancy contract in Indonesia', *Developing Economies*, **33**(4), 1995, pp. 374–396.

171. Michael Lyne, Michael Roth and Betsy Troutt, 'Land Rental Markets in sub-Saharan Africa: Institutional Change in Customary Tenure', in Roger Rose, Carolyn Tanner and Margot A. Bellamy (Eds), *Issues in Agricultural Competitiveness: Markets and Policies*, IAAE Occasional Paper No. 7, International Association of Agricultural Economists, Dartmouth Publishing Co., Ltd., Aldershot, UK, 1997, pp. 59, 60 and 65.

Bruce (1989)[172] noted that wealthier farmers with draft animals could no longer rent in land from other households that lacked oxen following the draconian enforcement of reform legislation that prohibited tenancy arrangements in the communal areas of Ethiopia. In Senegal, he observed that landholders were unwilling to observe the custom of "loaning" land after laws enacted in 1964 recognized the right to continued occupancy of those individuals cultivating land. Similarly, Roth (1993)[173] reports that respondents surveyed in Somalia's lower Shebelle region were wary of renting land out owing to legal provisions banning transactions and frequent disputes involving tenants who refused to return land at end of the agreed term.[174]

Underutilization of land because of prohibitions on rental also has been observed in the Barolong and Eastern Bangwaketse regions of Botswana and in Nigeria, Ghana, Cameroon and Zanzibar.

Many of the agrarian codes that originally forbade land rental have been amended to permit it, and to remove it from the list of grounds for expropriation of land in those instances in which a regime of expropriation still is in effect (examples include those of Mexico, Peru, Honduras and El Salvador). In the case of Honduras, the Agrarian Reform Institute developed (in 1993) model rental contracts and distributed them widely to peasant farmers throughout the country.

In addition to land rentals, which usually are short-term, there is an important role in agriculture for long-term private leases of land. The reasoning in support of them is much the same as that for land rentals, and their longer length encourages the tenant to undertake activities to improve the land.[175] Extensive discussion of the nature of legislation which would create the instrument of agricultural leases is found in a recent report developed for the Estonian Government by an FAO team.[176]

Attempts to control the levels of land rents generally fail. Under even modest inflation, the controlled rents can get seriously out of date and therefore, sooner or later, the presence of the controls tends to discourage landowners from renting out land. Once again, a consequence is that more land is likely to be left idle. Dissemination of model contracts and information about prevailing rental rates are more effective measures to head off attempts to exploit rental tenants.

In order that leasing arrangements serve the goals of promoting sustainable livelihoods and more equitable access to land, they should demonstrate the following characteristics:[177]

- equitability and fairness between the parties
- transparency
- preservation of the legal interests in property of both parties

---

172. J. W. Bruce, 'The Variety of Reform: A Review of Recent Experience with Land Reform and the Reform of Land Tenure, with Particular Reference to the African Experience', Land Tenure Center, University of Wisconsin, Madison, WI, USA, 1989.

173. M. Roth, 'Somalia Land Policies and Tenure Impacts: The Case of the Lower Shebelle', in T. J. Bassett and D. E. Crummy (Eds), *Land in African Agrarian Systems*, University of Wisconsin Press, Madison, WI, USA, 1993.

174. M. Lyne, *et al.*, *op. cit.*, p. 59.

175. The effect of short-term land rentals on the amount of investment in soil conservation has been the subject of discussion by researchers. In the case of Rwanda, convincing evidence has been shown to the effect that such investments are lower on rented land (Daniel C. Clay and Thomas Reardon, 'Determinants of Farm-Level Conservation Investments in Rwanda', in R. Rose, C. Tanner and M. A. Bellamy (Eds), *op. cit.*, 1997, pp. 212, 215).

176. W. Valletta, S. H. Keith and R. D. Norton, *The Introduction of Agricultural Land Leasing into Estonia*, TCP/EST/5612, Food and Agriculture Organization of the United Nations, Rome, March 1998.

177. N. Ravenscroft, 2001, p. 7.

- simplicity
- low transaction costs
- certainty
- sustainability
- promotion of the leasehold sector as a means of promoting flexibility in the market
- a minimum of state regulation and intervention

Model leasehold contracts should be designed with these characteristics in mind.

5.8.5 Share Tenancy Contracts

To some, sharecropping carries connotations of exploitation, but in fact it is the predominant form of land rental in developing countries. If some key features of reality are assumed away – lack of uncertainty, perfect access to credit and insurance for the rural poor – it has been demonstrated theoretically that sharecropping always is inefficient – it always results in lower output, and lower labor input on the land, than pure cash rental contracts. However, when the analysis is made more realistic, 'with risk aversion and uncertainty, a share contract provides the possibility of partly insuring the tenant against fluctuations in output', and 'a limit on the working capital available to the tenant (or to landlord and tenant) because of imperfection in the credit market, can lead to the adoption of a share contract as the optimal solution to the bargaining problem [between landlord and tenant]'.[178] Empirical tests have been made regarding the motivation for sharecropping contracts. Statistical tests performed on a sample of Tunisian farmers indicated that the credit constraint, rather than risk aversion, was the major factor

underlying the choice of a share tenancy contract.[179]

Research on land rental markets in Ethiopia found that the negative efficiency effect of share tenancy is small and that rental markets as a whole improve the efficiency with which productive resources are allocated in agriculture:

Our research indicates that land rental markets have two opposing effects on agricultural efficiency: a positive land reallocation effect and a negative incentive effect. The positive reallocation effect results from better matching between land, labor and oxen power. The negative effect arises from the standard inefficiency argument according to which sharecropping reduces labor effort. Of the two effects, the first effect is by far the largest. Although we find evidence of a negative incentive effect for sharecropping, the magnitude of the efficiency loss is quite small and the effect on output and profit is negligible. On the whole, land rental markets in present-day rural Ethiopia perform a beneficial function and should not be discouraged.[180]

In sharecropping (share tenancy) systems, it is not necessary to make policy interventions to avoid the danger of rental rates getting out of date, since the payment is a share of the output. Binswanger *et al.* warn against attempting to regulate tenancy contracts: 'Historically, land reform that resulted in establishing owner-operated farms appears to have been a far more successful way of addressing the equity question'.[181] On the other hand, *in some cases of highly unequal distributions of landholdings and a preva-*

178. Reprinted from J. Behrman and T. N. Srinivasan (Eds), *Handbook of Development Economics*, Vol. 3B, H. P. Binswanger, K. Deininger and G. Feder, 'Power, distortions, revolt and reform in agricultural land relations', pp. 2713 and 2714, Copyright (1995), with permission from Elsevier.

179. *Op. cit.*, p. 2714.

180. Marcel Fafchamps, 'Land Rental Markets and Agricultural Efficiency in Ethiopia', mimeo, Department of Economics, Oxford University, Oxford, UK, August 23, 2000, p. 2.

181. Reprinted from J. Behrman and T. N. Srinivasan (Eds), *Handbook in Development Economics*, Vol. 3B, H. P. Binswanger, K. Deininger and G. Feder, 'Power, distortions, revolt and reform in agricultural land relations', p. 2716, Copyright (1995), with permission from Elsevier.

lence of sharecropping arrangements, it appears that legislating the division of the crop may improve both efficiency and equity. The results for an agricultural tenancy reform of this nature in the Indian state of West Bengal were summarized as follows:

> A primary goal of the reform was to change the division of output between landlords and tenants in favor of tenants. As a result . . . the division of output on many tenancies changed from an even split to a 70–30 split in favor of tenants. By 1982 the reform had reached about half the state's sharecroppers, and over the next decade West Bengal achieved a breakthrough in agricultural growth. Banerjee and Ghatak (1996)[182] estimate that more than a third of West Bengal's growth in agricultural production during 1981–92 was due to the tenancy reform. Thus the land tenancy reform not only redistributed income to the poor, it also changed the incentives of the poor to be productive by changing the tenancy contracts under which they supplied labor. In this way, redistribution substantially increased the income generated by the poor.[183]

Share tenancy is increasingly recognized as a useful system for access to land, in part because landowners often are a source of credit for the sharecroppers. This kind of contract also reduces risk to the landowner, in the face of possibilities of crop failure, *vis-à-vis* the option of paying wage labor. In summary:

> Sharecrop tenancy seems to offer a package of incentives, insurance and interlinkage to other

markets which is often best for both parties when compared to the relevant alternatives. . . . This observation is consistent with the conclusion that cultivation under share tenancy is not less efficient or less productive than other forms of tenancy. . . . We now have an abundance of research which demonstrates, both in theory and practice, that while a particular form of tenancy may not be efficient in an economically perfect world, in many instances the type of contract which prevails is optimal in the real world.[184]

5.8.6 Policies Regarding the Sale of Agricultural Land

Among policy makers, sales of agricultural land have generated as much concern as rentals. A principal concern is that beneficiaries of agrarian reform may try to sell their plots, and the pattern of landholdings could become skewed once again. Independently of agrarian reform, the issue is that many smallholders may be forced to sell their land when they confront periods of temporary economic hardship, in order to provide subsistence to their families, and as that occurs inevitably landholdings will become concentrated over time. Platteau states the case:

> It appears to be a well-accepted proposition among economists today that the establishment of a land market in a land-scarce environment entails considerable efficiency gains while at the same time involving significant costs in terms of equity. Thus, Binswanger, McIntire and Udry (1989)[185] have . . . concluded that 'while both the land rental and

182. Abhijit V. Banerjee and Maitreesh Ghatak, 'Empowerment and Efficiency: The Economics of Tenancy Reform', Massachusetts Institute of Technology, Cambridge, MA, USA, 1996.
183. Karla Hoff, 'Comment on "Political Economy of Alleviating Poverty: Theory and Institutions", by Timothy Besley', in Michael Bruno and Boris Pleskovic (Eds), *Annual World Bank Conference on Development Economics, 1996*, The World Bank, Washington, DC, USA, 1997, p. 142.
184. Susana Lastarria-Cornhiel and Jolyne Melmed-Sanjak, with assistance from Beverly R. Phillips, 'Land Tenancy in Asia, Africa and Latin America: A Look at the Past and a View to the Future', Working Paper No. 27, Land Tenure Center, University of Wisconsin–Madison, Madison, WI, USA, April, 1999. This paper is a thorough review of theory and experience regarding share tenancy.
185. H. Binswanger, J. McIntire and C. Udry, 'Production Relations in Semi-Arid African Agriculture', in P. Bardan (Ed.), *The Economic Theory of Agrarian Institutions*, Clarendon Press, Oxford, UK, 1989.

sales markets promote the efficient allocation of factor endowments, land sales are likely to exacerbate landholding inequality'....[186]

In other words, a bad harvest, or a serious health problem, is more likely to force a small landowner to sell than a large landowner, because of wealth differences. The cumulative effect of such 'distress sales' may result in a greater concentration of landholdings, in part because long-term financing mechanisms do not generally exist that would permit the small farmers to repurchase land, or to avoid the distress sale in the first place. On the other hand, trying to restrict sales frequently leads to an even less favorable outcome:

Since these restrictions also prevent some transfers of land from worse to better farmers or managers, there is likely to be some efficiency loss.... Restrictions on the rights of land reform beneficiaries or settlers on State-owned land to sell the land also reduce their access to credit. Often new owners are forbidden to mortgage their land during an initial probation period. Since that period coincides with the establishment phase, when their need for credit is most urgent, the efficiency losses may be considerable.[187]

A more recent and complete formulation about the issue of restrictions on land sales is the following:

Even temporary restrictions on land sales can be counterproductive, however, because they prevent landholders from accessing credit when it is most needed.... Precluding beneficiaries of land reform from renting or selling their land is likely to prevent adjustments that reflect the settlers' abilities and could, if combined with restrictions on rentals, cause large tracts

of land to be underutilized. The goal of preventing small landowners from selling out in response to temporary shocks would be better served by ensuring that they have access to output and credit markets and technical assistance and by providing them with safety nets during disasters to avoid distress sales.[188]

Some observers are concerned that permitting land sales may reduce aggregate economic efficiency in the sector because of the possibility that it may on average shift land from small farms, which generally are more efficient per hectare, to larger operations. Yet, in many other instances a sale will enhance efficiency, for the same reason that land rental does: the land is placed in the hands of a person with a greater interest in or ability to work it. In cases like the above-mentioned agrarian reform co-operative in Puno, Peru, the efficiency of the operation would have been enhanced by an ability to sell off part of the holding. In any event, the main objection to permitting untrammeled sales of land has concerned their effects on equity, by leading to a greater concentration of landholdings.

Direct bans on land sales, and requirements for administrative approval of such sales, have the effect of destroying the market value of the land. In the final analysis, *the notion of private property is almost inseparable from the right to sell it or rent it*, and so restrictions on transfers of land ownership are difficult to justify in the context of a market economy. In practice, the most effective ways to prevent a concentration (or reconcentration) of landholdings in the form of large estates is to improve the access to land of lower-income rural families (a topic discussed below) and to improve rural credit systems so that small farmers have access to additional resources in difficult times.

In some circumstances, *prohibitions on land sales can themselves exacerbate inequality*. This point was illustrated by Feder and Noronha:

---

186. J.-P. Platteau, 1992, p. 193.
187. H. P. Binswanger, *et al.*, 1995, p. 2725.
188. K. Deininger and H. Binswanger, 1999, p. 264.

Increased inequality, the fear of which is often cited as a reason for prohibiting land sales and mortgaging, could in fact be an important consequence of the prohibition. Inequality also arises where governments recognize sales only by people from a particular group (chiefs, for instance) or where transactions in land involve complex procedures with uncertain results. . . . Inequality is a consequence of prohibition for two other reasons: First, those who know the law (usually the wealthier and better-off) can use the system to their own advantage. Second, they are protected in land transactions because their status ensures that no action would be taken to dispossess them. . . . in Ghana, 'many chiefs have benefitted as a result of their control over land . . . to acquire bank notes, tractors'.[189] . . . Similar consequences have occurred in Botswana. . . .[190]

In addition to these problems, prohibitions on the sales of land generate their own costs (transaction costs), in the form of measures taken to circumvent the ban. Hence, the operational question is, *are there classes of restrictions on land markets which would help to reduce the equity and efficiency losses, while not creating significant losses to society of other kinds?*

A common kind of restriction which, in principle, would not lead to additional transaction costs is the aforementioned establishment of a *first right of refusal in favor of a target group* (e.g. village members, land reform beneficiaries and low-income households in general). For this kind of provision to be effective, it needs to be accompanied by effective mechanisms for financing purchases of land by those who wish to exercise the first right, in favor of purchase. As discussed below, such mechanisms generally require government subsidies. An extreme case of this option is an outright ban on sales outside of the defined group. However, such bans are likely to lead to a fall in the value of the land for such groups, and correspondingly to a drop in its collateral value.

A ban on land sales of limited duration (say, for three to five years) may be justified for recipients of land through public programs. In this way, they will gain experience in working the land before making a decision to sell. Such a measure also may forestall the temptation to turn a quick profit by selling land which had been received at a subsidized price, but it also can be evaded through the device of unregistered sales contracts, and so it may be difficult to enforce.

An indirect way of restricting the land sales market is to impose *ceilings on land ownership*. Such ceilings have been in effect in many countries, including South Korea, Bangladesh, Pakistan, Peru, Honduras and El Salvador.[191] There are obvious possibilities for evading such a ceiling by making artificial sub-divisions of land among relatives and close friends, but such evasions can be carried out only to a certain extent, and so a ceiling may result in less concentration of land than otherwise would be the case. Binswanger *et al.* have pointed out that:

> . . . several studies do credit land ownership ceilings with a major role in preventing new large consolidations after land reform (Cain,

189. This quote is from Andrew Shepherd, 'Agrarian Change in Northern Ghana: Public Investment, Capitalist Farming and Famine', in Judith Heyer, Pepe Roberts and Gavin J. Williams (Eds) *Rural Development in Tropical Africa*, St. Martin's Press, New York, 1981, p. 177.

190. G. Feder and R. Noronha, 1987, p. 156.

191. Landholding ceilings should be clearly distinguished from policy-imposed limits on the size of parcels of State lands which are distributed or sold to cultivators. In conditions of land scarcity, such policy limits on land distributed always should be present, regardless of whether there exists a ceiling on landholdings, so that land reform is not abused by the granting of very large plots to a favored few. After the adjudication of government land to private parties, then the recipients of the land would be free to enlarge their farms if they were able to do so, up to the landholding ceiling if it exists.

1981[192]; Mahmood, 1990[193]). . . . Ceilings imposed after a land reform that results in fairly homogeneous holdings might be effective and less distortionary in preventing massive reconcentration of land.[194]

In the final analysis, any restriction on factor markets runs the danger of reducing efficiency in the sector, that is, harming the sector's growth prospects. For this reason, in some cases in which landholding ceilings are in effect, provisions have been legislated to permit exceptions to the ceiling provided that the landowner is willing to pay a substantial special levy (a one-time payment). Such payments should go into a fund whose uses are restricted to agricultural and rural development – perhaps a fund that supports market-assisted land reform, or a land fund.

5.8.7 Transforming Collective Farms: Corporations, Private Co-operatives and Individual Farms

There is no dominant model for converting collective farms, or State production co-operatives, into entrepreneurial structures. Differing approaches have been advocated. For example, in spite of the creation of large numbers of farm corporations in the former Soviet Union and Eastern Europe, the position of the World Bank for the agricultural transition in that region has been opposed to that trend:

. . . farms are poorly suited to the corporate form. Most corporate farms in North America, for example, are family farms incorporated for tax purposes, not companies with many shareholders. Secondary markets in shares of farm corporations are virtually unheard of. Corporatizing collective and State farms therefore creates farm structures with no counterpart in market economies and no ready mechanism for their evolution and reorganization, since share trading on secondary markets is unlikely to develop.

The reorganization of farmholdings should concentrate on establishing and documenting individual ownership of land and nonland assets and on creating markets through which owners can adjust farm size and capital intensity.[195]

These observations may have their relevance, but it should be pointed out that there are thousands of companies throughout the developing world whose shares are privately held and traded. In many parts of Africa and some countries of Latin American and the Caribbean, stock markets do not exist, and so the requirement that there exist a secondary market for shares cannot be fulfilled, neither for industrial corporations nor for agricultural ones. Shares are traded without the benefit of an organized market, and the mere fact of their issuance creates a pressure group interested in monitoring the management of the firm. In addition, corporate farms do exist in North America and, regardless of the motive for their founding, the corporate structure, once again, creates shareholder interest in the firm's performance. An example of a successful corporate farm in the developing world is Cítricos de Chiriquí, a citrus grower and processor in Panama. Initially, this was a State corporation, and it was privatized in the same form, before Panama had a stock market.

Another advantage of the corporate form is that it is particularly amenable to privatization of State enterprises in participatory ways, as mentioned in Chapter 3. Accordingly, State-owned

192. M. Cain, 'Risk and insurance: perspectives on fertility and agrarian change in India and Bangladesh', *Population and Development Review*, **7**, 1981.
193. M. Mahmood, 'The change in land distribution in the Punjab – empirical application of an exogenous–endogenous model for agrarian sector analysis', *The Pakistan Development Review*, **29**, 1990.
194. H. P. Binswanger, *et al.*, 1995, p. 2726.
195. The World Bank, 1997, pp. 58–59.

sugar mills (with cane-growing land) in El Salvador and grain silos in Honduras have become corporations with different classes of stockholders.

Most importantly, *after long experience of working together, however meager the economic results may have been, the members of some collectives may be reluctant to parcel out the entity's assets in the form of individual property.*[196] In these circumstances, the corporate structure can serve as a useful, private form of property, and by providing humane economic conditions for departure of those who are least interested in participating, it can promote a more consolidated, effective ownership over time. Corporations with full property rights are well positioned to secure credit and also have the option of leasing out some of their land, under long-term contracts, or selling some of it, options which were not available to the former collective farms and agrarian reform production co-operatives.

In Belarus as well as in Russia, the corporate, or joint-stock, form of organization is a principal option for members of former collective farms, along with the alternatives of production co-operatives, service co-operatives and individual farms. In Moldova, priority has been given to the creation of joint-stock companies but with land being treated as personal property, and to the formation of corporate-type enterprises for agro-processing and supplying inputs to agriculture.[197]

However, it should be recognized that often the managers of former collective farms become presidents of the joint stock companies, and this prospect creates incentives for the latter form of organization. In contrast to the cases of Belarus and Moldova, in Armenia and Georgia much more emphasis has been placed on land allocations to families.

In reality, in Eastern Europe and the former Soviet Union as well as elsewhere, the choice may not be so much one of a corporate form versus individual property, but rather one of a corporate form, based in commercial law, versus private co-operatives that often are constituted under the aegis of agrarian legislation.

Unless the legislation is very clear about the use of property in private co-operatives as collateral, those enterprises may have more difficulty in securing production credit than corporations would. Similarly, the legislation supporting the co-operatives needs to make it clear that creditors will be given a priority claim on assets in the event of voluntary dissolution or bankruptcy of the co-operative. On the other hand, for farm corporations, it is essential that the legislation clarify the rights to sell shareholdings, even though a first right of refusal may be granted to other members. 'If . . . the government in response to domestic pressures fails to support the freedom of individual exit as a natural extension of the share process, the newly registered shareholding structures will remain largely frozen in their scale and mode of operation'.[198]

An example of legislation that strengthens the process of transformation is found in the case of Bulgaria:

> The Bulgarian government is giving support to the formation of voluntary cooperatives, and legislation has been introduced covering, for example, the rights of participants in a cooperative and how land and assets can be reclaimed in the event of a member leaving.[199]

196. This was noted for the case of Bulgaria by A. Buckwell and S. Davidova, 'Potential implications for productivity of land reform in Bulgaria', *Food Policy*, **18**(6), December 1993, p. 505.

197. For comments on the cases of Belarus and Moldova, see G. H. Peters, 'Agricultural economics: an educational and research agenda for nations in transition', *Agricultural Economics*, **12**(3), September 1995, citing Victor Krestovsky and Sergey Chertan, respectively, pp. 207 and 209.

198. Reprinted from *Agricultural Economics*, **13**(1), K. Brooks and Z. Lerman, 'Restructuring of traditional farms and new land relations in Russia', p. 24, Copyright (1995), with permission from Elsevier.

199. M. Hristova and N. Maddock, 'Private agriculture in Eastern Europe', *Food Policy*, **18**(6), December 1993, p. 461.

Finally, flexibility in applying the options for transforming collective farms and respect for the members' wishes are important. As noted above, it is likely that corporate or co-operative structures would not be applied to all of a farm's assets, but rather only to some of them. This has usually been the case in Estonia. In addition, in Russia:

... agriculture will probably evolve toward a mix of various farming structures, ranging from [private cooperatives] through associations to family farms. The unifying feature of all these structures will be their foundation on private ownership of land and assets, which may be used collectively or individually, and free choice of the specific organizational form through voluntary regrouping of members with their land and asset shares.[200]

In the former Eastern Germany, a similar eclecticism has prevailed in regard to organizational forms, as collective farms 'became private enterprises under the ownership of groups of farmers (existing as limited-liability shareowners, joint-stock companies, and registered agricultural cooperatives)'.[201]

As the World Bank acknowledged in another statement in the same publication cited above:

... over sixty years of nonprivate farming in parts of the NIS has instilled a view that land is not a commodity like any other. . . . This has created considerable resistance to change. . . . One of the few specific mechanisms that has been implemented (on a pilot scale in Nizhniy Novgorod, Russia) is the internal auction. After an initial period of share distribution, public education, and asset valuation, participants bid their shares in auctions against the

farm's real assets. The farm is then liquidated, and the new enterprises created through the auction are registered. By mid-1995 sixty-eight farm enterprises had gone through this process. Out of five farms in the earliest stage of the program (1993–94), twenty collective [corporate and/or cooperative] enterprises, seventeen family farms, and six individual businesses were created. This is a promising beginning. . . . Whatever mechanism of initial privatization is adopted, the critical need is for freely functioning land markets.[202]

In the context of Africa, the World Bank has stated in another publication that 'Titles could also be provided to groups for collective ownership'.[203] Under such schemes, the collectivity is more efficient if each individual's economic participation (shareholding) is clearly defined in such a way that incentives for work and entrepreneurial behavior are provided. Moving in that direction usually leads to corporate structures or private co-operatives, although not always in the initial stage of transforming a system of customary rights.

The main lesson from Central America regarding the options for transforming collective farms is that *the members of the collectivity themselves should make the decision on their future form of asset ownership, provided that each option includes some form of private property and that the membership is fully informed regarding the nature and implications of each option*. Some choose to divide up the assets entirely into individual holdings, but many have chosen a mixed form, with some assets owned individually and others jointly through a corporation-like structure.

In broader terms, for developing *a strategy for transforming the ownership of the assets of collective farms*, there appear to be three optional paths for their evolution:

---

200. Reprinted from *Agricultural Economics*, **13**(1), K. Brooks and Z. Lerman, 'Restructuring of traditional farms and new land relations in Russia', p. 24, Copyright (1995), with permission from Elsevier.
201. G. H. Peters, 1995, citing Werner Schubert, p. 212.
202. The World Bank, 1997, p. 59.
203. The World Bank, 1989 (cited in Platteau, 1992, p. 296).

(i) *A breakup of all of the lands of the unit into individual, privately owned plots*. Under this alternative, which is commonly undertaken, complementary measures have to be designed to distribute the non-land assets of the unit (machinery, buildings, animals, etc.) among the membership. Animals and some equipment can be auctioned to the most interested farmers, but it is difficult to apply this procedure to major buildings without creating significant inequalities among the members of the former collective and disrupting their patterns of production. As previously noted, a common procedure followed in Bulgaria, Estonia and a number of other countries has been to provide all former members with shares which can be transformed into joint holdings in the non-land assets, normally in the form of joint stock companies (corporations),[204] but alternatively as private co-operatives. Members had the option of selling their shares to other members before they were exchanged for ownership in the assets. Many of the former members used the assets that were acquired in this way to form agricultural service enterprises, often jointly owned. In cases such as Mexico, Honduras and El Salvador, the decision to break up the unit or keep its lands together as one enterprise was left to the members of the farm, by law.

(ii) *Implementation of option (i) for individually tilled plots, along with the conversion of former communal lands into jointly held assets*, along the lines sketched out above for machinery and buildings.

(iii) *Conversion of all or part of the collective farm (land plus built-up assets) into a corporate structure or a private co-operative, titling the assets in the name of the enterprise*. In the corporate case, shares in the entire unit, land plus other kinds of capital, are distributed to members, who become the shareholders in a corporation. This kind of transformation can overcome problems of lack of full ownership and weak incentives. Under the new structure, the members will have a long-term interest in maintaining and increasing the value of their shares, which should encourage their interest in investing both labor and financial capital in the unit. The shares must be saleable in order for them to have economic value but, if desired, first rights of refusal can be given to the remaining shareholders when one of them decides to dispose of shares. Another advantage of 'corporatization' in this sense is that it can give members who are no longer interested in participating in the unit a viable exit opportunity, through sale of their shares (provided that the relevant legislation provides adequate support to this option). Under a collective regime, members who leave – and desertion of production co-operatives has plagued the agrarian reform in Honduras and other countries – do so without receiving any cumulative benefit for their time spent working in the unit. Members usually depart the production co-operative as poor as they joined it.

Forming a corporation or private co-operative which owns part of the farm's assets, option (ii) above, will be the preferred avenue in many cases. It has occurred on a widespread basis in Eastern Europe, both for whole collective farms and for parts of them. However, the new entity may be undercapitalized, and so it may be advisable to take the decision to sell some of the shares to outside investors. An option in this regard would be to create two classes of shares, each with its own set of rights and responsibilities.

These three basic options were presented to members of the production co-operatives of the

---

204. This was accomplished under the so-called 'agricultural reform' (as opposed to land reform) in Estonia. See Ministry of Agriculture, *Long-Term Strategy for Sustainable Development of the Agricultural Sector*, Ministry of Agriculture, Tallinn, Estonia, 1997, Chapter 3.

agrarian reform sector in Honduras, under a program titled 'Reconversion of the Agrarian Reform Sector'. Shortly thereafter, three similar options were presented to reform sector farmers in El Salvador in the early 1990s, under the name of the program of 'The New Options'. In both cases, it was stressed that the members of each production unit had to make their own decision on how they wished to proceed. In the majority of the cases, the members of the production unit chose option (ii), i.e. the mixed option.

Under any of the above options, members may wish to build on their experience of working together and take the additional step of forming a service co-operative, or second-degree co-operative. Such a co-operative can be limited to specified tasks, such as product marketing, or input purchase, or obtaining bank loans. It also can be made general, so that it can eventually undertake several activities. However, service co-operatives tend to be most successful when confined to one or very few activities, such as milk processing and marketing. It is vital to make the operating rules of the co-operative as simple and clear as possible at the outset.

Private production co-operatives will continue to be a relevant option, especially in Eastern Europe and the former Soviet Union, mainly because of the nature of the prior experience of the farmers. In Bulgaria, where both the corporate and co-operative option have been offered, it appears that most farmers have opted for the latter.[205] On the other hand, in Russia, 'virtually all farm enterprises registered in new shareholding forms and distributed shares in land and assets to their members and employees'.[206] A pattern similar to that of Russia was reported in Ukraine.[207]

5.8.8 Transition Issues in the CIS and Eastern Europe

A related lesson from Eastern Europe and the former Soviet Union is that *patience must be exercised with the process of transformation.* In the case of Ukraine, Peter Sabluk has commented that 'new forms of managerial arrangements in which share ownership is allied to self-governing mechanisms for the control of activity, and in which group initiative is rewarded according to results, were only slowly emerging'.[208] Similarly, 'Hungary, the veteran of privatization in Eastern Europe, managed during 1991 to transfer only 10% of all fixed assets [to private ownership], and is planning to privatize an additional 80% only over a 5-year period. Even that, however, is a short time span compared with the generation which Professor Lugachov estimated as being the period needed for the gradual cultivation of a spirit of entrepreneurship'.[209]

In Russia:

Reform is mainly proceeding through the reorganization of existing collective and State farms, in response to government decrees, directives, the setting of targets and of timetables for action, and a slowly emerging body of law. Nearly half have been reorganized into sub-cooperatives, collections of private farms, or closed joint stock associations. However, most of the units continue to rely on the old

205. A. Buckwell and S. Davidova, 1993, p. 497.
206. Reprinted from *Agricultural Economics*, **13**(1), K. Brooks and Z. Lerman, 'Restructuring of traditional farms and new land relations in Russia', p. 23, Copyright (1995), with permission from Elsevier.
207. Z. Lerman, K. Brooks and C. Csaki, 'Restructuring of traditional farms and new land relations in Ukraine', *Agricultural Economics*, **13**(1), October 1995, p. 37.
208. Reprinted from *Agricultural Economics*, **12**(3), G. H. Peters, 'Agricultural economics: an educational and research agenda for nations in transition', p. 199, Copyright (1995), with permission from Elsevier.
209. Reprinted from *Agricultural Economics*, **12**(3), G. H. Peters, 'Agricultural economics: an educational and research agenda for nations in transition', p. 205 (citing M. I. Lugachov), Copyright (1995), with permission from Elsevier.

kolkhoz or sovkhoz [collective farms or State farms] structure for the purchase of inputs and sales of output owing to the absence of organized markets. As yet there are few signs of improvement in land use practices.[210]

I. Lukinov of Ukraine has stressed that 'creation of more efficient economic structures is far from being a matter of passing a limited number of legal acts which do no more than remove the old system. It is a longer evolutionary process of forming new proprietorships, new institutions, new external trade relationships and a great deal of development from below'.[211]

The experience of Kazakhstan illustrates well the difficulties which can arise in the early years of a de-collectivization process, before adequate substitutes are found for the multiplicity of services that were offered to their members by collective farms:

Land reform has been the most [rapidly implemented] of Kazakhstan's economic reform packages. . . . instead of the 2500 State and collective farms in existence in 1991, at present there are more than 62 000 individual farming units, 8754 production cooperatives, 1169 business partnerships, 578 joint stock companies, and only 89 State enterprises. . . . The share of nonstate enterprises is 93.9 percent of all agricultural lands, 94.9 percent of arable land, and 91 percent of livestock and poultry. . . .

The positive changes in Kazakhstan's land allocation to individual farmers fail to reveal the huge losses and crises that resulted. . . . In reality, the reforms turned many efficiently

functioning large public farming enterprises into numerous small farms, most of which are not viable because they lack machinery and working capital and cannot adapt to market conditions. As a result, the owners of property and land shares again joined together and established production cooperatives on the basis of joint ownership. Further, social tensions in rural areas have dramatically increased because the social protection afforded by the previous system has weakened.

The areas under agricultural crops are decreasing. Crops covered only 21.8 million hectares in 1998, compared with 35.2 million hectares in 1990.[212]

Replacing a collective farm system entails much more than reassigning land rights. It requires developing a new agricultural extension system, new marketing channels, new modalities of input purchase, new sources of working capital, and new forms of social services in rural areas.

Lerman has documented a significant difference between the countries of Eastern and Central Europe (ECE) and the CIS in transforming the economic structures of co-operative production units:

. . . the new market sounding names [joint-stock societies, limited liability partnerships, etc.] more often than not hide an internal structure which is basically unchanged since the Soviet times. Survey data for the CIS (Russia, Ukraine, Moldova) reveal persistence of traditional management and organization features:

---

210. Reprinted from *Agricultural Economics*, **12**(3), G. H. Peters, 'Agricultural economics: an educational and research agenda for nations in transition', p. 196 (citing C. Csaki and S. Johnson), Copyright (1995), with permission from Elsevier.
211. Reprinted from *Agricultural Economics*, **12**(3), G. H. Peters, 'Agricultural economics: an educational and research agenda for nations in transition', p. 201, Copyright (1995), with permission from Elsevier.
212. Reprinted from *Food Policy*, **25**(6), A. Baydildina, A. Akshinbay, M. Bayetova, L. Mkrytichyan, A. Haliepesova and D. Ataev, 'Agricultural policy reforms and food security in Kazakhstan and Turkmenistan', pp. 734 and 738, Copyright (2000), with permission from Elsevier. However, much land cropped in the Soviet Union was unsuitable for agriculture and should have been left as pasture. The reduction of cropped areas is not always bad.

the restructured farms retain a strong central management apparatus, and the functional subdivisions have only token autonomy beyond general production planning. Specifically, finances and labor relations are handled by the central management, and not by the functional units. The majority of member-workers in large-scale farms in CIS report that nothing has really changed in their farm enterprise as a result of restructuring. Even farms restructured as part of international donor projects in CIS often strikingly resemble their collective predecessors. Interesting changes of farm organization are observed in Moldova and Azerbaijan, where many large farms are breaking up into independent multi-family units that occupy an intermediate position between individual farms and former collectives.

In ECE, there appears to be a more significant departure from the old collective-management pattern. Although no comprehensive data are available, case studies suggest that in Hungary, the Czech Republic, and Lithuania many of the large farms have transformed into market-driven corporations. In Romania, at least some of the large farms are new associations or cooperatives created voluntarily by individual landowners after the completion of land privatization. The large corporate or cooperative farms in ECE are now often forced to operate under hard budget constraints, with a real threat of bankruptcy proceedings in case of default. In CIS, neither budget constraints nor bankruptcy laws are enforced.[213]

Education and dissemination of information can speed up this process of development. In Russia, survey results indicated that information has been a major bottleneck to farm restructuring, and that the desire to restructure by the membership is not enough.[214] As in most areas of agricultural development, the factor of human capital is the most important one, and developing managerial capabilities and other skills appropriate for a market economy takes time.

Land tenure regimes are still in flux throughout Eastern Europe and the former Soviet Union, with significant differences between those two sub-regions. In most countries of the region, land markets are still very limited in their scope. The degree of change in land tenure regimes in the region, and the obstacles to further change, have been summarized very well by Renee Giovarelli and David Bledsoe. They highlight issues that are relevant to other regions of the world as well:

In the Western CIS countries, except Moldova, much of the agricultural land has been privatized under a 'land share' system, in which a large majority of private owners (former members of the state and collective farm system) still hold their rights in common, with some form of right to partition land in kind (as yet unexercised). In those countries, the right to a land share has little value because there is little chance to exercise meaningful control over that land share. [This experience illustrates the point made above about the need to make the shareholdings saleable in the privatized or restructured cooperatives.]

In the Transcaucasus, land was distributed and farms restructured at the same time. Private farmers have title to their land. Balkan privatization is advanced. . . . In the EU accession countries, the privatization issues are related to the restitution process primarily [which has] slowed down the privatization process.

During the privatization process, some countries have been slow to privatize state land, and instead lease out that land. . . . *An ongoing concern with leasing of state-owned land is that it is often leased at very low rent levels, thus undercutting the development of private market rents.* . . .

Lack of farm reorganization is an impediment to market development in the four

213. Z. Lerman, 1999, pp. 7–8.
214. K. Brooks and Z. Lerman, 1995, pp. 23–24.

Western CIS countries [Belarus, Russia, Ukraine, Moldova] and many of the EU accession countries that restituted agricultural land to its former owners. . . . *Providing a legal and policy framework in which individual farmers can adjust farm size to respond to market signals is crucial.* The policy and legal framework should not only allow, but also encourage farm reorganization into units of whatever size is chosen by farmers. . . .

Purchase and sale transactions will not occur in the Western CIS countries unless legislation clearly allows such transactions, permission of local bureaucrats is not needed, and procedures for notarization and registration are simple and affordable. . . . *Land-based lending will not occur until there is an active agricultural land market* and for Russia, Ukraine, and Belarus major changes in legislation and political will would have to occur first. . . . In fact, mortgage lending to any great extent will not occur until: (1) an active land market exists and agricultural land has market value; and (2) foreclosure procedures are reasonably quick and effective. . . .

The Western CIS countries have seriously under-functioning registration systems. . . .[215]

5.8.9 Policies to Promote Co-operative Forms in Agriculture

The diversity and persistence of associative forms of cultivation throughout the world underscores the fact that agriculture is a sector which, more than most, is characterized by co-operation among producing agents. Labor exchanges, for example, are traditionally practiced on farms in all continents. As Platteau has expressed it:

. . . even when individuals have to perform the same task (for example, preparation of the field, sowing or harvesting) and there are no labor indivisibilities . . . team work may be

more productive than 'isolated' production. This is because of the morale-boosting effects of group work: there are indeed good grounds for believing that individuals derive a positive utility from working in company, especially if the tasks to be performed are arduous and toilsome (an effect completely ignored by economic theory), and that team work increases the speed of operations and the intensity of effort. That these two effects may play an important role is probably evident for all those who had the opportunity to see large groups working in the fields of Africa, generally under the stimulus of pace-setting songs sung in unison by all the team members. In fact, the prevalence of reciprocal work groups throughout the whole African continent – e.g. the 'ploughing companies' (Kampani) or the kafo work groups in Lesotho and the Gambia, respectively – may to a large extent result from the significance attached to such efforts in the African cultures. . . .[216]

These forms of collaboration are not confined to Africa. In South Korea, traditionally rice has been transplanted in teams formed of community members, usually accompanied by musicians with traditional instruments and a banner-bearer. In frontier agriculture of the United States, community collaboration in 'barn raising' for each family was a firmly rooted tradition.

Service co-operatives, as noted earlier, are also useful forms of co-operation, especially in light of the economies of scale in input purchase, storage, marketing and provision of machinery services. Farmers' associations also play important roles in many countries, for example, in facilitating the provision of farm advisory services, veterinary services and other services for their members.

Given the widespread desire for and usefulness of co-operation in agriculture, a constructive role for policy is to support its spontaneous manifestations and to direct it into the most productive

---

215. Renee Giovarelli and David Bledsoe, 'Land Reform in Eastern Europe', report prepared for the FAO, The Rural Development Institute, Seattle, WA, USA, October 2001, pp. iii–iv [emphasis added].

216. J.-P. Platteau, 1992, p. 271.

channels. *Modes of co-operation also need to be given juridical and institutional support, and in some instances transitional finance to initiate the new co-operative structures.* In this regard, in many countries it would be appropriate to develop not only a strategy for land tenure but also a strategy for reinforcing modes of co-operation within the sector, particularly in regard to farm services. For land tenure, a long-term view is necessary, particularly when the issues include the evolution of communal and collective land rights. More than one stage may be required in the development of a new system of these rights, and new administrative structures associated with land rights have to be created and consolidated along the way.

It is difficult to form effective and independent corporate structures or other entrepreneurially oriented co-operative units in agriculture without appropriate legal structures and a clear definition of land rights and factor markets. This lesson emerges from Zhang Xioashan's recent review of Chinese experience with market-oriented agricultural production co-operatives:

In terms of the legal framework, there is no cooperative law in China and, in company law or legislation concerning other economic organizations, there are no articles or chapters specifically about cooperatives. Since the regulations or charters issued by the ministries concerned do not have the authority of law, cooperatives cannot be formed and operated under the law, nor can they be protected by the law. To a large extent, the survival and development of rural cooperatives depends on the policies issued and implemented by local governments, i.e. they are dependent on people, not law. . . . It was probably for this reason that rural cooperatives used to invite local leaders to take the position of director-general or chairman, and such a phenomenon is now seen in [other associative forms]. This inevitably

has a great influence on the independence of cooperatives.

. . . in China in the past, the centrally planned economic system, together with a household registration system that separated urban from rural residents, hindered the free movement of production factors. In addition, the markets for these factors are not yet mature, the foundation on which to build free combination of resources is still weak and it would be unfavorable for cooperatives to grow independently. For instance, since there is no sound market for land transactions, if a cooperative wants to purchase the use right to rural collectively owned land, it has to obtain permission from the many government institutions concerned. . . . Another case regards the credit supply. In China . . . banks and credit cooperatives, to a certain extent, are still affiliated to government and have not become independent financial organizations. . . .

For these reasons, at this stage of Chinese development, while a sound market system has not yet been established, the production factors market is still imperfect and the legal foundation defective, it is a rational decision for cooperatives to seek support and protection from government. Cooperatives that are completely self-organized and have no government support have to pay high transaction costs and spend a lot of time and effort doing business on their own. . . . The protector–protected relationship is useful and beneficial for both sides. . . . As the market economy develops and cooperatives grow, the costs of direct transactions between cooperatives and the market will decline. When the costs of government protection (the loss of a degree of independence) equal, or even exceed, the benefits of being protected (reduced transactions costs), the protector–protected relationship between rural cooperatives and the State might come to an end.[217]

---

217. Zhang Xioashan, 'Cooperatives, communities and the state: the recent development of Chinese rural cooperatives in transition', *Land Reform*, Food and Agriculture Organization of the United Nations, Rome, 1999/1–2, pp. 103–104.

As in other areas of economic policy, an adequate legislative foundation is a prerequisite – although not a sufficient condition – for making progress.

5.9 IMPROVING ACCESS TO LAND FOR THE POOR AND FOR WOMEN

5.9.1 Land Market Mechanisms and the Rural Poor

The main barrier to increased land ownership on the part of the rural poor is imperfections in financial markets, i.e. lack of mortgage finance. A barrier to their retaining the land is the lack of adequate mechanisms for insuring against risk. Associated with the financing problem there is another barrier in the form of the price of land. It has been pointed out by Binswanger *et al.* (1995, pp. 2710–2711) that the price of land is always greater than its expected returns (in the form of a discounted stream of annual returns) because it has other sources of value: a guarantee for obtaining credit (even for uses unrelated to the land), and prestige which can translate into political influence. Hence, for a farmer who wants land for purposes of production, purchasing it at market prices with financing often is not a viable option, for the annual cost of servicing the mortgage may be greater than the annual profits from using it. In addition, if the land were to be fully mortgaged, then it could not be used as collateral for production credit to obtain modern farm inputs.

Under most circumstances, a poor rural family without financial resources would not be able to purchase land. This fact accounts for the widespread recourse to State land development schemes, under which new land is made available to settlers at little or no cost. Along with the historical concentration of landholdings in many countries, it also explains why so many expropriative agrarian reforms have been carried out.

Can access to land for poor families be improved in the context of land markets? Attempts are underway all over the world to promote access to land for the rural poor via market mechanisms. Some of them are promising, but to keep matters in perspective it should be noted that to date 'there are few experiences where land markets have been used successfully for small peasants' land purchases. Lack of capital and unfavorable access to credit hamper the possibilities of land purchasing for these peasant groups'.[218]

Basically, there are nine possible approaches to improving access to land in a market setting, and they can be implemented individually or in concert. These are described in the following sections.

5.9.1.1 A Land Fund or Land Bank

A land fund or bank operates in land markets, facilitating the purchase of holdings that are for sale, often in the form of smaller parcels, by qualifying low-income families. It operates on the basis of voluntary sales by landowners. John Strasma has pointed out that land funds should not be government entities, nor should they become directly involved in the purchase and sale of land.[219] Direct governmental involvement in a land fund generates political pressures on it to offer an attractive price to large landholders that are selling and virtually makes its situation untenable. A temporary land bank that was created in El Salvador, to help honor the commitments of land that were made to ex-combatants at the end of the civil war, was government-owned and directly bought and sold land. Its President admitted to this author that he encountered difficult pressures of this nature.

Land funds are most effective if they operate as second-storey institutions, refinancing the land transaction. In this manner, they can operate

218. Land Tenure Service, FAO, 2001, p. 51.
219. John Strasma, 'Analysis of Land Markets and Land Banks', in Instituto Nacional Agrario and Land Tenure Center of the University of Wisconsin, *Seminar on Access to Land, Tenure Security and Investment in the Productive Use of Natural Resources*, Tegucigalpa, Honduras, 1990, in particular, pp. 32 and 33.

through a wide variety of intermediaries, including banks, non-bank financial institutions, NGOs, and even large landowners who decide to sub-divide their land and sell it. Non-financial institutions would have to post bonds as guarantees of their performance.[220]

A necessary condition for a land bank to function successfully is a source of subsidy. For the reasons mentioned above, low-income rural families cannot be expected to purchase land at market prices, and therefore a subsidy element must be planned in the operation of the land bank, and its source of financing correspondingly identified. Principal options include revenues from sale of State lands, revenues from land taxes, the regular government budget and contributions by international development organizations. The subsidy is translated into lower mortgage payments for those who acquire the land, or lower down payments or both. It is considered important that recipients of land pay something for its acquisition, if nothing else so that they may value the land psychologically, and also to reduce the temptation to realize windfall gains by selling the land received through such a program. Deininger and Julian May have commented that, in the context of market-assisted land reform in South Africa, requiring a contribution from beneficiaries does not appear to discriminate against the poor.[221]

So far, there are few cases of land funds in operation. The FAO has noted that 'Land funds experiences in Ecuador, Colombia, Brazil and Guatemala are presently trying to develop support strategies to facilitate land purchasing and land cultivation by the rural poor'.[222] Honduras is initiating an effort in this direction also, with external funding for the subsidy component.

Gustavo Gordillo de Anda has analyzed the experiences with land funds in Guatemala (FONTIERRA) and El Salvador (Banco de Tierras). His findings reinforce the points that contributions by beneficiaries are important, and therefore these instruments probably cannot assist the very poorest families in rural areas; that community organization is an important prerequisite for the functioning of land funds; that titles to the properties should be completely clear before they are transferred to the beneficiaries; and that these kinds of programs can be expensive:

Lessons learned from FONTIERRA: In principle . . . not all poor farmers are eligible to be beneficiaries of this program. Only those who fulfill certain characteristics (certain level of capitalization at the moment of entering the program, substantial degree of organization in community groups, experience with profitable crops) qualify to participate in the program.

. . . the program appears to respond to a specific type of farmer who is not necessarily the poorest, which suggests the necessity of other types of programs to meet the needs of the poorest rural population. . . . Owing to the relatively high cost of the program, it is clear that its potential coverage is limited. . . .

Lessons learned from the Banco de Tierras in El Salvador: . . . those [farmers] who lack resources and access to support networks would find it difficult to take advantage of the distributions of land . . . . From the perspective of property rights, the transfer of lands without previously 'cleaning up' their legal status creates serious difficulties at the moment of subdividing and titling the new plots in favor of the beneficiaries. . . . [which] illustrates the

220. Many of the operating concepts of a land bank can be found in John Strasma, Ricardo Arias, Magdalena García, Daniel Meza, René Soler and Rafael Umaña, *Estudio del Diseño Conceptual de un Fondo de Tierras en Honduras*, prepared for the Government of Honduras, Proyecto APAH, Tegucigalpa, Honduras, 1993.
221. Klaus Deininger and Julian May, 'Is there Scope for Growth with Equity? The Case of Land Reform in South Africa', mimeo, The World Bank, Washington, DC, USA, January 2000, p. 18.
222. Land Tenure Service, FAO, 2001, p. 51.

importance of strengthening the institutions linked to land administration (cadastre, property registries). . . .[223]

5.9.1.2 Market-Assisted Land Reform

Market-assisted or negotiated land reform is similar to the operation of a land fund, but it encompasses a broader range of options. The subsidy for land purchase may be given directly to poor families, instead of being channeled through a financial institution in the form of lower mortgage payments. In addition, the community may participate in selection of the lands to be sold, and in carrying out the negotiations with the owners. Land sales are voluntary but clearly there exists the potential for an element of local pressure to sell to be placed on some landowners. Another feature of market-assisted land reform is that beneficiaries may be required to submit a proposal for development of the farm to be purchased, and credits or grants may be given for investments on the farm, as well as for its purchase.

Deininger (1999) has written a detailed but preliminary review of three experiences in negotiated land reform, in Colombia, Brazil and South Africa. In the Colombian case, up to 70% of the value of the land purchased may be subsidized – 20% in cash and 50% in bonds. In South Africa, a grant of up to R15 000 is provided for the purchase, although consideration is being given to increasing that amount substantially. Special emphasis is placed in Colombia on the development of a viable farming project proposal, and it is a condition of selection of the beneficiaries. In both Colombia and Brazil, the subsidy element of the land purchase is channeled through a commercial bank, as in the case of land funds.

The purposes of market-assisted land reform go beyond simply redistributing land. Deininger

has drawn out some valuable lessons from these early experiences of the three countries:

(a) land reform through negotiation can only succeed if measures are taken to make the market for land sales and rental more transparent and fluid;
(b) productive projects are a core element of market-assisted land reform [since it should be] designed to establish economically viable and productive projects at a socially justifiable cost rather than to transfer assets;
(c) the only way to achieve effective coordination of the various entities involved in this process is through demand-driven and decentralized implementation;
(d) the long-run success of land reform is likely to depend critically on getting the private sector involved in implementation. . . .

Other strategic findings in Deininger's research concern the importance of:

co-financing . . . the land purchase through a private financial intermediary which, because it shares in the risk of default, will have an incentive to assess the economic feasibility of the proposed farming project [and the fact that] negotiated land reform is a complement, rather than a substitute for other forms of gaining access to land, especially land rental.[224]

The caution mentioned in Section 5.7, i.e. that any kind of land reform inevitably must proceed slowly because it depends on the development of institutional capacities at the local level, is applicable to market-assisted land reform. As noted above, in Colombia market-assisted land reform encountered a number of implementation problems, including difficulties in selecting beneficiaries, and effectively has come to a halt. Efforts to

---

223. Gustavo Gordillo de Anda, 'Un nuevo trato para el campo', paper presented at the *International Conference on Access to Land: Innovative Agrarian Reforms for Sustainability and Poverty Reduction*, Bonn, Germany, May 19–23, 2001, pp. 24–25 and 28–29 [author's translation].
224. K. Deininger, 1999, p. 666.

strengthen administrative capacities at the local level would appear to be essential to further applications of this approach.

5.9.1.3 *Improving Land Rental Markets*

As earlier comments in the chapter indicated, rental markets offer important opportunities for poor farm families to increase their access to land, and they also favor increased productivity in farming. The advantages of improving rental markets, and the need to provide assistance so that poor families can participate in those markets, have been stated clearly by Alain de Janvry, Elisabeth Sadoulet and Karen Macours:

Land rental markets offer vast unused opportunities to attack poverty and improve efficiency. Access to land in rental is cheaper than through the land sales market and less politically demanding than access to land through expropriation. Making land rental markets work requires 'assistance' on both the supply and the demand sides of the market. On the supply side, more land will be offered in rental if security of property rights is increased, obsolete land reform legislation that threatens property rights is voided, and reliable conflict resolution mechanisms are put into place. On the demand side, special grants need to be made to poor and young potential tenants to qualify them for the land rental market and make land rental as attractive as land buying under community assisted land reform. Innovative schemes can be explored to enhance access to rental for the rural poor, such as group rental, rental with option to buy, provision of a menu of alternative contracts from which to choose, and the design of contracts that offer provisions to compensate for the residual value of investments made by the tenant. . . .[225]

Improving the functioning of rental markets in these and other ways represents one of the potentially most valuable, and least explored, ways of increasing access to land for the rural poor and, as these three authors emphasize, if properly carried out it will not weaken tenure security for existing landholders. Given the value of land rental for low-income farmers, it is all the more important to ensure that existing customary rental rights are protected when traditional lands are subjected to a titling program.

5.9.1.4 *Lease-Purchase Contracts*

This kind of contract was mentioned above in the context of long-term leases of State lands, and they also can be applied to the programs of a land fund or market-assisted land reform or other programs designed to endow low-income families with land. The provisions would specify that after a family has worked the land for a given number of years, the cumulative lease payments would be credited (retroactively) to the down payment on purchase of the land, and remaining lease payments would be converted into mortgage payments. The need for a subsidy content in the land price still applies under this option.

5.9.1.5 *A Progressive, Area-Based Land Tax*

Policy makers concerned about the presence of idle land sometimes advocate a tax on land that is unused. However, a tax designed in that manner would invite problems in its implementation. As John Strasma observed in a University of Wisconsin seminar on land tenure in Honduras, in practice a punitive tax on idle land would encounter serious difficulties because it would be subjective, would be difficult to implement, and would be less effective than a tax applied to all land because the latter avoids the need to pass judgment on the degree of effective cultivation.[226]

225. Alain de Janvry, Elisabeth Sadoulet and Karen Macours, 'Land Policy and Administration: Lessons learned and new challenges for the Bank's development agenda', comments provided to the World Bank, Washington, DC, USA, March, 2001, p. 4.

226. John Strasma, 'Analysis of Land Taxation', in Instituto Nacional Agrario and University of Wisconsin Land Tenure Center, 1990, p. 42.

Assessments of the market value of rural lands are equally subjective in practice, and the arbitrariness of the assessments has been a major hurdle to the widespread implementation of a rural real estate tax. An alternative is a tax on all agricultural land, based on size of holding, and such a tax could be made progressive by exempting from taxation the first few hectares of each holding. The usefulness of this type of tax in stimulating an intensification of land use has been widely recognized; for example, in Bulgaria it was found that:

> There is . . . evidence that some owners will seek to hold land not for productive purposes but as an asset and a hedge against inflation. For this reason a land tax may be necessary to encourage the sale of such land.[227]

In Latin America:

> Because land taxes were low or difficult to collect, land was not farmed and was held for speculative purposes in Brazil and Costa Rica.[228]

Plus, in general:

> Heavy taxes on land . . . may induce more efficient utilization of existing assets and raise the level of output, while at the same time reducing inequality. The importance of such arguments can be determined only by a close examination of the situation in each particular country.[229]

The potential for a land tax to encourage a redistribution of land in favor of the rural poor has been noted by John Strasma and Rafael Celis:

> An annual tax on rural land. . . . is a fixed expense that raises the cost of holding property idle for speculation. Any significant tax [rate] is relatively onerous to an owner who earns little income from his land. If he invests and works to increase its productivity and income, the land tax does not increase, so the burden becomes relatively lighter. . . . The . . . tax should encourage owners to rent or sell land from which they derive little income. The increased supply of land will naturally depress land prices or make sellers offer better terms. As a result, the relatively poor will find it easier to buy or rent land with which they can raise their income by much more than the yearly tax.
>
> The land tax also is preferable to a tax on agricultural production or exports, from the government's point of view. Most tax incentives cut revenues. Land taxes can generate substantial revenues – yet they do not penalize the most productive, as do taxes on agricultural commodities or exports.[230]

Similar observations have been made by Mohan Rao:

> An interesting possibility is that the tax raises the yield of land and improves economic efficiency. This is based on an inverse relationship between farm (landholding) size and land yield that is generally observed in underdeveloped agriculture. The land tax may compel large farms to improve yields or to transfer – through sale or lease – their land to medium-sized and small farms. Whether such effects will occur will depend on whether the tax can modify the factors causing the lower yields. Credit constraints on smaller farmers and the noneco-

227. M. Hristova and N. Maddock, 1993, p. 461.
228. The World Bank, *Renewable Resource Management in Agriculture*, Operations Evaluation Department, The World Bank, Washington, DC, USA, 1989, p. 37.
229. Bird, Richard M., *Tax Policy and Economic Development*, p. 9, Copyright (1992), The Johns Hopkins University Press. Reprinted with permission of The John Hopkins University Press.
230. Reprinted by permission of Transaction Publishers. 'Land Taxation, the Poor, and Sustainable Development', in S. Annis, O. Arias, J. B. Nations, S. B. Cox, A. Umaña, K. Brandon, S. K. Tucker, J. D. Strasma and R. Celis (Eds), *Poverty, Natural Resources and Public Policy in Central America*, by John D. Strasma and Rafael Celis, pp. 149–150, 1992. Copyright (1992) by Transaction Publishers.

nomic behavior of larger ones are two factors that may respond to the tax.[231]

A skeptical view of the potential of a land tax to encourage a redistribution of land was put forth by Binswanger *et al.* Their first argument consists of citing a study for India which concludes that a land tax will increase the concentration of landholding, because in bad years the poor may be forced to sell land to meet the tax obligation.[232] However, that study did not take into account the option of exempting the first few hectares of each farm from payment of the tax, and therefore would not be applicable for such a variant. Their second objection is that 'dummy divisions of their holdings' by landowners would diminish the effectiveness of a progressive tax. There always would be an incentive for such evasion but if, say, the first 5 hectares are exempt from payment of the tax, it would be very costly to make sufficient fictitious divisions of a 500-hectare holding to substantially lighten the tax burden. Thus, while the effect of the tax could be attenuated somewhat by evasions, the tax would still be expected to operate in the direction of encouraging intensification of land use and some sales of parts of the less efficient large holdings.

In light of the struggles, sometimes violent, and policy distortions that have been associated with land reforms in many parts of the world, their third argument – 'it is not obvious why such an indirect approach would be politically more acceptable than direct redistribution of land' (p. 2724) – must be regarded as unsupported by empirical research. Finally, their fourth argument (*ibid.*) is that administering a land tax is likely to be very costly. A response to this argument is that *a per hectare tax (area tax) is much easier to apply than an* ad valorem *tax*, especially if relatively few categories of land are defined, in order to avoid disputes over the classification of a parcel. Cadastral surveys and land registration systems are pro-

ceeding in many countries, and local land taxes exist (usually at low rates) in a large number of countries, and so administratively the step to a land tax applied on a per hectare basis is not larger than that associated with any other kind of tax. It can be argued that such an area-based tax would in fact be easier to collect than an income tax in rural areas, where record-keeping is almost non-existent. The land tax could be made deductible from the income tax obligation, and so for most farmers it would become the amount of income tax paid, *de facto*.

In spite of their objections, Binswanger *et al.* concluded that 'flat or mildly progressive land taxes based on rough classification of holdings may still be useful for raising revenue and providing some modest incentives for owners to sell off poorly utilized land'.[233] In a more recent publication, they suggest a land tax could have wider benefits:

... governments can consider imposing a land tax and establishing land information systems. A land tax that is enforced at the municipal level not only could provide an incentive to large landowners to utilize their land more productively but could also make an important contribution to decentralization. On the one hand, a land tax is one of the few cases of a lump sum tax where – using asset, rather than production, values – the effective tax rate decreases as the income generated from land increases, thus encouraging more productive use of the resource. Several countries are currently experimenting with a land tax, either using a flat tax rate as in Nicaragua or basing land taxes on self-assessed values as in Chile. ... Land taxes have proven very useful in a wide range of urban contexts in developing countries and – if accompanied by appropriate institutions to help with accounting and implementation – should be feasible in rural ones as well.[234]

---

231. J. M. Rao, 'Taxing agriculture: instruments and incidence', *World Development*, **17**(6), 1989, p. 813.
232. H. Binswanger, *et al.*, 1995, p. 2724.
233. *Op. cit.*, p. 2724–2725.
234. K. Deininger and H. Binswanger, 1999, p. 265.

Strasma and Celis add that implementation of land tax would promote a clearer definition of property rights and, with exemptions for forested land, could be used to slow the rate of deforestation and encourage reforestation. They also note that such a tax could support the development of community infrastructure and help finance a land bank.[235]

They conclude that the main barriers to implementation of land taxes have been resistance by large landowners, lack of political will and indifference to the concept on the part of international development agencies, but that situation is changing, and 'the World Bank and some other aid agencies urged various countries to adopt land-tax reform as a tradeoff for lower export taxes and lower taxes on farm income'.[236]

Mahmood Hasan Khan, in a recent survey of agricultural taxation, makes the following observations about a land tax based on area, as opposed to one that is based on value of land:

The use of land area, with minor adjustments for differences in crops, soils, source of irrigation, as a tax base is convenient for administrative reasons. The major disadvantage is that land area, whether taxed on a flat or graduated rate, bears no relation to the land value as real estate or wealth. However, the assessment of 'fair' market or rental value of any property, particularly agricultural land, poses serious problems. . . . In most developing countries, the cadastral requirements for property taxation, even in urban areas, are difficult to meet. A block or group approach to agricultural land can overcome some of the valuation problems and reduce the cost of administration. Similarly, as an alternative to regular assessment it is possible to use an appropriate price index as a guide to changes in the value of land.

. . . there are a number of tax options that governments can use to mobilize additional resources, provided that they can address the political and administrative aspects of tax reforms. In most developing countries, these constraints have hampered the development of a rational and equitable tax regime affecting those in the agriculture sector who own or control large areas and have been the major beneficiaries of public investment, input subsidies and credit programs. . . . Governments have, however, started taking steps to reform their tax systems as part of their structural adjustment and economic reform programs. Both internal and external political and financial pressures are apparently encouraging them to overcome the existing political and administrative constraints.[237]

Khan's survey shows that the following countries are now collecting a land tax in rural areas, based on size of holdings rather than value of property: Ethiopia, Bangladesh, India, Malaysia, Pakistan, Sri Lanka and Egypt.

An area-based land tax has the potential to raise revenue for rural development and other programs, to strengthen the revenue base of local government, to encourage more productive use of agricultural land, to encourage sustainable forest management (through exemptions for forested land), and to provide incentives for a redistribution of large landholdings to smallholders. The latter can be facilitated by the operation of a land fund or market-assisted land reform which will subsidize the purchase of land by poor families. It is not possible to make a proper assessment of the contributions of an area-based land tax in these domains, since there has not yet been much cumulative experience with a systematic application of the approach – with few differences in category of land for tax rate purposes, exemptions for small farms, and no other loopholes. In light of its potential benefits, it is likely to be explored by an increasing number of countries. It is an approach

235. See the summary of their paper on p. 39 in the cited volume edited by Annis *et al.* (1992).

236. Reprinted by permission of Transaction Publishers. 'Land Taxation, the Poor, and Sustainable Development', in S. Annis (Ed.), *Poverty, Natural Resources and Public Poverty in Central America*, by John D. Strasma and Rafael Celis, p. 151, 1992. Copyright (1992) by Transaction Publishers.

237. M. H. Khan, 'Agricultural taxation in developing countries: a survey of issues and policy', Reprinted from *Agricultural Economics*, **24**(3), pp. 525 and 527, Copyright (2001), with permission from Elsevier.

worth pursuing more assiduously, but it must be stressed that its implementation requires considerable political will. The FAO's Land Tenure Service has commented on the lack of success of several experiences with land taxes, owing to the political factor. It also points out that in Brazil 'new regulations for a rural land tax ... were approved by the Congress in December 1996. With a stronger political support, the Government is now enforcing the application of the [land tax]'.[238]

Table 5.2 Productivity of land by farm type in Estonia (data from National Strategy for Sustainable Agricultural Development, Ministry of Agriculture, Tallinn, Estonia, 1997, Chapter 3, Table 3.9)

Type of farm	% of arable land in production[a]	% of gross agricultural output[b]
Household plots	22.2	34.0
Private farms	29.9	17.6
Farm enterprises	48.0	48.4

[a] As of January 1996.
[b] For 1995.

5.9.1.6 Titling Small Farms

In some cases, one of the barriers to property ownership for the rural poor has been legislation which prohibits the granting of title to small farms. In Honduras, for example, until the Law for Modernization and Development of the Agricultural Sector was passed in 1992, it was expressly forbidden to title farms with less than five hectares of land, even though the vast majority of farms in the country fell into that category. The result was to place smallholders in a category of second-class citizens, without access to the property rights enjoyed by their more fortunate neighbors.

In Estonia, the existing land reform legislation does not provide for titling the household plots, or former subsidiary plots of the State and collective farms. However, such plots are significantly more productive per hectare than either the large farm enterprises (direct successors to the State and collective farms) or the newly constituted private individual farms, as shown by the data presented in Table 5.2. Similar figures could be cited for many countries (see the example of Hungary above). The smallest farms (household plots) achieve their greater productivity in part by specializing in high-value crops, but that does not diminish their achievement. Failure to title them makes it more difficult for this class of farmers to emerge from poverty. It is sometimes argued that the household plots are too small to be technologically efficient, but they exist and they are economically efficient. They can be even more efficient with titling. Perhaps more importantly, titling them will set in motion a process of market dynamics which will lead to a gradual consolidation of the small units into larger ones, as some of their owners decide to leave for urban life, or as their heirs decide not to continue in agriculture. Without titling, and therefore without the prospect of realizing some economic benefits by selling the farm, such dynamics can hardly occur. In essence, it is a question of whether an 'optimal farm' size can be imposed on the private sector (in effect, by assuming the smallest farms will disappear without a full legal basis) or whether attaining that size should be left to market processes over time. It also is a question of designing poverty alleviation programs that increase the capacity of the poor to generate higher levels of income, and titling their *de facto* landholdings would be an important component of such programs.[239]

238. Land Tenure Service, FAO, 2001, p. 51.
239. The consequences of a policy bias against small farms have been underscored by Binswanger: 'Communist countries as well as many market economies have paid an enormous price for assuming – without much empirical evidence – that large farms are more efficient than small ones'. (Reproduced from H. P. Binswanger, 'Patterns of Rural Development: Painful Lessons', in J. van Zyl, J. Kirsten and H. P. Binswanger (Eds), *Agricultural Land Reform in South Africa: Policies, Markets and Mechanisms*, Oxford University Press, Cape Town, South Africa, 1996, p. 20, with permission of the Oxford University Press, Southern Africa.)

While titling small farms does not provide expanded access to land to poor families, it does improve the quality of their landholdings and consequently their economic status. In implementing such titling programs, the remarks made earlier about the possibility of subsidizing part of the costs of titling should be borne in mind.

5.9.1.7 Alternative Financial Mechanisms in Rural Areas

Worldwide, there is increasing interest in new modalities of lending to the rural poor, and of loan supervision for those clients (see Chapter 7). While the value of land ownership as collateral has been mentioned frequently, it should be recognized that in reality, in many countries, the social and political context makes it difficult for banks to foreclose on small farms. In any event, to pursue that option 'opens a huge avenue leading to landlessness and land concentration'.[240] Therefore, it is incumbent on lending institutions to develop other methods of securing loans – group lending is a common example. An alternative which deserves wider consideration is an approach in which a borrower assigns the right to a lender, in the event of default, to take over the former's land for a season or two, and reap the returns from cultivating it, rather than to seize the land. The lender could hire the borrower and his family as wage laborers for that period of time, thus providing them with an income equivalent to the minimum wage for part of the year. Equally, anyone else could be hired for that purpose. Obviously, the limits on loan size would have to be dimensioned with the potential harvest returns in mind, but such a procedure would represent an alternative way to provide a lending institution with the requisite guarantees without the threat of creating more landlessness among the rural poor.

In real estate law, this kind of arrangement for collateral is known as *antichresis*. It has foundations in Greek and Roman law. It has not always been regarded favorably in history; in an earlier era, the Catholic Church considered it to be usurious and inveighed against it. Originally, it had the legal status of a clause in standard real estate contracts, in which title to the assets was also pledged. In many modern European legal codes, it has been given the character of an independent contract. In the Italian code, it is a contract that conveys only the right of usufruct to the creditor. In the French, Spanish and German codes, it is also considered to implicitly convey the right of foreclosure if it is necessary to recover all of the credit. (Antichresis has not been accepted in any form in Swiss law.) Banks traditionally have been reluctant to enter into such contracts, not wishing to become involved in property administration. Nevertheless, it can be argued that contracting out the administration of a farm for a year or two is less onerous than foreclosure and sale of the property, particularly when there are doubts about whether the judicial and political systems would support foreclosure of small farms in low-income areas.

A form of antichresis has been implemented in Bangladesh, as a way to enable small farmers to provide security for loans:

> The most popular form of land mortgage found in the study area is transfer of user rights in land in exchange for cash, with the stipulation that such rights will revert back to the owner once repayment is completed. . . . A variant of this system is sometimes encountered in which the land reverts back to owner control automatically at the end of a stipulated period.[241]

Wider exploration of this mechanism for guaranteeing agricultural loans would appear to be

---

240. J.-P. Platteau, 1992, p. 127.
241. K. A. S. Murshid, 'Informal Credit Markets in Bangladesh Agriculture: Bane or Boon?', in G. H. Peters and B. F. Stanton (Eds), 1992, p. 660.

warranted, in the interest of both facilitating more lending to agriculture and protecting land rights of the poor.

5.9.1.8 Eliminating Subsidies that are Skewed in Favor of the Non-Poor

The ability of low-income rural families to acquire land can be enhanced by redirecting pricing policies and other policies so that they are targeted on the poor, as discussed earlier in Chapter 3. In practice, many agricultural policies and programs, from guaranteed crop prices to interest rate subsidies to agricultural extension, provide disproportionate benefits to the better-off farmers. While these farmers often are a vital part of the sector's productive structure, and should not have to face the macroeconomic distortions that frequently penalize agriculture (Chapter 4), there is no justification for targeting sectoral subsidies disproportionately in their favor. Effective action to redirect the benefits of special sector programs toward the poor would, in itself, improve their access to land through existing mechanisms. This point was made by Binswanger and Elgin,[242] as quoted in an article by De Klerk: 'a pre-condition for . . . land reform is the prior elimination of distortions favoring large farmers'.[243]

5.9.1.9 Distributing State-Owned Land to Low-Income Ruval Families

In countries where national and/or local governments hold title to significant amounts of agricultural land, one of the most powerful measures for improving the access of the poor to land is to adjudicate that land to the rural landless and to farmers with very small plots. As noted above, this kind of measure was one of the basic ingredients in the successful land reform in Taiwan. The land should be sold, rather than given away, but for the

reasons mentioned previously, its price should be subsidized. If the lands are owned by governments, then fiscal outlays would not be required to subsidize the land price. The size of the new plots should be well above subsistence level, but not so large as to reduce the number of beneficiaries significantly. The exact size will depend on agronomic conditions and other variables, but in Africa and Latin America, it normally would range from 5 to 20 hectares. In East Asia, the corresponding plot size would be smaller.

A principal operational difficulty that can arise is that State lands may already be occupied by squatters. In such a case, a policy option would be to require them to pay for the land they are occupying (with a mortgage), up to a specified ceiling of landholding size, and to place the lands above the ceiling amount in a pool for the program of redistribution. If the squatters have been on their land many years and are politically influential, the launching of such programs can face resistance. However, the potential disruption associated with changing their *de facto* degree of access to land would be considerably less than that attendant upon programs of expropriation of titled private lands, as pointed out above for the case of Honduras, and the government budget would benefit from a policy of requiring squatters to pay for the land that they use.[244] Production levels may well rise with such a program, given the well-known tendency of smaller farms to produce more output per hectare of land.

5.9.2 Improving Access to Land for Women

On average, households headed by women have lower incomes than those headed by males. The causes are well known: cultural factors leading to discrimination against women, including less willingness to educate girls in low-income households, institutional barriers, and legal restrictions that codify many of the cultural biases and give

242. H. P. Binswanger and M. Elgin, 'What Are the Prospects for Land Reform?', paper delivered at the XX International Conference of Agricultural Economists, Oak Brook, IL, USA, 1989.
243. M. J. de Klerk, 'Issues and Options for Land Reform in South Africa', in Csaba Csaki, *et al.*, 1992, p. 365.
244. In the late 1990s, Honduras instituted a requirement that squatters on State lands pay for them.

women less control over resources.[245] One of the most important ways to increase the earning power of both single and married women is to improve their access to land. That women's access is more limited than that of men cannot be doubted. Research by Rekha Mehra has concluded that:

> Few women in developing countries have secure and independent access to land. . . . traditional land tenure arrangements. . . . are regarded as relatively favorable to women. . . . Closer examination reveals, however, that tenure under traditional systems can be quite secure for men but not for women. . . . Where women have use rights to land, they are rarely free to act as independent agents; their rights tend to be restricted and use-specific. They cannot, for example, use their land for commercial purposes. . . . Women tend to face even greater difficulties in obtaining access to land under modern tenure systems. In fact, their land rights tend to deteriorate when governments institute land reform, land registration or resettlement schemes. Land registration programs throughout Africa (for example, in Ethiopia, Guinea-Bissau, Kenya and Zimbabwe) have failed to give women title to land even where they had customary access to land prior to registration. . . . In Latin America, women were similarly left out of the agrarian reform process of the 1960s and 1970s. . . . Recent reforms in China have also overlooked women. With the collapse of rural communes in the early 1980s, land was redistributed primarily among men, in effect reversing the 1947 agrarian reform which had given women

separate land deeds. . . . In India, for example, women have practically no right to inherit agricultural land. . . . Despite passage of the Hindu Succession Act (1956), which was intended to improve women's rights, the law has been interpreted so as to deny women access to agricultural land. In some Indian states, laws explicitly exclude widows and daughters from inheriting agricultural land. . . . Kenya's Succession Act (1972), intended to provide greater gender equality in inheritance, fails to do so because agricultural land was left under customary law, which denies women the right to inherit farmland. . . .[246]

A vicious circle has been created which tends to trap women in poverty:

> Worldwide, more women than men are poor, and the numbers of poor women are growing even faster than those for men, especially in the rural areas of developing nations. . . . As a consequence of their poverty, women are unable to acquire land even when laws permit. In Kenya, for example, recent laws do not prevent women from owning property, but most women cannot afford to acquire it. . . . Moreover, poverty prevents women from benefitting from some reforms; for example, women may be unable to take advantage of privatization that entails titling and registration because they cannot afford the costs of registration. . . .[247]

The difficulties that women face in obtaining access to land have been documented in many countries. Such difficulties are especially great

245. Although the fact of greater poverty among women is well known, and policies should aim to correct it, the extent of the phenomenon should not be over-estimated. Alain Marcoux has made some careful calculations that show that 'for those countries where adequate data are available, the average proportion of women among the poor appears to stand below 55 percent. . . . So, the gender bias in poverty does not reach the very high levels sometimes attributed to it. This does not mean that the bias is not real or not growing. It does seem globally real, although situations vary widely across countries and places'. From A. Marcoux, 'How much do we really know about the feminization of poverty?', *Brown Journal of World Affairs*, **V**(2), Summer/Fall 1998, pp. 187–194.

246. Rekha Mehra, 'Women's Land Rights and Sustainable Development', in R. Rose, C. Tanner and M. A. Bellamy (Eds), 1997, pp. 1, 2 and 5.

247. *Op. cit.*, p. 7.

where legal systems and social norms are like those of the Dominican Republic, where:

> several aspects of civil and agrarian law make it difficult for rural women in particular to gain access to land.... According to Dominican law, all property brought into and obtained during marriage becomes communal in nature ... but such marital property can only be legally administered (meaning sold, mortgaged or otherwise entered into a legal contract) by the husband.... Married women therefore have no rights even over land which they themselves may have inherited or purchased, and women in consensual unions – which are not recognized in the country's Civil Code – have even less claim on commonly held land and other property if their unions should dissolve. In the event of a husband's death, Dominican law gives hereditary rights to children, parents, and siblings of the deceased (in that order). Widows are eligible for up to one-quarter of their husband's property only if a will was drawn up before he died.[248]

Mehra's observation about the gender bias in agrarian reform finds echoes in many other countries. For example, in Colombia official records show that women represented only 11% of the agrarian reform beneficiaries during the first three decades of that program.[249]

Plus, in the Dominican Republic:

> The Agrarian Reform Law of this country stipulates that the head of the family unit, who generally is the man, is the one who has the right to benefit from the Agrarian Reform: it

Both customary and private land tenure systems frequently display gender bias, as illustrated in the case of Zambia:

The discrimination against women in both customary and privatized tenure is real. In customary tenure, women do not inherit land. In both patrilineal and matrilineal systems, the rights for women to land are not recognized. In privatized tenure [on State lands], women are required to meet some conditions not demanded from men before they are allocated land (V. R. N. Chinene, et al., 1998, p. 98).

does not recognize the woman as qualified to receive lands nor does it foresee her participation in the agricultural settlements managed by the [agrarian reform] cooperatives.[250]

Eve Crowley has reported that only 4 to 25%, approximately, of the beneficiaries of land reforms in the 1960s and 1970s in Latin America were women.[251]

The barriers to women's access to land are not always obvious but nonetheless pervasive, as pointed out by Crowley:

> Customary land institutions are familiar and convenient to rural women, reducing the transaction costs that prohibit recourse to formal land administration services. But this social and physical proximity can also be repressive. Government offices and land registries can introduce new principles, maintain public land records, and offer a neutral forum in which

248. Elizabeth Katz, 'Gender and Rural Development in the Dominican Republic', mimeo, prepared for The World Bank, Washington, DC, USA, November 2000, p. 4.
249. Elizabeth Katz, 'Gender and Rural Development in Colombia', mimeo, prepared for the World Bank, Washington, DC, USA, June 28, 2000, p. 3.
250. Beatriz B. Galán, 'Reglamentaciones jurídicas sobre el acceso a la tierra de la mujer rural en países de América Central y el Caribe', *Land Reform,* 2000/1, Food and Agriculture Organization of the United Nations, Rome, p. 81 [author's translation].
251. Eve Crowley, 'Empowering Women to Achieve Food Security: Land Rights', *Focus Note 6, A 2020 Vision for Food, Agriculture and the Environment*, International Food Policy Research Institute, Washington, DC, August 2001, p. 2.

women can effectively press their claims, but require transparent and consistent procedures and affordable transaction costs. In many countries, these institutions are inefficient, corrupt, time consuming and complex. Few women have the political connections, know-how, money, or physical proximity needed to secure land rights within them. . . . where women can legally purchase land, in practice only wealthier women and women's groups have the income to compete in the market. Nepotism, preferential treatment and complex, expensive procedural requirements restrict entry to land markets.[252]

Another kind of pervasive barrier to women is exemplified by the case of Uganda, where 'civil law . . . provides for equal rights in divorce – but customary law prevails in the division of conjugal property, and divorced women are unable to retain access to land'.[253]

Even in societies where the economic situation of rural women was more protected in the past, the forces of economic change may have placed them in a disadvantageous position:

In most societies, women have unequal access to, and control over, rural land and associated natural resources. In many cases, societies have protected the interests of women through customary law, religious law, and legislation in the past, but changing socio-economic conditions often result in the old rules failing to ensure that women have access to the resources needed to raise and care for families. Communities that now experience land shortages or rapidly increasing land values may be unable or reluctant to prevent male relatives from claiming land over which women, particularly widowed or single women, have rights.[254]

To overcome the legal, institutional and attitudinal barriers faced by women in trying to gain more access to land, five kinds of approaches are needed: *reforms in legal codes and regulations, targeted financial assistance for women, gender analysis in the design stage of projects and programs for rural development, training and awareness-raising among public officials, especially in land registries and land reform programs (market-assisted or otherwise), and campaigns of public education*. A single approach will not suffice, but rather all five kinds of programs needed to be implemented together.

In the case of Honduras, the earlier agrarian reform laws were amended specifically in order to enable women to become beneficiaries of agrarian reform. The previous agrarian reform legislation made it almost impossible for women to receive land through the agrarian reform process. Their access to land was governed by a separate article in the law which enabled a woman to receive land in the event that her husband died while the process of adjudicating land to him was underway. The newer Agricultural Modernization Law (1992) eliminated this dualism and replaced it with a single clause, applicable in equal measure to men and women, which defined their rights to receive land under the agrarian reform:

. . . the modifications introduced by the Agricultural Modernization Law in 1992 in this country eliminated the legal restrictions that impeded explicitly the participation women in the agrarian reform process.[255]

Within a year of the passage of the 1992 law, during which period the regulations were being drafted and approved, the President of Honduras presided over a ceremony at which 40 000 hectares of land were adjudicated to women, married and unmarried – a first in Honduran history.

---

252. E. Crowley, 2001, p. 2.
253. The World Bank, 'Engendering Development – through Gender Equality in Rights, Resources and Voice', World Bank policy research report, The World Bank, Washington, DC, USA, 2002, p. 16.
254. Land Tenure Service, FAO, 2001, p. 17.
255. B. B. Galán, 2000, p. 80 [author's translation].

Training and educational campaigns on gender issues are essential in order to be able to weaken the often deep-seated cultural resistance to a fairer treatment of women:

In many societies, improving access and security for women will require changes in policy and legislation, for example specifically recognizing the rights of a woman to hold land, and allowing a legal title to be issued in her name, either individually or jointly with her spouse. More importantly, it may require changes in cultural norms and practices. A country's laws may declare that men and women have equal rights to hold property and to inherit it, but if cultural norms and practices are in conflict with such laws, the rights of women are likely to be ignored.[256]

The FAO's Land Tenure Service has emphasized the importance of carrying out gender analysis at a very early stage in the development of projects and programs, a message that is directed as much at international institutions as at developing-country governments:

Within development projects, the design of land tenure components should incorporate gender analysis from the start to ensure that particular constraints faced by women are not overlooked. Attempting to incorporate gender considerations once the project objectives and design are in place often results in unproductively forcing gender issues into an inappropriate framework.[257]

In the words of Meinzen-Dick:

It seems time to shift the assumption – rather than assuming that registration and titling is a neutral process, and looking for the most 'efficient' process from the government's point of view, it is necessary to consider what will be done to remove barriers for poor and women.[258]

She recommends attention to the following specific points related to gender and poverty issues in land administration:

- Publicity and legal literacy programs (including mass media and working through various organizations . . .) to publicize the programs, and *make women aware of what is available to them*, including what safeguards for their rights (e.g. joint titling).
- Making the registration system available as close to the land in question as possible. This is especially important because women face particular travel restrictions.
- Reducing the up-front cash costs as much as possible, because women and poor households have the greatest difficulties in raising cash for such purposes.
- *Participatory processes for surveying the land*, so that local people are aware of the proposed boundaries, and can stake their claims.
- Provision for registration in the names of groups, rather than only individuals.
- Provision to recognize overlapping bundles of rights, e.g. for grazing on fallow agricultural fields, collecting tree products, etc. These overlapping bundles may not fit with western notions of 'ownership', but they often play a key role in the livelihoods of the poor.

I realize that this increases the cost of land registration systems, and makes transfers more complicated. However, without such provisions, land registration and titling programs increase the risk of elites, especially educated and/or urban elites, capturing resources . . . but with such provisions there is the possibility of strengthening the assets of the disadvantaged.[259]

---

256. Land Tenure Service, FAO, 2001, p. 18.
257. *Op. cit.*, p. 19.
258. R. Meinzen-Dick, 2001, p. 2.
259. *Op. cit.*, p. 3 [emphases added].

The gender analysis that is required to improve project design should be a thorough and carefully structured process. This section – and this chapter – are concluded with excerpts adapted from the FAO's guide to the questions that should be asked in a gender analysis for land tenure components of projects. The very nature of the questions sheds considerable light on the pervasiveness of gender bias.

5.9.2.1 Questions About Gender Policy Instruments

Family law:

- Do wives have equal status with husbands as regards marriage rights, grounds for divorce and separation, and marital property?
- In case of divorce or separation, do wives as well as husbands have rights to land and other property acquired during their marriage?
- In case of divorce or separation, are wives able to take with them the property they owned before entering marriage?

Inheritance law:

- Do statutory inheritance laws give daughters as well as sons inheritance rights to property, including landed property?
- Do widows have equal rights to the land and other property they and their husbands acquired during their marriage?

Privatization law:

- When land and other property are privatized, do women have the same opportunity as men to receive or buy property?
- Do privatization laws and/or regulations recognize only household heads as beneficiaries of these programs? If so, are women recognized as household heads?
- Do privatization laws and/or regulations recognize part-time workers (who may be mostly women) of State enterprises as eligible for receiving the property being privatized?

Land registration law:

- Do land registration and titling laws and regulations recognize only one household head per family as beneficiaries of these programs?
- Do registration and titling laws/regulations recognize the secondary rights of other household members to landed property?
- If land is considered family property, does legislation, particularly titling and registration regulations, establish property rights for all family members irrespective of gender or civil status?

Resource management law:

- Do resource management laws recognize the different roles women and men play in managing land and other natural resources?
- Do forest management laws take into account both women's and men's activities related to trees and other forest products?
- Are women's needs for water for both household and agricultural use explicitly considered in water management laws?
- Do laws that attempt to reform management of commons include the participation of women in management of natural resources?

5.9.2.2 Gender Analysis in Natural Resource Programs

- In programs that register and *title* land, are women's rights to land, including cultivation and other secondary rights, recognized?
- Do titling program laws and/or regulations target only one person per household?
- What are the cultural and ideological constraints, on the part of those implementing the program, encountered by women for acquiring land and rights to land?
- Where there is private *freehold ownership* of land, do customary norms and practices prevent women from acquiring land and other real property and/or from registering it in their own names?
- Do laws, regulations and practices (both commercial and cultural) allow women to enter into contracts without a man's (husband or related male) approval or signature?

• What constraints do women encounter to participating in land bank and land purchase programs, resettlement programs and market-driven land reform? Are these programs limited to one household member?

5.9.2.3 Other Aspects of Resource Management and Land Tenure Programs

• Do laws, norms or practices differ for women and men with regard to access to common-property resources such as forests, grazing lands and water resources?
• Do institutions that allocate community resources utilize gender, explicitly or implicitly, to determine who has access to these resources?
• Do privatization programs (including titling and registration programs) overlook and perhaps negate traditional property rights held by some persons such as women and junior men?
• In projects that deal with resource conservation or resource management, are women specifically consulted as to what resources, both individually and communally managed, need to be protected and how?[260]

DISCUSSION POINTS FOR CHAPTER 5

• Historically, concerns for both fairness (equity) and economic efficiency have motivated land tenure policies. The equity concern is basically about alleviating rural poverty. It has been found that both the initial distribution of land and the nature of land tenure regimes and policies have a direct bearing on the extent of rural poverty.
• Access to land is the most fundamental determinant of income-earning potential in rural areas in developing and transitional countries, but security of land rights is as important as access.
• Rights to land take many forms, and unrestricted private ownership is only one of them. In almost all customary systems of land man-

agement, rights of usufruct are defined but not ownership rights that would permit the owners to transfer the land.
• All over the world, the rights to individual parcels have been found to undergo an evolution as population pressure on the land increases. When the land is in abundance, the rights are *general* in that a family does not necessarily have access to the same plot year after year, but rather it is only guaranteed access to land somewhere in the village's domain. As population pressure increases, families become motivated to hold onto the land they have been working, out of uncertainty about the prospects of obtaining another piece of land which is equivalent in quantity and quality. Hence, over time rights to use the land tend to become specific to particular plots.
• Similarly, individual land rights tend to replace communal systems over time. Today in Africa, for example, individual rights are more prevalent than communal rights for most cropland.
• Traditional land rights can display great complexity and imagination in providing resource access to satisfy basic needs. Traditional systems have shown considerable flexibility in defining bundles of land rights, for example, allowing for grazing access at specified times of the year and under specified conditions.
• Private landownership also is rarely absolute but is subject to various conditions representing the rights of society or other individuals. Private ownership also can be considered to constitute a bundle of different kinds of rights over the land, including the right to use the land, the right to prevent others from using it, the right to derive income from it, the right to transmit it to heirs, etc.
• The great variety of forms in which land is held in the world today can be simplified into the following six basic types:

— *Open access lands.* In this kind of tenure no one can claim ownership to the land or resource and no one can be excluded from

---

260. Land Tenure Service, FAO, 2001, pp. 53–55.

access to it. It sometimes applies to forest lands or range lands. Marine resources usually are of the open-access type.

— *Communal lands* in *customary land regimes*. They are open to all members of the community, but there are community restrictions on their use and on access to them. Frequently, they are grazing lands.

— *Collective lands* are used for joint production by a group of farm families and are defined by a decision of central authorities. They can include individual plots and jointly worked plots. In most cases, the members of a collective farm have had no voice in deciding the form in which the land is to be held and exploited but rather the decision was made centrally.

— *Individual land rights under associative tenure*. These rights embrace individual plots in customary and collective tenure regimes.

— *Private land rights*. These rights include *ownership* (with varying degrees of restrictions), and other usufructuary rights in a market context such as rental, leasing and sharecropping. They may also be subordinated, temporarily and partially, to group decisions through *voluntary* co-operation in selected farming tasks or in the procurement of farm services. Ownership brings with it the right to dispose of the land according to the owner's wishes: in sale, leasing, rental, inheritance, and to encumber it with contingent claims such as a mortgage.

— *State lands*. In this case, property rights are assigned to an authority in the public sector, local or national.

• Tenure security requires clarity in defining a user's right to the land, and stability of that right over time. One kind of security guarantees the right to use a given plot of land and another, stronger kind, guarantees the right to transact the land, whether it be in rental, sale or another form of transaction.

• Tenure security can refer to protection of one's right to use the land, and can also mean the right of usufruct for a long period of time, long enough to stimulate investments in the land.

• Customary tenure systems vary widely. They usually supply tenure security in the sense of certainty of being able to use the land, and to some extent they occasionally guarantee rights to transfer the land, albeit usually with limitations. However, they normally cannot provide enough tenure security to permit the land to be used as collateral.

• Empirical research suggests a positive relation between land tenure security and farm performance. Studies have found that farms with full title to the land obtain more credit and invest more, and also that land values increase under titling.

• Research in several countries on the effect of farm size on productivity per hectare suggests a 'U-shaped' relationship, with the declining part of the 'U-curve' covering most of the relevant range of farm sizes. In other words, small farms produce more value per hectare than larger ones do for almost the entire range of relevant farm sizes.

• As noted, most traditional land tenure regimes do not permit a transfer of land right, although an increasing number do. As population pressure on the land increases, so does the pressure to permit transfers and also the tendency to formalize individualized land rights.

• A major concern in the transition from customary to formal systems of land rights is that those who are in command of better information about the new system can dispossess farmers who have held traditional rights of usufruct. Another hazard is that, when title to customary lands is vested in the State, farmers effectively can remain dispossessed from lands they traditionally had rights to use.

• Many land tenure systems include areas of common use. These 'commons' often are subject to over-grazing, as a result of the corresponding institutional incentives.

• This 'tragedy of the commons' tends to occur in open-access lands. Common property regimes with appropriate management controls, even if the system is customary, can diminish the possibility of over-exploitation. In addition, commons can be important resources for generating income for the poorer families in a community. However, the historical trend is clearly for common-property regimes to be converted into open-access regimes as population pressures increase, and therefore the trend toward environmental degradation of such lands increases over time in most cases.

- Providing community titles to common property lands and providing assistance in their management can help avoid land degradation. Group titles function best in small communities but still they may not constitute sufficient safeguards when new agricultural production or market opportunities open up.

- When group titles are issued, options that may be considered include providing to the community a 'first right of refusal' on the purchase of the land rights of an outgoing member of the community and, especially for former collective farms, issuing a form of shareholding in the community's assets, to encourage more entrepreneurial management of the lands.

- Collective and State farms have been created out of agrarian reform processes in many countries, except in East Asia where the trend was to provide land in individual, owner-operated plots. El Salvador has provided land to beneficiaries in both forms.

- As economic enterprises, most collective farms were born with a number of handicaps. Usually, their members were denied the right to mortgage their assets or to sell or rent out part of them. They also were legally denied the option of obtaining production finance from private banks, and all of their agricultural advisory services had to be supplied by the government. To offset these disadvantages, they usually have been heavily subsidized.

- From Latin America to Ethiopia to China, collective farms have largely failed everywhere, often with results far inferior to those on private landholdings. As a result, in many countries they are being transformed into private property of various forms.

- The State is a major or dominant landowner in many countries of Africa and the former Soviet

Union and in some cases in South East Asia. The reasons for this development are many, including historical, ideological and concerns for avoiding concentrations of landholding or speculation in land. However, as a practical matter the State has not generally proven to be a capable manager of land.

- To overcome the problems that typify State ownership of agricultural land, some countries have instituted a policy of providing land use rights through long-term, freely tradeable leases. The initial allocation of such leases merits careful study in order to ensure a transparent and equitable process.

- Land reform began in parts of Europe in the 18th century, based on willing sales of land to poor farmers, and in the 20th century it was transformed into a process based on coercion by the State. In general, coercive land reform has yielded disappointing results and has failed to reduce rural poverty. Exceptions include the East Asian land reforms of the 1950s, the first phase of land reform in Zimbabwe, and some aspects of the Philippine land reform of recent decades.

- The reasons for the disappointing record of coercive land reform include governments' inability or unwillingness to compensate former owners (thereby closing off the option of providing full title to the new beneficiaries), political motivations of land reform, the forms in which land has been adjudicated, failure to provide sufficient education and training to land reform beneficiaries, and failure to provide them with access to investment capital.

- Coercive land reform is inherently a disruptive process and can inflict long-term damage on the credibility of a country's institutions. It normally occurs in contexts of political upheaval or violent conflict. To lessen the disruptions and make land reform more effective in reducing poverty, new approaches are being tried out that introduce market elements in the land reform process, although early results suggest the process is a slow one owing to its demands on local administrative structures.

- Improving the functioning of land markets requires answering fundamental questions about what the nature of land rights should be, including whether customary land regimes should be protected and, if so, how, and what options should be provided to former members of collective farms.

- Experience has shown, especially in Africa, that customary systems can provide security of land tenure. Modern land registration systems are expensive and the requirements for introducing them are demanding, and so they are not always economically justified. However, as land values and population pressures increase, the need for even more secure land rights also increases. Some of the criteria for evaluating the effectiveness of traditional land tenure systems include whether women can hold land in their own right, the extent to which rental and sharecropping arrangements are permitted, the scope for inheriting land rights, and the degree to which traditional authorities are accountable to local people in their decisions about the allocation and protection of land rights.

- Thus, the question of whether to formalize land rights is a practical matter. One principal option is to give legal backing to customary systems of land rights as has been done, for example, in the new Rural Code of Niger. In all land tenure systems, strengthening local administrative capacities is a key to their effectiveness and durability.

- When a decision to shift to a modern land titling system is made, it should be borne in mind that the process is slow and it is important to provide protection in the process for renters and other holders of secondary rights. Equally, it is vital to carry out carefully local consultations on existing land rights before awarding the new titles. There is an *a priori* case for subsidizing part of the cost of registering new land titles. Strengthening institutions of dispute resolution at the local level often is a central element of any kind of program to improve a land tenure system.

- Prohibiting land ownership of small parcels, below a specified threshold in size, has proven to be counterproductive for reducing rural poverty.

- Land rental, in its various forms, is an important mechanism for helping reduce rural poverty through providing an affordable avenue of access to land for people on low incomes. It tends to make operated farm sizes more nearly equal and it leads to improvements in land productivity, because it transfers the use of the plot from a party who is less interested in or capable of working it to one who is more interested or capable. Finally, it reduces income risk for larger landholders by offering the renting-out option in periods when they are unable to work the land directly.

- For these reasons, prohibitions of land rental have proven counterproductive and have led to widespread evasion and to under-utilization of land. Equally, attempts to control the level of rents generally have failed.

- Sharecropping theoretically produces less efficient use of land than pure rental contracts do, but in light of the difficult access to credit that tenants frequently experience, in some circumstances it becomes the preferred mode of land utilization.

- Concerns have been raised that sales of farmland may lead to greater concentrations of landholdings over time. However, attempts to prohibit land sales usually have not worked well and some cases have increased inequality of landholdings, since those who are better connected may be better able to circumvent a ban. A more productive policy may be to provide greater access to credit for small farmers, and/or provide safety nets for them, so that they are not forced to sell land during periods of economic distress.

- For conversion of former collective farms into private property, there appear to be three principal options: private collective holding in the form of a joint-stock company or other corporate form, purely individual holdings, and a mixture of the two in which land is privately held and other assets are held jointly in corporate form. The corporate option, in which members possess shareholdings, offers stronger economic incentives for production than private co-operatives, although the latter have proven useful in the processing and marketing of some agricultural products.

- This process of conversion may take a long time because it requires training and other forms of support to former collective farms, in addition to passage of new laws and promulgation of new regulations.
- Access to land for the rural poor may be promoted in a market context through several mechanisms that may be instituted individually or jointly:

  — A land fund that finances the sub-division and sale of large properties and subsidizes their purchase by poor families.
  — Market-assisted land reform in which communities participate in the process and beneficiaries are given financial support for investments in newly acquired properties.
  — Improvements in land rental markets.
  — Institution of long-term leases and lease-purchase contracts for farmers on State lands and for beneficiaries of market-assisted land reform.
  — Institution of a progressive, area-based land tax.
  — Titling small farms.
  — Introducing alternative financial mechanisms such as antichresis or pledging of usufruct rights for limited periods.
  — Elimination of the many subsidies in agriculture whose incidence is skewed toward the non-poor.
  — Distributing State-owned lands to poor rural families.

- Women's access to land is more limited than that of men throughout the developing world, even under many traditional systems of land tenure. Frequently, laws discriminate explicitly against women in rights to hold and inherit land.
- To overcome the legal, institutional and attitudinal barriers faced by women in trying to gain more access to land, five kinds of approaches are needed: reforms in legal codes and regulations, targeted financial assistance for women, gender analysis in the design stage of projects and programs for rural development, training and awareness-raising among public officials, especially in land registries and land reform programs (market-assisted or otherwise), and campaigns of public education. A single approach will not suffice, but rather all five kinds of programs need to be implemented together.
- Specific measures that can materially assist in promoting access to land for women include: carrying out publicity and legal literacy programs to make women aware of what is available to them and what safeguards they have for their rights; making the registration facilities available as close to the land in question as possible; reducing the up-front cash costs; undertaking participatory processes for surveying the land so that local people are aware of the proposed boundaries and can stake their claims; allowing provision for group titling; and making provision to recognize overlapping bundles of rights, e.g. for grazing on fallow agricultural fields, collecting tree products, etc., which often play a key role in the livelihoods of the poor.

6

Water Management Policies in Agriculture[1]

6.1 Introduction   197
6.2 Policy Objectives for the Irrigation Sector   203
6.3 Strategic Planning for Irrigation as Part of
    Water Resource Management   207
6.4 Strategic Issues in Irrigation Development   211
    6.4.1 National Policies for Agricultural
          Development   211
    6.4.2 Intersectoral Water Policy   212
    6.4.3 Irrigation Rehabilitation versus
          New Systems   213
    6.4.4 Types of Irrigation Systems   216
6.5 Principal Policy Issues in the Irrigation
    Sector   223
    6.5.1 Instruments for Managing Water
          Demand   223
    6.5.2 The Pricing of Irrigation Water:
          Preliminary Considerations   226
    6.5.3 The Pricing of Irrigation Water:
          Conceptual Issues   227
    6.5.4 Irrigation Pricing Systems   232
    6.5.5 Water Rights Markets:
          An Introduction   236
    6.5.6 Issues in Water Rights Markets   239
    6.5.7 Potentials and Prerequisites of Water
          Rights Markets   247
6.6 Institutional and Process Issues in Water
    Management   248
    6.6.1 Sector-Wide Institutional Issues   248
    6.6.2 The Benefits of Local Water
          Management   255

6.6.3 Forms of Local Irrigation
      Management   260
6.6.4 Organizing WUAs   264
6.6.5 Gender Issues in Irrigation   268
6.7 Irrigation as a Tool of Rural Development   269
Discussion Points for Chapter 6   271

6.1 INTRODUCTION

As water is the font of life, irrigation has been the font of civilization. It underlay the rise of the first sedentary societies organized on a large scale, in Mesopotamia, Egypt, the Indus Valley and China. Irrigated agriculture appears to have been developed as early as the 7th century BC, on a small scale, in sites such as Jericho and the settlement of Çatal Hüyük, in what is now southern Turkey. Primitive methods of irrigation were also developed at approximately the same time by the early Sumerian people at the mouths of the Tigris and Euphrates Rivers. It was the Sumerians who were later to construct the first large-scale irrigation systems, between those two great rivers, to the north west of their original settlements. Historians agree that by approximately 3500 BC extensive irrigation systems were operated by several Sumerian city-states. The development of irrigation and water control measures in the other

1. Grateful acknowledgment is made of detailed suggestions on this chapter by Fernando Pizarro of the FAO and helpful discussion of some of its points with Alvaro Balcázar of the National University of Colombia.

Agricultural Development Policy Concepts and Experiences. R. D. Norton
© 2004 Food and Agriculture Organization of the United Nations
ISBNs: 0-470-85778-1 (HB)   0-470-85779-X (PB)   FAO Edition: 92-5-104875-4

great river basins mentioned above followed soon afterwards, within about 300 years.[2]

Irrigation also has been the basis of much of the agricultural growth in the 19th and 20th centuries, from the Western United States and the deserts of Northern Mexico and Peru to Mali and Sudan, to the Punjab and China and South East Asia. Flood control has been equally essential for agriculture in many places, especially in China and India but also in other locations such as the northern coastal valleys of Honduras. Drainage is important everywhere irrigation is practiced, and even without irrigation it is vital in Northern Europe, where winter's precipitation leaves vast quantities of standing water that must be removed from fields in order for cultivation to take place. In brief, in most parts of the world water control is a major component of agricultural technology and a basic determinant of the sector's prospects for expansion.

It has been estimated that '2.4 billion people depend on irrigated agriculture for jobs, food and income [and] over the next 30 years, an estimated 80 percent of the additional food supplies required to feed the world will depend on irrigation'.[3] In playing this fundamental role in food production, irrigation has become the world's largest user of fresh water, accounting for more than 80% of water use in Africa[4] and comparably high percentages in other developing regions of the world. For low-income countries as a whole, in 1992 irrigation accounted for 91% of water withdrawals, while for medium-income countries the corresponding figure was 69%.[5]

In the past, many irrigation strategies tended to treat water as an inexhaustible resource, and the emphasis was placed on the construction and financing of new systems to serve farmers. Now, the growing demands for water in all sectors have made it clear that water is a scarce resource, increasingly so, and former irrigation strategies are no longer viable in many areas. Ever larger numbers of countries are seeing their annual renewable water supplies fall below the critical level of $1000\,m^3$ per capita, below which they become a severe constraint on development prospects. Some of those countries, and their projected renewable water supplies per capita for the year 2000, in cubic meters, are as follows: Saudi Arabia (103), Libyan Arab Republic (108), United Arab Emirates (152), Yemen (155), Jordan (240), Israel (335), Kenya (436), Tunisia (445), Burundi (487) and Egypt (934).[6]

At least six other countries in Africa and the Middle East were projected to fall below the level of availability of $1000\,m^3$ by 2000. In some cases, the renewable supplies are being supplemented by mining groundwater and/or by desalinization of sea water, but the former is not sustainable and the latter is expensive, prohibitively so for agriculture.

The per capita availability of renewable water supplies is declining by more than 25% per decade in many cases. In some countries not yet at the critical level, those supplies are already below the level of $2000\,m^3$, which signifies serious problems in some of their regions, especially in drought years. More than 40 countries were expected to be in this situation by the year 2000,[7]

---

2. For a readable, yet reasonably thorough, summary of these early developments, see *The Age of God Kings*, Time-Life Books, Arlington, VA, USA, 1987.

3. FAO, 'Water Policies and Agriculture', in *The State of Food and Agriculture 1993*, Food and Agriculture Organization of the United Nations, Rome, 1993, p. 233, based on information from the International Irrigation Management Institute.

4. The World Bank, *A Strategy for Managing Water in the Middle East and North Africa*; The World Bank, Washington, DC, USA, 1994, p. 69, on the basis of estimates by the World Resources Institute and the World Bank.

5. World Bank, *World Development Report 1992*, The World Bank, Washington, DC, USA, 1992, p. 100, on the basis of data from the World Resources Institute.

6. FAO, 1993, p. 238. These figures include river flows from other countries, some of which may not be reliable sources in the future.

7. FAO, *Reforming Water Resources Policy*, FAO Irrigation and Drainage Paper No. 52, Food and Agriculture Organization of the United Nations, Rome, 1995, p. 7.

including Peru,[8] South Africa, Zimbabwe, Morocco and Iran.[9] By contrast, the *average* availability of water per capita in the world is much higher. By region, it was about the following levels (in m³) by the year 2000: Africa, 5100; Asia, 3300; Latin America, 28 300; Europe, 4100; North America, 17 500.[10] Endowments of fresh water supplies are not distributed spatially in a very even manner, neither within or between continents, nor within most countries. Since it is a product with low unit value, water is expensive to transport long distances from deficit to surplus areas. Therefore, most solutions to water problems have to be developed locally, specifically at the level of watersheds and groundwater basins.

Even in countries with relatively abundant supplies of water for the time being, irrigation is losing its former attractiveness for two other reasons: its productivity in terms of farm income has turned out to be substantially less than expected in many cases, and the cost of building irrigation systems has increased. In combination, these two factors mean that the economic returns to irrigation systems are often below levels that are considered acceptable.

It has been pointed out that the initial analysis of most African irrigation projects would not have looked promising if more realistic figures on expected yields, costs, scheduling of irrigation and other factors had been used in the calculations.[11] In addition, for a sample of irrigation projects analyzed *ex post* in Vietnam, Thailand, Myanmar and Bangladesh, it was found that their financial and economic returns were substantially reduced by below-target production levels and crop prices that turned out lower than expected. As a consequence, the sustainability of the

schemes was weakened by factors completely unrelated to requirements for operation and maintainance (O & M).[12]

Irrigation management also has been deficient in many instances. The FAO has concluded that, on a world scale, 'The overall performance of many irrigation projects has been disappointing because of poor scheme conception, inadequate construction and implementation or ineffective management'.[13] The International Irrigation Management Institute (IIMI) has underscored these problems:

... there is widespread dissatisfaction with the performance of irrigation projects. This is true whether performance is measured in terms of achieving planned targets or in terms of the production potential created by physical works. Sub-optimal performance can be observed in irrigation systems of all types and sizes, from the small farmer-managed systems in the hills of Nepal to the giant canal systems of India and Pakistan.

In most schemes around the developing world, more water is delivered per unit area than is required, leading to low irrigation efficiencies. According to a recent report, irrigation efficiencies are as low as 20–25 percent in Java, the Philippines and Thailand, and in Pakistan it is about 50 percent. ... In many irrigation schemes, the actual irrigated area is much less than the area commanded. Water deliveries rarely correspond in quantity and timing to crop requirements, resulting in low cropping intensities and low productivity.

Sharp inequities in water supplies to farmers in the head reaches of the irrigation system and

8. Inter-American Development Bank, 'Nuevas Corrientes en Manejo de Aguas', *El BID*, August 1997, p. 4.
9. (a) Ruth S. Meinzen-Dick and Mark W. Rosegrant, 'Managing Water Supply and Demand in Southern Africa', in Lawrence Haddad (Ed.), *Achieving Food Security in Southern Africa: New Challenges, New Opportunities*, International Food Policy Research Institute, Washington, DC, USA, 1997, Chapter 7, p. 204; (b) The World Bank, 1994, *op. cit.*, p. 69.
10. FAO, 1993, p. 237.
11. J. R. Moris and D. J. Thom, *Irrigation Development in Africa*, Westview Press, Boulder, CO, USA, 1991, p. 579.
12. E. B. Rice, *Paddy Irrigation and Water Management in Southeast Asia*, A World Bank Operations Evaluation Study, The World Bank, Washington, DC, USA, 1997, p. 52.
13. FAO, 1993, p. 233.

those located downstream are another manifestation of poor performance in many irrigation systems.[14]

In many irrigation systems, considerable amounts of water are wasted because they are withdrawn from the water source (e.g. well, reservoir, river, etc.) but only a portion of them are actually applied to the crops. For Latin America and the Caribbean as a whole, it has been calculated that this ratio of water applications to withdrawals averages 45%.[15]

System deterioration further reduces the efficiency of irrigation and is a common problem in irrigation facilities, especially in economies in transition. A clear illustration of this danger is provided by the experience of irrigation in Dashowuz Velayet, Turkmenistan:

Canal leakage is a major problem with delivery efficiencies ranging from 50 to 60 percent. . . . these low delivery efficiencies force excessive diversions in order to meet irrigation requirements. That increases the leakage and contributes to the high water table and, consequently, to the increasing soil salinity.

The ability to deliver precise amounts of water is also limited by the condition of the control gates. . . . However, many of the drive motors are inoperable or even missing. Many of the threaded shafts that control the gates indicate that the positions of the gates have not been changed in a long time. There is no evidence of lubrication and they are heavily rusted. A few of the shafts were even bent to the degree that they could not be adjusted. . . .

In the Yilany Etrap, it was reported that the entire area is served by pumping from the Shabat Canal. However, only 40 to 50% of the pumps are operational. Repairs are funded by the budget, but there are not enough funds for parts and maintenance. . . .

Increasing soil salinity caused by high water tables is the primary direct cause of low crop productivity. The primary on-farm water supply-related cause is excess application of irrigation water coupled with inadequate on-farm drainage systems. . . .

Problems with [the drain] collectors are primarily siltation and depth. Siltation is a continuing problem that reduces the capacity of the collectors and the collectors are too shallow to adequately receive water. . . . The World Bank also states that these drains have been neglected to the point that most open drains are silted and overgrown with weeds and the closed drains have not been flushed for years and almost all are reported to be non-functional.[16]

Deterioration of irrigation works also has been occurring widely in Kazakhstan:

It is estimated that irrigation systems in the country use 30–35% more water than irrigation systems growing similar crops in market oriented economies. In recent years irrigation system efficiency declined even further as most irrigation systems have deteriorated due to a lack of funds for maintenance and a breakdown in management.

Farm privatization and restructuring in the sector have created further problems. The rapidly increasing number of small farms has made irrigation management more complex involving financial, economic, environmental and institutional issues. State and district level water committees are facing increasing difficulties in terms of water fee collection and revenue generation. The inefficient use of the irrigation

14. International Irrigation Management Institute, 'The State of Irrigated Agriculture', in *25 Years of Improvement*, at the worldbank/cgiar website, 1998 [www.worldbank.org/cgiar].

15. Luis E. García, *Integrated Water Resources Management in Latin America and the Caribbean*, Sustainable Development Department, Environment Division, Inter-American Development Bank, Technical Study No. ENV-123, Washington, DC, USA, December 1998, p. 8.

16. Adrian O. Hutchens, 'Irrigation Management and Transfer Issues in Turkmenistan', prepared for Central Asia Mission, US Agency for International Development, April–May 1999, pp. 8–10.

systems has exacerbated local environmental problems with excessive water applications leading to water logging and salinity problems, as well as contributing to downstream environmental problems in the Aral and Caspian Seas.[17]

Centralized control of irrigation systems everywhere has tended to lead to system deterioration. A similar litany of deficiencies was reported for Andhra Pradesh in India, before the devolution of responsibilities for maintenance to the farmers themselves:

... the number of control structures at various levels of the systems is insufficient and communications systems are rudimentary. The most detrimental factor, however, has been the cumulative impact of chronic underfunding of maintenance over many years. This has resulted in the severe disrepair of most surface irrigation schemes: canals and drains are heavily silted, lined sections are damaged, drops are eroded and collapsing, many gates are inoperative, outlets are damaged, and shutters are missing. ... Irrigated yields and production per unit of water are well below their potential. ... Inequitable and unreliable water delivery from deteriorating systems has resulted in tail enders not getting water and yield shortfalls for other farmers ranging from 15% to 40%.[18]

Policy makers should be aware that irrigation projects can fail badly. Sometimes, it is more than a question of low returns on an investment. The Bura project in Kenya is one of the prime examples of irrigation gone very wrong. It was expected to be a showcase of agricultural development, and original estimates suggested that 100 000 ha or more could be irrigated along the Tana River, and then an early feasibility study suggested implementing the scheme on 18 000 ha. In the end, only 3900 ha were made operational, which meant that the high capital costs could not be recovered. There were more serious problems, however. As W. M. Adams has summarized them, they included:

... low yields, both of maize and cotton, and resulting low settler incomes. The chief cause of this has been poor and erratic water supply. ... Problems of maintenance, spare parts, fuel supply, and lack of trained operators and mechanics led to repeated breakdowns in 1983 and 1984, during which year water was unavailable for 25% of the time. ... siltation of structures and canals, and in places erosion of the main supply canals, have been significant. ... Malnutrition in children was a serious problem. ... the Health Center, finished in 1981, was not opened until 1983 due to budgetary constraints in the Ministry of Health. ... Disease incidence was high among settlers and their children during this period, particularly because of cerebral malaria. The construction of tenant houses fell behind schedule, and their cost rose, making the possibility of tenants being able to afford to repay those costs more remote. ... Furthermore, shoddy workmanship led to the collapse of many houses (34% in Village I in 1982, for example). ... There were also serious problems of fuelwood supply. ... In technical terms, it was clear that the poor soils of the project area were a serious constraint, resulting in low economic returns. ... Bura will continue to suffer operational deficits. Furthermore, tenant incomes are so low that it will not be possible to raise service charges to meet the shortfall.[19]

17. Sam Johnson, 'Economics of Water Users Associations: The Case of Maktaral Region, Southern Kazakhstan', prepared for Central Asia Mission, US Agency for International Development, November 1998, p. 4.
18. Keith Oblitas and J. Raymond Peter, *Transferring Irrigation Management to Farmers in Andhra Pradesh, India*, World Bank Technical Paper No. 449, The World Bank, Washington, DC, USA, 1999, pp. 5–6.
19. W. M. Adams, 'How beautiful is small? Scale, control and success in Kenyan irrigation', *World Development*, **18**(10), October, 1990, pp. 1317–1318.

Similarly disappointing stories could be recounted for other projects in various parts of the world. Thus, at a juncture in world history when the potential contribution of irrigation is ever more necessary and valuable, the barriers to the realization of that contribution have become formidable. For this reason, new emphases and approaches for irrigation strategies have emerged in the past decade or so, and a principal purpose of this chapter is to synthesize them and underscore the issues that are involved. A related purpose is to provide references to technical studies in the field that the interested reader can consult further.

While the problems of irrigation are difficult ones, the international consensus is that the problems can be overcome and that irrigation has an important role in the future. As the FAO has put it:

The mediocre performance of the irrigation sector is also contributing to many socio-economic and environmental problems, but these problems are neither inherent in the technology nor inevitable, as is sometimes argued.

Irrigation projects can contribute greatly to increased incomes and agricultural production compared with rainfed agriculture. In addition, irrigation is more reliable and allows for a wider and more diversified choice of cropping patterns as well as the production of higher-value crops. Irrigation's contribution to food security in China, Egypt, India, Morocco and Pakistan is widely recognized. For example, in India, 55 percent of agricultural output is from irrigated land. Moreover, average farm incomes have increased from 80 to 100 percent as a result of irrigation, while yields have doubled compared with those achieved under the former rainfed conditions; incremental labor days used per hectare have increased by 50 to 100 percent. In Mexico, half the value of agricultural production and two-thirds the value of agricultural exports [are] from the one-third of arable land that is irrigated.[20]

There is no other single technology or policy intervention in agriculture that promises benefits of the magnitudes that irrigation does, provided its potential can be realized.

In spite of the widespread emergence of problems in irrigation systems, there are a number of 'success stories' which bear out the FAO's optimism that, with proper policies, irrigation can improve its efficiency and continue to provide many economic and social benefits. The FAO has pointed to the experience of the Philippines as one example of how the management of irrigation systems can be reformed with positive results:

The National Irrigation Administration (NIA) of the Philippines is a good example of how a bureaucracy can, over time, transform its strategy and operating style. . . . The quality of the irrigation service provided by and for farmers has undoubtedly improved, system operating expenses have been reduced and the recurrent cost burden of the scheme on the national budget has been eliminated. . . . Recent research suggests that the reforms have made water supply more equitable.[21]

Sharma *et al.* have highlighted selected cases of successful irrigation management in Nigeria, Ethiopia and Mali:

Fadama irrigation in Nigeria: Fadama are river valley areas which are seasonally flooded or have high water tables for all, or a large part, of the year. . . . The projects made available motor pumps at subsidized prices, and drilled tubewells, which enabled farmers to greatly increase water use for dry season crop production. These systems each allowed irrigation on 1–2 hectares at a cost of US$350–700 per hectare. Individually owned and maintained, 70 percent of farmers in project areas use the pump systems, and the schemes have a 90 percent success rate. . . . with over 50 000 pumps operating nationwide, a strong service industry has developed to maintain them, ensuring sustainability.

Ethiopia's success with small scale participatory irrigation: . . . the project provided funds for rehabilitation and construction of over 4400 hectares of small-scale irrigation schemes (with average investment costs of $1200 per hectare) . . . and technical assistance support to the Irrigation Development Department at national and regional levels. . . . Over 40 water user groups voluntarily formed at the time of initiation then fully participated in scheme identification and construction. These groups have also assumed total responsibility for operation and maintenance. Income generation and growth have resulted in project areas, and 100 personnel have been provided with training.

Spectacular success in the Office du Niger: The Office du Niger (ON) in Mali began in 1932 as a French scheme to produce cotton and rice in a one million hectare area over a fifty year period. However, in 1982, the project was far from meeting its objectives: only 6 percent of the target area had been developed . . . infrastructure was poorly maintained, cotton production had been discontinued, average paddy yield was at a low point, and settlers were unhappy. However, after 10 years of careful [project] preparation and implementation, the ON is on its way to becoming a success. . . . between 1983 and 1994: average paddy yields have tripled, 10 000 ha of abandoned land have been rehabilitated, the settler population grew by 222 percent, and per capita production of paddy rose from 0.9 tons to 1.6 tons. . . . farmers were allowed to participate more fully with the ON (e.g. through the setting of fees and collection of payments). Additionally, the ON itself has been restructured and streamlined, and fee collection has reached a level of 97 percent (even with an increase of 43 percent [in rates] over the last two years).[22]

These excerpts suggest some of the factors that can contribute to success in irrigation schemes. They and other factors are reviewed systematically in this chapter.

6.2 POLICY OBJECTIVES FOR THE IRRIGATION SECTOR

As in other areas of policy, three overriding objectives may be established for the irrigation sector, namely, *efficiency*, *equity* and *sustainability*, but in some important respects their interpretation is different for this sector. As elsewhere, sustainability should be understood in its fiscal, institutional and environmental dimensions; in many irrigation experiences to date the institutional aspects have been particularly weak, both in governmental entities and users' organizations. The environmental sustainability dimension refers primarily to maintaining water and soil quality and a correct balance of water resources. Many thousands of hectares of irrigated land are lost each year as a consequence of waterlogging and the accumulation of salts in the soil. Natural habitats also can be lost if, for example, the construction of an irrigation system results in the drying up of swampy lands or the flooding of other areas. Public health is another dimension of the environmental concerns associated with irrigation. A principal risk is that some water supply projects can create conditions for the breeding of malaria and the spread of bilharzia (FAO, 1995, p. 21).[23]

In areas where groundwater extractions are significant, the sustainability objective refers

22. Narendra P. Sharma, Torbjorn Damhaug, David Grey and Valentina Okaru, *African Water Resources: Challenges and Opportunities for Sustainable Development*, World Bank Technical Paper No. 331, The World Bank, Washington, DC, USA, 1996, pp. 47–48.

23. On the other hand, there are important positive externalities associated with irrigation. It can improve rural water supply and sanitation and allow rural people to have easy access to water for domestic needs. Women are principal beneficiaries since normally it is they who carry out the arduous tasks of fetching and carrying household water, frequently over long distances (International Irrigation Management Institute, 1998).

to guaranteeing future generations' access to groundwater supplies. Unfortunately, the rates of use of groundwater are far from sustainable in many basins. For countries in the Middle East, groundwater extraction represented a small proportion of the renewal supply of such water in the 1960s but by 2000 extractions had reached a level of almost 180% of the renewable resource base in Jordan, 140% in Yemen and 99% in Tunisia.[24]

An alert over both the quantity and quality of available groundwater resources in many parts of the world was sounded by Jacob Burke and Marcus Moench:

> In many countries, groundwater extraction has increased exponentially with the spread of energized pumping technologies for irrigation in agriculture. In India, for example, the number of diesel and electrical pumps jumped from 87 000 in 1950 to 12.6 million in 1990. ... Expansion of pumping technology has often resulted in dramatic declines in the watertable in low-recharge areas. Competition among agriculture, domestic and commercial water users over access to limited groundwater resources is growing. Perhaps more seriously, pumping changes flow patterns, often resulting in migration of pollutants and low-quality water into aquifers. Pollution has also been caused by rapid growth in the use of agricultural chemicals and the all-too-common practice of discharging untreated industrial and domestic waster water directly into the ground. Furthermore, even minor spills of some industrial chemicals, such as organic solvents, can cause large-scale groundwater pollution. Once aquifers are polluted, their clean-up can be technically impossible or simply uneconomic.[25]

The equity objective for irrigation management refers to giving priority to the development of irrigation for and with small farmers and distributing water equitably *within* systems, so that no farmers are deprived of this essential resource. This means providing water to all farmers entitled to it, without any favoritism. 'Poorer farmers are often at the tail end of irrigation systems, where supplies are unreliable' (FAO, 1995, p. 20). Working toward greater equity also means ensuring that rural women are given prominent roles, not only as recipients of irrigation water but also as participants in system design and management. The needs and problems faced by women in farming households often differ from those of men, and accordingly irrigation systems require different designs in some respects in order to meet their needs. For example, women irrigators may not feel safe going out at night to open the irrigation gates. Women's participation, and that of farmers in general, also serves the efficiency objective.

Equity considerations deserve special emphasis because in practice they are often overlooked. Although they are frequently mentioned among the objectives of irrigation projects, policies for the actual distribution of water rarely are established in a way that is compatible with that objective.

Properly managed, irrigation can be an effective tool for reducing poverty. The International Programme for Technology Research in Irrigation and Drainage has pointed out that the benefits of irrigation include higher and more stable incomes for farmers, greater food security for the poor, more farm and non-farm rural employment opportunities, and enhanced supplies of domestic water that can improve health in low-income

---

24. *MENA/MED Water Initiative, Proceedings of the Regional Workshop on Sustainable Groundwater Management in the Middle East and North Africa, Summary Report*, Sana'a, Yemen, June 25–28, 2000, hosted by the National Water Resources Authority, Republic of Yemen and Co-Sponsored by the Swiss Development Cooperation and the World Bank, p. 4.

25. Jacob J. Burke and Marcus H. Moench, *Groundwater and Society: Resources, Tensions and Opportunities*, United Nations Department of Economic and Social Affairs and Institute for Social and Economic Transition, New York, 2000, p. 11.

households. However, the same study stresses that the realization of these benefits requires effective targeting of irrigation on land-poor families, including approaches that allow them to own irrigation systems and sell the water for a profit, ensuring selection of low-income families among the families for settlement schemes, using employment-intensive methods for building irrigation systems, institutional arrangements to ensure secure water supplies in periods of low water, and full compensation for cultivators who may be dislocated by the schemes. Equally, the use of participatory approaches in the design of irrigation schemes is vital, with emphasis on involvement of low-income households, and designs need to guarantee low costs of system operation.[26]

Increasing the efficiency of irrigation systems has two different meanings. Technically, it refers to the reduction of water losses. In a broader sense, it refers to increasing net economic returns for the systems' users, taking full account of externalities. Achieving it requires actions in distinct dimensions: technological, institutional and the dimension of the policy environment. However, irrigation differs from some other areas of agricultural policy in that water is not a sectoral resource. It is a unitary, mobile resource that may be used in all sectors of the economy and for different purposes. Hence, irrigation policies and programs cannot ignore the role that water plays in other parts of the economy and in other uses. It is now an internationally accepted principle that water management policies of all kinds must recognize that water is an economic good and that it has value in competing uses.

For these reasons, achieving greater 'efficiency' in irrigation in the broader sense may mean giving up water to other sectors where it has higher-value uses, even if sometimes that also implies reducing the value of agricultural output. As the value of those other uses is taken into account in irrigation planning, achieving greater efficiency within

typically will require measures to increase both system efficiency and on-farm efficiency. Such measures range from changes in system design to management improvements to economic incentives for better water use. Improvements in these areas in turn will lead to higher levels of agricultural output per unit of water used and/or reductions in total water used, and possibly to reductions in the costs of water delivery and of system management.

In line with the need to improve the efficiency of irrigation systems, in most parts of the world the emphasis in irrigation planning has shifted from augmenting supplies to managing demands. It also is an internationally accepted principle that irrigation management is a component of broader water management. It must take account of all uses in all sectors of the economy, by watershed or river basin, and irrigation development plans must be made consistent with overall frameworks for water policies. This holistic approach embraces considerations of water quality as well as quantity.

These interrelationships occur on the ***demand side*** of water balances. There are also basic linkages on the ***supply side*** which require holistic thinking about water management. The ***physical relationships*** on the supply side are complex and pervasive. Surface and groundwater characteristics need to be analyzed jointly in developing strategies for water management, since usually there are hydrological linkages, sometimes complex, between aquifers and surface flows. Typically, surface flows filter down to aquifers in part but movement may occur in both directions from flowing aquifers. Utilization of one of these sources is likely to affect the other, and so strategies for joint or 'conjunctive' use need to be developed. More broadly, it should be recognized in strategies that use of water in one part of a watershed or basin can affect both the quantity and quality of water in other parts.

---

26. International Programme for Technology and Research in Irrigation and Drainage (IPTRID), *Poverty Reduction and Irrigated Agriculture*, Issues Paper No. 1, Food and Agriculture Organization of the United Nations, Rome, January 1999, pp. 7 and 15–16.

The necessity of a holistic approach to water management derives in part from the diversity of its uses and their competition:

*'Water provides **four types of important economic benefits: commodity benefits, waste assimilation benefits, esthetic and recreational benefits, and fish and wildlife habitats**. Individuals derive commodity benefits from water by using it for drinking, cooking and sanitation. Farms, businesses and industries obtain commodity benefits by using water in productive activities. These commodity benefits represent private good uses of water which are rivals in consumption (e.g. one person's or industry's water use precludes or prevents its use by others)'. (From FAO, 1993, p. 259 [emphasis added].) [A special case is hydropower production, which generates benefits from water without consuming it, although it may introduce some constraints on other uses.]*

It may be added that the waste assimilation uses can preclude the commodity uses, and both of those uses can diminish water's availability or usefulness for fish and wildlife habitats and for esthetic and recreational purposes.

There also are **economic externalities and social linkages** on the supply side of a watershed. These linkages as well as the physical ones have been mentioned by John Dixon and William Easter in stating the case for integrated management of watersheds:

(1) The watershed is a *functional region* established by physical relationships.
(2) The watershed approach is logical for evaluating the *biophysical linkages* of upland and downstream activities because within the watershed they are linked through the hydrologic cycle. . . .

(3) The watershed approach is *holistic*, which enables planners and managers to consider many facets of resource development.
(4) Land-use activities and upland disturbances often result in a chain of *environmental impacts* that can be readily examined within the watershed context.
(5) The watershed approach has a *strong economic logic*. Many of the externalities involved with alternative land-management practices on an individual farm are internalized when the watershed is managed as a unit.
(6) The watershed provides a framework for analyzing the effects of *human interactions* with the environment. The environmental impacts within the watershed operate as a feedback loop for changes in the social system.
(7) The watershed approach can be *integrated with or be part of programs* including forestry, soil conservation, rural and community development, and farming systems.[27]

There are a few parts of the world, mainly in parts of Latin America and West-Central Africa, where water is still abundant relative to demands for it, and the competition for water among sectors will not become manifest for several decades. In those cases, expanding irrigation supplies still may have a priority, but experience has shown that methods of system management are often the principal determinants of the effectiveness of the systems in raising rural incomes. Therefore, in both water-scarce and water-abundant environments, a principal operational concern is to improve irrigation management in the broadest sense, from watersheds to the farm level, and from the performance of public agencies to the roles of user groups and other community-level associations.

---

27. John A. Dixon and K. William Easter, 'Integrated Watershed Management: An Approach to Resource Management', in K. W. Easter, J. A. Dixon and M. Hufschmidt (Eds), *Water Resources Management*, Westview Press, Boulder, CO, USA, 1986, Chapter 1, p. 6 [emphasis in original].

In order to translate these broad objectives into concrete actions, the following eight operational *sub-objectives for the irrigation sector* may be established, taking into account both the needs of users and constraints of a financial and institutional nature:

(1) Improving the effectiveness of holistic planning for irrigation.
(2) Making irrigation more productive for its users, i.e. increasing economic returns per unit of water used.
(3) Improving the access of women and poor farm families to irrigation.
(4) Reducing physical water requirements per unit of output.
(5) Reducing the total costs for the provision of irrigation.
(6) Reducing the fiscal costs to the national budget.
(7) Improving the environmental sustainability of irrigation.
(8) Improving the institutional sustainability of irrigation systems.

The *means* for achieving these objectives are varied and include the following: direct administration of water fees and allocations at the irrigation system level; management of water allocations, quality levels and fees through watershed or river basin authorities; capacity building among users and managers; involvement of stakeholders in irrigation system design and operating decisions; decentralization of public sector management organs; codification of water rights and establishment of regulatory frameworks; developing markets in water rights; joint management of transboundary water basins; and public investments in water extraction and conveyance facilities. The choices among the means depend on the principal water issues faced by a country and on the particular socioeconomic, political, geographical and hydrological environment. As will be seen later, the more participatory the management of the systems, then generally the better their performance.[28]

6.3 STRATEGIC PLANNING FOR IRRIGATION AS PART OF WATER RESOURCE MANAGEMENT

Irrigation is a sector in which development of an overall strategy is essential before proceeding with significant investments or programs to improve system effectiveness. As the FAO has stated, 'National water politics are shifting from *projects* to *policies* – this trend is likely to continue and even accelerate'.[29] This imperative has been stated in many contexts, and it is applicable to watershed management as well as to irrigation management. For Africa, the idea has been expressed in the following way:

A systemic approach based on a river basin or watershed facilitates the development of water resources in an integrated, multi-sectoral framework taking into account the political economy. This approach requires sovereign countries and users of water to think holistically and to take local and sectoral actions which are consistent with long-term development goals. It promotes a more objective assessment of water resources and better understanding of the interconnectedness and interdependencies among various components, such as land, water and forests, of an ecosystem. Land use and changes in land use patterns, deforestation, degradation of vegetation, soil erosion, decreased river flows, climate change and development programs are all interrelated and have an impact on water utilization. A systemic approach also facilitates an understanding of interactions among economic sectors, such as agriculture, industry, energy, and [commerce], as well as the role of various stakeholders, such as farmers, local communities, and domestic users. Further, a systemic approach provides a regional perspective that transcends political boundaries.[30]

28. L. E. García (1998, pp. 39–46) has discussed some of these policy instruments.
29. FAO, 1993, p. 296 [emphasis in original].
30. N. P. Sharma, *et al.*, 1996, p. 61.

However, Sharma *et al.* have pointed out that adopting a more systemic approach does not imply greater centralization of decisions or greater control over water resources by a government. On the contrary, it means developing partnerships among all stakeholders, including local and national government entities, communities, irrigation users and other water users (*ibid.*).

For the Middle East and North Africa, it has been commented that:

> Clear statements of national water policy are required, and national water resource management plans, supported by appropriate regional and basin plans, should be consistent with the policy. Alternative water development strategies should be systematically evaluated, giving full consideration to the balance between supply and demand management practices and measures.[31]

In addition, at a worldwide level the role of the public sector and of national water strategies has been summarized in the following way:

> The primary roles of the public sector are to define and implement a strategy for managing water resources; to provide an appropriate legal, regulatory, and administrative framework; to guide intersectoral allocations; and to develop water resources in the public domain. Investments, policies, and regulations in one part of a river basin or in one sector affect activities throughout the basin. Thus, these decisions need to be formulated in the context of a broad strategy that takes the long-term view, incorporates assumptions about the actions and reactions of all participants in water management, and fully considers the ecosystems and socioeconomic structures that exist in a river basin.[32]

The FAO has emphasized the need for inter-institutional co-ordination in order to make strategic planning effective:

The institutional challenges to better water management were spelled out frankly by the Government of Yemen in its strategy to deal with the rapid depletion of the groundwater in the basin under the capital, Sana'a:

The action plan consists of a variety of institutional, physical, technical and financial measures which should be implemented as an effort to slow down the aquifer depletion. . . . It is fundamental to establish an adequate institutional framework for water sector planning and management. . . . The existing institutional set-up is fragmented and is not capable of addressing the issues faced by the Sana'a Basin. . . . Water legislation aimed at controlling unregulated water abstraction is awaiting government approval [and] should . . . be implemented with full zeal. . . . The variety of policy measures are by no means simple to implement. . . . Their implementation will require heroic efforts by the government and the people of the Sana'a Basin. . . . Technological innovation is necessary but not sufficient . . . This will be a joint effort with the farmers, and their cooperation is essential for success. Some of the proposed measures will ultimately affect the life of every rural family in the target area. Rapid and effective dissemination of information among the rural population will be especially difficult. Problems of village isolation, illiteracy, lack of mass communication media, and shortage or lack of qualified personnel all enormously increase the magnitude of the task. (From The Government of the Republic of Yemen, High Water Council, Water Resources Management Options in Sana'a Basin, *Final Report, Volume IX, Sana'a, Yemen, July 1992, pp. 53–55.)*

A more integrated and broader approach to water sector policies and issues is important because of water's special nature as a unitary resource. . . . Water use in one part of the

---

31. The World Bank, 1994, p. 53.
32. The World Bank, *Water Resources Management*, A World Bank Policy Paper, The World Bank, Washington, DC, USA, 1993, pp. 40–41.

system alters the resource base and affects water users in other parts. ... Governments generally tend to organize and administer water sector activities separately: one department is in charge of irrigation; another oversees water supply and sanitation; a third manages hydropower activities; a fourth supervises transportation; a fifth controls water quality; a sixth directs environmental policy; and so forth.

These fragmented bureaucracies make uncoordinated decisions, reflected in individual agency responsibilities that are independent of each other. Too often, government planners develop the same water source within an interdependent system for different and competing uses. ... This project-by-project, department-by-department and region-by-region approach is no longer adequate for addressing water issues.[33]

In addition, strategic planning for the water sector must take into account macroeconomic considerations:

Improving water resource management requires recognizing how the overall water sector is linked to the national economy. Equally important is understanding how alternative economic policy instruments influence water use across different economic sectors as well as between local, regional and national levels and among households, farms and firms. For too long, many water managers have failed to recognize the connection between macroeconomic policies and sectoral policies that are not aimed specifically at the water sector can have a strategic impact on resource allocation. ... A country's overall development strategy and use of macroeconomic policies – including fiscal, monetary and trade policies – directly and indirectly affect demand and investment in water-related activities. The most obvious example is government expenditures. ... on irrigation, flood control or dams.

A less apparent example is trade and exchange rate policy aimed at promoting exports and earning more foreign exchange. For example, as a result of currency depreciation, exports of high-value, water-consuming crops may increase. If additional policy changes reduce export taxes, farmers are provided with an even greater incentive to invest in export crops as well as in the necessary irrigation. ... [34]

The strategic and policy issues that must be faced and resolved for long-term irrigation planning may be grouped into the following three categories:

(i) *Strategic issues*, including macroeconomic policies and basic questions of the future scope of irrigation in light of other water needs and the externalities of irrigation (system development questions).

(ii) *Sectoral policy issues*, to create an appropriate enabling environment, through legal frameworks, regulations and regimes of economic incentives and disincentives.

(iii) *Institutional and process issues*, which concern strengthening both governmental and private institutions involved in water management and fostering the appropriate participation of water users and community groups in decision-making.

These three classes of issues are reviewed in Sections 6.4 through 6.6, respectively. Some of the risks of proceeding with projects without having formulated a comprehensive water strategy which takes into account intersectoral linkages and social effects are well illustrated by a World Bank review of one of its irrigation projects in Sudan:

The Roseires Irrigation Project. ... did not meet appraisal expectations. Only 184 000 ha [vs. 489 000 ha planned] were actually provided with irrigation infrastructure ... and the rate

33. FAO, 1993, pp. 248–250.
34. *Op. cit.*, p. 253.

of return was a disappointing 9%. The dam was constructed prematurely because the need for stored water was overestimated at appraisal. Also, insufficient attention was given at appraisal to the planning of investment packages for developing agriculture. The return could have been greater if the project had been part of a comprehensive plan for utilization of the Nile water resources, which could have facilitated coordination and optimization of irrigation, agriculture and power development. ... No concern was expressed about potential problems of yield sustainability arising from a possible decline in soil fertility, nor about the future prospects of nomads living in the proposed reservoir area. ... there is no indication of problems which the nomads or others may have faced in adapting to the changed environment.[35]

On the other hand, the challenge of preparing a solid and useful water strategy should not be underestimated. In relation to the Middle East and North Africa region, the World Bank has observed that long-term planning for managing resources at the regional and watershed levels often is weak, as is the incorporation of water resource considerations into national development strategies. The reasons for this circumstance include lack of clarity about goals, inadequate technical data on water resources, inadequate financial resources for resource management planning and insufficient personnel with the needed capabilities, and lack of a clear political commitment to better management of the vital resource that is water. In addition, mechanisms for public participation in strategy formulation usually are inadequate, and this further weakens the political commitment to seeking better solutions and the capability of implementing them.[36]

Many of the same observations could be made with respect to other regions of the world.

However, an encouraging example has been set by the participatory process for changing water management policies in Andhra Pradesh, where extensive public consultations were undertaken to identify problems and possible avenues of solutions. The outcomes of the consultation process made policy makers aware that fundamental structural changes were required for all aspects of the irrigation management system. The strategy for change emerged from the consultation process and was implemented on a large scale.[37]

In overall strategic terms, as commented on previously, in most cases the emphasis needs to shift from augmenting irrigation supplies to *demand management*, for which enhanced user participation is a particularly effective means. In the words of Sharma *et al.*, for Africa:

> Policy makers need to focus more on demand management and changing people's behavior through incentives, regulations, and education. They must also turn their attention to reallocating existing supplies to higher-value uses, reducing waste, promoting conservation and water-saving technologies, and facilitating more equitable access.[38]

Similar observations have been made for Latin America, to the effect that radical changes are needed in the way in which water is managed. The Inter-American Development Bank (IDB) has proposed that, instead of continuing to finance individual projects for augmenting water supplies, investment decisions should be made in a holistic context, taking into account the economic, social and environmental value of water resources. In addition, the IDB has stated that decisions on investment programs for water should be made in a participatory manner and should be supported by appropriate policies and legislation. Special attention needs to be given to creating and strengthening institutions for integrated man-

35. Operations Evaluation Department, The World Bank, *Renewable Resource Management in Agriculture, A World Bank Operations Evaluation Study*, The World Bank, Washington, DC, USA, 1989, p. 166. It should be noted that in general Sudan has some of the more successful irrigation projects in Africa.

36. See The World Bank, 1994, pp. 24–25.

37. K. Oblitas and J. R. Peter, 1999, p. 10.

38. N. P. Sharma, *et al.*, 1996, pp. xix–xx.

agement of water resources, which largely are lacking.[39]

Notwithstanding this necessary change of emphasis, productive opportunities to expand water supplies should not be ignored. Yet in the future the means for achieving that will be more varied than simply building new physical systems. They will include the modernization of water infrastructure, improvement of water management institutions and capabilities, measures to guarantee more efficient delivery and utilization of water, and reductions in system operating costs.[40]

6.4 STRATEGIC ISSUES IN IRRIGATION DEVELOPMENT

6.4.1 National Policies for Agricultural Development

A prerequisite for an irrigation development program is a national economic policy framework that is conducive to agricultural growth. Chapter 4 reviews the policy instruments that can play a role in such a framework. Whatever the choice of instruments, it is essential that policies be configured in a way that provides adequate price incentives at the farm level for agricultural growth. Moris and Thom warn against undertaking irrigation projects 'where there are poor agricultural pricing policies, ineffective marketing facilities, high transport costs, or the unavailability of required agricultural materials'.[41]

The powerful influence of macroeconomic policies on water use is illustrated by the case study for Yemen, presented in the annex to this work. The most important aquifers of that water-scarce country are being over-exploited, with the consequence that water tables are dropping rapidly. It is urgent to encourage water conservation, and even at the margin to favor rainfed agriculture over cultivation based on tubewells. Nevertheless, of the 20 policy instruments examined in the study, *all* of them turned out to be biased in the direction of providing incentives for overuse of water.

Although the exchange rate was not a significant factor in the case of Yemen, at least at the time of that study, in some circumstances it can be the most important factor in determining whether irrigation investments will be worthwhile. The profitability of cropping is one of the most basic considerations in deciding whether to expand an irrigation network, and the analysis of Chapter 4 has shown that it depends crucially on exchange rate policy. This factor is so important that, as a general rule, it can be said that *irrigation investments should not be undertaken if the exchange rate is significantly overvalued*. Trade policies skewed in favor of the industrial sector also can undermine the profitability of irrigation investments.

The consequences for irrigation of unfavorable agricultural pricing policies include not only lower incomes for farmers, and therefore lower project returns, but also the inability of farmers to contribute to system maintenance costs, inadequate maintenance and therefore increasing physical deficiencies, and even partial or total abandonment of schemes as soils deteriorate and farmers find more remunerative alternatives elsewhere. The ability to arrest negative environmental effects is weakened. Net returns to water decline below levels that are obtainable from irrigation under more favorable conditions, which already are low by the standards of other water-using sectors. Thus, from a national perspective, expanding irrigation under policies that are biased against agriculture leads to wastage of water and the capital invested in the systems, environmental damage and failure to improve sufficiently the economic conditions of farmers.

Other basic elements of national policies for agricultural development include appropriate land tenure policies, effective transfer of agricultural technology and effective rural financial institutions. As Moris and Thom have summarized it, irrigation does not pay without attractive crop prices, security of land tenure, availability of inputs and credit and reliable water supplies.[42]

39. Inter-American Development Bank, 1997, p. 4.
40. N. P. Sharma, *et al.*, 1996, p. xx.
41. J. R. Moris and D. J. Thom, 1991, p. 33.
42. *Op. cit.*, p. 42.

6.4.2 Intersectoral Water Policy

Elements of a national, intersectoral water policy can include plans for construction and operation of multi-sectoral water facilities, definition of the legal and regulatory framework for water rights, procedures for surveillance of water use and conflict resolution, incentives frameworks, definition of the role of the private sector in water supply and management, and environmental regulations and flood control provisions, among other considerations.

A basic requirement of a national water policy, and therefore of an irrigation strategy, is a *national water assessment*. Before proceeding with the design of irrigation projects, a prior decision has to be taken as to how much supplies of irrigation water can be expanded, or whether they can be expanded at all, in light of projected water balances by watershed and nationwide. Only after such a decision has been reached is it appropriate to investigate how to make additional irrigation viable in technical, economic and institutional terms.

A water assessment has to take into account all sources of water, including both groundwater and surface water and non-conventional sources such as desalinized and treated waste water, and all water uses, current and projected. Where there are competing uses, it is necessary to estimate the economic value of water in each category of use and the possibilities of water conservation in existing uses. An assessment also evaluates the waste-assimilation capacities of the water systems in light of the likely loads of waste materials that will be discharged into fresh water, and the economic costs and public health risks associated with poor water quality. As well as direct dumping of wastes into water, considerable amounts of pollution occur via seepage of wastes from cesspools and landfills and runoff of agricultural chemicals. Groundwater is particularly vulnerable to such pollution because natural cleansing mechanisms are practically non-existent and its contaminants disappear only extremely slowly, if at all. Aquifers in coastal areas are particularly vulnerable to saltwater intrusion as they are pumped down. Rivers are also vulnerable because pollution increases the costs of utilizing water to downstream users, and pollution may make rivers completely unsuitable for some uses. In addition, river-borne pollutants often flow into lakes, estuaries and coastal seabeds where they create other problems.

The case for water assessments has been made in the following way by the World Bank:

> Specific options for investment and development must consider the interrelations among different sources of water. Surface and groundwater resources are physically linked, so their management and development should also be linked. Land and water management activities as well as issues of quantity and quality need to be integrated within basins or watersheds, so that upstream and downstream linkages are recognized and activities in one part of the river basin take into account their impact on other parts. Investments in infrastructure may displace people and disturb ecosystems. Thus, water resource assessments need to consider these cross-sectoral implications.[43]

The possibilities of augmenting year-round access to water supplies through the construction of dams may be included in some water assessments. Dams have a history as long as large-scale irrigation systems, and between 1950 and the late 1980s approximately 35 000 large dams were constructed throughout the world.[44] However, there is now greater awareness of the difficult environmental and social issues that large dams often pose, and an international dialog is underway on developing new planning procedures for such dams and new criteria for their evaluation. (See

---

43. The World Bank, 1993, p. 42.
44. Tony Dorcey, Achim Steiner, Michael Acreman and Brett Orlando, *Large Dams: Learning from the Past, Looking at the Future*, IUCN – The World Conservation Union, IUCN, Gland, Switzerland and Cambridge, UK and The World Bank, Washington, DC, USA, 1997, p. 4.

the referenced IUCN – World Bank paper by Dorcey *et al.* on large dams (especially pp. 4–13, 19–20 and appendices A1 and A2) for a preliminary statement of planning issues and evaluation criteria.) Future projects for the construction of dams will face more demanding criteria for acceptance but some of them can be feasible if designed conscientiously with these criteria in mind. It is important for designers of dams to keep up to date with the ongoing international dialog on that subject.

In situations of water scarcity, the possibilities of recycling treated waste water and desalinization of sea water need to be examined in a water assessment. For many areas of the Middle East and North Africa it has been decided that recycling waste water, under proper controls, can contribute both to expanding the water supply and improving the environment. A substantial amount of irrigating with waste water is already camed out in several countries of that region, such as Israel, Jordan and Saudi Arabia.[45] In the western hemisphere, Mexico has been a leader in the use of treated waste water for irrigation, in its central plateau where competition for water is especially acute. There still are concerns about the quality of waste water and research is underway on efficient ways to improve it. Research in Burkina Faso has found that putting waste water in holding ponds in which water lettuce (*Pistia stratiotes*) grows can improve the water quality to a level where it can be used to irrigate market gardens.[46]

A full water assessment will indicate, among other things, the possibilities of developing new water supplies and their approximate costs, the trends in water demands, the possibilities for water conservation in each kind of use and options for improving the efficiency of existing irrigation systems, trends in water quality and classes of measures needed to maintain acceptable quality, and the likely direction and magnitudes of intersectoral water transfers in the future. It should also identify agronomic constraints on irrigation, environmental issues in addition to water quality (e.g. soil deterioration and degradation of natural habitats) and social issues that can accompany water development and water quality issues (e.g. public health concerns, resettlement of populations as a result of both dam construction and the opening of newly irrigated lands). It is only within this kind of framework that irrigation planning should go forward.

Given the increasing importance of policies for water management, the FAO has made a case for complementing a water assessment with a *water policy review* as well as an assessment. This organization considers that such a review should take place if any of a number of problems emerge, including difficulties in balancing water supplies with demands, deficiencies in the standards of service in supplying water, degradation of water quality, serious inefficiencies in water systems (including irrigation), financial shortfalls in the water sector, inadequacies in the institutions charged with managing water, and symptoms of conflict among water users.[47] The guidelines for developing such reviews are provided in the 1995 FAO publication.

6.4.3 Irrigation Rehabilitation versus New Systems

In view of the disappointing performance of many irrigation schemes to date, most reviews of investment strategies recommend giving priority to rehabilitation of irrigation areas rather than development of new areas. As the FAO has put it:

A prime opportunity for irrigation advancement – and indeed for development in general – lies in the enormous potential of the 237 million ha already operating [world-wide].

---

45. The World Bank, 1994, p. 28.
46. D. Kone, *Epuration des eaux usées para lagunage à microphytes et à macrophytes en Afrique de l'Ouest et du Centre-état des lieux, performances épuratoires et critères de dimensionnemente*, PhD Thesis, Ecole Polytechnique Fédérale de Lausanne, No. 2653, Lausanne, Switzerland, 2002 (reference supplied by the International Network on Participatory Irrigation Management (INPIM)).
47. FAO, 1995, pp. 12–15.

While the total value of irrigation investment in the development world today is about $1000 billion, the returns are well below the known potential. Many irrigation schemes need substantial investment to be completed, modernized or expanded. Although it is increasingly expensive, rehabilitation can yield high returns.[48]

However, the priority accorded to rehabilitation needs to be qualified in a number of respects. It may be more important to improve the institutional aspects of the system, or the policy environment, than to rehabilitate the physical structures, or those kinds of improvements may be a prerequisite for successful system rehabilitation. These concerns have been summarized by Moris and Thom, as follows:

Obviously, in those countries like Niger or Tanzania where the rate at which already developed irrigable land is going *out* of production exceeds the development of new irrigation, rehabilitation should take first priority. In recent years, this has indeed been the emphasis among most donors. However, experience with attempted rehabilitation shows the issue is not so clear-cut:

— Engineering considerations tend to predominate during rehabilitation, when in fact the greatest need may be for O & M modifications.
— The need for rehabilitation is usually linked to a lack of adequate maintenance procedures. Unless these can be instituted within the local system, physical reconstruction will effect only a temporary improvement.
— Where the main system has been allowed to badly deteriorate, the costs of reconstruction can be just as high as for the building of new schemes.

— The pyramiding of new loans on top of old ones creates a crushing financial burden beyond the support capacity of many schemes. . . .

Thus, while the balance of effort in Africa probably should be directed towards improvement of existing irrigation, it does not necessarily follow that physical reconstruction of these schemes under external loan financing is what is needed. A carefully done, case-by-case comparative analysis of O & M deficiencies which makes rehabilitation necessary within existing schemes would appear to be [a precondition] before effective remedial measures can be instituted.[49]

In cases in which physical rehabilitation would appear to have a role, it is important to evaluate the original engineering design and decide whether it functions sufficiently well to warrant rehabilitating it. Caution should be exercised in financing rehabilitation projects in which the design of the basic water delivery infrastructure is deficient. In this context, Willem Van Tuijl has effectively posed three options for decision makers: (i) changing the system's basic design – upgrading it, (ii) rehabilitating it according to the original design, and (iii) leaving it alone if the original design is seriously deficient and too costly to improve. He points out that:

Guidelines for determining when upgrading would be preferable to rehabilitation are difficult to provide. The decision would depend on local conditions, such as investment costs, anticipated on-farm irrigation technologies, the value of additional crop production, and the value of the water saved by applying improved technologies. . . . As part of project preparation for future rehabilitation and improvement projects, more diagnostic work (technical, agronomic and socio-economic) should be carried out to evaluate the existing systems and to determine the need for upgrading.[50]

---

48. FAO, 1993, pp. 290–292.
49. J. R. Moris and D. J. Thom, 1991, pp. 561–562.
50. Willem Van Tuijl, *Improving Water Use in Agriculture: Experiences in the Middle East and North Africa*, World Bank Technical Paper No. 201, The World Bank, Washington, DC, USA, 1993, pp. 21–22.

A common reason for deterioration of the production potential of irrigation systems is inadequate drainage systems and/or lack of corresponding monitoring measures. The marked deterioration of the drainage facilities in Turkmenistan was mentioned. The International Irrigation Management Institute has observed that:

> In China, for example, more than 930 000 ha of irrigated farmland have become unproductive since 1980, an average loss of some 116 000 ha per year. . . . It has been estimated that about 24 percent of the irrigated area worldwide is affected by salinization, though many observers regard this estimate as too high.[51]

Returning to the more general question, programs for rehabilitation generally should be based on a broad vision of the system's functioning, including its supporting policies, its management

---

Waterlogging and salinity are among the principal causes of decreasing production on many irrigated projects. Waterlogging is due to an excessive input of water into systems that have finite natural drainage capacities. After waterlogging has occurred, soil salinity increases because the irrigation water leaves dissolved solids in the soil. It is essential to monitor the water table levels from the beginning of a project in order to implement corrective measures before the soil damage has occurred. . . . in rainfed agriculture, surface drainage is required to prevent any temporary waterlogging and flooding of lowlands. In irrigated agriculture, artificial drainage is essential under most conditions. It is vital to minimize drainage requirements and costs by reducing the sources of excess water through improved system design and on-farm water practices. . . . (FAO, 1993, p. 287).

---

components and the role of farmers. While physical rehabilitation or upgrading often is needed, a new conception of how the system is managed and operated may also be required and may be more urgent. An early illustration of some components of this broader approach to irrigation rehabilitation was provided by World Bank proposals for Mexico:

> There is vast potential for increasing productivity in the majority of the [Irrigation] Districts, and six areas of action are recommended in which the Government's efforts should be concentrated: (a) completion of structures and on-farm works in existing systems; (b) realization of water savings through more economical use of water resources; (c) better training and management of extension and research services to achieve more rapid yield gains; (d) less restrictive Government policies with regard to desired crops, leading to more diversified production patterns and a higher value crop mix; (e) provision of sufficient funds and resources for adequate maintenance through increased water charges as well as by more active farmer participation; and (f) raising farmgate prices to a point as close as possible to border price levels to stimulate production and enable farmers to pay for higher water charges and a share of investments to improve systems.[52]

The narrow objective of rehabilitation or upgrading, which appears to emphasize physical reconstruction, should be replaced with the more comprehensive objective of *improvement of irrigation's overall efficiency and of water distribution*. Efficiency can be decomposed into its constituents of on-farm efficiency and system efficiency, and both depend on institutional and economic factors as well as the physical aspect. According to Van Tuijl, the requirements for improving the efficiency of water distribution

---

51. International Irrigation Management Institute, 1998.
52. Latin America and the Caribbean Regional Projects Department, The World Bank, *Mexico: Irrigation Subsector Survey – First Stage, Improvement of Operating Efficiencies in Existing Irrigation Systems,* Vol. I, Main Findings, Report No. 4516-ME, The World Bank, Washington, DC, USA, July 13, 1983, p. 7.

include improvements in policies for land tenure in the direction of providing greater security of land tenure rights, more appropriate levels of water charges and more realistic budget allocations for operation and maintenance (O & M), improvements in institutional aspects of management such as water users' associations and institutions for agricultural support services, and, where necessary, irrigation system rehabilitation and improvement.[53]

Improvements in the agricultural policy environment could be added to this list, in order to provide greater incentives for farmers to cultivate under irrigation and maintain the system.

6.4.4 Types of Irrigation Systems

In engineering terms, there are many types of irrigation systems, but the most common distinctions are full versus supplemental systems, modern versus traditional (informal) schemes, and large versus small schemes. More than one type of system may have a place in a national water strategy; accordingly, all types should be reviewed in a national water assessment.

The word informal refers to traditional practices such as recessional (*décrue*) irrigation following the receding of annual flood waters, sometimes enhanced by simple structures such as polders and river intakes, or small rainfall catchment structures. Spate irrigation, assisted by simple earthen diversion structures, is common in places like Yemen. In spite of the pace of construction of irrigation systems in the past, *traditional systems* still represent the dominant form of irrigation in some areas. For example, of all irrigated rice grown in the 19 principal rice-producing nations of Africa, as much as 72% has been produced via traditional practices.[54] In general terms, it cannot be said that informal or traditional irrigation systems[55] are necessarily

preferred to modern, engineered systems, or vice versa. An evaluation has to be made of the circumstances surrounding each case. What is needed is to study more widely the experiences with traditional irrigation, and to take into account objectively the option of expanding it for some areas when national irrigation strategies are designed. To date, international agencies have tended to ignore the potential which could be offered by modest improvements of the traditional systems, and this oversight should be corrected. Moris and Thom have explained the situation in the following terms:

> . . . in many African countries, irrigation systems are polarized between a few, larger-scale government schemes and a number of very small independent irrigators. That latter practice various 'traditional' techniques, employed with hardly any outside assistance. These days they are incorporating some modern (or 'introduced') equipment, notably small pumps, but their whole mode of financing and operation is very different from that employed on the large-scale, official schemes. . . . The project documentation which one finds in donors' files tends to represent these official schemes. . . . Farmers' own efforts to control water are usually on a very small scale. The purchase of a single pump may represent the culmination of a major effort among subsistence farmers. . . . At this extreme, few expatriate engineers would consider such water management practices as 'irrigation'. Nonetheless, they achieve the same objective as the much more expensive, imported technologies used on official schemes. . . . The tremendous differences between the two main types of irrigation . . . has inhibited any sharing of experience or assistance between them. It has proven very difficult for government agencies and

---

53. W. Van Tuijl, 1993, p. 4.
54. Calculated from figures presented in J. R. Moris and D. J. Thom, 1991, p. 41.
55. In some taxonomies, traditional irrigation systems fall under the rubric of small-scale systems. The FAO, for example, says 'Small-scale programs include a diversity of technologies such as water harvesting, well development, river offtakes and use of wetlands' (FAO, 1993, p. 287).

external donors to work with Africa's small-scale systems, though there are a few instances of partial success (in Senegal and Tanzania). The extreme duality which characterizes the

> *In some of the literature there is confusion between irrigation systems or schemes, on the one hand, and irrigation methods, on the other. In a very simple classification, irrigation methods can be put into two categories: surface flow and pressurized irrigation. Surface flow irrigation can be provided in many ways (basin, furrow, borders, etc.) whose common characteristic is that water is applied at a certain point on a parcel of land and from there it moves on the surface over the rest of the parcel. Until the development in the 20th century of pressurized techniques, surface flow was the only method used in history, and it still is the most widely used. Although it has marked disadvantages such as low water application efficiency, the need for land leveling, difficulties in applying the correct volumes of water in the right frequency, and a high demand for labor, it is expected that it will continue to be the most commonly used method by far.*
>
> *Pressurized irrigation, sometimes called micro-irrigation, can be divided into sprinkler and localized irrigation techniques, with the latter referring mainly to drip and micro-sprinkler options. When they are well designed and managed, both pressurized techniques allow higher water application efficiency than surface flow methods do. Localized irrigation can apply water and fertilizers on a daily basis according to crop needs, thus promoting higher crop yields and quality, and labor is saved as well. The disadvantages of pressurized methods include high investment costs, the necessity of energy, and the use of sophisticated components that are not always available. For these reasons, the use of pressurized irrigation normally is limited to high-value crops such as fruit trees and vegetables.*

irrigation sector in most sub-Saharan countries is unfortunate. It makes it unlikely that successful smaller projects will evolve into medium-scale operations which might combine high farmer involvement with economies of scale in water management.[56]

It is not necessary to travel to remote areas to see traditional irrigation practiced. Where the Niger River passes through Bamako and its environs, for example, many small farmers can be seen taking water from the river in gourds and plastic buckets to water vegetables planted a few feet from the river's edge. A small number of them have invested in pumps and hoses.

A process of development of an irrigation strategy which incorporates greater participation on the part of farmers may lead, in some cases, to greater emphasis on expansion and improvement of traditional irrigation systems. For strategic planning, the significance of traditional irrigation resides in the following: (a) its value should be recognized, particularly when designing projects (usually dams) that may result in the reduction or elimination of existing opportunities for its use; (b) technical possibilities of incremental improvements in traditional irrigation, for example, by supplying more pumps or building small polders, may exist and may be considered where land tenure and agronomic, economic and social conditions are appropriate; (c) in considering expansion or improvement of traditional irrigation, priority should be given to involving in the discussions the farmers who are presently practicing it, soliciting their ideas for how to increase its effectiveness and returns. The objective in this case is to improve traditional systems of irrigation without undermining the inherent strengths that caused them to be developed in the first place.

Supplemental irrigation is used to compensate for dry spells during the rainy season, or to prolong the season of water availability for crops. It usually is based on pumping, whether from surface or ground water. Its desirability is dictated

---

56. J. R. Moris and D. J. Thom, *op. cit.*, pp. 6–7.

by climatic conditions; in regions where the rainy season often is irregular, it can play a vital role in preventing severe damage to crops. Worldwide, most irrigation is supplemental to one degree or another, except in very arid climates and in greenhouses. *Supplemental irrigation can be critical not only for increasing the volumes of production but also for ensuring the quality of products like fruit and vegetables, for it enables farmers to control the timing of water deliveries to the plants.* In view of the increasing importance of product quality for export, and even for domestic markets in developing countries, this contribution of supplemental irrigation can be significant for increasing the incomes of farm households.

Commercial farmers have played an important role in developing private systems of supplemental irrigation. A noteworthy example is that of coffee farmers in Kenya. The case for supplemental irrigation has been stated in the following terms:

In any farming system dependent on rainfall where the mean total [precipitation] lies near the boundary required for successful cultivation, minor deficits . . . can have a dramatic impact on crop yields. . . . In policy terms, what is important is to realize that the potential benefits from irrigation may be just as great in supplementing rainfed cultivation as in supplying all plant water requirements under 'total' irrigation in a semiarid environment.[57]

Furthermore:

Commercial farmers in East and Southern Africa have usually found it necessary to develop supplemental irrigation in order to achieve reliable crop yields. If so, the same need probably exists within smallholder farming. Regularization of rainfed crop returns by stabilizing planting dates and eliminating the within season dry spells might represent a more desirable (and water conserving) objective than

'full' irrigation with its heavy water demands. The main problem is, of course, the high cost of present technologies for achieving this objective. . . . We do not yet have answers, but the need to pay more attention to partial irrigation seems obvious. . . .[58]

The experience of the Machakos District in Kenya provides an example in which farmers led the way in developing traditional irrigation:

As a technique for increasing water supply, water harvesting methods work in the Machakos District in Kenya [via] the building of terraces to control soil erosion and to slow runoff, thus increasing the supply of water in the crop root zone. These measures have increased crop yields and production. Perhaps the most significant feature of this achievement was that it resulted almost entirely from farmers' decisions to invest their own resources in improving the natural resource base and other aspects of farm enterprise. The government contribution – by no means trivial – was to improve farm-to-market roads, both within the region and between the region and Nairobi (N. P. Sharma, et al., 1996, p. 47).

Within the category of modern irrigation systems, the wisdom of *large-scale versus small-scale irrigation projects* has been the subject of considerable discussion in recent years, with the consensus opinion inclined in favor of smaller projects but not excluding the larger ones under favorable circumstances. Moris and Thom reason as follows:

If in Africa small-scale projects are not necessarily cheaper to build, they are nevertheless easier to withdraw from; managerial assistance by an NGO rather than the government is more feasible; field layouts can be more adapted to farmers' needs; and there is at least a theoreti-

---

57. J. R. Moris and D. J. Thom, 1991, pp. 16–17.
58. *Op. cit.*, p. 572.

cal possibility farmers will be more involved and consequently more committed. We recommend, therefore, a bias towards assisting small-scale projects and technologies. . . . This recommendation. . . . ignores the fact that schemes requiring large reservoirs or major canals are bound to be large-scale in nature [and] a pervasive opinion . . . that small projects are just as demanding of supervision and management as are large ones. While this may be true, the consultants drawn upon in this study were nearly unanimous that in Africa smaller, flexible projects on average outperform the large ones. . . . these arguments do not rule out experimentation within large systems to decentralize scheme functions and increase farmer participation, e.g. as the Dutch have attempted in the Office du Niger.[59]

The international action plan for irrigation that is guided by the FAO (IAP–WASAD) also has identified small-scale irrigation as one of its priority areas. The requirements for successful small-scale irrigation, as put forth in that plan, include adequate technical advice, more participatory approaches to management of systems, and strengthened and more accountable public sector institutions.[60]

Sharma *et al.* acknowledge the greater success rate of the smaller projects but point out that larger ones can be successful also:

While the majority of irrigation success has come with small and medium-scale schemes, this does not mean governments should completely ignore the value of large-scale irrigation projects. Through a coordinated effort, Nigeria has irrigated 70 000 hectares of land . . . and Sudan, through careful upstream management to control sedimentation, has operated its Sennar dam for seventy-six years with only a 56 percent loss in total capacity. . . . Factors which hold true for other development efforts also hold true for irrigation projects of such magnitude: the availability and extension of a comprehensive package of improved technological messages, liberalization of crop marketing and processing, land tenure security, improved roads, bureaucratic reform, government commitment, focus on well-defined goals (in this case, water management), partnership with the individual farmer-producers, and donor coordination.[61]

Adams' review of the issue of project size in Kenya concludes that *farmer participation and control* is more important to the success of the scheme than size itself, that schemes run by bureaucracies tend to perform poorly independently of their size.[62] This lesson would appear to be valid for other countries as well. The challenge of ensuring farmer control at the level of tertiary canals, and maintaining adequately their communications with upstream managers, may be greater in larger schemes but, with adequate definition of responsibilities at all levels, it is not insuperable. The technical and managerial problems of ensuring stability of water levels in secondary and tertiary canals also tend to be greater in the larger systems.[63]

Provided that the institutional, engineering and policy requirements for proper operation of a modern irrigation system can be satisfied, size should not be a barrier. Irrigation systems in several countries, including Mexico, Pakistan and

---

59. *Op. cit.*, pp. 562–563.
60. FAO, 1993, p. 287.
61. N. P. Sharma, *et al.*, 1996, p. 47.
62. W. M. Adams, 1990, p. 1320.
63. For those involved in the design of small-scale irrigation schemes in developing countries, detailed and practical guidelines can be found in: (a) F. M. Chancellor and J. M. Hide, *Smallholder Irrigation: Ways Forward, Guidelines for Achieving Appropriate System Design*, H. R. Wallingford Ltd, Report OD 136, Department for International Development (DFID), Wallingford, Oxon, UK, August 1997; (b) G. Cornish, *Modern Irrigation Technologies for Smallholders in Developing Countries*, ITDG Publishing, London, 1998.

India, as well as China, Nigeria and Sudan, confirm this conclusion. However, where management institutions for irrigation are in their infancy or are not well structured, and traditions of farmer participation are not well developed, experience would suggest smaller schemes are more likely to be successful, in light of the diverse kinds of challenges that have to be met in making irrigation work.

Hervé Plusquellec, Charles Burt and Hans Wolter have introduced an additional and important distinction in the irrigation typology, and in irrigation strategies as well, by suggesting different engineering approaches to the design of surface systems. They point out that actual efficiencies tend to run at 50–85% of design efficiencies, and propose that efficiency levels can be raised significantly through use of *modern design concepts*. Many irrigation systems routinely fail to achieve the basic goal of delivering water to farmers in the requested quantities and with the specified timing. They emphasize *the importance of reliability of water deliveries, and one precondition for this is maintaining stable water levels in main canals*. The statement of the problem by Plusquellec, Burt and Wolter is as follows:

> Many designs are difficult to manage under real conditions. Operating instructions are often conflicting and sometimes meaningless. Murray-Rust and Snellen,[64] studying the Maneungteung Irrigation Project in Indonesia, observe:

> > The system calls for bi-weekly assessment of demand for every tertiary block, and a re-adjustment of every gate in the system to meet the changed water distribution plan. This requires a very intensive data collection program and an efficient and effective information management system. Because it is

carried out in an environment of unpredictable water availability it becomes almost impossible to achieve even if there were a huge increase in number and skills of field staff.

> Another example is the Kirindi Oya Irrigation Project in Sri Lanka. . . . it takes up to four days to reach a new steady state after changing the flow at the headworks. In the upper reaches of the canal the steady state is reached soon and the water level fluctuations are low. But in the lower reaches of the canal the water level fluctuations of about one meter occur for up to four days. . . . If the discharge at the headworks is only changed once in a week, steady flow conditions are rarely achieved. . . .

> Some irrigation designs guarantee anarchy at the turnouts. When water delivery is erratic, water users lose respect for the rules and regulations governing water usage. These conditions lead to passive water user associations (WUAs) and tremendous damage to distributaries and turnouts. Reports judge the incidence of such damage to be as high as 80 percent, even in some Asian countries with long traditions of irrigation. Such anarchy is not a necessary part of an irrigation project, nor is it inherent to any culture. . . . The authors contend that inadequate design and operation is a much more significant factor in creating conflicts and disorder than the absence of irrigation tradition or social and legal norms.[65]

These authors propose systematic investigations as to why an irrigation system is not fulfilling its operational potential and, when appropriate, modification of the design to make the system simpler to operate and more effective. Their approach to system design is worth review-

64. D. H. Murray-Rust and W. B. Snellen, *Performance Assessment Diagnosis* (draft), International Irrigation Management Institute (IIMI), International Livestock Research Institute (ILRI) and Institute for Water Education (IHE), 1991.
65. Hervé Plusquellec, Charles Burt and Hans W. Wolter, *Modern Water Control in Irrigation: Concepts, Issues and Applications*, World Bank Technical Paper No. 246, Irrigation and Drainage Series, The World Bank, Washington, DC, USA, 1994, pp. 2–4.

ing at a policy level, in the course of developing irrigation strategies. It includes the following elements:

> A good design increases reliability, equity and flexibility of water delivery to farmers. It reduces conflicts among water users and between water users and the irrigation agency. It leads to lower operational and maintenance costs.
>
> *Extended gravity irrigation schemes with manually operated gates and control structures rarely work*, despite all efforts to improve irrigation management and the capacity of staff. The performance is sometimes inferior to systems without adjustable structures. Basically there are two options to improve irrigation performance: (a) simplification through proportional dividers, unadjustable gates and rigid scheduling, or (b) modernization through the application of hydraulic principles, automation, improved communication and decentralization. . . .
>
> A good design [produces] the simplest and most workable solution. A good design is user friendly and not necessarily synonymous with 'high costs', 'high maintenance', or 'complexity of operation'. . . . Some modernized irrigation projects have failed because of improper choice of control structures, incompatible components, and a design that was not based on realistic operation and maintenance plans. This has created the wrong impression that modern design concepts are not suitable for the environment of developing countries.[66]

According to these authors, the characteristics of modern irrigation design include robustness, good communications systems and 'social capital' in the sense that the users have a sufficient degree of mutual trust and participate in designing and supervising the water allocations:

> Each level is technically able to provide reliable, timely and equitable water delivery services to the next lower level. . . . An enforceable system is in place that defines the mutual obligations and creates confidence at each level that the next higher level will provide reliable, equitable and timely water delivery service. . . . Good communications systems exist to provide the necessary information, control and feedback on system status. . . .
>
> The hydraulic design is robust, in the sense that it will function well in spite of changing channel dimensions, siltation, and communications breakdowns. Automatic devices are used where appropriate to stabilize water levels in unsteady flow conditions. . . .
>
> Engineers do not dictate the terms of water delivery; rather, agricultural and social requirements are understood and satisfied at all levels and at all stages of the design and operation process within overall resource availability.[67]

The FAO properly cautions against overemphasis of the engineering aspects of irrigation systems,[68] but nonetheless the modern design approach appears to offer significant practical advantages. The basic point here is that *improved system design, generally in the direction of simplifications, can make the system operations both more efficient in water use and more equitable among irrigators*. There would appear to be a need for further training of engineers who are involved in the design of irrigation systems and scope for regularly seeking additional professional opinions on designs before they are implemented.

---

66. *Op. cit.*, pp. 5–6 [emphasis added].
67. *Op. cit.*, pp. 6–7.
68. 'Another important influence on water resource policy is societies' partiality for technical solutions. In most countries, water management is typically relegated to the engineering domain. Indeed, most water managers are engineers, who are trained to solve technical problems. As inadequate public policies are increasingly blamed for water-related problems, a strong case is emerging for emphasizing human behavior as an additional component of water systems' (FAO, 1993, p. 257).

An ironic illustration of the importance of appropriate system design is provided by the contrast between the functioning of the older and newer irrigation works in Egypt:

Despite a minimum of management [the] traditional irrigation system has a high overall efficiency. . . . The irrigation infrastructure in the newly reclaimed lands from the desert ('new lands') is designed along the same lines as the 'old lands', except that canals are lined. In the 'new lands', however, the lack of water management devices, night or buffer storage and the inability to recycle spills has led to very low efficiencies. The heavy losses have caused water logging in adjacent 'old lands'.[69]

The option of the modern design approach should be considered not only for new systems but also when reviewing systems that need rehabilitation. 'Whether to rehabilitate current projects to existing standards only or upgrade them to standards for (future) adoption of improved irrigation technologies on the farm is an issue that has not been sufficiently addressed in rehabilitation projects'.[70] Van Tuijl had in mind the alternative of pressurized irrigation, but the same statement could be made with respect to modern design improvements in canal and control infrastructure. A call for design modification by simplifying the infrastructure and operations requirements by converting to fixed and automatic controls that need less discretionary intervention was issued by E. B. Rice, for the systems he studied in south east Asia.[71]

Moris and Thom have underscored the need to *adapt the engineering to the local agronomic and socio-economic conditions*, which may place severe limits on system potentials: 'Why are so many irrigation projects in Africa designed and justified for double-cycle cropping when it is common knowledge that few projects can attain such cropping intensities? Why do field specifications continue to call for wire gabions in surroundings where people have a high incentive to steal the wire? . . . By parceling out specialized tasks within the project cycle, donors have insulated the specialists involved at the design phase from the necessity of learning from past mistakes'.[72]

While improved design of the infrastructure for water delivery and more effective system management can contribute to significantly better performance of irrigation, there are two problems that have proven rather intractable, worldwide: the tendency toward waterlogging and salinization, and the difficulty of maintaining level plots. When land is not level, irrigation efficiencies drop sharply, and problems of waterlogging can be exacerbated. It is recommended that special attention be paid to these two issues, in preparing both new investments and rehabilitation projects, and as a matter of basic orientations of irrigation strategies.

Appropriate system design is also critical from a gender perspective. To ensure gender issues are addressed in the design, it is important to carry out a gender analysis of the involved communities first, with special emphasis on identifying the agricultural and water-related tasks that women carry out. The design process should be participatory and women's groups should be consulted without the presence of men. Demonstration visits to functioning irrigation sites should include women, and it is important that participants in the scheme understand what will be the workload implications for both men and women.[73]

In closing this brief review of issues related to types of irrigation, it is worth reiterating the importance of adequate planning of use of groundwater and the option of conjunctive use of surface and groundwater in some locations, since in effect groundwater represents stored water that can be used during droughts. At the same time, it

69. W. Van Tuijl, 1993, p. 20.
70. *Op. cit.*, p. 30.
71. E. B. Rice, 1997, p. 5.
72. J. R. Moris and D. J. Thom, 1991, p. 154.
73. Useful recommendations on dealing with gender issues in irrigation design are found in the 1999 study by Chancellor, Hasnip and O'Neill, mentioned in the box on p. 223.

is important to be aware of *the hazards of using systems based on pumping in environments that cannot readily support them.* In the words of Moris and Thom:

> In Africa today there are probably more pumps that do not pump than there are those that do. ... Pumps, when employed by unskilled operators without benefit of adequate mechanics and parts, may not last through their second season – the actual experience of many ... during the first phase of a USAID-assisted Mali Project. ...
>
> .... Pumps appear to have been *very* problematic when introduced into the more remote setting where small schemes are often located. The sources of the difficulty are the following:

— The great vulnerability of 'orphan' equipment which cannot be supported within the immediate commercial environment;
— The fact that a pump breakdown may immediately threaten the associated production system;
— Poor maintenance, which leads to a high rate of breakdowns and rapid deterioration of equipment;
— Frequent problems associated with obtaining fuel or power to keep pumps operating;
— Farmers' inability to pay operating costs at the times needed;
— Difficulties caused by fluctuating water levels, which may exceed a pump's lift capacity; and
— The generally poor quality of backup services (mechanics, parts, assistance, etc.).[74]

A solution to this problem is for international agencies to support the development of pumps based on local materials, as has been done in Nicaragua with the development of the award-winning 'mecate' (jute rope) pumps that are now widely used throughout the country and other parts of Central America.

> *Longdale irrigation scheme [in Zimbabwe] received an electric pump from DANIDA in 1993. Because no information was given on service requirements, the pump received little attention and has recently begun to give problems. In 1998, the pump was out of action for a number of months due to failure of a rubber seal, which led to the loss of one season's crops. A replacement rubber seal could not be found in Zimbabwe or South Africa. Finally the seal was reconditioned in Masvingo, but this is only a temporary measure and a new seal will be needed in the near future* (F. Chancellor, N. Hasnip and D. O'Neill, Gender-Sensitive Irrigation Design, Guidance for Smallholder Irrigation Development, *H. R. Wallingford Ltd, Report OD 142 (Part 1), Department for International Development (DFID), Wallingford, Oxon, UK, December 1999, p. 32*).

6.5 PRINCIPAL POLICY ISSUES IN THE IRRIGATION SECTOR

6.5.1 Instruments for Managing Water Demand

In light of the increasing scarcity of water relative to its demand in many countries, and the growing recognition of its economic value in sectors other than agriculture, the management of irrigation demands has become a central issue in irrigation policy. Prices of irrigation water usually are too low to influence demands significantly, and therefore recourse is frequently made to other mechanisms for allocating and reallocating water. The allocational issue is fundamental for several purposes: ensuring that water is allocated to its most productive and socially desirable uses, ensuring that equity objectives are satisfied in the provision of water, achieving water conservation objectives and minimizing negative externalities (environmental damage). Such objectives would be given concrete expression in a national water strategy and the challenge for policy and projects is to effect an allocation of water that is consonant with them. The word 'reallocation' is used as

---

74. J. R. Moris and D. J. Thom, 1991, pp. 176 and 178.

well, because water demands, equity requirements and opportunities for productive uses change over time, and so the allocational mechanisms that are used need to have built-in flexibility and responsiveness.

There are three principal systems of demand management, or water allocation, in use today. As explained by Meinzen-Dick and Rosegrant, they are as follows:

- *Administrative allocation* of water includes publicly managed allocation of water ... through quantity distributions or administered water pricing schemes. Quantity-based administrative water allocation is the traditional mode of operation in most large developing country irrigation systems and is by far the most common mechanism in use at all levels in the developing world today.
- *User-based allocation systems* are controlled by users with a direct stake in the use of water, often operating within the confines of a predefined water right. Institutions undertaking this type of allocation are irrigation districts, groundwater districts, cooperatives, irrigator associations, village-based organizations, or more informally constituted user groups.
- *Market allocation of tradable water rights* attempts to structure economic incentives for water users, whether irrigation, industrial, or municipal users, to consider the full opportunity cost of water when making water use decisions. ...

These three allocation mechanisms correspond to the three sectors in which water resource management can take place: the public sector, the collective action sector, and the private sector. These differ, among other things, by the assignment of property [usufruct] rights: Property rights in the public sector are assigned to the State, in the collective action sector to groups, and in the private sector to individuals.

In practice, different allocation mechanisms often overlap. For example, there may be public allocation between sectors and within large-scale irrigation systems, with user-based allocation ... on tertiary distribution units, and market allocation of groundwater used conjunctively to supplement surface irrigation. ... No single allocation type will be optimal, or even appropriate, for all situations.[75]

Of the three mechanisms, market allocations are the least frequently used but there is growing interest in them (see below). Administrative allocations in practice do not place much discretion in the hands of the managers of irrigation systems. In reality, allocations of water are usually decided by system design (e.g. strict rotations) or farmers' requests (within system parameters).

While the above-mentioned three mechanisms for allocation cover most of actual practice, two others should be mentioned that are relevant in some circumstances:

- *Joint allocation by users and governmental agencies.* The best-known example of this mechanism is the set of participatory institutions created for managing water supplies in each of the six principal watersheds in France.
- *Individual allocational decisions made by owners of infrastructure.* The most obvious example of this mechanism is provided by the irrigation decisions made by owners of wells. It may seem a trivial mechanism but there are hundreds of thousands of such owners in each region of the world. In some cases, e.g. in Yemen and in Tamil Nadu, India,[76] some of their decisions have been in the direction of reallocation, to sell their water for non-agricultural uses (usually through informal market arrangements), while others have continued to allocate their pumped water strictly within the farm, guided to varying

---

75. R. S. Meinzen-Dick and M. W. Rosegrant, 1997, pp. 210–211.

76. The informal water markets in Tamil Nadu are described in M. W. Rosegrant, R. Gazmuri S. and S. N. Yadav, 'Water policy for efficient agricultural diversification: market-based approaches', *Food Policy*, **20**(3), June 1995, p. 207.

An important aspect of the French system is that water resources are managed at the level of the river basin. There are six river basin committees and six river basin financial agencies, whose territories closely correspond to the main river basins. They specialize in water resource management (planning and macromanagement), which they have performed efficiently for twenty-five years. The river basin committees facilitate co-ordination among all the parties involved in managing water resources. These committees have become the center for negotiations and policymaking at the river basin level. . . . The committees approve long-term (twenty to twenty-five year) schemes for developing water resources. Every five years they vote on action plans to improve water quality. In addition, they vote annually on two fees to be paid by water users within the river basin: one fee based on the level of water consumed and the other one on the level of pollution at each point source. . . . The committees are composed of 60 to 110 persons, who represent interested parties: the national administration, regional and local governments, industrial and agricultural groups, and citizens (The World Bank, 1993, p. 46).

degrees by an appreciation of the value of water in alternative uses.

Meinzen-Dick and Rosegrant have also summarized some of the justifications for and issues concerning administrative allocations of water by governments:

Heavy State involvement in water allocation has been justified based on the strategic importance of the resource, the scale of systems required to manage it, and the positive and negative externalities in its use. . . . The large-scale systems used to deliver much of the water

for irrigation and municipal needs lend themselves to natural monopolies and would be beyond the capacity of most communities or private firms to organize and fund. The positive externalities [and the] high individual costs of internalizing negative externalities such as deterioration of water quality from agricultural runoff, sewage, and industrial effluents, or deterioration in groundwater levels, provide further arguments for a strong State role. . . .

The State's role is particularly strong in intersectoral allocations, as the state is often the only institution that includes all users of water resources and has jurisdiction over all sectors of water use. . . .

While public allocation or regulation is clearly necessary at some levels, government-operated irrigation . . . [tends] to be expensive to operate and often [fails] to live up to expectations. . . .

Under public management the dominant incentive to comply is coercion – that is, setting regulations and using sanctions for those who break them. But this type of incentive is effective only if the State detects infractions and imposes penalties. In many cases the State lacks the local information and ability to penalize for infractions such as breaking water delivery structures or excessive water withdrawals. Enforcement is more effective where there are fewer points to monitor. It works better, for example, for main canals of large irrigation systems than it does for small-scale irrigation. . . . [77]

Market-based and user-run allocational mechanisms are gaining increasing favor throughout the world. There is a strong trend towards transferring operation and maintenance of irrigation systems to users, and by now considerable experience has been accumulated regarding such transfers.[78] These mechanisms are discussed in subsequent sections. However, first the principal

77. R. S. Meinzen-Dick and M. W. Rosegrant, 1997, pp. 211–212.

78. Many of the lessons of this experience are summarized in D. L. Vermillion and Juan A. Sagardoy, *Transfer of Irrigation Management Services, Guidelines*, FAO Drainage and Irrigation Papers No. 58, Food and Agriculture Organization of the United Nations, Rome, 1999.

concepts and issues concerning the pricing of irrigation water are reviewed. In irrigation experiences to date, pricing and allocation of water usually have not been closely related, unlike the situation of many other resources and almost all commodities.

6.5.2 The Pricing of Irrigation Water: Preliminary Considerations

The rules for pricing irrigation water vary widely, within and between countries. It appears that the rules for setting prices are neither systematic nor uniform.[79] The only consistent feature of irrigation prices is that they usually are well below the cost of supplying the water. As the World Bank has stated, tariffs for irrigation water are usually far below even the prices of municipal water, which are inadequate for covering costs, and most governments have not even put forth a policy of trying to recover irrigation system costs through water charges.[80]

A review of international experience in water pricing concluded that, for 13 developing countries, the recovery of operating and maintenance costs ranges from 20–30% in India and Pakistan to about 75% in Madagascar.[81]

In spite of the reluctance of many governments to charge more for irrigation water, experience

---

Once established, irrigation projects become some of the most heavily subsidized economic activities in the world. In the mid-1980s, Repetto estimated that the average subsidies to irrigation in six Asian countries covered 90 percent of the total operating and maintenance costs. Case studies indicate that irrigation fees are, on average, less than 8 percent of the value of benefits derived from irrigation (FAO, 1993, p. 232).

---

has shown that farmers may be willing to pay more for it *provided that* the supply is reliable. This is a major caveat that often is not satisfied in gravity-fed systems. It has been pointed out that:

> Farmers are noted for their reluctance to pay water rates for water from public irrigation schemes. It is interesting to note, however, that the same farmers are often willing to spend considerable amounts (per unit volume of water) to develop groundwater supply. The obvious conclusion is that farmers are willing to pay for water if it is reliable and somewhat flexible.[82]

The World Bank has noted the willingness of poor farmers to pay for irrigation, linking it to the reliability of the supply of water:

> Information on communal and private irrigation systems in various countries in Asia shows that even very poor farmers will pay high fees for good-quality and reliable irrigation services.

> • In Bangladesh, it is not uncommon for farmers to agree to pay 25 percent of their dry season irrigated rice crop to the owners of nearby tubewells who supply their water.
> • In Nepal . . . farmers contribute large amounts of cash and labor to pay the annual cost of operations and maintenance. For example, in six hill systems studied in detail, the average annual labor contribution was sixty-eight days per hectare. In one thirty-five-hectare system, annual labor contributions were approximately fifty days per hectare, while cash assessments averaged the equivalent of more than one month of labor.

> Although many of these farmers are very poor in an absolute sense, they are willing to

---

79. R. K. Sampath, 1992, p. 973.
80. The World Bank, 1994, p. 36.
81. Ariel Dinar and Ashok Subramanian, *Water Pricing Experiences: An International Perspective*, World Bank Technical Paper No. 386, The World Bank, Washington, DC, USA, 1997, p. 8.
82. H. Plusquellec, C. Burt and H. W. Wolter, 1994, p. 11.

pay for good-quality irrigation services that raise and stabilize their income. Thus, the critical issue is providing these poor farmers with reliable, profitable, and sustainable irrigation services.[83]

Thus, *a principal constraint to the implementation of higher water prices is the poor functioning of many irrigation systems.*

One of the principal questions about irrigation pricing is: To what degree would farmers alter their behavior if the price of water were significantly higher? Is there empirical evidence that they would take actions to conserve water or to apply it to higher-value crops? Rosegrant *et al.* have cited evidence, some of it indirect, in Nepal, Tamil Nadu, Jordan and Chile, which strongly indicates that farmers have responded to a higher price or opportunity cost of water by conserving water, improving irrigation efficiencies, and/or shifting their cropping patterns to higher-valued crops when the corresponding product markets are available. The results for the case of Chile are especially clear, after reforms that introduced markets for water:

> Reform significantly increased the scarcity value of water, and the area planted to fruits and vegetables, which require less water per value of output than field crops, increased during the period 1975–1982 by 206 600 hectares, replacing traditional crops and irrigated pastures. In addition, aggregate water use efficiency in Chilean agriculture increased by an estimated 22–26% between 1976 and 1992. . . .[84]

By the same token, the opposite result of wasteful use of water, or over-exploitation of aquifers, has been observed in cases in which water is unduly cheap to the user. The case of Yemen, presented in the annex to this book, is only one example.

In summary, it appears clear that irrigation charges generally are very low and that raising them would cause farmers to respond in the desired direction of greater efficiency, as well as

generating more fiscal revenues in most cases. Why, then, do the charges remain consistently so low? What kind of policies or institutional arrangements would tend to increase them? Should they be increased in all cases, or are there justifiable exceptions?

6.5.3 The Pricing of Irrigation Water: Conceptual Issues

To provide answers to the foregoing questions, it is useful to begin with a review of the motives for raising water prices and the logic of each institutional arrangement with respect to prices. The point of reference is that, for irrigation water, price does not play the normal role of equilibrating supply and demand, except in the case of markets for water rights, which so far have had limited applicability. Accordingly, in most cases, the justification for the level of this price must be different than that of balancing supply and demand.

Within a framework of overall policy objectives such as the discussion in Chapter 2 of this book, there are five fundamental reasons, or subsectoral objectives, for setting the price of irrigation water at an appropriate (which usually means higher) level. The first three reflect societal concerns for the utilization of a scarce resource and the last two, fiscal concerns. They are as follows:

(i) To stimulate the conservation of water.
(ii) To encourage the allocation of water to the most water-efficient crops, i.e. to its highest-value agricultural uses, or to non-agricultural uses if water is more productive there in net terms, after allowing for intersectoral conveyance costs, provided that the infrastructure exists for delivery of the water to new users. This kind of allocation would maximize the economic growth benefits of a scarce resource, although it should be pointed out that often the intersectoral infrastructure condition is not satisfied in irrigated areas in developing countries.

---

83. The World Bank, 1993, p. 50.
84. M. W. Rosegrant, R. Gazmuri S. and S. N. Yadav, 1995, p. 208.

(iii) To minimize the environmental problems attendant upon irrigation, especially those arising from excessive use of water.

(iv) To generate enough revenues to cover operating and maintenance costs of irrigation systems so that, among other things, it would not be necessary to invest in expensive rehabilitation projects.

(v) To recover the original investment costs of each system, in addition to providing revenues for O & M costs.

Daniel Bromley has argued that promoting economic efficiency should not be the relevant goal for irrigation pricing policy, but rather:

the purpose of the irrigation pricing regime should be to ensure that water allocation within an irrigation system (or community of irrigators) is optimal with respect to the efficient operation of the system as a domain of shared access to a scarce resource. . . . water pricing must be seen as part of a regime in which farmers are induced to contribute to a public good – improved water management – that benefits each of them. The principle of reciprocity requires that all individuals contribute to the public good exactly that amount that they would most prefer every member of the group to make.[85]

In practice, Bromley's prescription would lead to price levels that cover O & M costs but would not rise to the level of opportunity costs (efficiency prices) of water.

In some circumstances, there may be one argument for *not* raising the price of irrigation water, or for limiting its increases:

(vi) If irrigators represent poor rural households, a higher price of water may cause them real economic hardship. This is the *equity concern.*

Whatever the merits of this last argument, it is evident that *raising* the price of irrigation cannot serve the equity goal of reducing rural poverty in the short run. It may be argued that improved cost recovery would enable the government budget to build more irrigation projects in the future, although it is doubtful whether most farmers would believe that such a linkage is firm. A more persuasive argument is that higher water rates would lead to a more efficient allocation of that resource, creating more employment and higher incomes. Nevertheless, the basic case for raising irrigation fees derives from the efficiency objective and the policy principles of fiscal and environmental sustainability (see Chapter 2). Since irrigation systems are not sustainable without cost recovery, there is a persuasive argument that irrigation fees are not an appropriate policy instrument for addressing the needs of the rural poor and, as argued in other chapters of this book, development of better research and extension and rural financial systems are apt instruments for addressing rural poverty.

The first three objectives above are associated with ***pricing policy for demand management***, while the last two refer to ***pricing policy for cost recovery***. If there is to be an irrigation subsidy for purposes of poverty alleviation, it pertains to the latter category of policy, i.e. it is calculated as part of public budgetary policy and constitutes a deliberate decision not to recover all costs, either of investment or of O & M, or both, and to provide public funds in place of user contributions, at least to a degree. However, obviously a reduction of irrigation charges for this purpose would have collateral, negative effects on the degree of achievement of the other three objectives. In any event, it would be difficult to justify reducing them below a level which is adequate for funding operation and maintenance.

Care must be exercised in adducing other 'social goals', such as higher levels of agricultural production and lower urban food prices, as justifications for low irrigation rates, with the possible

---

85. Daniel W. Bromley, 'Property Regimes and Pricing Regimes in Water Resource Management', in Ariel Dinar (Ed.), *The Political Economy of Water Pricing Reforms*, Oxford University Press, New York, 2000, pp. 37 and 47.

exceptions of extreme poverty and important gender concerns. However, irrigated farms almost always generate incomes above those of the poorest rural strata. Promoting the goal of higher agricultural production through subsidized irrigation would appear to represent a move in the direction of greater efficiency, but the distortions caused by low irrigation prices cause inefficiencies that may well cancel the putative gains in production. Recall the experience of Chile mentioned above in which *higher* irrigation prices generated changes in the direction of higher-value crops. Similarly, concerns about the welfare of urban populations are most appropriately addressed through targeted subsidies for those groups, and not through low agricultural prices (see the discussion of policy objectives in Chapter 2 and the discussion of subsidies in Chapter 3).

Objective (ii) is in essence the efficiency objective as interpreted for the irrigation sub-sector. For the latter, pursuing that objective via pricing policy requires that the price of irrigation water reflects the (marginal) productivity of water in its most productive alternative use – its opportunity cost. Sampath (1992, p. 972) correctly points out that, according to the general theory of second-best, the absence of competition in supplying irrigation water to farmers means that setting the water price equal to its marginal cost (of supply) does not necessarily guarantee the most efficient economic outcome. Nonetheless, the economic gains from setting the water price at its opportunity cost can be very real and significant, as Robert Hearne and William Easter have concluded in their analysis of the experience of four watersheds in Chile.[86]

The objectives pursued in practice vary with the institutional character of irrigation prices. When those prices are established by governmental agencies, the primary reasons usually are fiscal, that is, objectives (iv) and (v) above. An enlightened policy may be motivated also by objectives (ii) and (iii) and, especially if water is scarce, objective (i). Farmers, whose primary concern is their income level, are not likely to concur in the importance of all these objectives. An increase in administered irrigation prices 'is correctly perceived by rights holders as expropriation of those rights, which would create capital losses in established farms'.[87] Frequently, however, they can be persuaded that recovering O & M costs (objective (iv)) is worthwhile, for the sake of keeping the system viable. Hence, dialogs between governments and farmers tend to center on the O & M factor.

Typically, the level of irrigation fees required to recover O & M costs is well below the value of water in alternative uses. Most research has found that the average and marginal water productivity are greater than the average and marginal costs of making water available.[88] Furthermore, fees set at a level to recover costs still are usually well below the opportunity cost of water.[89] For the case of the irrigated districts of Mexico, which cover about 2.8 million hectares, Ronald Cummings and Vahram Nercissiantz found that although the Mexican National Water Commission has a legal mandate to collect O & M costs for systems, water prices to irrigators are as low as 4% of the scarcity value of water.[90] *Therefore, experience suggests that a dialog that is developed around the objective of financing O & M activities is not likely*

86. Robert R. Hearne and K. William Easter, *Water Allocation and Water Markets: An Analysis of Gains from Trade in Chile*, World Bank Technical Paper No. 315, The World Bank, Washington, DC, USA, 1995, in particular, pp. 38–41.
87. M. W. Rosegrant and H. P. Binswanger, 'Markets in tradable rights: potential for efficiency gains in developing country water resource allocation', *World Development*, **22**(11), November 1994, p. 1619.
88. G. F. Rhodes, Jr and R. K. Sampath, 'Efficiency, equity and cost recovery implications of water pricing and allocation schemes in developing countries', *Canadian Journal of Agricultural Economics*, **36**(1), March 1988, p. 116.
89. The World Bank, 1993, p. 50.
90. R. G. Cummings and V. Nercissiantz, 'The use of water pricing as a means for enhancing water use efficiency in irrigation: case studies in Mexico and the United States', *Natural Resources Journal*, **32**, Fall 1993, pp. 739 and 745.

to raise the price (opportunity cost) of irrigation enough to promote the efficiency objective to a significant extent.

When irrigation prices are established by water users' associations (WUAs), again the primary concern is likely to be funding O & M activities. Raising revenues for additional irrigation projects in other areas, or promoting economic efficiency at the cost of paying a higher 'tax' on water, is not normally seen to be in their interest. They can become motivated to save water (objective (i)) *if* that means being able to expand their own irrigable area with the water saved, but sometimes limitations of the irrigation system itself preclude that option unless substantial further investments are undertaken.

When the price of irrigation water is established by water markets, it usually turns out to be higher than if established by governmental agencies or users associations, because of the aforementioned relation between the marginal value of water and the marginal cost of supplying it. Hence objectives (i) through (iii) are likely to be satisfied by water markets, and normally this means satisfying (iv) as well. Whether the price would be high enough to represent recovery of the investment cost (objective (v)) would depend on the magnitude of that cost. However, *there is a fundamental difference between a market-determined price for water and a decreed price: if the price rises because of demand for water, as transmitted through markets for water rights, then farmers can be beneficiaries of the price rise, by selling their water rights*. They are compensated for giving up those rights. Obviously, they would do so only if the resulting gains are greater than what they would earn by irrigating with the same amount of water.

When irrigation prices are increased by administrative decision, whether by public agencies or users' associations, then farmers are losers rather than beneficiaries in the cash-flow sense, although they may be net beneficiaries in the longer run if adequate pricing can lead to better O & M, and thus prevent deterioration of the system.

In the paradigm of the private owner of a source of irrigation water, typically a well (tubewell or borehole), the owner pays the full supply cost of the water – not through a price but through the investment outlay and the subsequent operating and maintenance costs. Hence, indirectly, objectives (iv) and (v) are served. Since these costs, when considered together in discounted annual form, are considerably higher than the annual O & M costs alone, production patterns under the individually owned water source will come closer to satisfying objectives (i), (ii) and (iii) than will production in most gravity-operated, publicly owned systems. It is for this reason that irrigation from wells is more likely to be associated with the production of higher-value crops, although of course there are many exceptions. The more expensive the well (which is mainly a function of the depth of the water table), then the more likely this relationship is to be true.

What other institutional arrangements would lend themselves to recovery of full investment costs in an irrigation system? The answer by now is evident: systems in which the users are its owners. The investment costs (or a policy-determined portion of them) are recovered through the proceeds of the sale of the system to the irrigators. The ownership refers not to the water itself, but to the infrastructure for irriga-

---

Irrigation systems owned jointly by users are still relatively rare but by no means unknown. Rosegrant et al. *mention an example in Nepal:*

The Chherlung Thulo Kulo irrigation system was developed by initially issuing altogether 50 shares to the 27 households. Shares were issued based on proportional contribution of farmers to the initial investment costs of the system, and entitled the holder to 1/50 of the total water delivered by the system. Demand and supply factors have led to market transfers of shares among farmers as well as a rise in the prices of the shares over time. . . . Opportunity to sell and buy shares has created a high opportunity cost of water and thus incentives to conserve water, leading to expansion of irrigated area and improvement in overall technical efficiency of water use (M. W. Rosegrant, R. Gazmuri S. and S. N. Yadav, 1995, p. 207).

tion and to the rights to use the water conveyed by it.

Recent reforms of the irrigation systems in Shaanxi Province, China, have included experiments with six different models of local responsibility for irrigation systems. One of them is the ownership model of 'joint stockholders':

> This model converts communal ownership to shares. Portions of the irrigation system are divided into shares and these are sold to farmers, local residents, irrigation agency staff, and other local officials. Property rights belong to the individuals but the operation of the system is collective. . . . Part of the funds from the sale of shares is used to improve laterals as well as to expand the service area. In addition to paying O & M costs, a percentage of the irrigation water fees are used to pay a return on the investment to share holders.[91]

Not only are pricing objectives (iv) and (v) attained in fully privatized systems, but objectives (i) and (ii) as well, and it is likely that objective (iii) is at least partly achieved, owing to the conservation of water. However, protection of the environment (and third-party interests) almost always requires at least a degree of regulation on the part of the government. In all countries, in the final analysis, the degree of protection afforded to the environment is basically a political decision.

In Chile, the establishment of tradable water rights has been accompanied by joint ownership of the irrigation systems by users, and their associations not only manage the infrastructure and supervise water allocations, but they also approve water transfers subject to specific conditions and constitute the initial (and usually definitive) forum for resolving conflicts.[92]

Other examples of privately owned irrigation systems include those found in Eastern and Southern Africa. It is considered that some of Africa's most efficient irrigation systems are those in the privately owned, estate sector, such as those that have been functioning in Zimbabwe's Hippo Valley and Triangle estates.[93]

Among other benefits, *user ownership of irrigation systems provides a more complete set of incentives for irrigators to make the financial contributions that are necessary for maintenance.* An owner of any structure (house, factory, commercial building, etc.) maintains it for two reasons: (i) to keep it in adequate operating condition and to avoid deterioration which would cause major outlays on repairs, and (ii) to sustain the possibility of realizing capital gains in the future. Groups of irrigators who are responsible for system maintenance but are not system owners do not have the second motive for contributing to a maintenance. For them, the first motive also may be weak if they believe that the government, as system owner, will step in and fund any necessary rehabilitation in the future. Transferring ownership of the system to its users removes any doubt on this last score and thus provides a complete set of motives for ensuring adequate maintenance.[94] The formation of a water users' association in itself does not provide complete incentives.

For these reasons, it would appear useful to direct more research and policy attention toward the possibilities of ownership of irrigation systems by the participating groups of farm households. In instances of older systems that have deteriorated, a prerequisite of the transfer would be the government's carrying out a rehabilitation program.

---

91. 'Six Irrigation Management Models from Guanzhong', *INPIM Newsletter*, International Network on Participatory Irrigation, Washington, DC, No. 11, March 2001, pp. 8–9.

92. *Ibid.*

93. J. R. Moris and D. J. Thom, 1991, p. 20.

94. It may be argued that, since the value of water use rights is usually capitalized in the value of land, this relationship alone provides the opportunity for the realization of capital gains associated with the irrigation infrastructure. However, without ownership of the infrastructure the water rights may not be completely secure, and equally the landowners may not be certain that the infrastructure will be fully maintained. On both counts, the capital gains motive is weaker without ownership of the infrastructure.

Notwithstanding the foregoing arguments, a cautionary note regarding privately owned irrigation systems has been sounded that is worth bearing in mind in designing systems and their transfer to the private sector, in order to take remedial measures:

While privatized small-scale irrigation systems are evident in many countries, the existence of market failures [in water] calls for strong regulatory institutions. Principal among the market failures is the presence of externalities caused by unsustainable use of water, such as excessive drawdown of aquifers. Private sector development is also affected when the supply of irrigated water is highly variable and/or systems are complex, which makes drawing up contracts to cover all contingencies difficult.[95]

6.5.4 Irrigation Pricing Systems

The effectiveness of irrigation pricing in achieving the objectives listed above, particularly (i) through (iii), depends not only on the level of the price but also to some extent on the kind of pricing system employed. Yacov Tsur and Ariel Dinar describe eight methods of pricing irrigation water: volumetric pricing, output pricing, input pricing, area pricing, tiered pricing, two-part tariff pricing, betterment levy pricing and water markets.[96] Volumetric pricing charges per unit of water used, output pricing per unit of output produced with the water, input pricing per unit of a complementary input used (such as fertilizer), and area pricing per hectare irrigated. Tiered pricing is volume-based pricing in which the rate increases as a threshold volume is exceeded. Two-part tariffs include volumetric pricing plus a fixed fee for access to irrigation.

Some methods are basically variants of others, for example, tiered pricing and two-part tariff pricing are forms of volumetric pricing. There are yet other variations in practice. In India, area charges can vary by crop or across seasons, according to the method of irrigation (flood, ridges or furrows), and charges that are paid whether or not water is used.[97]

In the early 1970s and again in 1980, the International Commission on Drainage and Irrigation (ICID) administered a worldwide questionnaire concerning use of irrigation water and pricing practices. The sample area covered by the questionnaires was 8.9 million hectares, with conditions representative of 12.2 million hectares. This latter amount is equivalent to about 5% of the world's total irrigated area. M. G. Bos and W. Wolters analyzed those data and found that area pricing, which is the easiest pricing method to administer, was used in more than 60% of the projects sampled. In another 25% of the projects, volumetric pricing was used, and in most of the remainder, a combination of area and volumetric pricing.[98]

Y. Tsur and A. Dinar have conducted a conceptual analysis of the optional pricing systems from a viewpoint of promoting economic efficiency. They hypothesize that efficiency is achieved when the price of water equals the marginal cost of supplying it, in spite of Sampath's reminder (mentioned above) that, given the actual conditions of irrigation systems, the general theory of second-best asserts that achieving this equality condition does not guarantee economic efficiency. They conclude that volumetric pricing is the most efficient in the absence of implementation costs, but that if such costs (mainly the

95. Ashok Subramanian, N. Vijay Jagannathan and Ruth Meinzen-Dick, 'User Organizations in Water Services', in A. Subramanian, N. V. Jagannathan and R. Meinzen-Dick (Eds), *User Organizations for Sustainable Water Services*, World Bank Technical Paper No. 354, The World Bank, Washington, DC, USA, 1997, p. 6.
96. Y. Tsur and A. Dinar, 'The relative efficiency and implementation costs of alternative methods of pricing irrigation water', *The World Bank Economic Review*, **11**(2), May 1997, pp. 243–262.
97. *Op. cit.*, p. 247.
98. M. G. Bos and W. Wolters, 'Water charges and irrigation efficiencies', *Irrigation and Drainage Systems*, **4**(3), August 1990, pp. 267–278.

costs of collecting information on amounts of water used) amount to 10% or more of water proceeds, then area pricing becomes more efficient.[99] They point out that water markets can be efficient in their sense, especially as the costs of gathering information are internalized by users, provided that the costs of the institutions and physical infrastructure necessary for such markets have already been incurred.

In another theoretical study, G. Rhodes and R. Sampath also found that volumetric pricing leads to greater efficiency than area pricing does, and that the latter is superior to attaching the water price to proxies represented by crop outputs or complementary inputs.[100]

Bos and Wolters did not find an empirical relation between the kind of irrigation charges and efficiency, probably because in the sample in general the level of charges was low in relation to net farm income. However, they found that, independently of the type of pricing, there was an increase in efficiency as the charges increased. In particular, it was found that water charges have the greatest impact on tertiary unit efficiency, and not so much by changing farmer behavior as making more funds available for operation and maintenance of the tertiary units.[101]

The FAO suggests that, as water scarcities increase, volumetric pricing schemes will be increasingly adopted, because they encourage savings of water, which area pricing does not. It also recommends consideration of two-part pricing, in which a fixed levy represents investment cost recovery and a variable levy (approximating the short-run marginal cost of supplying water) is attached to the volumes of water used, and that the water rights market constitutes the most appropriate scheme of all.[102]

Volumetric pricing may not be difficult to apply if the engineering design of the system is appropriate for that purpose. In Morocco and Tunisia, water deliveries are made to smallholders in the amounts desired, under a system of volumetric water charges. This has been possible by the construction of reliable and easily operated systems and also by programs of consolidating dispersed smaller holdings.[103]

The administration of volumetric charges can be simplified considerably by the procedure of *selling water in bulk to groups of farmers*, in prearranged amounts and timing. The farmers are responsible for allocating it among themselves and for levying upon themselves the requisite fees in order to pay for the deliveries. Usually, such groups of farmers are those occupying the tertiary commands of an irrigation system. An example of an irrigation system that has adopted this practice is the Majalgaon Irrigation Project in Maharashtra State in India:

> Under the new concept water will be sold in bulk to water user associations, each serving about 300 to 400 ha. The WUAs will get annual quotas and are responsible for the distribution of water and the maintenance of the water courses and field channels. They order the volume of water required for each irrigation turn. The internal water distribution will be essentially on rotation and proportional to the size of the holding, but other arrangements such as buying and selling of water are possible.[104]

Other examples have been cited to the same effect:

> wholesaling of water volumetrically from agencies to WUAs, who then collect charges from their members (with a portion of the fees going to support the organization) has shown

99. Y. Tsur and A. Dinar, 1997, p. 259.

100. G. F. Rhodes, Jr and R. K. Sampath, 1988, p. 116.

101. M. G. Bos and W. Wolters, 1990, p. 277.

102. FAO, 1993, pp. 272–275.

103. H. Plusquellec, C. Burt and H. W. Wolter, 1994, p. 81.

104. *Op. cit.*, p. 72.

> *[Area] rates are criticized because they do not include incentives for rationing water in line with willingness to pay [and to alter irrigation behavior accordingly]. Such schemes are, however, simple to administer and assure the supplier of adequate revenue. The high cost of installing and monitoring meters is suggested as being the main reason for continuing the [area] rate approach. This argument is convincing in cases where water is plentiful, supply costs are low and managers doubt the rationing effects of volumetric pricing. In other cases, water managers are turning to volumetric pricing to address water scarcity problems and the high costs of developing new supplies (FAO, 1993, p. 272).*

promise in pilot projects, for example, Mohini and Mula in India. . . .[105]

From these experiences, it appears probable that *the future directions of irrigation prices can be characterized by higher levels, applications in volumetric form, administration through bulk sales to WUAs and, for a small but increasing number of systems, determination by markets for water.* Changes in pricing practices may be slow, but are likely to accelerate in those localities where water is becoming more scarce.

There are three important caveats to the forecast of an upward trend in (real) irrigation prices. First, whatever the type of water pricing system, it needs to be borne in mind at all times that *reliable delivery of water is an essential condition in order to be able to raise its price.* If reliability cannot be guaranteed under an existing system design and/or administration, then those factors need to be modified before irrigation rates are raised.

The reliability factor has been mentioned before, but it is so important that its absence can

even affect farmers' willingness to dedicate their labor to O & M tasks:

> farmer follow-through on O & M below the tertiary turnout is highly dependent on the quality of the service the agency provides at the turnout, that is, whether farmers can rely on it to provide the allocated water supplies. . . . Agency inefficiency reflected in a variable water supply destroys farmers' confidence and weakens their commitment to O & M.[106]

Almost invariably, a necessary condition for improved reliability of irrigation supplies is local management of the revenues that result from water tariffs. One of the principal shortcomings of centralized systems of irrigation management is that irrigation fees are deposited in the National Treasury and O & M activities remain underfunded.

Therefore, strategies for cost recovery, whether through pricing irrigation or through user contributions of labor, are inexorably linked to issues of system effectiveness, both in terms of management and the quality of system design, and they are assisted by decentralization of financial management.

Secondly, while under appropriate circumstances it may be possible to set administered prices at a level high enough to fulfill the fiscal objectives of cost recovery, it is not realistic to expect such prices to be raised to the yet higher levels that correspond to water's opportunity costs. The reasons why this would be difficult are explained by Mateen Thobani:

> By raising the user price of water to reflect its true scarcity, or opportunity cost (that is, the price the marginal user is willing to pay), authorities hope to induce users to conserve water, making it possible to divert supplies to higher value uses. . . . In principle, if irrigation water near a city could be priced at what a

105. Ruth Meinzen-Dick, Meyra Mendoza, Loic Sadoulet, Ghada Abiad-Shields and Ashok Subramanian, 'Sustainable Water User Associations: Lessons from a Literature Review', in A. Subramanian, N. V. Jagannathan and R. Meinzen-Dick (Eds), 1997, p. 49.
106. E. B. Rice, 1997, p. 27.

water company would be willing to pay for the crude water (adjusting for conveyance costs), some farmers would give up farming, and others would switch to more efficient irrigation or grow less water-intensive crops. The higher charges would free up water that could be transferred to the water company for treatment and subsequent sale. They would also generate fiscal resources that could be used to improve the performance and maintenance of the existing infrastructure or to invest in new infrastructure.

Serious practical and political problems, however, have prevented any government from pricing water at its opportunity cost. Even if governments could find an inexpensive way to measure and monitor water flow, measuring the opportunity cost of water is difficult because it varies according to location, reliability, season, use, and water quality. . . .

The political problems are even more intractable. It is politically difficult to charge a farmer for water from a river that serves a town (and therefore has a high opportunity cost) a higher price than a farmer using water from a river that is not near a town. Similarly, it is difficult to charge profitable hydropower companies less than poor farmers [even though the former may return almost all the water to the river]. Strong farmer lobbies typically pressure politicians to keep water charges well below their opportunity cost.

Another problem in pricing irrigation water at its opportunity cost is that the price of land already embodies the price of water rights. In areas of low rainfall, irrigated land may sell for ten times the price of unirrigated land, reflecting the expectation that the owner of irrigated land will receive water at a low charge. If charges are later raised to reflect the opportunity cost of water, this land will be valued the same as unirrigated land, resulting in an effective expropriation of the farmer's assets. Although government actions frequently alter

the value of private assets, the sheer magnitude of asset expropriation implied, the numbers of people affected, and the socially disruptive aspects (in agricultural unemployment) of such a policy make it highly unlikely that opportunity cost pricing can be introduced within a reasonable time frame.

A unique problem affecting water pricing involves 'return flows'. When a farmer waters crops, only part of the water is absorbed by the plant. Depending on the efficiency of irrigation, a significant share of the water – the return flow – will seep underground. This water may enter an underground aquifer and be pumped up by another user, or it may even rejoin the river and be diverted into a canal. If water were priced volumetrically, according to what was received rather than what was actually consumed, farmers using inefficient irrigation (thereby inadvertently helping out downstream users) would pay too high a price.[107]

In principle, the issue of return flows can be handled by pricing water according to its consumptive [net] use (see Section 6.5.6 below), but the other objections would make it difficult, in most circumstances, to set administered prices at levels that approximate opportunity costs. To put such a policy into effect, it is necessary to have recourse to water markets. *The practical upward limit to administered prices is likely to be the level which corresponds to recovery of O & M costs plus, in some cases, part of the investment costs.* This conclusion confirms the appropriateness of the goal of water pricing posited by Daniel Bromley, i.e. that of improving water management, or improving the collective management of a public good.

Thirdly, gender considerations may put a brake on increases in irrigation water rates since rural women usually are less able to make cash outlays than men. This is a variant of the equity concern mentioned above in regard to irrigation pricing.

107. M. Thobani, 'Formal water markets: why, when and how to introduce tradable water rights', *The World Bank Research Observer*, **12**(2), August 1997, pp. 163–165.

In the words of Chancellor *et al.*, drawing on experiences in Zambia, Zimbabwe and South Africa:

> Despite the advantages that are expected, the current climate of 'user pays' has potential for increasing the disparity between men and women irrigators. . . . women have difficulty in asserting and retaining control over cash. Payment for water is therefore likely to be problematic for female-headed households, especially the *de facto* heads [who have] least control of the household funds. Withholding water from such households may start a rapid downward spiral in their ability to sustain livelihoods.[108]

The fourth caveat to raising water prices in agriculture is that irrigation pricing policy must also be established in the context of overall pricing policy for the sector, which is effected mainly through macroeconomic policy, as described in Chapter 4. *Inadequate incentives on the side of output prices will place an insuperable barrier on the road to raising irrigation prices.* This issue was so dominant for the projects in South East Asia studied by Rice that he recommended:

> *Abandon cost recovery.* The farmers who agreed to the terms of these schemes at startup are now paying substantial penalties because of the collapse of international and local rice prices. Their losses are reflected in consumer surpluses far larger than even full recovery of capital, as well as O & M costs, would provide. Imposing cost recovery on these paddy farmers is more likely to drive them out of farming than into diversification. . . .[109]

In brief, irrigation pricing policy, as well as long-term strategies for water development, has to be drawn up in a holistic context.

6.5.5 Water Rights Markets: An Introduction

Water rights markets, or systems of transferable water rights, have generated considerable interest in recent years in the literature on water management and economic development, in spite of the fact that formal water rights markets exist, at least until very recently, only in a few places, most notably in the Western United States,[110] Spain, Brazil, Australia, Mexico and Chile. The interest arises primarily out of the potential of such markets to foster efficient use of an increasingly scarce resource. This section of the chapter examines the nature of water rights markets, issues that need to be addressed, their advantages, and the minimum requirements for their adoption.

The enthusiasm for water markets is tempered in the minds of some practitioners. For example, in the opinion of the World Bank's Middle East and North Africa Regional Office:

> . . . market mechanisms are particularly problematic in water. . . . No doubt local water markets often operate successfully. But it is unrealistic to expect that a general reallocation between sectors or improvements in water quality can be effected through the market, at least for the foreseeable future. . . . Moreover, renewable water is a fugitive and variable resource associated with pervasive externalities. It is thus inherently difficult to manage, becomes embedded in complex institutional structures, and cannot be apportioned and regulated solely – or even significantly – by the market. Thus, government is inevitably required to: (1) establish the policy, legislative, and regulatory framework for managing water supply and demand; and (2) ensure that water services are provided, notably by constructing large-scale projects . . . for which economies of scale or social externalities preclude private supply.[111]

108. F. Chancellor, *et al.*, 1999, p. 42.
109. E. B. Rice, 1997, p. 60 [emphasis in original].
110. Mainly California and New Mexico, with some markets in Colorado and elsewhere and markets for temporary transfers of water in Wyoming.
111. The World Bank, 1994, pp. 21 and 23.

The proponents of water rights markets take as the point of departure the fact that such markets exist, albeit without official sanction, in many places. Examples of informal sales of agricultural water in Bangladesh, Yemen and Tamil Nadu, India, have already been mentioned. Experiences elsewhere can be cited as well. Rosegrant and Binswanger refer to extensive water markets in India as a whole and in Pakistan:

> Shah (1991)[112] estimates that as much as one-half of the gross area irrigated by tubewells in India belongs to buyers of water; Meinzen-Dick (1992)[113] and Chaudhry (1990)[114] document the rapid development of markets for groundwater in Pakistan. Water trading within surface water systems is also expanding. A recent survey in Pakistan found active water markets (trading or purchasing of water) in 70 percent of watercourses studied (Pakistan Water and Power Development Authority).[115]

In another context, Thobani points out that in Mexico informal water rights existed before the passage of the 1992 Water Law which established a system of codified rights, and that 'semi-formal' rights have long existed in Brazil and elsewhere.[116] The trend toward formalization of water rights as water becomes more scarce is a parallel to the growing formalization of land titles, as population pressure on land increases, which has been discussed in Chapter 5. 'When water is plentiful relative to demand, laws governing water use tend to be simple and enforced only casually. Where water is scarce, more elaborate institutional systems evolve'.[117]

Leaving water trading as an informal activity, instead of establishing a legal and institutional framework for water markets, has disadvantages:

In some cases these [informal] trades have not performed well and have resulted in an economically inefficient allocation of water. In parts of South Asia, wealthier farmers with deep wells charge neighboring smaller farmers a high ... price for water. As a result, crop output is lower than it would be if the water were priced at its opportunity cost – and income inequality is exacerbated. The opportunity to sell such a valuable resource also increases the exploitation of groundwater, which can deplete underground aquifers. ... Moreover, because such transactions are illegal, it is difficult to ... protect the aquifers.

... illegal markets may allow upstream users to sell more than they actually consume (because they may sell the return flow component of their water right), thereby infringing on the rights of third parties. In addition, the buyer lacks the security of an enforceable contract. Trades are therefore limited to spot sales or to sales for a single season, often between neighbors; longer-term trades are nonexistent, depriving potential investors or water companies [of] secure long-term access to water ... informal markets do not ... provide sufficient incentives or means for the creation of new infrastructure (M. Thobani, 1997, pp. 165–166).

---

112. T. Shah, 'Managing Conjunctive Water Use in Canal Commands: Analysis for Mahi Right Bank Canal, Gujarat', in R. Meinzen-Dick and M. Svendsen (Eds), *Future Directions for Indian Irrigation*, International Food Policy Research Institute, Washington, DC, USA, 1991.
113. Ruth Meinzen-Dick, 'Water Markets in Pakistan: Participation and Productivity', draft research report under USAID to Pakistan Grant No. 391-0492-G-00-1791-00 for the Ministry of Agriculture, mimeo, Government of Pakistan and International Food Policy Research Institute, Washington, DC, USA, 1992.
114. M. J. Chaudhry, 'The adoption of tubewell technology in Pakistan', *The Pakistan Development Review*, **29**, 1990, pp. 291–304.
115. M. W. Rosegrant and H. P. Binswanger, 1994, p. 1616.
116. M. Thobani, 1997, p. 165.
117. FAO, 1993, p. 268.

Water markets work on the basis of legally recognized and registered water use rights which are separate from land titles, an infrastructure which allows for reallocation of water, regulatory provisions for the protection of the public interest and third parties, institutions for contract enforcement and conflict resolution, and rules for apportioning shortages and surpluses of water. As in the case of other allocational mechanisms, responsibility must be assigned for operating and maintaining the water infrastructure.

The system may be defined so that only groups of irrigators may make sales of water to non-agricultural interests, but individuals can trade rights within the group. Alternatively, individuals may be free to make any kind of transaction. Sales of water rights may be made on a long-term basis, even permanent, and they may also be 'rented' on a short-term basis. In all cases, the price of the rights is freely negotiated between buyers and sellers. In Mexico, the water rights may take the form of concessions of up to 30 years. This period is long enough to encourage investments by the user, for example, in water conveyance structures (both on and off the farm), land leveling, drainage structures, and the planting of high-value but long-maturing fruit trees.

However, water rights markets do not function like most other kinds of markets. The conditions of competition – many buyers and sellers – usually are not fulfilled, and transactions require approval from various groups, from water user associations in Chile and Mexico to state governments in the Western United States. 'To an extent that is determined by state laws, water rights transactions must be approved by the state. This will typically involve an application for a transfer submitted to the office of the state engineer who examines the proposed transfer for any adverse effects on third parties. In some states, the transfer is publicized and any interested party can challenge the transfer at hearings conducted by the state engineer'.[118]

Although the concept of formally *tradable* water rights is relatively new, water rights themselves are an established institution in most parts of the world: 'it must be stressed that usufructuary rights to water already exist in most developing countries, either implicitly (through custom) or explicitly (through bodies of law and regulations). Establishment of transferable property rights is a matter of reforming or modifying existing water rights systems'.[119] The infrastructure for reallocation of water among sellers and buyers does not necessarily have to be extensive; if the quantities traded are large enough to warrant it, the buyer may invest in new conveyance facilities. This may happen when the buyer is a municipality or a large industry. Most commonly, transactions occur within the same watershed so that, for example, an upstream seller agrees to cease diverting a defined quantity of water from the river, leaving that amount available for the downstream buyer to take out of the watercourse. Many transactions occur among farmers in the same irrigation system. In these cases, the irrigation infrastructure must be flexible; for example, it would need gates instead of proportional flow dividers.

The main motivations for establishing systems of tradable water rights have been summarized by Thobani:

The potential to sell water rights makes them more valuable and provides an incentive for conserving water and reallocating it to higher-value uses. In this sense, the outcome is similar to that under opportunity-cost pricing. Tradable water rights also allow leasing of water (for a season, say) and spot sales; in fact, they facilitate such transactions. Finally, by allocating initial water rights, without charge, to existing users or holders of water rights, tradable water rights can circumvent the political problems associated with raising water prices and setting non-uniform charges. Governments can monitor operations and more effectively

118. R. G. Cummings and V. Nercissiantz, 1992, pp. 748–749. See also the illuminating case studies of water rights, though not necessarily markets for them, in Bryan Randolph Bruns and Ruth S. Meinzen-Dick, *Negotiating Water Rights*, International Food Policy Research Institute and Intermediate Technology Publications, London, 2000.
119. M. W. Rosegrant and H. P. Binswanger, 1994, p. 1615.

> *In Mexico water rights markets are nascent but growing:*
>
> *Since 1995, when the massive effort of registering water rights began in earnest, the National Water Commission has approved 517 transfers of water rights, resulting in a total annual volume of water rights transferred of about 160 million m³. Because total annual water use in Mexico is estimated to be over 200 billion m³, the amount of transfers approved so far is small. However, when water can be made available to meet demand through water markets, it reduces the need for constructing costly supply-oriented infrastructure and leads to a more rational and economically viable allocation of water resources (Karin E. Kemper and Douglas Olson, 'Water Pricing: The Dynamics of Institutional Change in Mexico and Ceara, Brazil', in A. Dinar (Ed.), 2000, Chapter 16, p. 352).*

enforce laws and regulations aimed at preventing the abuse of monopoly power, at ensuring that sales do not negatively affect the water available to third parties (that is, addressing the problems of return flow), and at protecting the environment.[120]

In order for these benefits to occur, a number of issues must be addressed in the course of designing a water rights market, including ensuring that equity concerns are satisfied. The principal issues are reviewed in the following section.

6.5.6 Issues in Water Rights Markets

6.5.6.1 Types of Water Rights

In the Western United States, water rights have been established under the doctrine of (*prior*) *appropriation*, or *first in time, first in right*, which is codified in the laws of the states of that region. The first right to water from a given source is established by usage, and refers to the annual quantities taken out by that pioneering user. The use must be 'beneficial', in keeping with established criteria, and failure to exercise 'beneficial' use for a defined period of time can lead to loss of the water right. Subsequent rights, or 'junior rights', may also be established, but in the event of a drought the more senior rights have priority in receiving water.[121] Informal water rights in developing countries also tend to be of this type.

The system of water rights which has been established for formal water markets in Mexico and Chile differs from the appropriation system. Called the *proportional rights* system, it confers rights to given shares of a river or other water source. Thus, in the event of unusually dry or wet years, all rights holders perceive reductions or increases in their water entitlements by the same proportion. (This provision for adjustments in quantities is stated formally in Chile and is left to the discretion of the National Water Commission in Mexico.) It has been contended that markets are easier to create for the system of proportional rights, because users have greater certainty of receiving water.[122] However, markets function under appropriative rights also, and the price reflects the degree of seniority of the right (i.e. the degree of certainty of obtaining water). The principal advantage of the proportional rights is precisely that they share more equitably the burden of dry years.

Just as there is a role for communal land rights in an economy that recognizes land markets (Chapter 5), there also is a role for communal water rights:

> It has been argued that establishment of tradable property rights in water is somehow antithetical to traditional community values, and inimical to communal management of water.

---

120. M. Thobani, 1997, p. 166.

121. An illuminating discussion of this doctrine and issues surrounding the transfer of water rights in the Western US is found in Allen V. Kneese and F. Lee Brown, *The Southwest under Stress: National Reource Development Issues in a Regional Setting*, Resources for the Future, Inc., The Johns Hopkins University Press, Baltimore, MD, USA, 1981, Chapter 4.

122. M. W. Rosegrant, R. Gazmuri S. and S. N. Yadav, 1995, pp. 210–211.

. . . Tradable property rights can, however, be assigned to communal groups or water-user associations as well as to individuals. Assignment of traditional property rights to communal groups should in fact enhance the control of these groups over water resources, insuring better access to water than is often the case with existing water user groups. Assignment of tradable property rights in water to communal groups may be more cost effective than assigning rights to individuals in instances where internalizing bargaining within the group reduces the information, contractual and enforcement costs relative to pair-wise bargaining by individuals.[123]

Whatever the form that water rights take, water markets require a system for formally titling and registering those rights, as land titles are registered. Their duration must also be recorded. Thobani recommends that:

The best way to ensure support for the [water markets] law is to assign rights to users, without charge, based on their historic usage. Although this approach may provide a windfall gain to some farmers, it acknowledges that the land price already reflects access to water at low prices and that the government is unlikely to recover directly the capital costs of investment in infrastructure. Because this procedure also rewards users that are taking more than their fair share of water, there may be merit in trying to rectify some of the most egregious wrongs. If the government were to try to use this opportunity to correct all such mistakes or to confiscate all illegally obtained rights, however, chances are good that the legislation will fail and the injustices will continue. . . . New and unallocated water rights should be sold at auction in an open and transparent manner.[124]

This last recommendation provides another parallel to policies for the land resource. Chapter 5 recommends auctions of title or leasehold to newly opened agricultural areas on State lands.

6.5.6.2 Return Flows and Third-Party Rights

Water rights may be defined as total diversion rights, consumptive use rights or nonconsumptive use rights. In the case of irrigation, there always are return flows to the water source (which in some cases is a joint aquifer–surface water system). Consumptive use is that portion of the water which is not returned. Nonconsumptive rights are used mainly for the generation of hydroelectric power.

In the case of irrigation, where farmers are entitled to specified *gross* releases of water from the system, it is vital to define whether the *transferable* rights are gross or net of return flows. That is, whether they refer to full diversion rights or consumptive use only. For farmers in the same area, growing mainly the same crops with the same kind of on-farm irrigation technology, it may not matter whether they buy the gross or net rights, as long as all transactions are treated in the same way. However, if a municipality or other entity purchases farmers' rights to water and takes the water out of the irrigation system's area, then there may not be a return flow at all, or it may be sharply diminished. This could cause a reduction in availability of water for downstream users, and hence the concern over third-party effects of water transfers that was mentioned in the box on p. 239 on informal markets for water rights.

The portion of irrigation water which represents consumptive use is difficult to measure with precision, for it depends on the nature of a farm's soils (including their slope), the kinds of crops grown, the type of irrigation technology, and climatic factors. Therefore the need to distinguish

---

123. M. W. Rosegrant and H. P. Binswanger, 1994, p. 1620. The first part of this passage also is found in M. W. Rosegrant, R. Gazmuri S. and S. N. Yadav, 1995, p. 217. A number of passages in the first publication are repeated in identical or virtually identical form in the second one, so hereafter reference will be made only to the first one in such cases.
124. M. Thobani, 1997, pp. 173–174.

between full diversion rights and consumptive use, in order to protect the access to water of third parties, would seem to place an insuperable barrier in the way of creating water rights markets. Nevertheless, pragmatic solutions have been implemented. One such solution consists of the *retention of rights to return flows by the irrigation authority* (usually the local irrigation district or the WUA). An explanation of this approach has been presented by Rosegrant and Binswanger:

> In most western US states, water rights are based on consumptive use, with protection of third-party rights to return flows. . . . This system protects prior rights to return flows, but significantly increases the transactions costs of water trading, because of the difficulty in measuring consumptive use and return flows. In practice, farmers often must demonstrate fallowing of land to release consumptive water use for sale. In the Northern Colorado Water Conservancy District (NCWCD), however, rights are proportional to streamflow and rights to return flow are retained by the district. Return flows are made available to water users at no charge, but no rights are established to these flows. Changes in patterns of return flows due to trades are therefore not actionable. *By defining away third-party rights to return flows, the NCWCD has greatly reduced transactions costs to trades, resulting in a very active water market.* . . . Chile and Mexico have in effect followed the NCWCD model by defining rights which are proportional to stream or canal flow. Rights to return flow do not exist. In Chile, return flows to neighboring areas may be used by the recipients without the need to establish a right of use. Use of this water, however, is contingent upon the flow of the main waterways and usage rates of the rights holder. There is no obligation to supply return flows and such flows are thus not permanent. . . . If initial

water rights can be allocated equitably . . . proportional rights with no rights to return flow should facilitate efficient allocation without compromising equity.[125]

A second pragmatic solution is to *restrict the scope for transfers of water out of agriculture* in circumstances in which the effects on return flows could be substantial and could pose problems:

> In Chile, there are two important river basins where additional protection to return flows has been employed: the Aconcagua River, in an area with a large proportion of high-valued crops; and the Elqui River, a small but significant river because it is located in a desert zone. . . . The Elqui River Water Users Association has dealt with this problem by limiting trades within upstream areas to farmer-to-farmer transactions (to retain all return flows within the basin), with agriculture–urban transactions authorized only in the downstream area.[126]

In other words, municipalities basically are limited to using water whose return flows otherwise would go to the sea, for the most part. Further protection of this nature is offered in Mexico and Chile by *laws that require approval of the local WUA for each trade* and prohibit both transfers that could damage irrigated agriculture and payments of compensation to third parties affected by transfers.[127] Appeals may be made in Mexico to the National Water Commission and in Chile to the National Water Authority, and further appeals may be pursued in the court system. However, the vague wording of the protection clause leaves considerable discretion in the hands of the water authorities and raises the risk of politicization of the implementation of the clause.

This second solution, of permitting intersectoral water trades only for waters in the lower reaches of the river, may be put into practice

125. M. W. Rosegrant and H. P. Binswanger, 1994, p. 1619 [emphasis added].
126. M. W. Rosegrant, R. Gazmuri S. and S. N. Yadav, 1995, p. 213.
127. *Op. cit.*, p. 214.

alongside the first one. A third solution consists of *developing and applying standardized or average rules of thumb for the arithmetic relation between total water diverted and return flows*:

New Mexico uses simpler and less costly procedures than California. The State Engineer's Office determines transferable water quantities utilizing standard formulae together with historical and secondary data. Reliance on standard transferable quantities for specific regions, soils and climates reduces the transactions costs incurred by applicants for hydrologic and engineering experts, saves staff time of water agencies, and creates more certainty in the transfer process. . . . An even simpler procedure would be to create a uniform presumption regarding consumptive use and return flows, which eliminates the need to determine consumptive use on a case-by-case basis. In Wyoming, the statute which authorizes temporary water transfers creates a presumption that 50 percent of diverted water is allocated to return flows, with the remainder considered to be the tradable quantity. Although attempts to rebut the presumption could be made, these would likely be infrequent if the presumption is a reasonable approximation. If, as is likely, a uniform state-wide presumption is not feasible due to different agroclimatic conditions, regional presumptions could be established. . . . An approach worth assessing in practice in river basins where return flows are significant would be a combination of the Elqui and New Mexico methods. . . . the key is to keep the transactions costs low while limiting return flow presumptions to the maximum that are genuinely produced, so as to preserve incentives for conservation and increase the gains from efficient market transfers of water.[128]

To illustrate this point, suppose that, in a given area, return flows can be reduced to 30% of full diversions by implementing water conservation measures in irrigation, but that the presumption embodied in the regulations is the maximum observed return flow of 50%. Then, a buyer of water who obtains rights to 50 units of consumptive use actually obtains rights to 100 units of full diversion. However, with sound water conservation practices, those rights could be converted into 70 units of consumptive use, thus expanding the irrigation capabilities associated with those rights.

Thobani makes a similar recommendation to use average measures of return flow:

Because of technical difficulties in calculating the return flow component on a case-by-case basis, this approach may not be appropriate for developing countries. But it may be possible to calculate averages that specify the volume of water consumed by a certain crop or activity. In those cases in which return flows are an issue, this published volume would become the limit on the amount that owners could sell to buyers. . . . This procedure would work for both surface water and groundwater. Even though the system has shortcomings, it would be a vast improvement over prohibiting all transfers or having no controls, as is the case with informal water markets.[129]

These examples and recommendations make it clear that workable solutions to the problem of return flows exist, and also that the most appropriate package of solutions is likely to differ from case to case.

6.5.6.3 Intersectoral Transfers of Water

As the foregoing example of the management of the Elqui River shows, concerns have been raised that the creation of water markets could be injurious to farming by encouraging the sale of water rights for non-agricultural uses, which generally have a higher productivity per unit of water and thus can afford to offer a higher price for the

---

128. M. W. Rosegrant, R. Gazmuri S. and S. N. Yadav, 1995, pp. 213–214.
129. M. Thobani, 1997, p. 175.

> *In Chile, the benefits of intersectoral water transfers have gone beyond financial compensation to farmers:*
>
> *For example, the city of La Serena was able to purchase 28 percent of its water rights from neighboring farmers, allowing the government to postpone the construction of a proposed dam. Similarly, the city of Arica, in the arid north, has been able to meet the needs of urban residents by leasing groundwater from farmers. . . . Changes in the structure of water markets create new opportunities for conserving water. When Santiago's municipal water company, EMOS, was notified that it could no longer receive new water rights without charge, the company initially sought to purchase additional water rights. When potential sellers demanded too high a price, EMOS decided instead to rehabilitate its aging pipe structure to reduce water leakages (M. Thobani, 1997, pp. 168–169).*

water. Mexico and Chile have put in place institutional safeguards against this possibility, in the form of a requirement for prior approvals of sales at different levels, as explained above. The safeguards appear to be effective. 'Chilean irrigators are also generally content with the codification of their traditional water-use rights. And since water-use rights are a tangible asset, which do not currently face a property tax, irrigators benefit from ownership of property rights even when the market for these rights is inactive'.[130]

The presence of water markets has been shown to foster a shift in agriculture toward higher-value crops, and when that occurs the difference in water productivity between sectors may not be as large as supposed:

> One of the most interesting results of this analysis is the relatively modest economic gains from intersectoral [water] trade in the Elqui

Valley. Although the value of water in municipal water supply is high, the value of water to profitable farmers is also high. . . . If these farmers are profitable, then the economic gains [i.e. to society] of the reallocation are small, even though the financial gain to the seller is large.[131]

In light of the Chilean and other experiences, perhaps some of the doubts of the World Bank's Middle East and North Africa Regional Office (quoted above) about the appropriateness of water markets as mechanisms for intersectoral reallocations of water, can be assuaged. It would appear that water markets, appropriately designed and implemented, can indeed be useful channels for such reallocations, *provided that appropriate institutional mechanisms for oversight of the trades exist.* Usually, it is the watershed, not the sector, that normally constitutes the limiting framework within which water transactions can take place, unless the public sector has independently made the (expensive) decision to invest in infrastructure for interbasin transfers.

However, there has been a successful experience in Spain with the transfer of $4 m^3/s$ from the Ebro Delta Irrigation District to the industrial area of Tarragona, some 80 km away and located in a different watershed. Fees paid by industrial users of water are invested in improving the Ebro Delta irrigation network. In Peru, there are water transfers from the Amazon Basin to the arid coastal region, through tunnels in the Andes. Until now, these transfers have not been market driven, but it has been suggested that water fees collected in the coastal area could finance development projects in the poorer Andean region.

6.5.6.4 The Issue of Transaction Costs

A principal concern about markets for water rights is that the transactions would prove to be too cumbersome, i.e. that the transaction costs would be too high. Clearly, there is a tradeoff

130. R. R. Hearne and K. W. Easter, 1995, p. 38.
131. *Ibid.*

between transaction costs and protection of third-party rights: the more elaborate the mechanisms of review and approval that are required for trades, in order to ensure that neither other individuals nor the environment suffers, then the higher the transaction costs. There is a general consensus that the procedures for water trades in California are weighted heavily in favor of protection of third-party rights and environmental concerns, and for that reason the water market there is not as active as in other western states of the United States.[132] In Mexico and Chile, the procedures for transactions are less burdensome and, accordingly, the market in water rights is more active.

There are 'transaction costs' associated with the reallocation of water in any system (see the box opposite), and they include measures to avoid conflicts and to compensate injured parties in the event of negative third-party effects. Earlier in this chapter, examples are mentioned of farmers going as far as damaging irrigation infrastructure when they feel their water rights have not been respected by administrative allocations. It is not clear that transaction costs, defined in the broad sense, are necessarily higher in water rights markets than under systems of administrative allocation of water.

Hearne and Easter have provided perhaps the only empirical estimate to date of the transaction costs of water trading, for two valleys in Chile. When the those costs are expressed as a share of the transactions price, they are as follows:[133]

Buyers, Elqui Valley	0.21
Sellers, Elqui Valley	0.02
Buyers, Limarí Valley	0.05
Sellers, Limarí Valley	0.02

> *Transaction costs arise whatever the process by which water is allocated. Transaction costs include: the cost of identifying profitable opportunities for transferring water; the costs of negotiating or administratively deciding on the water transfer; the cost of monitoring possible third-party effects and other externalities; the infrastructure cost of actually conveying the water and monitoring the transfers; and the infrastructure and institutional cost of monitoring, mitigating, or eliminating possible third-party effects and externalities (M. W. Rosegrant and H. W. Binswanger, 1994, p. 1617).*

In the Limarí Valley, these authors comment that transactions are more frequent because of the degree of development of its irrigation infrastructure and the strong organization of its WUAs.[134]

These transaction costs are not generally high and, more importantly, it is evident they can be reduced by development of appropriate physical and institutional infrastructure. From a policy viewpoint, the issue of transaction costs can be restated. Because *any* significant reallocation of water implies transaction costs, the more basic question is whether the water policy framework should be designed to permit flexibility in the allocation of this basic resource. In light of the growing scarcity of water, the benefits to farmers of water rights markets as opposed to administrative increases in water prices, and the other considerations presented in this chapter, it seems clear that the answer to this question should be 'yes'. Whether or not water rights markets should be implemented depends on whether their basic

---

132. 'In California, the transferable portion of the appropriative water right is limited to consumptive use, with protection of third-party rights to return flows. This system protects prior rights to return flows, but . . . significantly increases the transaction costs of water trading. . . . California's system for determining the tradable fraction of appropriative water rights in terms of consumptive use imposes a strong burden of proof on the prospective water seller for determination of how much water is tradable' (from M. W. Rosegrant, R. Gazmuri S. and S. N. Yadav, 1995, p. 211).

133. R. R. Hearne and K. W. Easter, 1995, p. 27.

134. *Ibid.*

requirements can be satisfied, as discussed later in this section.

6.5.6.5 The Role of Water User Associations in Water Rights Markets

Strong, well-informed water user associations are one of the prerequisites for a successful water rights market. Their role in approving and overseeing trades, and even (in Chile) in owning the irrigation systems, has been noted above (see the discussion in Section 6.5.3). Hearne and Easter concluded their empirical study of water markets in Chile by emphasizing the importance of the role of the WUAs in facilitating reallocations of water via the market, especially in the Limarí Valley that has an active market in water rights and in the Elqui Valley where intersectoral trading occurs.[135]

WUAs play a key role in resolving conflicts among water users. The more that such conflicts can be resolved locally, through the mediation of these associations, then the lower are the overall costs of water management. Water user associations are also responsible for undertaking and financing the operation and maintenance of the systems. This responsibility includes collecting fees from their membership. Accordingly, *the price of water rights established in markets is not its full opportunity cost but rather the opportunity cost net of O & M costs*. If the buyer is not a member of the WUA, then he/she must cover the O & M costs in the new use of the water, as well as the transaction costs.

6.5.6.6 Water Rights Markets and Environmental Protection

Concern about the environmental effects of water rights markets is logical, given that market values largely ignore environmental externalities. Nevertheless, measures to safeguard the environment can be built into the water codes which authorize

water rights markets. In Mexico, a water law established norms for protection of the environment by water users. The law's provisions include regulation that requires specification of minimal quality levels for water discharged to non-agricultural users and the National Water Commission has the right to impose controls on water use if there are severe water shortages, environmental damage or drawdown of aquifers.[136]

In principle, protection of the environment in regard to water use should be no more difficult under a transferable water rights system than under a system of administrative allocations, provided that an appropriate legal framework is in place:

> . . . the presence of privately held water-user rights does not necessarily reduce the possibility of proper environmental management of rivers. Water quality regulations need to be established and enforced irrespective of the water allocation system.[137]

In Chile, the water code is weaker than the Mexican one in regard to environmental protection, although it does require the national water authority to review major projects of water infrastructure for environmental and third-party effects before authorizing them. In other respects, water-related environmental protection is the province of the environmental code.

Experience with the environmental effects of water markets in developing countries is too brief to allow a judgment as to how the various protection measures are functioning. Nevertheless, the case of California is illustrative. That state has placed relatively tight restrictions on trading water rights, as noted above, in order to protect the environment and third-party rights. Notwithstanding this policy orientation, many Californian environmental groups have seen water rights markets as a means of satisfying the increasing demand for water without having recourse to the construction

135. *Op. cit.*, p. 39.
136. M. W. Rosegrant and H. P. Binswanger, 1994, p. 1620.
137. R. R. Hearne and K. W. Easter, 1995, p. 41.

of new water infrastructure, which they perceive as more injurious to the environment than water transfers.[138]

This experience in California tentatively suggests that not only can environmental protection be at least as strong under water rights markets as under administrative systems, but in fact it has the potential to be superior. Of course, in practice the result will depend not only on the legislative framework but also on the degree of commitment to environmental protection on the part of national and regional authorities and on the strength of the judiciary. These qualifications apply to any system of water allocation.

6.5.6.7 The Equity Issue for Water Rights Markets

Much as it is feared that land markets will lead to concentration of landholdings in the hands of a few, concerns have been expressed that markets in water rights will lead to semi-monopolistic concentrations of command over water. While concern merits analysis, it should be recognized from the outset that water markets are fundamentally different to land markets in that the former are characterized by much greater community (WUA) control over transactions. While this feature does not completely obviate the concern, it makes it much less likely that sales of water rights would result in large holdings by a few persons. In addition, it should be borne in mind that issuance of the initial water rights, if done in an equitable manner and free of charge, as recommended by the researchers cited in this chapter, endows poor families with a valuable asset. Even if they choose to sell the asset eventually, they will be better off than before the transferable rights were created.

Thobani has argued that water rights markets will help reduce poverty for several reasons:

Secure and tradable water rights reduce poverty in several ways. First, they allow scarce resources to be redeployed for more productive

Barbara van Koppen has observed that the presence of water rights markets can benefit poor rural families by increasing their access to irrigation supplies, which they sometimes are denied by the operation of prevailing local political power structures:

Poor farmers were well served. . . . in private irrigation (as buyers) if competitive private water markets developed, as seen in South Asia, on which poor farmers buy water. These competitive groundwater markets deliver good water services at low prices for millions of assetless smallholders. . . . Implications: Water markets, also those in conjunction with canal irrigation, can be more pro-poor if technology is available that is relatively small, but still provides excess water that the owners cannot use on their own land (B. Van Koppen, 'From Bucket to Basin: Managing River Basins to Alleviate Water Deprivation', International Water Management Institute, Colombo, Sri Lanka, 2000, pp. 10–11).

purposes, thus leading to increased output and employment. This occurred, for example, when farmers in Chile and Mexico sold their water rights to more productive farmers or cities. Second, tradable water rights encourage new investment in activities that require large quantities of water. An investment in a fruit farm is more likely to be attractive if the investor knows that water will not be transferred to a neighboring city in times of scarcity and that additional water can be purchased from farmers during water shortages. In Mexico, investors built a water-bottling plant after negotiating for the water rights from a farmer. Not only was the farmer better off, but the increased investment also generated additional employment.

. . . . Third, by empowering user groups to have a say on the issuance or transfer or water rights, secure and tradable rights help protect

---

138. M. W. Rosegrant, R. Gazmuri S. and S. N. Yadav, 1995, p. 215.

the poor. When water rights are granted without charge by public authorities, it is typically the rich and politically influential who have easier access to them, often at the expense of the poor. Fourth, secure and tradable water rights increase the value of the rights, which are often the most precious assets of poor farmers. In Mexico, many small farmers were able to take advantage of their ability to sell their water rights while still remaining on the land.

Additionally, by making it easier for cities to obtain water, such markets benefit the poor because they are the most likely urban residents to have been excluded from piped service. Chile provides almost universal coverage of piped water in urban areas. . . . Finally, because the transfer of water to higher-value uses occurs without confiscating water from less productive users (farmers) and without having to build new infrastructure, it is cheaper and fairer than alternatives, such as raising water charges substantially.[139]

6.5.7 Potentials and Prerequisites of Water Rights Markets

The foregoing discussion suggests that water rights markets can offer several kinds of benefits. In the words of Thobani:

Economic principles and lessons from experience suggest that formal enactment of tradable water rights permits rapid and voluntary changes in water allocation in response to changing demands, thereby improving water use. These formal water markets also increase user participation in allocating water and planning new investments, while allowing businesses to invest in activities that require assured access to water. The resulting increase in employment and income generation can help reduce poverty.[140]

Above all, water rights markets lead to pricing water at its full value (when O & M charges are taken into account also), thus enhancing the prospects for achieving objectives (i), (ii), (iii) and (v) listed in Section 6.5.3 above, without imposing administered price increases, which are equivalent to confiscating a basic right that many farmers hold, even if under customary or informal rules.

Although the initial implementation of systems of transferable water rights has occurred in Latin America, interest has been aroused in other parts of the world. For Sub-Saharan Africa, Sharma *et al.* have recommended that:

Sub-Saharan African countries should – whenever feasible and appropriate – develop market-like solutions. . . . The main economic instruments which could be considered are: (a) market-related and demand-oriented laws, policies, regulations, and incentive structures; (b) water markets with tradable water rights; (c) more extensive utilization of the private sector.[141]

Sampath makes a comparable recommendation for all areas of the world, with emphasis on Asia, in light of the growing demands for municipal and irrigation water associated with population increases. His conclusion is that it is hard to over-emphasize the need for better pricing of all water services and the encouragement of water rights markets.[142] The FAO has concluded that the work on water pricing systems and on forms of water markets has converged in the idea that systems transferable water rights best reflect the opportunity cost of water.[143]

While the potential benefits of water rights markets make them an attractive proposition, caution should be exercised in ensuring that all of their prerequisites are satisfied before they are implemented. Thobani has commented that:

139. M. Thobani, 1997, pp. 169–170.
140. *Op. cit.*, p. 177.
141. N. P. Sharma *et al.*, 1996, p. 60.
142. R. K. Sampath, 1992, p. 974.
143. FAO, 1993, pp. 274–275.

... tradable water rights are not a panacea, and an effective system is not easy to introduce. Chile's experience and the demonstrated superiority of markets over publicly administered means of resource allocation in general suggest that markets are preferable when water is scarce, when the infrastructure to effect transfers exists or can be cheaply developed, when there is a minimum institutional capacity to implement trades, and when there is political will to establish appropriate legislation.[144]

Some of the specific prerequisites of a water rights market have been summarized by Hearne and Easter:

(a) [in defining water rights] in areas where water supplies are highly variable, it is necessary to designate how water is allocated during times of scarcity;

(b) great care should be exercised in the initial allocation of water-use rights among users in order to make sure that all the rights are not captured by a few individuals;

(c) proper technology and institutions such as adjustable gates and effective water user associations can substantially reduce transaction costs. . . . ;

(d) [defining] a continuing role for water management authorities in enforcing rights and resolving conflicts.[145]

To this list may be added:

(e) Defining procedures for the treatment of return flows and third-party rights.

(f) Thoroughly disseminating information on how water rights markets work, to potential participants.[146]

In addition, probably the most basic requirement of all for introducing a system of water rights is:

(g) a clear and strong political commitment and the patience required to develop and implement the system.

It should be emphasized that most of these requirements pertain to administered irrigation systems as well, and that failure to fulfill them adequately has accounted for many of the observed problems in irrigation systems.

The use of water rights markets in developing countries is still in a very early stage. Therefore, the pace of adoption of water rights markets is likely to be slow, but close study of this option can be recommended when it appears that its requirements can be satisfied. In the longer run, the unavoidable necessity of better water management is likely to accelerate the pace of its adoption. In any given case, however, it is advisable to start slowly and assure the development of adequate institutional structures for monitoring and managing water rights transfers. Usually, it is preferable at the outset to limit transactions to farmers in the same valley, and only later open up the possibility of transactions with other users.

6.6 INSTITUTIONAL AND PROCESS ISSUES IN WATER MANAGEMENT

6.6.1 Sector-Wide Institutional Issues

Neither good irrigation technology nor adequate price incentives, nor the combination of them, is sufficient to ensure satisfactory performance of irrigation systems. The institutional context is at least as important, if not more so. The institutional aspect refers both to *how irrigation systems are managed, at different levels* and also to *the process over time by which those systems are developed and implemented*. It is the human dimension of the system: the way that the various actors are grouped into joint entities, the ways they work

---

144. M. Thobani, 1997, p. 177 [emphasis added].
145. R. R. Hearne and K. W. Easter, 1995, p. 41. For useful case studies in how water rights markets function in Brazil, Spain and the USA, see Manuel Mariño and Karin E. Kemper (Eds), *Institutional Frameworks in Successful Water Markets*, World Bank Technical Paper No. 427, The World Bank, Washington DC, USA, 1999.
146. 'Experience shows that it is essential to explain to users and other affected groups the advantages of formal property rights to water. A well designed information campaign can overcome the opposition to reform by powerful vested interests' (M. Thobani, 1997, p. 173).

together (or fail to), the roles both individuals and entities play, and the objectives they pursue:

> Establishing an institutional structure for allocating water is a fundamental role of social policy for any nation. The choice of structure is ultimately a compromise between the physical nature of the resource, human reactions to policies and competing social objectives. Not surprisingly, different cultures make tradeoffs based on the relative importance of their particular objectives. Countries try various means to balance economic efficiency (obtaining the highest value of output from a given resource base) and fairness (assuring equal treatment). Individual freedom, equity, popular participation, local control and orderly conflict resolution are other important objectives which societies must juggle when choosing a structure for water allocation.[147]

In the final analysis, an institutional structure is effective to the degree that it establishes clear and appropriate roles for all of the concerned individuals and groups and motivates them to play those roles well. Motivation may require incentives but above all it requires a belief in the correctness and legitimacy of their roles, and of the roles of the other institutions involved in water management. The importance of inspiring a belief that institutions and procedures are essentially well grounded and fair can hardly be overstated. Hence, the FAO's reference to 'human reactions' in the above quotation.

Individuals are motivated to work together for common ends by what is called good governance. This applies to irrigation management as much as in any field. 'Improvement in irrigation performance depends on good government, or governance. . . . There are four main elements of governance which can be considered at the national or the local level: the *legitimacy of government*; its *accountability*; its *competence*; and its *respect for human rights and the rule of law*'.[148]

The institutional context is so basic that it now is given priority in planning for irrigation, both at the project level and at the sector level. If measures to strengthen institutional capacity are not incorporated in project design from the beginning, institutional reform considerations will be shunted aside as project managers become involved in day-to-day concerns. Equally, there is a consensus that the first priority within an irrigation sector should be to strengthen its institutional capacity, which may include changing institutional forms in the directions of greater decentralization and devolution of ownership of systems as well as the institution of participatory mechanisms for water planning and management.

There is agreement that strengthening institutional capacities requires *wide participation* at all levels of decision making and at all stages in the process, including in policy formulation and project design: '. . . the process of water policy formulation, assessment and appraisal needs to include more open groups that are representative of political, technical, managerial and (most important) water user associations. These policy groups would be consulted before policy selection and then provide feedback and adjustment in the light of experience. . . . The goal is to identify a broader range of water policy options, to have less "policy by crisis" management and more resilience in the face of outside pressures'.[149]

Two fundamental reasons argue for broadening the participation in formulation of water policy and in development of irrigation systems, one related to societal objectives or values and the other related to practical concerns. They are *promoting fairness* and *improving the quality of policies and projects*. Those whose interests are affected are likely to have valuable insights regarding the functioning of both policies and projects and how they should be designed. In addition, their participation in project design and management will strengthen their commitment to making the system work well.

There is an additional reason which is very important in a practical sense: making the policies and irrigation systems as *robust* as possible in the face of evolving circumstances and unex-

147. FAO, 1993, p. 268.
148. FAO, 1993, p. 291 [emphasis added].
149. FAO, 1993, p. 296.

Good governance is crucial to the functioning of irrigation systems in the following ways:

Legitimacy. *When a new project is planned, are those living in the area consulted about the design of the scheme? Are there recognized representative groups of farmers, including women? Are the office holders elected and accountable to the members? Do these groups participate in decisions that affect them? . . .*

Accountability. *Are the financial plans of the irrigation scheme made public and arrangements made to explain them to farmers? Are there performance criteria with audit arrangements to ensure that officials adhere to the rules and, if they fail to perform satisfactorily, call them to account? Are officials responsive?*

Competence. *Can the professional staff prepare accurate budgets and effectively deliver services such as timely canal maintenance? Are there arrangements for training them or replacing them with competent officers if they fall short of their duties?*

Rule of law. *Is there a clear legal framework to . . . prevent overpumping of the aquifers? Is it enforced? Can pollution by industry or by saline water from upstream drainage projects be regulated? Are illegal extractions by farmers at the head of canals monitored and offenders charged by legal processes that are fair, timely, objective and without discrimination . . . ?*
(From FAO, 1993, p. 291.)

---

pected changes – the resilience factor referred to by the FAO. If systems and their regulatory frameworks are developed in part by water users and others who are affected by them, i.e. if they have legitimacy as described above, then the participants will strive to make them function well and to adapt them, as necessary, to unforeseen changes in circumstances. Thus, the institutional context has a direct bearing on the *sustainability* of irrigation schemes.

In practical terms, Vermillion and Sagardoy have summarized the case for irrigation management transfer (IMT) to user groups in the following way:

Efforts by governments to finance and manage irrigation systems, and to collect water fees from farmers, have generally not been very successful. Under-financing and mis-management of irrigation systems has led to rapid deterioration of infrastructure, shrinking service areas, inequitable water distribution and low agricultural productivity. Generally, governments hope that IMT will reduce the cost burden of irrigation on the government and will increase the productivity and profitability of irrigated agriculture enough to compensate for any irrigation cost increases to farmers.[150]

The other side of the coin of promoting participation and seeking consensus is *decentralization*, including within government structures themselves. The intrinsic value of this approach was expressed by the World Bank in its policy paper on water resources management:

Because of their limited financial and administrative resources, governments need to be selective in the responsibilities they assume for water resources. *The principle is that nothing should be done at a higher level of government that can be done satisfactorily at a lower level.* Thus, where local or private capabilities exist and where an appropriate regulatory system can be established, the Bank will support central government efforts to decentralize responsibilities to local governments and to transfer service delivery functions to the private sector, to financially autonomous corporations, and to community organizations such as water user associations.[151]

Some of the possible types of specific reforms to improve the autonomy and accountability of public irrigation agencies and increase farmer

---

150. D. L. Vermillion and Juan A. Sagardoy, 1999.
151. The World Bank, 1993, p. 15 [emphasis added]; see also The World Bank, 1994, p. 23.

involvement in decision making include the following options:

Possible reforms include shifting from a line department to a semi-independent or public utility mode, applying financial viability criteria to irrigation agencies, franchising rights to operate publicly constructed irrigation facilities, and strengthening accountability mechanisms such as providing for farmer oversight of operating agencies. Many of these reforms can be seen as introducing quasi-market incentives into the management of public irrigation systems.[152]

In a similar spirit, the World Bank's policy paper on water resources management emphasizes the importance of financial autonomy of irrigation agencies, along with a clear decision that financial shortfalls will not be made up by the central government:

The lessons of experience suggest that an important principle in restructuring public service agencies is their conversion into financially autonomous entities, with effective authority to charge and collect fees and with freedom to manage without political inference. Such entities need to work under a hard budget constraint that enhances incentives for efficiency and revenue generation. Of greatest importance, the hard budget constraint unlocks incentives to collect fees and to provide services that consumers and farmers want.[153]

At the same time that emphasis is being placed on greater autonomy of irrigation institutions and decentralization of decision making, the capacity of government institutions themselves needs to be improved in most cases. The FAO has pointed out that, in addition to the well-known 'market failures'[154] that characterize a resource such as water, 'government failures' are fairly widespread, in part for the following reasons:

'Products' are hard to define. The outputs of non-market activities are difficult to define in practice.... Flood control or amenity benefits of water storage reservoirs are examples. The internal goals ... of a public water agency as well as the agency's public aims provide the motivations ... for individual performance. Examples of counterproductive internal goals include budget maximization, expensive and inappropriate 'technical-fix' solutions and the outright non-performance of duties. In addition, agencies may adopt high-tech solutions, or 'technical quality', as goals in themselves. ... Finally, irrigation agency personnel may be persuaded, by gifts or other inducements, to violate operating rules for a favored few.[155]

Under optimal circumstances, public management of irrigation systems can perform well but those cases are the exceptions, as observed by Ruth Meinzen-Dick *et al.*:

Examples of irrigation systems that are performing well, for example, in Malaysia, demonstrate that good performance under State management is possible.... Many irrigation projects have been based on introducing technological innovations to improve system efficiency.... Yet without proper management such innovations fail to deliver the desired irrigation services. ...

Countries have generally entrusted the management of their irrigation systems to government agencies, on the assumption that they will have the capacity and motivation to achieve high performance standards. Heavy State involvement in irrigation has been justified

152. M. W. Rosegrant and H. P. Binswanger, 1994, p. 1614.
153. The World Bank, 1993, p. 55.
154. Natural monopolies created by economies of scale in water supply, lack of internalization of externalities by those who make economic decisions concerning water allocation.
155. FAO, 1993, p. 264 [emphasis in original].

based on the public goods characteristics of irrigation, notably the positive and negative externalities, strategic importance, and scale of systems. . . . In practice, State agencies cannot be omniscient and omnipotent, particularly in dealing with problems at the local level. Moreover, the private incentives of agency staff are often at odds with official objectives in irrigation management, leading to rent-seeking behavior. The result has been sub-optimal levels of system performance. . . .[156]

Meinzen-Dick *et al.* also comment on the difficulty of seeking solutions by administratively raising irrigation fees to adequate levels (a difficulty which is discussed in Section 6.5.3 above), and hence the need to look elsewhere for the means of improving system performance:

> . . . simply raising irrigation charges is politically unpopular, and does not provide the necessary incentives for agencies or farmers to improve irrigation system performance. The traditional economic solution of 'getting prices right' has been difficult to implement and of limited use in improving irrigation system performance. . . . Market solutions, such as tradable property rights, are being explored by policy makers and analysts, but the difficulties of specifying clear and enforceable property rights and the presence of high transactions costs and positive and negative externalities, along with other types of market failures in irrigation systems, have limited the effectiveness of this strategy. Therefore, institutional reforms to reduce costs while improving incentives for better performance of irrigation systems are essential.[157]

For similar reasons, the World Bank's Middle East and North Africa Regional Office has concluded that 'The weakness of many governmental agencies is a matter for serious concern, and is an issue that cannot be bypassed or avoided. . . . Since there is little evidence that *ad hoc* institutional change provides a satisfactory basis for effective management, institutional issues need to be tackled in an integrated and coherent manner . . .'.[158]

A strategy for improving agency performance in managing irrigation systems must include provision of better training to managers and introduction of mechanisms to make them accountable to the users as well as their bureaucratic superiors. Several different kinds of skills are needed, including planning and supervision in different areas, environmental management and liaison with stakeholders. Another critical element is:

> ensuring that agency staff have proper incentives to work with farmers. The strongest and longest-lasting incentives for agency staff to work with WUAs follow from linking budgets to user fees, and staff compensation and rewards to improvement in farmer services.[159]

A frequent cause of the lack of institutional capacity is the dispersion of authority among many agencies which inhibits the integrated approach planning and policies for water management which was stressed in an earlier section of this chapter. Often there is fragmentation of responsibilities for management of water resources among different sectors and institutions, along with over-reliance on the public sector provision of services.[160] Consequently, the FAO has made a call for integration of the insti-

156. R. Meinzen-Dick, *et al.*, 1997, pp. 11–13.
157. *Op. cit.*, p. 13.
158. The World Bank, 1994, p. 52.
159. A. Subramanian, N. V. Jagannathan and R. Meinzen-Dick, 'Executive Summary', in A. Subramanian, N. V. Jagannathan and R. Meinzen-Dick (Eds), 1997, pp. xii–xiii.
160. N. P. Sharma, *et al.*, 1996, p. xvi.

tutions charged with developing and implementing water policy, as noted in Section 6.3 above.

An exception to this need for integration is the requirement for *a clear separation between the roles of defining water policy and providing (or co-ordinating the provision of) water services*. Policy responsibilities often have been located in agencies charged with water supply and management, but doing so may create institutional conflicts of interest and at minimum usually leads to a dilution of the effectiveness of one of the functions. Sharma *et al.* have expressed this point cogently:

An important strategy is to promote separation of the regulatory ('gamekeeper') functions from the supply ('poacher') functions of water supply and wastewater collection and delivery in order to avoid conflicts of interest and ensure compliance. For regulation to be fully successful, good governance is important as are experienced and accountable public sector staff.

Separating regulation from supply is a first rule of water resources management.... However, there is no 'correct' solution for the institutional framework for effective water resources management; solutions evolve in an adaptive way in response to constitutional and cultural settings and to clear policy objectives and political commitment to policy enforcement.... Institutional development and reform are a long 'journey' which is generally embarked upon without clear knowledge of the eventual destination. Nevertheless, the separation of regulation functions from those of supply is an important principle. Some countries make water resources management the responsibility of an agriculture ministry, or they have a water ministry also responsible for municipal water supply and wastewater services; both preclude effective water resources *management*.[161]

Separation of administrative authority is also necessary for macro and micro allocations of water. The general principle is that macro management must be carried out by an entity different from the sectoral ones (such as those for hydropower, irrigation, municipal water supply, etc.). The case of Spain is illustrative of this principle. Since 1927, water resources management in Spain has been regulated by a central organ but managed by decentralized basin authorities. The centralized organ is the National Water Authority (under the Minister of the Environment) – this is responsible for national water policy and regulations. The basin authorities manage water resources in a holistic, participatory and decentralized manner, with financing shared between users and the State. Their role ends where water is delivered to different classes of users, and there the responsibility of sub-sectoral institutions begins (WUAs, municipalities, etc.).

In developing modes of participatory management which are appropriate for a country's cultural traditions, some countries may find it useful to explore adaptations of the French system of river basin committees that was mentioned previously, although the process of establishing them is not easy. Senegal, for example, has considered adoption of some elements of the French approach in its institutional framework:

[A] Water Resources Planning and Management Service ... [was] created, responsible for managing and allocating all water resources and for issuing abstraction permits. [A] Water Advisory Board ... [was] proposed as [a] steering committee for water resources policy making and management, with membership including representatives of consumers' associations, farmers' and livestock breeders' associations, industries, water and forestry associations, rural communities, mayors, the national water supply holding company ... as well as [government agencies].[162]

In the same vein, 'water parliaments' at the river basin level are being considered in Zimbabwe and South Africa (Sharma *et al.*, 1996, p. 50).

161. *Op. cit.*, p. 68 [emphasis in original].
162. *Op. cit.*, p. 50.

> *As provided by the [National Water Law of Mexico], the river basin councils have a key role to play in river basin planning and management.... At present, there is one functioning river basin council, Lerma–Chapala, which started operating in 1993. In addition, about 20 more basin councils are in various stages of development. Experience has demonstrated that the establishment of functional river basin councils is difficult and time-consuming. Organizing all the different water users into functioning groups that then elect representatives on the councils with adequate communication with both users and government officials has proven to be much more difficult than officials had contemplated. Experience has shown that basins with serious water scarcity and management problems have an easier time establishing councils (From K. E. Kemper and D. Olson, 2000, p. 350).*

Sharma *et al.* have summarized the institutional requirements for adequate management of water resources in the form of **awareness, capacity and management**. Awareness is a key link in the chain, and frequently it is the weakest link. Different kinds of awareness are needed:

- at the regional, sub-regional and basin level, to create a climate of mutual understanding and knowledge;
- at the national political level, to create commitment;
- at the executive level, as part of building capacity;
- among the public, to create society-wide stewardship.[163]

Building awareness is not necessarily an expensive undertaking but it requires a well-planned effort and persistence. Sharma and colleagues (1996, p. xx) have cited the 'success-stories in Namibia and Botswana, where water conserva-

tion has become a national ethic'. As another example of the importance of building awareness, they point out (1996, p. 53) that:

... participatory capacity is firmly embedded in the culture of many parts of [Africa] and is a valuable resource. However, building on indigenous skills and cultural traditions can only be successful where there is awareness of the problem and identification of solutions.

The same observation could be made with respect to all other regions of the world.

Building societal awareness and capacity and management skills at all relevant levels of the public service are necessary conditions for making irrigation systems function well, but they are not sufficient. As indicated, the public sector alone cannot do the full job; it faces the inherent limitation of operating in a costly manner, in institutional terms, especially at local levels, and it has great difficulty in gathering all of the relevant kinds of information for local undertakings. Hence, capacity and management skills have to be cultivated among users and elsewhere in the private sector. Subramanian *et al.* have underscored this requirement:

Policy makers have turned their attention to the potential of using water user groups to plan and manage water infrastructure because of the twin problems of the institutional costs of implementing water distribution rules and of planning and managing water infrastructure with incomplete information. The arguments advanced for supporting water user associations are that water users ... have far more complete information on local conditions and must therefore be included in the planning and management process. Furthermore, water users also have traditional norms and conventions that often may be far more effective than a top-down water bureaucracy in enforcing contracts among users of tertiary and secondary distribution systems.[164]

163. *Op. cit.*, p. 63.
164. A. Subramanian, N. V. Jagannathan and R. Meinzen-Dick, 1997, pp. 5–6.

6.6.2 The Benefits of Local Water Management

The main form through which local irrigation management is implemented is through water users' associations. The number of such associations has been growing rapidly in recent years, throughout the world. They have become fixtures of irrigation management in diverse places such as Indonesia, the Philippines, Thailand, Taiwan, Bangladesh, India, Pakistan, Nepal, Sri Lanka, Egypt, Morocco, Zimbabwe, Senegal, Cameroon, Mali, Nigeria, Kenya, Guyana, Argentina, Brazil, Colombia, Chile, Mexico, Peru, the Dominican Republic, and more than 20 other countries (Vermillion and Sagardoy, 1999). An international network exists, INPIM, to facilitate exchanges of information and experiences.[165] In the case of Indonesia, for example, 'the government had transferred more than 400 irrigation systems, covering 34 000 ha, to water user associations associations by 1992'.[166]

In India, the state of Andhra Pradesh has 'opted for a "Big Bang" approach', constituting 10 292 WUAs covering all of the state's irrigation systems, an area of 4.8 million hectares.[167]

Clearly, WUAs are responding to well-defined needs and in all probability are giving rise to significant benefits. However, before exploring their roles and the nature of those benefits, it is well to note that WUAs do not always improve system performance, and often their creation requires a conscious and sustained effort on the part of supporting institutions. It has been asserted that in general:

> it has been difficult to establish that communal approaches to water management have actually increased farm production and income. . . . In practice, turnover of irrigation systems [to users] has legitimized the transfer of the responsibilities for operations and manage-

ment to farmers, thereby reducing the costs of financially strapped public bureaucracies. The turnover of costs and responsibilities, however, has not been accompanied by improved access to water through the establishment and granting of tradable water rights, nor has it been accompanied by a clear demarcation of duties between the government irrigation bureaucracy and the private water user groups. *As a result, there has been no change in the fundamental incentives governing water use.*[168]

These caveats aside, some of the valid motivations for promoting the formation of water user associations have been mentioned throughout this chapter. The basic case for them has been stated by the World Bank in the following way:

> The participation of users in managing and maintaining water facilities and operations brings many benefits. Participation in planning, operating and maintaining irrigation works and facilities to supply water and sanitation services increases the likelihood that these will be well maintained and contribute to community cohesion and empowerment in ways that can spread to other development activities. . . . In addition, governments benefit directly . . . financial and management burdens on government that result from administering water allocation can be reduced through user participation in both urban and rural areas. . . .[169]

The FAO has concluded that the formation of WUAs offers several kinds of benefits:

> Studies throughout the world demonstrate that user participation in irrigation services improves access to information, reduces monitoring costs, establishes a sense of ownership among farmers and increases transparency as well as accountability in decision making.[170]

---

165. INPIM website: [http://www.inpim.org].
166. FAO, 1993, p. 293.
167. K. Oblitas and J. R. Peter, 1999, pp. 1 and 2.
168. M. W. Rosegrant and H. P. Binswanger, 1994, p. 1615 [emphasis added].
169. The World Bank, 1993, p. 55.
170. FAO, 1993, pp. 292–293.

To put it in another way, perhaps the main advantage of water user associations is that, properly organized, they can be effective control mechanisms for reducing or eliminating the 'free rider' problem, which in this case would consist of some farmers receiving irrigation water without making proportionate contributions to system management and maintenance.[171]

Mexico has advanced rapidly in turning over management of irrigation systems to WUAs, under a scheme in which the latter handle tertiary distribution, frequently over areas covering tens of thousands of hectares. Users also belong to corporations, whose shareholders sometimes are WUAs, which control and operate the main canals. While it is still too soon to evaluate the longer-term effects of that program, a survey-based evaluation of results in four irrigation districts was carried out after one or two years of experience, with the length of the experience depending on the date of the transfer. Some of the principal findings of this evaluation in Mexico were summarized by Cecilia Gorriz, Ashok Subramanian and José Simas as follows:

- About 80 percent of the surveyed farmers . . . said that with the transfer of the district to WUAs, there were improvements in water management, and in timely and adequate water delivery and maintenance of the irrigation systems.
- About 45 percent of the farmers believed that water fees were high.
- About 30 percent of farmers had problems of salinity primarily due to inadequate drainage systems and poor on-farm water use. . . .
- Most of the irrigation districts have reached financial self-sufficiency, but large variations

in water availability had affected revenues of the WUAs, which must endure financial difficulties in years of water shortages or heavy rainfall.
- For the most part. . . . users claimed that irrigation service had improved: structures were being better maintained even though they needed rehabilitation; there was interest in improving the on-farm irrigation systems; in improving the technological level of water distribution; and in technical assistance to users. The users greatly appreciated the training and technical assistance provided by the [National Water Commission] and the [Mexican Institute of Water Technology].

However, the qualifications to these positive results are illuminating and are likely to arise in other circumstances as well:

- The transfer of the four irrigation districts to the WUAs led to reductions in the number of staff causing dissatisfaction among National Water Commission personnel.
- While the farmers have obtained a greater leverage in water management than before the transfer, they have had to *substantially increase their contributions to O & M of the sub-system under their management and on-farm capital improvements*.[172]

The benefit of WUAs that is most commonly confirmed is a reduction in government fiscal outlays for managing irrigation schemes, in exchange for greater contributions by farmers of cash and labor. Frequently, the management by farmers is also more efficient so that the total cost of management is reduced, independently of

171. This point is brought out in Elinor Ostrom, Larry Schroeder, and Susan Wynne, *Institutional Incentives and Sustainable Development: Infrastructure Policies in Perspective*, Westview Press, Boulder, CO, USA, 1993, pp. 136–137.
172. Cecilia M. Gorriz, Ashok Subramanian and José Simas, *Irrigation Management Transfer in Mexico: Process and Progress*, World Bank Technical Paper No. 292, The World Bank, Washington, DC, USA, 1995, p. 38 [emphasis in original]. These findings are based on a survey carried out by the Colegio de Postgraduados, Mexico: *Diagnóstico sobre la Administración de los Módulos Operados por las Asociaciones de Usuarios*, Mexico City, Mexico, 1994.

The cost savings from local irrigation manage-ment were remarkable in an experience ana-lyzed in Chile:

In Chile, the State management of a 60 000 hectare irrigated area on Río Digullín involved five engineers, eight to ten technicians, fifteen to twenty trucks, and five bulldozers, compared to one engineer, two technicians, one secretary, and two trucks under farmer management of the same area. Because farmers work collabora-tively with engineers and technicians, they are fully aware of the 'true' costs of running the irrigation systems, and for this reason perceive that the water fee charges they pay, even if they are high, are 'believable' costs.... (R. Meinzen-Dick, et al., 1997, p. 90).

which entity is funding those costs. Meinzen-Dick *et al.* have documented this benefit extensively:

The most tangible and well-documented gain from farmers' involvement in irrigation is the reduction in government costs. These cost savings come from reduced administrative and operations costs as the number of staff fielded decreases, better project design, increased fee collection rates, and decrease in the destruction of facilities. Numerous country experiences support this claim. For example, Bagadion and Korten (1991)[173] estimated an annual savings to the Philippine Government amounting to US$12 per hectare from the contributions of the irrigation associations in terms of man-hours spent on management, maintenance, repair, and improvement activities; water dis-tribution and fee collection; and direct cash outlays for canal repairs and supplies and materials.[174]

The Senegalese experience provides another illustration of cost reductions and efficiency improvements when management of irrigation is transferred to farmers, albeit at the expense of higher costs to the latter:

Under agency management, irrigation fees and service quality were both low. The agency pro-vided maintenance and paid for electricity on an irregular basis, leading to highly unreliable irrigation services. Agency field staff were poorly supervised, and would therefore turn on pumps and leave. This resulted in over-pumping and system breakdowns. By contrast, WUAs provided more careful supervision of staff, reducing over-pumping and thereby cutting electricity costs by up to 50 percent. Other cost savings came from WUAs paying staff less than full civil service rates. Neverthe-less, because WUAs had to pay for full elec-tricity consumption along with maintenance and a fund for pump replacement, farmers' fees increased by two to four times....[175]

It is clear that under local management of irri-gation systems, at least a portion of the savings in costs to governments takes the form of increased costs for farmers. Presumably, the principal gain for farmers is increased reliability of service, but in almost all cases they end up paying more for the service after the transfer of responsibilities:

Although total cost reductions are possible, in practice, farmers' costs usually increase with the transfer of irrigation management respon-sibility to the WUAs.... Part of the reason is the removal of State subsidies with man-agement transfers. If irrigation service fees were below full O&M costs prior to transfer and WUAs are expected to assume full costs after transfer, farmers' contributions have to

173. B. U. Bagadion and F. F. Korten, 'Developing Irrigators' Organizations: A Learning Process Approach', in M. M. Cernea (Ed.), *Putting People First: Sociological Variables in Rural Development*, 2nd Edn, The World Bank, Washington, DC, USA, 1991.
174. R. Meinzen-Dick, *et al.*, 1997, pp. 88–89.
175. *Ibid.*

increase (unless efficiency gains are great enough to make up for the loss of State subsidies). For example, irrigation fees in Mexico increased fourfold to sixfold when WUAs took over and had to cover full O&M costs. Johnson (1993)[176] shows that expansion in local participation in Indonesian pump schemes resulted in water charges that were five to seven times higher than those imposed by the government, but the new fees still covered only 30 to 50 percent of pumping costs.[177]

For this reason, the financial viability of WUAs will depend on the magnitude of monetary and transaction costs that farmers have to assume, in relation to their income levels (Meinzen-Dick et al., 1997, p. 21).

After fiscal savings, the second most commonly observed benefit of WUAs is increased efficiency in the operation of the irrigation scheme. Meinzen-Dick et al. (1997, p. 87) have compiled several examples, as follows:

> Improved efficiency of water deliveries (reduced water losses) saved 25 to 30 percent of irrigation supplies after the WUA took control in Azua, the Dominican Republic. This, in turn, reduced the need for drainage and related investments. . . . In Nepal, water is more efficiently distributed in smaller farmer-controlled systems during the wet season than in larger systems. . . . More equitable water distribution has been a positive attribute of farmer-managed pump systems in Bangladesh . . . and of the introduction of WUAs in Gal Oya, Sri Lanka.

Part of the gain in efficiency is attributable to a lower rate of damage to facilities and to more prompt responses to instances of damage or degradation:

To provide adequate water to their members, WUAs are likely to pay more attention to maintaining the canals and headworks. Farmer members in Taiwan carry out routine patrolling and inspections to ensure the proper upkeep of systems. . . . Technicians hired by the users' association do regular 'policing' in Chile. Because they are in direct radio contact with farmers, problems can usually be corrected within an hour, compared to weeks of bureaucratic delay under agency management. . . . in Sri Lanka [it was] noted that farmers who had damaged facilities began taking greater care of them after they were organized into water groups with a common interest in the irrigation system. . . .[178]

These gains in efficiency are sometimes associated with more effective resolution of conflicts. Another observation regarding the Gal Oya project in Sri Lanka was that:

> [the users'] groups discussed their problems and communicated with the government irrigation department staff. This process has greatly improved communications between farmers and government officials. Conflict among farmers has declined substantially, and the improved system provides more water for farmers at the tail end of the system. Careful to separate their organizations from party politics, the farmers have also eased ethnic tensions. In one area cooperating farmers cleared a canal allowing 1000 hectares to be cultivated in the dry season, which had previously been left fallow.[179]

An example from Tunisia suggests that another benefit of WUAs can be greater flexibility in irrigated cropping patterns. Over time, this advantage can translate into higher farm incomes than

---

176. S. H. I. Johnson, *Can Farmers Afford to Use the Wells after Turnover? A Study of Pump Irrigation Turnover in Indonesia*, Short Report Series on Irrigation Management Transfer, No. 1, International Irrigation Management Institute, Colombo, Sri Lanka, 1993.
177. R. Meinzen-Dick, *et al.*, 1997, p. 89.
178. *Op. cit.*, pp. 87–88.
179. The World Bank, 1993, pp. 103–104.

Farmer-managed irrigation schemes have a long tradition in Nepal, where 70 percent of all irrigation is controlled by farmers. Nevertheless, the government had been heavily involved in developing new irrigation, with poor results. With a shift in approach, the government now promotes farmer management as a way to improve irrigation performance and to reduce the financial burden on the government of developing and operating irrigation systems. . . . The results are impressive. In the first two years of operation, forty-three surface subprojects of the [World Bank's] Irrigation Line of Credit were completed out of sixty-one subprojects processed . . . and eighty-one tubewells were drilled. . . . The project's success is due to the enthusiastic cooperation of farmers and the good dialogue between government officials and farmers. Having water user associations take ownership and responsibility for oversight improved the quality of construction, adding a much-needed element of transparency in the use of government resources. The associations created strong organizations that achieved good cost recovery and levied penalties on members who fail to abide by the rules. Many associations are also involved in other aspects of community development (The World Bank, 1993, p. 103).

Recognizing the financial burden and inefficiency of this situation, in the mid-1980s the government began to strengthen water users associations and to allow more involvement by the private sector. . . . The most success has come in the south, where associations now control practically all tubewell irrigation schemes, ranging in size from 50 to 200 hectares. The farmers are responsible for all O&M, including hiring the appropriate labor and paying for electricity. The associations are well structured technically and financially. While they perform routine repairs, the Government performs large repairs, receiving a small contribution from the associations. One notable achievement of involving user associations is that farmers have greater flexibility to respond to changes in market demand for different crops. Previous government control precluded much flexibility.[180]

Another example of the flexibility of cropping patterns facilitated by farmers' management of irrigation, and its benefits, is provided by the experience of the Mohini Water Distribution Cooperative (Society) in Gujarat, India:

[It] would not show a profit if is maintained the planned cropping patterns. . . . At present prices, the society makes a profit only if the major area is put under sugarcane. If the major area was under food grains, the society would make losses. The Mohini Society became a financial success because more than 85% of the area was put under sugarcane, instead of the prescribed 18%.[181]

Positive effects of WUAs on the process of project identification and design and on the quality of project construction also have been observed. In the case of small-scale irrigation projects in Ethiopia, 'Over 40 water user groups voluntarily formed at the time of initiation then fully participated in scheme identification and

otherwise would be the case. The Tunisian experience also illustrates how government and users' associations can share costs:

Water user associations have existed in Tunisia for most of this century, with the French colonial government introducing their legal basis in 1913. The Government of Tunisia reaffirmed the legal status of the associations by legislation enacted in 1975 and in 1987. During the 1970s, however, the government became increasingly involved in developing irrigation.

---

180. *Op. cit.*, p. 104.
181. R. Meinzen-Dick, *et al.*, 1997, p. 50.

construction. These groups have assumed total responsibility for operation and maintenance'.[182]

Results have become available from a systematic multi-country review of experiences with transfer of irrigation management, carried out three to five years after the transfer by the International Irrigation Management Institute (IIMI) in Sri Lanka, Colombia, Indonesia and India. Some of the principal findings of that work, which show that the effects of WUAs are not always unambiguously positive, have been summarized by Doug Vermillion, as follows:

- *Does irrigation management transfer (IMT) reduce government expenditure for operation and maintenance?* The answer is a definite 'yes,' with the qualification that IMT sometimes did not directly cause the reduction but at least generally supported a broader policy of reduction.
- *Does IMT result in improved quality of irrigation service to farmers?* In the four sample countries, IMT did not cause dramatic changes in irrigation intensity or in the adequacy or equity of water distribution during the first three to five years after IMT. There is evidence from the Colombian case that in pump schemes, irrigation delivery efficiency did improve after IMT. Farmers in all four countries reported improvements in communications and responsiveness to farmer needs by management staff after IMT. . . .
- *Does maintenance of irrigation infrastructure improve after IMT?* Results are mixed. In India and in run-of-river schemes in Colombia maintenance has improved. But in Sri Lanka, Indonesia and for expensive lift schemes in Colombia, it is apparent that some continuation of government subsidy or a more clear policy about rehabilitation is needed.

- *Does IMT result in higher agricultural productivity?* The results are mixed between and within countries, depending on many factors. In general, where changes do occur, they are not dramatic.
- *Does IMT result in higher economic productivity?* Again, results are mixed. It is evident that IMT has not undercut the profitability of irrigated agriculture. . . .
- *Do farmers pay more for irrigation after IMT?* Yes, they do pay more.[183]

In spite of mixed empirical results from the early years of these particular IMT experiences, in general the results are convincing enough that this approach has gathered considerable momentum in all parts of the world. The strength of this consensus was reflected in remarks by Hatusya Azumi of the World Bank's Economic Development Institute:

How important is participatory irrigation management (PIM)? I believe that participatory irrigation management is the single most important step that governments can take to improve the productivity and sustainability of irrigation systems. Let me repeat this statement: '*participatory irrigation management is the single most important step that governments can take to improve the productivity and sustainability of irrigation systems*'.[184]

6.6.3 Forms of Local Irrigation Management

Although WUAs are the main form of user management of irrigation water, other approaches are being explored. In the above-mentioned case of Shaanxi Province in China, the following additional forms are being utilized, in addition to WUAs and the model of user ownership of the system through share holdings:

---

182. N. P. Sharma, *et al.*, 1996, p. 48.
183. Doug Vermillion, 'Impacts of Irrigation Management Transfer: Results from IIMI's Research', *INPIM Newsletter*, No. 7, April 1998, p. 3.
184. Quoted in David Groenfeldt (Compiler), *Handbook on Participatory Irrigation Management*, The Economic Development Institute, The World Bank, Washington, DC, USA, April 1998, p. 2 [emphasis added].

Contracts. The government agency which owns the irrigation infrastructure, and has traditionally been responsible for its operation and maintenance, signs a contract with an individual (usually a local farmer, or former agency staff member). The agency retains property rights for the infrastructure within the Irrigation District. The rights and responsibility for managing the system, however, are transferred to the contractor. The contractor will operate and maintain the system for a fixed time period, usually 10 to 30 years. The contractor makes all water management decisions, and he is solely responsible for profits and losses of the contract. In most cases, the agency requires the contractor to invest a specified amount of money to line and improve the laterals and sub-laterals. They also require the contractor to deliver a fixed volume of water. The contractor must pay a penalty if he does not deliver the expected volume. The contractor determines the services fees that farmers pay for irrigation, within a range established by the agency. He must also collect the water fees and pass them on to the local irrigation agency. In turn, the agency pays him a management fee that was agreed upon by the contract.[185]

Lease. A lease is a slight modification of the contract system. It is generally applied only when the irrigation infrastructure is in relatively good condition. . . . without requiring the contractor to make a significant investment. . . .

Auction. The auction model is used extensively in Jinghuiqu District. . . . It is a variation on the contract model where the local irrigation agency 'pre-qualifies' three or four contractors to bid on the operation and maintenance contract for the lateral canal. . . .

Water supply companies. In contrast to the above models [including WUAs and joint stockholder companies] that are primarily focused on one or two laterals, water supply companies cover a branch or sub-branch and therefore serve all the laterals that take water off the branch. This can include up to 20 laterals. In most cases to date, water supply companies have used the joint stockholder model in order to raise the large investments that are required to improve the branch canal and its multiple laterals. Shares in the company are usually sold to farmers, staff of the irrigation station, and local government officials, with some restriction on the number of shares that any investor can hold. . . .[186]

For WUAs, a major question in defining their roles concerns the level in the system at which farmer responsibility for O & M ends and government responsibility starts, or continues. In small systems, farmers can be owners of the entire system, as noted previously, but this solution is more difficult to implement in larger systems, and so normally a division of labor is specified under which the government is responsible for main canals and the farmers for tertiary systems. The range of options is wide:

(1) *Government does everything.* In Malaysia, the Department of Irrigation and Drainage provides for the operation and maintenance of the main and secondary canals, while government sponsored farmers' organizations are responsible for providing water to individual farms. Farmers have no responsibility, and make no management decisions, about the water upstream from their outlets.

(2) *State dominates; users help.* The conventional management division in large irrigation systems is that the State takes responsibility for operation and maintenance of the headworks such as a dam or river diversion, and the main, secondary and larger tertiary canals, while farmers are responsible for managing water distribution and maintenance along the lowest level canals. Typically this entails farmer groups

---

185. This model is similar to the one of franchising rights to operate publicly constructed irrigation facilities noted by Rosegrant and Binswanger, 1994, and cited above.
186. 'Six Irrigation Managment Models from Guanzhong', *op. cit.*, 2001, pp. 8–9.

of between 10 and 50 farm families who are expected to work out sharing arrangements on their own.

(3) *Users dominate; State facilitates.* In some countries, associations of water users enter into contractual agreements with State water agencies for the provision of specific water services. In the case of Mexico, the National Water Commission manages the headworks and main canals, while legally recognized water users associations employ their own technical staff for the management of the secondary and tertiary levels of the canal networks. Farmers pay their associations for the water, and a small portion of that fee is passed on to the National Water Commission for its services.

(4) *Farmers do everything.* In the Hill regions of Nepal most of the irrigated area is in the hands of local communities who have constructed their own canal systems, generally tapping small stream flows. Similar examples of local, farmer-managed systems can be found in nearly every country where irrigation is important. . . .[187]

Meinzen-Dick *et al.* (1997, p. 58) propose a more succinct but somewhat finer categorization of the options for joint management:

• Full agency control
• Agency O & M, user input
• Shared management
• WUA O & M
• WUA ownership, agency regulation
• Full WUA control

As suggested by these quotations, the operational question is at what intermediate point the responsibility passes from one party to another. Another kind of solution can involve promoting *federations of WUAs.* In North Africa and the Middle East, for example:

In public irrigation schemes, water users associations (WUAs) are being promoted (e.g. in Tunisia and Morocco) and have been successfully introduced in many other parts of the world. Generally, their responsibilities have been limited to the O&M of tertiary systems. In some countries, water user associations are federated for O&M of larger canals and for participation in overall system management, and the potential for full transfer of smaller public schemes can often be considered.[188]

Forming federations of WUAs has been pursued vigorously in Argentina:

Traditional irrigation associations in Mendoza covered 100 to 500 hectares but were not large enough to meet the associated costs. Maintenance was insufficient, the administration was weak, and those at the head of the canal benefitted the most. The situation changed when the small associations merged into larger ones of between 5000 and 15000 hectares. Twenty-one new organizations were formed, covering 200000 hectares. Each organization is autonomous, raises its own budget, and issues its own regulations in accord with the recently enacted water law. The organization hires professional managers to deal with all administrative matters, such as water delivery, cost recovery, and maintenance. Administrative costs decreased with the decline in the number of associations. The larger organizations have increased the efficiency of conveyance by 10 percent through more efficient distribution.[189]

The link between forming federations of WUAs and achieving more equitable distribution of water has been brought out clearly in Rice's study of South East Asian systems:

The study shows that equitable treatment is less evident on the longer tertiaries and among

---

187. D. Groenfeldt, 1998, pp. 4–5.
188. The World Bank, 1994, p. 50.
189. The World Bank, 1993, p. 102.

tertiary systems on the same distributary. Headend command areas, rather than headend farmers, present the greatest challenge to fair distribution. At this level, associations and formal federations of primary WUAs can make a substantial difference. Note that at Lam Pao [in Thailand], as the associations of water user groups sharing the same secondary canals gain strength, the functions and prominence of the watercourse WUAs tend to diminish. This is predictable, because once the association of WUA leaders has determined an appropriate formula for sharing water or a cleaning schedule, meetings at the lower level can be dispensed with. The turnover of O & M responsibility from agencies to irrigators in coming years will have to focus on these *systems* of tertiaries.[190]

In broader terms:

... The allocation of functions between agencies and WUAs ... varies between levels of the system. A greater degree of agency control is generally found at higher levels of the system, with a greater WUA role at lower levels. However, the exact division of responsibility varies widely between countries. ... At the *main system level* users may have input in decision making ... but the agency retains a strong regulatory, ownership and O&M role. Some smaller systems ... recognize WUA ownership of even the main system. Shared management is frequently found at the *subsystem* or *distributary level*, either through planned sharing of O&M responsibility or through farmers informally taking on some tasks. Management transfer programs in larger systems have generally provided for eventual WUA ownership and O&M responsibility only at the subsystem or distributary level.

Below the lowest outlet, at the *watercourse level*, agency involvement is usually minimal. ... Unfortunately, unless the farmers have been adequately involved in the design and construction process, they often do not acknowledge ownership of or ongoing responsibility for the watercourses, and unless they have some degree of control over water deliveries from the main system, ownership at the watercourse level has little value.[191]

The general principle is that the turnover of responsibility doesn't depend on the volume of water but on the kind of user. If downstream of a given point the only use is irrigation, then that is the point where the responsibility of farmers would start, irrespective of the area irrigated and the chosen institutional form: a WUA, a federation of WUAs, etc. While in Asia often WUAs deal only with tertiary canals, in Latin America it is not uncommon for WUAs to manage an entire system downstream of the main canal's river intake. It is important to note that sometimes these systems cover 50 000 ha and even up to 100 000 ha. Institutional arrangements vary according to the characteristics of the transfer process. In Peru, every irrigation district has its own WUA (Junta de Usuarios) which in turn is divided into irrigators' commissions (Comisiones de Regantes), with each one managing a secondary canal. In Mexico, WUAs (Módulos de Riego) are in charge of secondary canals and they are brought together in federations (in corporate form) at the district level for the management of the main canal.

The continuing role of the State under any arrangement with WUAs is illustrated by the case of Chile:

Users' associations in Chile have been empowered and have taken on a wide range of functions, but the State's role remains clearly defined in performing adjudication functions, such as in cases where applications for water rights involve a natural water source, and in resolving highly controversial internal conflicts, as in reaching agreements regarding the allocation of water during extended dry spells. Even when the State is not called in on disputes, the

190. E. B. Rice, 1997, pp. 54–55 [emphasis in original].
191. R. Meinzen-Dick, *et al.*, 1997, pp. 58, 62 and 64 [emphasis in original].

Meinzen-Dick et al. comment also on the question of farmer ownership of irrigation systems, which was raised in the earlier section on water pricing:

Ownership *of irrigation system assets provides a clear combination of rights and responsibilities. The most important types of irrigation property include structures, equipment, water, and other assets (such as fish or trees). Ownership is based on investment in at least part of the capital costs, and implies a commitment to bearing the property's full recurrent costs. At the same time, it provides greater control over the property and the rights to earn income from it, which improves incentives for management. While in most cases the State claims ownership of both the facilities and the water rights, WUA ownership is found in many traditional farmer-managed systems. It is also incorporated into an increasing number of turnover programs. ....* **The assignment of property rights to WUAs as a management transfer strategy can increase local responsibility and incentives for system O&M.** *Where it is infeasible – for practical or political reasons – to give WUAs ownership, assigning clear rights (such as the right to exclude others or to make binding decisions) should be pursued as a way to strengthen WUAs and the effectiveness of system management (from R. Meinzen-Dick, et al., 1997, p. 61 [emphasis in original]).*

potential for State intervention is influential in persuading users to reach agreement.[192]

6.6.4 Organizing WUAs[193]

Although empirical results are not always clear-cut to date, prior reasoning as well as these many illustrative cases supports the thesis that WUAs, properly conceived and organized, can play a useful role in improving the performance of irri-

gation systems. For example, there is little doubt that placing responsibility for system management in the hands of farmers makes the systems more responsive to farmers' needs, and that they will be more flexible in adjusting cropping patterns to market signals than State agencies generally would be.

Effective participation by water users – meaningful decentralization – does not occur spontaneously. It requires planning and follow-through. The successful Philippine example of improving irrigation management, mentioned earlier in this chapter, was based in large measure on decisive steps to create institutional autonomy and promote effective participation:

The National Irrigation Administration (NIA) of the Philippines is a good example of how a bureaucracy can, over time, transform its strategy and operating style. Since the mid-1970s, the NIA has evolved from an agency that focused primarily on the design and construction of irrigation systems and told farmers about its key decisions into an agency that gives priority to the management and maintenance of irrigation systems, and that gives farmers, through the membership in irrigators' associations, the opportunity to participate in system management and to make key decisions about maintenance. ... In 1974, the NIA was established as a public corporation and ceased to be part of a government ministry. It was given a five-year lead-in period in which to become financially self-sufficient in terms of its operating budget. ... The NIA experience illustrates some important **preconditions for farmer participation:** teaming community organizers and engineers in order to integrate social and technical activities into one process; involving farmers in all project activities from the very beginning, thereby building up their organizational skills; modifying NIA policies and procedures that obstruct farmer participation; allowing enough time for farmers to mobilize

192. *Op. cit.*, p. 63.
193. Useful case studies on WUA experiences are found on the website of the International Water Management Institute: [www.iwmi.org].

and organize themselves before new construction activity.[194]

In order to obtain the full benefits of WUAs, it is important that they be organized at the stage of project conception and be given an opportunity to participate in the design of the project:

The tendency – best illustrated at Dau Tieng [in Vietnam] – is for foreign experts first to design and construct irrigation systems, next for donors and governments to finance them, and only then for attempts to be made to try to organize farmers to assist with tertiary development. Experience elsewhere confirms that this is an ineffective way to organize farmers to take over responsibility for the system. *Farmers should be organized first, or at least brought into the design and implementation processes*, and then persuaded to enter into agreements for partial financing, approval of designs, participation in construction, and management after completion of construction.

The government's failure to involve farmers in project design has had disastrous consequences in Bangladesh. Similar consequences are seen at Dau Tieng. First, the development of the tertiary systems by the provincial authorities has practically stalled, and there is no organized protest group to represent those farmers who remain unserved to exert pressure on the agencies to complete the job, to persuade village authorities to cede land essential for water passage, or to take over construction. Second, farmers in advantageous locations get early access to irrigation, which gives them the perception of an abundant supply. . . . This fosters habits, perceptions and relationships with officials that have hindered expansion of irrigated area.[195]

One of the challenges in making WUAs work well is clarifying future financial responsibilities for system rehabilitation, and another is developing an appropriate legal and institutional framework. Experience with WUAs in many parts of the world has led to the following consensus on the necessary conditions for their successful and sustainable functioning.[196]

6.6.4.1 Participation

• All affected stakeholders, including women and tenant farmers, should be included in a WUA.
• WUAs should build as much as possible on pre-existing patterns of co-operation (social capital). Governmental authorities should recognize the right of water users to organize themselves in the way they prefer and to establish their own local institutions.
• When WUAs are formed, their members should be made fully aware of their rights and responsibilities.
• WUAs should be involved in the design of new systems and in the supervision of their construction.
• More consideration should be given to establishing WUAs for drainage and flood control and not only irrigation.
• As many responsibilities as possible should be devolved to WUAs. Partial devolution has not proven very successful.
• Clear agreement should be reached on the criteria for allocating water in periods of shortages. For example, perennial crops and livestock may deserve priority, and the rights of poorer farmers and tailenders should be respected.
• For larger systems, as emphasized previously, the formation of federations of WUAs should be emphasized.

---

194. FAO, 1993, pp. 294–295 [emphasis added].
195. E. B. Rice, 1997, p. 55 [emphasis added].
196. The following summary is drawn in good measure from: (a) E. Ostrom, L. Schroeder and S. Wynne, 1993, pp. 224–225; (b) E. Ostrom, *Crafting Institutions for Self-Governing Irrigation Systems*, Institute for Contemporary Studies, San Francisco, CA, USA, 1992, Chapter 4; (c) R. Meinzen-Dick, *et al.*, 1997, pp. 65–66; (d) E. B. Rice, 1997, pp. 58–59; and (e) D. Vermillion and J. A. Sagardoy, 1999, pp. 12, 13 and 21.

6.6.4.2 Policy and Legal Frameworks

- WUAs should have legally defined water rights and contract definition of the obligation of the water authority to deliver specified quantities of water to them.
- Prior to transfer of existing systems to WUAs, a decision is required regarding the responsibility for rehabilitating or upgrading the system. Normally, government would assume that responsbility.
- The responsibilities for system rehabilitation after transfer, when WUA maintenance does not prove to be adequate over time, should be made completely clear. It is recommended that government make it clear that it will *not* fund rehabilitation in those cases.
- WUAs and governmental agencies should establish mechanisms for conflict resolution, among users and between users and the government.
- Mechanisms for supervision of the performance of WUAs also should exist, with graduated sanctions when needed. As an ultimate recourse, government should have the right to withhold water deliveries in cases in which WUAs are not fulfilling their responsibilities for O & M.
- In larger systems, control of tertiary gates should not be transferred to user associations below those gates. The gates often disappear in such cases and the policing costs for government are high.
- A service orientation and accountability should be inculcated in the agencies that work with WUAs and deliver them water in bulk. Such agencies should be financially autonomous.
- WUAs should be supported by strong water basin management organizations.

6.6.4.3 Physical Aspects

- The boundaries of the irrigation system, its components, and all member farms should be delineated fully.

- The control structures of existing irrigation systems may need to be modernized in the direction of less flexibility and discretionality in their operations, to minimize the possibilities of conflicts among users.

The question of the legal underpinnings of WUAs is fundamental. It has been stressed in the conclusions of Meinzen-Dick, Raby (see reference below) and Vermillion and Sagardoy. Since WUAs have 'taxation' and fiduciary responsibilities, their legal structure has to be carefully and clearly specified. A review of the Mexican experience with WUAs has reached similar conclusions, including the need for a strong commitment from government at the highest levels to forge new types of relationships with water users, a strong legal framework, and the assurance of financial viability through cost recovery by the water users' associations.[197]

In addition to strong legal foundations, another basic requirement for the success of WUAs is effective and sustained participation by users. In the 'Big Bang' approach pursued by Andhra Pradesh for the creation of WUAs, high priority was given to encouraging participation from the outset, and it was found that this approach permitted a tripling of water fees:

Two main conclusions were reached [by the Government of Andhra Pradesh] on the appropriate direction of the reform agenda:

- *Farmer empowerment and management should form the heart of the reforms.*
- *The process should be bold and comprehensive rather than incremental.*

. . . community outreach has been a continuous feature of the reform process, gathering further momentum as actions have been thought through and implemented from 1997 to the present. Extensive discussions have taken place

---

197. D. Groenfeldt and C. Gorriz, 'Participatory Irrigation Management in Mexico', in D. Groenfeldt (Compiler), 1998, p. 60.

across the State, including state-level large conventions, project (scheme) level workshops and smaller workshops involving NGOs at the district level. Farmer perspectives have been central to crystallizing the policy thrust and program direction. Efforts toward building understanding have been furthered since the formation of WUAs through: rural rallies in every district (30 000 to 50 000 persons each in July 1997) addressed by the Chief Minister; farmers conventions; district-level workshops with WUA presidents and Irrigation and Command Area Development Departments, Agriculture and Revenue Departments; and state-level conventions of all 10 292 WUA presidents held in April and December 1998. This massive public participation campaign has been instrumental in sustaining dialogue and fostering transparency throughout the process.

It is thus interesting to assess why Andhra Pradesh succeeded where other states have not, or have had to settle for more modest increases than Andhra Pradesh's tripling of rates. The principal reason is considered to be the *combination of a widespread consultation and public outreach process preceding the increase, and the presentation of the increase not as a single measure but as part of a package of measures which were seen by the rural communities and the political parties as overall beneficial to the farmers*.[198]

The management of irrigation is a complex and often conflictive matter, and so in principle it is not advisable for WUAs to embrace activities other than those strictly related to managing irrigation. On the other hand, in developing countries there is a general deficit of farmer organizations capable of dealing with issues such as marketing, technology transfer and financing, and so in some circumstances it is useful to build on an existing WUA structure to create a shared capacity for handling some of these other matters. Preconditions for this to work are that

farmers are highly motivated and that there is a clear separation between irrigation management and other development activities. In Latin America, there are a few examples of WUAs undertaking broader activities in this sense, and in the Philippines as well. Some of the advantages and disadvantages of this trend in the Philippines have been outlined by Namika Raby:

A number of irrigator associations (IAs), particularly in the reservoir and pump systems, have accumulated capital through their O&M contracts, obtained loans from the Philippines Land Bank, bought tractors, established rice mills, and undertaken the trading of rice. The conversion of the IA into a vehicle for channeling private and government based enterprise for the development of the rural sector is envisaged as the next phase in IA development. However, this has posed a dilemma in that the IAs as presently constituted are registered as non-profit, non-stock, non-sectarian organizations which then must be converted into for-profit organizations. Also, in order to qualify for loans, from the Land Bank for example, the IA must be constituted as a co-operative. There is apprehension on the part of the National Irrigation Administration that in their enthusiasm for business related activities, an IA might forget its original functions of carrying out O&M activities. However, for an IA to build up capital and become sustainable, it is logical that it become a co-operative. . . . If water is viewed as one input for agriculture, then there is a need for farmer organizations servicing other aspects of a viable agricultural program, e.g. marketing, credit, etc. . . . Allowing micro-organizations such as the current lateral-based IAs to evolve organically in response to demands from the environment, while focusing on a viable irrigator association for water delivery at the branch canal, may prove to be an alternative solution.[199]

---

198. Oblitas and Peter, 1999, pp. 11 and 13–14 [emphasis in original].
199. Namika Raby, 'Irrigation Management Transfer in the Philippines', in D. Groenfeldt (Compiler), 1998, pp. 67, 68 and 70.

Organizing WUAs and ensuring that they operate well is not a simple task, but they can make large improvements in the functioning of irrigation systems. The alternative of central management has a much higher rate of failure. In the words of Ostrom *et al.*:

> when the design, construction, operation and management of infrastructure are predominantly organized within a single, national government and largely financed by external funds, we can predict with some confidence the following results:
>
> - overinvestment in poorly designed and poorly constructed large-scale infrastructure facilities;
> - underinvestment in the operation and maintenance of these facilities;
> - rapid deterioration of infrastructure;
> - excessive investment in repair and rehabilitation of previously constructed facilities.[200]

6.6.5 Gender Issues in Irrigation

To date, only a few researchers have looked into gender issues in the management of irrigation systems. Generally, it has been found that those systems are managed entirely or almost entirely by men, even though women may represent a significant minority of the farmers and/or irrigation users. Women usually are more sensitive to issues of water quality for household use, but their voices are rarely heard in decisions about water management. In addition, frequently irrigating tasks are left more to the men, although not always.

The importance of involving women in the design of irrigation systems, taking into account their roles and responsibilities, was mentioned earlier in this chapter. Elena Bastidas, for example, reported that women in selected communities in Ecuador did not go out to irrigate at night, in part for fear of their safety and in part because of household responsibilities. They did not participate in village discussions of irrigation matters for social and cultural reasons, because traditionally that has been the domain of men. However, the more educated women tended to play a greater role in irrigation decisions.[201]

Van Koppen has underscored similar findings but has added that 'Evidence shows that irrigation agencies have played an important role. In the past, they persistently assumed that irrigation was men's business. In a number of cases, their actions have even undermined women's existing businesses and further polarized gender relations'.[202] She points out (p. 7) that inadequate access to water (water deprivation) is a central characteristic of rural poverty, and that 'neither State-sponsored and subsidized [water] development nor private investments can be termed pro-poor'.

At least some of the bias against women and the poor in irrigation design and management can be corrected by reducing the implicit gender bias in the actions of irrigation agencies, including those of international project teams:

> Early inclusion of resource-poor women and men in the local forums, at the interface of the project and the community, is pivotal for their improved access to water and for poverty alleviation. In a sense, the first step to become a rights holder is to be a member of the forum that negotiates rights. Water users' organizations for operation and maintenance evolve out of these early forums. Agencies strongly steer the composition of these forums. So, inclusion

---

200. E. Ostrom, L Schroeder and S. Wynne, 1993, pp. 218–219.
201. Elena P. Bastidas, 'Gender Issues and Women's Participation in Irrigated Agriculture: The Case of Two Private Irrigation Canals in Carchi, Ecuador', IWMI Research Report No. 31, International Water Management Institute, Colombo, Sri Lanka, 1999.
202. B. Van Koppen, IWMI, 2000, p. 2.

of the resource-poor depends primarily on agency efforts.[203]

In addition, she recommends 'Governments and international agencies should approve new water policies and programs . . . only after an ex-ante assessment indicates a positive impact on poor women's and men's water use, and should also monitor and evaluate the implementation'.[204]

In spite of cultural barriers against greater participation, the important point is that these barriers can be reduced, especially with the support of sponsoring agencies. Attitudes can change, sometimes in a short period of time. Indications along these lines were found among women farmers in Macedonia by Kitty Bentvelsen:

> Do women want to participate in WUA meetings and other activities initiated by the irrigation project? In general, the interviewed women were not aware what WUAs are meant to do. After explaining, women had different reactions. One woman, who was very annoyed with the bad irrigation supply during the past years, exclaimed: "I will be the first one to go to such a meeting!" The majority of women, however, started saying that attending meetings is a man's task. But during the discussion that followed, many of them would acknowledge the importance of [attending] future WUA meetings.[205]

6.7 IRRIGATION AS A TOOL OF RURAL DEVELOPMENT

The discussion so far in this chapter has concentrated on two main themes: (a) *achieving efficient* *and equitable allocations of water*, within and among sectors, in the face of increasing scarcity of that resource, and (b) *making irrigation systems work better, and sustainably*, through use of more appropriate technology, improvements in system management (including greater participation by users) and more appropriate economic and legal environments (including better incentives for more efficient use of water). Although the value of irrigation in reducing poverty has been mentioned, its role in promoting rural development has not been touched upon directly, except insofar as it is related to approaches designed to make access to irrigation more equitable. However, of all of the approaches designed to improve the lot of smallholders, irrigation undoubtedly ranks among those with the greatest potential benefits. As remarked of the Chilean experience, 'Ex-post evaluations have demonstrated that irrigation has been the investment with the greatest impact on productivity, employment and the incomes of small farmers'.[206] Equally, there is little doubt that irrigating agriculture generates multiplier effects that spread broader benefits through the participating areas, especially in cases in which their infrastructure services have been weakly developed. Among other things, those multiplier effects include 'accelerating the growth of transport, commerce and storage services, of input distribution, of technical advisory services, etc., [which] are an additional contribution attributable to new irrigation'.[207]

Fuller consideration of the role of irrigation for promoting improvements in the welfare of smallholders and rural development in general runs squarely into the question of the ability of

203. B. Van Koppen, 'Gendered Water and Land Rights in Rice Valley Improvement, Burkina Faso', in B. R. Bruns and R. S. Meinzen-Dick (Eds), 2000, p. 105.
204. Van Koppen, IMWI, 2000, p. 23.
205. Kitty Bentvelsen, 'PIM and Gender: Examples from Macedonia', *INPIM Newsletter*, International Network on Participatory Irrigation Management, No. 7, April, 1998, p. 9.
206. Jorge Echenique L., 'Utilización de Subsidios para el Fomento de la Irrigación', report prepared for the Latin America and Caribbean Office of the FAO, Santiago, Chile, November 1996, p. 49 [author's translation of the quotations from this article].
207. *Op. cit.*, p. 9.

the users to pay for irrigation services, which has been alluded to already. Moris and Thom state the issue clearly:

> ... smallholders cannot afford major capital improvements.... some element of subsidy is inevitable if a country intends to proceed in developing irrigation. The operational question is ... not whether to subsidize, but where, when and with what consequences?[208]

This issue also was raised in Vermillion's summary of the IMT research findings cited above. Therefore the question of subsidies for smallholder irrigation must be addressed explicitly, especially in light of the prevailing international emphasis on cost recovery in irrigation. The FAO has summarized this emphasis succinctly in stating that, for irrigation, 'the era of large direct and indirect subsidies is nearly over'.[209] On the one hand, there is the concern for the ability of users to pay, especially if they are poor farmers, which has been expressed not only by Moris and Thom but also in Rice's recommendation to 'abandon cost recovery' in situations of low returns to farmers, and in Van Koppens' recommendations in regard to poor irrigators. On the other hand, there is a legitimate concern over the inability of government budgets to continue underwriting irrigation costs. As stated by Sharma *et al.* for Africa, in words that apply to other continents as well:

> ... water use is highly subsidized.... Consequently, cost recovery remains low, increasing the burden on the central government to provide capital for maintaining existing systems and developing new infrastructure.... African countries should place greater emphasis on ... cost recovery.[210]

How are these tensions, between the contrasting objectives of promoting smallholder irrigation and restricting fiscal outlays, to be resolved?

A first step is to examine the nature of cost recovery in irrigation. As stated in the section on irrigation pricing policies, attempts to strengthen cost recovery rarely go beyond raising the funding required for O & M expenditures. Very occasionally a part of the capital costs may be recovered, but that is unusual, and full recovery of capital costs is not expected even in instances when ownership of irrigation systems is transferred to users. Recall that, in the box in the preceding section, Meinzen-Dick *et al.* have said that 'Ownership is based on investment in *at least part* of the capital costs' [emphasis added].

Thus, it is important to distinguish between subsidies for system development and subsidies for the operating costs of the irrigation service. The principle that the user should pay clearly applies to the latter costs, and hence those subsidies should be minimized if not eliminated. The reasons why have been spelled out throughout this chapter and may be summarized in three ways: (a) user financing of O & M costs gives farmers a vested interest in seeing that O & M is properly performed and that its costs are reduced as much as possible, (b) cost recovery reduces the fiscal burden on the government, freeing up funds for other development projects, and (c) it helps guarantee the sustainability of an irrigation system, since the availability of government funding cannot be counted on in the indefinite future.

System development costs are another matter, and here there are three basic reasons why subsidies may be considered to be indispensable:

(a) As noted, smallholders simply may not be able to pay the capital cost of irrigation, much as they are not able to pay the full price of agricultural land (Chapter 5), and the equity objective would indicate assisting them in that regard.

(b) When irrigation development costs are placed in the context of overall sectoral policy, and if arguments for generalized fiscal support for

208. J. R. Moris and D. J. Thom, 1991, p. 560.
209. FAO, 1993, p. 295.
210. N. P. Sharma, *et al.*, 1996, pp. xv and xix.

agriculture are accepted (Chapter 3, Section 3.2), then funding the construction of irrigation systems, especially for small-scale producers, may be seen as one of the most effective ways to provide that support. For example, if sector support is desired in order to offset the deleterious effects on domestic farm incomes of international agricultural subsidies, then funding irrigation development would be a non-distortive way to achieve that aim, as opposed to interventions in prices.

(c) The positive economic externalities in rural areas, noted by Echenique, may justify subsidization of system development. However, care must be taken in system design and management to ensure that negative environmental externalities do not outweigh the positive externalities. (Positive externalities in the form of flood control should not be overlooked either.)

Chile and Nicaragua provide recent illustrations of the use of subsidies for supporting the development of irrigation among smallholders. In Chile, the program originally applied only to commercial-scale farmers, and then it was modified to include a separate 'window' for financing the construction of small projects. The program was designed so that project identification could be 'demand driven', in the sense that farmers propose the projects and local consulting firms assist them to formulate the proposals. A special commission reviews the proposals quarterly, and they are judged on the basis of user contribution to project costs, the area to be irrigated, and the total cost per hectare irrigated.

During the first six years of operation of the program for smallholders, 56% of the proposals were selected, and the projects provided benefits to 43 000 producers.[211] The total amount of subsidy for construction costs was $134 million in that period, and the contribution of the beneficiaries to construction costs was $120 million. Echenique emphasizes that two elements con-

tributing to the success of the program were that 'the State provides the subsidy only when the works have been formally accepted as complete and therefore [it has been] guaranteed that the irrigation infrastructure is ready to provide services',[212] and that the program provides financing for the studies conducted in order to develop the proposals. For the former element, a key provision was the creation of bankable government certificates which allow the entity carrying out the construction to raise private financing for it.

A somewhat similar experience was initiated more recently in Nicaragua by the government, with the support of the Inter-American Development Bank, in the form of a 'National Rural Development Program' which funds small-scale projects identified by rural communities, again with the intermediary services of local consulting firms (or NGOs). In such cases, where the value of public subsidies for the development of poor rural communities is widely recognized, it would be difficult to argue that irrigation construction should be excluded from such subsidies, especially since irrigation is one of the most powerful tools for rural development. As in other irrigation experiences, the principal challenge lies in appropriate design of the operational mechanisms, in this case including the mechanisms for encouraging local project identification and participation in its management, for assigning priorities to proposals, and for establishing a process of review and approval of project proposals which is appropriate to their small scale. These challenges are practical but never are simple, and the solutions always must be adapted to the local context. However, that is the nature of the development challenge itself.

DISCUSSION POINTS FOR CHAPTER 6

• Irrigation is the world's largest user of fresh water, and the proportion of water supplies it uses is highest in lower-income countries. The

211. J. Echenique L., 1996, pp. 25 and 27.
212. *Op. cit.*, p. 48.

vast bulk of future increases in world food production will depend on irrigation.

- The per capita availability of renewable water supplies is declining by more than 25% per decade in many countries. More than 40 countries now experience serious chronic water shortages in at least some of their regions.
- Even in countries without serious water shortages, irrigation has seen its economic viability decline, as a result of lower-than-expected farm returns under irrigation and high construction costs of the systems.
- In addition, the design of irrigation schemes and their management has been deficient in many cases, with the consequence of degradation and even abandonment of some systems. Among functioning irrigation systems, most of them deliver more water than is needed to crops, and many suffer from inequitable distribution of water among the system's users.
- While the barriers to making irrigation work well are multiple and formidable, there is no other single technology or policy intervention in agriculture that promises benefits of the magnitudes that irrigation does, provided its potential can be realized. Throughout the world, there are many examples of successful irrigation schemes.
- The principal policy objectives for irrigation systems are efficiency, equity and sustainability. Efficiency has both technical and economic dimensions – efficient physical use of the scarce resource that is water, and economic viability. Equity refers to equal treatment of all irrigators in the system in terms of their access to water and to targeting new systems as much as possible on 'land-poor' rural families. Sustainability means avoiding depletion of groundwater and salinization and other forms of degradation of the soils.
- In light of the growing scarcity of water worldwide, policy emphasis is shifting from developing new sources of fresh water and water delivery systems to managing demands for water.
- Irrigation management is a component of the broader responsibility for all water management. Irrigation management needs to take place in a holistic context in which all uses in a watershed are considered.
- Numerous classes of optional policy instruments are available for implementing water management strategies, including the following: direct administration of water fees and allocations at the irrigation system level; management of water allocations, quality levels and fees through watershed or river basin authorities; capacity building among users and managers; involvement of stakeholders in irrigation system design and operation; decentralization of public sector management organs; codification of water rights and establishment of regulatory frameworks; developing markets in water rights; joint management of transboundary water basins; and public investments in water extraction and conveyance facilities.
- Adopting a more systemic approach does not imply greater centralization of decisions or greater control over water resources by a government. On the contrary, it means developing partnerships among all stakeholders, including local and national government entities, communities, irrigation users and other water users.
- A prerequisite for an irrigation development program is a national economic policy framework, both macroeconomic and sectoral, that is conducive to agricultural growth.
- A basic requirement of a national water policy, and therefore of an irrigation strategy, is a national water assessment. Before proceeding with the design of irrigation projects, a prior decision has to be taken as to how much supplies of irrigation water can be expanded, or whether they can be expanded at all, in light of projected water balances by watershed and nationwide.
- Most reviews of irrigation investment strategies recommend giving priority to rehabilitation of irrigation areas rather than development of new areas. However, the priority accorded to rehabilitation needs to be qualified in a number of respects. It may be more important to improve the institutional aspects of the system, or the policy environment, than to rehabilitate the physical structures.

- In cases in which physical rehabilitation would appear to have a role, it is important to evaluate the original engineering design and decide whether it functions sufficiently well to warrant rehabilitating it.
- There are many types of irrigation systems, but the most common distinctions are full versus supplemental systems, modern versus traditional (informal) schemes and large versus small schemes. More than one type of system may have a place in a national water strategy; accordingly, all types should be reviewed in a national water assessment.
- Supplemental irrigation is used to compensate for dry spells during the rainy season, or to prolong the season of water availability for crops. It usually is based on pumping, whether from surface or groundwater. Its desirability is dictated by climatic conditions; in regions where the rainy season often is irregular, it can play a vital role in preventing severe damage to crops. Supplemental irrigation can be critical not only for increasing the volumes of production but also for ensuring the quality of products like fruit and vegetables, for it enables farmers to control the timing of water deliveries to the plants.
- Small-scale irrigation projects generally have been more successful than larger ones, but it has been found that the degree of farmer participation in system management is a more important determinant of success than system size.
- The design itself of irrigation systems sometimes tends to generate conflict among users, especially in larger systems with discretionary water control elements, and sometimes it encourages over-application of water. Improved system design, generally in the direction of engineering simplifications, can make the system operations both more efficient in water use and more equitable among irrigators.
- Appropriate system design is also critical from a gender perspective. To ensure gender issues are addressed in the design, it is important to carry out a gender analysis of the involved communities first, with special emphasis on identifying the agricultural and water-related tasks that women carry out. The design process should be participatory and women's groups should be consulted without the presence of men.
- It is important to be aware of the hazards of using systems based on pumping in environments that cannot readily support them. In many countries, more pumps are out of order than are functioning, owing to the lack of a ready supply of replacement parts and training in maintenance.
- The methods of water demand management in use include administrative allocations of water (by far the most common mechanism in use in the developing world today), user-based allocation systems controlled by users, market allocation of tradable water rights, joint allocation by users and governmental agencies (as through watershed 'parliaments') and individual allocational decisions made by owners of infrastructure.
- The rules for pricing irrigation water vary widely, within and between countries. The rules for setting prices are neither systematic nor uniform. The only consistent feature of irrigation prices is that they usually are well below the cost of supplying the water. The recovery of operating costs through irrigation prices ranges from around 20% to a high of around 75%.
- Experience has shown that farmers may be willing to pay more for it *provided that* the supply is reliable. This is a major caveat that often is not satisfied in gravity-fed systems.
- For irrigation water, price does not play the normal role of equilibrating supply and demand, except in the case of markets for water rights, which so far have had limited applicability. Accordingly, in most cases, the justification for the level of this price must be different than that of balancing supply and demand.
- There are five fundamental reasons for setting the price of irrigation water at an appropriate (which usually means higher) level. The first three reflect societal concerns for the utilization of a scarce resource and the last two, fiscal concerns. They are as follows:
  (i) To stimulate the conservation of water.
  (ii) To encourage the allocation of water to the most water-efficient crops, or to non-agricultural uses if water is more productive

there in net terms, after allowing for inter-sectoral conveyance costs, provided that the infrastructure exists for delivery of the water to new users.

(iii) To minimize the environmental problems attendant upon irrigation, especially those arising from excessive use of water.

(iv) To generate enough revenues to cover operating and maintenance costs of irrigation systems so that, among other things, it would not be necessary to invest in expensive rehabilitation projects.

(v) To recover the original investment costs of each system, in addition to providing revenues for O & M costs.

• It has been suggested that the optimal level of irrigation pricing is that which the irrigators themselves judge adequate to maintain and operate their system, with each paying what he/she would prefer the others to pay.

• It may be difficult to raise irrigation charges when the producers are very poor (and female-headed households tend to figure among the poorest). However, since irrigation systems are not sustainable without cost recovery, there is a persuasive argument that irrigation fees are not an appropriate policy instrument for addressing the needs of the rural poor, and in any case irrigated farms almost always generate incomes above those of the poorest rural strata.

• The opportunity price of water in non-agricultural uses is usually substantially higher than even a price required to cover irrigation O & M costs. For this reason, when the price of irrigation water is established by water markets, it usually turns out to be higher than if established by governmental agencies or users' associations.

• There is a fundamental difference between a market-determined price for water and a decreed price: if the price rises because of demand for water, as transmitted through markets for water rights, then farmers can be beneficiaries of the price rise, by selling their water rights. Obviously, they will do so only if the resulting net annual income is greater than that attained by cultivating the land. On the other hand, administered price increases for irrigation water represent economic losses for them.

• Pricing based on the area irrigated is the most commonly used, but systems based on the volume of water are more efficient and are gaining ground. Improvements in irrigation design can assist the implementation of volumetric pricing.

• Markets in water rights have been implemented in some countries. They are not yet widespread but they have demonstrated significant advantages in promoting more efficient use of water without penalizing farmers economically. Their implementation requires appropriate legal and institutional frameworks to safeguard return flows and other aspects of third-party rights.

• Informal water rights markets tend to arise in conditions of water scarcity, and they have been found to exist in places as diverse as Brazil, Mexico, Bangladesh, India, Pakistan and Yemen. Legal recognition and formalization of such markets can improve their functioning and provide necessary protection to the environment and third parties.

• Water rights markets do not function like most other kinds of markets. The conditions of competition – many buyers and sellers – usually are not fulfilled, and transactions require approval from various groups, from water user associations in Chile and Mexico to state governments in the Western United States.

• Beneficial as they are, water rights markets are not a panacea and their implementation requires fulfillment of several prerequisites, including the following: designation of how water is allocated during times of scarcity; ensuring that the initial allocation of water rights is fair; having a water management technology that permits reallocations; having a strong water users' association; defining appropriate roles for regulatory authorities and regulating the treatment of third-party rights; disseminating information on how water rights markets work; and the political will to carry out the implementation of the new system.

• Establishing an appropriate institutional structure for allocating water is a fundamental role of social policy for any nation, and it is critical to the functioning of irrigation systems. Improvement in irrigation performance depends on good governance. There are four main elements of governance which can be considered at national and local levels: the legitimacy of government; its accountability; its competence; and its respect for human rights and the rule of law. There is agreement that strengthening institutional capacities requires wide participation at all levels of decision-making and at all stages in the process, including in policy formulation and project design.

- The financial autonomy of public irrigation agencies is important, as is greater training of their staff and improved communications with irrigators. Water allocation decisions should be devolved to the lowest levels possible, and that usually means leaving at least some allocational decisions in the hands of water user' associations.

- A problem with public sector water management in many countries is the fragmentation of responsibilities. An integrated institutional approach functions better except for the requirement for a clear separation between the roles of defining water policy and providing (or co-ordinating the provision of) water services.

- Putting local water management in the hands of water users' associations (WUAs) usually results in more efficient system management (lower costs) and a greater commitment to maintenance of the system – although sometimes necessary rehabilitation is deferred because of its costs. WUAs also perform a conflict-resolution role. However, devolving O & M responsibilities to WUAs almost always means farmers pay more for their irrigation.

- There are many thousands of water users' associations throughout the world and they are considered a *sine qua non* for the effective functioning of irrigation systems. Much experience has been accumulated on how to form them and train them, and how to define their relations with government agencies. Deciding upon government and WUA responsibilities for system rehabilitation is a crucial issue in the transfer of system management to users. In addition, it has been found more effective to form the WUAs

before systems are built and to involve the associations in the design of the systems.

- An important operational question is defining the division of responsibility between WUAs and government agencies. Experiences range from complete ownership and control by WUAs to complete ownership and control by government. The main options have been summarized as follows: full agency control; agency O & M, user input; shared management; WUA O & M; WUA ownership, agency regulation; full WUA control.

- The degree of control by WUAs is usually greater in smaller systems and in the distributaries or sub-systems of larger systems, while either government agencies or federations of WUAs typically control the operation of main channels in large systems.

- A useful operational rule is that control should pass to irrigators, in WUAs and possibly federations of WUAs, at the point in the system below which the only use of water is irrigation.

- Some important preconditions for farmer participation in irrigation management include the following: joining community organizers and engineers in teams in order to integrate social and technical activities; involving farmers in all project activities from the very beginning, thereby strengthening their organizational skills; modifying irrigation agency policies and procedures that hinder farmer participation; and allowing enough time for farmers to organize themselves before new construction activity.

- Other important guidelines for developing and supporting WUAs include the following: recognition that WUAs are stronger if they can build on existing patterns of co-operation; defining membership to include all stakeholders, including tenants and women; ensuring that benefits to farmers of participation outweigh the costs of participation; institution of a supportive policy and legal environment; ensuring that public irrigation agencies carry out their corresponding responsibilities effectively; and as mentioned earlier, clarifying the government's role in supporting the costs of system rehabilitation. Government assurances in this last regard often are critical for the formation and successful functioning of WUAs.

- Gender bias is widespread in both design and operation of irrigation system. Overcoming it requires not only sensitization of irrigators but also of irrigation agency staff. Women need to be involved in the planning of irrigation systems from the beginning, and system designs need to take account of the differences between agricultural tasks typically performed by women and men. Gender analyses need to be carried out before projects are designed, and women's groups need to be involved in design and management processes for the systems.

- Irrigation can play a strong force for raising the incomes of rural poor and of smallholders in general if it is targeted on those groups. Normally, a subsidy will be required for the construction cost of systems for smallholders, but not for meeting the annual O & M costs. Chile has implemented a program of irrigation investments for smallholders, through competitive bidding among proposals, in which construction costs are paid only when the system has been shown to be functioning as planned. With innovative programs of this nature, irrigation can play an important role in poverty reduction and rural development.

7

Policies for Agricultural and Rural Finance

7.1 The Role of Finance in Agricultural
   Development   277
   7.1.1 The Nature of the Problem   277
   7.1.2 Agricultural Credit and Rural
      Savings   279
   7.1.3 Financial Services in Rural Areas   284
   7.1.4 Characteristics of Rural Financial
      Markets   286
7.2 Policy Objectives for Rural Finance   287
   7.2.1 Policy Objectives: Production versus
      Income   287
   7.2.2 The Objectives of Poverty Alleviation
      and Gender Outreach   289
   7.2.3 Objectives for Rural Financial
      Institutions   293
   7.2.4 The Contributions of Microfinance   293
7.3 Keys to the Sustainability and Efficiency
   of Financial Intermediation   295
7.4 The Regulatory Framework for Rural
   Finance   300
   7.4.1 Regulatory and Institutional
      Challenges   300
   7.4.2 Collateral   302
   7.4.3 Contractual Relationships   305
   7.4.4 Interest Rate Regulations   306
   7.4.5 Bank Regulation for the Rural
      Sector   308
7.5 Structural Considerations for Rural
   Financial Institutions   317
   7.5.1 Local Institutional Structures   317
   7.5.2 Credit Co-operatives   318
   7.5.3 Rural Banks   322
   7.5.4 Apex Organizations   324
   7.5.5 Rediscount Lines and Bond
      Financing   326

7.5.6 Governance Issues   328
7.5.7 Gender Issues in Rural Finance   329
7.6 Approaches to Managing Rural Financial
   Institutions   330
   7.6.1 Autonomy   330
   7.6.2 Interest Rates and Lending Policy   331
   7.6.3 Other Incentives for Repayment   333
   7.6.4 Techniques of Savings Mobilization   334
   7.6.5 Prudential Management in Rural
      Financial Institutions   337
7.7 Macroeconomic Policy to Support Rural
   Financial Intermediation   340
   7.7.1 The Problem of Directed and
      Subsidized Credit   340
   7.7.2 Managed Deposit Rates   340
   7.7.3 Inflation and Interest Rates   341
   7.7.4 The Role of Donor and
      Government Finance   342
7.8 Elements of a Strategy for Developing
   Rural Finance   346
   7.8.1 The Financing Gap   346
   7.8.2 Elements of a New Approach   346
   7.8.3 The Context of Gender Policy in
      Microfinance   348
   7.8.4 Towards Banking for All Farmers   349
Discussion Points for Chapter 7   351

7.1 THE ROLE OF FINANCE IN AGRICULTURAL DEVELOPMENT

7.1.1 The Nature of the Problem

The antiquity of agricultural lending practices cannot be doubted. The second book of the Bible provided rules related to loans in kind: 'When a

Agricultural Development Policy Concepts and Experiences. R. D. Norton
© 2004 Food and Agriculture Organization of the United Nations
ISBNs: 0-470-85778-1 (HB)   0-470-85779-X (PB)   FAO Edition: 92-5-104875-4

man borrows a beast from his neighbor and it is injured or dies while its owner is not with it, the borrower shall make full restitution' (Exodus 22:14). The cuneiform texts of the early Sumerians described the penalties associated with default on loans: 'Landless peasants ... sometimes sold themselves as slaves simply for meals and a place to sleep. ... A man in desperate financial straits might turn over his entire family – himself included – to a creditor for an agreed-upon time in satisfaction of his debts'.[1]

While the early rules had to do with financial transactions between individuals, governments eventually came to be concerned about agricultural credit as a matter of policy. Ensuring that farmers receive sufficient credit has been taken as a serious challenge by virtually all governments in our era. The extent of policy influence on credit differs by country, but every government in the world has intervened in the rural financial sector.[2] Governments in the developing world have justified their measures by the perception of inadequate volumes of commercial bank lending to agriculture, and of excessive interest rates and limited amounts of loanable funds in the informal credit market.

For many decades now, those interventions in credit markets have tended to be direct, usually taking the form of directed allocations of loans, subsidized interest rates and State ownership of banks. 'In the later 1970s, for example, the central bank in Indonesia administered nearly 200 directed credit lines, many of which were aimed at agricultural activities, and most of which were subsidized. ... In Thailand ... during the 1970s and 1980s the government required all banks to lend an increasing percentage of their total loan portfolio to farmers'. In addition 'in several countries such as the Philippines, major segments of the rural financial system were attached to crop production programs. In other countries such as

Egypt and Brazil large subsidized credit efforts were justified on the basis of compensating farmers for other ... distortions in the economy, such as food price controls or over-valued foreign exchange rates'.[3]

Enough experience with such interventions has been accumulated that the results are now in: they have failed to meet their objectives and have become an unsustainable fiscal burden. As a result, the real amount of formal credit available to agriculture has declined in the last two decades in most regions of the developing world. How to satisfy the credit needs of a growing agriculture in a viable way has therefore become a central issue of agricultural development policy. The crisis in the traditional approach was well articulated by Jacob Yaron:

Generally, the past performance of State- or donor-sponsored rural finance operations has fallen substantially short of expectations. Many of the institutions established or supported primarily for delivering credit programs have not developed into self-sustained rural finance institutions. The programs have reached a minority of the rural population, often resulting in benefits in the form of negative [real] on-lending interest rates which become an unintended 'grant element,' captured by wealthy and influential farmers. The maintenance and continued operation of many of the credit programs has turned into an extremely costly drain on government budgets. ... Administrative interventions have retarded the promotion of efficient financial markets and have had an adverse impact on the development of other sectors of the economy, mainly by depriving them of loanable funds and increasing their borrowing costs. Many of the large rural financial institutions have been associated with heavy losses generated by either

---

1. *The Age of God-Kings*, Time-Life Books, Alexandria, VA, USA, 1987, p. 27.
2. J. Yaron, M. P. Benjamin, Jr and G. L. Piprek, *Rural Finance: Issues, Design and Best Practices*, Environmentally and Socially Sustainable Development Studies and Monographs Series, No. 14, The World Bank, Washington, DC, USA, 1997, p. 20.
3. Elizabeth Coffey, *Agricultural Finance: Getting the Policies Right*, Agricultural Finance Revisited No. 2, FAO and GTZ, Rome, June 1998, pp. 2–4.

inadequate indexation in a highly inflationary environment (Brazil, Mexico) or by dismal loan collection in a stable economy (India and Bangladesh).[4]

State-owned agricultural banks that masked their administrative shortcomings with repeated infusions of capital from the government budget have found that they can no longer count on funds from the national treasury indefinitely. Throughout the world, many of them have closed or scaled down their operations drastically. The closure of these banks has stranded large numbers of small- and medium-scale producers without access to institutional credit, even though many of them had solid credit histories. The loss of these financial relationships and the associated information which had accumulated over time represents a loss to the economy.[5]

At the other end of the spectrum, large numbers of small rural credit organizations that depended on donor funding have folded as the programs that sustained them came to an end. These difficulties experienced by agricultural credit institutions have given rise to a search for approaches that will be **sustainable** as well as ensuring that sufficient volumes of credit are available:

ongoing market reform and privatization have not yet produced appreciable improvements in the provision of agricultural support services. Nor have they increased farming profitability. If anything, small farmers often have less access to rural banking and institutional agricultural lending facilities than before. A major reason is the absence of an adequate rural and agricultural finance policy framework.[6]

A considerable amount has been learned in recent years about how to make small-scale financial institutions sustainable, and the number of successful microfinance organizations has increased rapidly. These institutions are responding to part of the need for production credit in agriculture, but their contribution in that regard remains small in comparison with the needs. The following is one of many commentaries along these lines:

In view of the difficult experience with agricultural credit, development aid shifted towards supporting microfinance institutions (MFIs). These institutions predominantly grant small and short-term loans to marginal clients. ... However, it becomes evident that microfinance institutions mostly focus on urban and peri-urban areas. In addition, they predominantly grant loans for non-agricultural purposes such as trading activities. Thus the financing requirements for on-farm production of small farm households remain largely unsatisfied.[7]

7.1.2 Agricultural Credit and Rural Savings

Agriculture is both more capital-intensive and more labor-intensive than the manufacturing sector in developing economies. The capital required per unit of output is twice or more as high in agriculture as it is in industry, on average. This basic fact is confirmed by the data in the capital matrices of input–output tables. Labor also is more intensively used per unit of output

4. Jacob Yaron, *Successful Rural Finance Institutions*, World Bank Discussion Paper No. 150, The World Bank, Washington, DC, USA, 1992, p. 3.
5. See the survey evidence on this point cited in Claudio Gonzalez-Vega, 'Servicios Financieros Rurales: Experiencias del Pasado, Enfoques del Presente', presented at the international seminar, *El Reto de América Latina para el Siglo XXI: Servicios Financieros en el Area Rural*, La Paz, Bolivia, November, 1998.
6. Brigitte Klein, Richard Meyer, Alfred Hannig, Jill Burnett and Michael Fiebig, *Better Practices in Agricultural Lending*, Agricultural Finance Revisited No. 3, FAO and GTZ, Rome, December 1999, p. 68.
7. Scheme for Agricultural Credit Development, *Report of the Eighth Technical Consultation*, FAO, the African Rural and Agricultural Credit Association, and the Central Bank of Nigeria, Abuja, Nigeria, March 8–10, 1999, p. 17.

in agriculture, as confirmed by the high share of the economically active population that depends on agriculture, relative to the sector's contribution to national product.

Logically this circumstance means that in agriculture either capital or labor must have a lower return than in industry, or that both of those factors do. In practice, it means both: wages as well as average rates of return to investments tend to be lower in agriculture than in other sectors. The lower wages can be explained in most cases by the relative abundance of labor and its difficulty of moving, in the short run, to remunerative non-agricultural occupations, since many of them have higher skill requirements. The lower returns to capital cannot be explained by its abundance in the sector. Investment capital is scarce in agriculture.

According to the traditional view that economic growth must be based on industrialization (see Chapter 1), agriculture simply lacks profitable opportunities for investment. Empirical evidence strongly suggests that this cannot be the whole story, however. Many farmers who borrow capital do so on the informal financial market, where they pay very high interest rates. If the productivity of capital were low throughout the sector, then all of those borrowers would have defaulted and informal lenders would have disappeared from the agricultural sector. In fact, in all countries' agricultural sectors there are numerous examples of entrepreneurs and product lines that have successfully expanded output through investments financed by borrowing. The low returns to capital may characterize mainly lending by formal financial institutions.

A plausible explanation for the apparently low average return to capital in the sector is that investment funds do not flow readily to the most productive uses, both because of the way credit institutions have been managed and because of the structure of markets. Funds injected into agriculture by public agencies have not necessarily gone to the uses with the highest returns. Private capital markets are segmented in rural areas and are imperfect in other ways. An econometric study of credit in Philippine agriculture found strong evidence of market segmentation, based on occupations. In practice the segmentation facilitates screening of borrowers and enforcement of contracts since, for example, some traders tend to be linked to farmers who are large rice producers.[8] While this kind of arrangement may be efficient for those particular lenders and borrowers, it poses problems for the development of the sector:

A rice trader is better equipped to evaluate the creditworthiness of rice farmers than corn farmers at a lower cost due to his occupational specialization. However, with the growing need for crop diversification because of environmental and risk concerns, it may become difficult for such specialized informal lenders to adequately service diversified farms. . . . It would also be difficult to introduce a formal credit institution into this type of segmented market. . . . Formal institutions would have to solve these borrower screening and contract enforcement problems in order to effectively compete with specialized lenders. . . . The well documented failure of the Philippines rural banking system in the early eighties was due in part to its inability to develop appropriate financial technologies to meet this challenge.[9]

Many of the reasons for the limited development of private financial intermediation in the sector are well known: imperfect information

8. G. Nagarajan, R. L. Meyer and L. J. Hushak, 'Segmentation in informal credit markets: the case of the Philippines', *Agricultural Economics*, **12**(2), August, 1995, p. 180.

9. *Ibid.* Other studies have documented the segmentation of rural financial markets. For cases in Africa, see 'Informal Financial Markets and Financial Intermediation in Four African Countries', *Findings: Africa Region*, No. 79, The World Bank, Washington, DC, January, 1997. This last reference summarizes work by Ernest Aryeetey, Hemamala Hettige, Machiko Nissanke and William Steel in *Financial Market Fragmentation and Reforms in Sub-Saharan Africa*, Discussion Paper No. 356, The World Bank, Washington, DC, USA, 1996.

about borrowers and projects, as illustrated by the Philippines example, lack of adequate collateral (landholdings without full title, for example), asymmetric information on the part of borrowers and lenders about crop yield expectations and variability and repayment capacity, covariant yield risk[10] and price risk, and so forth. A consensus also is emerging that inappropriate policies for the rural financial sector are another reason for its low state of development. Exploration of this problem and ways to improve those policies is a main theme of this chapter.

It is indisputable that the generally low level of human capital formation in the rural sector is also an explanation for its low returns to investment capital: the two forms of capital are complementary. Nevertheless, there have been many successful experiences with reforms in rural finance in recent years, in all regions of the world, that have addressed the weaknesses of existing formal systems of rural financial intermediation. These experiences suggest that it is possible, through appropriate policies and programs, to improve the allocation of capital in rural areas, yielding the results of higher returns to capital and higher incomes for its users.

Investment capital takes the forms of equity and debt. In addition, in agriculture human capital can be converted into physical capital by on-farm effort, as in building irrigation channels and fences by hand. However, many forms of productive capital cannot be created in an artesanal way, and rural families in developing countries typically do not have much financial equity (savings deposits) that can be put into major investments, although they may have a capacity to save. Nor is corporate farming, based on share capital, very prevalent in developing agriculture. In fact, in all sectors in almost all economies, including in the more industrial countries, equity

capital plays a much less important role than borrowing in the financing of investments. As noted by Joseph Stiglitz, 'in most countries equity is a trivial source of new finance'.[11] Hence, more efficient and sustainable lending mechanisms can contribute in a central way to agricultural development.

Institutions and mechanisms for mobilizing financial savings also are essential for the financing of agricultural and rural development. They contribute to the sustainability of rural financial intermediation and provide needed financial services to the rural population. The capacity of low-income rural households to save has often been underestimated and used as a justification for the approach of channeling credit to farmers, rather than that of building viable rural financial institutions. As commented by Robert Vogel, as early as 1979–1981 a successful USAID-supported project in Peru demonstrated the rural savings potential:

This project [BANCOOP] shows that savings can be mobilized in rural areas of low-income countries when the proper incentives are present.... There is a myth ... that most of the rural population has no savings. If this were true, the rural poor would have become extinct long ago with the onset of the first emergency, and small farmers would have starved while waiting for the next harvest if they failed to save some of the previous harvest. The rural poor, more than anyone else, must have a liquid reserve to meet emergencies. Credit, usually from informal sources, can sometimes supplement this liquid reserve, but credit is available only to those who have actual or potential savings. Even the moneylender will not lend to someone with no accumulated or potential surplus, and friends and relatives, as well as

10. The large variability of crop yields increases the probability that an individual agricultural borrower will default or request a loan rescheduling, but for banks an even greater concern is that the yields of all crops and all farms in a given area tend to fluctuate together because of weather variations. This is *covariant* behavior of yields.

11. Joseph Stiglitz, 'The Role of the Financial System in Development', paper presented to the Fourth Annual World Bank Conference on Development in Latin America and the Caribbean, titled *Banks and Capital Markets: Sound Financial Systems for the 21st Century*, San Salvador, El Salvador, June 28–30, 1998, p. 3.

savings and credit societies, usually require the ability to reciprocate . . .[12]

Marguerite Robinson has observed that Indonesian banks' massive savings mobilization from 1986 onward has destroyed the myths that mobilizing rural savings is difficult in developing countries.[13] She points out that institutional savings provide numerous benefits to households, including the following:

- *Liquidity*. Rapid access to at least some financial savings is considered essential by many households in monetized or partially monetized economies. . . . people save for emergencies and for investment opportunities, which may arise at any time . . .
- *Returns on deposits*. Positive real returns on deposits are typically not available at low risk outside financial institutions. . . .
- *Savings for consumption*. Households with uneven income streams (from agriculture, fishing and enterprises with seasonal variations) can save for consumption during low-income periods. . . . Households also tend to save for other kinds of investment, such as children's education, house construction, and electrification. . . .
- *Savings for social and religious purposes and for consumer durables. . . .*
- *Savings for retirement, ill health, or disability. . . .*
- *Savings to build credit ratings and as collateral. . . .*

Many of the benefits gained from institutional savings by households are also applicable to enterprises [which] tend to have a high demand for liquidity. . . . Deposits mobilized in conjunction with commercial credit programs enable . . . financial institutions [to become sustainable]. As of December 31, 1991, BRI's[14] KUPEDES program had 1.8 million loans outstanding that were fully financed by . . . bank deposits from 8.6 million savings accounts. . . . KUPEDES supplies an increasing amount of the large demand for local credit at commercial interest rates.[15]

In addition to increasing the pool of loanable funds and directly benefitting rural households, savings mobilization by rural financial institutions generates other beneficial effects. Vogel has summarized some of them in the following way:

Income Redistribution

Policies that improve savings opportunities can do far more to redistribute income toward

> *Another experience that revealed the latent potential to mobilize savings in rural areas occurred in the Dominican Republic in the 1980s:*
>
> *The Banco Agrícola in the Dominican Republic began to offer passbook savings services in 1984 because it was in serious financial difficulty and urgently needed funds. By 1987 deposits had increased more than twentyfold. Although 60 percent of the depositors were previous borrowers from the institution, the rest were a new clientele who demanded only a safe and convenient store for liquidity (from the* World Bank, World Development Report 1989, *Washington, DC, USA, 1989, p. 119).*

12. Robert C. Vogel, 'Savings Mobilization: The Forgotten Half of Rural Finance', in D. W. Adams, D. H. Graham and J. D. Von Pischke (Eds), *Undermining Rural Development with Cheap Credit*, Westview Press, Boulder, CO, USA, 1984, pp. 249–250.
13. Marguerite S. Robinson, 'Savings Mobilization and Microenterprise Finance: The Indonesian Experience', in María Otero and Elisabeth Rhyne (Eds), *The New World of Microenterprise Finance: Building Healthy Financial Institutions for the Poor*, Kumarian Press, West Hartford, CT, USA, 1994, p. 30.
14. Bank Rakyat Indonesia. The main clients of the highly successful KUPEDES program are small- and medium-scale savers and borrowers.
15. M. S. Robinson, 1994, pp. 35–38.

the rural poor than projects based on low-interest-rate lending. Low interest rates create an excess demand for credit, thereby forcing financial institutions to ration credit away from small borrowers without traditional collateral who are perceived to be risky and costly to serve. . . . Such rationing consists not only of loan refusals but also of transaction costs that can easily exceed interest costs for small borrowers. . . .

Resource Allocation

Effective savings mobilization by financial intermediaries draws resources away from unproductive investments, especially inflation hedges, as the opportunity is provided to make deposits that earn positive real rates of interest. . . . These resources can be on-lent by financial intermediaries for those activities that promise the highest rate of return. . . .

Financial Institutions

The positive effect of savings mobilization on financial institutions is the third argument in favor of savings mobilization. Financial institutions neglecting savings mobilization are incomplete institutions. They not only fail to provide adequate services for rural savers, but they also make themselves less viable, as can be seen most clearly in the high rates of delinquency and default that plague most agricultural development banks. . . . When financial institutions deal with clients only as borrowers, they forgo useful information about the savings behavior of these clients that could help to refine estimations of their creditworthiness. Furthermore, [in the case of local financial institutions] borrowers are more likely to repay promptly and lenders to take responsibility for loan recovery when they know that resources come from neighbors rather than from some distant government agency or international donor. . . .

Incentives

Savings mobilization provides appropriate incentives and discipline not only for rural financial markets and institutions but also for governments and international donors. . . . financial institutions are likely to have little interest in savings mobilization or loan recovery when cheap funds are available through government loans, central bank rediscounts, or loans from international donors. It is largely ignored that the volume of resources that can be obtained through effective programs of savings mobilization and loan recovery is potentially far greater than the most optimistic estimates of the amount of subsidized loans and grants available from governments and international donors. . . . Emphasis on savings mobilization is also incompatible with programs of low-interest-rate lending because financial institutions cannot be expected to mobilize savings and on-lend them at interest rates that cover neither interest payments to depositors nor administrative costs. It has sometimes been alleged that government officials use subsidized lending as a means to distribute patronage. . . . If true, this provides another reason for imposing the discipline of savings mobilization. . . .[16]

Rural finance overlaps with the field of microfinance, and much of the ferment and creative evolution in financial systems in recent years has taken place in the context of microfinance institutions. Frequently, their borrowers are more urban than rural (BancoSol in Bolivia), and in rural areas they may represent shopkeepers and traders as much as or more than farmers (Grameen Bank in Bangladesh), but nevertheless microfinance institutions can contribute significantly to agricultural development.[17] They contribute directly, in the form of production loans, and indirectly by supporting the agricultural marketing sector. Sections throughout in this

---

16. R. C. Vogel, 1984, pp. 249–252.
17. Microfinance has lent significant amounts to farmers in Indonesia, Cambodia, Thailand, Albania and Mali, among other countries.

chapter are devoted to questions related strictly to agricultural finance, but equally much of the discussion about developing rural financial systems follows the experience of microfinance institutions.

In regard to the development of microfinance in general, María Otero and Elisabeth Rhyne have written:

> Savings mobilization is an indispensable ingredient . . . as important as credit. . . . When there is no institution available, poor people tend to save in other than financial form, such as in small livestock or jewelry. . . . The rise of financial institutions that specialize in reaching the poor opens a window for defining micro-enterprise finance as part of a broader financial system. It also forces a change in focus – from creating good projects to creating healthy financial institutions for the poor.[18]

These considerations underscore the importance of increasing the opportunities for saving in financial form and of improving techniques of managing credit, so that the sector – and the rural economy in general – may realize more fully their potentials for productive investments.

7.1.3 Financial Services in Rural Areas

Under the traditional policy conception, the only role of credit in agriculture was to increase output. Credit was viewed as a production input, one that is necessary in order to acquire other inputs, and it was thought it had to originate largely from outside the sector. The need for financial services for poor rural families was ignored.[19] In light of the widespread failure of programs of directed credit in agriculture, that view is being abandoned in favor of a broader vision of the role of financial services in rural areas.

The principal orientation in rural finance reform is shifting away from a sole emphasis on the channeling of production credit to agriculture to one of strengthening rural financial intermediation in general. In addition to the above-mentioned financial services required by rural households, Dale Adams has commented on the need for mechanisms of financial transfers (for example, for sending payments to a child studying in a city and for receiving remittances), for long-term finance for fixed investments, and for mechanisms for more efficient allocation of investment funds among competing alternatives. Within rural areas there are opportunities for reallocating funds from households that save to those that invest.

Stuart Rutherford has observed recently that:

> Financial services allow people to reallocate expenditure across time. This means simply that if you don't have the ability to pay for things *now*, out of current income, you can pay for them out of *past* income or *future* income, or some combination of both . . . The poor need [this facility] no less than other groups of people. Indeed, *they may need it more*. This is not just because their incomes are uncertain and irregular (which is often true), but because the absolute amounts of cash they deal with are very small. As a result, anything more than the tiniest expenditures will require sums of money greater than they have with them at the time. . . .[20]

From the viewpoint of rural financial intermediaries, an institution that can provide a variety of needed services will have additional sources of fee income, and it can earn greater loyalty from its clientele, thereby increasing the possibilities for a high rate of loan recovery.

18. María Otero and Elisabeth Rhyne, 'Introduction', in M. Otero and E. Rhyne (Eds), 1994, pp. 4–5.
19. Manfred Zeller, Gertrud Schreider, Joachim von Braun and Franz Heidhues, *Rural Finance for Food Security for the Poor*, Food Policy Review 4, International Food Policy Research Institute, Washington, DC, USA, 1997, p. 1.
20. Stuart Rutherford, 'Raising the Curtain on the "Microfinancial Services Era"', *Focus*, Note No. 15, Consultative Group to Assist the Poorest (CGAP), Washington, DC, USA, May 2000, pp. 3–4.

The gamut of institutions that provide loans and some of these other financial services is wide, including commercial banks, investment banks, credit unions or co-operatives, small rotating savings and credit associations (ROSCAs), NGOs, input suppliers, agroprocessors and marketing agents, retail businesses, friends and neighbors and moneylenders, among others.[21] On the basis of studies in five Asian countries, the World Bank has underscored the diversity of the informal financial sector and its importance and operational advantages, pointing out that professional moneylenders account for only a small proportional of total informal credit.[22]

Additional testimony to the cost-effectiveness and sustainability of informal, indigenous financial institutions is found in a landmark study of monetary and banking policy in developing countries by Maxwell J. Fry:

[There are] four characteristics that explain why indigenous banks exhibit lower transaction costs than modern banks. First, indigenous bankers know their clients better than commercial banks. This reduces information costs. Second, administrative costs are lower for indigenous banks than for modern banks because their employees are paid less (and are less educated), the establishment is less elaborate, and the paperwork simpler. . . . Third, indigenous bank interest rates are not regulated

and can therefore adjust fully to market forces. Nonprice competition [for loans] is thereby kept down to an optimum level. Fourth, indigenous banks are not subject to the reserve requirements that are imposed on modern banks.[23]

Informal credit can be productive as well:

In the broadest, albeit necessarily incomplete, survey of indigenous financial institutions in developing countries, Wai (1977, p. 301)[24] reports that 55–60 percent of the demand for noninstitutional credit is for purely productive purposes, a finding that differs from the commonly held belief that high-interest informal lending is invariably used to finance consumption expenditure.[25]

Some of the informal financial institutions are not capable of taking deposits or offering transfer services over long distances. In addition, their lending practices are based more on knowledge of borrowers than on collateral, a characteristic that restricts their scope for expansion at the same time that it improves their effectiveness in managing risk.

Nevertheless, these institutions clearly are vital to the agricultural sector and to microenterprise in general. An appropriate policy framework for the rural financial sector must reach beyond banking institutions *per se* and facilitate the participation of many kinds of agents.

21. A more extensive list of both formal and informal financial institutions is found in Richard L. Meyer and Geetha Nagarajan, 'An Assessment of the Role of Informal Finance in the Development Process', in G. H. Peters and B. F. Stanton (Eds), *Sustainable Agricultural Development: The Role of International Cooperation*, Proceedings of the 21st International Conference of Agricultural Economists, Tokyo, 1991, Dartmouth Publishing Company, Aldershot, UK, 1992, p. 646. See also the comprehensive treatment in Joanna Ledgerwood, *Microfinance Handbook: An Institutional and Financial Perspective*, The World Bank, Washington, DC, USA, 1999.
22. The World Bank, 1989, pp. 112–113.
23. Maxwell J. Fry, *Money, Interest and Banking in Economic Development*, 2nd Edn, The Johns Hopkins University Press, Baltimore, MD, USA, 1995, p. 346. In this case, Fry cites work by Thomas A. Timberg and C. V. Aiyar, 'Informal credit markets in India', *Economic Development and Cultural Change*, 33(1), October 1984, pp. 43–59.
24. U Tun Wai, 'A Revisit to Interest Rates Outside the Organized Money Markets of Underdeveloped Countries', *Banca Nazionale del Lavoro Quarterly Review*, No. 122, September, 1977, pp. 291–312.
25. M. J. Fry, 1995, p. 345.

> *Sometimes the potential for non-financial institutions to provide loans to farmers is inhibited by the existing regulatory framework:*
>
> *. . . dealers in feed, fertilizer, insecticides and machinery. . . . often . . . are willing to extend credit without collateral. However, because they lack the deposit base of banks, they must be able to borrow themselves in order to offer credit. In a well-functioning system of secured transactions, such credit sellers could use their inventories and accounts receivable to secure loans from the formal sector to extend more credit. But herbicide dealers in Bulgaria, equipment dealers in Uruguay and Argentina, and insecticide and fertilizer dealers in Bangladesh have all reported that they have been unable to do so. The absence of a good framework for secured transactions can break any possible link between rural supplier credits and urban formal sector lending.*
>
> *In this way defects in the secured transactions system reduce the availability of credit to those who borrow in small amounts or who cannot provide land as collateral (from J. Yaron, M. P. Benjamin and G. L. Piprek, 1997, p. 57).*

7.1.4 Characteristics of Rural Financial Markets

Rural populations in developing countries are inherently more difficult to provide with financial services than urban populations are. They are spatially dispersed and transportation costs are high relative to incomes, leading to problems of access. Loan sizes tend to be small on average, leading to higher unit costs of processing loans. Literacy rates are lower than in urban areas, sometimes generating suspicion of paperwork and creating problems of eligibility in signing contracts. Documented credit histories are essentially non-existent.

Agricultural activities, which are pursued by a large share of rural populations, are subject to climatic risk and risk of price fluctuations to a much greater extent than urban economic activities are. Accordingly, income levels are more subject to fluctuations, as well as being lower, than those in urban areas. Rural borrowers are less likely to have tangible, documented collateral than urban borrowers are. Karla Hoff and Joseph Stiglitz have summarized some of the idiosyncracies of rural financial markets in the following way:

> Rural credit markets do not seem to work like classical competitive markets are supposed to work. Interest rates charged by moneylenders may exceed 75 percent per year, and in some periods credit is unavailable at any price. . . . neither the traditional monopoly [by moneylenders] nor the perfect markets view can explain other features of rural credit markets which are at least as important and equally puzzling as high interest rates:
>
> - The formal and informal sectors coexist, despite the fact that formal interest rates are substantially below those charged in the informal sector.
> - Interest rates may not equilibrate credit supply and demand: there may be credit rationing, and in periods of bad harvests, lending may be unavailable at any price.
> - Credit markets are segmented. Interest rates of lenders in different areas vary by more than plausibly can be accounted for by differences in the likelihood of default.
> - There is a limited number of commercial lenders in the informal sector, despite the high rates charged.
> - In the informal sector interlinkages between credit transactions and transactions in other markets are common.
> - Formal lenders tend to specialize in areas where farmers have land titles. . . .

The new views of rural credit markets are based on the following three observations:

(1) Borrowers differ in the likelihood that they will default, and it is costly to determine the extent of that risk for each borrower. This is conventionally known as the *screening* problem.
(2) It is costly to ensure that borrowers take those actions which make repayment most likely. This is the *incentives* problem.
(3) It is difficult to compel repayment. This is the *enforcement* problem.

The new view holds that it is the markets' responses to these three problems, singly or in combination, that explain many of the observed features of rural credit markets, and that they must therefore inform the policy perspectives for designing specific interventions.[26]

All of these factors are reasons why many commercial banks prefer the less challenging task of lending to the industrial and service sectors, and to urban consumers, rather than the uncertainties and difficulties of lending to agriculture. They also explain the above-mentioned tendency toward segmentation of rural financial markets. By the same token, the presence of these factors constitutes a clear signal that the most appropriate institutional designs and management approaches must be different for the case of rural financial intermediation, as suggested by Hoff and Stiglitz. Rural areas also offer advantages that can facilitate the work of a financial intermediary. Perhaps the chief asset is the stability and strength of social relations at the community level. Many innovative financial intermediaries utilize this asset to help sustain high rates of loan recovery. Another asset is the very multiplicity of producers, as they effectively constitute a large sample that allows a clear assessment of costs of production and their possible ranges of variation, at least for principal crops and livestock products. In an industrial sector with one or two firms, evaluating the expected costs of production for a new factory embodying a new technology may be largely a speculative exercise, given the lack of experience to go by.

Major strides have been taken in adapting financial institutions to the challenges of the rural environment. Advances can be found in all regions of the world. However, the share of total rural credit provided through these sustainable formal institutions is still quite small, and so the road ahead is still long. 'There is no single formula for a successful rural financial institution. The most appropriate modes of operation will be determined by the needs and socio-economic characteristics of the target clientele, as well as by the physical, economic and regulatory environment'.[27]

Another cautionary note is that most of the innovations in institutional structures and lending technologies so far have concerned ways to serve low-income clients in rural areas, through the above-mentioned emphasis on microfinance. While this is an important development, less attention has been paid to the situation of the medium-scale farmers whose customary sources of credit have dried up with the shrinkage or disappearance of the portfolios of State-owned agricultural banks. Continuing innovation and adaptation of other experiences will be needed to meet the full challenge of financial intermediation in rural areas, for all categories of borrowers and savers.

7.2 POLICY OBJECTIVES FOR RURAL FINANCE

7.2.1 Policy Objectives: Production versus Income

The traditional approach of directed, subsidized credit for agriculture, whether offered through State banks or via rediscount lines through commercial banks, has been closely tied to production of principal crops and livestock products. Mexico provided a classic example. Every year, the chief agricultural experts in the central Ministries meticulously calculated the acreage expected to be planted by crop throughout the country, with emphasis on the irrigation districts, and on that basis estimated the required amounts of inputs and the associated credit needs, as well as the required timing of delivery of the funds, using standardized crop-by-crop formulas based on actual practice. The corresponding amounts of

---

26. K. Hoff and J. E. Stiglitz, 'Introduction: imperfect information and rural credit markets – puzzles and policy perspectives', *The World Bank Economic Review*, **4**(3), 1990; reprinted in *From the World Bank Journals, Selected Readings*, The World Bank, Washington, DC, USA, 1995, pp. 269–272.
27. J. Yaron, M. P. Benjamin and G. L. Piprek, 1997, p. 7.

loanable funds were then made available to farmers through State-owned banking institutions. It was a task taken seriously at high levels in the government and monitored closely from a political perspective.[28]

Many other countries have followed similar procedures. Although some of this credit may have substituted for private sources of funding, it cannot be denied that there was a net output effect, that production of the targeted crops would have been lower in the absence of the considerable amounts of funding channeled to the sector in this way.[29] Assessments of this traditional approach to agricultural credit are presented throughout this chapter; the point to make here is that its aim was increasing *agricultural output*, not necessarily *agricultural income*, much less *rural income*. Although *directed credit* may have fulfilled part of its objective of increasing the production of selected crops, it has generally had low private and social returns. Its overall effectiveness in promoting agricultural development has been undermined by the following factors:

(a) In authorizing the loans, no selection criteria were used that measured the profitability of the investment. (Most of them were short-term loans.) Nor was a determination made of the comparative advantage or economic profitability of the products to which the credit was directed. Indeed, in Mexico and Central America, most of the directed credit has gone to grain production, and studies have shown that most of those products do not have a comparative advantage in that region.

(b) Rates of interest were subsidized, and so it was possible to cover them with low-return investments of the funds.

(c) Loan repayment requirements usually were lax so that the average real rate of interest paid, *ex post*, was even lower, often negative. This meant that part of the credit became a grant which helped sustain the production of crops whose acreage otherwise would have reduced because of their lack of profitability in the absence of that subsidy.

By delaying the transition out of uneconomic crops into other lines of production, the directed, subsidized credit programs often have acted as drags on the sector's growth rate.

In contrast to this way of administering credit, an approach of rationing loanable funds through market interest rates, accompanied by sound portfolio management which generates high rates of repayment, ensures that the funds go to more productive uses. This latter mode of operation fits the policy objective of increased sector income rather than increased output of selected products. It is argued in Chapter 2 of this book that income is a more appropriate objective than production.

Many of the profitable investments in rural areas are found in non-farming pursuits, especially in the marketing of outputs and inputs. The well-known Grameen Bank in Bangladesh has made a high percentage of its loans to rural women who market agricultural produce and handicrafts: 91% of its borrowers were women in the late 1980s.[30] Later, its lending for rural cell phones became so extensive that this activity was spun off into a separate enterprise.

---

28. In the early 1970s, this author worked in an agricultural planning office in what was then the Ministry of the Presidency in Mexico, and one of the principal responsibilities of the office chief each year was this determination of agricultural credit needs and the justification of the budget proposal in the Ministry of Finance. See, for example, Secretaría de la Presidencia, Dirección Coordinadora de la Programación Económica y Social, *Sector Agropecuario: Aspectos Metodológicos de la Programación*, Mexico City, Mexico, 1976, p. 123.

29. 'The additionality of directed credit programs for agriculture cannot be quantified, but in the short run the programs often resulted in increased investment and seasonal credit that benefit agriculture' (J. Yaron, M. P. Benjamin and G. L. Piprek, 1997, p. 22).

30. J. Yaron, 1992, p. 76.

Rural households, especially those with lower incomes, tend to have diversified sources of income and so financial facilities which best serve their needs are not directed solely to agricultural lending.

A loan portfolio that embraces diverse uses of the funds, with an important role for non-farming uses, is less risky for a financial institution than one that specializes in agricultural loans. The successful Bancafé in Honduras started as an institution that lent only to coffee farmers but it expanded and consolidated its financial position by diversifying its portfolio significantly, so that eventually coffee loans came to represent somewhat less than half of its total.[31]

Thus the primary objective of reforms of the rural financial sector in developing countries has come to be, justifiably, supporting the expansion of *rural income*.

7.2.2 The Objectives of Poverty Alleviation and Gender Outreach

A principal feature of the innovative institutions in rural finance is that they deal mainly with low-income clientele. Accordingly, their average loan size is small. In 1995, the initial average loan size of BancoSol in Bolivia, which has a loan portfolio in excess of US$40 million and more than 70 000 loans on its books, was about US$108.[32] The average outstanding loan sizes in the early 1990s for other well-known rural finance institutions were as follows: Badan Kredit Kecamatan (BKK) in Indonesia, US$26; Bank Rakyat Indonesia's Unit Desa, US$290; Grameen Bank in Bangladesh, US$150.[33] In Cambodia, a rela-

tively new microfinance institution, ACLEDA, which gained formal status of a bank in 2000, had 60 000 clients as of the end of 1998 with an average loan balance of $150.[34]

These and myriad other rural financial institutions are now reaching large numbers of borrowers – many of them women – whose low income levels would have excluded them from access to formal finance two decades ago. However, Dominique van de Walle questioned the effectiveness of targeting the poor in most existing rural finance programs:

Evidence that women and the illiterate feature prominently among participants in African schemes is given as proof that the schemes are pro-poor. The literature examining the many Bangladeshi schemes also takes it on faith that the target groups of women and the functionally landless represent the poor in Bangladesh. . . . But there are reasons to question this assumption. It has been found that what constitutes poverty by cultural or social standards is often far from identical to what constitutes poverty as defined by objective criteria. . . . One study . . . in Bangladesh . . . found that perfect targeting to the landless entailed sizable leakage to the nonpoor and imperfect coverage of the poor: some of the poor owned land, while some of the rich (including teachers, doctors, and shopkeepers) did not (Ravaillon and Sen, 1994)[35]. . . .

Do targeted schemes reach the poor? Design features undoubtedly help promote targeting to the poor. Given loan sizes and the fact that programs such as the Grameen Bank impose

---

31. Source: conversation with the president of Bancafé, Tegucigalpa, Honduras 1992.
32. Greg Chen, 'The Challenge of Growth for Microfinance Institutions: The BancoSol Experience', *Focus*, Note No. 6, The Consultative Group to Assist the Poorest (CGAP), Washington, DC, USA, March 1997, p. 4.
33. J. Yaron, 1992, p. 78. More recent estimates show higher average loan balances, for BRI $494, and for BancoSol $530 (Robert Peck Christen, 'Issues in the Regulation and Supervision of Microfinance', in Rachel Rock and Maria Otero (Eds), *From Margin to Mainstream: The Regulation and Supervision of Microfinance*, Accion International, Somerville, Massachusetts, January 1997, Chapter II, p. 36).
34. Source: conversation with ACLEDA management, Phnom Penh, Cambodia, April 2000.
35. M. Ravallion and B. Sen, 'Impacts of land-based targeting on rural poverty: further results for Bangladesh', *World Development*, **22**(6), 1994, pp. 823–838.

participation costs, the rich will surely have better alternatives. But the poorest of the poor are not being reached extensively by micro-credit schemes. . . .[36]

A more optimistic view of the ability of appropriately designed rural financial institutions to reach the poor is provided by Robert Christen, Elisabeth Rhyne, Robert Vogel and Cressida McKean. They examined 11 microfinance institutions in the developing world and found that the average outstanding loan balances (by institution), which are taken as proxies for the incomes of borrowers, 'cluster in the range of US$200 to US$400 . . . with several well below that level'.[37] These authors conclude that:

> The study demonstrates that among high-performing programs there is no clear trade-off between reaching the poor and reaching large numbers of people. Several very large programs (BKK, Grameen) have among the smallest loan sizes. Mixed programs, which serve a range of clients, not just those of a given loan size, have successfully reached very poor clients. It is scale, not exclusive focus, that determines whether significant outreach to the poorest will occur.[38]

In reviewing Kenya's Juhudi Credit Scheme and other experiences with NGO-funded rural credit, Albert Kimanthi Mutua distinguished between a welfare orientation and development aims in targeting credit schemes on the poor:

> Welfare-oriented NGOs traditionally have approached development from a very broad perspective. Usually, welfare programs focus on alleviating poverty by providing a number of free or subsidized services. Sustainability-focused programs assume that they are providing a service that poor people want and for which they are willing to pay. When welfare NGOs begin to operate credit programs, their general orientation tends to make them focus on selecting the neediest of clients – the poorest of the poor – rather than on delivering credit efficiently. . . .
>
> Kenyan NGOs are struggling with the question of who the target beneficiaries of NGO credit programs should be: the very poorest of people or poor people who are already entrepreneurs? . . . When welfare criteria are used to select beneficiaries, the credit programs end up with borrowers who are not entrepreneurs. . . . [and they] would most likely divert loan money to serve more urgent needs.
>
> NGOs argue for below-market interest rates because they believe that the poorest people cannot afford higher ones. But studies in Kenya and elsewhere have shown that poor entrepreneurs are more concerned with convenience in borrowing than with the price they pay for these services. The biggest obstacle for NGOs in making the transition to a [sustainable] finance-based system thus appears to be the perceptions of the NGOs rather than reality.[39]

Detailed statistical evidence on actual targeting by rural finance institutions was provided by David Hulme and Paul Mosley in a study of 13 microfinance institutions in seven

---

36. Dominique van de Walle, 'Comment on "Rural Finance in Africa: Institutional Developments and Access for the Poor", by Ernest Aryeetey', *Annual World Bank Conference on Development Economics 1996*, The World Bank, Washington, DC, USA, 1997, p. 183.
37. R. P. Christen, E. Rhyne, R. C. Vogel and C. McKean, 'Maximizing the Outreach of Microenterprise Finance: An Analysis of Successful Microfinance Programs', Evaluation of USAID Program and Operations Assessment Report No. 10 (PN-ABS-519), as cited in Mohini Malhotra, 'Maximizing the Outreach of Microenterprise Finance: The Emerging Lessons of Successful Programs', *Focus*, Note No. 2, The Consultative Group to Assist the Poorest, Washington, DC, USA, October 1995, p. 2.
38. *Ibid.*
39. Albert Kimanthi Mutua, 'The Juhudi Credit Scheme: From a Traditional Integrated Method to a Financial Systems Approach', in M. Otero and E. Rhyne (Eds), 1994, pp. 270–271.

Table 7.1 Benefits for Clients of Selected Microfinance Institutions

Institution	Share of borrowers below poverty line (%)	Average increase in borrower income as % of that of the control group	
		Whole sample	Below poverty line
BancoSol, Bolivia	29	270	101
BRI Unit Desa, Indonesia	7	544	112
BKK, Indonesia	38	216	110
KURK, Indonesia	29	—[a]	—[a]
Grameen Bank, Bangladesh	Vast majority	131	126
BRAC, Bangladesh	Vast majority	143	134
TRDEP, Bangladesh	Vast majority	138	133
PTCCs, Sri Lanka	52	157	123
KREP Juhudi, Kenya	—[a]	133	103
RRBs, India	44	202	133
KIE-ISP, Kenya	0	125	—[a]
Mudzi Fund, Malawi	Vast majority	117	101
SACA, Malawi	7	175	103

[a] Not available.

countries.[40] They found that for most of the studied institutions a significant share, even 'vast majority', of the borrowers were below the poverty line but that the beneficial effect of the borrowing was proportionately much less for the poorest borrowers than it was for those above the poverty line.

Some of the results of the Hulme–Mosley study are shown in Table 7.1. The control group was carefully defined as persons whose applications for loans at those institutions had been approved but whose loans had not yet been disbursed. This procedure helped reduce potential differences in socioeconomic characteristics between the control group and the sampled group, for each institution's clients. The number of total borrowers varied widely by institution, from 223 in the Mudzi Fund of Malawi to 12 000 000 in the RRBs of India. However, seven of them had at least 400 000 borrowers and another three had less than that but at least 25 000.

Three results that emerge clearly from Table 7.1 are the following:

(a) Most of the institutions were able to reach sizeable numbers of very poor households.
(b) The benefits of access to loans were substantial on average, for the whole sample.
(c) The benefits of access to loans were marginal for the very poor.

The authors point out that some of the very poor even experienced a reduction in incomes as a result of the borrowing, because of increased indebtedness without a corresponding increase in ability to service it. The very poor were very much more likely to borrow for consumption purposes, than those above the poverty line. However:

Despite the overall tendency of better-off clients to enjoy larger income impacts from microcredit, some borrowers below the poverty line achieved substantial increases in income from their loans. Preliminary analysis ... indicates that these particular poor clients borrowed for relatively low-risk capital investments such as small irrigation, high-yielding

40. David Hulme and Paul Mosley, *Finance Against Poverty*, Routledge, London, 1996. This study is summarized by Paul Mosley in 'Financial Sustainability, Targeting the Poorest, and Income Impact: Are There Tradeoffs for Microfinance Institutions?', *Focus*, Note No. 5, The Consultative Group to Assist the Poorest, Washington, DC, USA, December 1996.

seeds in rainfed areas, and new carpet-weaving looms.[41]

In Bangladesh, many of the poorest of the poor were incorporated sustainably into the microfinance network by first providing them with food aid and training, along with very small amounts of credit. To qualify for this program, known by the initials IGVGD (Income Generation for Vulnerable Groups Development), households had to satisfy the following criteria:

• Be headed by widows or abandoned women
• Own less than half an acre of land
• Earn less than $6 per month

After an 18-month period of food aid was terminated, two-thirds of the participants became regular clients of microfinance institutions. This program was designed and administered by BRAC (the Bangladesh Rural Advancement Committee), an organization that always has emphasized training and other inputs as much as credit for the goal of rural poverty alleviation.[42]

On the basis of these studies and experiences, it seems clear that microfinance institutions can reach significant numbers of very poor households, but that special care must be taken with this group to ensure that they actually benefit from the borrowed funds. Poverty alleviation can be a feasible objective of rural financial institutions but the challenge is not easy to meet. It is easier to generate benefits for rural households of moderate and moderately low incomes than it is for those below the poverty line. Under the present state of knowledge of approaches to rural finance, poverty alleviation would not necessarily be the main objective of all rural financial intermediaries, but it could be a principal or subsidiary objective of

many of them. It is important to bear in mind that in programs oriented toward the poor women tend to form a large share of the borrowers. It also is true that the very poor can benefit indirectly from rural finance programs via the increased employment opportunities that are created as a result.

Nevertheless, in many contexts microfinance cannot be regarded as the only, or even the primary, instrument of poverty alleviation. Poor families may not be able to repay loans, and investments in schools and productive infrastructure may be more effective in raising their standards of living, and to make them eventually eligible for microcredit.

Shahidur Kandker gives the following perspective:

Microcredit programs are not a viable option for many people because such programs require skills, such as accounting ability, that many people in target groups lack. Credit-based interventions are best targeted to those among the poor who can productively use microcredit to become or remain self-employed, while public works programs are best targeted to the ultrapoor who lack the skills to benefit from microcredit.

Microcredit, which finances self-employment activities that are performed at home, is particularly well suited to the needs of rural women, who are restricted by social custom from working outside the home. Many women lack entrepreneurial skills to become self-employed, however. For these women – who cannot participate in microcredit programs because they lack skills and cannot participate in the wage market because of social restrictions – literacy promotion and training are necessary so that they can benefit from microcredit.[43]

41. Paul Mosley, 'Financial Sustainability: Targeting the Poorest, and Income Impact: Are There Trade-offs for Microfinance Institutions?' *Focus*, Note No. 5, The Consultative Group to Assist the Poorest, Washington, DC, USA, December 1996, p. 4.

42. This experience is described in Syed Hashemi, with Maya Tudor and Zakir Hossain, 'Linking Microfinance and Safety Net Programs to Include the Poorest: The Case of IGVGD in Bangladesh', *Focus*, Note No. 21, Consultative Group to Assist the Poorest, Washington, DC, USA, May 2001.

43. Shahidur R. Khandker, *Fighting Poverty with Microcredit: Experience in Bangladesh*, Oxford University Press, New York, 1998, p. 143.

7.2.3 Objectives for Rural Financial Institutions

It is clear that in some measure financial institutions that function well can serve the two national policy objectives of generating more rural income and alleviating poverty. In order to fulfill this role, the institutions need to keep in focus their own objectives. Above all, they need to become *sustainable*, or otherwise their benefits to the rural population will be transitory, and perhaps their problems will damage the prospects of other emerging rural financial institutions. Sustainability can be defined in two basic ways: *eliminating dependence on donated or subsidized funds* and *achieving profitability*. Both are important and are indispensable for sustainability in the long run.

For credit programs that start out by being dependent on subsidized donations of funds and strive to achieve long-run sustainability, Elisabeth Rhyne and María Otero define four levels of self-sufficiency:

The lowest level, level one, is associated with traditional, highly subsidized programs. At this level, grants or soft loans cover operating expenses and establish a revolving loan fund. When programs are heavily subsidized and performing poorly, however, the value of the loan fund erodes quickly through delinquency and inflation. . . .

At level two, programs raise funds by borrowing on terms near, but still below, market rates. Interest income covers the cost of funds and a portion of operating expenses, but grants are still required to finance some aspects of operations. . . .

At level three, most subsidy is eliminated, but programs find it difficult to eradicate a persistent dependence on some element of subsidy. This is the level associated with most of the well-known credit programs, and it is probably necessary to reach at least this point in order to achieve large-scale operations. . . . The Grameen Bank, for example, retains two kinds of subsidy: Its cost of capital is several points below market, and it receives income from soft loan funds placed on deposit. . . . The Badan Kredit Kecamatan (BKK) program has eliminated subsidy from its branch network but requires some grant support for supervision. . . .

The final level of self-sufficiency, level four, is reached when the program is fully financed from the savings of its clients and funds raised at commercial rates from formal financial institutions. Fees and interest income cover the real cost of funds, loan loss reserves, operations, and inflation. The only major microenterprise programs to have reached this level are those of the credit union movement in certain countries and the BRI Unit Desa system in Indonesia.[44]

Additional objectives for an institution that can be important for achieving national policy objectives include increasing the number of clients (outreach), increasing the number of poor clients, and improving both the quality and variety of financial services offered. However, care must be taken to ensure that pursuit of these objectives does not undermine the financial viability, or sustainability, of the institutions.

7.2.4 The Contributions of Microfinance

In addition to their ability to reach the poor, including in some instances the very poor as commented upon above, microfinance institutions have opened up a new world of possibilities for small-scale entrepreneurs throughout the developing world. The approaches followed by these institutions and their capabilities have evolved rapidly in recent years, toward a broader base of clientele and improved lending technologies that increase the sustainability of the operations.[45] By the same token, there has been movement toward

44. Elisabeth Rhyne and María Otero, 'Financial Services for Microenterprises: Principles and Institutions', in M. Otero and E. Rhyne (Eds), 1994, pp. 17–18.
45. See Rachel Rock, 'Introduction', in R. Rock and M. Otero (Eds), 1997, Chapter 1, p. 3.

treating borrowers as *commercial clients* rather than *beneficiaries* of assistance programs.[46] Although 'there have been many more failures than successes', it is now clear that 'there is an increasing number of well-documented, innovative successes in settings as diverse as rural Bangladesh, urban Bolivia, and rural Mali. This is in stark contrast to the records of State-run specialized financial institutions, which have received large amounts of funding over the past few decades but have failed in terms of both financial sustainability and outreach to the poor'.[47]

Khandker carried out one of the most rigorous evaluations to date of microfinance programs, for the case of three programs in Bangladesh: Grameen Bank, BRAC, and a rural development program with a credit component known as RD-12. He evaluated them against the opportunity cost of the subsidized funds they used and against alternative kinds of programs such as investments in infrastructure.

He comments that:

> so-called outreach indicators, such as the extent of program coverage, for example, do not reveal whether program participation benefits the poor and, if so, how and at what cost. ***Repeated borrowing at high loan recovery rates may not indicate that participants benefit from microcredit programs.*** In fact, since many microcredit borrowers have no alternative sources of finance, the very low dropout rate among members with a low loan default rate may signal the dependency of the participants on the program itself. Even worse, repeated borrowers may use other sources of lending, such as informal lenders, to remain in good standing with a microlender.... To establish the cost-effectiveness of supporting microcredit programs for donors and governments, research must show that the income and other gains generated by microcredit programs are greater than those generated by alternative uses

of subsidized funds currently allocated to microcredit programs....[48]

Some of his conclusions are as follows, including a strong indication of a gender effect, of the greater benefits of lending to rural women than to rural men:

> The objective of these [microfinance] programs is to help promote self-employment for the unemployed poor and for women in order to reduce poverty. Sustained poverty reduction requires actions and policies that help improve both the productive and human capital of the poor. Policy interventions must be well targeted if benefits are to reach only the poor. In Bangladesh agricultural growth policies, which increased farm production and income, failed to improve either the physical or the human capital of the poor because their growth impact was neither broad-based nor technology-neutral. Targeted antipoverty measures without a credit component ... have smoothed consumption for the poor who depend on wage income but they have failed to enhance their human and physical capital.
>
> In contrast, microcredit programs have been able to reach the poor and enhance both their productive and their human capital by generating self-employment. These programs promote human capital development through literacy and social awareness programs and by targeting women.... How effective microcredit programs are in reducing poverty and reaching the poor is an important policy question that merits careful ... evaluation....
>
> Microfinance reduces poverty by increasing per capita consumption among program participants and their families. Annual household consumption expenditure increases Tk 18 for every Tk 100 of additional borrowing by women and Tk 11 for every additional Tk 100 of borrowing by men.... Poverty reduction

---

46. J. Ledgerwood, 1999, p. 5.
47. *Op. cit.,* p. 4.
48. S. R. Khandker, 1998, pp. 146–47 [emphasis added].

> 'The success of microfinance has destroyed three commonly held myths in rural finance: that the poor are not creditworthy, that women represent greater credit risks than men, and that the poor do not save' (*S. R. Khandker, 1998, p. 150*).

estimates based on consumption impacts of credit show that about 5 percent of program participants can lift themselves out of poverty each year by participating in and borrowing from microfinance programs.

Microcredit programs also help smooth consumption, as well as the seasonality of labor supply. . . . Targeted credit also improves the nutritional status of children. The nutritional impact of credit is especially large for girls, and the impact is larger for loans made to women. . . .

Women have proved to be excellent credit risks, with loan default rates of only 3 percent – significantly lower than the 10 percent default rate for men. Women have clearly benefitted from microcredit programs. Program participation has enhanced women's productive means by increasing their access to cash income generation from market-oriented activities and by increasing their ownership of nonland assets. These improvements should enhance women's empowerment within the household, influencing their own and their children's consumption and other measures of welfare (such as schooling). . . .

Microfinance loans are well targeted. Large farmers . . . received more than 82 percent, small and medium-size farmers received just 13 percent, and poor and marginal farmers received only 5 percent of total loans disbursed by formal banks. In contrast, landless and marginal farmers received 72 percent, small and medium-size farmers received 24 percent, and large farmers received only 4 percent of microfinance loans.

Microcredit reduces poverty but so do other antipoverty programs. . . . In terms of their effect on per capita consumption, Grameen Bank and infrastructure development projects appear more cost-effective than other programs, including BRAC, RD-12, agricultural development banks, and targeted food programs. Because different types of programs reach different types of beneficiaries, however, the higher cost-effectiveness of these programs may not indicate that resources should be reallocated from other programs.[49]

In sum, well managed microfinance programs can provide real benefits, often to groups in the population that are hard to reach through other kinds of programs. The challenge is to make the microcredit programs cost-effective and sustainable.

7.3 KEYS TO THE SUSTAINABILITY AND EFFICIENCY OF FINANCIAL INTERMEDIATION

In light of the largely disappointing experiences with programs of agricultural credit in the past few decades, the objectives of *institutional sustainability* and *efficiency* have acquired overriding importance for rural financial institutions. Little benefit can be provided to rural households over the longer run from credit institutions that prove not to be sustainable, or self-sufficient. The operational efficiency of an institution contributes to its sustainability and it also permits the institution to reach greater numbers of clients.

Several kinds of institutional strategies contribute to the sustainability of rural financial intermediaries. At the level of the institution itself, the main determinants of sustainability are the following, some of which are interrelated:

(a) *Savings mobilization.* The advantages of a strategy of savings mobilization have been mentioned already. A basic consideration is

---

49. Author's note: BRAC and RD-12 place more emphasis on supplying training and other inputs to the participants, and so some of their benefits may show up only with a time lag.

that an institution that does not generate its own sources of loanable funds is likely to find, sooner or later, that outside sources are not reliable. In addition, as noted above, awareness by borrowers that their own communities' funds are at risk is likely to enhance their commitment to repayment. Funds from governments and donor agencies are often regarded as quasi-grants by recipients. Most microfinance institutions do not initiate operations with a capacity to manage savings accounts, but development of that capacity is a key to their long-run viability.[50]

> *Forced savings programs in which credit clients are required to 'deposit' a certain percentage of the loan received with the lending institution have been common among microfinance programs for several years and have, in many cases, improved the clients' ability to save regularly. Evidence strongly suggests that if appropriate services are available, the poor will voluntarily save in large proportions. Voluntary savings programs are critical for two reasons: as potentially the largest and most immediately available source of finance for microcredit programs and as a much needed and demanded financial service for the poor (from Rachel Rock, 'Introduction', in R. Rock and M. Otero (Eds), From Margin to Mainstream: The Regulation and Supervision of Microfinance, Monograph Series No. 11, Accion International, Somerville, Massachusetts January 1997, p. 7).*

(b) ***Minimal dependence on subsidies.*** Grants or low-interest rediscount loans from governments and donor agencies often substitute for savings mobilization by financial institutions, but even in cases in which they complement

savings deposits, experience has shown conclusively that the greater the use of such external funds, then the less sustainable is the institution. In part, this is due to the fact that access to low-cost sources of funds is likely to weaken the institution's drive for operational efficiency. In addition, for the reason mentioned above, the use of such funds may change borrower behavior in the direction of lower repayment rates. It also may encourage the institution to offer below-market interest rates on loans, a policy that creates other pitfalls.

(c) ***Setting interest rates at market levels.*** The arguments against subsidized interest rates also are powerful:

(i) They usually mean low real deposit rates, which weaken savings mobilization, unless they are offset by significant subsidies, which generate their own problems.

(ii) They erode the capital base of the financial institution, progressively diminishing its capacity to serve its clientele.

(iii) They encourage lending for low-return activities, weakening the possibilities for the credit program to generate increases in sector income.

(iv) Since low real interest rates cannot be used to screen applicants or projects, rationing of credit tends to be carried out on the basis of non-economic criteria.

(v) Small-scale loans entail higher administrative costs per unit lent, and higher interest rates are needed to cover those costs.

(vi) Agricultural loans are risky on average, and for the sake of the sustainability of the viability of the financial institution a risk premium should be built into the interest rate for institutions oriented toward agricultural lending.

---

50. The advantages of savings mobilization in this context are seen primarily for rural financial institutions and the rural sector. In the aggregate, the quantitative evidence on the response of private savings to interest rates in developing is unclear. However, that may be due to the fact that only recently have most developing countries moved toward financial liberalization. For a summary and discussion of this evidence, see P. R. Masson, T. Bayoumi and H. Samiei, 'International evidence on the determinants of private saving', *The World Bank Economic Review*, **12**(3), September 1998, pp. 483–501.

With reference to rural financial institutions oriented toward the small-scale borrowers, a CGAP note has said:

Experience around the world has shown that microentrepreneurs do not need subsidies and that microlenders cannot afford to subsidize borrowers. Low income entrepreneurs want rapid and continued access to financial services, rather than subsidies. Most microenterprise clients see the 'market interest rate' as the rate charged by the money lender or curb market, which is often double the interest rate charged by microlending institutions. Subsidies often send the signal to borrowers that the money comes from government or donors who regard the poor as objects of charity, and borrowers see this as a signal not to repay. Few low income entrepreneurs end up benefitting from subsidized programs, because these programs fail before they reach significant numbers. Efficient financial intermediaries need to charge high rates to cover the costs of making small loans (from Nancy Barry, 'The Missing Links: Financial Systems That Work for the Majority', Focus, Note No. 3, The Consultative Group to Assist the Poorest, Washington, DC, USA, October 1995, p. 3).

Dale Adams, one of the early advocates of using market interest rates for agricultural lending, summarized the arguments against subsidized interest rates in the following words:

Interest rates are critical in determining the performance of financial markets, and cheap-credit policies are a major reason for the poor performance of rural financial markets in low-income countries. They destroy the incentives for rural households to save in financial form and seriously distort the way lenders allocate loans.[51]

(d) *Sound governance.* Institutional autonomy is a basic requirement. In its absence, strong pressures often are applied to make loans on political criteria. An analysis of the the loan portfolio of the Honduran Government-owned agricultural development bank, BANADESA, revealed that the largest loans, commonly referred to by the bank's staff as 'political loans', had lower repayment rates than smaller loans.[52] This pattern has been experienced in other countries as well.[53] Sound governance also means adequate institutional arrangements to avoid conflicts of interest in making loan decisions and to ensure accountability. To satisfy governance norms, both *appropriate institutional structures* and *training* are needed.

(e) *Capable management.* Selecting the management well and training the managers and staff of a rural financial institution are keys to its effectiveness and to its viability as well. Training can be expensive but its benefits justify the costs. An evaluation of the Grameen Bank, for example, commented: 'Staff training includes an intensive 6 months of mostly fieldwork, with some time spent in the classroom. Training is also provided for borrowers and center chiefs. Grameen Bank's success is at least partially due to its intensive training program . . .'.[54]

Keeping administrative costs low is one of the determinants of the sustainability of a financial institution, and proper training

---

51. Dale W. Adams, 'Are the Arguments for Cheap Agricultural Credit Sound?', in D. W. Adams, D. H. Graham and J. D. Von Pischke (Eds), 1984, p. 75.

52. Secretaría de Recursos Naturales, Grupo Técnico de Trabajo sobre el Sector Financiero Agrícola, *Las Políticas y la Estructura del Sector Financiero Agrícola*, Tegucigalpa, Honduras, 1990.

53. A similar experience in Costa Rica is recounted in Robert C. Vogel, 'The Effect of Subsidized Agricultural Credit on Income Distribution in Costa Rica', in D. W. Adams, D. H. Graham and J. D. Von Pischke (Eds), 1984, Chapter 11.

54. J. Yaron, 1992, p. 111.

helps guide staff in that direction. The approach to managing the institution and its loan portfolio also must be appropriate to the rural milieu. Very different approaches are required for small-scale rural borrowers than for large industrial borrowers. This topic is addressed in Section 7.6 below.

(f) *A market orientation in designing its financial services.*[55] Most of the innovative work applied to rural institutions in recent years has concerned *policies for lending* and *techniques of savings mobilization*, largely for microfinance institutions. In both areas, new approaches consistent with rural markets have been tested on a large scale and, while there are no formulas that can be applied in all contexts, there are guidelines that are generally applicable and approaches that may be adapted to different circumstances. For example, 'In the early 1980s in Indonesia, BRI staff asked villagers why they did not make use of . . . the national savings program administered by Bank Indonesia, the central bank. . . . The replies were nearly unanimous. People from one end of the country to the other responded that [it] permitted withdrawals only twice a month, and that the restriction on withdrawals was unacceptable'.[56] It is now recognized that the creation of savings instruments and other financial services in developing countries must be based on a careful assessment of clients' preferences.

In a study of four financial NGOs in the Gambia it was concluded that institutional sustainability of a financial program depends on its ability to mobilize deposits through offering attractive interest rates, covering operating costs without subsidies, diversifying portfolios to reduce risks that arise from covariance of borrowers' incomes, relying on local loan approval committees, developing effective substitutes for physical collateral, and maintaining high rates of loan recovery.[57]

The viability of a financial institution is not entirely in its own hands, even with the best institutional strategies. Both its operating policies and its balance sheet are affected by national policies which influence the profitability of agricultural production and the conditions of operation for financial institutions. The two main policy determinants of the viability of rural financial institutions are the following:

(a) *The national regulatory framework for finance.* Appropriate regulations are required in a variety of areas, including collateral, bank supervision, interest rates and contracts. In some cases, as for interest rates, an absence of regulation may be preferable to existing specifications. Inappropriate policies, on bank supervision and movable collateral for example, not only fail to encourage the growth of rural financial intermediation but often they have the unintended effect of actually inhibiting it, in commercial banks as well as in microfinance institutions.

(b) *The national economic policy framework.* The types of macroeconomic policies discussed in Chapters 2 and 4 that are unfavorable for agricultural development also tend to restrict producers' access to credit. In this respect, sound policy begins with the exchange rate. As commented by Yaron *et al.*:

Exchange rate distortions are particularly damaging to rural financial markets. Exchange rate pegs that do not reflect macro-

55. This principle is emphasized by Elisabeth Rhyne and María Otero, 1994.
56. Marguerite S. Robinson, 1994, p. 38.
57. Douglas H. Graham, Geetha Nagarajan and Korotoumou Quattara, 'Financial Liberalization, Bank Restructuring and the Implications for Non-Bank Intermediaries in the Financial Markets of Africa: Lessons from The Gambia', in Roger Rose, Carolyn Tanner and Margot A. Bellamy (Eds), *Issues in Agricultural Competitiveness*, Occasional Paper No. 7, International Association of Agricultural Economists, Dartmouth Publishing Company, Aldershot, UK, 1997, p. 253.

economic fundamentals distort foreign price signals, causing financial markets to channel excessive resources to inefficient sectors and insufficient resources to sectors with a comparative advantage. The liquidity and solvency of rural financial institutions may be undermined if they base credit decisions on relative prices that are later realigned significantly.[58]

Yaron *et al.* have identified eight principal ways in which macroeconomic policies often are biased against the rural sector and against the development of rural financial institutions. Drawing in part on the work of Schiff and Valdés (1992) mentioned in Chapter 4, the central observations of these authors are the following:

The performances of financial and real goods markets are closely interrelated. Because competitive financial markets are guided by price signals, distortions in prices for real goods lead to the misallocation of resources by financial markets. For years most developing countries subjected their rural sectors to heavy taxation measures. The eight pillars of urban-based policies have had a devastating effect on the profitability of agricultural and nonfarm enterprises. . . .

Eight pillars of urban-biased policies. Urban-biased government policies and public investment priorities pervade development efforts. This approach to development is usually linked to the goal of rapid industrialization as well as to political pressure for low food prices from a vocal urban population. Here are the eight pillars:

(1) Overvalued exchange rates.
(2) Low, controlled, and seasonally invariant prices for agricultural products.

(3) High effective rates of protection to domestic industry, the outputs of which are used as agricultural inputs.
(4) Disproportionately high budgetary allocations for urban areas over rural infrastructure . . .
(5) Disproportionately high investment in human resources in urban over rural areas (health and education).
(6) Usury laws (that rule out the loans that are most common in rural areas: small, risky, and high-cost loans).
(7) Underdeveloped legal and regulatory provisions regarding land titling and collateral for typical rural assets (land, crops, and farm implements) relative to urban assets (cars, durables, and homes).
(8) Excessive taxes on agricultural exports.

This approach has stifled agricultural and rural development in most developing countries for several decades. . . .[59]

In the end, as several authors have emphasized, achieving sustainability requires that a microfinance program must become *profitable* without subsidies and at the prevailing cost of borrowing loanable funds, in order to avoid decapitalizing itself:

A basic condition for sustainability is financial efficiency, that is, the ability to break even given the cost of lending. A sustainable program operates in such a way that the cost of making a loan – the cost of funds plus administrative and loan default costs – is equal to or less than the price (that is, the interest rate) it charges borrowers. [Alternatively] where subsidized funds are used, a program must be economically viable, in the sense that it breaks even at the opportunity cost of the subsidized funds.[60]

---

58. J. Yaron, M. P. Benjamin and G. L. Piprek, 1997, pp. 47–48.
59. *Op. cit.,* p. 49.
60. S. R. Khandker, 1998, p. 84.

7.4 THE REGULATORY FRAMEWORK FOR RURAL FINANCE

7.4.1 Regulatory and Institutional Challenges

Lending operations are intertemporal in nature and uncertain, based on a promise to pay in the future. Everything else being equal, the more certain lenders can be of repayment, then the greater will be the supply of loanable funds. The degree of certainty in turn depends on the institutional and legal environment, for example, on how rapidly and cheaply contractual agreements can be enforced. It also depends on the lending technologies utilized. The supply of funds also depends in part on the profitability of lending, and of financial intermediation. While profits are influenced by an institution's own efficiency as an intermediary, they also are determined by aspects of the existing regulatory framework, such as laws and regulations concerning levels of interest rates.

One of the most important strategies for promoting the development of rural finance is strengthening the regulatory framework, in order to give both lenders and depositors greater security and incentives. Conversely, an inappropriate set of laws and regulations can greatly inhibit the expansion of both rural deposit facilities and rural loans.

The sustainability requirements of sound governance and capable management for a microfinance institution are mirrored in the risks presented by such institutions, to their clients, to regulatory agencies and to themselves. Managing those risks successfully is the challenge faced by regulatory and supervisory authorities. Shari Berenbach and Craig Churchill have described those risks clearly and have differentiated them from those that characterize commercial bank operations, in words that reveal the diversity and complexity of the management challenges for microfinance:

> Microfinance institutions hold many risk features in common with other financial institutions. For example, MFIs and commercial banks are both vulnerable to liquidity problems brought on by a mismatch of maturities, term structure and/or currencies.

On the other hand, many risk features of commercial banks are not directly applicable to microfinance. For example, commercial banks are vulnerable to concentration of risk when a large loan to a single borrower can put at risk the total bank's capital or when multiple loans are exposed to the risk of a related enterprise group. Insider lending is another area of concern for commercial banks when managers or owners can use their influence to extract unsound loans of significant size. Yet, because of the volume of transactions and their very small size, such issues do not present significant risks to microfinance institutions.

There are four principal areas of risk that are specific to microfinance institutions: ownership and governance, management, portfolio, and new industry.

Ownership and Governance Risk. . . . Ownership and governance risk occur if the owners and directors of the MFI do not have the capacity to provide *adequate management oversight*. This is an issue because of the nature of institutions and individuals who typically own an MFI, or who are on its board. . . . the directors of a non-profit organization may not have the skills and experience to govern a formal financial institution. . . . *Ownership and organizational structure:* In some regulated microfinance institutions, there are unclear ownership and organizational arrangements involving the nongovernmental organization or public institutions that played a role establishing the newly formed regulated MFI. High risk scenarios can develop. If an NGO, which is funded with public resources and does not have owners, oversees the management and determines the policies of the regulated financial intermediary, the social mission may take priority over financial objectives. . . . While MFIs may be capable of raising the initial capital requirements from their funding shareholders, these owners may lack *financial depth* or flexibility to respond to additional calls for capital as may be needed.

Management Risks. The management risks that apply to microfinance portfolios are generated by the specific service delivery methods required to serve this market . . . *Decentralized operational systems:* . . . a decentralized organizational structure . . . is central to microfinance service methodologies. Such decentralized operating methods present management challenges in any industry. Poor telecommunications and transportation infrastructure can compound this challenge. Furthermore, decentralized operating methods generate an environment that can be subject to fraudulent practices if internal controls are not sufficient. *Management efficiency:* Microfinance institutions offer a high volume, repetitive service that operates on low returns per loan. If a branch or unit falls short in the project loan volumes, profits can quickly turn to losses. . . . The quality of management to ensure brisk and timely services is essential to the financial success of microfinance portfolios. *Management information:* The backbone of an MFI's management is its management information system. While this is true of all financial institutions, the decentralized operating methods, the high volume of short-term loans, the rapid portfolio turnover, and the requirement for efficiency service delivery make accurate and current portfolio information essential for effective MFI management. . . . A weak management information system might delay the follow-up on delinquent loans and could quickly undermine the quality of a microfinance portfolio.

Portfolio Risk. The basic features of the products and services that are appropriate for the microfinance market contribute to a different set of portfolio risks than are usually encountered by commercial lending institutions. *Unsecured lending:* Most microlending is unsecured in traditional terms. . . . The non-traditional approaches employed in microfinance are usually as effective as traditional collateral, but economic shocks could expose the institution if these approaches break down in a crisis. *Delinquency management:* Some

MFIs have undergone significant swings in the on-time repayment of their portfolio. Although the MFI may hold delinquency levels low for extended periods, precipitous changes in delinquency rates may arise. . . . Since operating costs are so high relative to the size of the portfolio, temporary delinquency problems become serious more quickly than in traditional banking. . . . *Sector or geographic concentration risk:* Unlike the risk concentration typically borne by commercial banks, where an individual large loan or loans to a related enterprise group may put the bank at risk, MFIs may be subject to risk if many clients come from a single geographic area or market segment that is vulnerable to economic dislocations. . . . It is important to point out, however, that this risk is largely theoretical. There are very few examples to date of sector or geographic concentration risk affecting microfinance portfolios.[61]

New Industry Risk. A number of the risks that face the microfinance industry stem from the fact that its techniques are relatively new and untested. *Adequate professional experience:* There are few professionals with prior banking experience that are also directly familiar with microfinance methods. . . . Because . . . many MFIs are unable to offer attractive compensation packages, MFIs may have difficulty attracting capable management talent. *Growth management:* MFIs . . . can experience dramatic growth in their initial years of operation. . . . Sustaining such growth rates presents management with the challenges of developing a trained cadre of employees, implementing standard policies and procedures, and maintaining portfolio quality. *New products, services and methodologies:* While this industry has made considerable advances in the design of appropriate microfinance products and services, the field remains young and relatively untested. It is difficult to assess whether a new product, service or methodology is an ill con-

---

61. Author's note: The depression in Nicaragua's agricultural sector in 2001, led by a collapse in coffee prices, caused many MFIs affiliated with the federation ASOMIF to suffer losses and consequently they largely withdrew from lending to agriculture.

ceived deviation from an existing methodology or is a breakthrough in new services for the market.... *Young   institutions:* ....   there remains much to learn about how these institutions behave in a crisis. What is the institutional learning curve?[62]

The following sections of this chapter deal with a number of issues that are related to meeting these challenges successfully.

7.4.2 Collateral

The security of a lender's position can be pursued with and without tangible collateral, i.e. as secured or unsecured lending. The latter requires more knowledge of the borrower, or the ability to penalize borrowers in arrears. Since it usually is less costly to assess tangible collateral than to acquire all of the necessary information about an unsecured borrower, lenders can provide larger and cheaper loans on the basis of tangible guarantees.[63]

The success of unsecured lending also depends in good measure on the development of *social collateral*. A kind of social collateral has been created by microfinance institutions that have fostered the formation of groups of borrowers in which a member guarantees the loans of all of the others. As pointed out by Khandker (citing Besley and Coate, 1995)[64]:

> ... group-based lending is a necessary but not sufficient condition for better repayment or better functioning of a group. For effective functioning of groups, the group method has to create social collateral that imposes certain disciplinary actions on group members. A group-based method may fail to enforce eligibility criteria, for example, if the entire group colludes in doing so. To prevent this problem,

all microcredit programs use a larger community-based organization to ensure that the group meets the eligibility criteria.[65]

Appropriate government policies can facilitate both secured and unsecured lending. One of the greatest hurdles to the expansion of secured lending in rural areas of developing countries is the lack of formal title to land which could be used as collateral, as discussed in Chapter 5. From the viewpoint of financial transactions, there are two aspects of a land title which are crucial: confirming the ownership of the property, and registering all liens against it. The latter is essential so that lenders may *perfect their claims*, or establish publicly the priority of them. For this reason, it is important both to title agricultural lands and to develop effective *systems of land registry*.

Land titles are of little value for financial purposes unless the judicial system permits a rapid settlement of claims in cases of default. Lengthy, costly or uncertain judicial procedures diminish very considerably the value of collateral and therefore reduce incentives to lend. By the same token, procedures for settling claims that are swift and decisive act as incentives for borrowers to repay loans, thus reducing the probabilities of default.

For the most part, these considerations are relevant for medium- and large-scale farms. In many, though not all, countries, it is difficult for lenders to foreclose on smallholders' property, even when titles are registered.[66] Sometimes, there are legal restrictions against doing so for properties below a certain threshold size. Often, the court system is hesitant to foreclose on low-income farmers, and lenders themselves may share that reluctance. Advocates of financial development point out that the legal ability to

62. Shari Berenbach and Craig Churchill, *Regulation and Supervision of Microfinance Institutions*, The Microfinance Network, Occasional Paper No. 1, Washington, DC, 1997, pp. 19–24 [emphasis added].
63. J. Yaron, M. P. Benjamin and G. L. Piprek, 1997, p. 54.
64. T. Besley and S. Coate, 'Group lending, repayment incentives and social collateral', *Journal of Development Economics*, **46**(1), 1995, pp. 1–18.
65. S. R. Khandker, 1998, p. 31.
66. 'It is not cost effective for the Grameen Bank to exercise foreclosure' (J. Yaron, M. P. Benjamin and G. L. Piprek, 1997, p. 125).

foreclose is rarely exercised but maintain that its presence still enhances smallholders' access to credit. It is likely that the reluctance to allow foreclosure will continue to prevail in many countries, out of understandable social concerns in addition to the cost of foreclosure in relation to the value of the asset.

Under these circumstances, an alternative method of providing land as security for a loan, to date relatively underexploited, is **antichresis**. Under antichresis, the borrower agrees to cede control of his or her land, in the event of default, until the loan is repaid out of its harvests or a specified period of time has lapsed. To ease the burden on the smallholder family, the creditor may agree to hire the borrower's family to work the land during this period. In many developing countries, this specification would be more acceptable than one under which a smallholder family risked losing its land. Examples of this approach are found in Bangladesh.[67] See Chapter 5, Section 5.9, for a further discussion of this. For antichresis to become operational, enabling legislation must be enacted, and in most developing countries it does not yet exist.

In addition to use of land as collateral, options for lenders include the use of other forms of collateral as well as undertaking unsecured lending. Among the other forms of agricultural collateral, crops and livestock figure prominently. Liens against future harvests are common instruments although, because of yield risk, they obviously do not confer the same degree of potential security that land liens do. It is relatively common for agro-processors and exporters to lend to farmers against a pledge of the future harvest. In the case of livestock, the risks include not only the possibility of decimation of a herd through disease but also the possibility that the borrower could sell the crop or animals without advising the buyer that a lien exists. (In Colombia a major hindrance to greater participation by small farmers in long-term marketing contracts with agro-processors and modern marketing channels is their tendency to break the agreement if a better price is offered in the short term.) Hence, crops and livestock are imperfect forms of collateral, but when complemented by knowledge of the borrower a lending institution may choose to utilize them. Again, enabling legislation is required. In some countries, under existing legislation neither commodities nor other movable property can serve as collateral.[68]

Another form of collateral is **crops in storage**. Obviously, this instrument is applicable only to non-perishable crops, primarily grains but also others such as cotton and coffee. When crops are deposited in a registered place of storage, their owners can receive documents known as **certificates of grain deposit** or **warehouse receipts**. A certificate of deposit can then be used as collateral for a bank loan. One of the most important uses of such loans is to permit a farmer to wait until prices experience a seasonal upturn before selling the crop, since prices always reach their cyclical low point at harvest. A system of certificates of grain deposit also requires specific legislation. The requirements for **bonded warehouses** have to be established, along with **grading standards** for the crops to be stored and rules that permit banks to accept this form of collateral.[69] The enabling legislation should be broad enough to cover rotating inventories and changes in form of the

---

67. See K. A. S. Murshid, 'Informal Credit Markets in Bangladesh Agriculture: Bane or Boon?', in G. H. Peters and B. F. Stanton (Eds), *Sustainable Agricultural Development: The Role of International Cooperation*, Proceedings of the XXI International Conference of Agricultural Economists, Dartmouth Publishing Company, Aldershot, UK, 1992, p. 660.

68. (a) J. Yaron, M. P. Benjamin and G. L. Piprek, 1997, p. 55; (b) Heywood Fleisig, 'The Right to Borrow: Legal and Regulatory Barriers That Limit Access to Credit by Small Farms and Businesses', *Viewpoint*, Note No. 44, The World Bank, Washington, DC, USA, April 1995, p. 3.

69. Patience and persistence are sometimes required to implement new regulations. This author participated in a team that developed and implemented, among other policy reforms, a system of certificates of deposit for grains in Honduras. After the detailed proposal was approved by the Minister of Natural Resources (Agriculture) and the decree drafted, it took two and a half years to obtain the necessary additional approvals from government agencies in order to make the decree effective!

commodity as it is processed, as indicated in the adjacent box.

Another potentially viable form of collateral is accounts receivable and loans. Agricultural input suppliers and rural retailers frequently are sources of lending to farmers, but their capacity to lend may be constrained by their ability to borrow against their accounts receivables. Similarly, a village moneylender could expand operations if it were possible to use his/her portfolio of loans as collateral – if a secondary market existed in such paper.

As pointed out by Yaron et al.:

> Compared to the bank, the store owner knows more about the customers and can more safely select good risks. However, in most developing countries problems in the framework for secured transactions prevent banks from creating, perfecting, and enforcing security interests in accounts receivable. These problems limit the access to credit of store owners and choke off a potentially promising source of rural credit. The same story, with small variations, can be told for all rural credit sellers and non-bank lenders (1997, p. 59).

Heywood Fleisig adds:

> When dealers and nonbank lenders can refinance the credit they offer, the supply of such credit expands and its cost falls. But expanding such financing requires a secured transactions law that permits easy, inexpensive public registration of security interests in accounts receivable or in chattel paper, and inexpensive transfer of these accounts if the borrower (the dealer or nonbank lender) defaults. Otherwise, these dealers and nonbank lenders will find it impossible to raise enough money to fund the loans they otherwise could profitably make.[70]

In most developing countries, rules governing collateral need to be broadened in order to

Legal obstacles against use of movable collateral can represent a formidable hurdle for the expansion of agricultural finance. For example:

Few countries have [legal] provisions for continuation in proceeds that would permit the security interest to be maintained as the collateral is transformed. A lender with a security interest in wool in a warehouse will lose [it] when the wool is sold. Sometimes high costs prohibit certain transactions. In Uruguay it costs 6 percent of the amount of the instrument to register a pledge; in Russia it costs 3 percent. Such registration fees alone, calculated at an annual rate, will exceed the interest rate on short-term loans for storing farm inventory. . . . In Argentina and Bolivia, something that does not yet exist cannot be the object of a loan. Consequently, farmers cannot get credit against the eggs from their poultry, the milk from their cattle, or the wine from their grapes. In Peru a rotating inventory requires redefinition of the loan, so fruit extract in warehouses cannot serve as collateral but fish meal, stored in containers of fixed sizes, can. For similar reasons, wheat in an Argentine silo cannot secure a loan but sugar in a warehouse can. These problems are fatal to lending because the lender knows that in the event of default the borrower can claim that the underlying contract has no legal foundation. These legal problems have no valid basis in policy (from J. Yaron, M. P. Benjamin and G. L. Piprek, 1997, pp. 55 and 57).

encourage increases in the supply of agricultural credit from private sources. However, formulating new legislation in this area is a delicate task that must be approached with care because difficult issues can arise concerning lenders' rights versus borrowers' rights. When movable assets are used as collateral, obviously liens can be placed on them in the event of defaults. However, precisely because the assets are movable, legislation to facilitate their collateralization tends to allow

70. H. Fleisig, 1995, p. 3.

creditors to act quickly in seizing the assets. It can be argued that such a provision is necessary in order to protect creditors' interests and that without it loans will not be made against movable assets. However, the other side of the coin is that, in the event of delinquency, possibly temporary, farmers stand to lose assets which are vital to earning their livelihood or their family's welfare, such as livestock, stored grain, farm tools and tractors. The issue is complicated by the fact that poor farmers in remote areas may not fully understand the implications of a credit agreement based on movable assets as collateral, and those farmers are unlikely to be able to use the court system for redress.

One possible solution would be to create special rural tribunals that could promptly rule on financial disputes, at a minimal or zero cost to low-income litigants. Whatever the approach adopted to this issue, legislation that legalizes movable collateral can be an important tool for promoting the growth of agricultural finance.

Subsidized crop insurance has proven disappointing and financially unsustainable wherever it has been tried throughout the developing world. Luz María Bassoco, Celso Cartas and this present author carried out a quantitative assessment of the effects of the crop insurance program in Mexico and arrived at the following conclusions:

> it is clear that there is a sizeable net social loss as a result of the subsidized insurance. . . . If the crop insurance program were not subsidized, there would still be a loss in net social welfare compared to the no-insurance situation. This loss arises because the insurance is compulsory, and in aggregate the administration costs exceed the value of the risk-reduction benefits conferred by the program.[71]

Other concerns with crop insurance are the potential for moral hazard – that a farmer may not take enough measures to prevent crop loss because the harvest is insured – and the difficulty of establishing an objective basis for measuring the degree of crop loss.

Private, unsubsidized crop insurance has been launched in a few developing countries such as Mexico and Honduras. To date, it has been profitable for the insurance companies involved, a sign of sustainability. However, it is limited to catastrophic losses resulting from drought, hurricanes, torrential rains, very low temperatures, hailstorms, and other extreme natural phenomena. A farmer that lost, say, 30% of the crop from insufficient or irregular rains would not be able to receive compensation from this kind of insurance. The possibility of developing more refined kinds of insurance is inhibited by an asymmetric information problem, in that farmers know their own history of yields better than an insurance company does.

Government guarantees of loans have, if anything, fared worse than subsidized crop insurance. They are characterized by two principal problems: moral hazard (inducing lenders to make loans they would otherwise have considered too risky) and operational shortcomings. In Honduras, for example, it was found that the loan guarantees advanced by the National Agrarian Institute, for loans made to beneficiaries of the Agrarian Reform by the National Agricultural Development Bank, were in fact never paid to that bank for the many defaults that occurred.[72] This problem contributed significantly to the bank's decapitalization. Consequently, one of the provisions of the Honduran *Agricultural Modernization Law* of 1992 prohibited such guarantees.

7.4.3 Contractual Relationships

As mentioned previously, a judicial system that ensures unambiguous and rapid enforcement of

---

71. Luz María Bassoco, Celso Cartas and Roger D. Norton, 'Sectoral Analysis of the Benefits of Subsidized Insurance in Mexico', in Peter Hazell, Carlos Pomareda and Alberto Valdés, *Crop Insurance for Agricultural Development: Issues and Experience*, The Johns Hopkins University Press, Baltimore, MD, USA, 1986, pp. 141–142.
72. Secretaría de Recursos Naturales, Honduras, 1990.

contracts is one of the factors that can most encourage the growth of private finance in rural areas, along with an appropriate framework for collateral. The **strength of contractual relationships**, of course, is a very basic concern and has implications in many realms of the economy other than financial transactions. As an illustration of its role, a common hindrance to development of non-traditional agricultural exports is the occasional lack of compliance on the part of some export brokers with the terms of their contracts with growers, and the lack of affordable legal recourse for the growers, especially the small-scale ones.

In addition to guaranteeing the reliability of contracts, it also is important to ensure that requirements for contracts not be specified in a way that unintentionally excludes many of the poor. For example, a requirement that the farmer read and assent to the agreement, and sign it, may discriminate against illiterate persons. In such circumstances, legal provision could be made for witnesses to read the contract to the signer and serve as co-signers, and for signature to take the form of a thumbprint.

7.4.4 Interest Rate Regulations

Laws against high interest rates – usury laws – also date back to biblical times. In fact, the Koran forbids interest entirely. Islamic banking has found substitutes for interest, but usury laws persist in many non-Islamic countries. However well-intentioned they are, in cases in which portfolio risk is relatively high, as in agriculture, such laws have the effect of reducing the supply of private credit. A more effective strategy is to develop a regulatory framework that encourages the provision of credit, instead of discouraging it, so that interest rates will decline as a consequence of the workings of supply and demand.

Fleisig has put the problem in the following way:

Many countries limit the interest rate charged on loans to protect unwary borrowers from unscrupulous lenders. In practice, however, high interest rates are often justifiable. Many of the costs associated with a loan are fixed, and these fixed costs represent a higher percentage of a small loan than of a larger loan. Since operators of small farms and businesses are more likely to want a small loan than are operators of large farms and businesses, private lenders charge higher interest rates to small operators than to larger ones. The lender would ordinarily hope to recover these larger costs with some combination of higher interest rates and an up-front processing charge. In addition, the costs of monitoring small, unsecured loans are higher than those of monitoring larger, secured loans. A lender with a small, unsecured loan must regularly inspect the borrower's business premises to be confident that the business is still solid, while secured lenders know that they have the right to take property with some value even if the business is not solid. Finally, the risk associated with small loans that are unsecured can be higher than the risk associated with large loans that are secured. To compensate for the higher costs and risks of small, unsecured loans, private lenders could be expected to charge higher interest rates.[73]

Maxwell Fry has pointed that *gross* borrowing costs are not necessarily higher on the informal market than with formal financial institutions, once all pertinent factors are taken into account:

Indigenous financial institutions normally charge higher explicit loan rates than modern financial institutions. Nevertheless, it is far from clear that gross costs of borrowing from indigenous banks are greater than the gross costs of borrowing from modern banks. Zia Ahmed (1982, p. 135)[74] presents a unique comparative study of borrowing costs, based

73. H. Fleisig, 1995, p. 2.
74. Zia U. Ahmed, *Transactions Costs in Rural Financial Markets in Bangladesh*, PhD Thesis, University of Virginia, Charlottesville, VA, USA, 1982.

on survey results. He finds that 84 percent of credit in Bangladesh is supplied by indigenous moneylenders. He estimates that the gross cost of borrowing from such moneylenders averages 86 percent a year. The gross cost of borrowing from commercial banks, however, averages 108 percent in rural Bangladesh. There are sizable noninterest costs of bank borrowing, which include travel, entertainment, bribes, and the opportunity cost of time involved in securing the loan.[75]

The Bolivian microfinance program PRODEM (Fundación para la Promoción y Desarrollo de la Microempresa), which spawned the world's first private bank that caters to micro-entrepreneurs, BancoSol, has charged annual interest rates that are more than 20 percentage points higher than commercial rates.[76] One of the keys to the success of the Bank Rakyat Indonesia's Unit Desa system, 'a rare example of a successful financial institution that profitably reaches the enterprising poor', is a 'nonrestrictive interest rate policy. . . . Without the freedom and will to set rates on saving and lending, the Unit Desa system would not have become self-sustaining and profitable'.[77]

Legislative or administrative ceilings on interest rates also limit the ability of institutions to mobilize savings. Since the margins of financial intermediation tend to be relatively high in developing countries, especially in microfinance institutions,[78] the need to set deposit rates high enough to attract savings tends to force up lending rates, and sometimes they bump against legislated ceilings before reaching appropriate levels. The result is either weak performance in savings mobilization or a high rate of failure of rural financial intermediaries.

> As agricultural development itself proceeds, interest rates faced by farmers may decline. Evidence to this effect has been found in India:
>
> In a relatively more prosperous district like Burdwan in West Bengal . . . the average rural interest rate for different classes (such as casual laborers, tenants, and agricultural laborers) varied between 36 and 84 percent per annum, while in a relatively poorer district like Nadia . . . the average rural interest rates varied between 72 and 120 percent per annum (from Subrata Ghatak, 'On interregional variations in rural interest rates', Journal of Developing Areas, 18(4), October 1983, p. 32; quoted in Hoff and Stiglitz, 1995, pp. 279–280).

Two other considerations that support a removal of ceilings on interest rates for agricultural lending are that (a) often farmers would pay higher rates for borrowing on the informal, uncontrolled market than they would for any kind of institutional loan, and (b) rates of return to short-term capital in agriculture – which is what the vast bulk of lending is for – are very high. The author conducted an econometric analysis of loans to 78 farms in Colombia, separating fixed capital investments from working capital, and breaking down the latter into labor and other inputs. The results indicated moderate returns to fixed investments, in the range of 7 to 15%, while the returns to working capital ranged as high as 100%. Independent calculations based on crop budgets for different levels of technology of production, suggested that the working capital required to purchase modern inputs yields returns typically in the range of 60

---

75. M. J. Fry, 1995, p. 348.
76. Amy J. Glosser, 'The Creation of BancoSol in Bolivia', in M. Otero and E. Rhyne (Eds), 1994, p. 237.
77. James J. Boomgard and Kenneth J. Angell, 'Bank Rakyat Indonesia's Unit Desa System: Achievements and Replicability', in M. Otero and E. Rhyne (Eds), 1994, pp. 225–226.
78. In the case of BancoSol, 'administrative costs . . . are a very large percentage of its total costs of operation when compared with the rest of the industry. Administrative costs are more than 80 percent of the total, and the industry standard is about 20 percent. The huge difference between these percentages stems from the high costs associated with microlending. As the number of clients increases relative to the number of credit officers, efficiency will improve although costs will always be high' (A. Glosser, 1994, p. 247).

to 80% and sometimes higher, depending on the crop, but never less than 42%.[79]

These results are intuitively understandable, for they suggest that once the basic infrastructure of a farm is in place, high returns are associated with access to the working capital that is necessary to buy operating inputs and make the farm function. They also explain why farmers all over the world are willing to borrow at very high curb market rates in order to purchase seed and fertilizer and other basic inputs. However, they raise a question about how to finance long-term investments in agriculture; this issue is discussed in Section 7.8 below.

Thus, interest rate restrictions, on both deposits and loans, can be prejudicial from the viewpoint of agricultural borrowers. The same is true of legislation that impedes the development of credit rating systems and wider forms of collateral.[80]

7.4.5 Bank Regulation for the Rural Sector

7.4.5.1 Prudential and Non-Prudential Regulation and Supervision

Bank regulations basically are designed to protect shareholders and, above all, depositors. They also have the purpose of protecting the integrity of the financial system as a whole. 'Problems in one bank can quickly spread through the entire financial system. Bank failures have monetary and macroeconomic consequences, disrupt the payments system, and lead to disintermediation (which decreases the mobilization of resources and the availability of finance for investment)'.[81] The capacity for adequate bank regulation is a pervasive weak point in the developing world:

Bank supervisors in many developing countries . . . pay relatively little attention to the prudential aspects of financial monitoring. For example, in many countries supervisors make no independent assessment of the quality of assets and give scant regard to accounting procedures and management controls. Together with macroeconomic instability and the lack of adequate legislation, this is one of the main causes of bank insolvency.[82]

The distinction between prudential and non-prudential regulation and supervision is vital. Christen and Rosenberg have provided a useful discussion of this difference:

We refer to some requirements as non-prudential, not because they are insubstantial, but because *they do not involve the financial authority in vouching for or assuming any responsibility for the soundness of the 'regulated' institution*. Examples of such requirements include:

- Registration and legal chartering of licensed entities
- Disclosure of ownership or control
- Reporting or publication of financial statements; norms for content and presentation of such statements; accounting and audit standards
- Transparent disclosure of interest rates to consumers
- External audits
- Submission of names of borrowers and status of their loans (on time? late? by how much?) to a central credit information bureau
- Interest rate limits

---

79. The World Bank, Latin America and the Caribbean Regional Office, Projects Department, *Colombia: Rural Financial Markets Sector Study, Annex E*, Report No. 5860-CO, The World Bank, Washington, D. C., February 13, 1986.

80. Gonzalez-Vega has stated that much of the existing financial legislation was developed with the aim of protecting the interests of borrowers but perversely has had the opposite effect, since it limits interest rates, types of collateral, and so forth. See his paper 'El Papel del Estado en la Promoción de Servicios Financieros Rurales', presented to the international seminar *El Reto de América Latina para el Siglo XXI: Servicios Financieros en el Area Rural*, La Paz, Bolivia, November 1998.

81. The World Bank, 1989, p. 91.

82. *Ibid.*

Depending on the combination of elements, a package of non-prudential regulation could be painless, or burdensome in the extreme. But these requirements don't involve the government taking a position on the financial soundness of an institution. They don't embroil the government in any accountability, explicit or implicit, as an insurer of depositors' losses in the event of failure. Some kinds of non-prudential regulation don't even have to be lodged in the agency supervising financial institutions. . . .[83]

7.4.5.2 Regulations for Preventing Crises and for Minimizing Damage

Michael Fiebig has presented a useful classification of regulation into the categories of *preventive regulations* (designed to prevent crises) and *protective regulation* (for dealing with crises once they have occurred), and within the former category *entry requirements* for financial institutions and *ongoing requirements*.[84] He explains how regulatory issues are different for agriculture than for other sectors. His description of the regulations provides many insights and lessons for financial institutions that are involved or wish to become involved in the agricultural sector, and not only for regulators. They are the basis for some of the conclusions in the final part of this chapter, and therefore they are quoted here at length.

The principal regulations in each category are the following, as described in Fiebig's words [with emphasis added]:

(a) *Preventive regulations – entry requirements:*

Minimum capital requirements: . . . to ensure that sufficient capital is available to absorb financial shocks. Capital requirements also should be designed to shield the institution from becoming a captive of bad debtors. . . .

[They] are a commitment of the owners' own risk resources which may be lost in the event that the bank makes bad loans. . . . Minimum capital requirements vary substantially from country to country. . . . Low entry capital proposals for institutions that target microfinance operations range worldwide from US$25 000 to US$250 000. Concentration at the lower end and lower entry requirements are found in Africa and Southeast Asia. . . . Considerations in this regard should balance the necessity to provide for strong owners with substantial capital at stake as well as a safety net for the . . . institution on the one hand, and a non-restrictive entry opportunity on the other. *In the context of agricultural lending it is important to bear in mind that for innovative lending, which extends the limits of traditional formal financial intermediation, a strong equity base should be required.* Very low entry capital requirements are unlikely to create strong enough institutions that can weather external shocks and business downturns. Very low entry barriers can also potentially overburden the supervisory institution with myriad small institutions.

As an example, there are 2420 People's Credit Banks (Bank Perkreditan Rakyat, BPR) in Indonesia today, out of which many have difficulties in competing with commercial banks in their rural, periurban and urban target areas. This has resulted in . . . 37% non-performing loans in the loan portfolio of these institutions. In order to provide a disincentive for the establishment of new BPRs, minimum capital requirements have recently been increased from US$7100 to US$71 000.

Ownership requirements: . . . intended to promote strong owners. They are to provide that all owners operate in the best interests of the institution (mission compatibility), and second, that members of governing bodies make all effort to be fully informed about the

83. Robert Peck Christen and Richard Rosenberg, 'The Rush To Regulate: Legal Frameworks in Microfinance', Occasional Paper No. 4, Consultative Group to Assist the Poorest (CGAP), Washington, DC, USA, March 2000, p. 9 [emphasis in original].
84. Michael Fiebig, *Prudential Regulation and Supervision for Agricultural Finance*, Agricultural Finance Revisited No. 5, FAO and GTZ, Rome, 2001, pp. 26–43.

institution's activities and performance (internal regulation). . . .

Not only do good owners bring a good financial background but also prove to have an unclouded strategic concept for the institution. The absence of owners committed to financial performance, or a majority of socially oriented owners who intend to stress outreach at the expense of sustainability may prove dangerous.

Governance structure and institutional type: Many regulators require a formal financial intermediary to be a shareholding company in order to ensure owners with capital at stake and an incentive for active monitoring. . . .

Feasibility studies: To assess the suitability of the new entrant into the formal financial market, customarily a detailed feasibility study is required. Detailed institutional information and a comprehensive business plan are components. . . .

(b) *Preventive regulations – ongoing requirements:*

Capital to asset ratios and loan portfolio classification: . . . a key instrument of banking regulation throughout the world. . . . assets need to be sufficiently backed by a financial institution's equity in order to be able to cushion the risks of loss. Risks . . . such as reductions in the value of assets in the loan portfolio are to be covered by specific and general loan loss provision. Capital requirements address unpredictable changes in the economic or competitive environment. . . .

The definition of capital becomes difficult in case of heavy donor involvement. Is it advisable to account for donor grants as capital? One of the basic considerations behind the capital-to-asset ratio is to ensure that sufficient capital is available to cushion the risks from the asset side of a financial institution's balance sheet. But heavy capital involvement also ensures that owners whose capital is at stake keep strong control over the business. Donors, however, are rather lenient owners, that usually do not have, or do not process in an adequate and timely manner, first-hand

information provided to them. Accordingly, *donor grants should be valued at a lower ratio than other capital sources. The same applies to equity provided by governments directly or through state guarantees.*

. . . in agricultural development banks the amount of equity is often heavily overstated or misleading, since many banks do not account for loan loss provisions in a proper manner.

The institutional form of savings and credit cooperatives poses specific obstacles in calculating capital. Member shares are redeemable, and the extraction of shares by many members can potentially become a threat to the equity base. Accordingly, the World Council of Credit Unions (WOCCU) recommends that only institutional capital is taken into account when calculating capital adequacy. . . .

Financial institutions with a loan portfolio concentrated in the agricultural sector can quickly accumulate arrears in bad agricultural years. . . . if agricultural producers concentrate on the same lines of products, or the volatility of yields of the different products is interrelated, arrears may be accompanied by a decrease in deposit base. In this circumstance, having a strong capital position at the outset is essential. . . .

But how then can the capital-to-asset ratio be adjusted for the conditions of agricultural finance in developing countries? For developing countries a higher capital adequacy ratio than 8% has been proposed to cushion the specific risks of narrow and volatile financial systems and for microfinance providers in particular to buffer the danger of rapidly deteriorating short-term credit portfolios. . . .

[There are] three options for adjusting capital-to-asset ratios to high-risk environments. . . . First, the overall ratio can be set higher. Second, one can adjust the risk weighting of those assets that carry a higher degree of risk. Third, one can adjust the capital definition.

Loans to agricultural producers may be generally classified as higher risk, requiring a higher degree of capital coverage. . . . Another option to be considered would be the risk-weighting of loan portfolio assets according to past repayment performance. ... *The result of assigning agricultural loans to a higher risk category would result in an increase in costs of agricultural lending, as it decreases the possible financial leverage of a financial institution.* While this may appear not desirable from a developmental perspective, the risk-based view of a regulator may deem this appropriate.

[However] the process of [loan] classification creates additional costs, which have to be weighed against the benefits of differentiation. Also, these regulations form part of the agenda of supervisors, and may enlarge their workload (and thus costs) substantially.

Liquidity management: Liquidity problems are often early signals of bank failure. Many bank regulators ask for various liquidity ratios of supervised institutions. ... Financial institutions that mobilize savings and lend in rural contexts face very particular liquidity issues. Regions dominated by agricultural production face seasonal cash flow fluctuations. ... A broad coverage of diverse regions and/or access to a liquidity pool can help mitigate the liquidity risks. In any case, a sophisticated liquidity management system needs to be in place to prompt due action on the part of senior management. In addition, agricultural lenders that hold a substantial exposure to foreign exchange risks due to international refinancing should be required to keep higher liquidity levels.

These factors indicate that banks with substantial agricultural loan portfolios may need higher liquidity ratios. Access to interregional liquidity pools and sufficient refinancing opportunities in cases of liquidity crunches are of utmost importance for agricultural lenders.

Credit risk management: Documentation and collateral requirements are part of requiring appropriate credit risk management. ... In rural financial markets, collateral in many cases may be successfully substituted or complemented with co-signing, group joint liability arrangements and/or the pledging of non-traditional banking collateral such as movable assets. ... *the capacity to manage agricultural lending methodologies requires management ...*

capacities as well as sophisticated management information systems.

... While it can be argued that sectoral diversification is one of the key risk management tools for successful agricultural lenders ... setting fixed percentages [of loans by sector] seems neither necessary nor opportune. ...

In many countries, insider lending or lending to staff members and owners of a financial institution is restricted. This is an important part of limiting the opportunity for fraud and corruption within a financial institution. ...

Provisioning and write-off policies: Provisioning requirements are designed to ensure that the real value of a loan portfolio is reflected in the balance sheet. This implies that an appropriate part of the loan amount and interest outstanding is written off when recovery is improbable. ... Apart from individual loan provisioning, many countries require a global provisioning of usually 1–3%, which is supposed to reflect the residual credit risk remaining even for healthy loan portfolios.

Loan provisioning is usually based on factors such as collateral values, guarantees, repayment track record and days past due. ... In collateral-based lending it may appear more appropriate to use only the amount of the loan not covered by the collateral as a basis for provisioning. However, in weak legal environments, lack of enforcement possibilities or lengthy legal processes towards collateral realization substantially decrease the actual net present value of collateral pledged. *In most developing countries, collateral fulfills more an incentive and hostage function than providing a substitute for loan repayment.* ...

For agricultural loans, which usually have longer terms than microcredit and often have lump sum repayment installments, the provisioning according to days past due ... is obviously inappropriate. For example, a two-year loan to a farmer, with a lump sum repayment at the end of the period requires an evaluation before the amount becomes overdue. Also, late payment of loans for agricultural production may well be due to a belated harvest season, an argument urban traders cannot put forward.

The default risk in agriculture is not necessarily altered by a few days of lateness. Laxness in agricultural lending is not advocated, but a certain degree of flexibility on the part of the lender is required to deal with the external shocks characteristic of the agricultural sector.

Product restrictions: Many countries use restrictions on the range of products offered as a measure to reduce vulnerability of financial institutions. Often, where a specific regulatory framework for small financial intermediaries has been created, these entities cannot mobilize deposits from the public right from the start. . . . *For agricultural lenders it has to be ensured that a diversity of financial products does not lead to an accumulation of risks.* . . . In the Bolivian example, Private Financial Funds (PFFs) are not allowed to offer credit cards or foreign exchange services. . . . offering new products often requires a separate licence from the supervisory institution. Requiring detailed and well-founded feasibility studies (including demand estimations) for operating new products appears to be a prudent approach. . . . It can however impose excessive paperwork, which stifles innovation.

Credit information bureaus: . . . In rural contexts, meager or inconsistent national identification systems may pose a serious problem to the effectiveness of this regulatory instrument. Borrowers may then easily use different names to avoid documentation of their past repayment behavior. In addition, if non-formal competitors, such as NGOs, work in the same areas, identification of multiple borrowing becomes difficult. . . .

Branching: In many countries, specialized supervisory agencies' approval is necessary prior to opening a new branch. Sometimes hours of operation are also set. The background of these regulations is to establish a level playing field. However, [they] may prove limiting if branches are required to be opened on a full-time basis, or in any case need to be in a solidly constructed building. *Mobile banking units and part-time branches are important tools to decrease the operational costs implied by rural financial intermediation.* . . .

Branching regulations may include a restriction to certain geographic areas. For example, the Indonesian and Philippine Rural Banks are confined to a municipality or subdistrict. The Municipal Savings Banks in Peru have also long been restricted to operating in one municipality only. . . . [Such regulations] may severely restrict the opportunities for portfolio diversification and lead to a greater sensitivity to external shocks. Diversified agricultural lending may prove difficult under these circumstances.

Loan documentation requirements: In many cases, external regulation specifies the documents each loan folder should contain. *In agricultural lending, and more generally in rural lending, as in microfinance the documentation required often proves excessive and/or irrelevant.* In these specific financial businesses, prudent loan decisions are much more oriented at a character-based assessment, or they even leave most of the loan decisions to self-selecting joint liability groups. Often, this is a major regulatory challenge for microfinance and rural finance. . . .

Reporting requirements: Reporting requirements are the foundation for supervision. Usually, these requirements comprise topics and sets of data to be provided on a daily, weekly, monthly or yearly basis. Reports on loan portfolios are often required on a loan by loan basis. For institutions with a high number of small loans in their portfolios, this reporting requirement can be burdensome unless full computerization of loan operations and a direct electronic data connection to the supervisory institution is available.

In rural areas, reporting portfolio status or other data to the supervisory agency on a daily basis is expensive if not impossible. Lack of infrastructure, i.e. roads, telephones and computers . . . may make daily reporting difficult. . . . There are different ways of tackling this problem. One can be the relaxation of centralized reporting requirements with delegation of data collection to the financial institution itself. External supervisors could then check consolidated data and, on a random on-site basis, the internal reporting to financial institutions' regional offices. Also, reporting requirements may well be tiered according to size of the loan and type of loan, triggering a more detailed reporting of larger loans and loans granted. . . .[85]

(c) *Protective regulation:*

Deposit insurance: Deposit insurance schemes ensure that depositors' claims will still be served once a financial institution has gone bankrupt. This reflects the prime regulatory aim of protecting depositors. . . . *A major problem of deposit insurance schemes is adverse incentives.* . . . an extended coverage of potential losses not only builds confidence, but also provides a disincentive for market control of the financial institution. . . . [And] if premiums are not set on a risk-assessment basis, lower risk institutions implicitly subsidize high-risk institutions. . . .

Lender of last resort: If severe liquidity problems occur in a financial institution, which do not reflect a fundamental solvency problem, lenders of last resort step in. The central bank implicitly or explicitly plays this role. Distinguishing financial institutions with liquidity problems from insolvent institutions, however, is a difficult task, which in the case of an agricultural lender may coincide with high political pressure to rescue the institution. . . . Decision-makers need to balance reasons for institutional guarantees . . . with the regulatory aim of ensuring a competitive market structure, which implies that inefficient institutions cannot be sustained in the long run.

Commercial bank lending for agriculture, in spite of its unique features that Fiebig has identified, falls under the normal procedures for bank regulation supervision in all countries. As important as bank regulation is, the procedures designed to carry it out are not always relevant for microfinance and rural finance, since most lenders that cater to small-scale borrowers begin operations as lending institutions only. 'In most countries, 85 percent of microfinance institutions are not financial intermediaries – i.e. they are lenders only, and do not take deposits from the public. There is probably no strong reason for public prudential oversight of such microfinance institutions, since protection of depositors is usually viewed as the principal rationale for such oversight'.[86]

Even some institutions that accept deposits need not necessarily be brought under the umbrella of formal bank supervision:

Many NGOs require savings deposits as a condition to granting loans. Such deposits should probably be thought of as part of the cost of the loan, rather than as true financial intermediation requiring public intervention to protect the depositor. Likewise, public supervision is probably neither needed nor practical for community-level groups which accept voluntary deposits from a few dozen members, who know each other and who control the group's lending decisions.[87]

When financial intermediaries begin to accept voluntary savings deposits, as they must do eventually to ensure their long-term sustainability, then usually they begin subject to regulation and oversight. However, it has been found that the usual procedures for that purpose are difficult to apply to microfinance institutions, and a number of countries have begun to develop separate regulations and supervisory procedures for those institutions – or to delay the implementation of prudential regulations for such institutions.

---

85. Fiebig's study also includes commentary on regulatory requirements concerning the topics of internal auditing, risk identification mechanisms, staff qualification and institutional changes.
86. R. Rock, M. Otero and R. Rosenberg, 'Regulation and Supervision of Microfinance Institutions: Stabilizing a New Financial Market.' *Focus*, Note No. 4, The Consultative Group to Assist the Poorest (CGAP), Washington, DC, August 1996, p. 1.
87. *Op. cit.,* p. 4.

Echoing in part what Fiebig has asserted for regulation of agricultural lending, Rock *et al.* define some of the ways in which supervisory procedures should differ for microfinance institutions:

• Requirements for elaborate documentation should be avoided, since microfinance institutions try to keep their costs down by, in part, simplifying their documentation.

> *The argument that small-scale savings deposits in microfinance institutions should be permitted* **without prudential supervision** *has been made forcefully in the following terms:*
>
> In our view, it is a serious mistake to prohibit deposit-taking by community-based organizations just because they are too small or remote to be supervised effectively. *Kate McKee of USAID has pointed out that such a policy is often tantamount to telling people in those communities that if high-quality (i.e. effectively supervised) deposit services can't be delivered to them, then they should have no deposit services at all. Especially in rural areas, 'unsupervisable' deposit takers may be the only ones willing and able to operate in a given locality. Clients are often well aware that such organizations are risky, but continue to use them because the other available savings options are even riskier – cash in the house can be stolen, livestock can die of disease or be unsaleable when you need the cash, and so on. In such cases, the unsupervised community-based depository may well be the least risky option available to a saver. We do her a great disservice if we remove this option because of our paternalistic judgment that it is not safe enough for her (from R. P. Christen and R. Rosenberg, 2000, p. 11 [emphasis in original]).*

• The normal requirement of high provisioning (reserves against losses) for uncollateralized loans should not be applied to these institutions, since the majority of their loans do not require collateral in the traditional, tangible sense, but instead have found effective substitutes.
• Case-by-case reviews of their portfolios are not recommended and lower-cost supervisory methodologies should be used.
• On the other hand, capital requirements should be proportionately higher, at least until the institution has established a solid track record for many years.[88]
• Likewise, liquidity requirements should be kept high, especially since the majority of liabilities (deposits) are short-term in nature.

Christen has argued that microfinance institutions need to be held to a higher standard than commercial banks, not only in regard to capital and liquidity requirements but also in terms of quality of assets and earnings:

> [The] argument that microfinance institutions ought to be allowed to operate within reduced standards since they serve a public purpose is wrong. When one considers that many micro-credit institutions seek a regulatory umbrella in order to mobilize savings from the same low-income client group they serve with credit, this position becomes even more dangerous. If anything, we should hold microfinance institutions to a higher level of scrutiny than traditional commercial banks, particularly if they intend to capture the deposits of low-income clients who can scarcely afford to lose their limited savings due to poor management of the microfinance institution.
> Earnings, asset quality and capital adequacy go to the heart of the purpose of setting stand-

---

88. The case of BancoSol illustrates this point well: 'BancoSol's equity represents from 37 to 46 percent of its total assets; the industry average is 7 percent. This indicates that BancoSol is very well capitalized but also that it is not efficient at using its equity investment to leverage debt. BancoSol intends to reduce the proportion of equity investment to assets to roughly 20 percent, which is still well above the industry average. This will allow BancoSol to leverage debt, but not to the extent that the bank assumes unnecessary risk' (A. Glosser, 1994, p. 245).

ards. They represent the overall financial performance and the risk to depositors or investors from the institution's activities.[89]

Evaluating asset quality in small financial institutions in turn depends on examining the appropriate variables:

Protecting the microloan portfolio is essential to the long-term viability of regulated financial intermediation for the microenterprise sector. Regulators face three main challenges in evaluating asset quality: allowing for the value of personal guarantees, addressing the concentration of large numbers of small, short-term loans within certain sectors, and adapting documentation requirements . . . to capture appropriate client information while minimizing administrative costs.[90]

These considerations imply that *regulations and supervisory procedures should be substantially different for microfinance institutions than for commercial banks.* Accordingly, it is often recommended that a separate supervisory unit be established for microfinance. It may be a distinct department of the supervisory authority (superintendency of banks), or it may be an entirely separate unit – dependent, however, on the ultimate monetary authority, the central bank. Although it may be a separate unit, it needs to be endowed with full authority to impose sanctions, including the liquidation of intermediaries when necessary.

The timing of the creation of a regulatory system may be important. Christen and Rosenberg have argued that premature prudential regulation of microfinance institutions can be worse than no such regulation at all. Other observers have suggested that the creation of prudential regulations can encourage the growth of sound microfinance institutions, by permitting them to

accept deposits on a larger scale and have greater access to borrowed funds, as well as providing them supervision. The central concerns of Christen and Rosenberg are as follows:

- Costs of regulation. . . . the costs of such regulation tend to be higher than is generally recognized, not only in terms of cash costs of the supervisor and the supervised, but also in terms of stifling innovation and outreach. . . . the time-consuming process of legislating new windows can distract the attention of MFIs, donors and governments from other important issues [including appropriate supervision of commercial banks].
- Absence of entities eligible for supervision. *The fundamental question, which seldom seems to get enough attention, is whether the country has MFIs that are suitable for licensing* but cannot use an existing window. . . . *To qualify as a safe depository for . . . commercial-cost money, an MFI should be profitable enough not only to cover its costs today, but also to pay the full commercial costs of the money its license will allow it to leverage, in addition to generating a surplus to fund growth and perhaps give a return sufficient to attract high-quality investors.*

We believe that it is irresponsible to license MFIs that have not demonstrated their ability to operate at this level of sustainability. How can a supervisor vouch for the soundness of an institution unless she is reasonably sure it can operate profitably with the new sources of money it wants access to?
- Inappropriate regulation. Drawing microfinance institutions into the regulatory framework can result in the imposition of unduly onerous regulations, including restrictions on interest rates that they may charge. Once MFIs are on a "country's political radar screen", the high interest rates they

89. Robert Peck Christen, 'Issues in the Regulation and Supervision of Microfinance', in R. Rock and M. Otero (Eds), *From Margin to Mainstream: The Regulation and Supervision of Microfinance,* Accion International, Monograph Series No. 11, Somerville Massachusetts, January, 1997, Chapter II, pp. 34 and 35.
90. Rachel Rock, 'Introduction', in R. Rock and M. Otero (Eds), 1997, Chapter I, pp. 24 and 26.

need to charge to become sustainable may be prohibited.[91]

These authors have argued that regulation of microfinance institutions should be postponed until a country demonstrates adequate capacity to supervise its commercial banks, which are far more important to the health of a nation's financial system. While agreeing that eventually regulation of small financial institutions is essential, they are skeptical that regulation can improve the operations of such institutions, contending that only an institution's managers can achieve that. Five of their basic conclusions are the following:

- Microfinance is unlikely to achieve anything like its potential unless it can be done in licensed environments. Therefore, prudential regulation and supervision of microfinance is a topic that will unquestionably need to be addressed.
- Nevertheless, in most developing countries today the absence of special licensing regimes for MFIs is not the binding constraint to the development of microfinance.
- Rather, the bottleneck is usually the scarcity of MFIs that are not dependent on continuing availability of subsidies, and that *can operate profitably enough to be able to pay a commercial cost for a large proportion of their funds without decapitalizing themselves*.
- The creation of special regulatory windows for MFIs is probably premature in countries where there is not a critical mass of licensable MFIs.
- More attention needs to be paid to reforming regulations that make it difficult to do microfinance under existing forms of bank or finance company licences.[92]

Some groupings of non-bank financial intermediaries, such as associations of credit unions, sometimes establish their own supervisory organ. However, in doing so usually the implicit error is made of combining into the same unit the supervision function and the function of promoting development of the intermediaries. Logically, the two functions are in conflict: the same unit should not be expected to be both judge and advocate for the members of the association.

Again in the words of Christen and Rosenberg:

... we use 'self-supervision' to refer *exclusively* to arrangements under which the primary responsibility for monitoring and enforcing prudential norms lies with a body that is controlled by the organizations to be supervised – usually a member-controlled federation of MFIs. Here at last is a point on which experience appears to justify a categorical conclusion. *In poor countries, self-supervision of financial intermediaries has been tried dozens of times and has repeatedly proven ineffective*, even in cases where donors provided heavy technical assistance.[93]

Regulation can also affect the environment in which microfinance institutions develop. A frequent concern is that many NGOs operate non-sustainable credit programs, with highly subsidized interest rates and lax loan enforcement. This practice siphons off potential clients from sound MFIs and adversely affects the development of an appropriate credit management mentality in the client pool. It can be particularly damaging in environments where there has been a history of State development banks operating in the same non-sustainable way. The unsustainable NGOs can weaken the quality of the portfo-

---

91. R. P. Christen and R. Rosenberg, 2000, pp. 10, 12 and 15 [emphasis in original].
92. *Op. cit.*, p. 23 [emphasis in original].
93. *Op. cit.*, p. 20 [emphasis in original]. These brief excerpts do not do justice to Christen's and Rosenberg's thoughtful paper. It is recommended, along with Fiebig (2001), for anyone interested in the regulation of microfinance institutions, and in innovative ways in which such institutions can collaborate with other kinds of financial entities. Equally recommended are Berenbach and Churchill (1997) and Rock and Otero (1997), along with the valuable examples in C. Churchill (Ed.), *Regulation and Supervision of Microfinance Institutions: Case Studies*, Occasional Paper No. 2, The MicroFinance Network, Washington, DC, USA, 1997.

lios of sustainable institutions when borrowers can use loans from the former to stay current with loans from the latter. Christen and Rosenberg are relaxed about this threat, asserting that 'many of today's best MFIs have grown rapidly in the midst of competition from other microcredit programs with unsustainable interest and default levels'.[94] This may be true in many cases, but in the end it is an empirical matter, and in some countries (Nicaragua and Honduras, for example) the number of unsustainable credit NGOs is high enough that they are exerting at least some debilitating effect on the potential market for solid MFIs. Better regulation can correct this problem in cases where it is warranted.

In short, microfinance requires its own regulatory and supervisory system which, while strict and vigilant, is adapted to the unique characteristics of the institutions it controls and monitors. In developing and extending regulatory systems, new institutional specifications may be required, and the role of the private sector should not be overlooked:

> ... regulatory structures that provide considerable discretion ... can breed corruption. In this connection it is useful to have more than one agency engage in monitoring. Corruption aside, all monitoring is fallible. Considering the large costs associated with allowing insolvent institutions to operate, one way of reducing the likelihood of that occurring is to have more than one independent monitor. ...
>
> In principle, the monitors have supervisors. But often the supervisors are not well informed. If there is more than one monitoring agency, a system of peer monitoring can be employed; each monitoring agency in effect monitors not only the financial institutions but also each other. ... Reformers who ignore the central importance of information and control may look at organizational charts and suggest streamlining them to end the allegedly wasteful

duplication. Such reform efforts may, from this perspective, be fundamentally misguided.

Governments can take advantage of resources and incentives in the private sector to stretch its regulatory reach and make its monitoring more effective. The ... suggestion that the government should focus on regulating such variables as net worth or capital, which it can observe at relatively low cost, falls into this category.[95]

In the final analysis, as important as financial regulation is, it is necessary to be aware that regulation alone cannot guarantee a sound financial system. As Fiebig has put it:

> It is important to bear in mind that while financial institutions do benefit from an appropriate external regulatory regime, *there is not much evidence that the existence of a regulatory jurisdiction makes institutions stronger and less prone to external (or internal fraud) shocks.* Discussants on banking regulation generally agree the major role of ensuring safety, soundness and profitability of a financial institution remains with managers and owners.[96]

7.5 STRUCTURAL CONSIDERATIONS FOR RURAL FINANCIAL INSTITUTIONS

7.5.1 Local Institutional Structures

The structure of rural financial institutions displays great variety. One of the main distinctions already has been noted, that of institutions that only lend versus those that take deposits as well. While most informal rural financial intermediaries commence operations as purely lending entities, particularly those supported by NGOs, it is important to provide them with incentives and support to facilitate the eventual transition to institutions that can mobilize savings, for the sake of their long-run viability.

94. R. P. Christen and R. Rosenberg, 2000, p. 10.
95. Joseph Stiglitz, 'The Role of the State in Financial Markets', *Proceedings of the World Bank Annual Conference on Development Economics, 1993,* The World Bank, Washington, DC, USA, 1994, pp. 37–38.
96. M. Feibig, 2001, p. 22 [emphasis added].

Some small-scale financial institutions, including those of the greatest antiquity, are born with both a deposit-taking and lending capacity. Diverse cultures throughout the world have incubated small mutualist financial groups that are increasingly referred to by the generic term 'rotating savings and credit associations' (ROSCAs). Their size is restricted, ranging from a handful of people to about 60 at a maximum; 12 and 24 persons are common groupings. Normally, they are organized among persons who are brought into daily contact for other reasons, as in the case of neighbors or employees of the same firm, but they can be groupings for former schoolmates or be organized around any other factor of social cohesion. The members of a ROSCA make regular, usually monthly, contributions to a central fund, and withdrawals from the fund obey fixed rules. A common variant is that one member, in rotation, takes home all of the monies in the fund each month. Other arrangements allow for lending portions of the fund. In Korea, this kind of organization is known as *kye*; in Mexico, *tanda*; in Bolivia, *pasanaku*; in Egypt, *gamaiayh*; in Mozambique, *xitique*; in Ghana, *susu*.[97] Although simple, their structures have certain elements of sophistication. For example, in Korea different positions in a *kye* correspond to net lenders or net borrowers, and likewise the implicit interest rate (±) varies by position. According to their desires or needs, Koreans can look for a *kye* with for example, the number 9 position open, or the number 2, etc. In Korea at least, most members of *kye* are women.

ROSCAs work on the basis of peer pressure to enforce compliance with the rules, although occasions in which it is necessary to apply that pressure vigorously are by no means unknown. In their original form, they are not replicable on a larger scale. However, the principles that drive a ROSCA are found in some other credit organizations, including credit cooperatives and credit unions, and they have been applied in the suc-cessful lending technologies of institutions such as the Grameen Bank.

Although their structure prevents their attaining a significant scale of operations, ROSCAs illustrate the ability of societies to develop their own institutions in response to the need for a place to deposit savings and the need for short-term credit. The existence of an indigenous matrix of financial relationships and institutions also underscores the need to adapt new financial institutions to the local context.

In a study of village finance in the Gambia, Nagarajan, Meyer and Graham analyzed the role played by two international NGOs and found that one of them debilitated local, traditional financial institutions (*kafos*) which provided insurance (a fund for contingencies) and savings mobilization as well as lending. These indigenous institutions also arranged for labor sharing and other joint community activities. Their interest rates were at market levels. One of the NGOs carefully adapted its operations in order to complement the role of the *kafos*, while the other one entered the villages with a prior agenda of targeted and subsidized credit. The authors concluded that:

> the targeted loans ... at subsidized interest rates using external funds seem to negatively influence *kafos* by eroding their carefully built institutional safeguards. As a result, since [this NGO] only partially substitutes for *kafo* activities, it may well reduce aggregate village welfare because the gains realized by some may not offset the losses of others.[98]

7.5.2 Credit Co-operatives

Among local financial institutions that reach beyond a small group of persons, credit co-operatives (credit unions) are the most widespread type, apart from small-scale, donor-supported lending organizations. The structure

---

97. These names and more in other countries are found in The World Bank, 1989, p. 114.
98. G. Nagarajan, R. L. Meyer and D. H. Graham, 'Institutional Design for Financial Intermediation by NGOs: Implications for Indigenous Self-Help Village Groups in The Gambia', in R. Rose, C. Tanner and M. A. Bellamy (Eds), 1997, p. 274.

of credit co-operatives is fundamentally different from that of banks. Depositor–borrowers are owners of credit co-operatives, and many of the co-operatives do not have a base of capital. In addition, except in cases in which they are federated, they do not have recourse to borrowing from other financial institutions to tide them over liquidity shortages. Banks, in contrast, have owners who contribute share capital, and their more formal structure allows the access to them interbank loan market.

One of the advantages of credit co-operatives is the ease of forming them, as they do not have to comply with the minimum capital requirements and other regulations that pertain to banks. (The disadvantage of this situation is that they are not subject to prudential regulation, as banks are.) Another advantage is that the cost of their administration can be very low; sometimes they even rely on volunteer labor from their members. As a result of the heightened participation of members, credit co-operatives tend to be more 'user-friendly' than many banks, and the cost of their services generally is low. However, banks increasingly are emphasizing improvements in the quality of their services.

More formally, the reasons for the success of credit co-operatives lie in the facts that (a) they have a *long-term interaction* with their members, which fosters a sense of identification with the institution and responsibility on the part of the latter, and (b) even in larger co-operatives, *loan monitoring is carried out by peers* who often are members of the same community.[99]

In practice, the record of credit co-operatives has been uneven and their performance in many countries has been disappointing, often characterized by a high rate of failure. 'Developing-country credit unions have shown a strong tendency toward disruptive boom-and-bust instability. In more than a few poor countries, the majority of all credit unions are insolvent – that is, unable to pay off all depositors'.[100] Ledgerwood has made a similar observation: 'many [credit union] systems do not function as well as their basic philosophy would lead one to expect. At the same time, they are very difficult partners for foreign institutions'.[101] This outcome can be traced directly to their institutional structure (as Feibig also observed in the comments cited above):

... credit cooperatives have not performed well even in countries where they belong to the formal financial structure (i.e. are regulated) because their governance rules and ownership structure are inconsistent with their financial health. Credit cooperatives can only lend to their owners which for all practical purposes means that credit cooperatives have no capital. In case of bankruptcy, owners can capture their equity investment by defaulting on their loans. In the absence of capital, financial intermediaries are unconcerned about risk. Moreover, credit cooperatives can be easily controlled by individuals with negligible investments in them. These governance rules and property rights are perverse because individuals in control of these organizations are residual claimants only of their profits but not of their losses (i.e. borrower domination). .... Examples of ... practices adopted by borrower-dominated intermediaries are weak collection of loans and low interest rates.[102]

In a comprehensive study of the experience of credit unions in the developing world, John H. Magill summarized some of their major weaknesses and constraints as follows:

99. Abhijit V. Banerjee, Timothy Besley and Timothy Guinnane, 'Thy Neighbor's Keeper: The Design of a Credit Cooperative with Theory and a Test', *The Quarterly Journal of Economics*, 1994, **109**(2), pp. 491–515.
100. R. P. Christen and R. Rosenberg, 2000, p. 14.
101. J. Ledgerwood, 1999, p. 103.
102. The World Bank, Latin America and the Caribbean Region, Central America Department, Sector Leadership Group, *El Salvador Rural Development Study*, Yellow Cover Draft, Report No. 1625ES, Washington, DC, January 23, 1997, p. 14.

Credit unions tend to be small. . . . [and hence they] could not assume the risk involved in developing or implementing specialized programs designed to reach large numbers of small-scale enterprises.

Because of basic inadequacies in credit union financial and interest rate policies in most countries – particularly the precedence of credit over savings – credit unions are not generating capital rapidly enough to meet member demand. . . .

Increased credit union participation in enterprise lending is also limited by the fact that most are conservative, highly traditional organizations that do not have a modern growth- and service-oriented philosophy. . . .

Internal credit union policies and operating procedures need modernizing if credit unions are to significantly expand their role in small-scale enterprise lending. In particular, poor delinquency control and weak portfolio management capabilities limit the ability of many credit unions to expand loan portfolios or add new services. Management, operational systems, and even basic accounting systems need improvement, particularly in smaller credit unions.[103]

Substitute 'farm' for 'enterprise' in this passage and the problems of relying on credit unions for agricultural finance, in their present state of development, can be seen clearly.

The problems of credit co-operatives can be compounded, rather than lessened, by financial support from the government:

In many developing countries cooperatives operate under a government department that supports them with funds, technical assistance, and policy guidance. Government support is attractive to the cooperatives' managers because it allows lending to expand quickly, but it weakens the incentives of cooperative members to provide their own finance. When loans are made according to government directive, lenders may find it difficult to collect. Such loans are often seen as grants and hence as resources that can be spent on consumption . . .

Moreover, the goals of government and cooperatives can differ greatly: governments often view cooperatives as instruments for the conduct of broader policy. In Africa, for example, a ministry wished to use the cooperative credit system to channel low-interest funds from foreign donors to targeted programs. When the ministry's plan was presented to the cooperative, the director declined because he felt that the funds would never be recouped by his institution. The director was told to reconsider or resign. The plan went into effect, repayment rates were extremely low, and other cooperative lending programs were undermined. . . .

Similarly, the support of foreign donors can be a mixed blessing. Cooperatives may seem a suitable channel for development funds, but they often end up with heavy liabilities and a bad collection record. . . . [104]

Nevertheless, under appropriate institutional designs and an appropriate policy environment, some experiences of credit co-operatives have been successful. The opportunities afforded by credit co-operatives and insights into approaches which can make them function well were presented by the World Bank in reviewing the experience of Togo:

Despite the difficulties, cooperatives are a good way of increasing access to financial services. Their costs are often low because they use volunteer labor and because they can reduce risk through group accountability and local sanction. Where governments have been more concerned with the viability of cooperatives than with social objectives – and *where interest rate*

---

103. John H. Magill, 'Credit Unions: A Formal-Sector Alternative for Financing Microenterprise Development', in M. Otero and E. Rhyne (Eds), 1994, p. 149.
104. The World Bank, 1989, pp. 117–118.

Both typical weaknesses and potential strengths of credit co-operatives are illustrated by the experience of the World Council of Credit Unions in reforming credit co-operatives in Guatemala:

Before 1988 . . . their main purpose was to provide cheap rural credit. [They] were financed by subsidized external credit and by compulsory, zero-interest share deposits from members. Because loans were issued at below-market interest rates, members were in effect penalized for saving and rewarded for borrowing. The credit unions had serious operational problems (management information systems were underdeveloped, the credit unions carried a large volume of nonearning assets, the loan delinquency rate was about 20 percent, and loan loss reserves were underestimated by more than 50 percent). Liquidity reserves were so low (about 3 percent) that the credit unions could not always honor cash withdrawals from members. . . .

The World Council for Credit Unions, funded by the US Agency for International Development, implemented an institutional development program for the credit unions between 1987 and 1994. . . . By 1994 . . . deposits had grown from 24 percent of assets . . . to 55 percent. . . . The loan delinquency rate had decreased to 8 percent of the loan portfolio. . . . Several factors contributed to this turnaround:

- *. . . use of a business plan that incorporated institutional development, financial stabilization, savings mobilization, and credit administration. Firm financial targets were set and an effective management information system was put in place. Interest rates on deposits and loans were increased.*
- *. . . An agreement of participation was signed by all parties to demonstrate their commitment.*
- *. . . financial assistance for the stabilization process was provided to the credit unions in the form of non-interest bearing one-year loans. The loan principal was placed in high-yielding Guatemalan investments, and the interest earned went to the credit unions to offset non-performing assets*

(from J. Yaron, M. P. Benjamin and G. L. Piprek, 1997, p. 74).

---

restrictions have been relatively modest – co-operatives have flourished and the supply of financial services has broadened. In Togo, for example, savings in the credit union system grew by 25 percent a year and loans by 33 percent a year during 1977–1986. Members elect a board of directors, which decides on interest rates, dividends on shares, and lending policies. The credit unions are federated, and they jointly manage a **central fund**, invest in low-risk financial instruments, and mediate transfers between member unions with surplus funds and those with deficits. . . . **they have access to broader financial markets through their federated structure**.[105]

Magill feels that if the credit union movement overcomes the weaknesses he describes, they can play a stronger role in financing development:

Credit unions throughout the developing world, and particularly in countries facing high inflation rates, have begun to modernize their savings services. Regular savings accounts and deposits are being offered in addition to the traditional share savings accounts. Credit unions are paying interest . . . closer to market rates on these new savings instruments. . . . These changes will make credit unions more viable financial institutions. . . .

Modernization of credit unions also should focus on improving their financial products:

- Developing a broader range of savings services, including long-term savings instruments, with different interest rates and maturities.
- Developing an expanded range of loan services designed to meet the needs of a wider variety of members.

---

105. The World Bank, 1989, p. 118 [emphasis added].

• Pushing aggressively forward to offer quasi-transaction loan and share draft accounts to meet the transaction needs of their members.[106]

In summary, the credit co-operative model can be made more viable through improved financial management and policies that offset some of their structural weaknesses in the areas of governance and ownership: a realistic interest rate policy, creation of a federation of credit unions which includes a central fund, and a policy by governments and donors of abstaining from directing subsidized credit through the co-operatives. It also is important to subject credit co-operatives to adequate supervision, which they frequently do not have, and, as indicated above, this requires separating the supervisory authority from the entity that provides technical assistance for the development of the co-operatives. Regulations need to be implemented to prevent the institutions' directors from having special access to loans without passing through the normal process of application and qualification. Many credit unions have shipwrecked on the shoals of struggles by influential individuals over access to funds.

7.5.3 Rural Banks

International experience in the microfinance field emphasizes the need for credit programs to move to higher levels of self-sufficiency, essentially to become banks. However, in the meantime most of them are limited in scale and in financial terms their clients are small in scale. The credit needs of many farms, even those in the small- to medium-size range, usually exceed the lending limits or capabilities of most microfinance institutions. In this regard, millions of farms find themselves in the same situation that growing non-farm micro-enterprises are in, i.e. of being too large to be clients of these institutions and yet too small to borrow from commercial banks:

Too big for credit programs that provide only short-term working capital, yet still too small to meet the minimum loan amounts or collateral requirements of formal financial institutions, these enterprises find their growth curtailed by lack of credit available in amounts and terms that meet their expanding needs. Growing microenterprises find that 'their success has made them too steep a risk for both the informal and formal finance sectors. They are lost in the gray area, a true structural gap in which thriving businesses stagnate, their potential for generating further income and employment curtailed'.[107]

In many countries, the credit requirements of farmers have become even more urgent as State-owned agricultural development banks have been closed. *As useful as the growing microfinance movement is for meeting the financial needs of many rural families, it does not represent a complete solution for financing agricultural development.* A more complete response to this need can be developed by complementing the microfinance approach with the creation of private banks that emphasize lending to agriculture. A few examples exist – Bancafé and Banco del Occidente in Honduras and the Banco Ganadero in Colombia may be mentioned – but they are not very numerous. As noted earlier, no commercial bank can afford to specialize entirely in agricultural lending, owing to its greater riskiness, but it is possible to devote a significantly larger share of the portfolio to agriculture than most commercial banks do.

For a bank to emphasize agricultural operations and still be viable it needs to be large enough to cover geographically dispersed areas, in order

---

106. J. H. Magill, 1994, pp. 146 and 152.
107. H. P. Martínez, 'The gray area in microenterprise development', *Grassroots Development*, **14**(2), 1990, p. 33, cited in Larry R. Reed and David R. Befus, 'Transformation Lending: Helping Microenterprises Become Small Businesses', in M. Otero and E. Rhyne (Eds), 1994, p. 185.

to reduce covariant risk. Other requirements include:[108]

- Implementation of a policy of flexible interest rates, so that the higher risk in agricultural lending can be offset by higher interest margins. In the same spirit, link interest rates to a farmer's credit history.
- Adoption of some of the techniques of microfinance institutions for selecting clients and creating incentives for repayment. The formation of groups of farmers, making them jointly responsible for repayments, or making lending to some depend on the prior repayment record of others. Even more applicable for the case of agriculture may be a policy of progressive lending, explicitly making eligibility for future loans dependent on the timeliness of repayment of the present one.
- Decentralization of loan decision-making as much as possible to rural branches, even forming loan committees composed of local citizens, to take advantage of local knowledge of production conditions and clients.
- Development of modern information systems so that loan officers can follow up with clients if loan payments are even a day late.
- Consideration of innovative forms of collateral, such as antichresis and movable property, as mentioned above.
- Improvement of staff training and provision of incentives to loan officers for good client selection and loan recovery rates.
- Provision of additional basic financial services to the rural population.

In addition to efforts to encourage better lending practices for agriculture, it is becoming apparent that *an appropriate regulatory framework is needed for agricultural finance, and not only for microfinance*, as suggested by the observations of Michael Fiebig cited in Section 7.4 above. This issue is discussed further in the concluding section of this chapter.

In most cases, it is not realistic to expect to persuade existing commercial banks to take on greater volumes of agricultural lending, in place of their relatively more secure investments in urban properties and government bonds. An alternative is to create separate, agriculturally oriented banks, but the initial capital requirements for the formation of a bank may be daunting for even a large group of interested farmers. Both Bancafé and the Banco Ganadero were transformed from government institutions into private ones. In the case of the latter, some of the shares were purchased by Colombian ranchers over a period of many years via a special levy on livestock herds.

Another alternative would be to seek external sources of funding in order to found such institutions and begin operations, as most of the microfinance programs have done, with an agreement that share capital would be purchased over time by local farmers and other private domestic investors. In such a case, the path followed to self-sufficiency could be very similar to that followed by those programs. The difference is that the clientele would be of varying sizes, in financial terms, and the average loan size would be larger than is the case for the microfinance programs. Another difference would be relatively greater emphasis on agricultural lending.

Additionally, rural microfinance institutions can expand their operations through alliances with existing banks:

Another alternative to special licensing of MFIs is to have unlicensed MFIs take advantage of someone else's license. NGO MFIs have teamed up with existing banks or credit unions, in effect using the latter institution's license to increase services to the NGO's target clientele. The NGO Freedom from Hunger has such arrangements with financial cooperatives or rural banks in Burkina Faso, Ghana, Mali, Madagascar, and the Philippines.

---

108. See B. Klein, R. L. Meyer, A. Hannig, J. Burnett and M. Fiebig (1999) for a very useful discussion of recommended practices for agricultural lending.

... There is room for a lot of creativity in structuring a bank/NGO relationship, including some options that could preserve the NGO's control over the lending, including methodology, loan sizes/terms, and the choice of clients.[109]

7.5.4 Apex Organizations

Yet another option is to develop a system of small-scale rural financial institutions, all linked through a central fund or 'apex organization', or second-tier or second-storey institution. Individually, the small-scale member institutions might be exposed to higher covariant risk in their portfolios, but in principle collectively they could disperse risk. Examples exist in Costa Rica, where the apex organization is a US-based non-profit organization known as FINCA (Foundation for International Community Assistance), and in Colombia, where it is a local entity known as AGS (Asociación de Grupos Solidarios de Colombia). Apex organizations for microfinance programs can exist regardless of the structure of the programs of local affiliates. Apex organizations are regarded as 'brokers or wholesalers between banks and NGO-based programs'.[110]

The concept of second-tier institutions, like that of financial intermediation in general, has undergone a rapid evolution in recent years. They used to be conceived as sources of subsidized, directed credit; the Fondo Financiero Agropecuario in Colombia was an example of that approach. Now, their role is to assist with *liquidity management* of their affiliates, provide access to the interbank market, and sometimes also to provide technical assistance in financial management. Given that individual rural credit unions or 'mini-banks' are likely to be exposed to high covariant risk, their sustainability can be enhanced through linkages to a second-tier institution.

The new role of second-tier institutions has been summarized in the following way:

A second-tier institution is a financial intermediary or network that provides financial and institutional support services to retail intermediaries. ... the second-tier institution needs to be autonomous, and free from political interference; have the capabilities to mobilize funding; know the retail institutions intimately; and be able to motivate retail institutions while being tough in the enforcement of standards and eligibility criteria for support.[111]

While an apex organization must have some supervisory control over its affiliates, with enforcement powers, it should not attempt to carry out banking supervision *per se*. This would put it in the untenable position of being a participant in the system and judge of it as well. In fact, the combination of supervision and technical assistance roles in a central organization has contributed to some of the problems experienced by credit unions.

Another critical issue concerns the ownership of the apex organization. The affiliates may be shareholders in it, but if collectively they hold a majority of shares, then it is virtually impossible for the organization to play the leadership role for the system that it should play. This issue has been raised in connection with the AGS in Colombia:

The governing structure of AGS, including the board, is in the hands of the affiliates' executive directors. This structure has been instrumental in ensuring that the services provided by AGS respond to the needs of the membership. ...

Nonetheless, the fact that the directors are making decisions that directly affect their own organizations has had a negative impact on the objectivity of the process. The governing

---

109. R. P. Christen and R. Rosenberg, 2000, pp. 18–19. Options for a bank/NGO relationship are discussed further in that paper.
110. E. Rhyne and M. Otero, 1994, p. 22.
111. N. Barry, 1995, pp. 4–5.

structure of AGS, like that of many membership organizations, is weak because the beneficiaries of the services provided by AGS are the same agencies that control the institution. The executive directors are both judge and jury. Policy decisions seldom take into account a larger vision beyond the parochial concerns of the members. A typical example is the strong opposition of some board members to increasing the interest rates charged by AGS and implementing fees for its services. The growth of the organization and its long-term sustainability have been sacrificed in favor of cheap money and free services for the membership.[112]

To help ensure good governance, such a second-tier institution can and should be owned at least in part by private investors, rather than by the microfinance institutions linked to it; they may be commercial banks for whom the small-scale intermediaries offer cost-effective ways to penetrate rural areas. A system structured in this way obviously bears certain similarities to a bank with many rural branches. A difference is that a federated system can have very low-cost local branches, to the extreme of some that are managed out of private homes and are open for business only two days a week. The possible role of such systems is becoming increasingly recognized:

> . . . the most promising of recent strategies for financial market development is the linkage of member-controlled financial institutions with a liberalized banking and cooperative sector.[113]

Carefully designed to avoid the ownership problem and others, apex organizations can play a useful role in strengthening a group of small financial institutions. However, to date their performance has not always lived up to their potential. Ledgerwood has provided a summary of their potentials and weaknesses:

An apex institution can:

- Provide a mechanism for more efficient allocation of resources by increasing the pool of borrowers and savers beyond the primary unit.
- Conduct market research and product development for the benefit of its primary institutions.
- Offer innovative sources of funds, such as guarantee funds or access to a line of credit from external sources.
- Serve as a source of technical assistance for improving operations, including the development of management information systems and training courses.
- Act as an advocate in policy dialogue for MFIs.

The experience of apex institutions has been mixed. Apex institutions that focus on providing funds to retail MFIs, often at subsidized rates, have found limited retail capacity to absorb those funds. What MFIs most often need is not additional funding sources but institutional capacity building. Furthermore, by providing wholesale funds in the marketplace, apex institutions remove the incentive for retail MFIs to mobilize deposits.

There are other potential weaknesses of apex institutions:

- Vision and governance issues are made more complex by the number of parties involved.
- The level of commitment to expansion and self-sufficiency may vary among the members, affecting the pace of expansion and the ultimate scale achieved. The commitment to market-oriented operating principles may also vary, affecting the ability of the group to operate in an unsubsidized fashion.

112. Arelis Gómez Alfonso, with Nan Borton and Carlos Castello, 'The Association of Solidarity Groups of Colombia: Governance and Services', in M. Otero and E. Rhyne (Eds), 1994, p. 260.
113. M. Zeller, G. Schneider, J. von Braun and F. Heidhues, 1997, p. 4.

- Differential growth rates among the partners can also strain the relationship, especially when their needs for resources and technical support widen dramatically.
- Monitoring and supervision are essential to good performance but are made difficult by the number of partners. If there are weaknesses in financial reporting and management at the primary level, these can adversely affect the second-tier operation.
- Unless both the primary institutions and the apex are efficient, the ultimate cost to the client can be high. There needs to be constant attention to issues of productivity and performance.

While there are many disadvantages to apex institutions, if structured appropriately and set up with clear and market-oriented objectives, they can add value and aid in the development of microfinance.

For the most part, microfinance apex institutions provide more than just liquidity in the market. Usually the apex is set up when everyone agrees that there is a drastic shortage of retail capacity. The advertised objective of the apex is to foment the development of stronger retailers capable of reaching a much more substantial portion of the microfinance clientele.[114]

To define their roles more precisely, apex institutions may be particularly useful in the following situations (Von Pischke, 1996):[115]

- When they are lenders of last resort, not necessarily in situations of crisis but on the basis of cost. Retail lenders regard an apex institution as a good source of expensive funds and presumably use them sparingly and only for highly important and profitable programs.
- When the apex institution fills in seasonally. Agriculturally based retail lenders might

want to borrow seasonally as a means of managing cash flow. Again, the funds should not have to be provided by the apex institution at anything less than commercial rates.
- When the apex institution becomes a shareholder in the retail MFIs, in the expectation that this would provide an attractive overall return. In this case, the apex institution would expect to add value by providing expertise and oversight as well as funds in the form of equity and quite possibly debt.
- When retail lenders are not permitted to take deposits. Apex institutions could then play a useful role if they added value through their terms and conditions and behaved commercially.

An apex institution appears to be most successful when a critical mass of strong MFI retailers already exists and when it is focused on working with existing formal financial institutions that are 'downscaling' to meet the demands of low-income clients.[116]

7.5.5 Rediscount Lines and Bond Financing

Second-tier institutions that are devoted exclusively to on-lending government and donor funds to retail financial intermediaries are known as rediscount lines. They have gone out of fashion in recent years but they still are used when weaknesses in financial systems constitute severe obstacles to development of certain sectors or areas of the country. For example, the World Bank has supported a rediscount line for investments outside of the Maputo area in Mozambique, since almost all bank investment was concentrated in that area, and rediscount lines in Honduras and Nicaragua, mostly for agriculture. In Estonia, it agreed to the government's con-

114. R. Rosenberg, 'Comment on DevFinance Network May 15, 1996', Internet discussion group [devfinance@lists.acs.ohio-state.edu].
115. J. D. Von Pischke, 'Comment on DevFinance Net-work, May 14 1996', Internet discussion group [devfinance@lists.acs.ohio-state.edu].
116. J. Ledgerwood, 1999, pp. 106–109.

tinuation of a rediscount line for agriculture during the transition to a market economy, and also in Nicaragua it offered to channel subsidized funding to banks that agreed to establish branches in rural areas (without much success).

Rediscount lines tend to specify a limited number of economic activities for which their funds may be used, e.g. agriculture or housing. Their critics point out that directed credit may go to sub-optimal uses when viewed from an economy-wide standpoint, and that the purposes of directed credit may be subverted by the fungibility of funds. For example, loans to a farm family may permit it to construct a house in an urban area, by freeing up some of its own resources. Usually, rediscount lines are subsidized for two reasons: (1) an incentive must be offered to financial intermediaries to induce them to lend for purposes that they would otherwise not lend for, and (2) the target sector or group is often judged to need subsidized funding in order to develop. However, subsidization of the target clientele does not necessarily have to be part of the design of a rediscount line.

While the criticisms of rediscount lines are valid, it is undeniable, as pointed out earlier, that the collapse of public sector agricultural banking has left many medium and small-scale farms in a large financial void. This will not be filled until microfinance institutions mature and grow considerably more and/or new types of rural banks emerge. Neither of those solutions is likely to emerge soon in most developing countries. The need for rediscounted funds is most acute for long-term investments, such as reforestation, private irrigation systems, fruit trees and livestock development. To the extent that commercial banks lend for agriculture, almost all of it goes for short-term working capital. To be effective, a rediscount line should be directed toward supporting production lines that have a clear comparative advantage (often this is not the case), and its terms have to be sufficiently attractive to financial intermediaries.

A major barrier to the effective operation of rediscount lines can be a lack of capacity in the retail financial system to absorb more funds. Often banks are fully lent-up in terms of their capital adequacy ratios, especially in light of the attractive terms offered by government paper, and MFIs do not have significant capacity to expand lending. Thus, the core problem may be the identification of an appropriate financial intermediary. Associations of producers have sometimes tried to play that role, but the guarantees they can offer are usually weak, unless they can purchase a guarantee bond from a bank, but that may be expensive and in some countries, impossible. Thus, while there is a certain inexorable logic to the (transitory) need for rediscount lines to foster the development of key sectors in many developing economies, this need often leads back to the need to strengthen existing financial intermediaries at the retail level.

The bond market is rarely tapped for agricultural development but it can represent a financial resource when the circumstances are appropriate. At the end of the 1990s, El Salvador successfully floated bonds on the international market to support a program of renovating coffee plantations. However, the conditions for the success of this program were very strict. They included the following:

(1) A solid history of domestic loan repayment by most coffee farmers.
(2) A proven technology for raising coffee yields.
(3) A relatively homogeneous client group.
(4) A well-organized national federation of coffee farmers that was able to deliver the technology package for renovation and select the farmers for participation.
(5) An acceptable credit rating by the Government of El Salvador, which guaranteed the bonds, on world financial markets.[117]

Conditions such as these rarely can be satisfied, but when they can, the international bond market represents an under-utilized financial resource to support agricultural development.

---

117. Source: conversation with Carlos Fuentes of the Ministry of Agriculture of El Salvador, 2001.

7.5.6 Governance Issues

At the system-wide level, good governance requires appropriate mechanisms of financial supervision and also of technical assistance by a central fund to which local financial intermediaries are linked. At the level of a local institution, governance concerns establishing *accountability* and *transparency* in the operations of the institution and in the relations between shareholders, depositors, managers and directors. Ensuring solid governance may be the most important factor in the success of rural financial institutions.[118]

Good governance allows each actor the necessary freedom to act in the best interests of the institution while at the same time responding to the other legitimate interests involved in the institution. When an institution works well in these senses, it can meet the following challenges, as described by Max Clarkson and Michael Deck:[119]

(1) How does the business preserve its vision?
(2) How does it balance growth, risk and profitability?
(3) How does it establish a governance system that holds management accountable without undermining its independence?

In a concrete sense, governance eliminates conflicts of interest and establishes relationships of trust among depositors, managers, shareholders and members of a board of directors, so that confidence is generated that influential individuals will not abuse their relationship with the institution to obtain access to its funds in an irregular manner. Abuses of this nature, or allegations thereof, have undermined many rural credit cooperatives over the years.

There do not exist precise guidelines in the governance area for rural financial institutions. However, to help avoid the problem of insiders' access to funds and to increase accountability,

Clarkson and Deck (1997, p. 6) outline the basic responsibilities of a board, as follows:

(1) *Fiduciary.* The board has the responsibility to safeguard the interests of all the institution's stakeholders. As such, the board serves as a check and balance to provide confidence to the [institution's] investors, staff, customers, and other key stakeholders that the managers will operate in the best interests of the institution.
(2) *Strategic.* The board participates in the organization's long-term strategy by critically considering the principal risks to which the organization is exposed, and approving plans presented by the management. The board does not generate [the institution's] strategy, but instead reviews management's business plans in light of the institution's mission, and approves them accordingly.
(3) *Supervisory.* The board delegates the authority for operations to the management through the Chief Executive Officer. The board supervises management in the execution of the approved strategic plan and evaluates the performance of management in the context of the goals and time frame outlined in the plan.
(4) *Management Development.* The board supervises the selection, evaluation and compensation of the senior management team. . . .

They also point out that 'it is not necessary for board members to be shareholders. In fact, it may be preferable if some of the members are independent . . .'. Above all, 'board members should not receive any personal or material gain other than the approved remuneration. The board must have common and clear objectives. It is important that board members do not have political agendas that could influence the direction of the organization'.

---

118. J. Yaron, M. P. Benjamin and G. L. Piprek, 1997, p. 7.
119. M. Clarkson and M. Deck, 'Effective Governance for Microfinance Institutions', in Craig Churchill (Ed.), *Establishing a Microfinance Industry*, Microfinance Network, Washington, DC, USA, 1997, p. 6.

In addition, their review touches on the issue mentioned above of having stakeholders who have invested capital in the institution, as opposed to the pure co-operative model (Clarkson and Deck, 1997, p. 7):

> One of the rationales for changing institutional forms, from an NGO into a regulated financial intermediary, is that for-profit institutions have capital that is owned by someone who will be upset if their capital is dissipated. Once the institution has shareholders who have something to lose, then there are clear lines of accountability between the owners and the board members.

Finally, they stress that programs of *training* for such institutions should include board members as well as managers.

7.5.7 Gender Issues in Rural Finance

The research of Khandker (1998) cited above supports the increasing consensus that women are better credit risks for rural finance and that they also make better use of borrowed funds for improving household welfare. Microfinance lenders have been quick to target women in their programs:

> One important achievement of the microfinance movement has been its relative success in deliberately reaching out to poor women living in diverse socioeconomic environments. Of the nearly 90 thousand village bank members worldwide that have received loans from the Foundation for International Community Assistance (FINCA), 95 percent are women. The Association for Social Advancement (ASA), one of the most prominent microfinance institutions in Bangladesh, has provided US$200 million exclusively to women borrowers. In Malawi, 95 percent of loans provided by the Malawi Mudzi Fund go to women borrowers. Since 1979, Women's World Banking has made more than 200000 loans to low-income women around the world. Literally hundreds of similar examples can be found in Asia, Africa and Latin America.[120]

Nevertheless, in spite of these advances rural lending still goes predominantly to males. The FAO provides several empirical examples about the gender distribution of rural lending; although they are somewhat out of date, the pattern is still approximately the same today in many countries:

> A 1990 study of credit schemes in Kenya, Malawi, Sierra Leone, Zambia and Zimbabwe found that women received less than 10% of credit directed to smallholders and only 1% of total agricultural credit.
>
> In the Indian government's Integrated Rural Development Programs, although the proportion of credit directed at women increased, by 1989 women only received 20% of the program's credit.
>
> In a survey of Kenya only 3% of women farmers surveyed had obtained credit from a commercial bank compared to 14% of male farmers. Similarly in Nigeria the ratio was 5% to 14%.
>
> Under Chile's Institute for Agricultural Development small farmer credit program women were only 12% of borrowers in 1992.[121]

Among the reasons adduced by the FAO for this pro-male bias in lending, in spite of the better productivity of funds lent to women on average, are the following:

---

120. Manohar Sharma, 'Empowering Women To Achieve Food Security: Microfinance', *A 2020 Vision for Food, Agriculture, and the Environment*, Focus 6, Policy Brief 10 of 12, International Food Policy Institute, Washington, DC, USA, August 2001, p. 1.

121. FAO, *SEAGA Macro Manual*, draft, Food and Agriculture Organization of the United Nations, Rome, July 2001, Module 13.

- Credit is often concentrated on particular types of activity such as agricultural production rather than the whole chain of activities including processing, storage and trading where women often predominate.
- Credit is seldom provided for consumption activities despite the fact that women are more likely than men to borrow small amounts on a more frequent basis to meet short term family consumption needs.
- Credit is sometimes channeled through rural organizations to which women lack membership access.
- Credit delivery sometimes forms part of an overall input and extension package aimed towards more progressive commercial farmers.
- Collateral loan requirements disadvantage women who often lack legal ownership of resources they use, especially in the case of land tenure.
- Cultural barriers may limit women's engagement in the formal sector, restrict their mobility and interactions with men and hence their access to financial institutions.
- Women's lack of literacy and numeracy may limit their knowledge of credit availability and make it difficult for them to follow application procedures.[122]

Although these barriers may appear formidable, the experience of many microfinance institutions has shown that they can be overcome and, indeed, that women make up the vast majority of a microfinance institution's clients in a number of cases. Purposive policy can assist in overcoming the barriers through steps such as the following:[123]

(1) Provide materials and training to raise awareness in financial institutions of the value and importance of targeting women clients to a greater degree, and to strengthen their capability of improving outreach to this group, for both savings and lending.

(2) Train women farmers in basic literacy, numeracy, cash flow management and the requirements of credit programs. In some cases, subsidies can be provided to credit programs for training components along these lines.

(3) Remove legal and regulatory constraints that may limit women's access to credit and savings facilities, such as requirements for a 'head of household' to authorize borrowing contracts and savings deposits.

(4) Reform land tenure laws to strengthen women's title to land, which often serves as a source of collateral.

(5) In legislating the use of movable property as collateral (see Section 7.4 above), include jewelry and other household items that are likely to be owned by women. In many countries, the greatest possibilities for expanding the coverage of financial institutions lies in the potential of women, especially rural women, to become clients. Experience suggests that a financial institution's performance indicators are likely to improve to the extent that women represent a greater proportion of its clientele.

7.6 APPROACHES TO MANAGING RURAL FINANCIAL INSTITUTIONS[124]

7.6.1 Autonomy

For government financial intermediaries, management considerations begin with the essential requirement of autonomy. Government interference in the operations of these intermediaries, for political reasons and in the name of advancing social objectives, has caused huge difficulties in

---

122. *Ibid.*
123. Several of these options are adapted from FAO, 2001.
124. This section provides a brief review of management issues. For the reader interested in more detail on setting up and managing microfinance institutions in particular, very useful references are B. Klein, *et al.* (1999), J. Ledgerwood (1999) and Robert Peck Christen, *Banking Services for the Poor: Managing for Financial Success*, Accion International, Somerset, MA, USA, and Washington, DC, USA, February 1997.

their operations and has made them unsustainable. Yaron *et al.* state the case clearly:

> The financial performance of virtually all government-owned rural financial institutions has usually been extremely poor. Most rural financial institutions have remained highly subsidy-dependent. In India arrears as a proportion of amounts due and overdue hover at around 50 percent in most states. The recovery rate of Mexico's BANRURAL was around 25 percent in the late 1980s (ignoring recoveries from the loss-making national agricultural insurance company). Recoveries for the Small-Scale Agricultural Credit Agency in Malawi plummeted from almost 90 percent during the most recent elections; the agency was subsequently declared insolvent. Inflation eroded the real value of the equity of government-owed rural financial institutions throughout Latin America during the 1980s because of poor loan collection and agricultural on-lending rates that failed to keep pace with inflation.
>
> The economic cost of this dismal performance has been enormous and has often put macroeconomic stability at risk. For example, agricultural credit subsidies totaled 2.2 percent of Brazil's GDP in 1980, and 1.7 percent of Mexico's GDP in 1986. . . . In several cases, the subsidies could not even be measured because of poor accounting practices. . . .
>
> *The reason for the poor performance is evident: the interventions invariably have been, and generally still are, characterized by a lack of managerial autonomy for the rural financial institutions and by poor operating procedures.*[125]

As a result of these problems, most government-owned rural financial institutions are playing a much smaller role in the sector than they were in the 1970s. There are a few exceptions. They are those government institutions that have attained the greatest degree of operational autonomy, such as the Bada Kredit Kecamatan (BKK) and the Bank Rakyat Indonesia (BRI) in Indonesia and the Bank for Agriculture and Agricultural Co-operatives (BAAC) in Thailand.

7.6.2 Interest Rates and Lending Policies

Interest rate policies have been discussed throughout this chapter. Suffice it to say here that the management of rural financial institutions should follow a flexible, market-oriented interest rate policy and not attempt to subsidize rates for borrowers. Excessively high rates bring with them the danger of adverse selection of clients, as mentioned previously, but lending rates should be high enough to cover deposit rates plus intermediation margins (including profits), and modest provisioning in the form of a risk reserve, especially at the beginning of operations. Establishing the most appropriate set of rates may take time, depending on the results of pilot efforts in savings mobilization. After the introduction of savings deposits in an institution:

> interest rates on loans may have to be changed to ensure that the spread between interest rates for loans and deposits is sufficient to cover all costs and return a profit. . . . adjusting the interest rates requires some experimentation.[126]

However:

> Microenterprise programs can charge much more than formal financial institutions and still underprice informal-sector alternatives. Moreover, studies have shown that microenterprise borrowers are far more sensitive to the availability and convenience of credit than to the

125. J. Yaron, M. P. Benjamin and G. L. Piprek, 1997, pp. 25–26 [emphasis added].
126. Joyita Mukherjee, 'Introducing Savings in Microcredit Institutions: When and How?', *Focus*, Note No. 8, The Consultative Group to Assist the Poorest, Washington, DC, USA, April 1997, pp. 2–3. [This note is a synopsis of Marguerite S. Robinson, 'Introducing Savings Mobilization in Microfinance Programs: When and How?', Harvard Institute for International Development, Cambridge, MA, USA, 1995 or 1996.]

interest rate (Christen, 1989[127]). The nonfinancial transaction costs borrowers normally face dwarf interest costs.[128]

Innovative lending policies have permitted the authorization of unsecured loans. They include *criteria for borrower eligibility*, *incentives for repayment* and *techniques of monitoring borrower behavior*. These new policies represent responses to the three basic problems faced by lenders in rural credit markets, as described in Section 7.1 above: *screening* potential borrowers, providing *incentives* for compliance with the conditions of the loans, and *enforcing* repayment obligations.

Group lending is at the heart of the new lending techniques. In part, this approach is designed to overcome the common problem of asymmetric information between lenders and borrowers, for screening potential clients, but it also deals with the problems of incentives and enforcement. 'If clients have better information about each other's investments than the lender has about them, and the clients can engage in cooperative behavior, [then] interlinked contracts based on mutual guarantees can result in better loan terms for the clients with no reduction in expected income for the lenders. This is so because the co-guarantees can lead to higher effort levels and lower default rates on loans'.[129]

The Grameen Bank of Bangladesh pioneered a scheme in which borrowers are asked to form groups of five persons, each of whom will guarantee the loans of all of the others in the group. In the first stage, loans are made only to two of the five, and extending loans to others in the group depends on the performance of the first two borrowers. A similar approach was developed even earlier in Mexico for small-scale rural credit programs, under the name of *grupos solidarios.* An extreme case of the approach of mutual liability is found in Albania, where 'if a borrower fails to repay a loan, the line of credit to the borrower's entire village may be suspended'.[130]

Among other distinctive features, these group approaches may be considered to substitute *social collateral*, or *peer monitoring*, for more tangible forms of collateral. There are two alternative ways to implement this concept:

The two most common means of providing group accountability are (a) joint and several liability and (b) limited liability. Joint and several liability encourages extremely careful selection of members because any member can be held liable for the defaults of others. It may, however, deter the comparatively wealthy from joining the group, since they have more to lose. In rural Zimbabwe, schemes based on joint and several liability worked well in times of average production but fared worse than other schemes in the same area in times of drought and low production. The threat of default led farmers to withhold repayment and hope for a general amnesty, since they would be, in any event, accountable for other members' debts.

Group lending schemes based on limited liability are more common. In Malawi and Nepal borrowers are required to put part of their loans in a fund that would be forfeited if any member defaulted. If all members repay their loans, these deposits are returned. This practice has resulted in a good record of repayment. In Malawi, where 10 percent of loans was held as security, 97 percent of seasonal credit disbursed between 1969 and 1985 was recovered. In Nepal's Small Farmer Development Program, which required security deposits of 5 percent, the repayment rate in 1984 was 88 percent. These rates compare favorably with other small-borrower credit programs.[131]

---

127. Robert Peck Christen, 'What Microenterprise Credit Programs Can Learn from Moneylenders', Discussion Paper Series No. 4, Accion International, Cambridge, MA, USA, 1989.
128. E. Rhyne and M. Otero, 1994, p. 20.
129. J. Yaron, M. P. Benjamin and G. L. Piprek, 1997, p. 78.
130. *Op. cit.*, p. 72.
131. The World Bank, 1989, pp. 116–117.

The Grameen Bank's approach to group lending is somewhat different:

When should groups be used?

— *When communities have strong group cohesiveness.*
— *When . . . start-up transaction costs are high.*
— *When individuals can obtain information about each other at a lower cost than the bank can.*
— *When individuals do not have collateral.*

Guidelines for more effective use of groups:

— *Focus on small, homogeneous groups that will assume some responsibility for supervising their funds.*
— *Impose penalties (such as prohibiting access to further loans while an individual is in default) and provide incentives (such as promising a higher future loan amount to those who repay on time).*
— *Institute sequential lending (to allow groups to screen out bad risks).*

Risks associated with using groups:

— *Risk of poor records and lack of contract enforcement.*
— *Risk of corruption and control by a powerful nucleus or leader within the group.*
— *Risk of covariance because of similar production activities.*
— *Risk of spending excessive time and money to form viable groups.*
— *Risk that departure of a group leader may jeopardize group viability.*
— *Risk of potential free-riders. (This risk can be reduced if the group can impose penalties on individuals.)*

(from J. Yaron, M. P. Benjamin and G. L. Piprek, 1997, p. 108).

The bank's customers, who are restricted to the very poor, are organized into five-person groups, and each group member must establish a regular pattern of weekly saving before seeking a loan. The first two borrowers in a group must make several regular weekly payments on their loans before other group members can borrow. . . . the Grameen Bank has experienced excellent loan recovery. As of February 1987 about 97 percent of loans had been recovered within one year after disbursement and almost 99 percent within two years.[132]

A cautionary note is in order about the degree of local participation in the way in which the schemes are organized:

Groups often have been created at the initiative of governments or private development agencies. This top-down approach means that a scheme can be extended rapidly, but it may undercut the force of local sanction (*ibid.*).

The Juhudi Credit Scheme in Kenya organizes clients into associations of thirty members known as KIWAs. 'Savings begin before lending. After eight weeks of uninterrupted savings by each KIWA, eighteen members qualify for loans. The remaining twelve members in the KIWA qualify for loans after the first eighteen repay at least four installments without fail, and all KIWA members continue contributing toward the group savings fund without interruption. Once members repay current loans, they qualify for repeat loans, provided the other KIWA members continue to meet their obligations'.[133]

7.6.3 Other Incentives for Repayment

A related lending policy, also practiced by the Grameen Bank as well as other institutions, is stepped lending: starting borrowers with very small loans and gradually increasing the amounts as a function of the borrower's repayment record.

---

132. *Op. cit.*, p. 117.
133. A. K. Mutua, 1994, p. 272.

In the KUPEDES program of the BRI Unit Desa system, future loan eligibility is precisely established according to repayment performance on the current loan. If all payments are made on time, then the borrower is eligible for a 100% increase in the loan amount. If the final payment is on time but one or two installments are late, then the allowable increase in loan amount is 50%. If the final payment is on time but three or more installments are late, then no increase in the amount is permitted, although a new loan may be taken out. If the final payment is late but paid within two months of the due date, then any new loan must be 50% smaller than the current one. If the final payment is more than two months late, then no new loan is allowed.[134]

Another approach is to give the borrower who repays the entire loan early a rebate on part of the accumulated interest paid.

Additional techniques to enhance loan recovery rates include the following approaches, which are especially suited to the situation of poor borrowers:

(1) Requiring frequent payments on loan balances, often weekly. This helps maintain the loan repayment among the client's priorities. [However, this is an obvious example of a technique that is not applicable for the majority of agricultural loans.]

(2) Monitor frequently the situation of the client, for example, with weekly visits to his or her farm or place of work.

(3) Immediate visits to the borrower in cases of delays in payments. This requires an information system which gives updated information to the day, as is the case with BancoSol in Bolivia.

(4) Locating lenders 'close to the borrowers' places of work'.[135]

An alternative establishing an office close to the borrowers is to employ mobile banking:

The National Agricultural Bank of Morocco doubled its banking network by opening seasonal banking windows in existing local offices of the Ministry of Agriculture. Several rural financial intermediaries have employed agents who regularly visit villages by motorcycle or on foot to provide financial services.[136]

A useful role that technical assistance for rural financial institutions can play is to make them aware of these optional approaches and also to assist them to develop adequate *management information systems* so that problems in the loan portfolio and in lending policies are detected on a timely basis.

7.6.4 Techniques of Savings Mobilization

In development policies as practiced, emphasis on the mobilization of rural savings deposits is a fairly recent occurrence, dating basically from the 1980s. As commented upon earlier, from the viewpoint of rural households sometimes access to reliable savings facilities can be more urgent than access to loans:

Often the need for safe, liquid, and remunerative savings facilities takes precedence over the need for credit, because saving improves clients' ability to smooth consumption with their own resources, and allows them to avoid having to carry the burden of debt repayments during income downswings.[137]

Marguerite Robinson has cited the experience of an Indonesian villager with regard to savings:

---

134. J. J. Boomgard and K. J. Angell, 1994, p. 214.
135. Mohini Malhotra and James Fox, 'Maximizing the Outreach of Microenterprise Finance: The Emerging Lessons of Successful Programs', *Focus*, Note No. 2, The Consultative Group to Assist the Poorest, Washington DC, USA, October 1995, p. 2.
136. J. Yaron, M. P. Benjamin and G. L. Piprek, 1997, p. 78.
137. J. Yaron, M. P. Benjamin and G. L. Piprek, 1997, p. 77.

I used to save in goats, but goats take a lot of work. Now the shepherds are all in school and the parents have jobs. Now we have no time to save in goats. We prefer to save in the bank.[138]

Nevertheless, as noted earlier, most externally funded, small-scale rural financial institutions commence their operations with lending only. If they are to achieve sustainability, eventually they have to confront the issue of making a transition to a savings institution as well. The rationale for giving priority to savings deposits has been summarized in Joyita Mukherjee's synopsis of work by Marguerite Robinson:

Locally mobilized voluntary savings is potentially the largest and the most immediately available source of finance for some microcredit institutions. Another important reason for undertaking the institutional mobilization of voluntary savings is the vast unmet demand for institutional savings services at the local levels of developing countries.[139]

However, when participation in savings schemes goes beyond a very small local circle, then the issue of the potential need for bank supervision has to be weighed, although as mentioned above there are arguments against introducing supervision requirements too quickly:

For the protection of their clients, especially depositors, financial institutions that mobilize voluntary savings should come under *government supervision*. This, of course, requires a government that is willing to modify its banking supervision so that the rules for microcredit institutions are appropriate for their activities, and to ensure that the supervisory body is able to monitor these institutions effectively. . . . Before mobilizing savings, *a microcredit institution should have demonstrated consistently good management of its own funds*. In other words, it should be financially solvent

with a high rate of loan recovery, earning attractive returns. This established record is important because *in many countries low-income people who have entrusted their savings to small unsupervised financial institutions have lost their lifetime savings.*[140]

Robinson and Mukherjee underscore the importance for savings mobilization of sound macroeconomic policy and an appropriate legal and regulatory environment, including sufficient capacity to supervise deposit-taking institutions. In addition, they draw together several important lessons about how the introduction of savings deposits into lending institutions changes the nature and operations of those institutions. Those lessons are best presented in their words:

Adding voluntary savings to a microcredit program will fundamentally change the program

The institution should be prepared for these changes and should not believe that adding savings is like adding 'just another product'. In countries with large, unmet demand for savings services, microfinance institutions that offer loans and well-designed savings instruments have *many more deposit accounts than loans*. At BRI's local banking system, there are about six times as many deposit accounts as loans. At Bank Dagang Bali, the ratio of savings accounts to loan accounts is over 30 to 1. This pattern occurs primarily because most microfinance clients want to save all the time, while most want to borrow only some of the time.

The introduction of voluntary savings services thus implies the addition of many new customers – which in turn means *increases in staff, management, offices, systems, communications, staff training, security*. . . . interest rates on loans may have to be changed to ensure that the spread between interest rates for loans and deposits is sufficient to cover all costs and to return a profit.

138. M. S. Robinson, 1994, pp. 31–32.
139. J. Mukherjee, 1997, p. 1.
140. *Ibid.* [emphasis added].

Compulsory savings and voluntary savings are incompatible

The requirement for compulsory savings [to qualify for loans] and the mobilization of voluntary savings reflect two very different philosophies. The former assumes that the poor must be taught to save, and that they need to learn financial discipline. The latter assumes that the poor already save, and that what is required are institutions and services appropriate to their needs. Microfinance clients may not feel comfortable putting voluntary savings . . . in compulsory savings accounts, or even in other accounts with the same institution. They know that they cannot withdraw the compulsory savings until their loan is repaid . . . and they fear that they may also not have easy access, *de facto*, to their voluntary savings.
. . . . The lesson learned in mobilizing savings is to train the staff, not the clients!

Products should be designed and priced together

An institution aiming at full self-sufficiency must . . . set a spread between loan and deposit rates that enables institutional profitability. . . . adjusting the interest rates requires some experimentation. . . . a savings instrument that features quick and easy access (liquidity) and is in high demand can be labor-intensive to manage. It is, therefore, costly to a microfinance institution, especially if there are a large number of very small accounts. . . . Labor and other non-financial costs must be considered when setting interest rates on deposits. These costs are difficult to determine in advance, so pilot tests are needed to estimate costs accurately.
The introduction of voluntary savings will also require *some changes on the lending side*. For example, if a microfinance institution offers only group loans, it should consider introducing individual loans to its portfolio.
. . . . *limits on loan sizes will have to be increased when deposit services are added*. . . . Larger savers tend to qualify for and want larger loans. Borrowers who are forced out by the institution's loan limits before they can qualify for loans from commercial banks find

themselves in difficulty. In contrast, micro-finance institutions that help long-term borrowers to obtain larger loans and recommend them to other banks when they qualify will continue to have the client's goodwill, and at least some of their savings. Bank Dagang Bali retains its good borrowers, offering them increasingly larger loans as their enterprises grow. Eventually some of these borrowers find better loan terms elsewhere, but usually they remain savers in Bank Dagang Bali.

Deposit instruments should be appropriate for local demand

. . . . savings services [must] meet local demands for security, convenience of location, and a choice of instruments with different mixes of liquidity and returns. In the case of BRI, the same banking system that mobilized US$17 million in its first ten years of operation (1973–1983), mobilized US$3 billion from 1984 to 1996. . . . Deposit instruments were designed specifically to meet different types of local demand.

There is a substantial need to develop human resources

Managing a financial intermediary is more complex than managing a credit operation, especially since the size of the organization tends to increase rapidly. Training for staff and management becomes an urgent need. For commercial banks venturing into the micro-finance market, the tendency to down-load commercial and corporate banking instruments, spreads, training, and attitudes must be assiduously avoided. Bank staff must also learn to treat poor clients with respect, a lesson that comes hard to some bankers.

New marketing strategies will have to be developed

BRI's most liquid savings account (called SIMPEDES) featuring both interest and lotteries was an instant success because extensive research had been done on what features customers wanted in a liquid instrument and why they wanted these. Moreover, BRI conducted market research to determine what kind of lottery prizes were popular, what kind of bank book was wanted, and what kinds of publicity were effective. The results were excellent. . . .

Careful attention must be paid to sequencing

The following steps are relevant to many micro-credit institutions that are planning to introduce voluntary savings:

(1) Enhance the knowledge of the institution's board and managers on the experience of other microfinance institutions with regard to voluntary savings mobilization.
(2) Carry out market research and train staff selected for the pilot phase.
(3) Conduct and evaluate a pilot project (a crucial step because, until the extent of the demand and costs of different products, including labor, are known, only temporary interest rates can be set).
(4) Where necessary, second pilots should be carried out and evaluated. . . .

. . . . instituting a voluntary savings program is a prime illustration of 'haste makes waste'. *A microfinance institution that does it the wrong way will lose the trust of its clients, and eventually its own viability.*[141]

7.6.5 Prudential Management in Rural Financial Institutions

Prudential management includes both safeguarding against risk in the loan portfolio and keeping intermediation costs under control to ensure the institution's viability. The relatively high covariant risk associated with agricultural lending has to be recognized and addressed in order to assure the long-term sustainability of a financial institu-tion. Of course, careful screening of borrowers and close monitoring of them help to reduce the risk. However, additional measures are needed.

A direct response to the risk issue is simply to require higher proportions of *capitalization* for rural financial institutions – though the absolute amounts may be small when the institutions are small.

. . . most countries have relatively brief experi-ence with microfinance: in the absence of decades of empirical data about microfinance institutions' performance, regulators may wish to begin cautiously in fixing leverage ratios [for capital]. Second, because microfinance institutions operate with relatively high costs and lending rates, a given percentage of non-performing portfolio will decapitalize a microfinance institution faster than it would a commercial bank. Taking all these factors into account, several analysts . . . suggested an initial capital/asset ratio of about 20 percent for microfinance institutions, subject to down-ward adjustment as the institution and the industry gain experience.[142]

Because of the covariant risk associated with agricultural production and incomes in a given area, a higher capital/asset ratio than the Basel Convention's norm of 8% also can be suggested for banks that cater to medium- and large-size farms, although it may not have to go as high as 20%, owing to the wider use of tangible collateral and lower unit lending costs. However, policies like these on capital requirements have not been imple-mented in very many instances to date.

Another tool of prudential management is *liq-uidity*. For the above-mentioned reasons, institu-tions dedicated to microfinance and agricultural lending can maintain higher than usual levels of internal liquidity or ensure that ready access to additional liquidity exists, through a central fund

141. *Op. cit.*, pp. 2–5 [emphasis added]. Similar conclusions are drawn from four case studies of rural financial institutions in Asia and Latin America in the paper by Joyita Mukherjee and Sylvia Wisniwski, 'Savings Mobilization Strategies: Lessons from Four Experiences', *Focus*, Note No. 13, The Consultative Group to Assist the Poorest, Washington, DC, USA, 1998, pp. 2, 4–6.
142. R. Rock, M. Otero and R. Rosenberg, 1996, p. 2.

(in the case of small-scale financial institutions) or access to the interbank market (for commercial banks):

> For prudential liquidity management, each of the [four micro-finance institutions analyzed][143] has established an internal liquidity pool or is linked to the liquidity pool of a partner organization. . . . The internal liquidity transfer price is set high enough to encourage savings mobilization. Empirical evidence from the Bank Rakyat Indonesia and the Banco Caja Social shows that an internal liquidity price close to the interbank rate is conducive for savings mobilization.[144]

Again, the same considerations would apply, to a lesser extent, to commercial banks that are principally oriented toward lending to agriculture. *An institution's own auditing procedures* can contribute to prudential management. 'In the absence of effective supervision and the lack of a reliable deposit insurance system, internal auditing often takes on a more important role than external supervisors'.[145]

It should not be overlooked that one of the most basic safeguards a rural financial institution can institute is *portfolio diversification*. As mentioned previously, such a strategy has been carried out successfully by Bancafé in Honduras, although it still lends a higher proportion of its portfolio to agriculture than other commercial banks in the country do. The same may be said of Banco Ganadero, which has lent relatively more to medium- and large-scale ranchers and farmers than Bancafé does. Generally, it is not recommended that the proportion of a portfolio devoted to agricultural production exceed 40 or 50% of total assets. However, those levels are significantly greater than the agricultural proportion of most commercial banks' portfolios, which usually are in the 10 to 20% range at most.

On the cost side of financial intermediation, control of salaries is the most important consideration for microfinance institutions, owing to their high unit costs of lending. In the study of 11 microfinance institutions by Christen *et al.* (summarized in the *Focus* note by Malhotra and Fox, 1995) it is pointed out that 'programs paying lower salaries were more profitable than those that paid more [and they] used local personnel to staff their operations, which gave them a distinct cost advantage' (p. 3).

An effective way to control costs is demonstrated by the *unit desa* operations of the Bank Rakyat Indonesia. After sweeping reforms in 1983:

> Each unit became a separate profit center. . . . The income statement of the *unit desa* recorded transactions such as interest paid on excess funds borrowed from the district-level branch offices to meet liquidity needs, and interest received on any excess liquidity maintained as deposits with the branch office.
>
> . . . Direct responsibility for loan approvals and repayments rested with *unit desa* staff, particularly loan officers. . . . an incentive bonus that distributed 10 percent of a *unit desa*'s annual profits among its staff was introduced. . . . Internal supervision and audit capacities were also strengthened. The number of internal supervisors/auditors was increased from one per six *unit desas* to one per four. A standard audit manual provided simple and clear guidelines on supervision rules. Most importantly perhaps, supervisors, auditors and *unit desa* managers and staff underwent a periodic training program, over three years, on reporting and supervision techniques. Greater supervision led to early detection of problems and early remedy.[146]

---

143. The Bank for Agriculture and Agricultural Cooperatives (BAAC), Thailand, the Banco Caja Social, Colombia, the Bank Rakyat, Indonesia, and the Rural Bank of Panabo, Philippines.
144. J. Mukherjee and S. Wisniwski, 1998, pp. 5–6.
145. *Op. cit.*, p. 6.
146. J. Mukherjee, 'State-owned Development Banks in Micro-Finance', *Focus*, Note No. 10, The Consultative Group to Assist the Poorest, Washington, DC, USA, August 1997, p. 4.

Fiebig (2001) has summarized basic prudential management norms in terms of a series of basic approaches for effective internal control of a rural financial institution, elaborating on guidelines laid down by the Basle Committee[147] [emphasis added]:

Substantial institutional changes that led to BRI Unit Desa's success:

(1) Major reorganization of BRI management at all levels from head office to the unit banks.

(2) High priority accorded at the head office to the management of the unit banking system.

(3) Extensive reorganization and training of staff throughout the country.

(4) Establishment of a system of promotion and development of promotion criteria that reflect new expectations for performance.

(5) Fundamental revision of bookkeeping, audit and supervision systems, which permitted the establishment of the unit banks as independent profit centers (rather than branch windows) and made accountability and a sustained anticorruption drive possible.

(6) Opening of new unit banks and relocation of others to areas with high demand.

(7) Attention to learning about rural financial markets and emphasis on using this information to avoid potential problems with moral hazard and adverse selection.

(8) Crucial improvements in communications and computerization facilities.

(9) Overhaul of BRI's public relations.

(10) Implementation of an effective unit bank staff incentive system rewarding good performance.

(Source: M. S. Robinson (1994), cited in M. Fiebig (1999).)

Management oversight and control culture: Board of Directors should decide on and monitor overall business strategies, organizational structure, policies and major risks run by the bank. . . . boards should recognize [risks] . . . define acceptable risk levels as well as risk management policies. . . . Senior management, in turn, has the main responsibility for implementing strategies and policies regarding agricultural lending. . . . It also should develop a valid control process as well as maintain an organizational structure with clear responsibilities, authority and reporting requirements between . . . levels. *In rural intermediaries, the delegation of responsibilities together with the setting of internal off-site and on-site control policies is crucial.*

Reform of the internal regulation of development banks has been a particularly challenging part of development bank reform. But also in other institutional types, such as for example NGOs, the change of attitude towards enhanced internal control is often problematic. . . . While small NGO-type institutions may well survive and prosper without explicit focus on internal regulatory issues, a medium-sized NGO can be severely struck by fraud, lack of management oversight and weak Board control.

Risk recognition and assessment: . . . . Management information systems in financial institutions involved in agricultural lending should provide the applicable data to manage the client-specific and external risks of the agricultural sector. Recognition of risks and their assessment needs to involve branch and credit officer levels and cannot stop at headquarter/aggregate data levels.

Control activities and segregation of duties: Systems of checks and balances between different organizational layers form basic control activities. . . . Internal auditors need to be operationally independent to carry out their assigned tasks in a prudent manner. . . .

---

147. Basle Committee on Banking Supervision, 'Framework for Internal Control Systems in Banking Organizations', Basle, Switzerland, 1998.

Information and communication: . . . . Information generation and communication of information obtained is a severe problem in rural contexts. . . . In decentralized institutional structures, management information systems and internal control mechanisms need to be designed taking this into account.

Monitoring activities and correcting deficiencies: . . . . Management information systems that collect all risk relevant data are of little use as long as they are not used as part of proactive management, which should range from fraud detection to the active management of portfolio diversification. . . .

Incentive schemes: Positive control incentives should complement the control mechanisms in the form of staff motivational and remuneration incentive systems. Incentive systems for staff are crucial for the well-being of a financial institution. . . . The behavior of State-owned banks is particularly difficult to change as compensation is very rarely based on performance.

7.7 MACROECONOMIC POLICY TO SUPPORT RURAL FINANCIAL INTERMEDIATION

7.7.1 The Problem of Directed and Subsidized Credit

In setting the context of monetary and banking policy, it is worth recapitulating and extending the arguments concerning directed [selective] credit and subsidized interest rates. Fry has made the following observations:

Selective credit policies invariably produce the opposite results to those intended. That governments continue to pursue them is explained largely by the political pressures exerted by vested interests created by these policies in the first place. The Asian developing countries pursuing selective credit policies most vigorously – Bangladesh, India and Nepal – tend to be those recording the lowest rates of economic growth. Selective credit policies . . . have simply failed to improve domestic resource mobilization and allocation.[148]

Selective credit policies go hand in hand with interest rate policies. The defects of low interest rate policies are compounded by selective credit policies that differentiate and subsidize . . . particular categories of borrowers. The most obvious disadvantage of selective credit policies is that capital rather than labor is subsidized. . . . *Abandoning directed credit programs* must constitute one of the first steps of any sensible financial development program. *If the government is too weak to take this step, it may well be too weak to implement any well-conceived macroeconomic policy.*[149]

Even the more successful rural financial intermediaries in the developing world have had difficulties with directed credit programs:

BAAC (of Thailand) and BRI (of Indonesia) continue to implement mandated government credit programs with poor results. The losses from these operations are either covered by government funds in the case of BAAC or absorbed by the profits generated by the *unit desas* in BRI. These transfers represent a loss of resources and inhibit profitability in both cases.[150]

These lessons are very basic ones. Nevertheless, an argument for some occasional and transitory exceptions is presented above (under 'rediscount lines') and later in this section.

7.7.2 Managed Deposit Rates

Fry makes an argument for government intervention with regard to interest rates, but not in the

---

148. M. J. Fry, 1995, p. 448.
149. *Op. cit.*, p. 469 [emphasis added].
150. J. Mukherjee and S. Wisniwski, 1998, p. 4.

direction of subsidizing them. His argument is that the banking sector in most developing countries is not fully competitive and therefore an approximation of a competitive outcome for deposit rates may have to be imposed by policy:

Holding interest rates low by administrative fiat to encourage investment does not work because it discourages financial saving, the source of investible funds. With a **cartelized banking industry**, abolishing interest rate ceilings is unlikely to achieve the optimum result either. Indeed, it may not even produce higher real institutional interest rates. In the light of the recent experiences of Argentina, Chile, Sri Lanka, Turkey and Uruguay with excessively high real loan rates of interest, continued government intervention in the determination of institutional interest rates may well be the best interim policy until price stability is achieved and bank supervision and competition within the financial sector are adequate. . . . an administered interest rate policy should select an appropriate objective and not be hampered by pursuit of several incompatible aims.

Possibly the most appropriate objectives of interest rate policy in developing countries are the efficient mobilization and allocation of domestic resources. Efficiency of both the mobilization and allocation of saving is maximized when institutional interest rates are set at their free-market equilibrium levels. . . . *Since most financial systems in developing countries are oligopolistic, the competitive solution may well have to be imposed.* Provided adequate bank supervision is in place, there is probably little point in attempting to set loan rates of interest; they can so easily be evaded through [obligatory] compensating deposits. Instead, the monetary authorities might set deposit rates at levels that approximate the competitive free-market rates. Specifically, banks might be required to offer indexed six- or 12-month deposits with a modest (perhaps 3 percent) real return.

One way of stimulating rather than simulating a competitive interest rate solution is for the government to issue treasury bills and bonds with attractive yields. Typically, government securities in developing countries are held only by financial institutions as part of their required liquidity ratios. . . . Yields on government securities are so low that voluntary holdings are nonexistent. The appropriate interest rate structure, however, applies equally to the government and the private sector. *Indeed, if the government is unwilling to compete in this way, financial development is doomed.*[151]

South Korea in the late 1960s pioneered this approach to deposit interest rates, in fact carrying it so far that inverse margins (deposit rates above loan rates) were purposely created, and the Government compensated the banks for their consequent losses. The aim was to induce households to become accustomed to depositing their savings in banks, where they would be accessible to industry, instead of in informal financial structures or savings in kind.

7.7.3 Inflation and Interest Rates

The injunctions presented throughout this chapter against ceilings or subsidies on loan interest rates are widely applicable in economies with low or moderate inflation. However, under conditions of high inflation, and when drastic stabilization programs are being implemented, other economists have argued persuasively that releasing controls on interest rates should follow, not precede, the strengthening of bank supervision capacity and the achievement of relative price stability. The principal reason is that high interest rates during high and volatile inflation can give rise, to an unusual degree, to the problem of **adverse selection**, in which some of the less risky borrowers refuse to take out loans at high interest rates and lenders are left with a higher proportion of the more risky borrowers. If inflation is being brought down sharply, then a significant

---

151. M. J. Fry, 1995, pp. 467–468 [emphasis added].

share of these latter borrowers may end up defaulting, thus endangering the banking system.[152]

It also is the case that most farmers cannot afford to pay positive real interest rates when inflation is persistently of the order of 40–60% per year or more, as has been the case in Turkey for an extended period of time. The reason is that maintaining a policy of positive real interest rates for lending under high inflation causes total input costs, including interest, to rise geometrically, while output prices rise at a slower rate. Because a finance charge is attached to inputs, effectively the inflation on input prices is charged twice to the producer. An oversimplified example of the phenomenon may be stated as follows: (a) if inflation is 50% per annum and real interest rates are maintained at 12%, then the nominal interest rate will be 62%; (b) if, for simplicity, all inputs are financed for a year, then input costs rise by 50 + 62 = 112% each year, while output prices rise by only 50%.[153] In addition to this problem, João Sayad has raised the concern that input prices may increase more rapidly than some output prices during high inflation, making it even riskier to borrow to finance production if real interest rates are positive.[154]

These caveats about interest rates during high inflation and while stabilization programs are in progress basically refer to the timing of interest rate reforms. Under some circumstances, a deregulation of interest rates may have to be postponed until inflation is reduced. However, this consideration does not undermine the basic contention that over the longer run loan rates controlled at an artificially low level are inimical to the development of sustainable rural financial systems.

The bottom line is that 'legal limits on loan interest rates, if enforced, will usually make commercially viable microfinance impossible',[155] and microfinance represents an increasingly large share of agricultural finance.

Much of the commentary on interest rates from analysts of agriculture and the rural sector tends to concentrate on *lending* rates. In a period of transition out of high inflation rates, policy directed toward *deposit* rates can be more effective, as Fry has indicated. Fry's analysis also suggests that, in cases in which stabilization programs are being implemented in order to bring down inflation rates, a policy of reducing the rate of growth of the money supply combined with increases in deposit rates can also stimulate real growth. 'The simulations of my model suggest that stabilization policies raising the time deposit rate of interest are superior to policies relying solely on control over the nominal money supply. When the deposit rate is fixed below its equilibrium level, higher deposit rates can *raise* the rate of economic growth by increasing credit availability in real terms while slower monetary growth lowers the inflation rate'.[156]

7.7.4 The Role of Donor and Government Finance

Where do the considerations presented in this chapter leave donors who wish to make cash contributions to the development of rural finance? While there is clearly a role for funding technical assistance and advocacy efforts in developing rural financial intermediaries, do the lessons of experience suggest that no external funding should be contributed to them? Not necessarily:

152. See, for example, D. Villanueva and A. Mirakhor, 'Strategies for financial reforms: interest rate policies, stabilization, and bank supervision in developing countries', *IMF Staff Papers*, **37**(3), September 1990, pp. 509–536.
153. A more detailed illustration of this issue is found in the appendix to Roger D. Norton, *Agricultural Issues in Structural Adjustment Programs*, FAO Economic and Social Development Paper No. 66, Food and Agriculture Organization of the United Nations, Rome, 1987.
154. J. Sayad, 'Rural Credit and Positive Real Rates of Interest: Brazil's Experience with Rapid Inflation', in D. W. Adams, D. H. Graham and J. D. Von Pischke (Eds), 1984, pp. 146–160.
155. R. Rock, M. Otero and R. Rosenberg, 1996, p. 1.
156. M. J. Fry, 1995, p. 227.

Not even an extremely far-reaching and successful financial sector reform would eliminate the need for specific efforts to initiate and support target-group-oriented financial institutions. There does not seem to be a "market mechanism" which would induce established financial institutions to start providing financial services to the lower-income target groups as soon as a financial sector reform has taken place. . . . In the medium to long term [banks] will have other strategic options . . . and experience indicates that they prefer these other options and are reluctant to serve the poorer segments of the population if they are not given specific incentives and technical assistance. . . .[157]

There are four valid roles for external funding: (a) contributions to local intermediaries' pool of loanable funds, limited in relation to the amount of deposits mobilized, (b) start-up funding, especially if an institution begins life as a lender only and later acquires a capacity to manage savings deposits, (c) selected special incentives to offset higher costs of providing financial services to target groups, as in the case of subsidizing commercial bank branches in rural areas, and (d) filling the gap in long-term funding.

The first role can take the form of contributions to the fund of an apex organization in a system of 'mini-banks'. However, in such cases it is important to limit the amount of on-lending to the affiliates and to tie it to their own deposit mobilization. Experience has shown that a strict ceiling of this nature is important to maintain the integrity of the loan repayment response and the loan collection effort. Otherwise, the external support often generates the perception that the donor funding does not have to be repaid, and in turn that weakens the financial discipline of both borrowers and financial institutions.

Nancy Barry has made a case for limited financial support, specifically during a start-up period:

If financial intermediaries are to move to the significant volumes in microfinancing that are needed to make these programs sustainable, early operations will need to be subsidized. NGOs and other specialized financial intermediaries, which are not in the position to cross-subsidize microenterprise lending while they build lending volumes, will need some form of institutional subsidy for a period of five to seven years. Specialized institutions that meet performance standards need capitalization and low-cost, long-term funds, preferably repayable in local currency, while they expand their volumes to sustainable levels.[158]

The reasoning behind this argument is that unit administrative costs will be very high in the early stages of operation of a microfinance institution, owing to lack of economies of scale and the cost of learning, and that without a source of subsidy it would have to charge interest rates so high that they might discourage potential clients (or foster an adverse selection of clients). By the same token, it would be important to tie such funding to fulfillment of annual performance targets, and make the phase-out of the funding unmistakably clear from the outset.

Seed capital of this nature can perhaps be most effective when granted to a second-tier institution, or central fund, that oversees and supports a network of credit co-operatives or mini-banks. In addition, as noted, a case may be made for providing subsidies to encourage the opening of branches of rural financial institutions in remote areas, including commercial banks, in order to offset the higher unit administrative costs associated with their smaller scale of operations.[159] In such cases, funding formulas have to be specified

---

157. J. Yaron, M. P. Benjamin and G. L. Piprek, 1997, p. 99, citing J. P. Krahnen and R. Schmidt, *Development Finance as Institution Building: A New Approach to Poverty-Oriented Banking*, Westview Press, Boulder, CO, USA, 1995, p. 8.
158. N. Barry, 1995, p. 3.
159. This suggestion has been made by Yaron, Benjamin and Piprek (1997, p. 39), and it has been prescribed in World Bank funding for commercial banks in rural areas of Nicaragua.

in advance to ensure that the assistance is tied to unavoidably higher administrative costs and that it does not, for example, compensate for lax loan recuperation. Another area in which external funding can play a crucial role is in training the managers and directors of rural financial institutions, including providing them with study tours to review successful experiences in other countries.

The argument for selective government support has been stated in the following ways:

> Government interventions in rural finance . . . should be based on the principle of removing the causes of market failure in a cost-effective way. Ultimately, they should facilitate the effective working of market forces. This may include grants or subsidies for information generation and institution and capacity building, providing seed capital for capital enhancement of new rural financial intermediaries, *providing rural financial intermediaries with access to refinancing facilities, in particular, to be used for term lending*. Subsidies and grants, however, should always be transparent and temporary and provide incentives to strengthen the role of private sector operators in rural financial markets.[160]

Plus:

> in some areas public intervention to provide long-term resources might still be justified. Firstly, during transition periods of establishing efficient capital markets or periods of political and economic instability public involvement is needed. This involvement might be *in the form of long-term credit lines*, lending quotas or additional capital injections to rural financial institutions. However, it should be emphasized that, as the experience with public short-term finance in the past has shown, the private and voluntary mobilization of appropriate funding sources must not be discouraged or undermined.[161]

The conceptual argument for government intervention on grounds of market imperfections has been articulated by Stiglitz in a nuanced manner:

> There *is* a role for the state in financial markets; it is a role motivated by pervasive market failures. In most of the rapidly growing economies of East Asia government has taken an active role in creating financial institutions, in regulating them, and in directing credit, both in ways that enhance the stability of the economy and the solvency of the financial institutions and in ways that enhance growth prospects. Although limitations on markets are greater in developing countries, so too, many would argue, are limitations on government. It is important to design government policies that are attentive to those limitations.[162]

The financial crisis in East Asia led to widespread questioning of government attempts to mold the development of financial markets. In response to assertions of the need to intervene in cases of market failure, the general point has been made that 'Direct measurement of the existence and extent of market failure is difficult and uncommon'.[163]

Given the poor track record of government interventions in financial markets, the thrust of policy at present has to lie on the side of encouraging private efforts at savings mobilization and innovative approaches to financial intermediation, aided by technical assistance and by relatively modest and temporary direct support.

It is clear that injections of cash into the rural financial system have to be limited in quantities and associated with very carefully designed

160. E. Coffey, 1998, p. 46 [emphasis added].
161. Thorsten Giehler, *Sources of Funds for Agricultural Lending*, Agricultural Finance Revisited No. 4, FAO and GTZ, Rome, December 1999, pp. 76–77 [emphasis added].
162. J. Stiglitz, 1994, p. 50 [emphasis in original].
163. J. Yaron, M. P. Benjamin and G. L. Piprek, 1997, p. 38.

programs for development of financial institutions. Nonetheless, public interventions on a larger scale can be directed toward ameliorating the conditions that cause rural financial markets to be inefficient:

Is there, then, any role for public policy? Greenwald and Stiglitz (1986)[164] have recently shown that markets with imperfect information give rise to externality-like effects, for which government intervention may be most successful. In the context of credit markets, one externality is the reduction in information costs brought about by development in other markets. Examples are land titling and commercialization in goods markets. More generally, *government expenditure on rural infrastructure that reduces farmers' risks will likely reduce the importance of information asymmetries, improve the level of competition, and therefore reduce the distortions in rural credit markets*.

Another type of externality may reside in institutions which facilitate the overcoming of informational problems in rural credit markets. One such institution is that of small-scale peer monitoring. . . . Individuals form a small group which is jointly liable for the debts of each member. The group thus has incentives to undertake the burden of selection, monitoring and enforcement that would otherwise fall on the lender. . . . There is, however, an externality in this institutional innovation. An individual who bears the initial cost of organizing such an institution is providing a form of social capital from which all members of the group will benefit. As is well known, when this type of externality arises there will be an undersupply

of the socially beneficial service, and there is therefore a role for the government to help organize and act as a catalyst in the formation of such institutions. . . . there are notable successes when the government has acted in this way.[165]

As noted, the Grameen Bank, of which the Bangladeshi Government owns 25%, has invested considerable effort in organizing these kinds of groups. External donors have played the organizing role in other circumstances as well.

The above-mentioned donor-supported rediscount lines in special cases constitute the exception to Fry's argument about rediscount lines. Reforestation and sustainable forest management, for example, are activities that rarely can find the needed long-term financing from domestic banks, even though they may be highly profitable and consistent with a country's comparative advantage.

Vogel and Adams have made a spirited argument against the idea of rediscount lines, correctly pointing out that the emphasis should be on the development of viable institutions for financial intermediation,[166] but in practice the gaps in availability of medium- and long-term financing are so critical that second-storey financial windows will continue to be used while financial markets develop. As Mark Wenner has recommended for the Inter-American Development Bank:

In order to improve the availability of long-term finance, the Bank should prepare operations that . . . provide temporary access to external funds through second-tier banks to compensate for the lack of funding sources. . . .[167]

164. B. Greenwald and J. Stiglitz, 'Externalities in economies with imperfect information and incomplete markets', *Quarterly Journal of Economics*, May 1986, pp. 229–264.
165. K. Hoff and J. Stiglitz, 1995, pp. 282–283.
166. Robert C. Vogel and Dale W. Adams, 'Old and New Paradigms in Development Finance: Should Directed Credit Be Resurrected?', CAER II Discussion Paper No. 2, Harvard Institute for International Development, Cambridge, MA, USA, April 1997.
167. Mark Wenner, *Rural Finance Strategy*, Sector Strategy and Policy Papers Series, Sustainable Development Department, Inter-American Development Bank, Washington, DC, USA, December 2001, p. 20.

It is clear that such rediscount lines should be temporary, but it is equally clear that there are externalities to be reaped by inducing otherwise conservative commercial banks in developing countries to explore new areas of lending. In many countries, an argument could be made along these lines for a rediscount line for reforestation projects or fruit trees, for a period of five to ten years, after which the banks should have acquired sufficient experience in that field. By the same token, rediscount lines cannot be justified for well-established activities that already have been receiving commercial lending, e.g. production of grains and beef fattening in many countries. Too often, rediscount lines are used for this kind of purpose.

7.8 ELEMENTS OF A STRATEGY FOR DEVELOPING RURAL FINANCE

7.8.1 The Financing Gap

Greater access to financial services of various kinds can be a crucial factor in catalyzing the efforts of rural households to break out of poverty and stagnation and place themselves on a self-sustaining path of improving incomes and well-being. Yet, the availability of formal rural finance has decreased relative to the demand for it in the last ten or fifteen years, as official rural lending institutions have seen their activities shrink drastically. However, even at their height of activity, when they were flush with government subsidies, they were not reaching a very significant share of the rural population:

In Mexico, notwithstanding a network of more than 500 agricultural bank branches and billions of US dollars in directed credit programs through both a State agricultural development bank and nationalized commercial banks, a recent World Bank study found that formal credit reached only 8 percent of rural enterprises, and direct government loans reached less than one percent of rural enterprises.[168]

The traditional approach hasn't worked. Channeling credit to agriculture proved not to be a sustainable solution. The concept of a State development bank for the sector has not proven durable. Nor have the accompanying monetary and regulatory policies been propitious for rural lending. 'The three most damaging interventions are excessive reserve requirements, a large volume of directed credit programs, and subsidized interest rates or interest rate ceilings'.[169] In addition, as the work of Michael Fiebig has demonstrated, most of the main elements of frameworks for bank regulation are inappropriate for agricultural finance.

7.8.2 Elements of a New Approach

As the material in this chapter has shown, new approaches have evolved in response to the failure of the old ones. In considerable measure, the new approaches will be implemented by NGOs and the private sector without government intervention, but the experiences of Bangladesh, Thailand, Indonesia and other countries have shown that there can be a useful supporting role for government in rural microfinance. The main unresolved question is how to strengthen the mechanisms for *agricultural* finance.

It also is clear that favorable macroeconomic policies and policies for sector development are crucial for the sustainability of agricultural finance. Profitability of the sector's production is a basic requirement for the viability of financial approaches and institutions in the sector.

There is a consensus that the new set of policies for promoting rural finance includes the following principal elements:[170]

168. J. Yaron, M. P. Benjamin and G. L. Piprek, 1997, p. 25, citing results from Rodrigo Chaves and Susana Sánchez, 'Mexico: Rural Financial Markets', Report 14599-ME, The World Bank, Latin America and the Caribbean Department, Natural Resources and Poverty Division, Washington, DC, USA, 1995.
169. *Op. cit.*, pp. 50–51.
170. Most of these recommendations, and others, are summarized in M. Wenner, 2001, pp. 10–16.

(1) An appropriate overall legal and regulatory environment, especially concerning interest rates, bank supervision capacities, more secure property rights and the legislative framework for contracts and collateral. Among other benefits, such a framework will promote lending to agriculture on the part of non-bank intermediaries such as input suppliers and marketing agents.

(2) Transitory, selective subsidies for microfinance and rural finance institutions that show sufficient management capabilities and appropriate governance structures to assist them in attaining the scale and capacity required for sustainability and in targeting the poor.

(3) Emphasis on savings mobilization by rural financial institutions, large and small.

(4) Use of new techniques of lending on the basis of intangible collateral, to extend the outreach of rural financial institutions to poor households. In some cases, the techniques are applied by institutions that deal largely with the poor, and in other cases, they are used by specialized units of commercial banks.

(5) Greater attention to gender issues in the design of rural finance programs.

(6) In programs of technical assistance and funding for the sector, greater attention to the structure of rural financial institutions, with emphasis on governance issues and, in some cases, the role of second-tier institutions. Creation of supporting institutions such as credit bureaus to improve financial information.

(7) Emphasis on training farmers and rural households in financial management.[171]

The bulk of the innovation in recent years has concerned *methods of reaching low-income households*, thus addressing a crucial need of long standing that was not adequately satisfied by the traditional approach. It must be emphasized that, although a great deal has been learned and the new approaches are very promising, the reach of formal rural finance is still very slight in comparison with the needs for it. The learning and implementation process must continue, with constant adaptations of experiences that have proven successful in one context or another.

One promising development that has not been stressed in this chapter so far is the increasing *interest on the part of commercial banks in microfinance*. Some lend directly to low-income clients (the Centenary Bank in Uganda, the Multi-Credit Bank in Panama, Bancosol and the Caja de Ahorro y Crédito Los Andes in Bolivia, and the Banco del Occidente in Honduras), others have independent or semi-independent units that deal with such lending (Banco del Desarrollo in Chile, the Unit Desa of the Bank Rakyat Indonesia, the Social Enterprise Program of the Bank of Nova Scotia in Guyana, and the Institute of Private Enterprise Development of the Demerara Bank in Guyana), and yet others lend to microclients indirectly through NGOs (Banco Wiese in Peru).[172] This is a trend that appears to be accelerating as commercial banks observe the success of microfinance institutions in capturing new segments of the market.

What has not been stressed in the literature is the need for adaptation of bank regulatory frameworks to the special conditions of agricultural finance. As mentioned above, Fiebig's monograph has made it very clear that virtually all elements of bank regulation and supervision require modification for the agricultural sector in developing countries, including capital requirements, portfolio classification rules, liquidity rules, documentation and reporting requirements, and branch regulations. It now is widely recog-

---

171. On this topic, see the monograph by Jennifer Heney, *Enhancing Farmers' Financial Management Skills*, Agricultural Finance Revisited No. 6, FAO and GTZ, Rome, August 2000.

172. Most of these references are taken from Mayada Baydas, Douglas Graham and Liza Valenzuela, 'Commercial Banks in Microfinance: New Actors in the Microfinance World', Development Alternatives, Inc., Bethesda, MD, USA, summarized in *Focus*, Note No. 12, The Consultative Group to Assist the Poorest, Washington, DC, USA, July 1998. Others are drawn from this author's experience.

nized that microfinance institutions need their own regime of regulation and supervision. What has not been sufficiently recognized is that *agricultural and rural banking institutions, microfinance or otherwise, need their own regulatory and supervision framework* as well. In many countries, laws have been proposed or passed to create a special regulatory regime for microfinance, but comparable steps have not been taken for agricultural finance. The implications of the need for a new regulatory framework are developed further below.

7.8.3 The Context of Gender Policy in Microfinance

Microfinance has an important *gender dimension*. The research of Khandker in Bangladesh showed clearly that women are better loan risks and also that women make better use of borrowed resources in terms of improving household welfare. This finding accords with anecdotal evidence from many other places, and provides a rationale for special support for those microfinance institutions that cater largely to women. Lending to women can also help change their status in rural areas. An evaluation of a women's credit program in Ecuador commented that:

> Women who participated in the program went through a learning process that strengthened their leadership and organizational skills. Other qualitative changes that were observed were improvements in the women's self-worth. . . . In addition, women's increased financial acuity prompted favorable changes in male partners' attitudes and level of respect. For example, members of the credit associations were able to purchase inputs, livestock, and vaccinations that helped them improve animal husbandry, and change that was appreciated by men as a positive contribution of women to the household economy.[173]

Sharma has summarized the justifications for targeting women in microfinance programs in the following words:

> Women's status, household welfare and microfinance interact in the following ways:
>
> • A woman's status in a household is linked to how well she can enforce command over available resources. . . .
> • Newly financed microenterprises open up an important social platform for women to interact with markets and other social institutions outside the household, enabling them to gain useful knowledge and social capital. . . .
> • Women's preferences regarding household business management and household consumption goals differ from men's, particularly in societies with severe gender bias. In such situations, placing additional resources in the hands of women is not a mere equalizer: it also materially affects both the quality of investments financed by the microfinance programs and how extra income is spent. . . .
> • Women are thought to make better borrowers than men. . . .
> • Loans are not simple handouts. If microfinance programs are designed to cover all costs. . . . development goals related to women's empowerment and improved household welfare are self-financing and no subsidies are required.

Several suggestions for operational policy measures to assist in the promotion of women's participation in rural financial programs have been made in Section 7.5 above. While these policy orientations can be effective in enhancing the role of women in microcredit programs, it is well to bear in mind Sharma's summary of evidence on empowerment and cautionary words in that regard:

---

173. Fundación de la Mujer Campesina (FUNDELAM), 'The Socioeconomic Impact of Credit Programs on Rural Women: A Study in Carchi, Ecuador', Report-in-Brief, PROWID, International Center for Research on Women (ICRW) and the Centre for Development and Population Activities (CEDPA), Washington, DC, USA, 1999, p. 2.

One widely cited study that made special efforts to construct measures of empowerment incorporating client perspectives is based on a 1996 survey of 1300 married Bangladeshi women members of the leading microfinance institutions, Grameen Bank and the Bangladesh Rural Advancement Committee (BRAC). The study found that married women participating in these credit programs scored higher than nonparticipating women on a number of empowerment indicators such as involvement in major family decision making, participation in public action, physical mobility, political and legal awareness, and the ability to make small and large purchases.

However, empirical studies point out that positive gender effects cannot always be taken for granted or are not as large as might be supposed. Many women, lacking skills and confidence, lean on their husbands to make use of their loans. A 1995 study in Bangladesh indicated that 'while 94 percent of Grameen Bank's borrowers are female, only 37 percent of them are able to exercise control over loan use . . .'.[174]

Sharma's conclusion about the overall policy context is an important one:

Ultimately, women's empowerment requires fundamental changes in society that call for more direct policy instruments. New policies should renegotiate property rights, replace rules sustaining gender inequality, and improved access to and quality of education. Fundamental change of this scale can hardly be worked out easily or quickly, especially in countries where gender bias has been a norm for centuries. Over the short run, microfinance programs provide a handy, potentially cost-effective, and politically feasible tool for moving towards gender equality. Group-based activities by women have served as important catalysts of change in Asia and Africa. The scale of change they ultimately catalyze will depend, however, on how seriously other social reforms bearing on women's empowerment are pursued.[175]

7.8.4 Towards Banking for All Farmers

The needs of the rural poor deserve priority, but that concern should not detract from concomitant attempts to strengthen financial mechanisms for the many medium-scale farmers in the developing world. Although on grounds of social equity they may not have deserved the subsidies that were channeled largely to them through programs of directed credit in the past, their financing requirements are legitimate and they play a vital role in the growth of agriculture. Services to them as well as to many lower-income farmers can be provided by commercial banks that are oriented to agriculture. The examples of Bancafé in Honduras and the Banco Gandero in Colombia have been mentioned. Another example is the Caja de Crédito Agrícola-Gandero in Ecuador. The Banco del Occidente in Honduras has a long tradition of lending to small farmers, most of them without tangible collateral, in the western parts of Honduras. As noted, these banks do not, and should not, lend the largest portion of their portfolio to agriculture but nonetheless, at up to half of the portfolio, they are lending much more to the sector than the average commercial bank does.

To foster an agricultural orientation in a commercial bank, a special commitment is required by the board of directors and managers. Many commercial banks do not have the **expertise to evaluate agricultural lending projects and clients**, and given the riskiness of the sector, in contrast to more of the certain earnings from investments in government bonds and urban real estate, they often are not interested in acquiring that expertise. Nevertheless, the examples cited have demonstrated the potential for agriculturally oriented banks to be profitable and sustainable, given appropriate operating policies and a supportive regulatory framework.

A policy instrument that can be utilized to foster the development of such banks is the

---

174. M. Sharma, 2001, p. 2.
175. *Ibid.*

privatization of State-owned development banks. In carrying out privatization, governments can encourage the creation of agricultural commercial banks by selling some of the shares to farmers and farmers' associations. In the case of Banco Ganadero, a government holding of over 80% of the shares was reduced to less than 20% over a period of about a decade through the mechanism of an agreed levy per head of livestock that was invested in shares of the bank, in the name of the ranchers. In cases in which the financial wealth in the sector is not sufficient to make such a conversion of ownership feasible, an alternative is to use government *seed capital* funding to capitalize a privatized bank, in the name of producers' associations (of small, medium and large farmers), assuming that other sources of private capital can be tapped as well and producers can repay the seed funding over time.

However, *in order to promote a more complete development of agricultural banking, two needs are critical: (a) the development of a separate regulatory framework for agricultural finance, and (b) a policy decision on how the higher cost of agricultural banking, which results from the higher risk, is to be apportioned*. Once it is acknowledged that agricultural loan portfolios require different prudential and non-prudential norms, and liquidity management for both rural saving and lending also requires a different treatment, it follows that a separate regulatory framework for agricultural banks is needed. Liquidity requirements and many other regulations apply to an entire financial institution, not to part of its portfolio, and so *a special financial regime is required under which private agricultural banks would be licensed*. An operational decision would have to be taken as to what percentage of its portfolio must be in agriculture before it falls under the new regime, but the principle is clear.

In the absence of such a regime, most commercial banks in developing countries have found it too risky to become significantly involved in agriculture. In the past, the response was the creation of State agricultural banks, instead of reforming the regulatory regime. Today, a more appropriate response would be the creation of a regulatory regime that would encourage the development of private agricultural finance.

As Fiebig (2001) has noted, *recognition of the higher capital and liquidity requirements for agricultural lending implies that it necessarily is more expensive*. Pretending that agricultural finance can function adequately under the same regulatory requirements as urban-oriented finance does has the effect of discouraging private banks from lending to agriculture. There are three options for covering this higher financial cost in agriculture: (a) through higher interest rates, i.e. the clients pay the cost, (b) through subsidies to cover the risk premium, i.e. taxpayers pay the additional cost, or (c) a combination of the two approaches. If a decision is made to authorize a subsidy element for agricultural banking, care must be exercised not to create moral hazard. In other words, the subsidy element would have to be small in relation to the total cost of lending, so that it does not prejudice efforts at loan recovery.

The arguments mentioned in Chapters 1–3 of this book, that agriculture is not simply another sector and its growth generates exceptional benefits for the rest of the economy, can be read as arguments for a partial subsidy of an agricultural banking system, through a special risk fund. Equally, it can be argued that producers should pay part or all of the risk premium. It is a decision that would have to be made according to the circumstances of each country.

A crucial need of the sector which is not well serviced by financial markets in developing countries is for *long-term finance*. Many investments in the sector require this kind of finance, from irrigation pumps and channels and small silos to barns and livestock to tree crops and soil conservation works. It was mentioned earlier in this chapter that returns to fixed investments in isolation appear to be much lower than those associated with working capital in the sector. A possible approach that merits exploration, by private agricultural banks, is the packaging of long-term and short-term finance, so that disbursements are made for both purposes over a defined period of years, and the interest rate is calculated for the package as a whole. In addition, arguments have been mentioned for donor support and government support for long-term lending in the sector. International bond issues may be considered as well, if the financial institution is sufficiently solid.

As an example of how this issue has been dealt with in Africa,

> In general, savings are loanable funds of a short-term nature. The main sources of medium and long-term agricultural lending funds used to come in the past from equity and international and government loans, that were often made available at concessionary terms. The Cooperative Bank of Kenya has set up a subsidiary merchant bank as a specific institutional set-up for its agricultural term lending. This bank mobilizes funds in the form of fixed-term deposits as well as by issuing bonds. Similar arrangements are found in India. The Land Bank of South Africa, as a strong financial institution, on the other hand, obtains its medium and long-term loanable funds mainly from issuing promissory notes and long-term debentures, government loans and a reallocation of general reserves from retained earnings. The African Development Bank Fund provides international loans for development purposes. . . .[176]

Rediscount lines through commercial channels are already implicit in the approach of land funds or land banks that are being implemented in several countries (see Chapter 5). To the extent that rediscount lines are used, to ensure their viability it is essential to target them on sub-sectors that represent a country's comparative advantage, and not necessarily those that are applying the greatest amount of political pressure for access to new financing.

It should also be noted that there are opportunities to attract more domestic investors into agriculture through securitization of assets such as livestock herds, crops in the field, private irrigation systems, forward contracts, and so forth. The national agricultural commodity exchange in Colombia (*Bolsa Nacional Agropecuaria*), initially conceived as a commodity exchange, has become a major source of finance for the sector though measures like this.

For adapting and implementing all of the new approaches, the most essential need is for **technical assistance and training**, of various forms and at all levels, from clients to directors of institutions. This is the area in which finance supplied by donors and governments can have the greatest impact on the development of the rural financial sector in the long run. The experience of BRAC in Bangladesh, cited above, is one of many that show the value of accompanying credit programs with training of clients. Along with the establishment of an appropriate regulatory framework for agricultural finance, this emphasis accords with the new vision of the government's role in the sector: that of *facilitating* the development of rural financial systems, rather than supplying credit for production.

DISCUSSION POINTS FOR CHAPTER 7

- Past policies for agricultural credit, which were based on the creation and maintenance of State agricultural development banks, control and subsidization of interest rates on loans, and use of directed credit allocations by crop on a massive scale, have failed in all regions of the world. The fiscal cost was high, benefits usually were regressive in their incidence over farm households, and the development of more viable rural financial institutions was discouraged.
- With the failure of this approach, the supply of credit to agriculture has been reduced significantly in real terms in most developing countries. Small- and medium-scale farmers have suffered most from this credit reduction.
- New approaches to agricultural finance are being developed that emphasize the creation of *sustainable* financial institutions.
- Returns to investment in developing agriculture can be very high, but market imperfections of

various kinds have prevented greater flows of financial capital into agricultural investments.
- One of the keys to sustainable agricultural finance is mobilization of greater amounts of savings in rural areas. The capacity and desire to save in financial form exists in rural areas in all countries, but financial institutions have not always developed the most appropriate instruments for attracting savings, nor have monetary policies always been supportive of this aim.
- Informal finance can be productive in agriculture but its scale is limited, often by restrictions in the regulatory framework which prevent, for example, the securitization of suppliers' credits.
- Agricultural credit markets are characterized by geographical dispersion, high risk, market segmentation, asymmetric information between lenders and borrowers, and the frequent presence of excess demand for credit at prevailing interest rates.
- Three major problems that are widespread in agricultural credit markets in developing countries are: (1) the *screening* problem – borrowers differ in the likelihood that they will default, and it is costly to determine the extent of that risk for each borrower, (2) the *incentives* problem – it is costly to ensure that borrowers take those actions which make repayment most likely, and (3) the *enforcement* problem – it is difficult to compel repayment. Sustainable rural financial approaches and institutions need to find ways to deal with all three of these problems.
- The past approach of allocating credit among crops and livestock products by governmental fiat usually led to unprofitable lending and encouraged investment in lines of production that were not necessarily in a country's comparative advantage.
- The approach of microfinance emphasizes reaching low-income clients and it attempts to achieve sustainability through the use of innovative lending techniques and, in many cases, through local savings mobilization. So far, it has had a greater impact in urban and peri-urban areas and, within rural areas, in supporting non-agricultural activities, but in some cases it also has expanded lending to small-holders. However, by the same token it is clear

that microfinance alone, as valuable as it is, will not fulfill the need of agriculture for greater access to finance.
- Microfinance programs do not necessarily reach the very poorest households but empirical evidence shows that they do reach significant numbers of borrowers below the poverty line.
- One of the important contributions of microfinance is increasing income levels of rural women. They have generally proven to be more dependable borrowers, and an additional unit of spending power in a woman's hands tends to have greater benefits for household nutrition and education than an additional unit in the hands of men in the same household.
- Sustainability for a financial institution requires, eventually, eliminating dependence on donated or subsidized funds and achieving profitability at a commercial cost of capital. Usually, sustainability in these senses is reached in stages.
- Other key strategies for attaining institutional sustainability include setting interest rates at market levels, developing sound governance and capable management of financial institutions, and designing the institutions' financial services with a market orientation.
- At the policy level, supporting the efforts of rural financial institutions to achieve sustainability requires appropriate regulatory frameworks, for agricultural finance as well as microfinance, and a national economic policy framework that is favorable to agricultural development.
- The risks faced by microfinance institutions differ in some important respects from those faced by commercial banks. For example, they are less exposed to risks of insider lending and of concentration of the portfolio in a handful of large loans. On the other hand, they face significant risks in the areas of ownership and governance and management and risks arising from the nature of their portfolios and the fact that they are a relatively new kind of activity.
- Microfinance institutions depend to a considerable extent on developing social collateral

since most of their clients cannot offer physical collateral as loan guarantees.

- The provision of agricultural finance can be expanded, through both banks and non-bank financial intermediaries, by policies and legislation that widen the forms of collateral that is legally allowed. Examples include pledging usufruct rights to a plot of land for a specified period (antichresis), crops in storage and in processing, other forms of movable collateral, and accounts receivable and loans extended. In many developing countries, this is an area that has not received sufficient emphasis.

- Strengthening the regime of contract enforcement also helps encourage more agricultural finance. This is often a major weak point in developing economies.

- Ceilings on interest rates in the formal financial sector usually are counterproductive. Non-interest costs of obtaining a loan often exceed the interest costs, and in any case lending in the informal sector comes at a very high interest cost. Capping interest rates has the effect of discouraging lending to the sectors with interest controls and inhibiting the possibilities for savings mobilization. It also fosters use of non-economic criteria for allocating financial resources and tends to direct part of those resources to low-productivity uses.

- Regulation of financial institutions can be divided into *prudential* and *non-prudential* types. The former involves the regulatory authority in guaranteeing the soundness of the regulated institutions and therefore protecting depositors. The latter involves many requirements for starting up financial institutions and accounting and audit procedures, and reporting and disclosure obligations. Both kinds are important tools for supervising the growth of the financial sector.

- Another classification characterizes regulations as *preventive* (avoiding crises) and *protective* (handling crises once they have occurred). The former class of regulations can be further divided into *entry requirements* and *ongoing requirements*.

- Among the entry requirements, a common error is to set minimum capital levels too low.

Especially in agriculture, lending institutions need a strong equity base. Ownership requirements are important for ensuring that the owners bring to the institution a clear vision of its aims and context, as well as financial strength.

- Agricultural loans generally should be assigned a higher risk classification. This increases the cost of lending to agriculture but is necessary to ensure the sustainability of the financial institutions that lend to the sector. Likewise, agricultural portfolios need to be backed with higher liquidity ratios.

- Traditional bank regulations governing the nature of collateral are not always applicable to agriculture, especially when social capital (group lending) is utilized. Equally, loan documentation and reporting requirements, repayment intervals, and criteria for past-due loans all may have to be different for agricultural lending and for microfinance. Branch restrictions developed for commercial banks also may not be applicable for rural and small-scale lending operations.

- While many financial regulations need to be different for microfinance and agricultural portfolios, it has been argued that small-scale deposit-taking at a local level need not be brought under the umbrella of prudential regulation, because supervision could not be effective and the result would be to prohibit access to deposit facilities for many poor families.

- Equally, it has been argued that the creation of a special regulatory window should wait until there are enough microfinance institutions that are suitable for licensing, because regulation itself has a significant cost, both for the authorities and for the regulated entities.

- Rotating credit funds have been in existence for centuries, if not millennia, and they have generally functioned well at a local level on a small scale. However, the expansion of this concept into credit co-operatives tends to create institutions that are dominated by borrowers and not by those who have invested capital in the institutions. Hence, their financial performance has tended to be weak.

- Credit co-operatives can be made into stronger institutions with appropriate reforms that focus on ownership and governance, interest rate policies, and the types of financial services offered.

- To satisfy the financial needs of many small- and medium-scale farmers, greater emphasis could be placed on encouraging rural banks, or agricultural banks, a few of which exist already. These banks do not lend exclusively to agriculture, because the risk would be too high, but they devote a significantly higher share of their portfolio to lending in the sector. To be viable, such banks need to be willing to compensate for agricultural risks with higher interest rates, adopt some of the lending techniques of microfinance institutions, decentralize their operations as much as possible to take advantage of local knowledge of production conditions and clients, use innovative forms of capital, and improve their information systems and staff training.

- Apex organizations, or second-storey institutions that support systems of agricultural banks or microfinance institutions, can play a valuable role in liquidity management, obtaining access to additional funds, and offering technical advice. However, they suffer from a number of weaknesses. Their presence can complicate the system's vision and cloud governance issues, and differential growth rates among the primary-level institutions can strain the system. Therefore, apex institutions and the associated systems need to be designed and monitored with care.

- Another option for guaranteeing the long-term viability of small-scale financial institutions is for them to affiliate with larger banks. Increasingly, banks are purchasing stakes in those institutions.

- The financial instrument of rediscount lines has been justly criticized for directing lending to inappropriate lines of production and low-productivity uses, but nevertheless international development institutions continue to support it in special cases. There are major gaps in funding for agricultural development, especially for long-term capital needs, and rediscount lines can play a useful role in filling those gaps on a transitional basis.

- International bond markets represent another potential source of financing for agricultural development, in cases of well-organized associations of producers for particular products, but the requirements for tapping those markets are demanding.

- For both financial systems and individual financial entities, governance issues are critical. They basically concern the establishment of relationships of accountability and transparency.

- The few successful government financial institutions that function in rural areas are invariably characterized by a great deal of operational autonomy.

- Some microfinance institutions direct most of their lending to women, since they often are better financial clients on average and tend to make better use of borrowed funds than male clients do. However, in most financial institutions there still is a pro-male bias, attributable in part to the ways in which lending institutions are accustomed to work and the legal nature of collateral requirements. Legal reforms, training and awareness building can help overcome the barriers to more lending to women.

- Innovative lending policies have permitted the authorization of unsecured loans. They include criteria for borrower eligibility, incentives for repayment and techniques of monitoring borrower behavior. These new policies represent responses to the three basic problems faced by lenders in rural credit markets, of screening potential borrowers, providing incentives for compliance with loans, and enforcing repayment obligations.

- The new lending policies include various forms of group guarantees for loans. Limited liability schemes, in which all borrowers in a group make deposits into a fund which will be returned if all loans are repaid, appear to be more workable than requiring all members to be totally liable for the loans of all other members. Any form of group lending requires social cohesiveness and sufficient information about the other members of the group.

- Just as new lending techniques have been developed by microfinance institutions, new savings mobilization techniques have emerged as well.

However, introduction of deposit instruments into a financial institution greatly expands its workload and responsibilities. It fundamentally changes a financial institution and is a more demanding step than sometimes is supposed.

- Prudential management of a microfinance institution requires attention to capital adequacy, the role of a board of directors, liquidity levels, management information systems and procedures for recognizing risk, internal auditing procedures, performance incentives for staff, and portfolio diversification, and it also requires efforts to control costs.

- It has been argued that *deposit* rates are generally too low in developing countries, that they often reflect oligopolistic market positions of banks, and therefore that regulatory authorities should raise them instead of endorsing the banks' positions on these rates. This would enhance a country's prospects for financial savings mobilization.

- Highly inflationary conditions present a special case in which deregulation of lending rates may have to be deferred until inflation is reduced, since agriculture cannot pay high nominal interest rates. Under inflation and high interest rates, total costs of production rise much more rapidly than output prices.

- Donor finance can play a useful role, especially when it is linked to technical advice, is used for transitional purposes (such as starting up institutions), and subsidizes access of target populations to finance. It also can fill the gaps that exist in regard to availability of long-term loans in agriculture and forestry. Donor contributions to the financial system should be accompanied by other support for reducing market imperfections through, for example, land titling and the development of transport infrastructure.

- Commodity exchanges can represent large sources of finance for production and marketing.

- Developing agricultural finance requires application of the lessons learned in microfinance and also the development of a regulatory framework unique to lending for agriculture. The normal bank regulatory requirements for capital adequacy, liquidity, loan classification and monitoring, reporting and branch establishment are not applicable to much of agricultural lending. For these reasons, commercial banks typically are not very interested in agricultural lending. In the past, the policy response was to create State agricultural development banks. A better policy may be to create a regulatory environment that can make agricultural lending more attractive to private financial institutions.

- To encourage more private agricultural banking, a policy decision also needs to be made in regard to who pays the higher risk of agricultural lending, borrowers (farmers) or taxpayers, or a combination of the two. If a decision is made in this area and a special regulatory regime is instituted for agricultural lending, it can be expected that the volume of agricultural finance may increase instead of continuing its downward trend in relation to needs.

- As valuable as lending to rural women is, it needs to be recognized that often women do not control the funds that are lent to them. To enable women to participate more fully in managing finances, and their economic destiny, reforms are also needed in basic areas such as property rights legislation and educational policies.

8

Policies for Agricultural Technology[1]

8.1 Introduction: The Role and Context of
Agricultural Technology   357
  8.1.1 The Role of Research and Extension   357
  8.1.2 The Policy Context   360
  8.1.3 Institutional Considerations   361
8.2 Issues in Agricultural Research   363
  8.2.1 Research Capacity and Effectiveness   363
  8.2.2 The Appropriateness of Technology   365
  8.2.3 Gender in Agricultural Research   368
  8.2.4 The Public and Private Sectors in
    Research   371
  8.2.5 Pest Management   372
8.3 Issues in Agricultural Extension   375
  8.3.1 The Historical Development of
    Extension in LDCs   375
  8.3.2 The Rationale for Public Extension
    Services   377
  8.3.3 The Performance of Public
    Agricultural Extension   381
  8.3.4 Gender in Agricultural Extension   384
  8.3.5 The Challenge of HIV/AIDS for
    Agricultural Extension   385
  8.3.6 Towards a New Paradigm for
    Agricultural Extension?   387
  8.3.7 Trade in Agricultural Technology   388
8.4 New Directions in Agricultural Research   390
  8.4.1 Identification and Implementation of
    the Research Agenda   391
  8.4.2 Management and Institutional
    Structures for Agricultural Research 396
  8.4.3 The Financing of Agricultural
    Research   398
  8.4.4 Agricultural Research and Poverty
    Alleviation   399
  8.4.5 Gender Approaches in Agricultural
    Research   402
8.5 New Approaches to Agricultural
Extension   404
  8.5.1 Alternatives for Agricultural
    Extension Systems   404
  8.5.2 Promoting a Client Orientation in
    Extension Systems   409
  8.5.3 Gender Approaches in Agricultural
    Extension   413
  8.5.4 Responding to the Challenge of
    HIV/AIDS   414
  8.5.5 Illustrations of New Trends in
    Agricultural Extension   416
  8.5.6 A Synthesis of New Approaches to
    Agricultural Extension   417
Discussion Points for Chapter 8   420

8.1 INTRODUCTION: THE ROLE AND CONTEXT OF AGRICULTURAL TECHNOLOGY

8.1.1 The Role of Research and Extension

Agriculture has two ways to increase its output: expanding the land area under cultivation and improving the yields on cultivated land. If agricultural growth is taken to mean increases in the farming incomes of rural families, then a third

---

1. The author is grateful to the staff of FAO/SDRE for their very helpful comments on an earlier draft of this chapter.

Agricultural Development Policy Concepts and Experiences. R. D. Norton
© 2004 Food and Agriculture Organization of the United Nations
ISBNs: 0-470-85778-1 (HB)   0-470-85779-X (PB)   FAO Edition: 92-5-104875-4

way can be added, i.e. shifting the product composition to higher-value products. There are no other ways.

For decades it has been commented that, globally, the possibilities for expanding the land area under cultivation are diminishing steadily. Pursuing that route increasingly risks environmental degradation in many parts of the world as forests are cut down and soils on slopes are eroded. Therefore, the only viable options that remain are increasing yields and changing the product composition. However, while shifting to higher-valued crops and livestock products is a sound strategy from the viewpoint of farmers, it does not help to increase food supplies in the aggregate. For that aim, the only course left is to increase yields. In addition, for many poor farmers who do not have adequate access to diversified markets or cannot fulfill other requirements for shifting to higher-valued crops, increased yields are also the only route to higher incomes.

Increasing agricultural productivity is all the more urgent because the majority of the developing world's poor are found in rural areas, and the sector's average productivity is actually declining in many low-income countries. Dina Umali-Deininger has expressed the dilemma in the following words:

> Rapidly growing populations have unleashed a spiraling demand for food, while the food-producing capacity in many nations is increasingly constrained by both diminishing opportunities to bring new land into production and by the declining productivity of over-cultivated areas caused by natural resource degradation. . . . At the same time, the significant majority of the poor continue to depend on agriculture for their livelihood. Of the 720 million poor identified by the World Bank . . . 75 percent live in rural areas. Thus increasing farmers' incomes through improved productivity is an important

element in agricultural development and poverty reduction strategies.[2]

Yields can be increased dramatically by the application of irrigation. Adopting irrigation requires the training of farmers and the provision of extension services over a sustained period of time, but it can achieve substantial increases in yields without new agricultural research. Yet the possibilities for expanding irrigated areas also are limited in most parts of the world, and indeed many existing irrigated areas are suffering from salinization, waterlogging and other problems that reduce productivity, as discussed in Chapter 6. Therefore, while all possible efforts should be made to improve irrigation management and expand irrigated areas where feasible, irrigation alone cannot be expected to provide the physical basis for the world's needed increases in agricultural production in the future. The burden of meeting this challenge falls principally on the systems for development and transfer of improved agricultural technology – on agricultural research and extension. It falls equally on systems for education of farming families; indeed, some observers contend that education is the most important factor for improving productivity.

The declining productivity mentioned by Umali-Deininger is not an isolated phenomenon. Lilyan Fulginiti and Richard Perrin reviewed cross-country studies of changes in agricultural productivity and carried out their own estimates via alternative methodological approaches. They commented that all developed countries have experienced productivity increases while most low-income countries have seen their agricultural productivity decline, even those in which farmers have widely adopted Green Revolution varieties of wheat and rice.[3] They went on to ask whether measurement problems had biased these results. On the basis of their own analyses of the data sets, they concluded that declining productivity is

---

2. D. Umali-Deininger, 'Public and private agricultural extension: partners or rivals?', *The World Bank Research Observer*, **12**(2), August 1997, p. 203.

3. L. E. Fulginiti and R. K. Perrin, 'Agricultural productivity in developing countries', *Agricultural Economics*, **19**(1–2), September 1998, p. 45.

a real phenomenon and that unfavorable agricultural pricing policies may be a major cause of it:

> The most significant result . . . is that agricultural productivity in these [18] countries seems to have receded at an average rate of 1–2%, and this result is robust with respect to measurement techniques. . . . This result was not uniform across countries. Chile and Colombia consistently show gains in productivity across the methods employed. Ghana, Ivory Coast, Zambia, Pakistan, Thailand and Korea show productivity losses in all three methods. . . . We conclude that the phenomenon of negative productivity trends indicated by previous studies has not been an artifact of the analytical methods used, since the general results are supported by a variety of methods. The diversity of performance across countries, however, opens the possibility of discovering what factors contribute to productivity improvement in these countries. In other studies, we did find that countries that tax agriculture most heavily had the most negative rates of productivity change, consistent with previous results suggesting that price policies may be one of the important contributing factors.[4]

The overall picture is a mixed one. Improvements in cereal yields were found in some districts of 13 African countries in a recent period by William Masters, Touba Bedingar and James Oehmke:

> Using case studies, we show that by the late 1980s numerous techniques produced by research were being adopted, and are now producing high levels of social gain. These include new varieties, whose principal feature is often early maturity for drought escape, as well as new management techniques aimed at moisture

retention and soil fertility. This type of technical change is very different from that which produced the Green revolution in Asia and Latin America, where greater moisture availability made short stature and fertilizer responsiveness the keys to higher yield.[5]

While there are bright spots, at best the performance of agricultural productivity in less developed countries (LDCs) generally has not been encouraging. It is clear that the challenge faced by agricultural technology systems in those countries is great, and it will probably become greater in the future. The challenge is exacerbated by a secular downward trend in funding for agricultural research in developing countries in the last two decades. Total international assistance to agriculture fell from about $12 billion a year to about $10 billion a year during the 1980s, and the agricultural share of development assistance also declined. The same trend, which continued in the 1990s, has characterized funding for agricultural research.[6]

> *A great deal of evidence shows very severe under-investment in agricultural research in developing countries. The economic returns from past agricultural research for developing countries are very high and the potential benefits from additional research by far exceed the expected costs. (From Per Pinstrup-Andersen, 'Is Research a Global Public Good?', Entwicklung + Ländlicher Raum, 34 Jahrgang, Heft 2/2000, reprinted in the* Research Themes *series of the International Food Policy Research Institute, Washington, DC, USA, 2000, p. 3.)*

For Latin America and the Caribbean, it was also found that spending on agricultural research

4. *Op. cit.*, pp. 49–50.
5. W. A. Masters, T. Bedingar and J. F. Oehmke, 'The impact of agricultural research in Africa: aggregate and case study evidence', *Agricultural Economics*, **19**(1–2), September 1998, p. 81.
6. L. S. Hardin, 'Whence international agricultural research?', *Food Policy*, **19**(6), December 1994, p. 564.

declined by about 13% between the beginning of the 1980s and the start of the 1990s, even though demands on agricultural research are increasing.[7]

From a viewpoint of economic efficiency – the true basis for growth – the justification for the role of agricultural research and extension arises not directly from their contribution to the increased physical levels of production, but rather from the economic rate of return associated with expenditure on these activities. In this regard, quantitative studies have consistently shown high rates of economic return to research and extension, and so the observed declines in their funding would not appear to be supported by economic considerations.

8.1.2 The Policy Context

The pace of productivity improvement in agriculture is not only determined by administrative allocations of budgetary funds to research. As indicated by the suggestion by Fulginiti and Perrin, the rate of agricultural technical progress is not immune to the policy and institutional context and other underlying factors. The earliest exponents of the theory of induced technical change were Vernon Ruttan and Yujiro Hayami, whose work is cited in Chapter 9 of this volume. One of their main conclusions is that over the long run technical progress tends to be driven largely by the same influences that shape a country's comparative advantage: relative factor endowments (and hence relative factor prices). They also conclude that the most effective innovations are those that are consistent with a country's relative factor endowments, a conclu-

sion illustrated by the above-cited findings of Masters *et al.* for recent agricultural research in Africa.[8]

Rubén Echevarría has pointed out that technical progress is essential for a country to be able to realize its inherent comparative advantage through international trade.[9] Taking advantage of trade possibilities requires adaptability in cropping patterns and product quality. Therefore, international trade imposes a requirement for agricultural research to be more flexible. Trade opens new opportunities for production that can be seized only if the stock of available technologies of production is adequate. In the words of Echevarría:

> the product mix in the agricultural sector is subject to rapid changes through trade, so that research organizations must be positioned to react to these changes [and] with growing urbanization, the share of the retail price of agricultural commodities captured by the farm sector is declining, requiring more research focus on post-harvest processing and marketing.[10]

Besides trade policy, other macroeconomic policies are critical to the success of technology development and transfer particularly, as noted, those that affect farm prices and real returns to farming. Dennis Purcell and Jock Anderson have commented that a supportive policy environment is essential for both research and extension to contribute to higher productivity:

> Investment in technology development and its dissemination cannot be expected to increase

---

7. Rubén G. Echevarría, Eduardo J. Trigo and Derek Byerlee, *Cambio institutional y alternativas de financiación de la investigación agropecuaria en América Latina*, Inter-American Development Bank, Washington, DC, USA, August 1996, p. 3.
8. See the discussion in Chapter 9, where reference is made to Vernon W. Ruttan and Yujiro Hayami, 'Induced Innovation Model of Agricultural Development', in C. K. Eicher and J. M. Staatz (Eds), *International Agricultural Development*, 3rd Edn, The Johns Hopkins University Press, Baltimore, MD, USA, 1998, pp. 163–178.
9. R. Echevarría, 'Agricultural research policy issues in Latin America: an overview', *World Development*, **26**(6), 1998, p. 1107.
10. *Ibid.*

productivity unless stakeholders operate in an otherwise conducive environment. Appropriate macroeconomic and sectoral policies, favorable market opportunities, access to resources, inputs and credit are all necessary to realize the full potential of new technology.[11]

In the African context, Mywish Maredia, Derek Byerlee and Peter Pee wrote:

The adoption of improved varieties of food crops that respond to the use of purchased inputs, such as improved seed and fertilizer, is strongly conditioned by the policies that affect input supply and prices, and market infrastructure. High rates of return to agricultural research are difficult to sustain in an environment where inputs are not accessible or affordable to farmers. This is exemplified by the post-structural adjustment experiences of Malawi and Zambia, where there has been a substantial dis-adoption of improved maize seed and fertilizer technology following the market liberalization of the early to mid-1990s. . . .[12]

In an era in which the private sector is playing an increasingly important role in both research and extension, as discussed below, an aspect of policy that has special relevance to the incentives for agricultural research is that of intellectual property rights (IPRs). A lack of well defined IPRs weakens incentives for privately funded research at the same time that it strengthens the case for internationally funded public research:

Unlike in much of industry, critical agricultural technologies (principally new seed varieties) are not well protected by IPRs, either globally or

nationally. Therefore private investors do not provide enough R & D, especially for technologies applicable in the poorest countries, where information and market problems add to those of weak IPRs. The potential international spillovers that discourage private investors also enhance the economic effectiveness of international collective efforts in agricultural R & D. . . .[13]

8.1.3 Institutional Considerations

Although a favorable policy framework and adequate funding levels may be necessary conditions for the success of agricultural research and extension, a consensus has emerged that profound institutional changes are needed in these fields. Summarizing that consensus, the present author has written that:

In order to reverse these trends and put agricultural productivity back on an upward course, agricultural research must undergo extensive institutional transformation. This is the first point to be addressed in bringing about a transformation in production. One of the main challenges is to find a viable way to involve non-governmental organizations (universities, foundations, producers' associations) . . . and private companies in the research process. A second challenge is to orient research more effectively toward the clients' (producers') needs, by more closely involving farmers in decisions regarding research strategies. Transformations of this nature are already underway . . . but they will have to be accelerated. If they are successful, it will be much easier to convince lending institutions to redirect their attention to agricultural research, as

11. Dennis L. Purcell and Jock R. Anderson, *Agricultural Research and Extension: Achievements and Problems in National Systems*, Operations Evaluation Study, The World Bank, Washington, DC, USA, 1997, p. 2. See also D. Birkhaeuser, R. E. Evenson and G. Feder, 'The economic impact of agricultural extension: a review', *Economic Development and Cultural Change*, **39**(3), April 1991, p. 611.

12. M. K. Maredia, D. Byerlee and P. Pee, 'Impacts of food crop improvement research: evidence from sub-Saharan Africa', *Food Policy*, **25**(5), October 2000, p. 554.

13. The World Bank, *Knowledge for Development*, World Development Report 1998/99, The World Bank, Washington, DC, USA, 1999, p. 37.

in the past. But, national political leadership is also essential in this sense.[14]

Likewise, the modalities of agricultural extension are coming under heavy pressure to change along similar lines, giving a greater role to the private sector and NGOs, partly because of the need to reduce the fiscal costs of large extension staffs.[15] M. Kalim Qamar has written about new trends that are changing the basic orientation of agricultural extension:

The very definition, scope and technical focus of agricultural extension are now under scrutiny. The question being raised is why should extension services focus exclusively on the transfer of agricultural technology, which is not only a passive function but also utilizes a top-down approach? The result is that more emphasis is now being placed on human resources development, i.e. on developing the problem-solving and decision-making capabilities of farmers. . . . There is a drive to decentralize agricultural extension, and a number of countries have disbanded the top-down, multilayer organizational structures . . . The modality of using both public and non-public institutions for delivering extension services to farming communities, called the pluralistic extension system, is gaining popularity. . . . The old practice of delivering the same message to all farmers using the same extension methodology is gradually being replaced by client-focused approaches.[16]

In short, in recent years it has been accepted that traditional ways of carrying out agricultural research and extension are no longer satisfactory; that in spite of their apparently high returns, these systems can perform better under new approaches that lead to different institutional arrangements and that respond to a new operational philosophy. Dissatisfaction with former ways of doing things has arisen mainly from three concerns:

- tighter fiscal budgets;
- a perception that not all research and extension programs have been efficient;
- a mandate to devote proportionately more resources to seeking ways to increase the productivity of low-income farmers.

Perhaps the principal limitation of the past generation of systems of technology development and transfer is that they are best suited to conditions of approximate agricultural homogeneity – in which large numbers of farmers face similar growing conditions. These were conditions that favored the spread of the Green Revolution. Those approaches to technology development and transfer tended to be centralized administratively and were based on the implicit assumption that uniform technological recipes could be developed by scientists and subsequently delivered down a chain of command to farmers, almost as if they were factory workers. Under the circumstances of agro-ecological heterogeneity that characterize most groups of low-income tillers, the applicability of a centralized, top-down approach is limited. The new push for institutional change in agricultural technology systems strives to incorporate adequate feedback from farmers themselves, both on the nature of problems they face and on possible lines of solution.

This topic of institutional transformation of the agricultural knowledge system is the subject of much of the remainder of this chapter. As well as influencing the speed of productivity improvement for the sector as a whole, institutional

14. Roger D. Norton, 'Critical Issues Facing Agriculture on the Eve of the Twenty-First Century', in *Towards the Formation of an Inter-American Strategy for Agriculture*, IICA, San José, Costa Rica, 2000, p. 291.

15. See R. Picciotto and J. R. Anderson, 'Reconsidering agricultural extension', *World Bank Research Observer*, **12**(2), August 1997, p. 254.

16. M. K. Qamar, 'Agricultural extension at the turn of the millennium: trends and challenges', in M. K. Qamar (Ed.), *Human Resources in Agricultural and Rural Development*, Food and Agriculture Organization of the United Nations, Rome, 2000, pp. 159–160 and 162.

considerations play an important role in helping research and extension contribute to the alleviation of rural poverty.

8.2 ISSUES IN AGRICULTURAL RESEARCH

8.2.1 Research Capacity and Effectiveness

Mohinder Mudahar, Robert Jolly and Jitendra Srivastava point out that in most situations four kinds of research may be distinguished. In their words, they are as follows:

- *Basic research*, which creates new scientific knowledge to achieve new understanding but with no immediate commercial application.
- *Strategic research*, which provides knowledge and techniques to solve specific problems that have a wider applicability.
- *Applied research*, which develops new technologies and tangible inventions by adapting basic and strategic research to solve specific field problems.
- *Adaptive research*, which involves selecting and evaluating technological innovations to assess their performance in a particular agricultural system and adjusting technologies to fit specific environmental conditions.[17]

They also point out that basic research falls mostly in the domain of the public sector (because of externalities that make it a public good), whereas private sector participation is more likely to occur in applied and adaptive

research. It is in these last two types of research that most of the ferment regarding institutional approaches is occurring. Private sector research does not necessarily lead to private ownership of the results; it can be carried out with public funding. Given the resource limitations in developing countries, many would argue that their research systems, both public and private, should concentrate on applied and adaptive research, building on international findings as much as possible. Thus, *the type of agricultural research* to be undertaken within a country is one of the first issues that requires an answer.

In regard to *economic rates of return to agricultural research* as a whole, Mudahar *et al.* cite a review of studies by Evenson and Westphal that reported the following mean rates of return: Africa, 41% (10 studies); Latin America, 46% (36 studies); and Asia, 35% (35 studies).[18]

Estimates of the linkage between research effort and agricultural productivity have been made as well. Purcell and Anderson[19] collated studies of this linkage for both developed and developing countries, although they found only two studies for the latter. For India over the period 1965–1987, Evenson and Mark Rosegrant[20] estimated that the elasticity of total factor productivity in the crop sector with respect to the public sector's research 'stock variable' lay in the range 0.05 to 0.07. In other words, a 14 to 20% increase in research effort would generate a 1% increase in productivity per year. For 22 countries in Sub-Saharan Africa over the period 1971 to 1986, Thirtle, Hadley and Townsend[21] derived less optimistic results, calculating the elasticity of

17. Mohinder S. Mudahar, Robert W. Jolly and Jitendra P. Srivastava, *Transforming Agricultural Research Systems in Transition Economies: The Case of Russia*, World Bank Discussion Paper No. 396, The World Bank, Washington, DC, USA, 1998, pp. 61–62.
18. Robert E. Evenson and Larry E. Westphal, *Technological Change and Technology Strategy*, Center Paper No. 503, Economic Growth Center, Yale University, New Haven, CT, USA, 1995 (cited in Mudahar *et al.*, 1998, p. 7).
19. D. L. Purcell and J. R. Anderson, 1997, p. 116.
20. R. E. Evenson and M. W. Rosegrant, *Total Factor Productivity and Sources of Long-Term Growth in Indian Agriculture*, Environment and Production Technology Division Discussion Paper No. 7, International Food Policy Research Institute, Washington, DC, USA, 1995.
21. C. Thirtle, D. Hadley and R. Townsend, 'Policy-induced innovation in Sub-Saharan African agriculture: a multilateral Malmquist productivity index approach', *Development Policy Review*, 13(4), 1995, pp. 323–348.

total factor productivity in all agriculture with respect to public sector research to be 0.02, which implies that a 50% increase in research effort would be required to attain a 1% increase in productivity per year. Among other considerations, such estimates are sensitive to initial scale, to the relation between the size of the research effort and the size of the agricultural sector. They also assume no improvements in research efficiency with the existing capacity. Therefore, these estimates are only suggestive of the relationship between research and productivity.

A more recent analysis of the contributions of agricultural research was carried out for India by Evenson, Pray and Rosegrant.[22] They concluded that (p. 63) the returns to additional public investments in agricultural research reach almost 60% in each of the periods 1956–1965, 1966–1976 and 1977–1987. In regard to the contributions of research to total factor productivity, they estimate (p. 59) that public research accounted for about 29% of the growth in total factor productivity over the entire sample period, with the rest accounted for by increased use of inputs and by private research, extension, literacy and markets.

Strong contributions of agricultural research in Africa also have been observed:

In recent years, as a result of the growing donor pressure to demonstrate impacts of agricultural research, several studies have been conducted to document impacts and estimate rates of return to research investment in Africa. These studies provide evidence of the increasing availability of improved varieties of major food crops to farmers in Africa, increased food production in regions where adoption has occurred, and positive returns to research investment, indicating that agricultural research in Africa has had productivity increasing impacts on its agriculture. The widespread adoption of improved maize, wheat and rice varieties is especially noteworthy, covering more than 50% of the area under these cereal crops by the early 1990s.

As a result of this growing evidence, the impacts of agricultural research in Africa can no longer be denied. The generation and diffusion of improved, higher-yielding maize open-pollinating varieties in Western Africa and hybrids in Eastern and Southern Africa, higher yielding wheat in Eastern and Southern Africa, hybrid sorghum in Sudan, semi-dwarf rice for irrigated regions in West Africa, early maturing cowpeas in West Africa, and disease-resistant potatoes in Eastern and Central African highlands are now cited as outstanding success stories of technological change in food crop production in sub-Saharan Africa.[23]

For Indian agriculture, Evenson *et al.* observe that private research accounts for about half the expenditure that public research does, and that it is concentrated on those crops in which hybrid seeds play a significant role (p. 18). In addition to the private sector, other non-governmental institutions in developing countries have the capacity to participate in agricultural research, including farmers' organizations (especially in the applied and adaptive modes) and universities and specialized institutes (in all modes, including basic research). *Lack of co-ordination between governmental and non-governmental research institutions* has been identified as a pervasive problem. In their review of World Bank-supported agricultural research, Purcell and Anderson identified this issue as a principal one, along with several others:

The net result of investment [in agricultural research] has been an improved human resource base (albeit with some mismatches between available and needed skills); a substantially expanded research infrastructure in facilities and equipment, combined with doubts about the appropriateness of some investments; improved links with external research entities; advances in agency coordination within

22. Robert E. Evenson, Carl E. Pray and Mark W. Rosegrant, *Agricultural Research and Productivity Growth in India*, Research Report 109, International Food Policy Research Institute, Washington, DC, USA, 1999.
23. M. K. Maredia, D. Byerlee and P. Pee, 2000, p. 554.

national agricultural research systems (NARS), but inadequate attention to involvement of academic institutions; mixed results in improving research-extension farmer linkages; weak development of the incentive structure for researchers; and, despite considerable emphasis in the second half of the review period, slow progress in improving the efficiency of resource allocation in NARS agencies.[24]

They felt that in some cases an emphasis on financing an expansion of research capacity detracted from the need to improve the efficiency of the research effort per unit of expenditure. Thus, while *research funding problem* is a central issue in almost all developing countries, improving the *efficiency of research programs* is another critical issue that is encountered in all regions of the world.

When a given level of research expenditure is spread over increased numbers of research staff, efficiency suffers, and the problem is exacerbated by the declines in real levels of expenditure. Donor agencies often try to take up the slack, but this creates an *issue of sustainability*. These points have been made cogently for African agricultural research by Philip Pardey, Johannes Roseboom and Nienke Beintema, as follows:

Sub-Saharan African countries made some progress in developing their agricultural research systems during the past three decades. Particularly the development of research staff has been impressive in terms of numbers (a sixfold increase if South Africa is excluded), declining reliance on expatriates (from roughly 90% expatriates in 1961 to 11% in 1991), and improvements in education levels (65% of the

researchers held a postgraduate degree in 1991). . . .

Developments in agricultural research expenditures were considerably less positive. After reasonable growth during the 1960s and early 1970s, growth in expenditures basically stopped in the late 1970s. . . . Donor support has clearly increased in importance. Its share in the financing of agricultural research increased from 34% in 1986 to 43% in 1991. While increased donor support somewhat compensated for declining government funding, it is unlikely that such high levels of support can continue indefinitely.

Many of the developments of the past decade in personnel, expenditures, and sources of support for public sector R & D in Africa are clearly not sustainable.[25]

8.2.2 The Appropriateness of Technology

In the end, various kinds of diagnoses of research systems and their effects underscore one central message: in the words of Charles Antholt, it is *'the importance of getting the technology right'*, and the message is valid 'whether [the technology] is evolved over time by farmers themselves, borrowed directly from other parts of the world, or borrowed and then locally adapted'.[26] Reversing the decline in research budgets may be an integral part of any reform to the system, but how to ensure the appropriateness of the technologies developed is the single greatest, and most enduring, issue faced by agricultural research systems. What is right for a few farmers may not be right for the majority.

It is well known that low-income farmers may value risk avoidance more than income gains.[27]

---

24. D. L. Purcell and J. R. Anderson, 1997, pp. 7–8.

25. P. G. Pardey, J. Roseboom and N. M. Beintema, 'Investments in African research', *World Development*, **25**(3), 1997, p. 421.

26. Charles H. Antholt, 'Agricultural Extension in the Twenty-First Century', in C. K. Eicher and J. M. Staatz (Eds), 1998, Chapter 22, p. 360 [emphasis added to first quote].

27. Many quantitative estimates of this tradeoff have been made. A discussion of a risk aversion parameter measuring the tradeoff between increments in farm income and the standard deviation of income is contained in Chapter 5 of Peter B. R. Hazell and Roger D. Norton, *Mathematical Programming for Economic Analysis in Agriculture*, Macmillan Publishing Company, New York, 1986.

Risk aversion goals can be pursued by many means, including reducing the height of stalks of grains, accelerating the crop maturation process, reducing the dependence on purchased inputs (to reduce financial risk), increasing resistance to pests, and various other ways. Of course, farmers themselves have developed many traditional modes of reducing risk, including intercropping,[28] crop diversification, and holding widely scattered plots. Hence, the 'appropriateness' of new agricultural technology has to be evaluated not only in the dimension of yield increases or even in terms of increases in net income per hectare.

As well as risk aversion, other factors that bear on the appropriateness of technology include its environmental implications (its sustainability), gender considerations, and compatibility with market requirements and with the needs of agroindustrial processes (product quality considerations). These considerations bear directly on the issue of *defining priorities for research programs* and criteria for selection of the programs' targets for outputs. They imply that research goals and the varietal selection process cannot be guided only by narrow criteria of physical yields.

Above all, research programs need to respond to different kinds of farmers and farming conditions. A technology that may be appropriate for large-scale farms with fertile, level land and ready access to production finance may not be as appropriate for small farms on hillsides and with no collateral. The technologies used on small farms vary enormously even within a district. Purcell and Anderson have commented on the urgency of making research more relevant to the situation of smallholders, especially in demanding agro-

ecological conditions.[29] In the same context, they derive the corollary that research needs to be guided more by farmer demand:

Demand-driven research should involve the intended beneficiaries (farmers and other industry stakeholders) in its design and evaluation. The expansion of on-farm adaptive research encourages beneficiary involvement, but this has not always occurred in projects and often only in a limited way. . . . Researchers must be made aware of the circumstances of farmers, whether through direct interaction with farming communities or their representatives, significant reliance on intermediaries in public or private extension systems, or a combination of these approaches. . . . Regardless of the methods used, this interaction has to be an integral part of the research process.[30]

The logical implications and relevance of this imperative have engendered an approach called 'participatory technology development', in which researchers and farmers become full partners in the process of research and technology dissemination. This approach is based on the recognition that:

scientists alone cannot generate site-specific technologies for the wide diversity of conditions of resource-poor farmers throughout the world, or even within one country. . . . the knowledge and skills of farmers in, for example, influencing soil fertility or managing pests and diseases, will play a key role in developing appropriate technologies.[31]

---

28. Intercropping is not an isolated phenomenon and it can be enormously complex. In a quantitative study of Nigerian agriculture that used extensive farm survey data, the author found that farmers in northern Nigeria typically combined three to five crops in intercropping practices and many combined a larger number, up to nine crops. (These findings were reported in R. D. Norton, 'Pricing Policy Analyses for Nigerian Agriculture', report submitted to the Western Africa Regional Office of the World Bank, Washington, DC, USA, September 1983.)

29. D. L. Purcell and J. R. Anderson, 1997, p. 13.

30. *Ibid.*

31. L. van Veldhuizen, A. Waters-Bayer and H. de Zeeuw, 1997, p. 4.

> *Participatory technology development should strengthen the capacity of farmers and rural communities to analyze ongoing processes and to develop relevant, feasible and useful innovations. . . . The process of technology development is closely linked with a process of social change . . . the planning and assessment obliges the participants to take account of their situation and the responsibilities of different people in the community. . . . (Laurens van Veldhuizen, Ann Waters-Bayer and Henk de Zeeuw,* Developing Technology with Farmers: A Trainer's Guide for Participatory Learning, *Zed Books Ltd, London, 1997, p. 4).*

How to organize such collaboration with farmers and rural communities is a major issue facing national agricultural research systems. This issue has been dealt with successfully through the mechanism of a Local Agricultural Research Committees (known by the Spanish acronym CIAL) in Latin America. Originally organized in Colombia's Cauca Valley by the international research institute CIAT, they have spread to seven other countries (Honduras, Ecuador, Bolivia, Brazil, Nicaragua, Venezuela and El Salvador). Some of the keys to this dissemination have been training trainers, both farmers and researchers, sensitizing national research and extension institutions to avoid trying to deliver technology messages to farmers in top-down fashion, as opposed to having farmers participate in developing them, giving true control over key aspects of the research process to farmers, and endowing each CIAL with a small fund for financing the inputs into the research. Governmental institutions sometimes cannot give cash to particular groups of farmers, and so NGOs play a critical role in the process in this regard as well as working with farmers on research issues. In addition, it was found that the CIALs take root most successfully in localities where the degree of farmer organization is already strong.[32]

The ability of farmers to contribute fruitfully to the research process has been well illustrated not only by the CIALs but also by experiences in Rwanda, Zimbabwe and elsewhere. This ability was exemplified by experiments in which women farmers were asked to make their own varietal selections in the test plots and then their selections were compared with those of researchers:

> Scientists at the Institut des Sciences Agronomiques in Rwanda and at the Centro Internacional de Agricultura Tropical in Colombia collaborated with local women farmers to breed improved bean varieties. The two or three varieties considered by the breeders to have the most potential had achieved only modest increases in yields. The women farmers were invited to examine more than 20 bean varieties at the research stations and to take home and grow the two or three they thought most promising. They planted the new varieties using their own methods of experimentation.
>
> Although the women's criteria were not confined to yield, the breeders' primary measure for ranking, their selections outperformed those of the bean breeders by 60 to 90 percent. The farmers were still cultivating their choices six months later.[33]

Van Veldhuizen, Waters-Bayer and de Zeeuw have pointed out that participatory technology development represents a response to a different ***goal of the development process*** which is more all-encompassing than the usual goal of increasing incomes, or increasing incomes of the poorest groups in the population. It is the goal of helping rural people gain more control over the direction their lives take.[34]

---

32. These lessons and others are reported in Jacqueline A. Ashby, Ann R. Braun, Teresa Gracia, María del Pilar Guerrero, Luis Alfredo Hernández, Carlos Arturo Quirós and José Ignacio Roa, *Investing in Farmers as Researchers, Experience with Local Agricultural Research Committees in Latin America*, International Center for Tropical Agriculture (CIAT), Cali, Colombia, May 2000, in particular, pp. 90–121.
33. The World Bank, 1999, p. 38.
34. L. Van Veldhuizen, A. Waters-Bayer and H. de Zeeuw, 1997, p. 4.

The role of participatory research in developing and enriching human capabilities also has been made by Jürgen Hagmann, Edward Chuma and Oliver Gundani on the basis of a research experience in Zimbabwe:

The integration of formal research into the participatory technology development process enabled both farmers and researchers to jointly develop technologies and have the benefits in terms of data (researchers and policy makers) and a deeper understanding of processes (farmers and researchers).
A very important effect of the process of farmer experimentation, although difficult to quantify, was the gain in confidence and pride of people who have been looked down upon as being helpless peasants. This human factor is the starting point for sustainable bottom-up development.[35]

Working directly with rural communities to strengthen their capacity to articulate their agricultural knowledge and carry out adaptive research on their own farms contributes to enhancing their control over the circumstances of their lives. These authors have raised the issue of defining the goal of development, as a prerequisite for establishing the kind of agricultural research program that is needed, as well as under-scoring the technical contributions of participatory research.

For Latin America and the Caribbean, Echevarría has commented on the slow response of agricultural research systems to the new challenges, and the need for better institutional incentives for the kinds of research required:

National research organizations are being called upon to broaden their agendas and to give greater attention to concerns of poverty alleviation, environmental degradation and resource management. In addition, agricultural technologies are becoming more management intensive, both through the substitution of improved information for environmentally harmful chemicals (e.g. integrated pest management) and through demands on all sectors of society to reduce costs in order to increase competitiveness.
. . . advances in molecular biology and information technology have opened new avenues for agricultural research which can reduce the costs of developing improved technologies. These technologies, however, call for substantial initial investments in human and physical capacity. Given the global trend to privatize knowledge, more public investment is required in basic sciences as a prerequisite for the generation of future streams of technology.
Because government activities in the agricultural sector in most countries of the region have shrunk and because the private sector is not 'filling the gap' (Pray and Umali-Deininger, 1998),[36] the move toward a more poverty-oriented and environmentally sound research agenda has been very slow. Institutional structures to break down the separation of research by commodity and discipline, and incentive systems to develop accountability in terms of impacts at the farm level, are needed to handle these new demands on research systems.[37]

While more appropriate institutional forms for agricultural research need to be developed in each country's own context, one general lesson is that research systems should be decentralized. This topic is taken up in Section 8.4 below.

8.2.3 Gender in Agricultural Research

The relevance of much of existing agricultural research can also be questioned from a gender

35. Jürgen Hagmann, Edward Chuma and Oliver Gundani, 'Integrating formal research into a participatory process', *ILEIA Newsletter*, Center for Research and Information on Low External Input and Sustainable Agriculture, Leusden, The Netherlands, **11**(2), 1995, p. 13. See also Jean-Marie Diop, Marga de Jong, Peter Laban and Henk de Zeeuw, 'Building Capacity in Participatory Approaches,' PTD Working Paper 4, ILEIA, Leusden, The Netherlands, 2001.
36. C. Pray and D. Umali-Deininger, 'Private sector investment in R&D: will it fill the gap?', *World Development*, **26**(6), 1998, pp. 1127–1148.
37. R. G. Echevarría, 1998, p. 1107.

viewpoint. Even when researchers do not pursue a top-down approach, the farmers they consult are likely to be male farmers – in spite of the above-mentioned example of the efficacy of the participation of women farmers in research in Colombia and Rwanda. Often, this is putatively justified on the ground that they are interacting with heads of households. However, in male-headed households women often have significant agricultural responsibilities, and in addition a significant number of rural households are headed by females. For example, in the Dominican Republic female-headed households represent about 22% of the total in rural areas.[38] In Colombia, 'between 1973 and 1985 the share of women in the rural economically active population rose from 14 to 32%'.[39]

This gap in research programs is increasingly recognized as a limitation both for women's development and for the improvement of household welfare in general in rural areas. Thelma Paris, Hilary Feldstein and Guadalupe Duron have summarized the problem in the following words:

> More than twenty years of experience with research and development has shown that technology is not neutral. Women are vital to food security and family well-being and their need for labor-saving and income-generating technologies is acute. However, most research and development programs from the 1970s through the mid-1990s only partly recognized women's contributions to the development process and the effect of the process on them. As a result, new technologies often had detrimental consequences not only to the economic security and social status of women and their families but

also to these programs' and projects' ability to meet national and regional development objectives.

Women's work, particularly in rural areas, is arduous and time consuming. Women and children carrying heavy loads of wood and water, and women pounding grain, are familiar images. Increasingly, though, girls are also headed to school, studying science, and contributing to technology development. Three areas of technology research and adaptation can make substantial contributions to rural women's well-being and empowerment: agricultural production and post-harvest processing, information technology and energy.[40]

Most existing research modalities will have to be modified if they are to start to meet the needs of rural female producers. Improvements in household technologies, almost always ignored by research and extension systems, can play a valuable role in increasing household income, by releasing women's time for more agricultural work. Pareena Lawrence, John Sanders and Sunder Ramaswamy analyzed the effect of both agricultural and household technologies on household incomes in rural Burkina Faso.[41] Since there is very little rigorous empirical evidence regarding the gender effects and total effects of introducing new household technologies, their study merits a close review.

In Burkina Faso, as in many countries, women typically have their own private agricultural plots, and both from cultivating the plots and working off-farm they have their own sources of income. The authors also cite empirical evidence that women are often paid by their spouses for various

38. Elizabeth Katz, 'Gender and Rural Development in the Dominican Republic', note prepared for the World Bank, Washington, DC, USA, mimeo, November 2000, p. 2.
39. Elizabeth Katz, 'Gender and Rural Development in Colombia', note prepared for the World Bank, Washington, DC, USA, mimeo, June 28, 2000, p. 2.
40. Thelma R. Paris, Hilary Sims Feldstein and Guadalupe Duron, 'Empowering Women To Achieve Food Security: Technology', *A 2020 Vision for Food, Agriculture, and the Environment*, Focus Note No. 6, Policy Brief 5 of 12, International Food Policy Research Institute, Washington, DC, USA, August 2001, p. 1.
41. P. G. Lawrence, J. H. Sanders and S. Ramaswamy, 'The impact of agricultural and household technologies on women: a conceptual and quantitative analysis in Burkina Faso', *Agricultural Economics*, **20**(3), May 1999, pp. 203–214.

tasks such as supplying fuelwood and cultivating rice fields. Their analysis was carried out under three alternative assumptions about decision-making within the household on labor allocation and wages: exploitative (male controlled; paying women the traditional wage regardless of their marginal productivity), bargaining (between spouses) and altruistic (males paying females at least their marginal productivity). They point out that in reality most household decision-making is some variant of the bargaining mode,[42] but by covering the entire spectrum of options their results are more robust. In their analysis, the choice of household decision-making mode affects women's incomes but does not affect total household income. The household technologies considered include improved fuelwood stoves, pestles with steel tips, parboiled sorghum, and wells with pumps that are closer to the village. The new agricultural technologies include application of moderate amounts of inorganic fertilizer and pesticides and use of new cultivars of cotton and maize.

Their results (pp. 211–213) show that adoption of the new agricultural technologies alone increases farm household income from the family plot by 26% (animal-traction households) to 58% for hand-traction households, and that adoption of both the agricultural and household technologies leads to further increases in household income of 11 to 12%. For women, the effects of new agricultural technology vary markedly with the decision-making mode, as expected. The resultant increase in their incomes ranges from zero, under the exploitative model, to 25–60% under the other decision-making modes, depending on the household's form of field traction. However, the effects on women's incomes of the introduction of new household technologies (with adoption of new agricultural technology) were rather constant over types of decision-making. Women's incomes increased by a further amount of 30–38% as a result of the improvement in household technologies alone.

Although these results refer only to one case, they suggest the potential value of emphasizing improvements in household technologies as well as production technologies, especially from a viewpoint of promoting greater gender equity.

It is frequently remarked that women-headed rural households are slower to adopt new agricultural technologies than those headed by men. This phenomenon warrants close analysis in order to facilitate better design of adoption strategies, and understanding it is central to improving the status of rural women through technological advances. Cheryl Doss and Michael Morris have analyzed this question with data from a national survey of maize growers in Ghana, and they also comment that 'A wealth of case study evidence suggests that female-headed households are less likely to adopt new technologies than male-headed households', citing evidence from Malawi and Zambia in particular.[43] Their empirical conclusions for Ghana are that gender itself does not determine adoption rates, but rather land ownership, ability to hire labor, education, contact with extension services, and market access are the main determining factors and are the reasons why male-headed households display greater adoption rates:

after we control for a farmer's age and level of education, access to land and labor, contact with the extension service, and market access, there is no significant association between the gender of the farmer and the probability of adopting modern varieties or fertilizer.... Failure to control for gender-linked factors can lead to misleading conclusions about the importance of gender *per se* as an explanatory factor.

---

42. *Op. cit.*, p. 209. There is some evidence from Mali that sometimes male-dominated household decisions make women worse off with the introduction of new field technologies, since they are paid less than their previous returns from labor on their own private plots (*op. cit.*, p. 208).

43. C. R. Doss and M. L. Morris, 'How does gender affect the adoption of agricultural innovations? The case of improved maize technology in Ghana', *Agricultural Economics*, **25**(1), June 2001, p. 32.

Given the experiences with women as clients of rural financial institutions that are mentioned in Chapter 7 of this volume, it is reasonable to expect that women also would be good farm managers. The Doss and Morris study concludes, for that particular case at least, that women are as amenable to implementing new technologies as men are. It is other kinds of constraints that may inhibit them from doing so as rapidly as men.

8.2.4 The Public and Private Sectors in Research

The revolution in agricultural research that has sprung from advances in molecular biology has brought into sharper focus the issue of *the role of public and private sectors* in that research. The justification for the traditional role of the public sector arises from the public-good nature of much of agricultural research: once an innovation emerges from the laboratory, farmers cannot be excluded from participating in its benefits, and the fact that one farmer reaps the benefits of a research innovation does not reduce the amount of the innovation available to other farmers. Obviously, this characterization has never applied to many input-specific innovations, such as improved agrochemicals and hybrid seeds. Nevertheless, even the basic research that underpins that class of innovation can satisfy the conditions of a public good in some cases. Research that leads to improvements in self-pollinating crop varieties and to discoveries of better methods of crop management in the field and after harvest has always fallen in the domain of 'public goods'. In these cases, since a private company cannot appropriate a share of the economic benefits of the research, the private sector will have little interest in funding it.

Advances in biotechnology have opened greater possibilities for research results that are appropriable by private firms, i.e. that are not public goods. These results include a wide variety of hybrids that have desired agronomic characteristics and qualities preferred by agro-industry and consumers, plus plants with desirable characteristics that may be triggered by the application of certain chemicals in the field. While the fact that products of research laboratories may be appropriable acts as an incentive for private investment in research, the private sector may not respond in ways that are consistent with development aims of poorer countries:

Clearly defined and enforceable intellectual property rights are essential for private-sector research and development of new biotechnology products. However, overly broad patents may grant excessive market power to patent holders, reducing their incentives to provide socially desirable levels of production or investment in innovation. Unduly broad patents and/or overly restrictive licensing of academic inventions will diminish the capacity for new entrants to compete. . . .

Developed countries should not be overzealous in their enforcement of intellectual property rights in developing countries. First, excessive fees will encourage cheating and, second, undue emphasis on IPR protection may conflict with other goals, such as promotion of free trade. Consideration should be given to establishing two-tiered pricing systems for intellectual property rights, with developing countries paying lower prices.[44]

Pinstrup-Andersen argues that the amount of private research will be less than socially optimal on two grounds: it is difficult to enforce patent rights in developing agriculture, and even if they could be enforced, a significant share of the research benefits accrue to consumers through lower food prices, and therefore the entity generating the research can never capture its full economic benefit. In his words:

---

44. David Zilberman, Cherisa Yarkin and Amir Heiman, 'Agricultural Biotechnology: Economic and International Implications', in *Food Security, Diversification and Resource Management: Refocusing the Role of Agriculture?* G. H. Peters and J. von Braun (Eds), Proceedings of the 23rd International Conference of Agricultural Economists, International Association of Agricultural Economists, Ashgate Publishing Ltd, Aldershot, UK, 1999, p. 160.

neither computer chips patented by Intel nor Round Up Ready soybean seed patented by Monsanto are public goods. . . . The intellectual property rights are clearly defined. However, enforcement of these rights is likely to be much more difficult in the case of biological technology such as improved seed because, contrary to computer chips, seeds multiply and the farmer may use his own seed in future planting without paying the original owner, e.g. Monsanto. Although farmers may enter into contracts with seed companies agreeing not to use their own seed, such contracts are difficult to enforce. . . .

But even if the private research agency could enforce property rights, for example through hybrid seed or built-in gene switches [triggered by

applying chemicals], research investments by the private sector would be less than socially optimal. The reason is that . . . consumers, would benefit through lower prices. Since the private research agency does not have the right to tax consumers, the benefits derived by the farmers will set limits for how much the agency can capture.[45]

Much of the discussion in this section has been concerned with the sector-wide, or aggregate, effects of agricultural research and their implications for the structure of research systems. Recently, another class of issue has been raised: *whether agricultural research has contributed to poverty alleviation in developing countries*, and what can be done to make it more effective in that regard. The issue is related to the earlier one of how to make research more relevant to the needs of smallholder farmers.

Some empirical evidence on this issue for the case of India indicates that agricultural research has made a contribution to poverty reduction, but that a change in research priorities will be needed for it to continue to contribute to that aim and also promote overall growth in the sector. Past gains in poverty reduction have come through expansion of the irrigated area, but now the returns to additional irrigation are declining, and some rainfed areas may be showing the greatest returns to agricultural research.[46]

However, skeptical notes have been sounded about the effectiveness of trying to base research priorities on poverty alleviation goals.[47] This issue is also taken up again later in this chapter.

8.2.5 Pest Management

Montague Yudelman, Annu Ratta and David Nygaard have collated the available, very approximate aggregate data on crop damages attributable to pests of all kinds, including pathogens and

---

Pinstrup-Andersen suggests new modes of international co-operation with the private sector:

Such a situation calls for publicly funded agricultural research. Strong national agricultural research systems (NARS) focused on solving problems facing poor farmers and consumers are likely to make major contributions to both efficiency and equity goals. These contributions would be significantly enhanced through innovative partnerships with private sector research agencies in which non-exclusive rights to processes and traits are transferred from the patent holder to NARS for restricted use in research to develop technology for eco-regions and commodities of little or no commercial interest to the patent holder. The private research agency holding the patents would, in turn, expect to improve public relations and develop new markets as poor farmers who benefit from the technology become customers (from P. Pinstrup-Andersen, 2000, pp. 3–4).

---

45. P. Pinstrup-Andersen, 2000, p. 2.
46. S. Fan, P. Hazell and T. Haque, 'Targeting public investments by agro-ecological zone to achieve growth and poverty alleviation goals in rural India', *Food Policy*, **25**(4), August 2000, pp. 426–427 (see also the further discussion of this point in Section 8.4).
47. See D. Byerlee, 'Targeting poverty alleviation in priority setting for agricultural research', *Food Policy*, **25**(4), August 2000, p. 442.

weeds. They conclude that crop losses from these causes are somewhere between one-third and one-half of production in the world, and higher in developing countries than in developed countries. Insects appear to be the largest cause of crop losses, and pathogens are next, and then weeds.[48]

As a consequence of this disturbing finding, these authors offer a number of reflections and recommendations, including the following:

The possibilities of rising costs of production and the slowing down of yield increases from the existing technology make it timely to reconsider some of the options and priorities for raising productivity and increasing future food supplies. Any such review should include examining whether higher priority should be given to reducing wasteful and unnecessary crop losses and protecting crops from pests. . . . Theoretically, higher priority for improving crop protection would be warranted up to the threshold where the marginal costs of reducing losses are equal to the marginal costs of expanding a comparable volume of production by other means. . . .

One of the obstacles to formulating any such strategy is that the current state of knowledge about actual losses from pests and the gains from improved pest management leaves much to be desired. . . .

The concept of integrated pest management has gained strong support among environmentalists as well as among agriculturalists. . . . there is still no agreed-upon definition of IPM. However, in its broadest sense IPM involves moving from a chemically based treatment of pests to a biologically based treatment. At present, most systems of crop protection in developing countries, outside of traditional agriculture, are chemically based (especially for cotton, export crops and rice). . . .

The promotion of IPM requires an effective, easily managed method that can be introduced on a large enough scale to offer what chemical pesticides currently deliver: insurance against pest damages and acceptability among smallholders who can ill afford any losses. For this to happen, international development agencies, governments and others will have to make major commitments to IPM, including the commitment of resources to develop and promote this form of management. This will involve both *acquiring new knowledge about improving pest management through research, as well as disseminating information that is already known.* It will also involve educating and organizing producers so that they can apply that knowledge. This will be no easy task. The experience in Indonesia and elsewhere points to the importance of sustained government commitment and support and the introduction of innovative approaches for persuading small-scale, risk averse producers to adopt new approaches to pest management.[49]

In fact, chemical treatment of pests on a regular basis is often an uneconomic strategy for farmers. Gains in yields may be offset by the cost of pesticides, and over time pests may develop resistance to the chemicals. As far back as the 1960s, growing resistance to chemicals made it necessary to eventually treat cotton as many as 30 to 40 times per growing cycle in North-East Mexico, and for that reason cotton cultivation was abandoned in that area.

Use of pesticides on rice in Asia also has been found to be frequently uneconomic:

On-farm experiments and examination of farmers' yields do not suggest any positive yield or profitability response to insecticide applications. . . . Herdt *et al.* (1984)[50] concluded that

---

48. Montague Yudelman, Annu Ratta and David Nygaard, *Pest Management and Food Production: Looking to the Future*, Food, Agriculture and the Environment Discussion Paper 25, International Food Policy Research Institute, Washington, DC, USA, 1998, p. 8.

49. *Op. cit.*, pp. 40–42 [emphasis added].

50. R. W. Herdt, L. Castillo and S. Jayasuriya, 'The economics of insect control in the Philippines', in *Judicious and Efficient Use of Pesticides on Rice*, Proceedings of the FAO/IRRI Workshop, International Rice Research Institute, Los Baños, Laguna, Philippines, 1984.

the expected returns to rice production are lower for farmers applying insecticides on a prophylactic basis rather than not applying insecticides at all. This result was validated by on-farm trials of alternative pest control practices conducted by Litsinger (1989)[51] and Waibel (1986)[52]. Both Litsinger and Waibel observed no significant yield differences between the insecticide treated and untreated plots in more than half the cases.
... Rola and Pingali[53] found, for lowland tropical rice systems, natural control to be the economically dominant pest management strategy. Natural control, in association with varietal resistance, proved to be consistently more profitable in an average year than pro-phylactic treatment.... The dominance of natural control becomes even greater when the health costs of pesticide exposure are accounted for.[54]

The magnitude of the pest problem has raised questions about the priorities for international agricultural research, as exemplified in the following excerpt from a letter by Wightman:

As an integrated pest management (IPM) specialist of some 30 years' standing, I was pleased to have been able to attend the discussion on pest management held at IFPRI in September. ... we learned that 50 percent of the world's food production is lost to pests. So why is the CGIAR focusing (1) on soil and water research, a topic that has, for the most part, remained impervious to progress during the life and in the context of the CGIAR, and (2) on biotechnology/breeding, where progress can

normally only be measured in 1 or 2 percent increments per year? The CGIAR will only be able to demonstrate impact if it attacks the most important constraints to production (which differ considerably from crop to crop) and leaves the peripheral issues to location-specific enterprises, where there is a history of success.[55]

Equally, national agricultural research systems should re-evaluate the priority they assign to research on pest problems and integrated pest management, as opposed to plant breeding and other kinds of research.

The FAO's Farmer Field Schools (FFSs) have made progress in promoting integrated pest management and soil fertility management through a participatory research and extension mode. Farmer Field Schools have been established in more than 40 countries in Asia, Africa and Latin America. Their aim is to formulate locally viable approaches as a result of combining prior scientific information with the results of farmers' own experiments:

The concept behind an FFS is that groups of farmers meet on a regular basis in a field to do practical structured learning exercises that allow them to combine local knowledge with scientific ecological approaches.... All courses are very hands-on, practical and field-based, with few or no lectures and using the field itself as a teacher....
The extension officer's role has evolved from that of a primary knowledge source to that of the facilitator of knowledge creation.... The FFS methods have transformed farmers

---

51. J. A. Litsinger, 'Second generation insect pest problems on high yielding rices', *Tropical Pest Management*, **35**, 1989, pp. 235–242.

52. H. Waible, 'The economics of integrated pest control in irrigated rice: a case study from the Philippines', in *Crop Protection Monographs*, Springer, Berlin, 1986.

53. A. C. Rola and P. L. Pingali, *Pesticides, Rice Productivity, and Farmers' Health – An Economic Assessment*, World Resources Institute and International Rice Research Institute, Los Baños, Laguna, Philippines, 1993.

54. P. L. Pingali and R. V. Gerpacio, 'Living with reduced insecticide use for tropical rice in Asia', *Food Policy*, **22**(2), April 1997, pp. 112–113.

55. John A. Wightman, letter in *News and Views, A 2020 Vision for Food, Agriculture and the Environment*, International Food Policy Research Institute, Washington, DC, USA, November 1998, p. 7.

from recipients of information to generators and manipulators of local data. . . .[56]

In all of the experiences of the Farmer Field Schools, emphasis has been placed on developing a research agenda attuned to local needs in each area. 'If farmers feel that they are getting a "national" curriculum they may avoid the FFS'.[57] The Farm Field School approach has evolved into a program for creating a capacity in each community to develop its own methods of IPM. The goals of this 'Community IPM' are to foster conditions in which farmers:

- act upon their own initiative and analysis;
- identify and solve relevant problems;
- conduct their own local IPM programs that include research and educational activities;
- elicit the support of local institutions;
- establish or adapt local organizations that enhance the influence of farmers in local decision making;
- employ problem solving and decision-making processes that are open and egalitarian;
- create opportunities for all farmers in their communities to develop themselves and/or benefit from their IPM activities;
- promote a sustainable agricultural system.[58]

The success of the FFS experience in Indonesia in widening the horizons of farmers and encouraging their initiative was confirmed by a government official in charge of agricultural services:

Experience in Gerung[59] has shown that IPM farmer alumni are inclined to conduct field studies. For example, IPM trained farmers studied the effectiveness of SP 36, conducted variety trials, analyzed the effects of defoliation, and tested various planting distances and their influence on yields. Alumni conducted demonstrations for themselves and others on the ability of plants to compensate for damage caused by pests. There are other things that you can see regarding IPM alumni. They are creative, dynamic, and have taken on the leadership for developing a sustainable approach to agriculture.

(Ir. L. L. Noverdi Bross, Head, Provincial Agriculture Service[60])

The FFS program in general, and the Community IPM approach in particular, represent a successful application of the concept of participatory research and extension. This theme is discussed further in other parts of this chapter.

8.3 ISSUES IN AGRICULTURAL EXTENSION

8.3.1 The Historical Development of Extension in LDCs

The approach to agricultural extension in developing countries has changed considerably over the past five decades and still is undergoing evolution. Cogent summaries of the historical development of extension in these countries have been provided by Antholt (1998) and Picciotto and Anderson (1997). In brief outline, that history runs as follows:

Fifty years ago agricultural extension organizations in developing countries mirrored the administrative traditions of former colonial powers. . . . Like other agricultural support

---

56. Kevin D. Gallagher, 'Community study programmes for integrated production and pest management: Farmer Field Schools', in M. K. Qamar, 2000, p. 62.
57. John Pontius, Russell Dilts and Andrew Bartlett (Eds), *Ten Years of Building Community: From Farmer Field Schools to Community IPM*, FAO Community IPM Program, Jakarta, Indonesia, 2000, p. 36.
58. *Op. cit.*, p. 39.
59. Gerung is a Sub-District of West Lombok District on the island of Lombok in West Nusa Tenggara Province, Indonesia.
60. Quoted in J. Pontius, R. Dilts and A. Bartlett (Eds), 2000, p. 45.

services, extension services were geared to producing and marketing export commodities. . . . extension programs often relied on the proposition that farming productivity was held back not so much by technology and economic constraints as by farmer apathy, inadequate social arrangements, and lack of local leadership (Picciotto and Anderson, 1997, pp. 249–250). . . . there was a high degree of confidence in the ability of Western agricultural technology to solve the needs of the 'hungry, poor and ignorant' in the developing world. . . . The problem of developing agriculture was seen as one of accelerating the rate of growth of agricultural output and productivity via what became to be known as the 'diffusion model' of agricultural development. . . . In that model the process was conceived as a hierarchical, unidirectional process; it provided to traditional agricultures new technology, usually from the West, which was delivered to farmers by extension workers in departments of agriculture (Antholt, 1998, p. 355).

In the 1950s and the early 1960s, the agricultural extension service tended to be subordinated to multipurpose rural development programs. Extension agents carried out a variety of functions, ranging from credit delivery and input distribution to sundry coordination duties. And because extension agents were among the few government officials available at the village level, they were often asked to undertake clerical, statistical, or even political chores. Typically the service had only weak connections to agricultural research. Looking back, the rural development movement was the victim of a poor enabling environment for agricultural development. Eventually it fell into disfavor as lack of profitable technical packages and an overly broad agenda led to a thin spread of resources, excessive administrative costs, and slow agricultural production growth (Picciotto and Anderson, 1997, p. 250).

. . . the results of village-level studies in the 1950s and early 1960s documented that peasant farmers were 'poor but efficient', and that lack of profitable technology was a major cause of stagnation. Schultz's pioneering book, *Transforming Traditional Agriculture* (1964), drew on these studies to challenge the extension/diffusion model and urged developing countries and donors to shift their resources from extension to building agricultural research capacity. . . . The diffusion model obscures the fact that farmers are innovators, not just passive receptacles of information.

Unfortunately . . . these legacies . . . generally reinforced the limited, linear, and sequential view of how information and knowledge need to be developed and made accessible to farmers – that is, from basic science to applied science to technological innovations to farmer recommendations. . . . in the early 1970s, after the first flush of the Green Revolution, there was a sense among many agriculturalists that there was a backlog of technology yet to be moved to farmers. It therefore followed that it was necessary to increase the intervention capacity of extension through more staff, more training, more buildings, more motorcycles, etc. . . . The aim of the T & V approach was to reform the **management** of extension systems and turn a cadre of poorly supervised, poorly motivated, and poorly trained field agents into effective technology transfer agents, through fortnightly training of agents, who then made regular visits to farmers, conveying clear extension messages (Antholt, 1998, pp. 355–356).

Yet the degree to which the remarkable food production gains of the green revolution can be attributed to any particular mechanism, such as T & V, has long been disputed. . . . This said, T & V has dominated extension in South Asia and Africa for more than two decades, partly because of the strong support offered by the World Bank (Picciotto and Anderson, 1997, pp. 250–251).

Today, extension is viewed much more broadly, as indicated by comments earlier in this chapter. In addition to being carried out through a variety of approaches and institutions, it is seen to be part of a broader 'agricultural knowledge and information system for rural development (AKIS/RD)', of which other principal components are agricultural research and agricultural education. In this view, knowledge generation and dissemination do not proceed linearly but rather are interactive and are the result of joint efforts among different kinds of participants. The system needs to generate mutual learning and exchanges of information in order for the sector to advance at a satisfactory pace. However, the starting point for designing improvements in the system is a full recognition of its shortcomings that still exist throughout the developing world:[61]

- Farmers' needs do not sufficiently drive the orientation of research and extension, and labor market requirements are not adequately translated into curriculum design in agricultural training institutions. . . .
- The know-how and technologies that are produced by AKIS/RDs, even when relevant, are not widely taken up by farmers, suggesting a lack of effective transfer. Concerns over cost-effectiveness mean that public research and extension services have trouble ensuring their financial sustainability.
- Public decision-makers are often unaware of the actual results achieved and the long-term resource allocations needed. . . .
- In many settings, the quality of human capital in AKIS/RDs is low, suggesting that investments in human capital formation are inadequate and that the training and educational institutions themselves are insufficiently responsive to changing demands.
- A lack of systematic collaboration among educators, researchers, extension staff and farmers has limited the effectiveness and relevance of support services to the rural sector.

The responses to these deficiencies are diverse and are developed by different kinds of institutions. The diversity is likely to be a permanent feature of the institutional landscape in agriculture. 'It can . . . be reasonably argued that no single approach best suits extension development in all circumstances. . . .'.[62]

8.3.2 The Rationale for Public Extension Services

Agricultural extensionists are *intermediaries* between farmers, on the one hand, and researchers, input and credit suppliers, marketing agents and other agents that intervene in agriculture, on the other hand. Thus, fulfillment of their role requires management of two-way flows of information and communications skills as well as technical knowledge. Often, their main role is *to stimulate a learning process* in which both they and farmers participate. John Farrington has listed the four principal functions of agricultural extension:

- *Diagnosis* of farmers' socio-economic and agro-ecological conditions and of their opportunities and constraints.
- *Message transfer* through training courses and mass media, and through direct contact between extension agent and farmer or indirect contact involving intermediaries, such as 'contact farmers' or voluntary organizations. Messages may comprise advice, awareness creation, skill development and education.
- *Feedback* to researchers on farmers' reactions to new technology to refine future research agenda.
- *Development of linkages* with researchers, government planners, NGOs, farmers' organizations, banks, and the private commercial sector. In remote areas, extension agents have taken on a number of input supply functions directly.[63]

---

61. FAO and The World Bank, *Agricultural Knowledge and Information Systems for Rural Development (AKIS/RD), Strategic Vision and Guiding Principles*, Food and Agriculture Organization of the United Nations, Rome, 2000, pp. 7–8.
62. William M. Rivera, 'Agricultural and Rural Extension Worldwide: Options for Institutional Reform in Developing Countries', paper prepared for the Extension, Education and Communication Service, Draft, Food and Agriculture Organization of the United Nations, Rome, October 2001, p. 9.
63. J. Farrington, 'The changing public role in agricultural extension', *Food Policy*, **20**(6), December 1995, p. 537.

It can be added that the messages delivered by extension agents can include information on public-sector programs in which farmers may be eligible to participate. In an era when increasing emphasis is being given to policy measures involving direct support to farmers, as opposed to market interventions, extension agents can inform farmers of the nature of those measures, and also provide feedback to the government that is useful in designing such measures. In addition, the HIV/AIDS crisis is imposing special demands on extension agents in many countries, as discussed later in this chapter.

In regard to the purely technical role, the information transmitted to farmers by extension activities takes two forms: that which is embodied in physical inputs (machinery, seeds, etc.), and pure information that is not embodied in goods. Umali-Deininger has classified pure information into four categories:

- *Cultural and production techniques*, such as timing for planting and harvesting, use of inputs, animal husbandry and livestock health, crop protection and farm-building design.
- *Farm management*, such as record-keeping, financial and organizational management, and legal issues.
- *Marketing and processing information*, such as prices, market options, storage procedures, packing techniques, transport and international standards for quality and purity.
- *Community development*, such as the organization of farmers' associations.[64]

Extension services traditionally concentrated on providing the first and last types of information. Increasingly, it is necessary to provide the second and third types as well. Improving productivity requires attention to farm management and not just cultivation techniques. Adapting to more open trade regimes and shifting to higher-value crops requires timely access to marketing and processing information, and the ability to translate it into actions on the farm. This circumstance points to the importance of basic education for increasing production and incomes on farms.

The rationale for public provision of these functions has been the public-good nature of information about many agricultural technologies. The classical justification for the role of the public sector in disseminating such information is that it is likely to filter from one farmer to another. It cannot be provided only to one person, excluding others from access to it. In addition, the value of the information is not diminished by an increase in the number of recipients of it. There are specialized kinds of information that are marketed by combining them with the sale of inputs, but many kinds of knowledge about agricultural technology cannot be marketed in that way.

In her seminal paper, Umali-Deininger has refined the public-good argument concerning agricultural extension. She applies the principles of *rivalry* and *excludability*. 'Rivalry (or subtractability) applies when one person's use or consumption of a good or service reduces the supply available to others. . . . Excludability applies when only those who have paid for the product or service benefit from it' (Umali-Deininger, 1997, p. 208). Most goods are rival in the sense that one person's purchase of a good makes it unavailable for others. Services can exhibit different degrees of rivalry, including decreasing rivalry over time as, for example, information provided to a group gradually leaks out to others.

Purely private goods are those that are both rival and excludable; *purely public goods* are neither. A tractor is an example of the former, and mass communication of agricultural information is an example of the latter. However, many kinds of agricultural information, including those that extension agents are required to deliver, fall somewhere in between these extremes. Umali-Deininger suggests that the modalities of delivering information need to be carefully designed in recognition of its inherently public

---

64. D. Umali-Deininger, 1997, pp. 206–207.

Fulfilling the role of an extension worker requires special aptitudes. Miguel Angel Núñez offers the following list of characteristics of the 'new extension worker':

- *Be a native of the area in which he or she works and have family links there.*
- *Be familiar with the cultural values of the area.*
- *Have knowledge of forms of mass education.*
- *Know agro-ecological techniques.*
- *Have experience in participatory undertakings at the community level.*
- *Have experience in training.*
- *Have a commitment from the organizations that sponsor her/him to continue to disseminate the training process throughout the region.*

(Miguel Angel Núñez, 'La extensión agrícola en el marco del desarrollo sustentable', Políticas Agrícolas, IV(1), 1999, p. 61 [author's translation].)

and private characteristics, otherwise they will fail to perform to expectations. To clarify the area between purely private and purely public goods, she proposes use of the concepts of **toll goods** and **common pool goods**. In her own words:

Toll goods are excludable, but not rival; for example, the supply of information provided by a private extension consultant exclusively to a group of farmers is not reduced by the addition of another member to the group. . . . **Common pool goods** are those that are rival but not excludable; in other words, other people cannot be stopped from using them. For example, the purchase of high-yielding self-pollinating seeds such as rice and wheat reduces the supply of such seeds, but their ease of replicability makes exclusion difficult and costly in the long run. Farmers do not buy rice and wheat seeds every season, because they can set aside part of their harvest for planting the next crop (Umali-Deininger, p. 208 [emphasis added]).

This approach to analyzing the information flows relevant to agricultural technology leads Umali-Deininger to some conclusions about the role of extension services. Her principal conclusions are the following:

Information designed to improve existing cultural and production practices, farm management, or marketing and processing techniques and provided by traditional agricultural extension approaches is a toll good in the short term. . . . But the diffusive nature of nonexcludable information transforms it into a public good quickly. . . . Thus, how quickly information is diffused determines whether the private sector has an incentive to provide it. If the information diffuses easily, the possibilities of charging for it are limited, and private firms will have little or no incentive to provide such services . . . delivery of nonexcludable information will remain the responsibility of the public sector or of private nonprofit agencies. . . .

As farm operations become more commercialized and agricultural technology more specialized, the corresponding extension services need to support these activities also become highly specialized. Such specialization leads to exclusivity of the information and, therefore, the extension activity. For example, the results of a soil analysis or the development of computer programs to facilitate farm operations are location- and client-specific. Such information may not be useful to other farmers. . . . Asymmetric information problems, however, increase the difficulty of assuring quality. Unless the private fee-for-service extension industry can effectively police itself to ensure the quality of the information communicated, public intervention will be necessary to enforce quality standards and legal contracts. . . .

Medium- and large-scale producers can spread the cost [of private extension services], resulting in lower per-unit costs and higher rates of return. Consequently, the larger the farm operations, the greater the potential demand for 'fee-for-service' extension.

. . . because the value of their marketable output is low, resulting in higher per-unit costs, small-scale farmers typically find it less attractive or profitable to 'purchase' the extension service. Subsistence farmers have limited, if any, incentive to pay for extension services.

Government policies can greatly affect the demand for extension services, through their (direct and indirect) influence on commodity prices and aggregate demand. High (direct and indirect) taxes on agriculture reduce farmers' incentives to adopt improved technologies. . . . The allocation and level of public expenditure on rural roads, markets, and irrigation infrastructure, for example, influence the development potential of particular localities and thus the return on investments in technologies that enhance productivity. Public expenditures on education, especially in rural areas, have a strong influence on the capacity of farmers and consumers to absorb new information.

A major implication of the shift in the classification of information from a 'free good' to a 'purchased good' is that the demand for paid agricultural extension services will originate almost exclusively from market-oriented farming operations and particularly from medium- and large-scale farmers. . . . Conversely, private-for-profit firms will tend to neglect areas composed of more marginal farmers. . . . (Umali-Deininger, pp. 210–211 and 215–216).

This forceful conclusion refers to *unsubsidized extension services*. Umali-Deininger adds the qualification that small farmers may be able to purchase extension services in groups. Her broad recommendation regarding the financing role of the public sector in extension, including the option of subsidization of privately supplied services, is as follows:

When should extension be funded by the public sector? Where extension delivers public goods and information with high externalities, such as environmental or conservation-related information, complete privatization is neither desirable nor feasible. Two other arguments could justify public subsidization of extension to small farmers: first, *when small farmers may be*

unaware of the benefits of improved technologies and unable to afford them; and second, when small subsistence farmers may derive considerable non-monetary benefits (including better nutrition and health) from adopting new technology (p. 217 [emphasis added]).

This conclusion is part of an increasing consensus that there is a role in agricultural extension both for the public sector and for unsubsidized private services, as well as for public services that are offered at a cost to clients. Umali-Deininger's analysis marks an important step forward in understanding the role of public and private sectors in agricultural extension. While her framework provides the first truly systematic basis for dealing with these questions, her analysis can be made more realistic by including the role of uncertainty and other factors that impinge heavily on farm decisions. Consequently, the outlook for private extension services may be brighter than she suggests, albeit with subsidies for small-scale farmers to enable them to purchase those services.

Even if the information provided by extension proves to be non-excludable as time passes, farmers may wish to purchase it for any of the following reasons:

(1) There is a *premium on the timeliness and quality of information*. Farmers may not wish to wait to obtain information from other farmers through the diffusion process, and they may be fear that second-hand information is not as accurate as it was originally. Thus they may be willing to pay for information that is more timely and of higher quality.

(2) As a corollary to the timeliness issue, *access to information can raise barriers to entry for potential competitors*. Even if information is non-excludable eventually, those who gain first access to it may be able to sew up markets, thus effectively excluding competitors even if the latter obtain the same information. This consideration can be particularly relevant to non-traditional products. For example, in Honduras low-income rural women in the village of Sabana Grande have

recently specialized in eggs and certain kinds of flowers, shipping container-loads of both products to nearby Tegucigalpa each month. Those who try to emulate them in the market for flowers would face lower prices, since the women from Sabana Grande are now meeting the bulk of the demand for certain varieties.

(3) *Uncertainty* is omnipresent in agriculture. Entering into a contract for extension services may represent a form of insurance against the eventuality of infestations of other problems that require a rapid response, even if the recipe for response eventually filters out to other farmers. A farmer who is on the verge of losing his or her crop needs a rapid solution and may not be concerned that other farmers may be able to obtain the same information at no cost in due course.

These considerations imply that the potential market for commercial agricultural extension is greater than Umali-Deininger's analysis implies. However, income constraints and seasonal cash shortages limit the ability of poor farmers to pay for extension services. If poverty alleviation is a national goal, then a public subsidy to poor farmers to enable them to obtain extension services may be justified. The modality of granting such a subsidy almost always will require organization of poor farmers into groups of beneficiaries, but the case for the subsidy can be strong on grounds of poverty alleviation alone.

8.3.3 The Performance of Public Agricultural Extension

In practice, public agricultural extension services often have proven disappointing. Funding and management deficiencies have led to the frequently observed syndrome of extension agents spending more time in the office than on farms, and linkages between extension and research services generally have been weak. In the words of William Rivera, 'In many low-income developing countries, agricultural and rural extension is in disarray, a fact that bodes ill for countries that must now accommodate the new paradigm that is increasingly being shaped by global trends toward market-driven and highly competitive agribusiness enterprises'.[65]

However, this has not always been the case. In the late 1970s the present author visited a village in The Republic of Korea where rice farmers had recently begun to cultivate tobacco, under plastic sheds. Sitting with them under the plastic, I asked where they had learned to grow tobacco. 'From the government extension agent', was the reply. I asked how often he came by. 'Oh, he comes every day', they said. Looking at their watches, they added 'He will be here in a half hour'.

In contrast, in 1972 a Mexican Government colleague and I calculated that the average Mexican farmer saw an extension agent once every forty years.

A different kind of issue was observed in Honduras: a marked bias in the provision of public extension services in favor of large-scale farmers. In 1990 Gilberto Galvéz et al.[66] carried out a large-scale survey of economic and social characteristics of Honduran farms. In the area of agricultural extension, the survey asked whether services were provided in a timely fashion or not, for timeliness is of the essence when an infestation is affecting a crop. It also asked whether those services were of good, fair or poor quality. The results were tabulated in two groups, i.e. one for the extension service provided by the Ministry of Natural Resources (MNR) and the other for the service of the National Agrarian Institute (INA) for the agrarian reform sector. Of the smallest farms receiving extension from the MNR (defined for this purpose as those with less than 10 ha), 39% said the extension service was timely and of good quality. Of the largest farms (50 ha

65. W. M. Rivera, 2001, p. 1.
66. Gilberto Gálvez, Miguel Colindres, Tulio Mariano González and Juan Carlos Castaldi, *Honduras: Caracterización de los Productores de Granos Básicos,* Secretaría de Recursos Naturales, Tegucigalpa, Honduras, November 1990.

or more) who were clients of MNR, 72.7% responded that way. The bias in favor of large farms was even greater for clients of INA's extension service. Of the smallest farms,[67] only 20.2% said the extension service was timely and of good quality, whereas 81.7% of the largest farms responded that way.

Since public extension services in Honduras have been free, as in virtually all developing countries, this bias in favor of larger farms represents a *regressive subsidy* in public expenditure. Anecdotal evidence suggests the bias occurs in other countries as well.

One of the most serious systemic problems with public agricultural extension services is the lack of adequate incentives for the agent to serve the client well. The client or customer is the farmer, and most extension services have not had a strong orientation of 'serving the customer'. This has translated into lack of timeliness of the services, lack of responsiveness to the farmer's own problems, which may be different from those envisaged by researchers, and in the worst cases a total lack of attention to the majority of the farmers.

This absence of a customer orientation is largely attributable to the incentive system under which extension agents operate. They are not paid by their clients as a function of the adequacy of services rendered. Their income comes from a large bureaucracy which has a limited ability to monitor the quality of the services they have provided. Appointments may be at least partly political and indifferent field performance often is not punished; indeed, an agent who does not meet clients' needs may continue to receive promotions. Clearly, this critique is not true of all extension systems, nor of all agents in the weaker systems, but it is relevant to many situations. It points to the need to restructure the incentives aspect of extension systems, as well as strengthening the links between extension and research and making other reforms.

The opposite side of the accountability coin is *expecting the beneficiaries of extension to be responsible for some of the support*, even if it is only a proportion of total costs. This is important for three reasons. First, it gives the beneficiaries ownership and drawing rights on the services. Secondly, it takes some of the financial pressure off of the central government and therefore responds to the issue of financial sustainability. Lastly, if ownership and responsibility rest with clients, the basis for more demand-driven, responsive service is established. An example of this has been the National Farmers' Association of Zimbabwe.[68]

The most widely adopted approach to agricultural extension in recent decades has been the

> The most important policy initiative that is needed in extension is to shift the primary focus of power and responsibility for extension to the clients. To borrow Robert Chambers' phrase, we need to 'put the farmers first'. . . . There is abundant evidence that the 'normal' incentive system facing government employees, even under the most enlightened circumstances, puts a premium on not making a mistake and on length of service but not necessarily on service to clients, particularly small farmers. This is not acceptable and does not have to be taken as inevitable. Sims and Leonard found that the most important determinant of extension success is the strength of farmer organization (C. Antholt, 1998, pp. 360–361; referring to Holly Sims and David Leonard, 'The Political Economy of the Development and Transfer of Agricultural Technologies', in Making the Link: Agricultural Research and Technology Transfer in Developing Countries, D. Kaimowitz (Ed.), Westview Press, Boulder, CO, USA, 1990).

---

67. In dealing with reform sector co-operatives, farm size was defined by dividing the total area of the co-operative by the number of members.

68. K. Amanor and J. Farrington, 'NGOs and Agricultural Technology Development', in *Agricultural Extension: Worldwide Institutional Evolution and Forces for Change*, William Rivera and Dan Gustafson (Eds), Elsevier, New York, 1991.

Training and Visit (T & V) system, first introduced by Daniel Benor in Turkey in 1967. This system has been credited with many of the successes of extension in the 1970s, 1980s and 1990s, and it also has been held responsible for some of the failures. In reviewing the system, it should be borne in mind that:

> The aim of the T & V approach was to reform the *management* of extension systems and turn a cadre of poorly supervised, poorly motivated and poorly trained field agents into effective technology transfer agents through fortnightly training of agents, who then made regular visits to farmers, conveying clear extension messages.[69]

Indeed, a review of World Bank extension projects found that in practice the T & V system exhibited strengths in the management area, although they were not always fully realized:

> many of the organizational principles included in the T & V model are internalized in the majority of 'good' extension services and are unquestionably sound: programming of activities; the technology focus; continuous staff training; program supervision; close research–extension links; and farmer feedback to allow adaption of technology to farmer circumstances. Unfortunately, many of these principles were not adequately developed in the projects.[70]

In spite of the potential advantages of the system, the consensus of expert opinion emphasizes its shortcomings and thus leans against putting into practice more widely, at least without substantial modifications. Antholt summarized its limitations as follows:

- The model was rigid and was often not appropriate, given the variation in cultural, historical and institutional factors among and within countries.

- Problems of recurrent cost funding, lack of appropriate technology and deficiencies of staff quality threatened long-term sustainability of extension programs.
- The T & V concept of using a contact farmer as the primary recipient of extension visits (for subsequent transfer of technology to other farmers) was not very effective and was often replaced with farmer groups.
- A top-down approach to delivering extension messages was often based on standard packages of recommendations that ignored the heterogeneity among farmers.[71]

Picciotto and Anderson highlighted an overlapping but somewhat different set of deficiencies in T & V, on the basis of a World Bank evaluation:

- Ninety percent of the projects faced budgetary constraints, in part because almost half did not evince strong borrower or implementing agency ownership.
- More than half of the projects suffered from inadequate extension messages resulting from research weaknesses or poor linkages between extension and research.
- Twenty-five percent of the projects were hindered by the low education level of frontline staff.
- The training programs of more than half of the projects did not give the frontline staff sufficient practical knowledge.
- Almost 40 percent of the projects suffered from inadequate adaptation to local conditions.

T & V's hierarchically organized and strictly programmed method of agricultural extension presumes the availability of a sustained flow of research innovations coupled with the ability of implementing agencies to secure, retain and motivate good technical staff. Where both of these elements were available, T & V may well

---

69. C. H. Antholt., 1998, p. 356.
70. D. L. Purcell and J. R. Anderson, 1997, p. 5.
71. C. H. Antholt, 1998, p. 357.

have accelerated the spread of new agricultural technologies on a rewarding scale. Where the initial conditions were not suitable – for instance, because farming conditions were highly differentiated, the research pipeline was empty, and either a disciplined organization or adequate skills, or both, were lacking – T & V proved poorly adapted to the challenge.[72]

In T & V's defense, it can be said that many of these deficiencies can characterize any public extension service. However, to the extent that is so, then the very concept of a public extension service is open to question, in spite of the above-mentioned arguments that extension is a public good. Certainly, the top-down approach of T & V has proven to be a central weakness, especially in regard to meeting the needs of small farmers located in heterogeneous agro-economic conditions.

8.3.4 Gender in Agricultural Extension

As in agricultural research, agricultural extension has been slow to recognize differences in gender

Average % of Agricultural Extension Resources Allocated to Programs for Women Farmers:	
Worldwide	5%
Near East	9%
Africa	7%
Latin America	5%
Asia and Europe	3%
North America	1%

Source: FAO, Agricultural Extension and Women Farmers in the 1980s, Food and Agriculture Organization of the United Nations, Rome, 1993 (cited in FAO, 2001).

needs in farm production and to develop different approaches for women clients. This is beginning to change in some places, but the problem is still very widespread. As Kalim Qamar has expressed it:

> The agricultural technology systems, until quite recently, were almost totally focusing on male clientele. The right to improve agricultural technology was considered by and large, belonging to men only. Women were just not included on the list of clientele in spite of the fact that they made very significant contributions to agricultural production. No surprise, most extension programs ignored them . . . [and] still . . . few extension agencies take into account the needs of farming women.[73]

In the words of the FAO's SEAGA macro manual:

> in many countries there are severe socio-economic and gender biases in the delivery of extension services. . . . Although problems are country specific some of the more common difficulties are:

> • Extension services are geared towards male recipients who often do not discuss production decisions or share extension knowledge with their female partners. This is particularly true in Africa . . . [and of] the contact farmer approach as used in training and visit programs, with the focal contact farmers expected to pass on their knowledge to other farmers. A study in Tanzania showed that women are seldom chosen as contact farmers and that female headed farms derived fewer benefits from this system than other types of farm households.[74] . . .

---

72. R. Piccioto and J. R. Anderson, 1997, p. 252.
73. M. Kalim Qamar, 'Effective Information Systems for Technology Transfer: Challenges of Transformation for Conventional Agricultural Extension Services', in *Agricultural Research and Extension Interface in Asia*, Asian Productivity Organization, Tokyo, Japan, 1999, p. 52.
74. J. M. Due, N. Mollel and V. Malone, 'Does the T & V system reach female-headed families? Some evidence from Tanzania', *Agricultural Administration and Extension*, **26**, 1987, pp. 209*ff*.

- Services tend to be devoted to progressive farmers who own land and who are willing and able to obtain credit and invest in inputs and technological innovation.... This often excludes resource-poor farmers and the landless, including women. A study by Swanson, Farmer and Bahal (1990)[75] found that less than a quarter of the world's extension resources are designated for subsistence farmers, only 2% is aimed at landless producers and only 7% to rural youth and young farmers. Thus, while resource poor and subsistence farmers make up 75–80% of the world's farmers they are allocated only about one third of extension time and resources.
- Attitudes of extension personnel can be a barrier to effective... extension services. A study of extension in Africa found commonly held beliefs that women are not significant contributors to agricultural production, that they are always busy with household chores, that they are hard to reach and that they resist innovation....
- In many countries women face barriers to tertiary agricultural education. As a result, there are few female extension officers....
- The timing of extension visits often conflicts with women's household chores such as meal preparation.
- Women lack access to membership in rural organizations that often serve to channel or provide training and extension activities.[76]

In light of this pervasive gender bias in agricultural extension, the FAO points out that:

Extension services that neglect the needs of women and resource-poor farmers risk low returns and a failure to achieve development objectives such as food security, sustainable agricultural growth and poverty alleviation. Closing the gap between the existing and potential productivity levels of female farmers may be one of the most important ways of promoting ... overall agricultural development. For example, in Kenya following a nationwide information campaign targeted at women under a national extension project, yields of corn increased by 28%, beans by 80% and potatoes by 84%.[77]

8.3.5 The Challenge of HIV/AIDS for Agricultural Extension

At the outset of the HIV/AIDS epidemic, the prevailing attitude was that agricultural extension approaches did not have to be modified in any way, that it was a problem to be dealt with through other institutions in each country. This attitude, however, is necessarily changing in light of the gravity of the problem and the havoc it is wreaking on rural societies. Qamar has described the nature of the challenge from several perspectives:[78]

Until recently, HIV/AIDS was considered mainly as a health issue, and all the programs for combatting the epidemic were based on health and medical sciences.... However, views are changing fast. The adverse effects of HIV/AIDS on development institutions and their programs in Africa have forced the health and non-health development agencies alike to approach the problem from an entirely different angle. The HIV epidemic is now being considered as an important cross-sectoral

75. B. E. Swanson, B. J. Farmer and R. Bahal, 'The Current Status of Extension Worldwide', in *Report of the Global Consultation on Agricultural Extension*, Food and Agriculture Organization of the United Nations, Rome, 1990.
76. FAO, *SEAGA Macro Manual*, Food and Agriculture Organization of the United Nations, Rome, Draft, July 2001, module 12.
77. *Ibid.*
78. M. K. Qamar, 'The HIV/AIDS epidemic: an unusual challenge to agricultural extension services in sub-Saharan Africa', *The Journal of Agricultural Education and Extension*, 8(1), December 2001 (selections from pp. 2–5 are cited in the rest of this section).

developmental issue bearing far reaching implications for policies and programming, both for the governments and international development agencies.

The loss of breadwinners due to the epidemic is leading to increased poverty and food insecurity among affected families in sub-Saharan Africa. Also professionals and other categories of skilled labor have not been spared by the epidemic. The main consequence of this calamity in many affected countries is the reversal of the social and economic progress made during the last few decades, coupled with the serious negative impact both on households and relevant organizations and institutions. This is especially true for small-holder agriculture.... An enormous cost burden has been imposed on households and organizations due to diversion of resources to health care, loss of both skilled and unskilled labor, funeral costs, costs of recruiting and replacing staff, and reduction in productivity due to losses of human resources.

Both subsistence and commercial agriculture have been affected by AIDS significantly in the way of decline of crop yields, increase in pests and diseases, and decline in the variety of crops grown in case of subsistence farming.... Major financial and social crises have been created in the agro-industry due to protracted morbidity and mortality and loss of skilled and experienced labor....

He points out that extension services themselves are directly affected because staff members incur higher risks than others as a result of the frequency of their visits to areas affected by HIV/AIDS, and their need to attend to sick family members and neighbors. The epidemic has undermined morale in many extension services. In addition, the costs to extension organizations have increased because of outlays on treatment of affected staff, funerals and insurance. In Uganda, it is estimated that 20 to 50% of extension staff time has been lost because of the direct and indirect effects of the epidemic.

Qamar adds that the effect on farming populations has been devastating in many cases:

The epidemic is changing the traditional composition of the clientele for extension services. In the areas of high HIV prevalence, the category of healthy and able-bodied men, women and youth, in the late adolescence to middle age range, is the one that has been most affected through high levels of morbidity and mortality. One finds more women and children now engaged in farming due to prolonged illness and/or death of their spouses, parents, guardians and older members of the family. Paradoxically, the struggle for feeding a large number of children left behind by their parents who have died young, has forced many very old persons back into farming who had retired from active farming long ago. The emerging target population for extension services increasingly includes more physically weak, sick, and elderly persons, widows and young orphans. These newcomers, who even though they are exposed to farming due to living in rural areas, have relatively less experience in agronomic practices ... and have limited physical and technical capacities for the use of heavy tools, farm machinery and animal-drawn farm equipment.

And the technical messages traditionally conveyed by extension agents are losing relevance:

there are now applications for agricultural credit from orphan- and widow-headed households, which are often not eligible according to the existing criteria for the approval of credit applications. The extension staff who, in general, are supposed to support the applications ... feel lost in the absence of the new criteria needed for this new clientele ...

The notoriously persistent denial and 'conspiracy of silence' about HIV/AIDS, common among rural communities, is gradually giving way to relative openness.... The farmers' questions are no longer limited to farming. There are so many queries related to HIV/AIDS. However, the extension staff, who know little about the epidemic and have not received any special training in this subject, feel helpless and embarrassed in front of the farmers. They are not in a position to offer any useful information or meaningful advice.

Meeting this challenge in the areas affected by the epidemic obviously requires substantial modifications in the way agricultural extension is conceived and carried out. As Qamar has put it, 'Presently there are no extension programs and strategies to improve agricultural skills of inexperienced young farmers including a large number of women and orphans who have suddenly become clientele of the services. The notoriously weak linkages between extension, research and other relevant agencies are no help in addressing the need for developing new technologies and equipment suitable for the new situation' (*op. cit.*, p. 6).

8.3.6 Towards a New Paradigm for Agricultural Extension?

A process of developing new methods of agricultural extension is underway in all regions of the world. In addition to concerns about gender bias, dissatisfaction with the past ways of carrying out extension is widespread, and it has given rise to a quest for a better approach. The reasons for this search for a 'new paradigm' have been summarized by Umali-Deininger in the following way:

Three major developments have brought about a rethinking of the appropriate channel for delivering agricultural extension. First and most important, *fiscal crises and economy-wide budget cutbacks*, often associated with structural adjustment programs, have forced governments to make sharp reductions in the budgets of public extension programs. Fiscal sustainability and cost-effectiveness have become the priority concerns.

Second, *the poor performance of some public extension programs*, as reflected in the slow adoption of extension messages, has spurred the search for alternative approaches to improve extension services. . . .

Third, *agriculture's dependence on more specialized knowledge and technologies has changed the economic character of the services delivered by the extension system*. The institutionalization of mechanisms that permit the seller to appropriate the returns from new inventions and new species of plants . . . has improved the private for-profit sector's incentives to provide extension services. The growing commercialization of agriculture and increased competition in domestic and international markets have further strengthened the economic incentives for farmers and other rural entrepreneurs to treat extension as another purchased input. . . .

In the search for a new paradigm of the agricultural extension system, developing countries are wrestling with several questions: What are the appropriate roles for the public and private sectors? Can the private sector deliver services more efficiently? What are the welfare implications for small-scale farmers and the rural poor?[79]

In the context of sub-Saharan Africa, the Neuchatel Group has observed that the environment for agricultural extension is changing in the following ways:

The aims of official development assistance are becoming more focused [on] reducing poverty and social inequalities, the sustainable use of natural resources, and participatory development. . . .
Many developing countries are at various stages in the process of economic liberalization, decentralization and privatization. . . .
New actors are becoming involved in extension activities. There are today four types of actors in agricultural extension: public agencies, private service providers, producer organizations . . . and non-governmental organizations. . . .
Public spending on extension is shrinking. Policies to bring down public deficits in most

---

79. D. Umali-Deininger, 1997, pp. 204 and 206 [emphasis added].

developing countries have led to expenditure ceilings on agricultural extension and the introduction of fee-based schemes.[80]

In summarizing the challenges that require new approaches for agricultural extension, Picciotto and Anderson stressed *administrative constraints* in managing a large system in addition to fiscal limitations. They add that:

the perception of agriculture's potential and constraints has changed. In many situations *the dissemination of standard packages of inputs and practices is no longer relevant, if indeed it ever was. . . . What is increasingly required is an approach that can generate custom-made, environmentally friendly solutions based on the farmers' involvement . . .*

the spread of education and modern communications and the rise of commercial farming have created opportunities for alliances among the public, private and voluntary sectors. More open and liberalized agricultural markets are bringing the knowledge and skills and private agribusiness to farmers without involving public-sector intermediaries. In both more- and less-developed countries, *farmer-led approaches to extension are spreading*, while farmers' associations, cooperatives and self-help agencies are contributing handsomely to the diffusion of modern technology.

According to Tendler (1997),[81] informal performance contracts between Brazilian farmers and extension agents have increased the commitment of extension workers, improved the customization of advice, and increased productivity. In Indonesia integrated pest management programs and the FAO's Farmer Field Schools show the value of turning farmers into extension agents and extension agents into farmers. . . .[82]

To this list of issues should be added the concern that rural poverty continues to be widespread in developing countries and that effective modalities of agricultural extension for reaching the poorer farmers need to be developed and, where they already exist, supported and reinforced.

Rivera has commented that 'there will have to be diverse extension systems to meet disparate needs' and that the new extension approaches 'will become more *purpose-specific, target-specific, and need-specific*'.[83]

8.3.7 Trade in Agricultural Technology

One of the principal means of transferring agricultural technology is through imports of inputs and capital goods that embody a new technology, including seeds, agrochemicals and machinery. Given the rapid evolution of agricultural technologies, developing countries clearly benefit from importation of technologies developed elsewhere, sometimes adapting them, while at other times using them directly. Against these benefits, a counterweight of concerns has been raised about the importation of inferior seeds and chemical products and the lack of adequate information for farmers on the quality of products they are receiving. There are two contrasting approaches for dealing with this concern: (i) restricting imports until the products in question are tested, and (ii) allowing them to be imported but disseminating test results, as information to buyers, as the results become available. In the latter case, usually not all products will be tested – sometimes only a small portion of them – and

80. Neuchatel Group, *Common Framework on Agricultural Extension*, Paris, 1999, selections from pp. 7–9. The Neuchatel Group consists of representatives of eight bilateral economic co-operation agencies and five international agencies, and it was formed for the purpose of forging a common approach to agricultural extension in sub-Saharan Africa.
81. Judith Tendler, *Good Government in the Tropics*, The Johns Hopkins University Press, Baltimore, MD, USA, 1997.
82. R. Picciotto and J. R. Anderson (1997), pp. 254–255 [emphasis added].
83. W. M. Rivera, 2001, pp. 11–12 [emphasis in original].

test results may lag adoption by farmers by a long period of time, even several years.

However, a policy of restricting imports until they can be tested runs a serious risk of slowing significantly a country's rate of technical progress in agriculture. The dangers of the restrictive approach have been pointed out clearly by David Gisselquist and Jean-Marie Grether:

Agriculture has become a high-technology field, with rapid advances in crop and livestock genetics, pest and livestock management, and machinery.... The issue of access to foreign technologies is particularly important because, as in other high-technology fields, agricultural technology is now international. Leading countries continually borrow and build on research from other countries.... Whatever the source, most new technologies reach farmers through marketed inputs.... Many developing countries lag behind, in part because of self-imposed barriers to the introduction of private agricultural technologies.

Access to new technologies may come either through multiple channels, with regulations focusing on [negative] externalities, or through a single channel, with regulations based on performance tests.... In industrial countries (and in some developing countries) governments maintain liberal trade regimes for foreign and domestic inputs, allowing multiple channels for the introduction of new technologies.... Governments regulate inputs to limit externalities (for example, by not approving dangerous pesticides), but otherwise allow companies to market new technologies, trusting that farmers and companies interacting through markets be able to choose those that are most efficient. ... This liberal approach to technology transfer is appropriate for agriculture, a field in which local conditions are critical in shaping the impact of new technologies. ...

In contrast to the market-friendly regimes common in industrial countries, many devel-

oping and transition countries strictly limit market access for new agricultural technologies. Restrictions are most common and problematic for seeds, but they also may interfere with machinery, fertilizers, low-risk pesticides, feed mixes and other items. Many developing countries maintain positive lists of allowed inputs, even those for which externalities are not a serious concern. For example, many countries list allowed plant varieties, and some also list allowed models of machinery, compositions of fertilizers, and feed mixes based on official performance tests.... Positive lists are far more restrictive than negative lists, which allow anything not listed.

For most inputs, environmental or public health concerns can be addressed with negative lists. For example, instead of approving each feed mix, governments can simply ban or limit potentially dangerous components. Although positive lists are standard across all countries for pesticides and veterinary medicines, countries differ in the conditions for registering new products.... With regulations and policies that make it difficult for companies to enter and to operate, many developing countries effectively block almost all transfers of private technologies for seeds and other major categories of agricultural inputs....

A single channel system severely constrains the flow of new technologies. In many developing countries with single-channel systems farmers are offered an average of less than one new seed variety a year for each major crop, while farmers in countries with multiple channels may see dozens of new varieties each year for a single major or minor crop or vegetable. Even where private companies are able to operate, regulatory costs limit the transfer of private technologies. Regulatory costs are particularly troublesome in small markets – small countries or minor crops – where companies may judge that registering a new technology is not worth the effort, leaving farmers with no access.[84]

84. D. Gisselquist and J.-M. Grether, 'An argument for deregulating the transfer of agricultural technologies to developing countries', *The World Bank Economic Review*, **14**(1), January 2000, excerpts from pp. 112–117.

These authors conclude that, on balance, *the risks of the restrictive approach to trade in technology are greater than those of the liberal approach*. When the latter approach is adopted, some of its risks can be ameliorated by accelerating a program of testing for informational purposes, and not for purposes of control. The testing does not necessarily have to be carried out by the government itself, although normally the government would underwrite its costs. Universities and producers' organizations often have the capability to do the testing and disseminate the findings.

8.4 NEW DIRECTIONS IN AGRICULTURAL RESEARCH

Agricultural research systems are in evolution throughout the developing world, in response to the issues mentioned earlier in this chapter. In large part, the evolution is a response to budgetary stringency, but other factors are at work as well. The role of government is undergoing a re-assessment nearly everywhere. Research systems also are being pressed to respond to other concerns such as poverty alleviation, and this requires new ways of doing things.

In spite of the general success of African agricultural research that was noted earlier in this chapter, weaknesses are apparent in the area of yield improvements as well:

> despite the introduction of new varieties, there have been less than expected impacts on yields, especially in the case of crops grown under adverse conditions without the use of other external inputs ... returns to research (and extension) investments are reported to be quite high, but represent variable performance across countries and crops. Food crops grown in regions dominated by commercial farmers and under more favorable conditions using improved management practices generate higher returns than crops grown in regions where the main change is only the shift to new

varieties. Results also reflect considerable variability by country as a result of agroclimatic factors and the policy environment, which can affect the supply of seeds and other inputs, and the continuity and stability of research investments.[85]

Mudahar *et al.* have summarized 'some of the major changes occurring in agricultural research systems around the world' in the following terms:

- increased emphasis on the cost effectiveness of agricultural research, often requiring reductions in staff and streamlining of bureaucracies;
- resource commitments based on anticipated applied research outcomes;
- increased involvement by users in decision-making....
- more access to research resources by those likely to benefit from its outcomes, including farmers, processing firms and seed producers....
- responsibility and substantial autonomy for management of research in main centers, with policy and funding bodies providing only overall guidelines on programs and outcomes....
- shift from basic to applied research while ensuring public-good linkage between basic and applied research....[86]

While funding levels and staff capabilities will always remain core issues for agricultural research, three other central issues are priority setting, research modalities and provisions for transfer of the research results. A definition of priorities is required for making the 'resource commitments based on anticipated applied research outcomes' mentioned by Mudahar *et al.* Adequate setting of priorities in turn requires *appropriate mechanisms for identification of the problems farmers face*. Research modalities include the way the research system is structured and the role of non-governmental, academic and

85. M. K. Maredia, D. Byerlee and P. Pee, 2000, p. 554.
86. M. S. Mudahar, R. W. Jolly and J. P. Srivastava, 1998, pp. 36–37.

voluntary institutions, and farmers themselves in the research process. Provisions for transfer of research findings depend in part on a specification of the role of extension agents and farmers in research, and they also include programs for production of basic seed, quality control in seed, and seed multiplication. (Quality control for seeds is a major issue in many developing countries. It is not uncommon to find that unscrupulous persons package and sell defective seed as certified seed, even to the point of counterfeiting the official seal of certification. It can be difficult to bring these persons to justice and stop these practices given the weakness of judicial systems. This is one of many examples of how governance issues can impede agricultural development.)

This present section begins with a review of the critical question of establishing agendas for agricultural research and then discusses research modalities, which largely involve issues of the institutional structures for research and the role of farmers. There follows a review of alternative approaches to financing agricultural research. Then, issues of the relevance of agricultural research to poor farmers and women farmers are discussed. Questions related to the transfer of research findings are touched upon partly in this section and addressed more directly in the following one, in the context of new directions in agricultural extension.

8.4.1 Identification and Implementation of the Research Agenda

There is an international consensus that generally agricultural research systems in developing countries suffer from insufficient funding levels, and in some cases from lack of sufficient qualified staff.[87] *An equally fundamental problem is that the research agenda has been science-driven*, rather than driven by the needs of farmers. It usually completely ignores the needs of women farmers. Agricultural research is all about science, and clearly its purpose is to apply science to the prob-

lems of the sector. Nevertheless, scientists are not always the best persons to identify the research priorities. The reasons for this include:

- The principal agronomic and varietal issues can vary widely across regions and districts of a country, and researchers are not always aware of them. Indeed, sometimes they vary almost from farm to farm, as a function of soil characteristics, slope, natural drainage qualities and other factors, including the farmer's own approach to cultivation.
- Researchers tend to view issues on a crop-by-crop basis, whereas producers tend to take a farming systems approach.
- Although they are aware of risk aversion considerations in farming, researchers tend to emphasize improvements in physical yields as the primary goal of their efforts, and they may assume a degree of access to inputs that many farmers do not have. In contrast, farmers are concerned about overall economic returns to the farm more than physical yields, the availability and affordability of inputs, risk aversion, access to markets, and other variables.
- When choosing among optional lines of research, scientists may sometimes succumb to the natural human tendency to select those that they are most familiar with or are most promising in professional terms, rather than the ones most relevant to actual farming conditions.

Qamar has posed the relevance issue in the following way:

If the technology transferred addresses the needs of its potential users, it has high probability of being adopted. The generation of such demand-driven technology is possible only if the research agenda is drawn on the basis of real-life field problems. One way to ensure this will be to follow a particpatory research approach where researchers, extension agents and farmers have a chance of expressing

---

87. Brazil is an example of a country that is a strong exception to this characterization. It is richly endowed with well-trained and adequately remunerated agricultural research scientists, as the result of decades of investment in their education at higher levels.

their observations and concerns. Otherwise, the luxury of purely academic research will remain a burden on the scant budgets of research institutions in developing countries.[88]

A research agenda can be made more relevant to farmers' needs, and thus more productive for the sector, if it is designed in collaboration with farmers. This is a way of responding to Antholt's observation about the central importance of the appropriateness of new technologies. This point is widely recognized, but putting it in practice is not simple, given the geographical dispersion of farmers, their frequently low education levels (which influences their ability to articulate their problems more than their ability to perceive them), and the centralizing tendency of bureaucratic institutions. In addition, it must be acknowledged that the point is more relevant to smallholders in heterogeneous farming conditions, often on hillsides, than to more commercial farmers.

> *The Valley of Jalapa in Nicaragua is a fertile valley with considerable agricultural potential. It suffers from extensive poverty but also has farms of moderate size. Discussions there, in January 2001, with farmers and extension agents revealed that the valley had received no benefits from the national agricultural research system. The research efforts had been concentrated in the more arid zones of the country, and the cereal varieties produced for those zones do not flourish in the more humid conditions of Jalapa. In subsequent discussions with the managers of the national agricultural research system, they agreed that Jalapa and similar areas had been neglected in the research programs to date, and that they should not be overlooked. Similar stories can be related about many agricultural zones throughout the developing world. Decentralization of research is the only viable way to respond to these concerns.*

Typical responses to this issue include incorporating representatives of farmers' associations on the boards of directors of research institutions, and carrying out field surveys of agronomic issues at regular intervals. In addition, in principle an extension service should be providing feedback to researchers on the most pressing cultivation problems, but such a mechanism usually does not function very well.

Nor is much gained from putting farmer representatives on boards of directors of national research agencies. These boards meet infrequently, and the main choices in research directions usually are determined well in advance of board meetings, by the researchers themselves, and filtered through the management of the agencies. Farmer representatives in any case usually are a minority on those boards. Furthermore, *discussing research priorities at a national level is not as useful as discussing them at a local level*. Field surveys of agricultural issues can be valuable inputs into the definition of a research agenda, but their worth depends heavily on the thoroughness and timeliness with which they are carried out, and on the procedures put in place to ensure that researchers internalize their findings. In practice, they usually represent only a marginal influence on the research program.

An approach that has proven useful is to *decentralize the research effort*, creating a number of local research centers in accordance with the variability of agronomic conditions and cropping patterns and allowing the local centers to formulate their own agendas. By itself, this approach tends to bring researchers in closer contact with farmers' concerns, but it is not likely to be sufficient. In spite of the importance of an appropriate degree of decentralization, some agricultural research systems have been consolidating geographically because of fiscal pressures. Clearly a balance has to be struck, but wherever possible it is important to establish local research centers.

Measures complementary to decentralization can include one or more of the following:

---

88. M. K. Qamar, 1999, p. 56.

- Making local farmers a majority on the boards of these research centers.
- Requiring farmers to make financial contributions, however modest, to the budget of the research program.
- Forming local research teams comprising researchers, extensionists, farmers, and sometimes additional persons such as agricultural economists and other kinds of development experts.
- Giving farmers a leading role in some aspects of the research process, with the support of facilitators, and giving them control of the directions of that process.

Taken together, farmers have a large reservoir of knowledge about cropping practices and even about varieties, as illustrated by the previously mentioned experiment in which women farmers in Colombia and Rwanda selected better-performing bean varieties than researchers did. In the words of Van Veldhuizen *et al.*, 'any organization involved in participatory technology development needs to realize that it does not have *the* answer to farmers' problems; it must be prepared to learn through interaction with farmers' (1997, p. 8).

Establishing research teams, in which researchers spend considerable amounts of time in the farmers' home villages, is at the heart of the participatory research process tested and advocated by these authors and others. It represents a way of tapping this reservoir of local knowledge, as well as directing the research toward the more pressing problems confronted by farmers.

In writing about 'early involvement of farmers in technology design and evaluation, use of farmer-focused criteria in economic evaluation and more participatory approaches', Sara Scherr has observed that:

A highly effective strategy has been the integration of research and extension functions in pilot field programs based on diagnosis–design–feedback–redesign with farmers. An

> *In Mali the Institut d'Economie Rurale has been deconcentrated to the regional level, and central and regional research users' commissions have been created. This has been done, with a view, on the one hand to strengthening the farmers' participatory process and, on the other hand, to coordinating initiatives, avoiding duplication of efforts and fostering the flow of information.* (*From Lawrence D. Smith*, Reform and Decentralization of Agricultural Services, *FAO Agricultural Policy and Economic Development Series No. 7, Policy Assistance Division, Food and Agriculture Organization of the United Nations, Rome, 2001, p. 119 (based on personal communication from R. Pantanali of the FAO).*)

other emerging approach is local farmer-led technology development with technical and scientific backstopping. . . .[89]

At the core of a process of participatory technology development is the fact that 'many farmers perform their own small experiments as part of a process of gradually changing their farming systems' (Van Veldhuizen *et al.*, 1997, p. 4). The outside members of the research team need to promote and support this process. This kind of approach is most valuable for smallholding farmers in diverse agronomic conditions. It does not obviate the need to also carry out research in a more traditional way, in research stations, especially research on varietal improvement. The two approaches are complementary. Traditional research programs are likely to place more emphasis on crop improvement (new varieties), although farmers can play a role here in defining priorities, and in varietal selection as shown by the example of women bean farmers. Participatory research is likely to emphasize crop management (cultivation techniques) and natural resource management, although it can help identify priorities for varietal research and also can develop new varieties directly.

89. S. Scherr, 'A downward spiral? Research evidence on the relationship between poverty and natural resource degradation', *Food Policy*, **25**(4), 2000, p. 494.

One of the earliest, and perhaps the boldest and most successful example of participatory research, for carrying out varietal selection and improving cultivation techniques, was begun in the early 1990s in the Department of Cauca Valley, Colombia, where farmers were given the initiative on local research (as mentioned briefly in Section 8.2 above). A Local Agricultural Research Committee (CIAL) consists of at least four persons elected by a community, and is supported by a facilitator and a small grant. In about nine years, the number of CIALs expanded to 249 in eight countries in Latin America, and many CIALs have branched into related enterprises, especially seed merchandising but also the construction of milling facilities and other processing capabilities.

The CIALs tap into a rich existing resource in the form of farmers' traditions of varietal and cultivation experimentation on their own plots, usually on a small section of them. They need support from outside institutions, particularly in the first two years, but if they survive that period they have been found to be fairly self-sustaining. The facilitators work with the CIALs on a weekly or bi-weekly basis, and they are evaluated by the CIALs themselves. The facilitators provide seed varieties to be evaluated and other information, and in some cases they are beginning to establish internet communication centers in the participating communities.[90]

Farmers quickly learn the basic concepts and language of agricultural research, and the CIALs increase their ability to interact with other institutions in the outside world. The CIALs also work on crop diversification and explore marketing options. In the area of IPM, their functioning is similar to that of the Farm Field Schools mentioned earlier. It is important that their experimental plots be small, especially at the beginning, to reduce their exposure to risk. While they need outside support, it has to be given with a light hand, always leaving the farmers in control of the process. Many CIALs have made remarkable progress in raising members' incomes. Some are operated by women farmers.

Other examples of participatory research and use of indigenous knowledge have been noted by the International Fund for Agricultural Development (IFAD):

> Worldwide, over US$300 million of pigeon pea, mostly grown by poor farmers, are lost yearly to pod-borer. In India ... at a farmers' meeting organized by an NGO ... an elder showed the defunct method of shaking larvae gently on to a plastic sheet and feeding them to chickens. ... By 1999 the method had spread to thousands of farmers....
>
> ICRISAT millets and [CIMMYT] maizes are crossed with landraces to suit local conditions and preferences, even at the cost of losing hybrid vigor in maize, as in the late 1990s by smallholders in Chiapas, Mexico.[91]

Requiring financial contributions from farmers for the research budget serves two aims: giving farmers a sense of ownership over the process, and thus a greater willingness to demand that research address their priority concerns, and helping defray part of the cost of research. The latter is becoming an increasingly important concern. Echevarría, Trigo and Byerlee also point out that such a system promotes equity in the sense that the principal beneficiaries of research also pay for part of it.[92] The approach has long been used by farmers in Mexico's irrigation districts, who are organized in *patronatos* for the purpose of financing research. In the Mexican

90. The comments on CIALs in these paragraphs are based on an interview with Jacqueline Ashby at CIAT in Cali, Colombia, and on the above-referenced publication by J. Ashby, *et al.*, 2000.

91. IFAD, *Rural Poverty Report 2001: The Challenge of Ending Rural Poverty*, Oxford University Press, Oxford, UK, 2001, p. 140.

92. Rubén G. Echevarría, Eduardo J. Trigo and Derek Byerlee, *Cambio institucional y alternativas de financiación de la investigación en América Latina*, Inter-American Development Bank, Washington, DC, USA, August 1996, p. 20. This document was also published in English by the World Bank under the title *Institutional Change and Effective Financing of Agricultural Research in Latin America*, World Bank Technical Paper No. 330, The World Bank, Washington, DC, USA, July 1996.

State of Sinaloa, they pay 0.16% of the value of production, thereby supporting about half of the local agricultural research budget. In Colombia, financial support to research is provided by producer associations for coffee, oil palm, cacao, rice, sugar and other crops, and the monies are raised through a surcharge on exports or local sales, depending on the crop. In Uruguay, producers pay 0.4% of the value of their production to support research, and those payments helped the national agricultural research system double its budget in five years.[93]

These kinds of approaches to bringing research closer to farmers and addressing their needs represent ways of *making research more accountable to farmers*, which is one of the principal thrusts of the new international consensus about agricultural research. With such accountability, *the research agenda becomes demand-driven, or client-oriented, rather than supply-driven*.

When farmers and extension agents become partners in the research itself, they become agents of dissemination of the research findings. This is part of the raison d'etre of the FAO's Farm Field Schools that have proven to be effective in Indonesia, Bolivia and many other countries. A participatory research project on Colombia's Atlantic coast, known as PBA and funded by the Dutch Government, involves poor farmers in varietal selection based on the results of applied biotechnology, in basic food crops such as plantain and cassava. The participating farmers have gained an additional economic benefit by going into the business of seed multiplication and sales. Through this kind of approach, the question of linkages between research and extension is solved directly. It is no longer necessary to develop elaborate mechanisms of institutional co-ordination that, in any case, have reduced effectiveness.

The design of a national research agenda should respond not only to clients' needs but also to opportunities for successful adaptation of research from elsewhere. Most developing countries do not have a capacity for addressing all of their important agricultural research questions, and thus *adaptation of research results from other countries and international institutions is necessary*. Linkages in the outward direction need to be strengthened, both to research systems of other countries and to regional and international research institutions. An example of successful collaboration of this nature is found in the work of the International Institute for Tropical Agriculture (IITA) in Nigeria in developing improved cassava varieties and disseminating them in many countries of Africa.[94]

When national agricultural research systems establish local research centers in different zones of a country, one of the priorities has to be a capacity for *laboratory analysis of soils*. The nature of the soils is one of the first and most basic questions put forward by farmers and extension agents and, unfortunately, many zones of developing countries are not endowed with a capacity to provide a response. Providing such a capacity is an integral part of the thrust of making agricultural research more relevant to the needs of farmers in heterogeneous conditions.

At the other end of the agro-economic chain, another frequently neglected priority is *research on post-harvest management techniques and processing and handling technologies*. In an era of globalization, when product quality considerations are increasingly important to successful marketing, neglect of these issues can seriously affect farmers' incomes. It no longer is sufficient to focus only on the quantity of agricultural output supplied. Responding to this challenge requires changes in research management as well as research procedures. Management has to develop mechanisms for staying in touch with evolving trends and requirements in the markets for agricultural products. In Xalapa, Mexico, a local university research institute has invested in developing adequate post-harvest management and packing procedures *before* working on varietal adaptation and crop management for non-traditional crops.

---

93. These examples are taken from R. G. Echevarría, E. J. Trigo and D. Byerlee (1996) and the author's own experience.

94. M. K. Maredia, D. Byerlee and P. Pee, 2000, p. 556.

An increasingly important aspect of the appropriateness of agricultural technology is its *environmental sustainability*. Technologies that are intensive in the use of chemical inputs create problems of contamination of soil and water, and they can lose effectiveness in pest control as pests mutate. Technologies intensive in water use run the risk of degrading irrigation systems over time with waterlogging and salinization, and depleting groundwater reserves. Some of these problems have emerged in the Indian and Pakistani Punjab, where the Green Revolution had some of its earliest successes, with the consequence that they have undercut productivity growth in those regions:

There are indications that the wheat–rice cropping system in the Indian Punjab was hurt by a steep decline in the water table, while rising water levels in the wheat–cotton zone led to severe waterlogging in [that] zone. Data from the Pakistan Punjab also confirm a serious problem of waterlogging and salinity, due in part to deterioration in the quality of tubewell water (reflected in a significant increase in residual carbonate and electroconductivity of groundwater). Soil quality in Pakistan (in terms of available soil organic matter and phosphorus) also deteriorated, particularly in the wheat–rice zone. . . .

In the wheat–rice system, resource degradation more than canceled the productivity-enhancing contributions of technological change, education, and infrastructure. The [resource degradation includes] the development of pest complexes due to inappropriate use of pesticides \and to monocropping of cereals.

. . . halting resource degradation will require that research systems, which have been oriented toward developing technologies based on packages of modern inputs, place more emphasis on input-efficient and environmentally friendly

practices. This will require considerable location-specific research on such themes as integrated pest and nutrient management and cropping systems. It will also require diversifying rotations to include legumes and the use of conservation tillage. Many such practices are information intensive and will require much greater information dissemination and extension efforts. Research systems in both [Indian and Pakistani Punjab] have shifted direction toward these new priorities in the 1990s.

The results of this study . . . raise *serious concerns about the long-term sustainability of intensive irrigated Green Revolution systems due to resource degradation*. For Pakistan, this study provides the first quantitative evidence of the impact of resource degradation, which is estimated to reduce productivity growth by one-third overall, and in the case of wheat–rice, to practically cancel the effect of technological change.[95]

The new orientations of agricultural research are moving far away from the earlier conception that agriculture is almost an industrial process of applying more material inputs to gain more output under uniform cropping conditions, and toward greater appreciation of the complexity and fragility of agricultural systems. Productivity improvement is now seen more as an adaptive process characterized by great diversity of approaches and requiring close communication and partnerships with those who ultimately make the technological decisions in the field – the farmers.

8.4.2 Management and Institutional Structures for Agricultural Research[96]

Effective decentralization of a national agricultural research system requires a change in man-

---

95. R. Murgai, M. Ali and D. Byerlee, 'Productivity growth and sustainability in post-green revolution agriculture: the case of the Indian and Pakistan Punjabs', *The World Bank Research Observer*, **16**(2), Fall 2001, pp. 204–205, 210 and 214 [emphasis added].

96. The main points in this section are explored more fully in D. Byerlee (1998), R. G. Echevarría, E. J. Trigo and D. Byerlee (1996), C. Pray and D. Umali-Deininger (1998), and C. Thirtle and R. G. Echevarría, 'Privatization and the roles of public and private institutions in agricultural research in sub-Saharan Africa', *Food Policy*, **19**(1), February 1994.

agement style, a willingness to grant sufficient autonomy to local research centers, and a new emphasis on providing liaison between head-quarters and the local centers. Equally, it requires *liaison between the work of international research centers and those of neighboring countries with similar agronomic conditions, and the local research centers within the country*. This kind of liaison is a role that usually is not well developed in existing research systems. The work of the local centers needs to be reviewed carefully on a con-tinuing basis, in order to understand the nature of the problems being addressed, and then the inven-tory of technologies developed outside of the country needs to be scanned in order to determine if relevant approaches have already been devel-oped elsewhere. In this sense, decentralization of a research system places greater weight on the liaison function among research efforts at differ-ent levels. The internet technology can greatly facilitate this liaison with minimal amounts of financial resources dedicated to it.

At the same time, taking into account the trend toward reduced budgets for research, agricultural research organizations need to become more cost-effective. There is a need for *a more entre-preneurial approach to the management of national agricultural research systems*, along with an emphasis on marketing the results of research, for full cost recovery, in those cases where the benefits are appropriable by the users. Where the choice is between a larger research staff and a higher quality staff, the latter should always be chosen. Equally, expenditures on research equip-ment and materials need to be adequate to support the efforts of the scientists. *Recruitment of good staff and provision of adequate personnel incentives should become central thrusts of re-search management*, along with a greater client orientation.

Greater efficiency is one of the main new thrusts with regard to research management. Achieving this requires change in the ways of working, and many research systems are reorga-nizing themselves to become less bureaucratic and more like private-sector organizations in their management styles.

A greater range of institutions is becoming involved in agricultural research, including universities, private companies, research founda-tions, ministries other than agriculture, non-governmental organizations, such as farmers' associations, and farming communities them-selves through mechanisms like the CIALs.

An increasingly common mechanism for the involvement of other institutions in agricultural research is competitive bidding for public re-search funds. Under this formula, universities, NGOs, producer associations and public agencies themselves compete for the available funding, on the basis of the quality of the research proposals submitted. In evaluating the proposals, quality can be interpreted to have several dimensions, including relevance to farmers' needs, cost-effectiveness of the proposed research, and demonstrated research capabilities. This mecha-nism can be a powerful tool for diversifying the institutions involved in research, and for provid-ing a stimulus for those institutions to improve their capabilities and efficiency. PRONATTA in Colombia, financed by the World Bank, is a successful experience of this nature in which the main decisions on allocation of research funding are made at the level of regions within the country by panels of experts.[97]

Nevertheless, this approach, while valuable, is not a complete solution in itself, since it is diffi-cult for an organization to invest in and sustain research infrastructure on the basis of occasional and uncertain research contracts. If the institu-tion has other means of supporting its most basic infrastructure, then participation in the bidding process offers a way to gain additional research experience and strengthen its capabilities in some areas. In other cases, single research projects, limited in duration, may be appropriate for some classes of problems.

Although the private sector is increasingly active in agricultural research in developing coun-tries, its role is complementary to that of the

97. Another example of the use of competitive funding for agricultural research is found in the experience of the Fund for Agricultural Research (FIA) in Chile, founded in 1981 (see R. G. Echevarría, E. J. Trigo and D. Byerlee, 1996, p. 17).

public sector, rather than competitive with it. Obstacles that limit a greater role for the private sector include the difficulty of capturing the financial returns to some kinds of research – owing to their public-good nature – and the small sizes of markets for some innovations. The private sector is unlikely to play a strong role in basic research and in situations where the legal framework for protection of intellectual property rights is not well developed. In addition, researchers may not see much productive potential in marginal lands that often are cultivated by poor farmers. Private sector enterprises tend to be most interested in research in agricultural equipment, then chemical products and biological products (in that order), with little interest in purely agronomic technology. However, farmers tend to be effective in research on the latter when organized and supported adequately.

Participatory agricultural research presents special requirements for management. A national participatory program is probably best coordinated through an institution devoted to rural development, or to supporting rural families in some manner, since successful efforts at facilitation are a key to making participatory research function well. Although this kind of research needs the technical support of scientific research organizations, it may not be appropriate to administer a participatory program through a national research system, since it is hard to resist the impetus to transmit the views and priorities of scientists in a top-down fashion. The risk is that attempts to encourage community-level research efforts may be overwhelmed by messages and guidance from the center.[98]

8.4.3 The Financing of Agricultural Research

Sources of financing for agricultural research, as well as the institutional makeup of research, need to be diversified. Public sector budgets alone will not be sufficient to support the required increases in research effort and quality. In addition to research financed purely by the private sector and research funding allocated through the mecha-

nism of competitive bidding, Echevarría *et al.* (1996) identified the following principal avenues of potential financing for research:

- Marketing of research products obtained by public research institutions
- Research in universities
- Establishment of endowed research foundations
- Financial contributions by producers (as exemplified above)

It should be added that NGOs are active in financing participatory research with farmers, where the financial requirements tend to be small in relation to the provision of facilitation services.

In regard to *the marketing of public-sector research*, Echevarría *et al.* mention (1996, p. 12) the case of EMBRAPA in Brazil, which now receives 8% of its budget through the sale of its research results. They also cite the case of Uruguay (p. 13), where agro-processors contributed $100000 to public sector research institutions toward the improvement of malt barley, with the result that both new varieties and improved management practices were developed.

Looking to universities for research efforts does not solve the financing problem, but with their existing staff capacity they can represent a cost-effective means of carrying out research, one that is generally under exploited in developing countries.

In the Honduran case mentioned in the box overleaf, the foundation has emphasized research in non-traditional crops and has supported trials by farmers. It has been successful in promoting the export of non-traditional crops in that country. Farmer associations are represented on its board of directors, along with the government and USAID. In the long run, the establishment of an appropriately endowed agricultural research foundation represents one of the best responses to the dilemma of financing the research programs. However, marketing of research efforts to the private sector and establishing arrangements for financial contributions

---

98. I am indebted to Jacqueline Ashby for discussion of this point.

from producers are also important elements of a solution.

Finally, a solution to this problem also requires the creation of a stronger base of support for research, both among farmers and among the citizenry in general. Generally, there is a very low awareness of the high returns to agricultural research and the public-good nature of many research products, even among members of national legislatures. There is a need for **strong and sustained campaigns of public information about the benefits of agricultural research, including lobbying efforts (supported by farmers) with governments and legislatures**. For the most part, national agricultural research systems have not been sufficiently attentive to the need for such campaigns. Without improving awareness and public support in regard to agricultural research, it will be difficult to solve the funding crisis in a lasting way.

---

*The avenue of **creating a research foundation** can provide a solid answer to the concern about the sustainability of agricultural research undertakings. Establishing a foundation and convincing donors to make contributions to its endowment requires a considerable effort but its payoff can be high for the development of the sector. USAID has played a leading role in Latin America in helping fund such foundations. Examples include the Jamaican Agricultural Development Foundation (JADC), the Honduran Foundation for Agricultural Research (FHIA) and the Agricultural Development Foundation (FUNDAGRO) of Ecuador (R. G. Echevarria, E. J. Trigo and D. Byerlee, 1996, pp. 18–19).*

8.4.4 Agricultural Research and Poverty Alleviation

The charge has been levied against agricultural research that its benefits flow primarily to larger-

scale, commercial farmers, and that it does less to reduce rural poverty. Empirical studies have tended to support this assertion. In large measure, this result stems from the relative factor endowments of larger- and smaller-scale farmers, especially in regard to farm size and soil quality, and their ability to purchase inputs, rather than from an explicit research strategy to favor the former. As Mitch Renkow has expressed it:

Almost by definition, the ultimate productivity impacts of improved agricultural technologies will be lower in marginal areas than in favored areas. And where 'marginality' is correlated with physical remoteness, inferior infrastructure or institutional inadequacies, the poorer availability and higher cost of complementary inputs tends to even further widen inter-regional disparities in the direct effects of new technologies. Furthermore, because the direct effects of new technologies are usually larger in favored areas, spillover effects operating through factor and product markets also tend to be larger when they emanate from favored areas.[99]

This kind of bias in favor of larger farms also reflects the traditional emphasis of agricultural research on (a) new varieties, as opposed to crop management and natural research management, (b) single commodities as opposed to farming systems, (c) varieties that are relatively intensive in modern inputs, in comparison with native varieties, and (d) top-down methods of developing and transmitting research findings. Therefore, the question remains as to whether a change in those emphases, towards different kinds of varieties, cropping systems and resource management, and in the direction of participatory research, could lead to greater benefits from research for poor farmers. On the other side of the debate, skeptics assert that the aggregate benefits to agricultural research will always be greater to the extent that more commercial farmers adopt the research findings, given their ability to extract greater

---

99. M. Renkow, 'Poverty, productivity and production environment: a review of the evidence', *Food Policy*, **24**(4), August 2000, pp. 475–476.

productivity (at the margin) from those findings. Hence, the debate is framed in terms of a trade-off between equity and efficiency.

To date, little direct evidence has been brought to bear on the debate. For example, Byerlee's doubts about using poverty alleviation goals to guide research strategies, mentioned earlier in this chapter, were based on a comparison of existing research programs (by crop), in terms of the incidence of their benefits. He asked whether shifting research resources away from some programs and towards others would provide greater benefits to the rural poor and concluded that it largely would not. However, in that study for Pakistan, he did not pose the options of different kinds of varietal research and, more importantly, participatory research with poor farmers.

Fan *et al.* analyzed the effects of high-yielding varieties and other public interventions on farmers in irrigated areas and rainfed zones in India. They concluded that technology improvements and better rural infrastructure have accelerated agricultural growth and helped alleviate poverty, but that these effects vary considerably between irrigated and rainfed zones, and among different types of rainfed conditions.[100] They argued (2000, p. 427), in apparent contradiction to Renkow's conclusions cited a few paragraphs above, that greater returns to public agricultural research and infrastructure investment are now being obtained in rainfed areas, including some apparently marginal areas. However, while this conclusion may be true of some of the areas in their sample, their statistical results do not fully support it.[101] It also should be noted that the authors studied only the effect of existing high-yielding varieties. The same caveats can be applied as were mentioned above for Byerlee's work: there was no analysis of the possible bene-fits for the poor of other kinds of research strategies. Thus, some principal questions are left unanswered about the potential for agricultural research to reduce rural poverty.

For the case of the Philippines, Keijiro Otsuka has argued that the main poverty-alleviation benefit of agricultural research comes through expanded aggregate output which reduces food prices for all, including the poor.[102] There are two problems with this argument. First, in a relatively open economy which is a price-taker, expansion of agricultural production may lead to reduction of imports or increase of exports without a change in domestic prices. Secondly, many rural poor have a small marketable surplus and thus are harmed by a drop in food prices. Indeed, even landless families and farmers whose plots are too small to generate a marketable surplus usually are beneficiaries of an increase in farm-gate prices, because it stimulates production and thus increases demand for rural labor. The study by Dean Schreiner and Magdalena García, cited in Chapter 4 of this volume, shows that the lowest income stratum in rural areas was decisively the greatest beneficiary of increased food prices in Honduras. In the end, the poverty-reducing effect of a change in food prices is an empirical question that depends in part on the numbers of rural landless compared with the numbers of farmers.

Otsuka makes the valid point (2000, pp. 459–460) that shifting the orientation of rice research toward less favored agricultural zones greatly complicates the research task with the result that the aggregate benefits, including for the poor, may be substantially reduced. In addition, he makes the following point (2000, p. 460), about making research consistent with an area's comparative advantage, that has wide applicability:

---

100. S. Fan, P. Hazell and T. Haque, 2000, p. 426.
101. Their results show that only in 6 of 13 rainfed zones is the poverty reduction effect of high-yielding varieties greater, per unit of expenditure, than in irrigated areas. In fact, in five of the zones that effect turns out to be zero. (In addition, their coefficients of the effect of high-yielding varieties were not statistically significant at the 5% level for three of the rainfed zones.) The intervention that has the largest poverty-reducing effect, by a wide margin, is the construction of rural roads in rainfed zones.
102. K. Otsuka, 'Role of agricultural research in poverty reduction: lessons from the Asian experience', *Food Policy*, **24**(4), August 2000, p. 447.

We do not argue that agricultural research should not try to develop new technologies for marginal areas. On the contrary, we argue that more resources should be allocated to research that generates appropriate technologies for such areas. We argue against rice research for unfavored areas, simply because the development of appropriate technologies can hardly be expected. We would like to suggest that the development of new technology for agroforestry, growing commercial trees, has high potential, because it is much more efficient than shifting cultivation. The development and wide adoption of new and more efficient agroforestry systems will both improve the incomes of poor farmers in marginal areas by increasing the efficiency of land use and contribute to partial restoration of forest environments. Yet, surprisingly, no international agricultural research center has conducted serious research on this promising technology. There might also be other crops and technologies particularly appropriate for marginal agricultural areas.[103]

Otsuka's suggestion is stated in a broader way by Hazell and Haddad, who point out *the importance of better technologies for natural research management on less-favored agricultural lands*. To the extent that poorer farmers are located on marginal lands, which often is the case, then improved natural resource management is central to increasing their economic productivity:

While some types of commodity improvement work seem vital for less-favored areas – improving drought tolerance, yield response to scarce plant nutrients, food nutrient content, pest and disease resistance, and livestock health and productivity – there is a growing consensus that major productivity improvements will first come from improved natural resource management (NRM) practices and technologies.[104]

This observation points toward an approach to reducing poverty by means of better technology that is based on agro-economic zones, rather than by attempting to identify target groups by income criteria, which is always more difficult in rural areas than in urban areas.

Clearly, the choice of strategy for research in relation to the rural poor depends very much on the context. As Renkow has summarized the debate:

Weighing the impact on poverty alleviation of alternative breeding and crop management research activities requires a careful assessment of where the poor live, what types of income-generating activities they engage in and the ways in which new agricultural technologies alter the returns to resources owned by household members. Available evidence does not support easy generalizations about the best means of improving the welfare of the poor in marginal areas. Instead, it reinforces the need for continual examination of alternative policies and investment strategies on a case by case basis. . . .

Undoubtedly agricultural research that specifically targets difficult production environments may represent the most pro-poor public investment available for some marginal areas. This is especially likely to be true for locations in which agricultural income shares of the poor are high, agronomic circumstances limit the adoption of technologies developed for other, more favorable production environments and prospects for research success are relatively high. However, in many situations, government investments in infrastructure and institutional

---

103. Crop diversification, including agro-forestry, plus better crop management by poor, hillside farmers, improvements that were achieved through participatory work with rural communities in Western Honduras, have been found to yield important gains in income levels of the poor, in efforts sponsored in that country by the FAO.

104. Peter Hazell and Lawrence Haddad, 'Agricultural Research and Poverty Reduction', Food, Agriculture, and the Environment Discussion Paper No. 34, International Food Policy Research Institute, Washington, DC, USA, August 2001, p. 27.

reform may well yield significantly larger and more rapid benefits to the poor in marginal areas than will investments in agricultural research targeted to those areas – especially where non-agricultural sources of income are relatively important.[105]

Where research has neglected the more difficult and heterogeneous agricultural environments, a case can be made for participatory agricultural research as a valuable instrument for poverty reduction. Co-operation with farmers in research can lead to better identification of varietal characteristics relevant for them, and to directing research toward their priorities among pest control, natural resource management, varietal improvement and post-harvest technologies. Farmers also can be skilled agents of varietal selection in their own agro-ecological contexts.

Ashby *et al.* have underscored the potential for participatory research to confer significant benefits on poor farmers:

With its emphasis on empowerment, the CIAL process is likely to have highly positive equity effects. In several cases very poor or marginalized groups normally left behind by development have taken up the process enthusiastically.[106]

The participatory approach can be intensive in terms of human resource requirements relative to the number of farmers benefitted, but when properly managed it has already proven to be promising in many cases in Africa, Latin America and, especially for pest management, in Asia.

In general, to make research more useful for poor farmers, it is necessary to promote *sustained dialogs between poor farmers, including women, and facilitators and agricultural scientists*, and not limit the collaboration to occasional visits by researchers to farms and villages. Many scientists fear that an involvement in 'development' will hinder their efforts at 'good science'. However,

the two goals should not be contradictory, and the latter should be seen as supporting the former. Otherwise, the argument for giving financial support to agricultural research becomes very much weaker.

8.4.5 Gender Approaches in Agricultural Research

Participatory approaches to agricultural research can be more beneficial if they give special emphasis to the involvement of women in the dialog. The case of Malawi is instructive about the benefits of taking into account the viewpoints of rural women in both research and extension:

throughout the 1980s and early 1990s uptake of . . . improved maize varieties was disappointing. . . . the vast numbers of farmers producing maize for home consumption were reluctant to adopt them for several reasons. They required expensive fertilizers and pesticides to grow successfully which women and poor farmers could not afford, they were not as drought resistant as local varieties of maize and hence posed a risk to food security, they were much harder to store and pound into meal so placed extra demands on women's scarce time and they did not have the favored taste characteristics of local maize which women knew their family preferred. . . .

In the late 1990s the Malawi Government . . . successfully re-oriented its research and extension activities to take into account the above problems. Research has successfully focused on developing an improved maize that has the taste, storage and pounding features of local maize and its uptake has prompted comments [that Malawi is experiencing a] 'delayed green revolution'.[107]

Although a gender orientation is still not widely used in agricultural research, the FAO's

---

105. M. Renkow, 2000, pp. 475–476.
106. J. A. Ashby, *et al.*, 2000, p. 140.
107. FAO, 2001, Module 12.

SEAGA Macro Manual (2001) cites other examples in which it has been successfully employed:

In Peru the International Potato Center is testing and improving staple food crops grown by women in Africa, such as the sweet potato, to find combinations of early maturity and high yield with a degree of drought tolerance. Such crops are often used by women during periods of famine and shortage and are eaten before the main harvest or when the staple harvest is poor.

In the Côte d'Ivoire the West African Rice Development Association (WARDA) has been conducting surveys to identify the preferences of men and women in adopting improved rice varieties. They found that while men prefer short statured high-yielding varieties, women are reluctant to grow these varieties due to the difficulties of harvesting them with infants on their back. In response WARDA has shifted its research emphasis towards the development of medium to tall statured varieties.

In Burkino Faso the study of gender dynamics of irrigation led to the adoption of appropriate technology that boosted production possibilities for women and the poor. Small dams for irrigation were introduced enabling the planting of trees for fruit and firewood as well as providing a convenient supply of domestic water. A training package was also introduced covering technical, organizational, credit and input uptake and marketing skills. This enables the technology to benefit all members of the community and household.[108]

As these examples show, *incorporating a gender focus in agricultural research is not difficult but it requires a sustained commitment on the part of research institutions*. An important starting point is the carrying out of a gender analysis of new and existing technologies. Another is the identification of rural women's activities. They vary by context but often include post-harvest and marketing activities, the cultivation of staples and/or vegetables, weeding field crops, breeding and raising small livestock, collection of water and fuelwood, and many other domestic chores. A participatory emphasis in the research program helps identify these activities and the ways in which women's labor can be made more productive. Of the 249 CIALs operating as of 2000, 7% were women-only and 37% were mixed in gender.[109]

Gender-sensitive research can generate household and field technologies that liberate women's time for more productive activities by reducing the labor demands of some of these tasks. The potential benefits of better household technologies were quantified for the case of Burkina Faso in the study by Lawrence *et al.* cited in Section 8.2 above. Research can also be directed towards improving the yields of crops that women typically produce and the efficiency of activities such as post-harvest management that they typically engage in.

It always is necessary to accompany agricultural research with infrastructure investments, programs of enhanced access to land, and other efforts directed towards improving the farmers' resource base, but this is especially true for women farmers. In studying factors that determine the rate of technology adoption among women farmers in Ghana, Doss and Morris concluded that:

On the whole, these results from Ghana suggest that technology adoption decisions depend primarily on access to resources rather than on gender *per se*. This conclusion should be interpreted with caution, however, because it does not necessarily mean that modern varieties and fertilizer are gender-neutral technologies. If adoption of modern varieties and/or fertilizer depend on access to land, labor or other resources, and if, in a particular context, men tend to have better access to these resources than women, then in that context, the technologies will not benefit men and women equally. Policy changes thus may be needed to

---

108. *Ibid.*
109. J. Ashby, *et al.*, 2000, p. 84.

increase women's access to the key resources; alternatively, it may be desirable to modify research efforts by deliberately targeting technologies that are particularly suited for the resources that are available to women. The bottom line is that it is important to examine both the technology itself and the physical and institutional context in which the technology is implemented. . . .[110]

On the one hand, this is a reminder of the value of a holistic approach to agricultural and rural development, and on the other hand it suggests the potential scope for re-orienting at least part of the effort of agricultural research in each country, to take better account of gender-related factors that influence the rate of adoption of new technologies.

Above all, a gender-sensitive research program requires a continuing effort to maintain adequate channels of communication with women farmers, and this requires changes both in the way research is carried out and in the ways that extension services are organized and operate.

8.5 NEW APPROACHES TO AGRICULTURAL EXTENSION

Although bright and dedicated extension agents can be found in every agricultural area, the weaknesses of existing extension systems are all too familiar throughout the developing world: extension messages that have little relevance, insufficient farming experience on the part of extension agents for them to be credible to farmers, lack of mechanisms for transmission of farmers' chief concerns to agricultural researchers, weak linkages between research and extension, poorly paid and poorly motivated extension agents, and insufficient supporting budgets with agents frequently sitting in offices in cities for lack of transport.

Farrington puts it more generously:

Numerous examples can be cited of successful public sector extension. . . . Yet, in many circumstances the picture is one of resources spread too thinly to be effective, inflexibility and inability to respond to the changing infrastructural and institutional contexts.[111]

While the pervasive fiscal crises of recent years in developing countries figure prominently among the causes of these problems, there are fundamental issues of management and structure of the extension systems, including incentives for performance and recruitment criteria. At this stage, it is clear that the earlier, centralized systems are no longer viable and that new approaches are needed.

In Section 8.3 above, it was noted that the environment for agricultural extension has changed drastically in recent years. Principal factors mentioned by various observers, in addition to the observed poor performance of many systems, include fiscal restrictions, the increasing involvement of the private sector, farmers' associations, community groups and NGOs, the effects of globalization on agriculture, changing priorities of donors, and, in some cases, the devastation wrought in farming communities by HIV/AIDS.

For these reasons, extension systems have been subject to intensive review and reform throughout the world. In addition, increasing attention is being paid to alternative means of enriching and transmitting knowledge about agricultural technologies.

8.5.1 Alternatives for Agricultural Extension Systems

The Neuchatel Group has enunciated and explained six guiding principles for developing new approaches to agricultural extension, in terms that are applicable to any region of the world:[112]

---

110. C. R. Doss and M. L. Morris, 2001, p. 39.
111. J. Farrington, 1995, p. 540.
112. Neuchatel Group, 1999, pp. 5 and 10–15.

- *A sound agricultural policy is indispensable.*
- *Extension consists of 'facilitation' as much if not more than 'technology transfer'.* Extension is too often merely seen as a vehicle for spreading scientific and technical progress and technology transfer. But this is a narrow and highly unsatisfactory definition. The dissemination of knowledge is not a one-way street from scientists to producers. Farmers' own knowledge must be collected, analyzed, capitalized on, propagated and disseminated. Producers need more than just technical information. There is rarely a 'one size fits all' solution to address the mix of technical, economic, commercial, social and environmental aspects that farming problems consist of. . . . producers must themselves be able to analyze the constraints, seek out and test solutions, and make choices from an array of existing service providers.

  The essence of agricultural extension is to facilitate interplay and nurture synergies within a total information system involving agricultural research, agricultural education and a vast complex of information-providing businesses. . . .

  Therefore, agricultural extension activity facilitates:
  — Direct exchanges between producers as a way of diagnosing problems, capitalizing on existing knowledge, exchanging experiences, disseminating proven improvements, and even fashioning common projects.
  — Relations between producers and service providers (including public extension services). Extension is advisory, not prescriptive. This requires extension workers to be 'actors in' not 'instruments of' extension. Trust must be established between the customer-small farmer and the advisor. Solid technical expertise remains essential but the abilities of extension workers must go beyond that. Extensionists must nowadays be adept in participatory techniques, and resourceful in drawing on a mix of communication methods and technologies. They must

think in terms of market opportunities, increasing producer incomes and total farm management . . .
- *Producers are clients, sponsors and stakeholders, rather than beneficiaries of extension.* Extension activities are more effective when farmers are directly involved in defining, managing and implementing them. When farmers fund or purchase training services the impact is significantly better than when they attend training entirely designed and funded by someone else.

  This happens when:
  — Farmer organizations manage their own technical services.
  — Producer groups and private . . . or public . . . service centers work together on a contract basis.
  — Producers can target funding on problem-solving for their specific needs.
- *Market demands create an impetus for a new relationship between farmers and private suppliers of goods and services.* A major theme in agricultural development is the gradual transition from low-productivity subsistence farming to specialized production based on comparative advantage. . . . Small farmers must produce a sufficient range of competitively priced outputs in the right quantity and quality at the right time. This move from subsistence to commercial farming is consumer, rather than producer-driven.

  Because input suppliers and produce buyers are business people, they must have their fingers on the pulse of demand and offer suitable products and services. Without inputs or markets, extension service recommendations are a dead letter. . . . Impartial and unbiased marketing and technical information are essential if producers are to be enabled to respond to market conditions. An extension activity which delivers that advice and facilitates balanced relations between producers and private business is a development-nurturing source of security for producers.
- *New perspectives are needed regarding public funding and private actors.* Public

funding of extension is essential, but that does not mean that public extension institutions should carry out or run extension services. . . . Governments may contract out some or all of the implementation to nongovernmental institutions (farmer organizations, specialized consultants, NGOs). . . . In order to do this effectively, governments must develop the capacity to monitor and evaluate activities they finance. . . . Having producers and private sector actors cofinance extension, either individually or through their professional organizations, can result in savings and the more efficient use of public resources.

- *Pluralism and decentralized activities require coordination and dialogue between actors.* Centralized and standardized national extension systems do not produce satisfactory results. No single approach or organization fits all. . . . To be effective, extension must be able to address change. Extension systems must be ultra-flexible to respond to new situations (opportunities or crises). Decentralizing guidance and decision-making bodies can facilitate that. . . . Producers should have a choice or a range of providers in terms of methods, quality of service and cost. . . . Nonetheless, the multiplicity of actors combined with decentralization make national coordination and consultation essential. National and local forums for dialogue and coordination between farmers and other stakeholders . . . are required to:
  — set common aims and frame policies;
  — harmonize working methods and tools;
  — capitalize on experiences and exchanges of information;
  — carry out follow-up and evaluation;
  — orchestrate activities and fairness in target groups;
  — achieve efficient deployment of public resources;
  — pool training and research facilities.

This dialogue must be equitable. Coordination must not become central control by another name.

A central conclusion that follows from these principles is that technology development and extension work must become more demand-driven:

There is now a growing consensus that to create a *demand-driven technology system* is to directly involve farmers in identifying problems, establishing priorities, and carrying out on-farm research and extension activities. . . . The creation of a balance between institutional "supply systems" and farmer-initiated demand-driven extension/technology systems should in many cases be the ultimate goal of countries eager to advance to higher stages of development and competitive power.[113]

While the responses to farmers' demands for information must be provided by a variety of institutions, Rivera (2001, p. 27) warns that *pluralism is not always the same as partnership.* The latter results in learning by all institutions involved in the process, and it is a relationship of equals. If the process is not structured properly, the existence of multiple service providers under government contract can simply be a way of implementing government mandates, without relinquishing central control over the process.

The approach of participatory technology development can be carried out relatively independently of existing extension systems, although the requirements of contact with research institutes and sustained sources of funding argue for some kind of institutionalization of the work, however loosely it may be structured. Farrington has summarized some of the ways in which a participatory approach can be applied:

- *Approaches based on farmer participation in diagnosis, testing and dissemination.* Normally organized with groups of farmers rather than individuals, these approaches recognize that researchers and extensionists are unlikely to capture the complexity, diversity and risk facing low-income farmers, that

---

113. W. M. Rivera, 2001, p. 12.

farmers' own knowledge is important, and that farmers themselves are best placed to interpret how relevant new technologies might be. These approaches demand the types of group organizing and support skills hitherto rarely found in public sector extension. Yet some types of organization (e.g. NGOs) have successfully supported the growth of cohesive membership organizations. . . . A simpler but potentially more powerful and increasingly popular approach is *to have farmers visit experiment stations* in order to select technologies appropriate to their circumstances, then provide feedback to researchers.

- *Farmer-to-farmer dissemination.* Less formal efforts based on many of the same principles, but not necessarily requiring group formation, have been used at least since the 1960s, when Oxfam sponsored farmer-to-farmer visits across Central American countries, and subsequently have been tried elsewhere, particularly in Southeast Asia.
- *'Para-professional' extensionists.* Some groups select one or more of their members to interact with public sector extensionists and researchers either across the board or on specific aspects of local farming systems. Whilst some initiatives assume that the para-professionals will do this largely on a voluntary basis, others link the provision of advice with input supply. Small farmers may pay for a package linking inputs and advice . . . payment for advice alone is largely restricted to commercial farming.[114]

The FAO's Farmer Field Schools, outlined at the end of Section 8.2 above, represent an effective means of involving farmers in both agricultural research and extension, as do experiences such as the CIALs in Latin America, the PBA program on the north coast of Colombia, and the thrust of participatory technology development pioneered in Zimbabwe, Malawi and other countries in Africa. *Through participatory experiences,* *farmers themselves become extension agents for their neighbors and nearby communities.* In fact, one of keys to the spread of CIALs has been training farmers as trainers who can go to other locations to explain and facilitate application of the approach.

Farrington also indicates that governments are responding to the availability of these 'non-traditional' kinds of extension in various ways:

First, government is tending to pull out village-level extension workers, partly because of financial pressures and partly because of farmers' growing capacity to reach higher into the technology generation and transfer system [through both their own organizations and NGOs] in order to draw down suitable technologies. Government is therefore scaling back, but the boundary will necessarily shift unevenly; where those producing commercial crops can readily obtain technical information from private sector input supply, processing and marketing organizations, the scaling back needs to be more extensive than among subsistence-level food crop producers.

Second, the number of organizations representing or working on behalf of the rural poor is increasing rapidly. Some government departments are beginning to provide technical support to them and to learn lessons through "feedback" from them. . . . Further, they need to provide an environment that will support the emergence and growth of such organizations. . . .

Third, there is a move toward providing the funds for low-income farmers to contract extension services from government departments and NGOs . . .[115]

One general lesson to emerge from such experiences is that there are many ways to disseminate agricultural technology, some of them more effective (and cost-effective) than formal extension services as they have been structured in the past. Another lesson is that the organization of farmers

---

114. J. Farrington, 1995, pp. 540–542 [emphasis added].
115. J. Farrington, 1995, pp. 542–543.

> *NGOs have begun to assume a greater role in agricultural extension, frequently focusing on areas that the government has neglected. One reason for their success has been their community-based focus. In West Africa, for example, the Se Servir de la Saison Séche en Savanne et el Sahel (the 6-S Program for the Savannah and the Sahel) promotes village organizations, helps groups establish community development programs, and provides funding and technical assistance for projects including village crafts, cereal banks, market gardening, soil conservation, and reforestation. With an annual budget of $1.25 million, 6-S is now operating in Burkina Faso, Mali and Senegal. Since its founding in 1976, it has established 2000 farmer organizations (averaging eighty members per group) in about 1000 villages. . . . In Northern Ghana the Agricultural Information Service, funded by the Presbyterian Agricultural Station at Langbensi, works with more than twenty church-based agricultural stations and coordinates with the government research station at Nyankpala. . . . (D. Umali-Deininger, 1997, pp. 214–215).*

at a local level is a key to successful extension, especially in the case of low-income tillers and women farmers. **Often NGOs are leaders in promoting community organization.**

While participatory research programs sometimes eliminate the need for a separate extension effort in those localities, extension still can add considerable value for the vast majority of farmers, and in some cases it can strengthen participatory research programs. In this regard, a third lesson is that NGOs are a very useful force for technology transfer. However, official policy generally has been slow to recognize this contribution, and accordingly it is not always as effective as it might be. In many countries, rural NGOs operate with complete autonomy, in isolation

even from one another. This circumstance gives them greater flexibility to work closely with rural communities but it has some disadvantages. One drawback is that approaches to crop management and resource management which prove successful are not shared among NGOs, or between them and public sector extension workers, and so the benefits of the approaches are more limited than they might otherwise be. Another shortcoming is that the NGOs' own extension workers do not receive the full benefits of the experience of the public sector extension system, and their linkages to the agricultural research system usually are weak. In other words, valuable as their contributions might be, they could be enhanced through greater co-ordination with other efforts. Sometimes, a single NGO is effectively 'trying to rediscover the wheel', in circumstances where other NGOs or public sector experts already have discovered it in the same country.

Co-ordination of rural NGOs is a delicate issue, since they understandably see great advantages in their autonomy. However, co-ordination carried out with a light hand could lead to improvements in the knowledge transmitted to rural populations. The public sector could usefully sponsor differ-ent kinds of forums in which NGO experts exchange information about experiences and lessons learned and attempts are made to collate the lessons and transmit them even more widely. Alternatively, the NGOs themselves can form umbrella organizations to carry out this function.[116] An example of this on an inter-country basis is found in Latin America, where eight NGOs, working in seven countries, formed the *Consorcio Latinoamericano para la Agroecología y Desarrollo* (CLADES) for the purpose of strengthening their efforts to disseminate agricultural technology to small farmers.[117]

The role of the internet as an informational tool should not be overlooked even in poor rural areas, although it has been little exploited for this purpose to date.

---

116. These conclusions emerged from a workshop with rural NGOs in Nicaragua, funded by USAID and led by the author, held in Managua in July of 2000.
117. D. Umali-Deininger, 1997, p. 215.

8.5.2 Promoting a Client Orientation in Extension Services

The frequent lack of a client orientation in extension services was remarked on earlier in this chapter. It is essential to improve the accountability of extension agents to farmers, so that their primary mission becomes one of *understanding and addressing the needs of farmers, including women farmers*. The responsibility of understanding the needs of farmers takes extension agents well beyond the realm of delivering extension 'messages' designed in the system's central office. Often, it requires that agents understand not only the agronomic conditions on a farm but also the farmer's constraints in terms of access to inputs and markets, and also the role of gender and community factors in shaping a farmer's decisions. The task of addressing farmers' needs also requires that agents provide feedback to agricultural researchers and maintain links with them to receive responses.

For this orientation to be effective, *the performance of extension agents needs to be evaluated by farmers themselves*, and not only by higher levels of a centralized organization. Putting farmers 'in the driver's seat' in this sense is necessary for achieving accountability to them. In the best extension systems, performance evaluations are based at least in part on feedback from farmers, but this element of the system often is weak. Such a requirement is the heart of the argument in favor of *private extension services for which farmers pay at least part of the cost, sometimes with the aid of transfers made by the government to farmers for that purpose.* Payment by users creates powerful incentives for extension agents to satisfy them rather than their superiors in a bureaucratic hierarchy. Along with payment goes *the farmers' right to select extension agents*, and to change them if their performance is found to be unsatisfactory, to fulfill the Neuchatel Group principle mentioned above that farmers be able to 'make choices from an array of service providers'.

Antholt has discussed the importance of incentives of this nature and the responsibility of the clients for helping create them:

> The opposite side of the accountability coin is expecting the beneficiaries of extension to be responsible for some of the support, even if it is only a proportion of total costs. This is important for three reasons. First, it gives beneficiaries ownership and drawing rights on the services. Second, it takes some of the financial pressure off the central government and therefore gets at the issue of financial sustainability. Lastly, if ownership and responsibility rest with clients, the basis for more demand-driven, responsive service is established. . . .[118]

In Estonia, the government funds farmers to hire private extension agents, and it also finances the role of farmers' associations for technology transfer, and the latter in turn contract with private extension firms. However, it retains a public extension service for poor farmers.[119] Private agricultural extension also is used in Hungary, Slovakia and the Czech Republic.[120] It is being developed in Azerbaijan.[121] In El Salvador, the national association of dairy farmers (APROLECHE) has, with funding from its members, contracted with a leading international extension expert in dairy management, and as a result members' milk yields increased very significantly in the 1990s.

---

118. C. Antholt, 1998, pp. 360–361.

119. M. K. Qamar, 2000, p. 161.

120. Geoffrey Adams, 'Extension advisory services in Central and Eastern Europe', in M. K. Qamar (Ed.), *Human Resources in Agricultural and Rural Development*, Food and Agriculture Organization of the United Nations, Rome, 2000, p. 12. Adams also points out that fully public extension services still are used in Albania, Bulgaria, Croatia, Poland and Romania.

121. John Lamers, Georg Dürr and Petra Feil, 'Developing a client-oriented, agricultural advisory system in Azerbaijan', in M. K. Qamar (Ed.), *Human Resources in Agricultural and Rural Development*, Food and Agriculture Organization of the United Nations, Rome, 2000, pp. 105–117.

Another scenario using this approach might be for extension departments to develop cooperative or contractual agreements with local bodies, as in China. Under such arrangements local organizations might take responsibility for provision of their own extension services, but the center would reimburse the local entities for some percentage of their costs. Alternatively, as in some areas of China or in Ecuador ... an arrangement to share in the output of the farming enterprise can be developed.

Another alternative is seen in Chile: contracting with private firms or NGOs for the provision of extension services. The government's role is to lay out the ground rules for service, select consultant firms through competitive bidding, evaluate performance, and subsidize the cost of the services. Consultants carry out the technical extension services, and farmers contract with the firm of their choice. ... there is cost-sharing between the government and farmers, the proportion of which is dependent on the amount of land owned by the farmer.[122]

In addition to strengthening the effectiveness of extension work in general, good farmer organization is a prerequisite for participation in schemes of providing vouchers to small farmers for the payment of part of the cost of extension services. Such schemes have been implemented in Costa Rica,[123] and in Nicaragua on a trial basis, and proposed formally for consideration in Honduras. They represent a way to bring small farmers into the market for extension services, and at the same time they help guarantee that a market exists for extension agents who might otherwise be fearful of the consequences of privatization of an extension system. (It is worth bearing in mind the circumstances under which farmers would be willing to pay for extension information, as reviewed in Section 8.3.2 above.)

A cautionary note needs to be sounded regarding the ability of low-income farmers to pay for

> *The importance of farmer organization comes up again in the context of payment for extension services:*
>
> *Provided farmers can overcome the difficulties of organizing into a group, farmers' associations can allow small farmers to pool their resources to purchase extension information that individual farmers may not be able to afford on their own (D. Umali-Deininger, 1997, p. 217).*

extension services. The principle that they should pay for part of the services is a valid one, but sometimes unrealistic expectations are created regarding how much they can pay. For example, a typical World Bank formula has been to ratchet up the percentage of extension costs paid by farmers, regardless of their income levels, in equal increments over five years, so that farmers pay 20% the first year, then 40%, etc., until they support 100% of the cost in the fifth year of the program. Although it has arithmetic neatness and beneficial effects for the fiscal budget, this formula has proven to be quite unrealistic for low-income farmers. It should be understood from the outset that poor farmers will be unable to pay the full cost of extension services for very many years, if ever.

At the same time, fiscal savings can be realized by *targeting the subsidy for extension services*, eliminating the regressive aspect of that subsidy that usually occurs under public extension services and which was mentioned earlier in this chapter. In this way, the cost of extension to the government can be reduced without requiring that poor farmers pay the full cost for it. When subsidized and targeted extension services are introduced, governments must decide to whom and how to target them.

Umali-Deininger (1997, pp. 213–214) provides other examples of farmer associations providing or contracting for extension services, in Argentina, the Central African Republic and

---

122. C. Antholt, 1998, p. 361.
123. M. K. Qamar, 2001, p. 160.

Zimbabwe. She also stresses the role of agribusiness enterprises as extension providers:

> During the 1970s dairy farmers in Argentina faced serious obstacles. Livestock there was unproductive; the milk supply was unstable and often of poor quality. These problems were mainly the result of poor animal nutrition and inadequate farm hygiene. The two largest dairy processors, Santa Fe-Córdoba United Cooperatives (SANCOR) and La Serenísima, whose own growth was jeopardized by the plight of the dairy farmers, launched extension programs to overcome these constraints. SANCOR created an extension department with eight regional offices, each managed by an agronomist assisted by middle-level technicians. Each office provided extension services to almost forty cooperatives and assisted small groups of farmers (usually six to fifteen) who met monthly to discuss a visited farm's progress and problems. SANCOR initially financed technical assistance for these small groups, but after thirty months, each group took on the cost of the professional agronomist. By 1990 SANCOR had 120 farmer groups participating in the program.[124]

In the end, the approach adopted must fit the circumstances, and different ways of combining private and public sector efforts can be developed. In the German State of Thuringia, for example, the government provides extension services in relation to public goods, i.e. on environmental issues and plant protection, and also to promote national goals, as in furthering the development of women farmers. Farm-related extension services are provided by private agents in both Thuringia and Saxony-Anhalt, but they are partially subsidized by the government in the latter.

In 2000, Thuringia decided to introduce partial reimbursement to farmers also; before that the vast majority of farms employing private extension advisors were large farms.[125] The Agricultural Advisory Service in Norway provides three categories of services: 'some services fully financed by government, some partially financed by government and another that receives no funds from government'.[126]

As illustrated by the Argentine experience, the agricultural marketing chain itself is a source of extension services. Increasingly, farmers both large and small are obliged to understand market requirements and to generally adopt more commercial modes of operation, including keeping records of production costs and cash flow. In this regard, another element that has been missing from most agricultural extension services is training in simple bookkeeping and principles of farm management. The goal of agricultural development is to make farmers more successful entrepreneurs, and without training in the basic tools of entrepreneurship it is difficult for them to advance. There is a growing consensus that extension work should include not only agronomic considerations but also basic principles of cost accounting and farm management. For example:

> Potential advisors must first learn to think and act in terms of farming systems and enterprises rather than activities. . . . Their technical ways of thinking need to be complemented by effective communication (listening is a key) and social skills.[127]

As noted above, the Neuchatel Group has endorsed the concept that agricultural extension services need to help farmers improve their linkages with input and output markets.

---

124. D. Umali-Deininger, 1997, p. 212.
125. Jochen Currle and Paul Schütz, 'Privatizing agricultural extension services in two new German federal states: necessary conditions emerging from experience', in M. K. Qamar (Ed.), *Human Resources in Agricultural and Rural Development*, Food and Agriculture Organization of the United Nations, Rome, 2000, pp. 131–140.
126. W. M. Rivera, 2001, p. 21.
127. J. Lamers, G. Dürr and P. Feil, 2000, pp. 110–111.

Umali-Deininger also has stated that the basic roles of extension include marketing advice and work on community development (see Section 8.3 above). Extension cannot be confined to technical issues of plant cultivation if it is to help develop rural sectors economically. This role requires further training of extension agents. In the context of Central and Eastern Europe, it has been observed that:

> The only practical way to provide advice to small mixed farms is through generalist advisors who can appreciate the needs of the whole family unit. Training specialists to become generalists has proved difficult, but has been successfully achieved in some countries (Estonia, Slovakia, Lithuania and Latvia).[128]

One of the most effective ways to promote a client orientation in extension work is through the use of participatory approaches. The basis of the Zimbabwean experience with a participatory extension approach (PEA) was described as follows:

> The concept [of] participatory innovation ... and extension is based on communication [through dialogue], farmer experimentation and strengthening of self-organizational capacities of rural communities. Encouragement of active participation and dialogue ... among all actors on the local level, for example, farmers and their institutions, extensionists and researchers, [is] the mainstay.
>
> Dialogue and farmer experimentation is being encouraged in an environment where a very powerful top-down extension service has considered farmers' knowledge to be backward and of no importance for nearly three generations, and where farmers have been conditioned to accept externally developed standardized technologies. ... the knowledge and understanding gained through this process strengthens farmers' confidence in their own solutions and increases their ability to choose options and to develop solutions appropriate for their specific ecological, economical and social-cultural conditions. ...[129]

Evison Moyo and Jürgen Hagmann have summarized the lessons for participatory extension from the Zimbabwean experience. Participatory extension necessarily involves participatory technology development as well. Their synthesis is as follows:

> In many countries, public sector extension services have been accepting that there is a need for participatory approaches to agricultural service delivery ever since the potential of such approaches was demonstrated by nongovernmental organizations. The acceptance and promotion of these approaches ... in hierarchical government bureaucracies, where they are often implemented by low-paid extension agents with low-level qualifications, has proved to be difficult. Many existing organizations will have to transform their approaches to extension from ones that are based on top-down teaching and a narrow orientation on production to ones that are people-centered, learning-oriented and participatory. ...
>
> Community-based extension, full community ownership of the process and joint learning are central to PEA. The characteristics of PEA include:
>
> • a focus on strengthening rural people's problem solving, planning and individual, as well as collective, management abilities. ...
> • equal partnership among farmers, researchers and extension agents who can all learn from each other and contribute their knowledge and skills;

---

128. G. Adams, 2000, p. 15.
129. J. Hagmann, E. Chuma and K. Murwira, 'Improving the output of agricultural extension and research through participatory innovation development and extension; experiences from Zimbabwe', *Journal of Agricultural Education and Extension*, 2(3), 1996, p. 16.

- promotion of farmers' capacity to adapt and develop new and appropriate technologies/innovations by encouraging them to learn through experimentation, building on their own knowledge and practices and blending these with new ideas in an action learning mode. . . .
- recognition that communities are not homogeneous but consist of various social groups with different and conflicting interests, powers and capabilities. The goal is to achieve equitable and sustainable development through the negotiation of interests among these groups and by providing space for the poor and marginalized in collective decision-making. . . .

PEA is far more than a participatory methodology and is distinctly different from [participatory rural appraisal], which is essentially a tool-box. PEA is a comprehensive, iterative learning process approach to rural innovation and problem solving that enhances governance and civil society in rural areas.[130]

In sum, *when carried out in its fullest spirit, PEA effectively merges with the participatory agricultural research approach.* It is more demanding to implement than traditional top-down extension systems at the outset, but it appears to provide an effective way to ignite processes of technical change in rural areas that have been bypassed by existing systems of technology delivery. It would appear to be particularly appropriate in communities of small farmers characterized by heterogeneity in farming conditions. Its implementation requires substantial institutional changes and therefore can proceed only with a strong commitment from higher levels of governmental institutions responsible for agricultural policy and services, and with an overriding focus on facilitation rather than delivering technological messages.

8.5.3 Gender Approaches in Agricultural Extension

A client orientation in extension will promote greater delivery of extension services to women. However, it is often necessary to make a special effort to recognize the pro-male biases in extension systems, in order to be able to overcome them. For example, prior to the reforms in research and extension in Malawi that were mentioned earlier in this chapter:

Extension work to promote the uptake and efficient production of . . . crops was mainly carried out by male extension officers and was directed to the more commercially successful and progressive farmers who were predominantly male. . . . Extension services formed part of a package made available to [them] through male dominated 'farmers' clubs'. . . . Despite the fact that over a third of farm households were female headed many women farmers were unable to benefit. Their lack of access to farmers' clubs and credit meant they were unable to adopt the packages and benefit from the supporting extension advice.[131]

The FAO presents several key actions that are necessary for the successful incorporation of gender sensitivity into agricultural extension work:

- Move towards a *demand-led extension system* where both male and female extension workers are trained in gender issues and participatory planning and are better able to identify women's and men's needs, constraints, priorities and opportunities and adapt extension packages accordingly. . . .
- Improve the links between extension and research *to ensure that local knowledge and practices are incorporated into research design.*

130. Evison Moyo and Jürgen Hagmann, 'Facilitating competence development to put learning process approaches into practice in rural extension', in M. K. Qamar (Ed.), *Human Resources in Agricultural and Rural Development*, Food and Agriculture Organization of the United Nations, Rome, 2000, pp. 143–146.
131. FAO, 2001, Module 12, Box 1.

- Broaden the range of extension activities to include local food crops, poultry and small ruminants, advice on labor saving devices for women's household chores and off-farm income earning activities.
- Schedule extension visits to ensure they do not clash with women's household responsibilities such as food preparation.
- Adapt training materials to meet women's level of literacy and numeracy.
- Adopt a group approach to extension provision to enable resource-poor farmers to pool resources. *Women's accessibility appears to be greater in group extension activities compared to their participation in the contact farmer approach*. . . .
- Introduction of gender-related analytic methods in the syllabus for the training of new and existing technical, extension and other field staff in rural development.
- Increase female enrolment in agricultural tertiary education and increase the number of female extension officers. . . .
- Introduce a system to monitor the extent to which extension services reach and benefit female and resource-poor farmers.[132]

As part of Malawi's reforms to its extension service:

Male extension workers have received specific guidance on working with women farmers and on-farm demonstrations now often use women farmers' fields. Women's participation at training sessions has increased tremendously since their own fields have been included in the training and demonstration program. In addition, extension workers have been instructed to include the following in their activities: low as well as high resource farmers; women farmers with both high and low resources; women as both household heads and wives (FAO, 2001, Module 12).

Participatory research and extension systems have already moved in these directions. It is important that these orientations be incorporated into the supervision guidelines that apply to private, subsidized extension services as well as to public extension services themselves.

8.5.4 Responding to the Challenge of HIV/AIDS

As indicated above, the spread of HIV/AIDS infections in many countries is not only having a devastating effect on families but also on rural services and agricultural production. Clearly, national health agencies are primarily responsible for co-ordinating the responses to the epidemic, but extension services also have to review their approaches in light of the changes in the rural labor force wrought by the disease and the effects on the services themselves:

Agricultural extension services cannot and should not be expected to put an end to HIV/AIDS. . . . However . . . the fact remains that these are the only organizations whose field staff are very familiar with rural life. They, therefore, can and should play a meaningful role in helping the farming communities for protection against AIDS. If they do not move fast, in collaboration with other relevant institutions, to properly respond to the increasing impact of HIV/AIDS on the overall farming situation in general and on their own weakening organization capacity in particular, the consequences could be disastrous and far reaching.[133]

Qamar outlines a number of possible strategies to face the challenge of HIV/AIDS in rural areas, including the following:

- Formulation of a national policy on AIDS and extension. . . .
- Preparation of extension staff [through] revision of pre-service and in-service training curricula. . . .

---

132. FAO, 2001, Module 12 [emphasis added].
133. M. K. Qamar, 2001, p. 6.

- Fast-track training of extension staff, through intensive orientation sessions of short duration [given] by health specialists, rural sociologists, and anthropologists [so that] extension staff should possess knowledge on the relationship between food security and HIV/AIDS, the main causes for the spread of HIV/AIDS, its visible signs, precautions to be taken in the handling of patients, ethical and privacy considerations, development of a healthy and constructive attitude towards sick persons, coping with the new clientele of extension, common fears about the epidemic which have no scientific basis, and on tactful strategies to discourage certain sexual practices embedded in culture that expedite the spread of HIV infection. . . .
- Revision of extension strategies and technical messages. Agricultural extension strategies, methods, and technical content, should all be revised and adjusted in light of the fact that large numbers of inexperienced men, women, youth, widows and orphans are being forced into farming due to the death of their traditional bread winners. Many of these persons are physically weak. They are not able to use heavy farm machinery and equipment, nor are they able to follow any cropping patterns requiring vigorous and frequent physical labor. . . . In Malawi, for example, the families affected by HIV/AIDS are giving up labor-demanding tobacco cultivation and post-harvest processing in favor of crops like cassava and sweet potatoes, which require less manual labor. . . .

  Introduction and/or strengthening of extension methodologies using a group approach that can be applied with relatively small number of extension staff in view of dwindling number of extension workers. Development of HIV/AIDS-oriented participatory, client-focused extension approaches

and technical messages in order to address specific extension and training needs of old and new clientele in terms of age, gender. . . . Involvement of rural youth in extension program planning and implementation since they constitute the sexually most active social group and therefore are hardest hit by AIDs. . . .
- Preparation of multi-media extension materials on HIV/AIDS. . . .[134]

Daphne Topouzis has also reported on the incidence and seriousness of the effects of HIV/AIDS on agricultural production:

Farming systems with fertile soils, abundant and well-distributed rainfall and a wide range of crops are less likely to be sensitive to labor loss than those with poor soils, little rainfall and a limited range of crops. . . . For example, Uganda's farming systems are less vulnerable to the HIV epidemic than the maize-based cropping systems of southern Africa.

. . . agricultural output in Zimbabwe has declined by nearly 20 percent among households affected by AIDS. Maize production by smallholder farmers and commercial farms has declined by 61 percent because of illness and death from AIDS. Cotton, vegetables, groundnut and sunflower crops have been cut by nearly half, and cattle farming has declined by almost one-third. . . .[135]

She emphasizes the need to strengthen existing multi-sectoral responses and the urgency of improving our understanding of the implications of the epidemic for training programs:

Multi-sectoral approaches to HIV/AIDS were widely adopted in the 1990s in recognition of the growing realization that the HIV epidemic was more than just a health problem and that

---

134. *Op. cit.*, pp. 7–8.
135. Daphne Topouzis, 'The impact of HIV on agriculture and rural development: implications for training institutions', in M. K. Qamar (Ed.), *Human Resources in Agricultural and Rural Development*, Food and Agriculture Organization of the United Nations, Rome, 2000, p. 94.

the intervention of ministries of health was not enough to stop the spread of the epidemic and mitigate its impact.

Multi-sectoral approaches in ministries of agriculture (MoAs) frequently have two components: the establishment of HIV/AIDS local points within the MoA; and information, education and communications activities for MoA staff and target groups. Both of these components have been primarily health-based. Information, education and communications activities have often been added on to training programs and projects but have rarely touched on the core areas of agricultural and rural development work. In other words, multi-sectoral responses have essentially consisted of AIDS-specific components that were implemented in relative isolation from the mainstream activities of MoAs. Similarly, the vast majority of donor projects in MoAs have not addressed the implications of HIV/AIDS for food and livelihood security. . . .

In view of this, the multi-sectoral response concept needs to be redefined within a developmental context in order to extend responses to HIV/AIDS beyond the health sector and into the core technical areas of agriculture and rural development. . . .[136]

She recommends addressing the following questions in order to improve the responses to HIV/AIDS:

• What are the key implications of increased young adult morbidity and mortality for training institutions and their training strategies?
• How can training programs be adjusted to reflect the developmental implications of HIV/AIDS? . . .
• What is needed to ensure that the technology developed and diffused by publicly financed agro-research institutions is relevant to the changing needs of rural producers and consumers in view of the HIV epidemic? . . .
• What changes are needed in training curricula and methodologies at the trainer and trainee levels?
• What are the required steps for facilitating the necessary adjustments to training curricula, methodologies and procedures?[137]

8.5.5 Illustrations of New Trends in Agricultural Extension

In short, a wide range of approaches to extension, characterized by a greater role for non-governmental entities and greater emphasis on farmer participation, has proven to be effective in developing countries. In the case of Thailand, as another example, the private sector has been credited with the successful dissemination of new varieties of cassava. A microcosm of these new trends is found in the case of Bangladesh, as presented by Antholt, citing work by Chowdhury and Gilbert:[138]

Bangladesh's experience with extension is of general interest . . . because the government has moved beyond T & V and developed a new national extension strategy that is less costly, more demand-driven, more decentralized, and relies heavily on NGOs. . . . The T & V approach was . . . consistent with the development thinking of the 1970s, in which the state played a central role in development and little attention was given to the possible contribution of NGOs in extension activities or to private firms in input delivery and marketing. . . . the T & V system had several major weaknesses, including its emphasis on delivering routine messages to farmers and failing to take farmers' constraints and priorities into account. Also, the use of contact farmers proved to be ineffective, and the program proved to be costly and was not

---

136. *Op. cit.*, p. 99.
137. *Op. cit.*, p. 100.
138. Mrinal K. Chowdhury and Elon H. Gilbert, 'Reforming Agricultural Extension in Bangladesh: Blending Greater Participation and Sustainability with Institutional Strengthening', Agricultural Research and Extension Network, Paper No. 61, Overseas Development Institute, London, 1996.

financially sustainable. Finally, the impact of T & V on agricultural production in Bangladesh was mixed. . . .

Bangladesh's new extension strategy includes the following reforms and institutional innovations: greater decentralization of authority and accountability from the center to the districts; use of groups of farmers rather than contact farmers; demand-driven extension methods and recommendations; broader participation of the private sector, including NGOs; a sharper focus on the disadvantaged, including women; and greater emphasis on financial sustainability.[139]

Antholt has provided a cogent summary of the new trends in approaches to agricultural extension, in the following words:

thinking about agricultural extension services needs to have conceptual horizons broader than the conventional public sector. It also follows that more attention needs to be given to financing.

The time for change is now, given the long gestation periods for institutional modernization. Below are some general parameters for the future that will provide useful guidance for contemporary policy changes and investment initiatives.

- Farmers need to come first. This means placing real ownership of, and accountability for, public extension organizations into the hands of the client community, particularly farmers (but agribusiness as well).
- Competition in the provision of extension services needs to be fostered through pluralism in the provision of extension services.
- Pluralism means redefining the roles of the public and private sectors in extension and, in particular, enhancing the role of the private sector through privatization, particularly of frontline extension.

- Mechanisms for public support – for example, vouchers, cost sharing, local taxes – need to be developed whereby farmers, farm organizations, and farming communities can draw on public resources to be used by them for extension services of their choice (public or private).
- Current public extension systems need to be downsized.[140]

As public extension systems are retrenched, emphasis needs to be placed on better recruitment standards, including a requirement for experience in farming. Similar criteria can be applied to the qualification of private extension firms for participation in publicly funded programs. A glaring management weakness of most extension systems in the past has been the lack of attention given to the need for women agents. Women play an important role in farming in all developing countries, they can be catalysts in community organization, and they can function effectively as agents of change. This weakness needs to be remedied as an urgent matter.

8.5.6 A Synthesis of New Approaches to Agricultural Extension

While the pervasive fiscal crises of recent years in developing countries are one of the causes of weaknesses in extension systems, there also have been fundamental concerns about the management and structure of the extension systems, including incentives for performance and recruitment criteria. In light of the demands on extension, it is clear that the earlier, centralized systems are no longer viable and new approaches are needed.

Recent experiences in many countries, and the ideas expressed in the literature on agricultural extension, are converging in a consensus about new modalities of extension. By now, the deficiencies of past approaches are very evident and have given rise to efforts to develop new orienta-

139. C. Antholt, 1998, pp. 364–365.
140. *Op. cit.*, pp. 365–366.

tions in many countries. As illustrated by the material presented in this chapter, the new approaches are diverse and can be characterized from various perspectives: a client orientation, participatory approaches, gender awareness, pluralism of providers, cost sharing, facilitation of producer linkages to markets and input suppliers, and other new orientations. In the final analysis, the structure and operating modalities of extension services will vary by country and by region within countries, but nevertheless there are common threads of thinking that inform the new approaches. Many of these threads derive from a few basic ideas that are gaining increasing acceptance. The main ideas and their corollaries can be presented in more structured form, as shown in the following.

8.5.6.1 Core Propositions for Re-orienting Agricultural Extension

The common threads in the new approaches to extension can be derived from a few core propositions, which are the following:

(1) *Farmers often can identify and characterize their problems better than advisors, can prioritize them, and possess at least some knowledge that is relevant to finding solutions.* The more heterogeneous the farming conditions, then the more applicable is this proposition.

From this basic realization, two other important ideas have taken root:

(2) *Extension programs should emphasize human resource development; strengthening the inherent capacities of farmers to solve their own problems and make appropriate farming decisions is the key to promoting agricultural and rural development.*

This fundamental proposition also follows from the basic aims of sustainable rural development which are to develop the capacity of rural families and communities to raise their standard of living through their own efforts. This capacity has several dimensions, including human capital, social capital, on-farm physical capital and local infrastructure.

Human resource development embraces both the human capital and the social capital aspects.

(3) *Governments alone are not able to provide fully adequate extension services*, in part because of proposition (1) above, which implies that farmers need to be participants in formulating solutions, and in part because of organizational and financial limitations in government agencies throughout the developing world. The centralization of many government services and processes of decision-making themselves impose severe limits on governments' abilities to interact with agricultural producers.

These three basic ideas, or core propositions, may be called the axioms of the new approaches to agricultural extension. Although they are related to one another, each one derives from an independent or partially independent basis. They may be called *the axiom on farmer knowledge, the axiom on capacity building for development, and the axiom on government limitations*.

8.5.6.2 Additional Propositions for Agricultural Extension

A number of other key facets of the new approaches to extension follow from the three core propositions or axioms. In abbreviated form, those other facets (or additional propositions for extension) are as follows:

A Policy implications of the axiom on farmer knowledge –

(4) *Extension services need more of a client orientation, and primary accountability to the client, who is the farmer*. Farmers are not passive recipients of a benefit from extension programs but rather they are stakeholders in the process. The messages that extension agents currently bring to the field do not always respond adequately to the needs of their clients.

(5) *Extension is a process of facilitation of the acquisition of knowledge and skills, more than*

a process of technology transfer. It facilitates direct contacts with other farmers, with researchers, with service providers and marketing agents, and with other economic and social agents in the rural milieu. This proposition reflects the reality that extension agents alone do not possess all of the experience and knowledge that is relevant to solving farmers' problems.[141]

(6) *Participatory approaches to extension are effective*. These are approaches that use local knowledge as much as possible, that use farmers as extension agents (and also researchers to an extent), and that work with groups of farmers rather than individual contact farmers for each area. They involve producers in problem identification, setting priorities among issues to be tackled, solving problems through analysis, and making choices. The participatory approach is a direct and logical consequence of accepting the axiom of farmer knowledge, and it is the surest way to guarantee the relevance of new technologies.

(7) *Incentives for extension workers need to be structured in a way that encourages them to emphasize satisfying their producer clients, rather than their superiors in an institutional hierarchy*. Even if individual extension agents are committed to working closely with their clients in order to understand and address their concerns, the institutional incentives they work under may not push them in that direction.

(8) *Decentralization of public extension services is likely to improve their effectiveness*, since it brings them closer to the clients, i.e. the producers. The more local is decision making on the provision of services, then the more capable it is of responding to clients' needs. This is an example of the principle of subsidiarity.

B Policy implications of the axiom on capacity building for development –

(9) *Extension services need to develop approaches that are suitable for rural women, who have been largely ignored by most extension work to date*. Development cannot occur to a meaningful extent for rural families if the potential of one of their main resources, women, is left untapped. Worldwide, only about 5% of the extension effort is directed toward women, and the percentage of farmers in developing countries who are women is very much higher.

(10) *Basic education makes extension much more productive*. Education is the single most important factor in economic development, and the benefits of educating women are especially strong.

(11) *Extension should facilitate not only the acquisition of crop cultivation skills, but also skills in farm management and accounting, marketing, dealing with credit institutions and input suppliers, community organization, and responding to the threat of HIV/AIDS*. In a globalizing world, in which agricultural production increasingly responds to consumer preferences, export possibilities and agro-processors' requirements, knowledge of cultivation techniques alone is insufficient for success as a producer.

C Policy implications of the axiom on government limitations –

(12) *Government funding of extension does not necessarily mean government provision of it. It is desirable to have multiple extension providers, competition among them should be encouraged, and producers should be in a position to evaluate them and choose among them*. Without steps to ensure competition, a public monopoly of extension might be replaced with a private monopoly. Providers can include NGOs, private extension agencies, input suppliers, export agents, agro-processors and universities, in addition to public sector agencies.

---

141. This proposition is central to the recommendations of the Neuchatel Group of donor agencies (see Neuchatel Group, 1999).

(13) *Mechanisms of support are needed so that poor producers may have access to extension services*. Different forms of support have been explored, including vouchers for the purchase of extension services, reimbursement of producers by governments for part of the cost of extension services, direct government payment to private providers after the delivery of the service has been verified, etc. The important point to recognize is that poor producers are unlikely to be able to pay for extension services in the foreseeable future, and that the public-good character of much of agricultural technology obliges governments to share extension costs with farmers.

(14) *Different forms of financing of extension need to be explored, including cost-sharing with producers who can afford it*. A financial contribution by producers puts them in a better position to judge the quality of the service and direct the service toward their own priorities. In addition, payment by the better-off producers reduces the element of a regressive subsidy that has been present in free public extension services.

(15) *A multiplicity of extension services requires mechanisms of co-ordination, especially among NGOs, without putting hindrances on their efforts*. Too often, each NGO goes its own way, unaware of what other NGOs and public agencies have been doing in the field of extension, and with what results.

(16) *An important role for government is the establishment of quality standards for extension providers and rules governing their provision of services*. Extension providers in effect have to be licensed to carry out their work.

D An additional implication of proposition (6) on participatory approaches –

(17) *Farmer and community organization play an important role in determining the effectiveness of extension services, and they should be encouraged by the extension effort itself*. This is especially true for women farmers. NGOs are particularly effective in promoting local organization.

Extension services throughout the world are moving in many of these directions. It bears reiterating that no single formula is appropriate for all or even most circumstances. The appropriate variants for each country and region have to be defined by the participants in the process in each case. Nevertheless, the foregoing propositions, or at least a sub-set of them, have been found to be relevant in virtually every case.

It should be reiterated, as remarked by the Neuchatel Group, that a sound overall agricultural policy framework, one that encourages agricultural growth, is a prerequisite for the success of agricultural extension efforts.

Finally, the importance of better education for rural populations cannot be emphasized too strongly, especially in light of the trends toward devolving responsibility for acquiring new knowledge to the farmers themselves. The receptivity of rural populations to new information, and their ability to assimilate and apply it, increases markedly with levels of education.

Education is the most important determinant of the ability of rural populations to improve their well-being. When there is a choice at the margin between allocating resources, for a given rural population, to agricultural extension and allocating them to basic literacy training, the decision inevitably has to be made in favor of the latter. Basic literacy opens the doors to many kinds of development that are otherwise impossible.

DISCUSSION POINTS FOR CHAPTER 8

• Three-quarters of poor people in the developing world live in rural areas, but increasing agricultural productivity is becoming more difficult because agriculture is encroaching upon more marginal lands and existing areas are suffering loss of productivity owing to soil and water degradation.

• In many cases, agricultural productivity in developing countries actually has declined, in spite of the high returns to investments in agricultural research. In part, this may be attributable to a sector-wide crop composition that does not coincide with a country's comparative advantage, but in part it also may reflect problems of soil and pest management at the farm level. Research also suggests that generally unfavorable agricultural price polices (including macroeconomic policies that affect prices) may be responsible for part of this trend.

- Funding for agricultural research in developing countries has declined in real terms, and there is a consensus that investment in this area is far below what it should be in light of the returns to these kinds of expenditures.
- The central challenge for agricultural research is developing results that are appropriate to farmers' needs. Collaboration with farmers in formulating research priorities helps meet this challenge.
- Agricultural technology is not neutral with respect to gender, but in practice usually the needs of women farmers are ignored in technology development and dissemination.
- Women and entire households can benefit not only from development of appropriate cultivation technologies but also from development of household technologies that lead to savings of women's time.
- Both the private and public sectors have important roles to play in agricultural research. The public-good nature of many research results limit the potential interest of the private sector in developing them, although biotechnology is opening more avenues for private sector participation. Clearly defined intellectual property rights help encourage the private sector to invest in agricultural research.
- It is estimated that between one-third and one-half of all of the crops in the world are lost to pest damage, and the proportion is higher in developing countries. In light of this finding, some experts recommend giving a higher priority in agricultural research programs to creating pest resistance.
- Chemical treatment of pests often is uneconomic as well as being damaging to the environment.
- The FAO's Farm Field Schools have proven to be an effective means through which farmers themselves participate in developing techniques of integrated pest management.
- Agricultural extension in developing countries was initiated with an aim of delivering technological messages from researchers to farmers, with little provision for feedback or for farmer participation in identifying the problems to be analyzed by researchers.
- The Training and Visit (T & V) system reformed the management of many extension systems but retained an essentially top-down approach to the relation between researchers and farmers.
- Farming technology sometimes is embodied in improved inputs but often it takes the form of pure information. Such information spans a continuum from a pure public good to a completely private good. The nature of a public good is that others cannot be denied access to it and the 'consumption' or use of it by one person does not reduce the quantity available for others. The public-good character, partially or totally, of much agricultural technology is the basic rationale for a public extension service, since the private sector cannot appropriate the benefits of its diffusion in those cases.
- However, even when information has some public good qualities, farmers may be willing to purchase it when there is a premium on timeliness of information, or when acting on that information (as in penetrating a market) can effectively exclude others from the same opportunity, or simply to reduce uncertainty about cropping risks.
- The public-good nature of much agricultural information constitutes an argument for subsidizing extension services, and the poverty of many farmers in the developing world also constitutes an argument for such a policy. The provision of technological information that has positive environmental externalities is another justification for subsidizing extension services. Nevertheless, many better-off farmers in developing countries can afford to pay for extension and are willing to when it is a private-good in nature or when one or more of the caveats in the preceding paragraph apply. Providing fully subsidized extension to all farmers constitutes a regressive subsidy.

- In many developing countries, the performance of agricultural extension services has been disappointing. There is a growing consensus that better performance can come about only by creating a greater client orientation in extension services, fostered by appropriate changes in incentive structures.
- Extension services also have been very slow in responding to the needs of women farmers. Worldwide, only 5% of extension resources are allocated to programs for women farmers, although they constitute a considerably higher proportion of all farmers and directing extension efforts to them has been shown to have significant benefits.
- The HIV/AIDS epidemic is devastating rural areas in many countries and also is depleting and demoralizing some extension services. In these cases, extension services need to disseminate agricultural technologies that are more appropriate for persons who are weak, very old and very young. They also are called upon to respond to pleas for information and guidance in dealing with the epidemic, and in general to assist communities in coping with the crisis.
- Extension services around the world are undergoing a re-evaluation and new modes of providing extension are being explored. The basic reasons for this change are government budget limitations, perceived deficiencies in existing extension services, the evolution of more specialized agricultural technology requiring different kinds of extension, the increasing importance of poverty alleviation as a policy objective, and the increasing importance for farmers of market linkages and sound farm management.
- International trade in seeds and inputs is an important means of improving a country's stock of agricultural technology. Some countries restrict the import of all such items until they are certified as safe and effective. However, this conservative approach runs the risk of slowing down the rate of technological progress in the country. A more liberal approach has been proposed in which testing of products is carried out *ex post* and test results are disseminated for purposes of information and not for control.
- As in the case of extension, agricultural research systems around the world also are undergoing a process of re-thinking. The reasons for this include disappointing rates of yield increase, budgetary cutbacks for research, evident declines of the quality of research staff in some cases, and lack of sufficient research on the main problems perceived by some farmers in each country, especially the poorer ones.
- A major issue for agricultural research is how to set the research agenda. Traditionally, research scientists have established it, but that approach runs the risk of missing the most important problems for many farmers. The issue is how to make research more relevant to real farming needs.
- Different approaches have been used to make research more relevant to farmers, including putting them on boards of directors of (local) research institutions, requiring financial contributions from them for research, involving them in multi-disciplinary research teams, and giving them a leading role in some phases of the research process.
- Farmers possess a considerable reservoir of knowledge about agro-climatic conditions and cropping systems in their areas, and they also usually have a tradition of experimenting with varietal development and cultivation techniques. Research methods that involve farmers in the process of scientific inquiry have been shown to be effective in producing results tailored to farmers' needs and in raising the incomes of poor farmers.
- Participatory processes often generate ancillary economic activities, such as seed marketing enterprises for small farmers. When participatory research is successful, the farmers themselves become agents of dissemination of the findings.
- Participatory research needs to be managed sensitively, and the role of facilitators is important. In addition, it requires small amounts of funds at the local level, under farmer control, to finance inputs into their research process.

- Other areas in which it is important to strengthen national agricultural research systems include the adaptation of research results from other countries and from international institutions, capacities for conducting local laboratory analyses of soils, research on post-harvest management techniques and methods of crop processing and handling, and the environmental sustainability of agricultural technologies.
- National agricultural research systems increasingly are taking an entrepreneurial approach to their management. Recruitment of good staff and provision of adequate incentives are central aspects of management of these systems.
- Agricultural research is carried out by a more diversified array of institutions than it used to be, including NGOs, producer associations, universities and private firms. One approach to involving them is to allocate research funds through a competitive bidding process in which research proposals compete annually against specified amounts of funding.
- The financing of agricultural research also is becoming more diversified. Principal sources of finance now include generating revenues by marketing research products, encouraging financial contributions from producers, and creating endowments to support research foundations.
- It has been debated whether agricultural research largely benefits the more commercial farmers, because these farmers tend to be located in agronomically more favored areas and because of the ways in which research priorities are set and the ways in which research is carried out.
- Poor farmers can benefit from research that emphasizes varieties that are not intensive in purchased inputs, farming systems and improved natural research management, and also from participatory methods of conducting research that focus the efforts on their main problems and tap into their own accumulated knowledge. Participatory methods of research constitute perhaps the main avenue for bringing the benefits of technological advances to poor farm families.

- Participatory research can be more effective if it gives special emphasis to involving women in the process. To do so requires a sustained commitment on gender issues on the part of research and facilitating institutions.
- It is sometimes observed that women farmers are slower to adopt technological innovations than men are, but a study of technology adoption in Ghana shows that this phenomenon is attributable to the fact that women farmers usually have less access to factors of production, and not to an inherent unwillingness to adopt new methods.
- Agricultural extension systems are undergoing significant change throughout the world, with emphasis on the role of extension agents as facilitators and the promotion of a two-way dialog between farmers and scientists on agricultural technology. Many extension systems are trying to implement the view that farmers are clients rather than beneficiaries of extension.
- The responsibilities of extension agents include, in addition to delivering technological advice, helping diagnose farm problems jointly with farmers, providing feedback to researchers, developing capacities for better farm enterprise administration, and developing linkages between farmers and NGOs, farm organizations, input suppliers and marketing channels.
- While there are solid arguments for public funding of part of extension services, that does not necessarily imply that public agencies always should carry it out. In public agencies, the performance incentives tend to lie on the side of satisfying institutional superiors, who may not always have reliable feedback from farmers. In contrast, in privatized extension systems, the incentives lie on the side of satisfying the clients. For this reason, some countries have developed ways to share extension costs between private and public sectors while leaving the delivery of the services in private hands. One of the extension principles developed by the Neuchatel Group of donors for Africa is that farmers should be in a position to choose from an array of extension providers.

- Good farmer organization is a prerequisite for small farmers to participate in many kinds of programs, including research and extension. Often, NGOs can play a valuable role in promoting better farmer organization.
- Extension systems need to make stronger efforts to meet the needs of women farmers, and take a number of specific steps in that direction, including hiring more women agents, promoting more female enrolment in agricultural tertiary education, schedule extension visits at times that do not conflict with rural women's household chores, broaden the range of crops and livestock products that extension activities are provided for in order to include those most relevant to women, and promote the formation of women's groups for interacting with extension services.
- A re-orientation of extension systems is required in places where the HIV/AIDS crisis has reached a significant scale. Better preparation of extension agents is needed for dealing with this, and national policies for extension in relation to the disease are required. The re-orientation should include the development of technological messages more appropriate to elderly persons, youths, widows and sick people, because the crisis has had a heavy impact on the age structure of rural populations in many areas. Extension agents also need to be better prepared to respond to community questions about coping with the crisis and reducing its rate of spread.
- The new approaches to extension attempt to put the farmer first, and public-sector extension systems are being downsized in most countries in favor of greater participation of the private sector, NGOs and farmers themselves. On pp. 418–420, this chapter presents a summary of the new orientations in the form of seventeen basic propositions for agricultural extension, and the basic assumptions, or axioms, from which they are derived.

9

Agricultural Development Strategies: Process and Structure

9.1 The Roles of an Agricultural Strategy   425
    9.1.1 A Strategy as a Vision and a
       Unifying Document   425
    9.1.2 A Balanced Package of Reforms   426
    9.1.3 The Educational Role of a Strategy   427
9.2 Participatory Processes for Developing
    Strategies   428
    9.2.1 Reasons for Encouraging
       Participation   428
    9.2.2 The Participants in a Process of
       Developing a Strategy   430
    9.2.3 Managing a Participatory Process   432
    9.2.4 Challenges and Risks of
       Participation   439
9.3 Structure and Consistency in a Strategy   442
    9.3.1 Shaping a Strategy   442
    9.3.2 Institutions and Human Capital   444
    9.3.3 Consistency Dimensions of a
       Strategy   446
9.4 Substantive Orientations of an
    Agricultural Strategy   450
    9.4.1 Lessons from Long-Term Models of
       Agricultural Development   451
    9.4.2 Strategic Orientations for Agriculture   455
    9.4.3 Approaches to Agricultural
       Development Policies   458
9.5 Rural Development and Poverty Alleviation   460
    9.5.1 Rural Development Projects   461
    9.5.2 Decentralizing Rural Development   462
    9.5.3 Economic Transfers to Support
       Agricultural and Rural Development   465
    9.5.4 Investment Priorities   468
    9.5.5 Frameworks for Rural
       Development Policies   469
9.6 Implementation of a Strategy   474

9.7 Concluding Observations   475
Discussion Points for Chapter 9   476

9.1 THE ROLES OF AN AGRICULTURAL STRATEGY

9.1.1 A Strategy as a Vision and a Unifying Document

A strategy is a consistency framework for bringing together diverse policy initiatives in an overarching logical structure that adopts a medium- or long-term view of the sector's prospects. Chapter 2 of this volume discusses the usefulness of a strategy for ensuring consistency among proposed policies, for linking them to national development goals, and for ensuring that the coverage of areas needing reform is sufficiently comprehensive.

A defining characteristic of a strategy is that it presents an integral approach to development issues. A viable and solid strategy, one that is capable of being implemented and that can move the sector forward, must have a solid conceptual framework, and its policy proposals must be developed according to rigorous technical criteria. If this is not the case, then the strategy may become a shopping list and forfeit much of its persuasive power. It should not be overlooked, however, that a successful strategy also is more than a technical document: it is a vision and a rallying cry. It is a vision in that it presents fresh

Agricultural Development Policy Concepts and Experiences. R. D. Norton
© 2004 Food and Agriculture Organization of the United Nations
ISBNs: 0-470-85778-1 (HB)   0-470-85779-X (PB)   FAO Edition: 92-5-104875-4

possibilities for the sector and a path to attain them. One of its principal roles is to show a feasible way to fulfill legitimate aspirations of the rural population. It is a rallying cry because, if successful, it can become a means of mobilizing support for the vision and its implementation. Agricultural sectors are populous and very diverse, and therefore a compelling vision of the sector's potentiality is necessary in order to mobilize support. By the same token, without widespread backing it cannot progress beyond the status of another technical study.

Support for a strategy should to be developed on at least three principal fronts:

- The producers – more broadly, rural families – on whose behalf it is formulated and without whose active assent and participation it cannot be fully implemented.
- The government, in its various manifestations, which must lead the effort of implementation. Governments are agglomerations of individuals with differing and sometimes conflicting viewpoints, and so achieving a sufficient level of consensus in government requires a multifaceted and sustained institutional dialog.
- International development agencies, whose agreement and financing are required for successful implementation of a strategy.

To the extent possible, a strategy should also strike a largely receptive chord with other segments of the population and with opinion leaders in general, even though many groups will be less involved in its development and therefore less interested in it. If it generates active opposition in influential sectors, gaining approval for it and implementing it are likely to become much more difficult.

This chapter reviews some of the main issues involved in the *process* of developing an agricultural or rural strategy, utilizing illustrations from the author's advisory work on strategy-building efforts in Honduras, Guyana, Estonia, Nicaragua, El Salvador, the Dominican Republic and Mozambique, plus information on experiences in other countries. The chapter then discusses the *substance* of strategic orientations of

The difficulties of carrying out policy reforms without having mobilized sufficient support were illustrated in the case of the Kenyan Cereal Sector Reform Program, which was part of a structural adjustment program:

... the inconsistency in implementation of the CSRP is substantially the result of neglect of the political dimension in the reform process, with no culture of reform being developed in the pre-reform period. Specifically, it is suggested that if the designers and managers of the CSRP had taken the trouble to provide adequate information on the expected benefits of the reforms, the process would have received more consistent support. This is evident from the behaviour of market actors, particularly the sifted maize millers and politicians with stakes in maize milling and the food distributive trade, who opposed reforms initially but later supported them. ... the absence of full backing by the President and Cabinet, and lack of understanding by all interest groups of the likely benefits and costs, did create uncertainties and conflicts of interest which turned out to be more apparent than real. These contributed to inconsistent implementation and avoidable conflict between donors and government. It is suggested that managing the politics of reform is a task that was neglected and should be explicitly incorporated in such programmes. (From Peter M. Lewa and Michael Hubbard, 'Kenya's Cereal Sector Reform Programme: managing the politics of reform', Food Policy, 20(6), December 1995, pp. 573–574.)

agricultural development policies, presents material on rural development programs and reviews criteria for allocating resources in such programs.

9.1.2 A Balanced Package of Reforms

As part of the effort of mobilizing support for reforms, a strategy can define tradeoffs and can attempt to balance gains and losses for different sub-sectors. Sometimes, it may be difficult to gain approval of a single reform proposal, since the stakeholders that stand to lose from it, even if they are a small minority, may be able to apply sufficient political pressure to derail it. However, if it is packaged with other reforms, those who stand to lose from one of the reforms may perceive enough benefits in the entire package that they will support it.

In the making of Honduran agricultural policy reforms of the early 1990s, farmers' support for free trade was obtained in part by proposing and implementing a system of price bands that automatically increased tariffs when world prices were exceptionally low (and decreased them when prices were high, to the benefit of consumers). Similarly, they agreed to dismantle the system of guaranteed crop prices in part because the measure was packaged together with a proposal to privatize State-owned grain silos in a way that enabled farmers to become part owners of the facilities. If free trade or freeing up prices had been presented alone, it is unlikely it would have gained enough backing to be approved in the circumstances of the time.

Other elements of that reform package contributed to its acceptance as a whole. Large-scale, commercial farmers perceived the possibility that a restructuring of the agrarian reform process could bring an end to policy inconsistencies on that topic and twenty years of land invasions and violent confrontations over land, and therefore they were willing to accept the elimination of the guaranteed price system from which they had been the major beneficiaries, and a removal of quantitative controls on imports, which also had benefitted some of them. The rural poor saw potential benefits from a more effective agrarian reform sector, an elimination of the prohibition of land rentals and from other elements of the package that were targeted on their income group, and accordingly they became willing to abandon the tactic of land invasions which, in any case, sometimes rebounded to their detriment.

It is valuable if *a fundamental thrust of a strategy is the creation of a level playing field in the policy arena and the elimination of special economic privileges*. This is desirable from a viewpoint of both economic efficiency and equity.[1] It can be difficult, however, to reach agreement in this regard. It may be easier to achieve a consensus on the need for uniform treatment of all economic actors when various interests are brought together to participate in formulating a strategy, instead of leaving them to negotiate individually with the government. When a broad range of issues is discussed in a forum with many participants, each player then can see that eliminating privileges for others requires reciprocity: its logic requires all privileges to be surrendered. Developing a widespread consensus on a reform package in this manner can be one of the more effective ways to repeal entrenched privileges, or at least reduce them significantly.

9.1.3 The Educational Role of a Strategy

One of the most basic roles of a strategy is to elevate the level of the national dialog on policies. As long as the policy dialog is conducted between special interests and government officials, the result tends to be a pattern of exceptions to the rule of uniform treatment. Such discussions tend to be conducted in narrow terms of benefits and losses for a particular group which might result from reforms, even though society as a whole might gain from them. The process of developing a strategy provides an opportunity to raise the dialog to the level of national development issues to be confronted and constraints to be overcome, and the gains to the nation that would flow from doing so. It can have a long-term educational value for the public that transcends the concrete benefits that arise from implementing key parts of it.

---

1. Hans Binswanger has underscored the equity aspect: 'Providing privileges or reducing competition in output, input, and credit markets is costly to consumers and taxpayers and ends up hurting small farmers and the rural poor, even if such an effect is unintended'. (H. Binswanger, 'Agriculture and Rural Development: Painful Lessons', in Carl K. Eicher and John M. Staatz (Eds), *International Agricultural Development*, 3rd Edn, The Johns Hopkins University Press, Baltimore, MD, USA, 1998, p. 298.)

Developing a strategy also is an educational process for the participants. It is emphasized later in this chapter that, to the extent possible, strategy formulation should be a participatory process. As well as respecting democratic values, such an approach strengthens the support for the strategy. It also improves its content because the participants can collectively represent the best national expertise on each of the problem areas. Yet, technical experts are not always familiar with all policy considerations in their fields. Often, they will have been working completely within an existing policy framework, and they may not be accustomed to thinking in terms of alternative frameworks, especially radically different alternatives. Hence **work on a strategy can provide a learning-by-doing process for all participants, in learning how to define the policy issues themselves in fruitful ways and how to pursue solutions in perhaps unaccustomed directions**.

The learning process can extend to providing participants with a greater understanding of the agendas of donor agencies, through dialogs with representatives of those agencies. The dialogs, in turn, can establish communications channels so that in the future donor agencies are able to work with a broader spectrum of leaders of society in formulating their programs for a country.

A draft strategy and the final version of the document may play different roles. The latter is the formal basis for an implementation program but the former, simply because it has the status of a draft, sometimes can address sensitive issues that policy makers have sidestepped. An example of this role is provided by the draft *National Development Strategy* for Guyana.[2] Among the issues it discussed openly was the option of privatizing the national sugar industry. At the time, the Cabinet was firmly opposed to that option, and in fact it had not been possible to discuss the question in any public forum. However, since the

> *In one of its many editorials commenting on the National Development Strategy in Guyana,* The Stabroek News, *an independent newspaper often regarded by Government as representing an opposition viewpoint, had this to say:*
>
> *The draft National Development Strategy should be required reading for our politicians, businessmen, trade unionists and academics. Containing useful and interesting ideas and discussions on every aspect of the economy and its processes it cannot help but raise the level of public debate which in so many areas is ill-informed and bereft of any theoretical or systemic framework. (From the* Stabroek News, *Georgetown, Guyana, March 6, 1997.)*

strategy was explicitly a preliminary and technical draft for public consultation, in the context of a comprehensive national review of economic options, the Ministry of Finance agreed to let stand a full analysis of the privatization option. For this reason, it can be valuable to label the first version as a technical draft and to underscore that fact, to free up the discussion from political constraints.

9.2 PARTICIPATORY PROCESSES FOR DEVELOPING STRATEGIES

9.2.1 Reasons for Encouraging Participation

In recent years, worldwide, there have been many efforts to promote the active participation of communities and citizens in development projects that have an effect on their well-being. Common examples include the approach of user participation in the management of irrigation systems, techniques of group lending in microfinance, the involvement of rural communities in programs for forest and watershed protection, and the organization of groups of farmers for receiv-

---

2. This document, first issued as a technical draft in 1996, covered all sectors of the economy, in six volumes, and was developed through the efforts of 23 sectoral working groups comprising experts from within and outside government. The work was suspended for more than a year because of an electoral crisis, and when it was re-convened (by the Minister of Finance), civil society played an even stronger role in the updating and revision of the draft. It was eventually submitted to Parliament, and during the process the government implemented many of the draft recommendations. The Carter Center played a facilitating role in the process.

ing technical assistance, purchasing inputs and obtaining better market access. In addition, the emphasis on participation has been extended to the area of social services, such as in community participation in the administration of some rural schools in El Salvador.[3]

In spite of this emphasis on participation, the vast majority of participatory undertakings have been aimed at fostering *participation in development projects and programs*, above all in specific localities. There have not been many efforts at using the participatory approach for formulating integral and specific proposals for *sector-wide policy packages and development strategies*. In a number of countries, activities have been undertaken for the formulation of national strategic visions for the longer term, in Mali, Ghana and Mozambique, for example. That type of experience can be very valuable and contributes to raising the national level of awareness about possible future development scenarios. It also can move political parties closer to a consensus on national development priorities, as has been the case in Mozambique, but eventually it needs to be complemented by concrete proposals for reforms to the national policy framework. In some cases, a vision document can be conceived of as a first step in developing a more specific strategy.

It may be asked, why should policy formulation be based on participatory processes? What is gained from it? Why cannot government alone take responsibility for policy formulation? The answer may be summarized in five main points:

(1) To improve the chances of attaining *a national consensus on policy reforms*. Consensuses on policy actions are never achieved 100%, but a participatory process can widen the area of agreement significantly. The greater the consensus, then the stronger the political support for change.

(2) To strengthen the *channels of national dialog*, thus empowering citizens to participate more effectively in the resolution of future policy issues. Frequently, these channels are underdeveloped or atrophied. This role of a participatory strategy represents *capacity building for civil society and the private sector* in regard to national development policy issues.

(3) To develop *more solid policies*. Experience has shown that representatives of the private sector and NGOs not only can play a role of supporting consensus policies, but that they are also capable of making fundamental contributions for improving the quality of the reforms, even to the point of assisting in the drafting of proposed legislation. As noted in World Bank reports:

> The majority of [World Bank] staff responsible for these CASs (Country Assistance Strategies) felt that the benefits of incorporating civil society participation in the process significantly outweighed the costs. They felt that participation in the CAS led to more informed development priorities for the country. . . .[4]

Plus, in the context of strategies for poverty reduction (PRSPs):

> Participatory analysis of the poverty impact of public expenditure can generate deeper understanding than analysis by officials and experts only.[5]

(4) To improve the *accountability and transparency of the policy making process*.
(5) To better *empower the country in international dialogs*, so that truly national priorities can

3. E. Jiménez and Y. Sawada, 'Do community-managed schools work? An evaluation of El Salvador's EDUCO program', *The World Bank Economic Review*, **13**(3), September 1999, pp. 415–441.

4. Maria Aycrigg, 'Participation and the World Bank: Success, Constraints, and Responses', Social Development Paper No. 29, Environmentally and Socially Sustainable Development, The World Bank, Washington, DC, USA, November 1998, p. 11.

5. S. Tikare, D. Youssef, P. Donnelly-Roark and P. Shah, 'Organizing Participatory Processes in the PRSP', Draft, The World Bank, Washington, DC, USA, April 2001, p. 7.

serve as guides for programs of international assistance, instead of allowing national policies to be defined implicitly by the sum of conditionalities attached to international loans and grants.

An axiom of democracy asserts that citizens should have a voice in the decisions that affect them. Developing countries sometimes lack a robust tradition of participation by civil society[6] in national issues, and lobbying is left to a few, economically influential interest groups. Putting into practice this axiom may require more active collaboration between the private sector and civic groups, on the one hand, and the executive branch, on the other hand.

9.2.2 The Participants in a Process of Developing a Strategy

The making of a successful agricultural strategy requires *(i) technical expertise on the relevant issues, (ii) intimate knowledge of a country's agriculture, and (iii) political leadership for a process of change*. The first requirement can be supplied by national and/or international experts, preferably in collaboration with representatives of farmers or rural communities. The second is provided by those representatives, and the third is usually supplied by the country's political leaders, although spokespersons for producers' groups can play leadership roles as well. In the Dominican Republic, for example, an influential producers' group, the Junta Agroempresarial Dominicana (JAD), has traditionally played all three of these roles in work on sectoral strategies.

In brief, the participation of representatives of farmers – or, more broadly speaking, of civil society – is important both for helping give a strategy the most productive orientations and for facilitating its implementation. Civil society can participate in two ways: by being a full partner in the effort of researching and drafting the strategy, or by being consulted during the process. The former is likely to produce a better strategy, but it requires a greater investment in organization and co-ordination of the undertaking, and there is a risk that in the end a consensus will not be attained. Once a commitment to involve rural society in the process is made, it is important *to ensure that those who participate are sufficiently representative* of a broad spectrum of that society. This includes women, the poor, and other traditionally disadvantaged groups. Exclusion of key groups can weaken the document and lead to controversy over its content after it is issued, which undermines its prospects for being implemented.

Participants do not necessarily have to be formal representatives of associations or other recognized entities. Sometimes, those who are in such positions feel obliged to defend entrenched positions. Depending on the circumstances, it may be possible to advance more rapidly in developing new and creative proposals jointly if the participants in the process are selected on the basis of their recognized status as thoughtful individuals who take a broad view of issues, rather than on the basis of formal organizational affiliations.

As noted, involving persons from outside official circles in a process of formulating a strategy can help strengthen civil society in countries where its voice has been weak in the national policy dialog. However, trying to involve political parties formally in the process may weaken the chances of reaching a consensus, since parties by definition have a vested interest in disagreeing with each other, i.e. in defining their respective platforms. In most cases, the process is more likely to lead to a consensus if the participants include a range of individuals whose own party affiliations span the political spectrum but who are not officials in party hierarchies.[7] In the Honduran case, the implicit representation of all political parties in the dialog helped ensure that the reform program was sustained when elections subse-

---

6. As used hereafter in this chapter, civil society includes the private sector (business sector), as well as NGOs, universities, citizens' associations and other groupings.
7. In an agricultural strategy exercise in the Dominican Republic, the UNDP commissioned chapters on various policy topics by national experts and then convened a national workshop in late 1994 to review them and

quently led to a change of the party in power. However, in Mozambique, it was decided to involve representatives of political parties in a visioning exercise, in view of the need to further the process of healing the wounds of a devastating civil war, and that experience was successful.

Sometimes, technical experts – including those from international agencies – are concerned that the participation of persons who do not have an extensive technical background will slow down the process and perhaps dilute the soundness of the recommendations. An answer to the first concern is that if the result is better, then it will have been worth the additional investment of time. An answer to the second concern is that it is incumbent on technical experts who feel they have better insights and policy prescriptions to explain them in accessible terms – and to demonstrate sufficient flexibility to accept suggestions for improvements when they come from non-technical sources and are not always expressed in technical language.[8] In any case, in virtually all developing countries, civil society will be able to contribute persons with strong technical qualifications in economics and other fields.

The risk of not reaching a consensus among members of civil society is a legitimate concern. However, if schisms exist in society on basic policy issues, the work on a strategy is as good a forum as any for trying to bridge them. It is relatively non-political and the presence of a wide range of participants including, usually, international advisors, effectively puts pressure on all participants to present constructive positions and to listen to new ideas. In the end, if agreements cannot be reached in some areas, an option is to present minority positions in a final report.

In the context of an agricultural strategy, experience has shown that major agro-industrial interests are likely to be diametrically opposed to farmers on issues such as import tariffs and product prices. Processors of grains and oilseeds usually favor low or zero tariffs on their raw materials, to the detriment of the country's farmers who produce those classes of goods or close substitutes for them. (The practice of waiving tariffs on donated food or concessional food aid exacerbates the problem.) For this reason, sometimes participation in an agricultural strategy is limited to representatives of primary producers, without involving the processing industries. The cost of doing this, of course, is a smaller domain of consensus on the resulting proposals, and perhaps lost opportunities to promote collaboration between farmers and processors on topics such as quality standards and the option of processors contracting with farmers for the supply of their raw materials. The decision would have to be made independently in each case, but this particular issue deserves consideration in advance of the launching of a participatory process for the agricultural sector.[9]

develop modifications. Participants in the workshop included representatives of national farmers' organizations, agro-industry, NGOs, government and the political parties. The discussions were animated and productive and the results were useful, but the goal of achieving a policy consensus including the political parties was not attained.

8. Getting experts to play this role is not always easy. As noted in one of the World Bank's assessment of participation efforts in its own programs, 'Participation efforts run counter to the Bank's "expert" culture' (M. Aycrigg, 1998, p. 27).

9. An interesting approach to participation was developed in El Salvador where, through the leadership of Mercedes Llort and with the support of the Inter-American Development Bank, the Agricultural and Agroindustrial Chamber of El Salvador convened four national forums on strategic issues in agricultural development and also commissioned technical studies on topics identified as priorities. National and international experts were invited to speak at the forums and the audience included a large number of representatives of farmers, agro-industries, government and international agencies. Then, a strategy document was drafted on the basis of the findings in the forums and the technical studies. It was issued as Cámara Agropecuaria y Agroindustrial de El Salvador (CAMAGRO), *Estrategia concertada de desarrollo agropecuario*, Convenio CAMAGRO/BID No. ATN/SF-5509-ES, Diálogo nacional sobre estrategias de desarrollo agropecuario, San Salvador, El Salvador, September 1998.

9.2.3 Managing a Participatory Process[10]

Each participatory process of strategy formulation will arise out of its own context and will have its own characteristics. The comments below reflect a preliminary assessment of some central issues and lessons that have been extracted from experiences with participatory strategy processes, but different issues may emerge in other contexts, and the lessons will have to be adapted to each situation. The comments here are offered to assist practitioners in anticipating issues and possible avenues of solution, without pretending to offer comprehensive or definitive recipes.

9.2.3.1 Ownership of the Process

A principal objective of participatory processes is to engender a sense of ownership of the product(s) on the part of the participants. In good measure, achieving this objective depends on how the process is carried out, on how fully involved the participants are. In some circumstances, it can also depend on how the process is initiated or, more precisely, who initiates it.

In many cases, the sponsor of a strategy-building exercise will be the government which, in the end, will be responsible for the bulk of the effort of implementing the policies in the strategy. However, in some political contexts members of civil society may feel that, by participating in a strategic exercise convened by the government, they could be co-opted into agreement with official positions on issues, and hence they may be reluctant to participate under those terms. This reaction is likely to be greater the sharper are the political schisms within the society and the closer are the next national elections. The seriousness of this risk has been underscored by recent experience in Uganda, where the government has been unusually supportive of participatory approaches:

Participation of civil society organizations in the Poverty Reduction Strategy Paper process, should not, however, been seen as an isolated event. Indeed over the last five years, the Government of Uganda has made an effort to involve the civil society organizations in policy planning and implementation of programs. . . . However, the lack of a clear framework for participatory planning between government and civil society is a cause of concern that depicts civil society organizations as clients of Government.[11]

Ideally, a partnership between government and civil society is needed, not least because the leading topical experts in a country can be found both within government ranks and outside them. The sharing of responsibility between leaders of government and civil society is something that has to be worked out in advance in each context. It is perhaps an easier task to manage for agricultural strategies than for economy-wide strategies, since the main participants from civil society would be farmers whose leadership qualities were already recognized in the sector. As mentioned earlier, care could be taken to ensure that all main political strands are represented among the farmer representatives who participate, and this may help reassure government that the strategy will not become a political platform for opposition parties.

As an alternative, work on a strategy may be sponsored by an international organization, but with that approach there is a risk that the strategy may be branded as a product of that organization, no matter how strong the efforts are to make the process genuinely participatory. *At issue is who will be 'champions' of the effort, who will take the lead in both drafting the document and lobbying for formal acceptance and implementation of it*. If the strategy is a programming document for an international organization, then it

---

10. The comments in this section and the previous one apply to the development of major policy statements as well as to strategies as a whole.

11. Zie Gariyo, 'Civil Society in the PRSP Process: The Uganda Experience', paper delivered as the workshop *Voices and Choices at the Macro Level: Participation in Country-Owned Poverty Reduction Strategies*, The World Bank, Washington, DC, USA, April 3–5, 2001, p. 3.

may be difficult for civil society in a developing country to become a true champion of it in this sense, rather than simply providing inputs into the document. As noted in a recent evaluation of participatory work in 189 World Bank projects:

> Sometimes meetings with stakeholders were little more than opportunities for the Bank to present and gain acceptance for its country programs, rather than to learn about local priorities. Participants were generally given too little feedback after they were consulted . . . possibly discouraging their vigorous and creative engagement in future consultations.[12]

In such circumstances, there is a danger that participants may come to feel that the purpose of the consultations was only to extract information from them.

The experience of Mali and other countries[13] points to the danger that governments may, in the end, tell international financial institutions (IFIs) what they would like to hear in order to qualify for debt relief and other assistance. The economic power of the IFIs relative to most host country governments is overwhelming and *it is not always realistic to expect a dialog of equals*. All of the heavy incentives come to bear on the side of accepting the IFI package of policies and moving forward with their implementation.

Among other observers, Kalthleen Selvaggio has commented on 'governments' desire to avoid a conflict with the IMF and World Bank which might threaten the flow of loans and debt reduction, as well as the fact that many government elites often subscribe to structural adjustment policies and even benefit materially from them'.[14]

This unbalanced relationship indicates a need for caution in tying conditionality to participation, or to the results of a participatory process. More fundamentally, it suggests *inherent limits to the extent to which IFIs can directly sponsor a participatory effort*. They can favor it, they can encourage it indirectly, but the route of direct sponsorship and tied incentives may run counter to the aim of fostering a sense of national ownership of a strategy.

It can be helpful if a programming document for an international agency is seen as subsidiary to, and derived from, the national strategy. In The Gambia, for example, it was decided that the national strategy, and not the PRSP, would be maintained as the guiding document and process for resolving policy issues.[15]

In the Guyanese case, the effort was convened by a neutral 'third party' (the Carter Center) which received financing for the effort from seven different international institutions and foundations. Even so, and in spite of the prolonged and intensive participation of more than two hundred representatives from civil society, alongside Guyanese government experts, one major political party made public assertions that the first draft was the work of the Carter Center. It was only through the second phase of work on the strategy, after a nearly two-year hiatus caused by a political and constitutional crisis, that it became accepted that it was a Guyanese product. Now, it is seen as thoroughly Guyanese, and the centerpiece in national policy discussions, by all Guyanese interested in policy issues.

Participatory efforts begin with an identification of the relevant stakeholders. *Who identifies and selects participants can be relevant to the*

---

12. Operations Evaluation Department, *Participation Process Review*, Executive Summary, The World Bank, Washington, DC, USA, October 27, 2000, p. 2.
13. Cited in Roger D. Norton, 'Development Cooperation Processes: Issues in Participation and Ownership', presented at the Development Co-operation Forum, the Carter Center, Atlanta, GA, USA, February 22, 2002. This paragraph and the following two are adapted from that paper.
14. Kathleen Selvaggio, 'From Debt to Poverty Eradication: What Role for Poverty Reduction Strategies?', CIDSE and Caritas Internationalis, Brussels and the Vatican City, June 2001, p. 24.
15. Abdou Touray, 'The Gambian Experience in Participatory Processes in Poverty Reduction Efforts', paper delivered at the workshop *Voices and Choices at the Macro Level: Participation in Country-Owned Poverty Reduction Strategies*, The World Bank, Washington, DC, USA, April 3–5, 2001, p. 2.

question of ownership. In both the Guyanese and Honduran cases, the participants were selected by international advisors. The technique was to call many civic leaders (leading farmers in Honduras) and ask them for recommendations for participants in such a process. The persons whose names were mentioned most frequently were then approached to participate. In Guyana, this mode of selection raised doubts among some of the participants, after they had organized themselves for the work. In Honduras, it did not prove to be a concern, but later in the Honduran process the participation had to be widened to incorporate *campesino* representatives who had felt left out of the first stage of the process. They had expressed forceful concerns about its apparent directions, albeit partly on the basis of inaccurate press reports, and after joining the process they made valuable contributions to the final shape of the reform proposals.

There are no set prescriptions for dealing with this issue of participant selection, but it is advisable to arrange in advance frank discussions of the sponsorship question between members of civil society, the government and the concerned international entities.

9.2.3.2 *The Drafting Process*

Often, much of the ownership issue can be resolved by the way in which the process is carried out. If first drafts are prepared by international advisors and submitted for review by civil society in brief consultations, it is unlikely that a sense of national ownership will be generated. Correspondingly, if the drafts are written by government officials, civil society is unlikely to feel ownership. *Effective participation in drafting a strategy is a key to ownership of the document.* This is a process that is necessarily time-consuming, especially if participants are expected to absorb some suggestions from advisors, revise them as needed, and make them their own. It is a much deeper form of involvement for participants than simply being consulted on drafts developed elsewhere.

Civil society representatives, especially those from rural areas, may not have much experience

in drafting tightly reasoned policy documents although they may have the most useful insights about the problems in their areas and possible means of addressing them. Therefore, advisors, national or international, may have to play a *facilitating role for the process* – and even a limited role in writing sections of a first draft in some cases. Special measures need to be taken to ensure that the drafts reflect the participants' concerns and many of their orientations toward solutions, and that they feel a sense of ownership of the proposals.

Illustrations of this kind of process may be worthwhile. In the first phase of the process in Honduras, farmer representatives met with two advisors a full day each week for brainstorming sessions with a blackboard. The sessions were organized by policy topic, such as land reform, agricultural finance, pricing policy, and so forth. The advisors took extensive notes of issues raised and optional avenues of solution. When there were sound economic reasons for resisting some proposals, forthright discussions were held. In such a process, in the end the participants have the final word but, when necessary, it is important that the advisors explain why some types of proposed solutions may create more problems than they solve. A genuine dialog establishes mutual confidence and paves the way for a more solid document. At the same time, the advisors had the responsibility of respecting the validity of the concerns that might underlie proposals by the farmers, and helping put on the table technical approaches (preferably with options) that responded to those concerns. As much as possible, rough consensuses were achieved in each session, and on that basis the advisors drafted in very preliminary form a chapter of the strategy for review at the next meeting.

There were no time limits on the reviews of the drafts. The next topic was not broached until the participants were fully satisfied with the draft presented to them and the changes they wanted to make in it. In addition, in subsequent meetings they were free to go back to any previous chapter and suggest further changes for discussion. It is vital to allow as much time as necessary for the process. In this way, the first Honduran strategy for the entire agricultural sector was drafted by the private sector in about four months.

If proposals are put forward that are economically unsound, and if the advisors choose not to respond directly to their proponents and end up ignoring their ideas in the drafting of the strategy, then there is a risk that some of the participants may feel *ex post* that the discussions were meaningless. The aim of such a process is *to move forward jointly, so that ideas from all participants contribute to the final formulation on each topic*, while suggestions that cannot be supported are perhaps weeded out along the way, to the degree possible, always on clear grounds.[16]

In a later phase of this Honduran work, the participation was broadened to include representatives of national *campesino* organizations. By then, the strategy had been transformed into draft legislation. In this round of discussions, it was found to be helpful to use a matrix to guide each day's discussions about the draft of a new law for agricultural policy. Each row in the matrix corresponded to a problem to be solved. The first column contained a short paragraph summarizing the nature of the problem. The second column contained the proposed solution in the draft law. The third column was left blank, to be filled in jointly by all participants in order to revise the draft law. At such a stage in the process, presenting a written technical draft can be a valuable aid in that it *concentrates the discussion on specific questions* and helps reduce the tendency that some participants may have to make speeches on tangential topics.

As part of this phase of the Honduran process, each day's agreement was recorded in the third column and the matrix was subsequently retyped and distributed prior to the next meeting. The first order of business in each session was to give all participants an opportunity to confirm that their consensus in the previous meeting was accurately reflected in the new wording. Sometimes, afterthoughts were brought up and earlier issues were revisited. Again, no restrictions were placed on the time needed for reviewing all issues. It was found that *presenting a revised matrix for review each day, before moving forward to the next problem area, was a transparent manner of confirming the results of the discussions, and it proved to be a useful confidence-builder*. The participants could see their own ideas taking form from meeting to meeting.[17]

In Nicaragua, a similar process for developing a national agricultural strategy with the private sector also involved interaction between advisors and farmers over an extended period, through task forces organized by topic. In this case, each task force developed a lengthy matrix in its area of policy at the outset. The three columns of each matrix corresponded to problems, existing approaches, and proposed new approaches. Once the matrices were completed, they were used to develop drafts of the strategy for review by the producers.

In Guyana and Mozambique, and also in the case of developing a national agricultural, forestry and fisheries strategy for Estonia, the national task forces drafted the chapters themselves. In Guyana and Estonia, they were subsequently reviewed by others and plenary sessions were held to discuss possible modifications. In these two cases, prior to the completion of the drafts, the advisors sat in on many of the working sessions, to develop lines of communication and common viewpoints with the participants as much as possible. This close and continuing interaction greatly eased the subsequent process of review of the drafts.

---

16. On the other hand, the procedure used in some group initiatives of having participants write down single words on 3 × 5 cards, and then pasting them on the wall for all to see and discuss, is not likely to be fruitful for producing a document with the degree of conceptual content and logical structure that a strategy normally requires.

17. In between the two phases of that work, there had been discussions with various ministries in the government and with international agencies. All told, twenty-five versions of the draft law were produced before sufficient consensus was reached to send it to the Honduran Congress. The process was lengthy, but each new draft signified a widening of the consensus and therefore was worthwhile.

9.2.3.3 Quality Control[18]

The quality control issue is at the heart of participatory efforts. If the document turns out to be weak on technical grounds, or irresponsible from a fiscal viewpoint, it will not be taken seriously. However, experiences have shown that civil society members usually are anxious to put together a sound, responsible document. *Capacity building is critical to ensure quality*. In this regard, often neutral technical assistance – by parties who do not represent official positions of donor agencies – can help when provided through a joint learning-by-doing process.

Often, capacity building consists of tapping into latent national talent. In every country there are experts familiar with issues in most sectors. The challenge sometimes is to familiarize them with broader policy frameworks and also with policy options that have been explored in other countries, more than entering into a teacher–student relationship, and to give them an opportunity to work on policy issues that they may not have worked on previously.

In spite of the advances of economics as a discipline, quality in sectoral policy work can reside in the eye of the beholder. The criteria for judging quality need to be as objective as possible. The internal consistency of a strategy and fiscal responsibility are the most basic criteria. In other areas, flexibility often is needed.

Qaulity can best be promoted through partnerships between participants and technical advisors that contribute to capacity building in policy work. Formal and informal training are basic tools of capacity building, but an equally powerful one is the approach of working side-by-side in teams, advisors and national counterparts together, over sustained periods of time. Efforts also should be made to explain technical jargon, as civil society representatives frequently have emphasized. There are few, if any, macroeconomic scenarios that cannot be presented in lay terms.

Whether at the sectoral or macro level, it bears emphasizing that *capacity building requires a sustained commitment over time on the part of technical advisors, national and international*. In the Honduran experience of agricultural policy reform, over a period of a year 80 all-day meetings were held with representatives of *campesino* organizations and large-scale producers in order to develop the reform packages, including both draft laws and draft implementing regulations. The participatory process in Guyana involved an advisory presence in more than a hundred meetings of civil society task forces over a period of several years. Perhaps the most salient characteristic of these meetings is that through them *civil society members actually drafted specific policy reforms*. The process was interactive between advisors and national counterparts, but the latter always had the last word and through the process they became authors of the reforms – and felt confident that they were the authors. This sense of ownership encouraged them to go forward and lobby for full acceptance of the strategy by society and international agencies and for its implementation by government.

Capacity building does not stop with the completion of one strategy document. As people move on to other occupations and even migrate abroad, capacity and the sense of commitment to the reforms can weaken. Capacity requires continuous nourishment. Providing financial endowments for independent 'think-tanks' is one way to ensure that sufficiently attractive incentives are offered to trained people to stay with the effort. There are several examples of productive policy think-tanks in developing countries, from the Thai Economic Development Institute to FUSADES (Fundación Salvadoreña de Desarrollo Económico y Social) in El Salvador. In this regard, it is important for international agencies to commit themselves to meeting the challenge of *institutionalization of civil society's capacity for policy work*.

In the final analysis, *one of most valuable roles for official development assistance is building*

---

18. This section is adapted from R. D. Norton, 2002.

national capacity for policy analysis and formulation. Such an undertaking is complementary to private investment flows – not a substitute for them. Investment flows respond to the quality of the policy environment more than anything else, and therefore expenditures on capacity building for policy work can have a significant influence on them.

9.2.3.4 Organization of a Participatory Process

A participatory process for developing a strategy requires a secretariat or co-ordinating committee to guide it. In most circumstances, it would be staffed by representatives of the participants, along with clerical staff, and ideally it would include a legal expert to assist with legislative issues and perhaps additional kinds of researchers to support the undertaking. If government is a sponsor of the work, it would be represented on the co-ordinating committee; in many cases it would chair it. One of the roles of this committee is to ensure the overall consistency of the strategy by reviewing the drafts of the chapters and indicating possible inconsistencies among them to the concerned working groups. Another role is to outline a tentative structure of the entire draft and enunciate a first version of the objectives and guiding principles, subject to modification by the full group. A co-ordinating committee has the responsibility of circulating these preliminary products to participants for discussion in the wider group, along with notes on issues and themes that might form part of an overriding vision that informs the entire strategy.

In both Mozambique and Guyana, a three-level structure was used to co-ordinate the effort. In Mozambique, an Executive Committee of four persons managed day-to-day logistics and a Committee of Counsellors of 14 persons was responsible for reviewing the quality and consistency of the drafts, and for receiving input from the local level throughout the country and disseminating drafts for comment. The actual drafting was carried out by 12 technical working groups or 'nucleos'.[19] In Guyana, a civil society committee of 35 persons guided the effort of the working groups, and that committee elected five co-chairpersons to manage the process.

If there is international sponsorship, international advisors may form part of the co-ordinating committee and/or topical working groups, as deemed appropriate. (In Mozambique, both the co-ordinating entities and the technical working groups were staffed only by Mozambicans.) In all cases, it is essential to clarify, by both words and actions, that the international advisors are not bringing an external policy agenda with them, but rather are present to serve as a sounding board for preliminary ideas, on the basis of their own international experience. If the advisors are staff members of international organizations, it may not be easy for them to play this kind of disinterested role and to stand back from the policy recommendations that their organization(s) may be advocating and may even be tying to loan conditionality. This question of how to erect a 'firewall' between technical advisors and those who are responsible for institutional positions has come to the fore in recent years. Equally, the international development community has begun to recognize the value of a 'third force' or neutral international advisory group that can play a supportive role for national teams, without trying to impose a set of recommendations, but so far no general solution has emerged.[20]

For international advisors participating in these kinds of processes, the bottom line has to

---

19. This group of civil society, known as Agenda 2025, was endorsed by the President and Prime Minister and the opposition parties contributed participants. It produced a detailed vision document covering all sectors of the economy and social issues in June of 2003.

20. A preliminary conference to explore this issue was organized by the Carter Center in 1997 and its conclusions were presented in *Toward a New Model of Development Cooperation: The National Development Strategy Process in Guyana*, Global Development Initiative, The Carter Center, Atlanta, GA, USA, May 1997.

be, as noted above, that local experts and civil society participants have the ultimate say. An advisor can try to convince them of a viewpoint on some issues, but he or she has to be prepared to yield in the end. Once a strategy is completed, it will be negotiated with other entities, including government and international agencies, and any truly unfounded recommendations probably will be vetted at that stage. The advisor's role is to *assist* in ensuring a better technical quality of the document, but it is not a role of being a guarantor of its technical quality.

The other key component of the organizational structure of the process is *a set of task forces, or working groups*. They are usually organized by policy topic. Several experiences have placed in high relief the role of task forces. They are the primary vehicle for facilitating participation on a continuing basis. They also are the organs that eventually produce drafts, and it is the task forces that assume ownership of the drafts in the first instance. It is essential to support the task forces, both logistically and substantively, so that they cohere and function sufficiently well to produce results and gain this sense of ownership.

Task forces can mobilize national talents that otherwise might remain outside the ambit of discussions of national policies. They can play a long-term role in national policy dialogs. When the task forces have developed a firm commitment to their results, they sometimes will undertake, on their own initiative, a sustained and forceful lobbying effect to ensure that the proposals are formally accepted and implemented.

A process of this nature is a very human undertaking. Experience shows that not all participants will work with the same degree of commitment, and some will fall by the wayside. (In the case of Guyana, 5 of the 23 task forces did not function well at the beginning of the effort, and eventually 4 of them had to be reconstituted.) Nevertheless, experience demonstrates that *from such processes and task forces leaders will eventually emerge, spokespersons for civil society* who will help push the process along and over the longer run will represent civil society in the dialog with government and international agencies over policies. In addition, often they will be sufficiently informed and

interested to participate is monitoring the policy implementation process.

Another management issue that sometimes arises is *whether the participants from civil society should be reimbursed for their time spent in the effort.* The argument in favor of such reimbursement is that it is a way to tap into the best national talent from outside government. Otherwise, some of the best experts, such as university professors who work as consultants, might not be available for the effort. Payments for time invested also can help sustain commitment to the effort. The argument against making such payments is that thereby the national participants become beholden to the sponsoring entity, whether it be government or an international agency, and accordingly they cease to be genuine representatives of civil society. At least such a perception could become current.

On these grounds, no reimbursement was made to the participants in the Honduran, Guyanese and Nicaraguan cases, except for providing lunches and snacks during meetings and, in the Honduran and Nicaraguan cases, reimbursing *campesinos* for their costs of transport to the capital city for the meetings. (Government officials, of course, continued to receive their salary while working on the effort.) Partly as a consequence of this decision, at times the level of commitment flagged in several of the Guyanese task forces, but they eventually completed their work. In Nicaragua (in 2001), participating producers were sent on a number of study tours to other countries. This provision proved to be a useful incentive for sustaining interest in the process and also was educational for the work on the strategy.

In the Estonian case, the principal participants were faculty members, who were contracted for the effort by the FAO, as well as government officials. In that case, it would have been virtually impossible to assemble a team of experienced national personnel without the contracting mechanism. Nevertheless, this question did not prevent the recommendations of the Estonian strategy from figuring prominently in national debates over agricultural policy, and it stimulated adoption of the policy approach of direct payments to producers instead of price controls (see

Chapter 4), new legislation in the area of land reform, and other policy initiatives.

In Guyana, the use of entirely volunteer experts in the second phase meant that the work took much longer than expected to complete. In the Honduran case, although the representatives of farmers were not compensated through the strategy project, the participating *campesino* leaders held paid positions in their respective associations and federations, and hence they did not have to make a personal financial sacrifice to contribute time to the effort. Thus, what can be achieved on a purely volunteer basis should not be overstated, although the principle of civic participation should be respected as much as possible for the sake of generating a sense of ownership of the product.

Participants in a process of this nature will not necessarily have had previous experience in participatory undertaking. Therefore, once it has been agreed who are the co-ordinators of the effort, it can be helpful to provide them with brief training on the basics of managing a process in a participatory manner. In societies without a well developed tradition of civil society activity, this can be quite useful. In the course of the training, it should be emphasized that communications among the participants can occur in various ways over the course of the project, and through both formal and informal channels. It is not necessary to leave all discussions to formal sessions. Informal gatherings can be organized as well, to enhance the group dynamics among the participants.

In terms of institutional strengthening, the effort of developing a strategy, with all of the policy analysis that it involves, is an invaluable exercise not only for civil society but also for the staffs of a ministry of agriculture and of other government agencies involved. Their understanding of policy issues, constraints and options can be greatly improved through the process, and they can gain a deeper appreciation of the thinking of other national participants.

If, for unavoidable reasons, it is impossible to organize a full partnership with rural society, then extensive consultations on a strategy should be carried out in at least two rounds: before the drafting process begins, and after a first full draft is available for discussion. Effective consultations are those that are well planned and structured. Asking the assembled individuals open-ended questions about their main concerns and recommendations can be useful at the beginning of a visioning exercise – as was done in Mozambique in virtually all districts of the country – but for developing more policy-specific documents it is helpful to structure the discussions around concrete issues. The first round of policy consultations can be based on a preliminary vision of principal issues, and the discussions can be mainly oriented toward what to do about those issues, without closing the door to the introduction of other concerns by the participants.

The effectiveness of later rounds of consultations can be enhanced if the draft document is discussed section by section in a structured manner, while the agreements reached on changes in it are recorded carefully. Likewise, it should be distributed to the participants well in advance of the consultations. However, even though these kinds of consultations are well planned and carried out, it should be recognized that they alone are not likely to be sufficient to produce a complete sense of ownership of the strategy on the part of civil society.[21] In most cases, the process necessarily will involve both direct participation, in the sense of drafting the strategy, and consultations in order to bring a wider circle of citizens into the process.

9.2.4 Challenges and Risks of Participation

By its nature, a process of this type faces several challenges and risks. Although it may be carefully structured, the process represents an attempt to catalyze civil society's own response so that it will

---

21. In Guyana, consultations were used as an adjunct to the task force approach. After the first draft was completed, it was taken to many towns and villages throughout the country, usually by the Minister of Finance himself, to solicit comments.

take on a more active role in national dialogs on development policies. Therefore, it cannot be predicted with any confidence where the process will end up. This simple, but basic, fact should be recognized explicitly at the outset.

Some principal challenges faced in participatory policy processes are summarized in the following paragraphs. Associated with each challenge is the risk that the challenge will not be successfully met.

(1) The first challenge is to *motivate a sufficient number of leading farmers and ranchers to commit themselves to the process* and dedicate the time required to bring it to a successful conclusion. Since both agricultural entrepreneurs and peasant farmers have many claims on their time, it is not easy to satisfy this requirement. In some circumstances, an additional factor that can complicate the challenge is a prevailing attitude of cynicism about the prospects of being able to effect true reforms in the policy framework. On the other hand, if the rural sector is experiencing a deep economic crisis, that circumstance alone may motivate people to try their hand at reforming policies through a participatory process.

(2) The second challenge, especially relevant for agricultural and rural strategies, is to *reach beyond the group of persons who reside in the capital city and attain a degree of geographical representativeness in the participation.*[22] The logistics of this challenge can be difficult in large countries and countries with poor communications and transport networks. Usually, it is necessary to pay transportation and lodging expenses for low-income persons who travel long distances to participate in the

effort. In Mozambique, the members of Agenda 2025 visited almost all of the more than 100 districts in the country to explain the effort, often traveling by small plane and small boat, and sometimes on horseback.

(3) The third challenge is to *overcome schisms that may be present in the participating group for reasons of political partisanship or socio-economic differences*, in order to be able to put together a consensus position that has the implicit support of all or most major groupings in rural society. Partisan differences may bring with them ideological discrepancies with regard to the role of the government in the economy and other basic issues, and so it may not be an easy task to overcome these differences and foster a true consensus on policy reforms. However, if a consensus can be achieved, it may represent a solid foundation for sustained co-operation on policy issues in the future.

(4) The fourth challenge is to create an environment in which *civil society representatives jointly go forward to lobby for sound economic policies that will benefit the entire sector*, instead of promoting narrow personal or sectarian interests. A fundamental reason for the success of a producers' group in Honduras, in developing a package of reform legislation, was their early agreement to propose uniform, sector-wide policies, setting aside their previous tendency to negotiate special privileges with the government of the day.[23] This is a basic issue of *quality control* for an effort of developing a strategy. The importance of a methodology for ensuring quality control of a participatory strategy was highlighted in the African context by Bodo Immink and Macaulay Olagoke, who put it in the form of

---

22. 'In national processes, such as the design of the Poverty Reduction Strategy, the government generally engages with organized civil society groups in the capital or major urban areas. However, national level civic engagement can also allow the government to reach a wider range of stakeholders and initiate a dialogue with smaller civil society organizations such as farmers associations, cooperatives, unions, chambers of commerce, women's groups. . . .' (S. Tikare *et al.*, 2001, p. 14).

23. This group now is known by the acronym CONPPAH and, twelve years after initiating their successful effort they still were playing a significant role in national policy deliberations.

a question, 'How do we ensure quality without being unparticipatory?'.[24]

(5) The fifth challenge is to *avoid creating a parallel process of dialog that may weaken existing channels of communication and social processes rather than strengthen them* (S. Tikare *et al.*, 2001, p. 26).

(6) The sixth challenge is to *avoid premature publicity about the participatory work in its drafting stage*. A proposed strategy needs to go through various drafts before it is mature, and the process of forging a consensus among the participants takes time. Although some publicity about the process can be useful, to help keep the public informed, premature publicity about the content of the strategy runs the risk of becoming a 'lightning rod' for sharp, sometimes partisan, criticisms of the tentative policy formulations, even in the national press. The danger is that such criticism early in the effort could fracture the emerging consensus among participants and even derail the entire effort. At a later stage in the process, wide publicity is essential in order to promote a national dialog about the policy recommendations, but if exposure comes before the participatory group coheres well, the effort can be undone.

A necessary condition for meeting this challenge is to create an atmosphere of mutual confidence among the participants and between them and the advisors. However, it is not a sufficient condition, because the work can be affected by currents and events beyond the control of the strategy effort and of the producers.

(7) The seventh challenge is to *gain acceptance for the principal policy recommendations from the government of the day, or the next government* if elections are due relatively soon. The

possibilities of success of the program in this sense can be increased by two factors: (a) achievement of a true consensus among the majority of the participating farmers or members of civil society, independently of their personal political affiliations, and (b) receiving support for the consensus on the part of international development agencies when recommendations are presented to the government. It has to be anticipated that this challenge will be a difficult one:

The biggest single constraint and challenge to the [World] Bank's ability to pursue participation across all its operation is [lack of] government commitment.[25]

The nature of some of these challenges has been summarized by David Brown and Rajesh Tandon in the following words:

The presence of a serious problem that does not respond to orthodox solutions can indicate an opportunity for collaborative strategies. Collaboration offers the opportunity to mobilize several resources of many different actors, for example, it can increase the stock of information and ideas. . . . [However, collaboration] requires bringing together persons with diverse interests, information, resources and power. Since collaboration means influencing each other mutually, these differences imply that there always exists the possibility of conflicts over objectives and means. . . .[26]

On the other hand, meeting these challenges successfully endows a private sector or civil society with a substantially enhanced capacity for playing a role in national policy dialogs and for making significant contributions to them.

24. Bodo Immink and Macaulay Olagoke (Eds), *Participatory Approaches in Africa: Concepts, Experiences and Challenges*, Proceedings of the Exchange Forum for Practitioners of Participatory Development Approaches in Africa, Uganda Catholic Social Training Center, Kampala, Uganda, July 1997, p. 6.
25. M. Aycrigg, 1998, p. 20.
26. L. David Brown and Rajesh Tandon, 'Multiparty Cooperation for Development in Asia', IDR Reports, Institute for Development Research, Boston, MA, USA, 1992, p. 30.

9.3 STRUCTURE AND CONSISTENCY IN A STRATEGY

9.3.1 Shaping a Strategy

While a creative and feasible vision about the future is an essential element of a strategy, the means of making the vision come true eventually need to be stated in sufficient detail and rigor as to constitute an implementable program. A vision can help consolidate the unity of a country, and promote agreement on future policy priorities, but alone it will not necessarily lead to the formulation of action plans and concrete measures. *Seeing the way forward simultaneously in broad terms and in specifics is also one of the major challenges in putting together a successful strategy*. Equally, a strategy should present a sense of priorities.

A useful step in the transition from a vision statement to a program of policy proposals is to identify principal obstacles to the fulfillment of the vision. Farmers and other rural dwellers will be aware of many of the principal problems that need to be overcome in order to realize the sector's potential. They may express their concerns in terms of low farmgate prices, perceived price-gouging by intermediaries, high costs of production, soil erosion, poorly functioning irrigation systems, lack of access to credit and banking facilities, and the like. A role of a strategy is to articulate those concerns in tractable ways, in concepts that can be addressed through policy actions. Low farmgate prices may be attributable to poor marketing infrastructure, but they may also arise from an overvalued exchange rate or an uneven tariff policy that lets food products in with low or zero tariffs while taxing the import of other goods at higher rates. It is the task of a strategy to diagnose accurately the sources of the farmers' concerns, and then to develop the corresponding policy remedies. *The logical links between diagnoses and solutions need to be made clear*.

Another worthwhile step is for the task forces to develop descriptions of the existing policy framework. Sometimes, they will have worked all or most of their professional lives in a given policy framework and may not find it easy to mentally step outside it and describe it and articulate alternatives. In this area, reviewing experiences of other developing countries, through advisors or through direct international meetings with counterparts, can be very helpful.

A document structure that provides stimulus to asking the right questions can facilitate the work considerably, especially when non-specialists are involved through a participatory process. It is particularly helpful to try to guide those who are developing the strategy towards the identification of principal constraints that need to be overcome, and issues that need to be addressed, in each area before they attempt to analyze alternative policy recommendations.

When a strategy is divided into several chapters by topic, it can also be useful to ask the writers of the chapters to translate the broad sectoral objectives into subsidiary objectives for each of their topics. In this way, the recommendations then can be presented in such a way that they fulfill the 'sub-objectives' and, in the process of doing so, overcome the major constraints that have prevented meeting those subsidiary objectives in the past. The statement of subsidiary objectives should be brief, typically one-half page to one and a half pages. If it is longer, a confusion of means and objectives tends to creep into the discussion.

Equally, it is helpful if each chapter contains relevant background material on its topic, plus a summary of the past and current policy orientations, so that the reforms are placed in an appropriate context. A good strategy should be much more than a set of recommendations – which are vulnerable to being quickly forgotten if presented without support. *A strategy should also become a major reference work for researchers and future policy analysts*. Among other things, this approach helps maintain the work on a rigorous level and also makes it more difficult for decision makers and politicians to ignore the strategy's proposals.

As an illustration of how the work may be organized, in some cases (Estonia and Guyana) it has been found useful to divide each chapter into the following sections, in order to give a logical sequence to the work of the task forces:

(1) Basic characteristics of the sector, sub-sector or topic under review.
(2) Review of past and present policies for the area or topic.
(3) Major issues and constraints to be dealt with.
(4) Specific (subsidiary) objectives for the area or topic.
(5) Policy recommendations and supporting technical justifications. (The policies are developed logically as ways to overcome the constraints and satisfy the objectives.)
(6) Appendices: recommended legislative reforms and recommended investment program (not applicable for all areas).

The introductory part of the strategy would state the overall objectives and interpret them in the prevailing context. In such a section, it can also be useful to state a set of *policy principles*, as illustrated earlier in Chapter 2 of this book.

Comprehensiveness in a strategy does not mean that it is necessary to reinvent the wheel in all areas. A strategy should incorporate worthwhile analyses and well-founded recommendations that have already been developed in other contexts. It is *an inclusive document but not necessarily original in all areas*. Indeed, incorporating previous work done by other groups can be one of the means of consolidating a societal consensus and mobilizing support for the strategy's recommendations.

In the agricultural sector, comprehensiveness also does not necessarily mean that all topics related to agriculture must be included in the strategy. The circumstances may dictate separate work, at a different time, on a national water strategy for example, or on a national forest management strategy. However, it does mean relatively complete coverage of all relevant policies within the scope of the strategy. It is not advisable to develop separate strategies for different agricultural products (grains, vegetables, root crops, livestock, etc.), since in the long run those kinds of outputs can be substitutes for each other in production, at least to a degree, and since they all respond to the same basic set of policies: to trade policy, agricultural financial policy, land tenure policy, and the like. Specific actions for

crops and livestock products can be handled at the level of programs within the framework of the strategy. (Chapter 2 discusses the distinction between policies, programs and projects.)

While the chances of a strategy being implemented depend on many circumstances, some of them difficult to predict, they also depend crucially on the conceptual strength of the document, on the respect it commands, as has been emphasized. However, this does not mean it should be an academically written and abstruse document but rather one whose logic is clear and convincing and whose empirical foundations are firm. Formal and quantitative analyses of issues may play a supporting role in developing a strategy; indeed it is essential that a strategy have rigorous underpinnings, but the document itself should be expressed in language understandable to a wide audience. Particularly important analyses can be presented in annexes if so desired.

There are many possible *dimensions or principal thrusts that help provide overall consistency to a strategy*. They are the structural girders or axes on which the substance of a strategy is erected. They may be based on historical, economic and social factors, agronomic and ecological considerations, and other relevant factors and considerations.

The choice of such a framework depends in large measure on the nature of the problems being addressed and the circumstances of each country and its agricultural sector. In the Estonian case, principal thrusts of the strategy included the historical imperative to continue to de-sovietize the agricultural sector, which lagged behind the rest of the economy in that regard, and the need to devise market-compatible compensatory measures for the extreme economic pain inflicted on agriculture by the country's adoption of an overvalued exchange rate and ultra-liberal trade policies. As a result of the abrupt decline in the agricultural economy, there was also an urgent need to redress emerging social problems in rural areas.

A logical framework that has been found useful for developing strategies in a number of circumstances is one that is based on the economic paradigm of the farmer (Chapter 2): *incentives for production, the agricultural resource base including security of land tenure, and access to inputs, markets and technologies of production.* Many other frameworks are viable so long as they lend themselves to clear descriptions of issues and development of implementable policy recommendations.

9.3.2 Institutions and Human Capital

Economic reasoning sometimes suggests that policy reform packages can be organized around the concept of correcting **market failures**. The concept of market failure is complementary to the framework of farmers' incentives, resources and access, since the effort to make improvements in each of those areas would involve correcting the corresponding market imperfections. In many cases, *policies and programs designed to improve incentives and correct market failures involve strengthening institutions.* For example, improving small producers' access to credit, and reducing the interest rates they pay at least somewhat, often requires development of new kinds of private rural credit institutions and changes in financial regulations. Likewise, an effort to improve access to land may require legislation that more clearly defines property rights or establishes new rules of the game for land tenure, and it also usually requires strengthening local land registries and sometimes requires new financing mechanisms for land purchase and improvements in rental markets as well. Agricultural technology delivery systems are undergoing profound institutional change throughout the developing world, as discussed in Chapter 8.

In most circumstances in developing countries, an agricultural strategy is incomplete if it does not review the *role of human capital*, which deserves emphasis of its own in a strategy, apart from the other basic resources. Theodore Schultz has put the matter in the following way:

The decisive factors of production in improving the welfare of poor people are not space,

An evaluation of the Chinese experience in agricultural development underscored the importance of incentives and institutional development:

The structure of incentives determines the economic outcome. The institutions – the rules of the economic game – structure the incentives. There is a strong tendency for political systems to develop institutions which result in weak incentives for agricultural productivity. The first move in the Chinese economic revolution was to institute the household responsibility system and to (slowly) reduce the role of the large state enterprises. It was clearly an important innovation, leading to increased productivity. The lesson is: get the institutions linking actions to consequences right. This is not as simple as it sounds. (From James D. Shaffer and Simei Wen, 'The Transformation from Low-Income Agricultural Economies', in G. H. Peters and D. D. Hedley (Eds), Agricultural Competitiveness: Market Forces and Policy Choice, Proceedings of the 22nd International Conference of Agricultural Economists, Dartmouth Publishing Co. Ltd, Aldershot, UK, 1995, p. 203.)

energy and cropland; the decisive factors are *the improvement in population quality and advances in knowledge.*[27]

A major constraint on technological improvement in many agricultural sectors is the low level of literacy of farmers and their lack of familiarity with basic concepts of cost accounting and business administration. In the context of a strategy, this issue merits review, along with options for increasing levels of schooling in rural areas and improving farmer literacy and numeracy. The Government of Mexico, for example, has recently decided that the human capital constraint is so dominant that it has initiated a program of financial transfers to rural families as a function of the rate of school attendance of each family's children.

---

27. Theodore W. Schultz, 'Investing in People', in C. K. Eicher and J. M. Staatz (Eds), 1998, p. 329 [emphasis in original].

Empirical evidence regarding the relation of literacy and farm productivity has been compiled in a number of developing countries. Thomas Pinckney studied the impact of education on productivity in coffee-growing villages in Kenya and Tanzania by analyzing survey data. He found that:

> Results for the two sites are striking. . . . Holding other inputs constant, output is more than 30 percent higher when agricultural decision-makers can add and subtract two-digit numbers and read and comprehend simple paragraphs.[28]

In short, *human capital is the most strategic factor for agricultural development* as new technologies emerge, markets demand higher quality and safer products, and as consumer requirements for quality and delivery timing change. Learning how to continuously access information and assimilate it is acquiring increasing importance in agriculture throughout the world. Equally, institutions and policies that facilitate such access are increasingly critical, but often Ministries of Agriculture are slow to recognize the importance of quality issues – for all types of farmers.

In countries in which land is scarce, another indicator worth monitoring is the land-to-labor ratio in agriculture. In the early stages of development, with high fertility ratios, the average farm size may decline unless new arable land is opened up. In addition, developing more arable land may come at the expense of forests. In establishing the framework for a strategy, it can be a valuable exercise to analyze population trends, including the rural–urban migration rate, to determine if the growth rate of the rural population is slowing and, if so, how soon it may be likely to begin to decline in absolute numbers, thus reversing the trend toward further fragmentation of farms or additional incursions into forested areas or marginal lands.

Both the logic of improving incentives and emphasis on human resource development point in the same direction of *institutional development*. Institutions are an integral part of a country's endowment of human and social capital. Virtually all fundamental strategic emphases have an institutional dimension. In the words of James Bonnen:

> Clearly a nation that cannot sustain long-term institution building and human capital improvement will never have a highly productive, industrialized agriculture. Just as clearly, long-term institution building and human capital accumulation must involve more than the research and education institutions. Physical capital and conventional input development are necessary for soil and water conservation, reclamation, and development; for long-term, intermediate, and short-term credit; for rural roads, mail service, and eventually electronic communications; and for the development of modern market institutions. . . .[29]

The role of institutions in the economy has been expressed as follows:

> Institutions can be regarded as arrangements among economic agents to attempt to decrease the uncertainty in exchange and ownership (North, 1990).[30] Ill-defined property rights

---

28. Thomas C. Pinckney, 'Does Education Increase Agricultural Productivity in Africa?', in Roger Rose, Carolyn Tanner and Margot A. Bellamy (Eds), *Issues in Agricultural Competitiveness: Markets and Policies*, IAAE Occasional Paper No. 7, International Association of Agricultural Economists and Dartmouth Publishing Company Limited, Aldershot, UK, 1997, p. 346.

29. James T. Bonnen, 'US Agricultural Development: Transforming Human Capital, Technology, and Institutions', in Bruce F. Johnston, Cassio Luiselli, Celso Cartas C. and Roger D. Norton (Eds), *US–Mexico Relations: Agriculture and Rural Development*, Stanford University Press, Stanford, CA, USA, 1987, p. 299.

30. Douglas C. North, 'Institutions and a Transactions-Cost Theory of Exchange', in J. E. Alt and K. A. Shepsle (Eds), *Perspective on a Positive Political Economy*, Cambridge University Press, Cambridge, UK, 1990.

induce opportunistic behavior to capture residual benefits, within and outside firms (Milgrom and Roberts, 1992).[31] High transaction costs and uncertainty result from such incomplete property rights. Hence institutional arrangements represent attempts to reduce the uncertainty in exchange, and lower transactions costs, by defining the rules of the game.[32]

In a rural economy, the scope of institutional development goes beyond reducing uncertainty and transaction costs. In many cases, it facilitates access that previously was completely lacking for a market or a service, at least for some groups of farmers. It can be said that in such cases it reduces transaction costs from infinity to a finite and more manageable level. Perhaps most fundamentally, *solid institutions open the doors to new knowledge for farmers and allow them to enter into reliable arrangements for future delivery of inputs and products and deferred payments*. If all transactions and payments have to be made on a spot basis, there is little scope for development beyond the village marketplace.

9.3.3 Consistency Dimensions of a Strategy

Another integrating dimension for a strategy is that of consistency and complementarity among the various policies. This is the dimension of *intra-sectoral consistency in policies*. As commented upon earlier, adequate access to farmland can require improvements in financial institutions, and in some cases the creation of new programs and institutions such as land funds. Regularization of land tenure in turn can be a pre-condition for the implementation of other policies and programs. For example, in order to carry out a program of transfers to small farmers

it is an unavoidable requirement that farmers' landholdings be registered in some form. Adequate availability of rural finance also is a necessary condition for many programs of technology transfer. On the side of crop production and marketing, a prerequisite for implementation of a program of certificates of grain deposits (crop liens) is the development and dissemination of agreed quality standards for grains. The list of examples could be extended considerably, but it is evident that these kinds of consistency requirements must be satisfied in order for a strategy's recommendations to be feasible.

When the agricultural sector is defined in a broad sense, additional consistency requirements enter the picture. In a number of countries in Latin America, existing land tenure laws are incompatible with sustainable management of national forests. In Panama, to cite only one instance, a rural family cannot aspire to tenure rights on State-owned lands as long as those lands are forested. However, if the family cuts down the trees and plants annual crops on the land, then provisional title to the land can be acquired under the agrarian reform law, and eventually that title can be converted to fee simple title.[33] Other policy linkages may be found between agricultural and rural environmental policy and between agricultural and water management policy.

Notwithstanding the importance of policy consistency between agriculture and natural resource management, the effort should avoid the temptation that arises in some kinds of strategizing to try to decide in a centralized way which crops can be grown in each class of soils. In the end, farmers are better placed than government to make those choices. Sometimes, the crop sown in a given area may be the second or third or fourth best for those soils, because the most appropriate crops are grown in even better soils elsewhere, and their

31. P. Milgrom and J. Roberts, *Economics, Organization and Management*, Prentice-Hall, London, 1992.
32. John C. Beghin and Marcel Fafchamps, 'Constitutions, Institutions and the Political Economy of Farm Policies: What Empirical Content?', in G. H. Peters and Douglas D. Hedley (Eds), 1995, p. 288.
33. These kinds of contradictions and other legal obstacles to sustainable rural development in Panama were reviewed in Roger D. Norton, 'Obstáculos jurídicos e institucionales al desarrollo sostenible del Darién', Panama City, Panama, April 1998 (study developed for the Inter-American Development Bank).

production from those other areas satisfies the market's needs. In both international and national production and marketing, the law of comparative advantage holds sway. What policy planners can usefully do in regard to the suitability of soils for given crops is to present workshops and training courses to inform farmers in a given region of cropping alternatives they may wish to consider and how such crops can be grown and marketed.

In addition to the intra-sectoral consistency dimension, the **inter-sectoral dimension** is basic as well. As explained above in Chapter 4, macroeconomic policies usually figure among the principal determinants of agriculture's growth path, and so a strategy for the sector cannot be presented without taking these linkages into account. Macroeconomic policy has such a pervasive effect that it is no exaggeration to say that it has a strong bearing on the type of society that is evolving. Sometimes it is felt that there is only one 'correct' set of macroeconomic policies, but this is not usually true. The composition of revenue and expenditure policy, the type of exchange rate policy, and the speed at which inflation is reduced all can vary, to mention only three examples. In a recent agricultural strategy for El Salvador, it was pointed out that the high degree of dependence on workers' remittances from abroad, and the policy of allowing them to appreciate the real exchange rate, was creating an economy and society dependent on the service sectors plus the bonded processing industry. Both agriculture and manufacturing *per se* were being permanently diminished in importance. That strategy also quantified the cost to society of the annual increments to rural–urban migration that resulted from the exchange rate bias against productive sectors, and it pointed out macroeconomic alternatives which would help correct that bias and promote the growth of both agriculture and manufacturing.[34]

The World Bank has commented on the relationship between macroeconomic and sectoral policy in the following words:

Reforms of specific sectoral policies in agriculture should not be divorced from reforms of economy-wide policies and development strategies that induce strong biases against agricultural production and exports. . . . many developing countries have discriminated against agriculture through high industrial protection and through inappropriate macroeconomic and exchange rate policies. . . . The linkage between sectoral and macroeconomic policies is usually so strong that it is best to carry out agricultural reforms in conjunction with reforms of general economic policies.

The most important priority in agriculture is to ensure that the profitability of farming is not artificially depressed because of either macroeconomic or sectoral policies.[35]

An agricultural strategy may have to take on the task of outlining viable alternatives for macroeconomic strategy, if they have not been brought into the public dialogue already. In addition, it should present clearly the interdependencies between macroeconomic and sectoral policies, and between macro policies and the sector's development prospects. Nevertheless, although a strategy should include both macro-sectoral linkages and a thorough analysis of agriculture's potential and policies for realizing that potential, it also is important to avoid adopting a partisan stance in favor of agriculture *vis-à-vis* the rest of the economy. If real agricultural prices have declined because of macroeconomic and trade policies, it may be appropriate to institute new policies to reverse that trend, but at the same time to discuss measures that could compensate the urban poor for the effects of rises in food prices. From an intersectoral perspective, the

34. Roger D. Norton and Amy L. Angel, *La Agricultura Salvadoreña: Políticas Económicas para un Macro Sector*, FUSADES, San Salvador, El Salvador, 1999.
35. The World Bank, *World Development Report 1986*, The World Bank, Washington, DC, USA, 1986, pp. 149–150.

guiding precept can be *to avoid policies that are biased against agriculture, but not to make agriculture a ward of the rest of the economy either.*[36] Gale Johnson has expressed the idea in the following language:

> Agriculture has important contributions to make to economic development, but must receive even handed treatment if the possible contributions are to be realized.[37]

Fortunately, there are many kinds of policies that foster agricultural development without impinging negatively on other sectors, as discussed throughout this volume.

A productive way to incorporate macroeconomic considerations in an analysis of the agricultural economy is through the *framework of incentives in the sector*. The net incentives for production are the result of many different policies, including taxes, tariffs, export rebates, trade controls and fiscal transfers, in addition to the exchange rate. Sometimes, the net incidence of these policies is uneven within the sector. In Nicaragua, for example, it was pointed out recently that the coffee sub-sector, which is a major contributor to both foreign exchange earnings and smallholder incomes, had been taxed more heavily than other sub-sectors in the economy, thus discouraging its development. A useful strategic exercise in these kinds of circumstances can be to analyze alternative combinations of tax and tariff instruments that would generate the same amount of fiscal revenue with fewer distortions. A more complete exercise would consist of a joint analysis of the incidence of all policy instruments in the sector and the identification of alternative combinations that would lead to *a uniform rate of effective protection* for all products.

Another basic dimension of planning documents is *time*. It can be worthwhile to develop projections of the sector's possible path of evolution over time, to illustrate its potential for generating income, employment, foreign exchange and so forth. However, such projections play a purely illustrative role and it is difficult to link them rigorously to the effects of policy reforms and their timing, although they can be useful in a pedagogical sense. A strategy's basic role is to present a vision of the sector that is valid for the medium term, without trying to forecast dates in which the recommendations will generate their effects. In this context, medium term signifies at least five years, and usually the appropriate time span is ten to fifteen years. A clear definition of priorities among policy actions usually is more useful than forecasts of the effects of reforms period by period in the future.

> ... *models themselves cannot provide the answers. This is especially true as attempts are made to build into the models the response of policy itself to changes in the economic environment. ... Such endogenous policy models might reveal some of the historical factors that accounted for policy shifts, but they seldom provide a sense of when the degrees of freedom for policy initiative are about to expand* (Peter C. Timmer, 'The Agricultural Transformation', in Carl K. Eicher and John M. Staatz (Eds), International Agricultural Development, 3rd Edn, The Johns Hopkins University Press, Baltimore, MD, USA, 1998, p. 132).

Finally, a major unifying axis of a strategy is its set of *policy goals or objectives*, and operational sub-objectives and the means to attain them. It is suggested in Chapter 2 that the most

---

36. In the wording of the Estonian strategy, '. . . policies shall be oriented toward ensuring that agriculture is not an economic parasite on the rest of the economy, and neither is it exploited by the other sectors in the economy' (Food and Agriculture Organization of the United Nations and the Estonian Agricultural University, *Estonia: Long-term Strategy for Sustainable Development of the Agriculture*, Tartu, Estonia, 1997, p. 9).

37. D. G. Johnson, 'Role of agriculture in economic development revisited', *Agricultural Economics*, **8**(4), June 1993, p. 421.

all-encompassing goals are real rural incomes and, within that category, real incomes of the rural poor. The strategic emphasis on rural poor is warranted from a national viewpoint as well as from a sectoral perspective, since in most developing countries the bulk of the poverty, including the extreme poverty, is found in rural areas. For Africa, for example, it has been noted that:

> Overall, household data show that while a large percentage of the urban population does poorly, the rural population does worse.[38]

Rosamund Naylor and Walter Falcon examined trends in poverty and demographics in the developing world and concluded that, in spite of continuing rural–urban migration:

> the poor in rural areas will continue to outnumber their urban counterparts well into the 21st century, even if they are less visible or less vocal politically.[39]

This emphasis on income leads directly to the subsidiary goal of productivity increases and, in many cases, to the goal of improvements in relative farmgate prices.

It should be pointed out that goals based on rural income and the income of the poor do not include food price stabilization as a primary objective, although it has its place in the pantheon of lesser objectives. Price stabilization often is accepted almost reflexively by some analysts as a fundamental policy goal. Peter Timmer, for example, in an otherwise illuminating piece, sug-

gests that 'food price stability' is an essential goal, on the basis that it enhances food security and attracts more investment into agriculture.[40]

The empirical and theoretical evidence on the harm caused by price fluctuations is still unclear. As pointed out by Stephen Jones:

> there is a lack of securely based consensus about the significance of, and how to define and measure, the costs of price variability for both food producers and consumers. Microeconomic modeling of the impact of missing markets for price insurance has suggested that the benefits in terms of allocative efficiency from price stabilization through conventional buffer stock policies are likely to be small relative to the distributional impact of these policies and the costs of intervention. On the other hand, this approach has been criticized for neglecting the dynamic impact of price instability on investment and the feedback from food price instability to the macroeconomy. . . .[41]

Unquestionably, extreme and frequent fluctuations in food prices can be detrimental to both food security and investment prospects, but the price stabilization goal has to be placed in an appropriate context. The first objection to a blanket acceptance of that goal is, price stability at what level of prices? Consistently low food prices (in relation to other prices in the economy) undercut both food security and investment possibilities. While both theoretical and empirical measures of economic well-being (utility functions) depend on both the expected value of farm

38. Kevin Cleaver, 'Rural Development, Poverty Reduction and Environmental Growth in Sub-Saharan Africa', *Findings*, No. 92, The World Bank, Washington, DC, USA, August 1997.

39. R. L. Naylor and W. P. Falcon, 'Is the locus of poverty changing?', *Food Policy*, **20**(6), December 1995, p. 517.

40. Peter C. Timmer, 'The Macroeconomics of Food and Agriculture', in C. K. Eicher and J. M. Staatz (Eds), 1998, pp. 204–206. Other illustrations may be cited of the practice of invoking food price stabilization as a principal policy goal. For example, 'Policy and institutional support to stabilize the prices of primary products of special interest to poor developing countries, especially in sub-Saharan Africa, should be initiated and firmly implemented'. (From D. J. Shaw, 'Conference report: Development economics and policy: a conference to celebrate the 85th birthday of H. W. Singer', *Food Policy*, **21**(6), December 1996, p. 562.)

41. S. Jones, 'Food market reform: the changing role of the state', *Food Policy*, **20**(6), December 1995, p. 556.

income and its variance,[42] and farmers have been shown to be risk averse, most farmers would take a significantly higher average product price at the cost of a degree of greater variability in it. The second principal objection is that food prices must fluctuate seasonally to some extent in order to attract investment in marketing and storage facilities, and without those kinds of investments the price fluctuations tend to become even more extreme under external shocks.

A third objection is that undue stress on food price stability can lead to recommendations for policy configurations that have distortive economic effects and have proven to be unwise from a viewpoint of governance in many countries, as in the case of crop support prices and other kinds of price interventions.[43] Finally, national policy makers often use the goal of food price stability as an argument for preventing needed increases in real farmgate prices.

In the words of one of the World Bank's *World Development Reports*:

Prices of agricultural commodities are expected to vary more than the prices of industrial products for three reasons: agricultural markets are vulnerable to climatic changes; the short-run responsiveness of supply and demand to changes in prices is usually less in the case of agricultural products than it is in industrial markets; and the output of most crops is necessarily seasonal. . . . The variability of agricultural commodity prices explains why governments in developing countries often try price stabilization schemes to protect farmers from large price falls and consumers from large price increases. When greater price

stability leads to greater income stability, farmers benefit from reduced risks. These benefits, however, are extremely hard to estimate in practice. . . . it is possible to overstate the benefits of stabilization. Farmers, for example, can lose rather than gain if incomes fluctuate because of variations in crop yields and outputs – stable prices can then destabilize incomes. . . . Moreover, farmers, consumers, traders, and industrial users can reduce the risks they face by diversifying their activities, by using capital markets, by storing products, and by sharing risks through purchase and sales contracts.

Stabilization is a particularly complex task for any government to undertake, and its costs can be very high.[44]

Extreme price fluctuations can be moderated by dismantling trade barriers. Again, as the World Bank has commented:[45]

Greater priority should be given to moderating stabilization and producer [price] objectives, to bringing about stability and predictability of the public policy regime, and to encouraging private sector operations.

9.4 SUBSTANTIVE ORIENTATIONS OF AN AGRICULTURAL STRATEGY

Both the selection of the integrating dimensions of a strategy and the development of the policy prescriptions depend on the circumstances of each country and the historical moment in which the strategy is formulated. However, it may be asked if there are universal guidelines for the

---

42. Path breaking work in the empirical measurement of farmers' utility functions was reported in R. R. Officer and A. N. Halter, 'Utility analysis in a practical setting', *American Journal of Agricultural Economics*, **55**, 1968, pp. 257–277.

43. For example, 'a public marketing agency can implement a price stabilization policy that requires active government intervention to defend a floor price for farmers and a ceiling price for consumers – a particularly useful role for a government agency in reducing extreme and unexpected seasonal price swings'. (From C. Peter Timmer, Walter P. Falcon and Scott R. Pearson, *Food Policy Analysis*, published for the World Bank by The Johns Hopkins University Press, Baltimore, MD, USA, 1983, p. 209.)

44. The World Bank, 1986, pp. 87–88 [emphasis added].

45. *Op. cit.*, p. 90.

basic orientations of a strategy, if international experience to date has not provided general lessons that are applicable to the specific content of a strategy. The answer is that some very general guidelines have indeed emerged, and they are sketched out in the following pages. Nevertheless, it must be stressed above all that a successful strategy is a pragmatic document, with a problem-solving orientation, and that simple rules, such as 'free up markets', 'diversify production' or 'invest in productivity', are not specific enough to give adequate guidance to the development of an implementable strategy. They do not indicate the policy instruments with which those orientations would be implemented.

In all countries, *the aim of an agricultural strategy is to generate sustainable growth that is widely shared*, but producing a successful strategy requires creativity in finding concrete and detailed solutions to specific problems that are viable in their context. A broad vision or general prescriptions alone will not suffice, and specificity on both policies and sequencing is required. (To cite only one example, it was pointed out earlier in Chapter 4 that in Central Asian countries the policy of freeing up input markets before output markets led to disastrous results for farm incomes.)

The point about specificity of policy recommendations was made in a review of cases of successful agricultural adjustment and transformation in Africa and Latin America. It came to the following conclusion:

success resides to a large extent in the nitty-gritty of the transition process. More than a rather simplistic 'privatization' strategy or 'export-led' strategy, empirical evidence suggests that successful transition implies a lot of attention to details, such as choice of wording

in legislation, identification of market 'niches' for specific products to improve coordination between supply and demand in terms of quality, form and delivery time, and various kinds of market arrangements and contracting. ... The most successful countries were the ones that not only applied ... sound principles but found innovative price and institutional mechanisms to implement the principles.[46]

9.4.1 Lessons from Long-Term Models of Agricultural Development

As mentioned in Chapter 1, early theorizing about the role of agriculture in economic development was cast in very aggregate terms and tried to draw applicable inferences from the observation of international patterns. One of those strands of thought was the 'dualistic model' in which the industrial sector was expected to be the engine of growth and to pull resources out of agriculture for that purpose.[47] However, strategic recommendations based on this model are based on a misinterpretation, for policy purposes, of the fact that agriculture's share of GDP inexorably shrinks over time. Part of Engel's Law – the fact that income elasticities of demand for food products in the aggregate are less than unity – will alone explain this trend. Contrary to what the early growth theories predicted, neglecting agricultural development has depressed the overall growth rate of economies, and the other side of the coin is that there has been a clear positive association between agricultural growth and the growth of total GDP, as pointed out in Chapter 1.

In the forthright words of Hans Binswanger:

The time is long overdue for declaring bankrupt the notion that urban development can solve rural poverty.[48]

46. Frédéric Martin, Sylvain Larivière and John M. Staatz, 'Success Stories of Adjustment: Results and Lessons from Africa and Latin America', in G. H. Peters and Douglas D. Hedley (Eds), *Agricultural Competitiveness: Market Forces and Policy Choice*, Proceedings of the Twenty-Second International Conference of Agricultural Economists, Dartmouth Publishing Company Limited, Aldershot, UK, 1995, p. 223 [emphasis added].
47. A leading example of the dual-economy approach is found in John C. H. Fei and Gustav Ranis, *Development of the Labor Surplus Economy: Theory and Policy*, Irwin Publishing Company, Homewood, IL, USA, 1964.
48. H. Binswanger, 1998, p. 290.

John Mellor and Bruce Johnston[49] tried to derive policy-relevant implications from a more comprehensive theory of agricultural development. Agreeing with other early models that agriculture's primary role includes releasing productive factors for industry, they also asserted that a successful agricultural growth strategy must be aimed at smallholders, and that government must play the key role through investment in human capital, technical innovation and farmer organization. While strong on the role of government, they passed over central issues such as production incentives, property rights, and the need to correct market imperfections, and their prescriptions remained at a general level.[50]

Timmer attempted to bring aggregate theorizing closer to the realm of policy decisions, observing that neither the Mellor–Johnston approach nor the laissez-faire, urban-driven approach provides adequate guidance to agricultural policy formulation, and that what is needed are 'carefully designed interventions into the prices determined in markets, not by leaving markets alone or by striving to reach the objectives through direct activities of government'.[51] He acknowledged the 'heavy analytic costs' of this 'price and marketing policy' but did not mention the governance issues that are often associated with many kinds of price interventions, nor the other disadvantages of attempting to control prices directly which are discussed above in Chapter 4.

As was demonstrated in Chapter 4, the weight of experience has come down on the side of *minimizing government interventions in product markets*.[52] Instead, effort and priority are more productively channeled to *improving the functioning of factor markets* in agriculture, particularly in the areas of education training (human capital) and for the factor markets related to land, water, credit and technology – the latter being another dimension of human capital. For this reason, the four most detailed chapters of these Guidelines are devoted to policy issues in those areas. In regard to education, which has been stressed previously in Chapter 8, Gale Johnson has made the following points:

The capacities of governments to intervene are legion, except in one area. Governments have seldom, if ever, adopted policies or programs designed to ease the costs of agricultural adjustment as economic growth occurs. . . . The governments of developing countries should learn from the failures of agricultural policies of the industrial countries. A major role for government action is to assist farm and rural people to adjust to declining employment opportunities in agriculture. *This means shifting the emphasis from commodity markets, where governments like to intervene, to that of making factor markets work more efficiently*. The welfare of farm people depends far more on the functioning of labor markets than on the markets for commodities, yet governments neglect activities appropriate to such markets, such as information and education. . . . the critical neglect is that of rural education. . . . For those who worry that assisting the farm adjustment process will result in inundating the cities, there is a very simple answer. If the coun-

49. (a) B. F. Johnston and J. W. Mellor, 'The role of agriculture in economic development', *American Economic Review*, **51**, 1961, pp. 566–593; (b) J. W. Mellor and B. F. Johnston, 'The world food equation: interrelations among development, employment and food consumption', *Journal of Economic Literature*, **22**, 1984, pp. 531–574.

50. A useful summary of this conceptual model and others developed at an aggregate level is found in C. P. Timmer, 'The Agricultural Transformation', in C. K. Eicher and J. M. Staatz (Eds), 1998, pp. 113–135.

51. *Op. cit.*, p. 132. See also C. P. Timmer, *Getting Prices Right: The Scope and Limits of Agricultural Price Policy*, Cornell University Press, Ithaca, NY, 1986.

52. This statement is not meant to understate the value of carefully crafted policies in areas such as product standards, market information systems, certificates of grain deposit, and programs of support for exports such as subsidies for trial shipments abroad of new lines of exports.

tryside is made an attractive place to live and work, through investment in rural infrastructure (schools, roads, electricity, communications), the flow of people to the city will not be of concern.[53]

The 'historical model' of Vernon Ruttan and Yujiro Hayami is one of the richest of that genre in policy implications. Their model emphasizes the fundamental role of technical innovation in agricultural development, and it also states that the nature of innovation is heavily influenced by relative factor prices and also by real prices of agricultural outputs. In an article first published in 1980, Ruttan made the case that technical change in agriculture has become the single most important factor in determining agricultural growth possibilities:

Prior to this century, almost all increase in food production was obtained by bringing new land into production. . . . By the end of this century almost all of the increase in world food production must come from higher yields – from increased output per hectare.[54]

Together, Ruttan and Hayami hypothesized that whether innovation tends to be land-saving or land-intensive depends crucially on the relative prices of land and other inputs:

There is clear evidence that technology can be developed to facilitate the substitution of relatively abundant (hence cheap) factors for relatively scarce (hence expensive) factors in the economy. The constraints imposed on agricultural development by an inelastic supply of land have, in economies such as those of Japan and Taiwan, been offset by the development of high-yielding crop varieties designed to facilitate the substitution of fertilizer for land. The constraints imposed by an inelastic supply of

labor, in countries such as the United States, Canada and Australia, have been offset by technical advances leading to the substitution of animal and mechanical power for labor.[55]

Ruttan and Hayami state that innovation is also influenced by factors other than relative factor prices and point to the role of institutions. *Historically, labor-saving technical change (e.g. mechanization) has tended to be in the hands of the private sector, whereas producing land-saving innovations (e.g. higher-yielding varieties), tends to be in the domain of the public sector.* This allocation of effort corresponds to the fact that the benefits of mechanization can be internalized, that is, captured by the firm producing the machinery, while the same is usually not true for the benefits of biological innovations. New plant varieties can be widely reproduced, and new cultivation techniques can be copied. These authors contend that the direction of public sector research also is endogenous to a degree, since the allocation of public funding to research tends to respond to perceived constraints in the sector and since basic scientists and applied researchers often collaborate in solving real-world problems.

In their view, institutional innovation responds in part to the same influences. Creating public sector institutions for agricultural research:

represents an example of a public sector institutional innovation designed to realize for society the potential gains from advances in agricultural technology. . . . It is unlikely that institutional changes will prove viable unless the benefits to society exceed the cost. Changes in market prices and technological opportunities introduce disequilibrium in existing institutional arrangements by creating profitable new opportunities for the institutional innovations.[56]

53. D. Gale Johnson, 1995, p. 19 [emphasis added].
54. Vernon W. Ruttan, 'Models of Agricultural Development', reprinted in C. K. Eicher and J. M. Staatz (Eds), 1998, p. 155.
55. Vernon W. Ruttan and Yujiro Hayami, 'Induced Innovation Model of Agricultural Development', in C. K. Eicher and J. M. Staatz (Eds), 1998, pp. 163–164.
56. *Op. cit.*, p. 172.

Things like the wages for agricultural researchers matter greatly. Ruttan and Hayami point out (1998, p. 169) that 'response of public sector research and extension programs to farmers' demand is likely to be greatest when the agricultural research system is highly decentralized' and urge flexibility in the role of the public sector in research and in institutional innovation in general:

> Profitable opportunities, however, do not necessarily lead to immediate institutional innovations. Usually the gains and losses from technical and institutional change are not distributed neutrally. There are, typically, vested interests that stand to lose and that oppose change. There are limits on the extent to which group behavior can be mobilized to achieve common group interests. ... the process of transforming institutions in response to technical and economic opportunities generally involves time lags, social and political stress, and in some cases disruption of social and political order. Economic growth ultimately depends on the flexibility and efficiency of society in transforming itself in response to technical and economic opportunities (1998, p. 172).

One of the implications of the Ruttan–Hayami work is that innovation should respect a country's relative resource endowments if it is to contribute effectively to agricultural development. A concomitant conclusion from their research is that inappropriate pricing policies, which reduce agricultural profitability and therefore reduce land prices, may give the wrong signals for technical change, encouraging labor-saving innovations (mechanization) when labor is abundant and it is land that is scarce. This perverse result can also occur as the result of tariff and financial policies that effectively subsidize capital in the form of machinery.

Another policy-relevant implication is that, since institutional innovation is driven partly by incentives and profitability, low incentives in the sector can create a vicious circle: agricultural growth is slow as a result of low incentives, but the lack of incentives also means that the innovations needed to accelerate the sector's growth are unlikely to be forthcoming. El Salvador provides an illustrative case, where one of the obstacles to improving agricultural education has been the lack of attractive prospects in agriculture for youths who are selecting their future careers, owing to the sharp and sustained decline in real agricultural prices that had been brought about by an appreciating real exchange rate.[57]

To try to break this vicious circle, public policy for agriculture needs to place firm emphasis on creating and sustaining a capacity for agricultural research and extension, and orienting it in directions that are consistent with the country's comparative advantage. In the words of Ruttan and Hayami:

> If the induced development model is valid – if alternative paths of technical change and productivity growth are available to developing countries – the issue of how to organize and manage the development and allocation of scientific and technical resources becomes the single most critical factor in the agricultural development process. It is not sufficient simply to build new agricultural research stations. In many developing countries existing research facilities are not employed at full capacity because they are staffed with research workers with limited scientific and technical training; because of inadequate financial, logistical, and administrative support; because of isolation from the main currents of scientific and technical innovation; and because of failure to develop a research strategy that relates research activity to the potential economic value of the new knowledge it is designed to generate (1998, p. 173).

Ruttan and Hayami also underscore the role of complementary sectoral and macroeconomic policies, especially those that influence prices:

---

57. Roger D. Norton, 'Perspectivas y opciones para la Escuela Nacional de Agricultura "Roberto Quiñónez"', report prepared for the Ministry of Agriculture and Livestock, San Salvador, El Salvador, 1998.

One of the most important . . . areas for public investment is the modernization of the marketing system through the establishment of the information and communication linkages necessary for the efficient functioning of factor and product markets. . . . An important element in the development of a more efficient marketing system is the removal of rigidities and distortions resulting from government policy itself – including the maintenance of overvalued currencies, artificially low rates of interest, and unfavorable factor and product prices for agriculture. . . . (1998, pp. 173–174).

9.4.2 Strategic Orientations for Agriculture

International experiences, supplemented by historical analyses of agricultural growth, provide a rich trove of lessons about agricultural development. The lessons concern both *economic variables, or growth factors*, that promote agricultural performance, and *approaches to development, or the ways in which the growth factors can strengthened*. The first strategic aspect refers to *what* conditions or factors need to be strengthened for agricultural growth performance to improve, and the second strategic aspect is *how* those conditions are strengthened. The 'how', or the approach selected, determines much of the effectiveness of policies in influencing growth factors. These two aspects of an agricultural strategy are reviewed in order.

Agricultural production is only one part of a chain of related activities, stretching from input supply and technology development through production to post-harvest management, marketing and processing. Therefore agriculture cannot prosper without forging solid linkages to markets, both domestic and international, and markets in turn are ever more demanding in terms of product quality and conditions of delivery. The primordial role of markets is as true for smallholder production as for larger farms. The International Fund for Agricultural Development (IFAD) has pointed out that most production for specialized organic markets comes from small farms, in part because of their higher endowment of family labor per unit of land and in part because they are less likely to have been using chemical-intensive production methods to start with.[58]

Markets go together with prices. Policy can play an indirect but key role in facilitating market access, and markets play a more direct role in influencing prices received by farmers. In order for farmers to respond to market opportunities and price incentives, they need to increase their capital of various types: human capital, as education and as technological knowledge, social capital, that is, effective organization of farmers and communities, institutional capital, and physical capital. Thus, in the broadest sense, *the most important factors for generating agricultural growth, without which growth prospects are very dim, are adequate market and pricing conditions and sufficient productive capital.* In the latter, *human capital* is the most essential for improving development prospects.

In regard to pricing conditions, for too long agricultural production incentives have been treated as secondary considerations or, worse yet, as something to be suppressed for the sake of consumer welfare. As the discussion in Chapter 4 and elsewhere in this book has shown, adequate real prices for farmers are an essential requirement for agricultural growth and for rural poverty alleviation. This conclusion is widely recognized but often forgotten, for example, for the sake of importing cheap food or for making imports in general cheaper by manipulating the exchange rate mechanism. Higher real farmgate prices generate gains in terms of poverty alleviation that outweigh the burden of high food prices to poor consumers, and in any case targeted subsidies can help reduce the impact of the latter.

Equally, policy has tended to ignore the importance of *product quality* in markets. It plays a central role not only in organic markets but

---

58. Octavio Damiani, 'Small Farmers and Organic Agriculture: Lessons Learned from Latin America', Office of Evaluation and Studies, International Fund for Agricultural Development, Rome, 2002.

increasingly in all markets. As discussed earlier in Chapter 4, there are three major dimensions to product quality: phytosanitary conditions, or freedom from plant diseases and pests, food safety conditions, which refer mainly to freedom from chemical residues, and consumer preferences as reflected in product taste, size, shape, color, uniformity and suitability for preparation in kitchens. Few developing countries have adequate institutional and policy frameworks to facilitate achieving adequate levels of product quality.

The most essential form of capital, *human capital*, extends *from basic literacy to acquiring better production technologies and marketing skills*. It extends to social capital, which at the local level is individuals' ability to work together, in various kinds of associations, to overcome barriers to access to inputs and output markets. *Institutional capital* refers to institutional capacities for supplying factors such as production finance and security of land tenure and the capacity for continuously developing agricultural technology. *Physical capital* also is vital, especially *access to agricultural land and endowments of infrastructure in key areas*.

These growth factors of market conditions, human capital, social capital, institutional capital and physical capital may be viewed from a policy perspective in the following ways, summarizing discussions of earlier chapters in a kind of checklist of principal issues for an agricultural strategy.

9.4.2.1 Market Development and Pricing Policies for Agricultural Growth

- *Market development policies* include a country's international trade negotiations for agriculture (bilateral as well as multilateral), policies to ensure compliance with phytosanitary norms and food safety standards, product grading (especially for grains), export promotion measures, market information and testing efforts, finance for marketing and storage, training of extension agents in product quality issues and organic production, and related efforts. An emphasis on production alone no longer is sufficient – it is now a question of production *for whom*. An earlier conception of marketing policies was that they consisted mostly of constructing wholesale markets in rural areas, but physical infrastructure is not generally the limiting factor – with the exception of rural roads. The private sector increasingly shoulders the burden of market development, but sensible policies can also assist in this area, and they are necessary to promote the food quality certifications needed for export. A firm and co-ordinated government commitment to agroexport policies can help overcome many of the barriers that block the entry of individual producers into international markets.

- Pricing policy should strive to *avoid declines in real agricultural prices*, and to reverse, at least partially, the large declines that may have occurred. However, attempts at direct control of prices are counterproductive and the main instruments here are macroeconomic and trade policies. *Appropriate macroeconomic policies* are an essential condition for successful agricultural development.[59]

- Pricing policy also should aim for *relatively uniform rates of effective protection* among products within the sector, and between sectors. This implies removing distortions induced both by non-competitive market behavior and by undesirable policy interventions. At the same time, the growth-creating effects of an *open trade regime* should not be underestimated.

- Implementing an open trade regime does not mean that obvious distortions in international markets have to be transmitted to the domestic economy. As Peter Timmer has pointed out, developing countries should not necessarily accept distorted international prices as the basis for relative prices in their economies. Consideration should be given to *systematic policies for*

---

59. 'The foreign exchange, trade, and taxation regime should not discriminate against agriculture but should tax it lightly, preferably using the same progressivity and instruments as it applies to the urban economy' (H. Binswanger, 1998, p. 298).

> *Sometimes, there is a misconception that opening up the trade regime means reducing the effective protection accorded to agriculture. This is not necessarily the case, especially if greater opening is accompanied by moves to deregulate the agricultural sector, eliminate unjustified tariff exemptions, and reduce industrial tariffs. This was the kind of opening of the agricultural economy that was promulgated in Honduran agriculture in the early 1990s (along with the introduction of price bands to moderate international price fluctuations). The lemma of those reforms, first articulated by Julio Paz, became 'moving from a controlled but unprotected agriculture to a decontrolled but more protected sector'. Before the reforms, the sector's effective protection rate had averaged about zero and was negative for some products, in contrast to an industrial protection rate averaging 100%. After the reforms, including a real exchange rate devaluation, effective protection for agriculture rose to modestly positive levels, and the sector's production responded vigorously.*

counteracting the effects of subsidies in exporting nations, such as countervailing tariffs that are scaled to match the effects of those subsidies (see Chapter 4), consistent with a country's comparative advantage. Alternatively, direct support to producers, in an amount that compensates for the price effects of the international subsidies, can be considered.

9.4.2.2 Human Capital Policies for Agricultural Growth

- It is almost impossible to exaggerate the importance of **rural education**, and creative efforts are needed not only to build schools but to train teachers and give students an incentive to attend. It may also be necessary, in some cases, to change the way in which such schools are managed.
- Extension and training programs in rural areas should incorporate elements of community and

producer organization, especially as regards women. The ability to work together co-operatively is a key to success in many development efforts, especially those that involve penetrating new markets. This is the factor of **rural social capital**.

- The capacity for **agricultural research and extension** needs to be strengthened very considerably in most developing countries, although the public sector does not always have to be the executing agent in these fields. In practical terms, this can require paying emoluments to researchers that are well above the public sector's salary scale, and recruiting researchers with stronger qualifications. Equally, it means more emphasis on decentralized and participatory approaches to both research and extension, as discussed in Chapter 8.
- Human capital also is manifest in institutions, and **strengthening institutions is vital to agricultural progress**. In the most general form, it requires inculcating respect for the rule of law, and for property rights in all forms. **Sound governance is a key to all aspects of agricultural development**. In a more specific form, institutional strengthening requires that they be made more **accountable and efficient**, and usually in developing agriculture this requires a greater degree of **decentralization of institutions** than now exists.

9.4.2.3 Physical Capital Policies for Agricultural Growth

- It is frequently necessary to make institutional and policy reforms to increase the access of poor families to land and to make land more productive by clarifying and buttressing **property rights**, including lease and rental rights. Chapter 5 provides illustrations of policy reforms that can increase **access to land** without disrupting the security of property rights.
- *Investments in rural physical infrastructure*, especially those related to *irrigation, transportation, electrification and communications*, form an essential part of the sector's base of capital, without which agriculture's growth

prospects will remain stunted.[60] All farmers are entrepreneurs, and an entrepreneur cannot be successful without these basic facilities. Although the importance of rural roads is widely recognized, sometimes insufficient priority is given to electricity supplies and telecommunications in rural areas.

Agriculture is increasingly becoming a sophisticated sector, even for smallholders. Product quality requirements are faced by ever larger numbers of producers, and demands are increasing for access to information on both technologies and markets, and awareness of how to respond to changing information. Agriculture is no longer viewed just as a matter of production, but as a holistic process ranging from technology development to solidifying linkages with markets. These new emphases are increasingly accompanied by requirements for decentralization of government services to make them more effective for producers, as discussed below. *The farmer is at the center of successful agricultural development strategies*:

. . . in the present era of globalization, it is clearer than ever that the producer, and not the government, is the agent of sustainable development. From this perspective, the most useful contributions policy can make in directing agriculture toward sustainable growth are:

- To improve farmers' capacities to understand and analyze their options, and to be able to deal with change, through education and specialized training.
- To improve their access to markets and to relevant information, through key infrastructure and information services.
- To improve the operation of institutions of major importance to the lives of rural

families (financial institutions, land registry offices, water management organizations, marketing cooperatives and others) through decentralization, institutional reform and staff training.[61]

At the same time, the importance of supportive macroeconomic policies is undeniable. It has been underscored in a recent report of the World Bank, referring in this case to the experience of Latin America and the Caribbean in the 1990s:

The liberalization of markets worsened things for smallholders by reducing their levels of protection at a time when international [agricultural] prices were, except in 1996, in their lowest levels in history (for example, for corn, wheat, coffee) and when many producers of import substituting products were not (and still are not) competitive in their national urban markets.[62]

9.4.3 Approaches to Agricultural Development Policies

The *ways* in which agricultural policies are conceived, designed and implemented are major determinants of their results. For example, centralized, command-economy approaches in relation to the growth factors mentioned above, from agricultural extension systems to land redistribution, have not proven effective. The lessons of experience point to five key approaches that help ensure the effectiveness of policy reforms:

- *Reducing distortions in product and factor markets*. This can mean appropriate regulation, as, for example, in the cases of privatization of agricultural marketing and processing facilities or the creation of water rights markets, and it

---

60. 'Government investments in education, rural electrification, agricultural extension, transportation, the communication infrastructure and other areas where externalities occur are required because the commercial market is not capable of delivering critical services' (T. L. Vollrath, 1994, p. 475).
61. Roger D. Norton, 'Critical Issues Facing Agriculture on the Eve of the 21st Century', in IICA, *Towards the Formation of an Inter-American Strategy for Agriculture*, San José, Costa Rica, 2000, p. 312.
62. The World Bank, March 2001, p. 16 [translated by the author].

also can mean removing unproductive regulations and controls.[63]

- **Reinforcing the legal strength of contractual relationships.**[64] This requires not only appropriate legal codes but also strengthened judicial organs, including in some cases the establishment of special rural tribunals. In the final analysis, fostering *relationships of mutual confidence* is crucial for penetration of new markets and obtaining production finance.

- According *special emphasis to women, small farmers and the rural poor* in the targeting of policies and programs. Such an emphasis is warranted not only from a viewpoint of equity but also to unleash the potential in the proven productive efficiency of small-holders.[65] As the material presented in the previous chapters illustrates, a special focus on gender in agricultural development is justified in almost all policy areas.

- **Decentralization and participation**: an approach of devolution of public services, with privatization where appropriate, and encouraging farmer and community participation in the design and implementation of programs and policies. Greater local participation has proven to be very effective in many areas, including water management, market-assisted land reform, agricultural research, agricultural extension and rural finance.

- Emphasis on creating *viable institutions*. This is a corollary of the emphasis on decentralization and participation, but it also implies attention to long-run financial viability and developing operating modalities that are sustainable. This last consideration is particularly relevant to

rural financial institutions. *Institution-building* in the sense of nurturing entities that serve the rural sector is a vital part of an agricultural development strategy, especially in regard to *water management entities, rural financial institutions and functioning land registries*, but the goal in all areas always should be the development of institutions that are self-sustaining and viable in the long run.

To enhance the concreteness of policy prescriptions in a strategy, they can be accompanied by recommendations for *legislative reform* where appropriate. Without *specificity in the proposals*, there is the danger that the reforms ultimately approved may differ in significant ways from what the strategy proposes. Equally, lack of specificity creates a loss of momentum in the reform process and thus a greater risk that the proposals may wither away instead of being approved and implemented.

It is evident that an agricultural strategy cannot be limited to issues that fall within a narrow conception of agricultural policy. As Robert Thompson has said:

> Macroeconomic policy, trade policy, factor market policy, and public investment policy (especially as it relates to education, research, and infrastructure) can all have a greater impact on agricultural development than narrow sectoral policy.[66]

Conceiving and carrying out policy reform is a demanding task in all circumstances but, in the words of Vernon Ruttan, 'it is imperative that the

63. 'All too often, government policies are themselves distortionary. The pricing of public goods, such as water, provides an example. Water may be provided at little or no cost to farmers and then be wasted. . . .' (T. L. Vollrath, 'The role of agriculture and its prerequisites in economic development', *Food Policy*, **19**(5), October 1994, p. 476).

64. 'A viable system of property rights and an effective legal system and judiciary to secure these rights must be established' (*ibid.*).

65. This is a theme stressed by many authors, perhaps beginning with the work of Mellor and Johnston mentioned above. Binswanger (1998, p. 298) has encapsulated the argument as follows: 'A strategy that promotes an open economy, employment intensiveness, and a small farmer orientation is both economically efficient and most likely to reduce poverty, both rural and urban'.

66. R. L. Thompson, 'Public policy for sustainable agriculture and rural equity', *Food Policy*, **23**(1), February 1998, p. 2.

poor countries design and implement more effective agricultural development strategies than in the past'.[67]

9.5 RURAL DEVELOPMENT AND POVERTY ALLEVIATION

The extent of the rural poverty problem in the developing world is well known, as illustrated by the comment of Naylor and Falcon on p. 360 that the rural poor will continue to outnumber the urban poor well into this century. In Latin America and the Caribbean, to mention one example, rural poverty worsened between 1986 and 1996, both in incidence and in absolute numbers. Plus, although the proportion of rural poor in the total population is expected to decline in that region over the next twenty years, the absolute numbers of rural poor will hardly change.[68] Similar situations prevail in Africa and large parts of Asia.

At the same time, rural development programs as originally conceived have been in decline in some international agencies:

There is . . . recognition of the decline of importance of rural development in national agendas and the decreased lending portfolio of the World Bank for rural development activities, despite the strategic importance of rural development and its potential to significantly reduce poverty.[69]

To some extent, broad rural development programs can be replaced by a series of specific initiatives to promote agriculture, such as market-assisted land reform or land funds (Chapter 5),

better management of irrigation systems (Chapter 6), and community participation in agricultural research and extension (Chapter 8). However, it is always valuable to coordinate policies and programs under a rural, spatial focus. Rural development can become another integrating dimension of an agricultural strategy and at the same time carry its policy prescriptions beyond the sector. It goes without saying that the linkages between agricultural and non-agricultural activities in rural areas are strong, and the latter constitute significant sources of employment and income for rural families:

rural development issues must be integrated into agricultural policy. . . . Only job creation can solve the problem of rural poverty. In many countries rural development policy is limited to agricultural policy; however, no country has ever solved the problem of rural poverty exclusively on the farm.[70]

In addition to the direct benefits of creating non-agricultural jobs in rural areas, taking into account off-farm employment opportunities for smallholders – the measure of the opportunity cost of their time – can be crucial for the design of agricultural production technologies that are acceptable to them. In the short run, this opportunity cost can be different for different members of a household. An integrated view of how a rural household functions, including the traditional divisions of labor by gender, is necessary in order to formulate realistic approaches to rural development.[71]

The Estonian strategy mentioned in these pages is effectively a rural development strategy.

---

67. V. W. Ruttan, 1998, p. 155.
68. The World Bank, *Plan de Acción para el Desarrollo Rural en América Latina y el Caribe*, Resumen del Informe, draft document presented at The City of Knowledge, Panama, March 2001, p. 13.
69. The World Bank, March 2001, p. 3.
70. R. L. Thompson, 1998, p. 4. On this point, see also T. Reardon, K. Stamoulis, M. E. Cruz, Al Balisca and J. Berdegué, 'Rural Non-Farm Income in Developing Countries: Importance and Policy Implications', special chapter in *State of Food and Agriculture 1998*, Food and Agriculture Organization of the United Nations, Rome, 1998.
71. The distributional effects of non-farm rural employment can differ widely among countries, in good part as a function of the availability of farmland. For example, 'In Egypt the poor (those in the lowest quintile)

One of its principal chapters deals with rural social issues and economic policies for the rural poor. A major concern was the very low standard of living of former collective farm members, mainly the elderly, whose only remaining productive assets are tiny household plots. Accordingly, the strategy recommended titling those plots, with no charge to the families concerned, and widening the net of retirement benefits to cover former collective farm members. That chapter also dealt with issues of child care for working parents in rural areas and recommended, among many other measures, the establishment of special industrial development zones throughout the rural sections of the country, a recommendation which has since been put into effect. These brief examples illustrate how non-agricultural issues can be relevant to an agricultural strategy.

It is beyond the scope of this present study to try to synthesize the literature on rural development, or the richness of experience in this field in agencies such as the FAO. Rather, observations are offered about illustrative rural development experiences and conceptual issues, and the link between agricultural development policies and rural development is explored. The orientations underlying a renewed approach to rural development projects are summarized, and a conceptual framework for guiding resource allocation in rural development is suggested.

9.5.1 Rural Development Projects

In past decades, funding for rural developments tended to be channeled into integrated rural development projects that provided infrastructure for basic community services as well as productive investments, in defined geographical areas. After a number of years of pursuing this approach, the results were generally regarded as unsatisfactory. One issue concerns the economic rate of return. It usually was quite low because the investment packages contained a social infrastructure component. Another concern was the lack of community participation in the design of the investment packages, and a related problem was the frequently weak co-ordination among the agencies of the central government that were charged with implementing the different types of investment in a given community.

Binswanger (1998) summarized the disappointing record in the past, and the main reasons for it, in the following way:

Many integrated rural development (IRD) projects in the seventies and eighties were unsuccessful, because they encountered the following problems (World Bank, 1987):[72]

Adverse policy environment. It quickly became apparent that many IRD projects, when pursued in an adverse policy environment for agriculture as a whole or for smallholders, amounted to pushing on a string, and could not succeed....

Lack of government commitment. Many governments did not provide the counterpart funding required for implementation of the programs....

Lack of appropriate technology. This proved to be an important problem in rainfed areas, especially in Africa, where there was little history of past commitment to agricultural research, or where colonial research efforts had decayed. Some integrated rural development projects included research components, but most failed to develop improved technologies. Also, many of these research components undermined the national agricultural research systems by depriving them of talented researchers.

receive almost 60 percent of their per capita income from nonfarm income. In Jordan the poor receive less than 20 percent of their income from nonfarm income. So nonfarm income decreases inequality in Egypt and increases it in Jordan'. (From Richard H. Adams, Jr, 'Nonfarm Income, Inequality, and Poverty in Rural Egypt and Jordan', Working Paper No. 2572 (abstract), The World Bank, Washington, DC, USA, March 28, 2001, pp. 1–2.)

72. The World Bank, *World Bank Experience with Rural Development: 1965–1987*, Operations Evaluation Study 6883, The World Bank, Washington, DC, USA, 1987.

Neglect of institutional development. Many IRD projects set up project coordination units, sometimes staffed by expatriates. However, this approach postponed the development of a local and district-level institutional capacity to plan, execute, and monitor rural development programs.

Lack of beneficiary participation. The programs were often designed in a top-down approach within which beneficiaries were not given authority for decision making or program execution. . . .

The complexity or coordination problem. It is ironic that complexity should have become the Achilles heel of rural development. After all, building rural roads, small-scale infrastructure, and providing agricultural extension are dramatically simpler tasks than the construction of large-scale irrigation infrastructure or ports. . . . The coordination problem emerged as a consequence of delegating subprogram execution to government bureaucracies or parastatals, which were typically highly centralized and had their own objectives. Many of them were out of touch with beneficiaries, who could much more easily have coordinated the relatively simple tasks at the local level. . . . Indeed, one might classify integrated rural development as the last bastion of central planning, swept away by reality like all other central planning schemes.[73]

9.5.2 Decentralizing Rural Development

Because of these concerns, among international agencies the emphasis in rural development has shifted to decentralized and participatory approaches, including **'demand-driven' investment programs** in rural areas. Increasingly, communities are asked to take the lead in defining or selecting the programs they will participate in, and the role of local governments is being put in higher relief. The distinguishing characteristic of demand-driven investments is that communities propose or choose them. Chile, for example, created an irrigation development fund that is devoted to financing projects proposed by communities or groups of farmers (see Chapter 6). In Paraiba, in Brazil's northeast, greater community participation and better co-ordination of central agencies' efforts have created successful rural development projects with World Bank funding.[74] In Nicaragua and Honduras, the Inter-American Development Bank has financed rural development projects which are entirely composed of small projects that are proposed by communities and screened at the level of county governments (rural municipalities). While in most cases it is impossible to estimate a rate of return on such proposed projects, for use as a screening criterion, the presumption is that communities will have a relatively accurate perception of the bottlenecks to their development, and hence of the investments needed for releasing them.

However, complementary kinds of criteria or filters also can be used for screening small projects, including cost guidelines (a maximum cost per kilometer of feeder road built or rehabilitated, for example), a restricted definition of the categories of investment that will be funded under the program, a requirement that a community contribute at least a specified minimum percentage of the cost of each project selected,[75] and a requirement of follow-up over time with the community in the case of training projects. In Paraiba, the formula included giving communities the right to propose projects and the right to elect half of the members of a committee that would make the final selection of one group of projects, leaving

---

73. H. Binswanger, 1998, pp. 292–293.

74. A summary of this experience and the reasons for its success is found in Maximiliano Cox, *Mejores prácticas en políticas y programas de desarrollo rural: implicancias para el caso chileno*, CEPAL, Naciones Unidas, Serie Desarrollo Productivo No. 86, Santiago, Chile, March 2001.

75. 'Successful community-driven projects have been those in which funds *were locally generated* to cover the ongoing operational costs of the project'. (From Operations Evaluation Department, 'Lessons on Community-Driven Development', Lessons and Practices No. 12, The World Bank, Washington, DC, USA, August 1, 2000, p. 2.)

the final choices on another group to municipal (county) authorities, and allowing State (provincial) leaders to select yet another group of projects.[76]

This approach represents a particular kind of *fiscal decentralization*. As well as empowering local communities (villages), it also strengthens the role of municipalities and, in some cases, NGOs (although care should be taken to ensure that NGOs do not come to dominate community decision making). Apart from supporting rural development more effectively, fiscal decentralization has powerful appeal on other grounds, and at the same time it presents complex issues to be dealt with if it is to be successful. A number of countries have legislated a requirement that a given share of central government revenues be transferred to local governments, but concerns have arisen about the capacity of local governments to manage the funds well and the effects of the transfers on the budgetary balance of the central government. Such transfers can also weaken the will of local governments to collect the taxes that fall under their jurisdiction. Nevertheless, the consensus appears to lie on the side of greater decentralization if it is properly handled. A primary requisite is a greater effort in training local officials.[77]

Colombia has successfully placed agricultural research and training expenditures in the hands of producers through the mechanism of 'parafiscal' funds for selected crops. Producers pay fees to support those funds and in turn decide how they are to be used. This model could be extended to wider classes of development expenditures at the local level, with participation of producers as well as local governments in the decision-making process. It should be noted that the parafiscal model has worked well in Colombia when the crops are relatively homogeneous and the producers well organized, as in the cases of coffee, oil palm and rice, but it has not worked so well for heterogeneous crops or producers, as in the case of the parafiscal fund for fruit and vegetables and the fund for artesanal raw sugar (*panela*).

The main points here are that rural development projects need to be conceived and carried out in a decentralized fashion,[78] but that there is no single approach that is universally applicable. The most appropriate approach will depend on the political and institutional circumstances of the country concerned. Equally, care must be taken to avoid domination of decision making by local leaders or elites.[79] The basic rationale for decentralization has been stated by Lawrence Smith in terms of accountability:

Decentralization of the public administration is expected to improve the system of incentives, which confronts suppliers of goods and services where provision is not dictated entirely by market forces. The closer the administration of the service is to the clients, then the greater the likelihood that decisions about which services should be provided, how much, where, and to whom, will be more responsive to the demands

76. M. Cox, 2001, p. 40.
77. Two good surveys of the issues involved in decentralization, one addressing mainly economic questions and the other mainly political and institutional issues, were published in Michael Bruno and Boris Pleskovic (Eds), *Annual World Bank Conference on Development Economics 1995*, The World Bank, Washington, DC, USA, 1996. The first one is by Vito Tanzi, 'Fiscal Federalism and Decentralization: A Review of Some Efficiency and Macroeconomic Aspects' (pp. 295–316), and the second is by Rudolf Hommes, 'Conflicts and Dilemmas of Decentralization' (pp. 331–349). Lawrence D. Smith provides a very useful discussion of many issues in decentralization, including definitions of different degrees and types of it, in his *Reform and Decentralization of Agricultural Services: A Policy Framework*, FAO Agricultural Policy and Economic Development Series No. 7, Policy Assistance Division, Food and Agriculture Organization of the United Nations, Rome, 2001.
78. Binswanger (1998, p. 294) mentions recent rural development projects in Mexico, Colombia and Brazil that have devolved decision-making authority to municipalities.
79. M. Cox, 2001, pp. 37–38.

of users. It is expected that the quality of service will improve, and that the efficiency of the provider will increase as a result of closer accountability to clients.[80]

Experiences with demand-driven rural investments illustrate some of the questions that need to be considered in designing decentralized rural development projects. One issue that frequently emerges is whether to use public monies (from an international loan) to make grants for on-farm investments. A decision usually is made to limit the scope of the investments to community-wide projects (e.g. rural roads, training and programs, rural market facilities, and the occasional small-scale irrigation project that has a sufficient number of beneficiaries), and to projects oriented toward co-operatives (e.g. supplying the initial capital for a women's marketing co-operative, or a women's co-operative for raising and selling iguanas). By defining the scope of the program in this way, both on-farm investments and some basic investments in infrastructure (electrification and installation of communications networks) are ruled out. This is not to deny the value of the kinds of investments that are funded by the programs, but rather to point out that an investment package developed through local community decision processes may have inherent limits in its coverage and therefore may need to be complemented by other kinds of actions.

While greater decentralization of decision-making is important, rural development needs to be supported at all levels of policy making, including by appropriate sectoral policies and macroeconomic policies that create growth incentives. A comprehensive approach is required. Appropriate sector policies are also needed to make decentralized approaches effective, including **the legal underpinnings for community management**:

Sound sector policies enable effective community-driven development. . . . Supportive sector policies need to include financial policies (including on community contributions), sector norms, technology options/standards, and laws supportive of community management and contracting of goods and services by communities themselves.[81]

Adopting a holistic approach in turn raises a fundamental issue of another kind:

A holistic approach means confronting with more resolve a theme historically viewed in many international agencies as taboo in the development process: politics and its relation to policy. It has become more apparent that *success is linked to the quality of governance and the political process*. Good political management of technical aspects and good technical management of political ones often form the basis for success and project effectiveness. Demand-driven approaches also are a response to create greater stakeholder involvement and hence greater social accountability in project implementation.[82]

In some circumstances, demand-driven investment programs have acquired partisan political overtones, as a presidential administration tries to direct the investments toward communities where its supporters are most numerous. Safeguards are required against such tendencies in order to protect the integrity of the programs and raise their effectiveness.

As salutary as decentralization of decision making can be, it also complicates the process of co-ordination among institutions and can make it more difficult to reach a consensus on the priorities for rural development expenditures. In the words of Lawrence Smith:

---

80. L. D. Smith, 2001, p. 44.

81. Phillipe Dongier, 'Community Driven Development Principles', Draft, The World Bank, Washington, DC, USA, January 7, 2000, p. 3 [emphasis in original].

82. World Bank Action Strategy, Consultation on Rural Development for Latin America and the Caribbean, *Summary Report of Conclusions and Proceedings*, CIDER (IICA), The City of Knowledge, Panama, April 2001, pp. 10–11.

Central government, local government, and CSOs [civil society organizations] may agree on goals, but disagree on priorities and strategies. This should not be an issue when provision [of services] is deconcentrated or delegated because, in theory, the central government is firmly in control. The problem emerges with devolution and partnerships. How should central government administrations deal with a variety of different priority rankings by local governments and CSOs claiming public support for their initiatives? What happens if local priority rankings differ from central government priorities established under a development plan? . . .

Central governments can handle these problems in several ways by:

(1) Introducing devolution gradually by giving membership in the elected local government assemblies to officers of deconcentrated units of the public administration, who can influence decisions.
(2) Transferring the appropriate priority functions outside the public administration to CSOs.
(3) Retaining the *management* of priority programs under central control while deconcentrating or delegating the *production* or delivery functions.
(4) Introducing conditionalities in the intergovernmental fiscal transfer. . . .[83]

In effect, the solution devised in the Inter-American Bank's rural development programs for Nicaragua and Honduras was to have the central government narrow the scope of decision-making for local governments, by defining the fields eligible for funding, and they also imposed technical criteria on the eligibility. Within that set of restrictions, priorities were defined locally.

9.5.3 Economic Transfers to Support Agricultural and Rural Development

The reach of rural investments can be broadened by implementation of complementary programs of *fiscal transfers to poor rural households*. Mexico has been more decisive than most developing countries in putting into effect programs of fiscal transfers to farmers, first through its own adaptation of the European Union's McSharry Plan (the PROCAMPO program, which mandated payments of $US100 per hectare to specified categories of active farms), and then through the above-mentioned program of direct transfers to rural households on the basis of children's attendance at school (the PROGRESA program). Estonia also adopted a variant of the McSharry Plan to ease the transition of its agricultural sector to a market economy.

Transfers to poor rural households are a category of subsidy that is justified on grounds of poverty alleviation (Chapter 2) and, properly conceived, it can help put farmers on the road to self-sustaining development. Examples might include payments to defray the costs of land titling (made upon completion of the titling process) and vouchers for poor farmers which can be exchanged for the purchase of agricultural inputs or extension services or for participation in specialized training programs.[84]

Transfers of this type may be considered to be instruments of agricultural policy, rather than rural development programs *per se*, but in any event they are complementary to other kinds of

---

83. L. D. Smith, 2001, pp. 21–22 [emphasis in original].
84. The concept of direct transfers targeted on smallholders for the purpose of increasing their productive capacity is a relatively new one. For example, a 1992 report of the Inter-American Development Bank listed four kinds of subsidy to rural families but did not include productive transfers. The types of subsidies mentioned were rural employment programs, school feeding programs, cash subsidies for subsistence to vulnerable groups (the aged, young children, pregnant and lactating mothers), and the distribution of food to nutritionally deficient groups in the population. (Gladys Aristizábal, Jorge Echenique, Ruy de Villalabos y Wolfram Fischer, 'Combatiendo la Pobreza Rural en América Latina y el Caribe: Una Nueva Estrategia de Desarrollo Rural', Banco Interamericano de Desarrollo, Washington, DC, USA, December 1992, p. 58.)

investments in rural areas. Worldwide there is a long history of transfers to poor farmers by providing some classes of agricultural inputs in physical form, particularly seeds, seedlings for trees, tools and agrochemicals. However, the next step, of empowering poor farm households to make their own choices of inputs, and to obtain them from established market channels, generally has not been taken. Direct transfers, especially for smallholders, take on added importance in light of the worldwide trend toward closure of State-owned development banks, and the consequent contraction of agricultural credit. To the extent that such credit programs were, in effect, disguised subsidies because of low loan recovery rates, then it would make sounder policy to replace them with direct transfer payments.

Programs of direct transfers facilitate the capitalization of smallholder farms and would represent neutral policy interventions with respect to cropping patterns. Therefore, in the frequent cases in which existing policy incentives are biased in favor of import substitute crops and against export crops, the implementation of direct transfers would reduce this bias, in effect compensating for the lack of programs of export incentives. In addition, when fiscal incentives are biased in favor of larger farms – another frequent occurrence – the use of direct transfers, with an upper limit on hectareage eligible in any farm, would tend to ameliorate that bias.

If both an area-based land tax, as discussed earlier in Chapter 5, and a program of direct transfers were implemented, the net effect would be progressive taxation, since the tax would exempt the first few hectares and the direct transfers would be subject to an upper limit on the cultivated area per farm that was eligible for the benefit. The effects are shown schematically in Figure 9.1, where on the horizontal axis the point 'a' represents the area exempt from taxation, point 'b' represents the upper limit on the area eligible for the transfers, and farmers with land areas greater than 'c' would be net tax payers. Farms smaller than 'c' would be net gainers.

The benefits of an area-based tax have been discussed above in Chapter 5. These include administrative simplicity, avoiding the problem

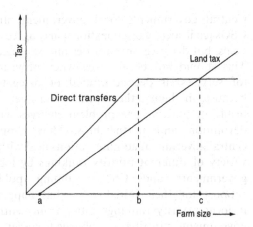

Figure 9.1 Schematic representation of the effects of direct transfers and land tax (see text for further details).

of cadastral valuations that are continuously out of date, and avoiding disincentives to on-farm investments. They could be deductible from income tax obligations, and for most farmers would effectively substitute for income taxes, since the latter are not paid very much in rural areas of developing countries.

The combination of the area-based tax and direct transfers would enable the sector to fulfill its fiscal revenue obligations at the same time that production incentives and rural income support were provided to small- and medium-scale farms. As noted, unlike most policy support measures in developing agriculture, direct transfers would be neutral with respect to choice of cropping pattern.

The administration of direct transfers raises a number of practical issues, and perhaps the principal one is the need to *decentralize the implementation of agricultural policies.* It would be impossible to administer a program of transfers to tens of thousands or more smallholders by operating directly out of a centralized ministry. For purposes of this kind of program, *greater autonomy should be assigned to regional offices of ministries of agriculture* as well as involving organs of local government.

The role of regional or local offices of a ministry of agriculture can be visualized as that of an *intermediary,* or a broker. A local representative of the ministry has the following basic tasks: (i) to get to know his/her clients, i.e. the farm families in his or her area, and their needs and potentials; (ii) to transmit to the center the main agricultural problems and issues in the area; (iii) to be familiar with the range of policy programs that may be implemented locally, and the criteria for qualification of the participants and the mechanisms for their execution; (iv) to create awareness in the rural population of the programs in which they may participate and, in this way, facilitate a matching of clients and programs; (v) monitor the implementation of programs and report to the center any design flaws that need correcting, as well as possible variants of existing programs or new programs that might be developed.

For local agents to play this role, the central institutions have to define carefully and clearly the options that are available, and to indicate in detail how they are to be implemented and monitored. Basically, *the role of the center is to define clearly the policy options* and issue a set of guidelines, one for each policy program or module and each kind of investment that may be implemented if the farmers demonstrate interest and meet the qualifying criteria. They can be as diverse as financial and technical assistance in establishing small-scale irrigation programs, assistance in community programs of watershed management, specialized training courses (for raising small livestock, cultivating non-traditional crops, small-scale agricultural processing, marketing, cost accounting, the rudiments of veterinary care, and the like), opportunities for titling land, vouchers to cover titling costs, opportunities for purchase of land with the financing of a land bank, vouchers for input purchase, vouchers for purchasing privatized extension services, transfers for school attendance, participation in programs of local infrastructure investment, etc. Emphasis should be placed on the role of women among the recipients of such transfers, and also opportunities should be sought to link financial transfers to environmentally sound agricultural practices.

Each program can have a descriptive brochure and an operational manual, developed by the center in co-ordination with local agents. The latter can assist in pilot tests of the usefulness of the materials. Looking at the implementation issue in this way, it is clear that *most of the operations of a ministry of agriculture need to be decentralized to local offices*, and the co-ordination between the center and the local agents needs to be strengthened. Yet, this is a topic that rarely receives priority in programs of institutional strengthening in agriculture, even in those that are oriented primarily to ministries of agriculture.[85]

An outstanding example of this approach to rural development can be found in the experience of the Province of Andalucía in Spain. An extensive set of brochures has been issued, covering not only agricultural topics but also rural tourism, the rural environment, forest management and other subjects.[86] Of course, in the Andalucian case the role of intermediaries is made easier by the fact that the range of programs on offer includes several that are funded by the European Union as well as those of the Government of Spain. Nevertheless, it is a useful example, not only of administrative decentralization but also of developing a wide range of actions for stimulating activities in the rural non-agricultural sectors. The central focus of the program is on developing the capabilities of the farm family members themselves, to some extent by capitalization of farm and non-farm activities but even more so through offering a range of carefully designed training programs in economic activities, agricultural and otherwise, that have potential in that region.

---

85. The aforementioned 1992 report of the Inter-American Development Bank on the subject of rural poverty and rural development commented: 'one of the organizational challenges of a system of programming, budgeting and management for rural development will be to attain optimal combinations among the decentralized levels of action for rural development projects and programs and the formulation of policies at a national level' (G. Aristizábal. J. Echenique, R. de Villalabos and W. Fischer, 1992, p. 41; author's translation).

86. These manuals were issued in 1996, under the umbrella title of *Guía de Desarrollo Rural*, by the Consejería de Agricultura y Pesca, Junta de Andalucía, Sevilla, Spain.

Specialized training programs for smallholder families should not overlook the possibilities of *linkages with agroindustry*. Various kinds of linkages between small farms and processing industries are possible, ranging from renting out farmland to the industrial companies and supplying them with labor, to entering into contracts to purchase inputs from the agroindustry (usually accompanied by supervision) and sell them products, to simply entering into contracts to provide them with the products.[87] The main stumbling block to successfully fulfilling such contracts can be lack of sufficient *quality control* on the products, as stressed earlier. It is in this area that training can have its greatest payoff.

9.5.4 Investment Priorities

The demand-driven or decentralized approach to rural development has widened the range of possible investments in rural areas. However, it can be argued that it should be complemented with a clear sense of *priorities for rural investment*. A case can be made that *a substantial share of government support should be channeled into a few essential kinds of infrastructure, instead of being dispersed over many fields*, and that those investments will enable rural families to make other kinds of choices that will raise their standard of living. International experience would appear to confirm that the priorities areas are four; i.e. education, transport, electricity and communications. If agroeconomic conditions are favorable to irrigation, then it should be added to the list as a fifth priority, in view of the large potential that it holds out for increased agricultural productivity.

A rural development strategy developed along these lines could include a phased set of infrastructure investments, in essence taking one area or district at a time and endowing it with adequate schools and roads, electricity supply, telecommunications systems and, where appropriate, irrigation systems. This kind of locality-specific package of investments would be accompanied by programs of special incentives for rural teachers, to attract the most qualified persons and to ensure their attendance at schools in remote locations, and transfers to rural families on the basis of regular school attendance of the children. The package would not exclude other rural development activities, especially those that impart specialized kinds of training to rural families, but it could be argued that focused investment packages deserve priority in government funding.

Infrastructure investments give rise to many kinds of benefits. Dominique van de Walle has argued that 'an important share of the benefits to the poor from rural roads cannot be measured in monetary terms' and proposes a methodology for selection of areas for road investments on the basis of three criteria: degree of existing road access, poverty and development potential.[88]

Basic infrastructure investments and rural development investments conceived and implemented in a decentralized manner can form the basic building blocks of an agricultural or rural development strategy. In addition, the creation of programs of transfers for low-income rural households can help give greater priority to the poverty alleviation objective. Pursuit of that objective can be made still more effective by strengthening microfinance institutions, clarifying the land tenure status of poor households, providing assistance for small-scale irrigation projects that are managed in a decentralized way, and developing mechanisms for better access to improved technology on the part of poor farmers. It is necessary to give specific emphasis to the rural poor in these and other ways because, as shown for the case of Honduras earlier in this book, many of the traditional, non-targeted agricultural policies and programs in fact favored the upper-income strata in rural areas.

---

87. Alexander Schejtman, *Agroindustria y pequeña agricultura: Alcances conceptuales para una política de estímulo a su articulación*, CEPAL, document No. LC/R.1660, July 29, 1996, Santiago, Chile, p. 22.
88. Dominique van de Walle, 'Choosing Rural Road Investments to Help Reduce Poverty', Working Paper No. 2458, The World Bank, Washington, DC, USA, October 2000.

As much as possible, investment planning should be made an integral part of the work on a strategy. Investments that follow from sector-wide priorities and policies are generally the most effective. An identification of investment needs can emerge from the same participatory process that gives rise to recommendations for policy reform. However, investment planning is a specialized process in its own right, and it may be necessary to incorporate the required investment planning expertise into the work on a strategy. The most that normally can be expected, without a special effort in the investment area, is a general listing of priority projects, with an indication of the nature and scope of each one. Quantifying the amount of investment required in each case usually requires a separate companion exercise.

Nonetheless, once the priorities in a strategy are presented, then approximate estimates of the amounts of investment needed can quickly be developed by first identifying the types and coverage of projects that are deemed priorities, e.g. hectares of irrigation rehabilitation, hectares of new irrigation, kilometers of feeder roads, kilometers of electrical power lines, hectares to be terraced for soil conservation, etc., and second applying average unit costs from past experience. In the case of demand-driven investments on a small scale, the amount of funding available determines the size of the portfolio.

Hence, the co-ordinators of a strategy should be encouraged to devote the small amount of additional effort required for the *companion exercise in investment planning*. If this step is taken, a strategy can provide a coherent overall framework for an investment program. It would represent a substantial improvement over the usual investment planning process, which is to compile a list of desired projects without any unifying logic or linkage to sector policy reforms. Obviously, all of the project ideas that emerge from a strategy would be subject to more careful review during the implementation process, including pre-feasibility and feasibility studies.

Above all, it should be remembered that a propitious and consistent policy environment is the single most important factor in attracting investment from outside the sector and in inducing

The possibilities and hurdles involved in linkages between smallholders and agroindustry are well illustrated by contrasting experiences with tomatoes in Chile and Peru:

The industry of tomato concentrates in Chile ships a large part of its production to the domestic market, for which reason the quality of the raw materials used is a central concern . . . since the quality of the final product depends on it. Since [this is] a product that demands great care, is employment-intensive, and does not have economies of scale . . . the industries contract for their supplies with small producers whose great advantage is having abundant family labor to assure the harvest. These producers receive technical assistance and training from the processing industry. . . .

. . . in the Valle of Ica [in Peru] it was found that the tomato paste factory had started out depending on small producers for a portion of its supplies but the difficulties and costs of supervision to assure that the farmers adapted to the enterprise's technical instructions caused it to abandon this practice, replacing it with renting in land and producing under its direct control and supervision (A. Schejtman, 1996, pp. 19 and 25).

farmers themselves to invest more in their operations.

9.5.5 Frameworks for Rural Development Policies

The foregoing discussion has touched on the main themes that characterize a successful rural development program. Such programs normally have a territorial dimension, in the sense that they comprise a set of activities to be carried out in specified geographical areas. However, a rural development program can be implemented in many areas throughout the length and breadth of a country. More importantly, *effective rural development programs concentrate on increasing the capacity of rural families to improve their economic status through their own efforts.* Rural development programs are more productive if they aim at reducing the causes of poverty rather than simply treating the symptoms of poverty. Sometimes, the symptoms are so acute, as in cases where people are undernourished, that urgent attention has to be given to ameliorating them. However, to prevent recurrence of the symptoms, the responsibility of development programs is to increase the abilities of rural families to produce more and work at more productive occupations. Approaching the problem in this way means implementing both programs aimed at individuals and families directly and programs aimed at improving the economic and institutional environment in which they live and work.

Five principal classes of constraints hold back the efforts of rural families to improve their lot, and on the basis of these constraints *five classes of rural development policies and programs may be designed* and priorities among them may be defined separately for each community. The classes of constraints or limitations represent insufficiencies in the following areas:[89]

(1) *Human capital*: low education levels and insufficient training in topics relevant to their work. 'Education increasingly involves lifetime skill acquisition for management and for acquiring and processing information'.[90]
(2) *Social capital*: low levels of community organization, producers' organization, etc., reflected in an inability to undertake productive efforts in co-operative or associative ways.
(3) *Physical infrastructure*: inadequate roads, communications facilities, energy supplies and irrigation water.
(4) *Institutional infrastructure*: weak agricultural research and extension, underdeveloped rural finance systems, inadequate rules and institutions for defining property rights and resolving conflicts over those rights, weak contract enforcement, and so forth.

(5) *Private physical capital*: insufficient land and on-farm irrigation infrastructure, and inadequate investment levels in livestock, tree crops and other productive capital. This represents the constraint of insufficient farm-level physical capital for the poor.

This perspective on the rural development problem is similar to the framework for agricultural development proposed earlier in this chapter. The main difference is that rural development efforts concentrate on the *rural family*, and on increasing its ability to control its world and improve its standard of living, whether through agriculture or other occupations. Agricultural development benefits from such an emphasis as well, but the orientation in agricultural strategies is more toward agricultural production *per se*.

Both human capital and private physical capital represent forms of capital owned by rural families themselves. Social capital represents the capacities of communities and other local groupings to work together. Building social capital is facilitated by higher levels of human capital and also by appropriate policies. Physical infrastructure and institutional infrastructure are forms of capital in the broader environment surrounding rural communities. Increasing their availability normally requires the assistance of national or regional efforts, as opposed to purely local efforts.

A rural development program cannot provide all of these forms of capital, but it can be linked to a national rural or agricultural strategy that highlights needs in the area of institutional infrastructure and develops policies for meeting those needs. *Rural development programs are especially appropriate vehicles for strengthening the first three forms of capital: human capital, social capital and physical infrastructure.* In addition, although international agencies are frequently reluctant to make loans or grants for private

89. Preliminary proposals along these lines were formulated for the Ministry of Agriculture and Forestry in Nicaragua with the support of the Inter-American Development Bank.
90. IFAD, *Rural Poverty Report 2001: The Challenge of Ending Rural Poverty*, Oxford University Press, Oxford, UK, 2001, p. 105.

physical capital for the rural poor, this form of capital can be a powerful instrument for poverty alleviation in the context of a rural development program. Since funding is made available for private physical capital when financial support is given for market-assisted land reform, which usually contains a subsidy element, it should be possible to extend the approach to other kinds of on-farm investments, provided the recipients are appropriately screened.

The endowments of water should be considered in the light of basic human subsistence needs, and not only for irrigation:

The poor share even less in farm water than in farmland, and suffer serious drinking-water shortages.[91]

This framework for rural investments can be applied to develop concrete recommendations by village or rural area. In a first phase of the work, diagnoses would be made as to the degree of adequacy of each kind of capital, and then in the second phase priorities would be defined to address the needs identified. For example, if most of the adults in a village were found to be illiterate, then human capital needs would figure among the priorities, and adult literacy programs could be a way of fulfilling that priority. *Each of the five forms of capital represents a path along which a rural community must advance in order to become self-sustaining in economic development*. Rural development – in the sense of strengthening the capacity for self-development of families and communities – cannot occur with advances along only some of the five paths. All five kinds of capital and infrastructure need to be created in adequate amounts and quality. In the area of physical infrastructure, for example, the first three priorities probably would be transport, communications and energy, in that order. For a village with access to the outside world only on foot part of the year, clearly a road would be a high priority investment. For human capital, the order of priorities would be functional literacy,

completion of primary school, and the acquisition of specialized skills through training courses, such as farm management, gender awareness, marketing and irrigation management. Without basic literacy, there would be little point in offering training in farm management, but once primary school had been completed, many options would open up for further development of human capital.

In this manner, priorities can be defined along each of the five paths, and the needs of each rural area can be assessed against those priorities. Policies and programs· can then be developed accordingly. Education might become a priority in some villages, while organizing groups of women for marketing and credit management (social capital) could be a priority in another one, depending on how far each village had already advanced along each of the paths. Putting in internet communications, with solar power for the energy source, might become a priority in other villages, as has been the case in parts of India.

Figure 9.2 illustrates three of the five paths, or axes, of rural development, showing the principal stages of progression along each of them. This kind of framework can also be used in the design of surveys for monitoring advances in rural development in selected districts or regions.

While rural families and communities themselves have to play a role in expanding their endowments of all five kinds of capital, outside institutions participate in the process as well. For developing *human capital*, central government, local government and NGOs all have roles to play. In some countries, NGOs have been the most innovative force in rural education. For strengthening *social capital*, local governments, NGOs and local associations make the most critical contributions, although national governments can direct some of their support to programs in this area.

For *physical infrastructure*, central and local governments have the main roles to play, but communities participate as well, often assisted by NGOs. Development of *institutional infrastruc-*

---

91. IFAD, 2001, p. 112.

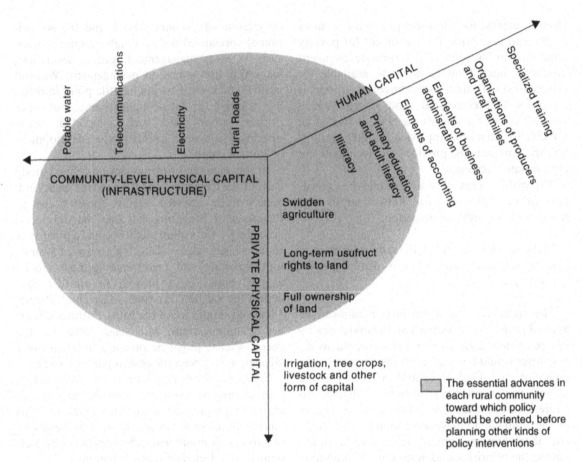

Figure 9.2   Basic dimensions of rural development: Stages of evolution of the capacity for development through own initiatives

ture depends crucially on appropriate national policies, but NGOs sometimes play a major role, as in the case of rural financial institutions. All of the foregoing four kinds of capital contribute vitally to expanding endowments of *private physical capital*, but in this area the efforts of families, communities and NGOs are also essential.

In the area of social capital, there is little doubt that the priority should be placed on removing gender disparities, on both equity and efficiency grounds. This issue can be addressed at the community level, through programs of education and awareness building and through economic empowerment of women, and through sector policies in the ways indicated in the foregoing

chapters. It also can be addressed through national policies and legislation regarding inheritance rights, eligibility for land under agrarian reform, domestic violence, women's health, and related topics.

Such policies and programs can have tangible results, as commented upon in a recent study of gender-oriented women's programs in China, which in any case has had a long history of enlightened national gender policies:

Recent empirical evidence from many countries has repeatedly shown that gender-focused public policy generates substantial social externalities, including improvement of child welfare (e.g. health, nutrition and education attainment) and reduction of gender bias and fertility rates.

... when combined with the income effects identified ... our analysis provides empirical evidence to support the claim that the gender-focused programs that emphasize promoting women's participation in both economic and social community activities can generate significant benefits. ...

We find support for the view that ... programs [of empowering rural Chinese women] can substantially increase the household incomes of participants but that some of this comes at the expense of negative income externalities for non-participants. Our results also suggest that the program is extremely successful at increasing participation rates within villages and that the program's income impacts depend sensitively on the ability to achieve such increases. The more successful the program is at increasing participation rates the greater both the positive impact on participants' incomes and the negative impact on non-participants' incomes, with the former substantially greater than the latter. In this sense, in the presence of the program, the gains from participation come from protecting oneself from these negative effects and from buying into the substantial income gains accruing from increased participation rates.

In conclusion, then, our results support the view that public policies geared towards increasing women's economic and social participation can generate substantial economic and social returns. ... Our results also lend support to the view that the range of gender-focused public policies implemented over the last few decades provides a complementary background, which contributes to the success of effectively implemented gender-focused programs.[92]

The IFAD has summarized the arguments in favor of giving priority to education of females in persuasive terms:

There are huge gaps between male and female educational access and literacy levels. These gaps are greater in rural areas, and greatest for the rural poor. Inequity helps cause inefficiency: female schooling does much more at the margin for income, poverty reduction, and child health and nutrition than extra male education. Women's lower adoption of agricultural innovations is due entirely to lower levels of education; at the same level, women farmers are as quick to adopt as men. Extra education raises household income more if it goes to females. Across Indian States in 1957–1991, the responsiveness of poverty to initial female literacy was higher than to any other initial condition. Mothers' education is also associated with better child health in many studies, often holding income constant.[93]

A framework based on the five paths – the five kinds of capital required for development – would have to be expressed in different concrete terms in each setting, and in a much more detailed form than has been traced out here. In whatever form it is utilized, it provides logical backing for the proposals in a rural development strategy and a way of integrating the disparate pieces of action that would comprise the strategy, so that they are mutually reinforcing. It also offers guidance for establishing investment priorities, on the basis of an assessment of the degree of advance of each community along the five paths.

To put this approach in a broader context, Gustavo Gordillo de Anda has properly stressed the importance of political transparency and a strong institutional framework which reduces the uncertainty associated with economic transactions in rural areas. Uncertainty – about whether

92. David Coady, Xinyi Dai and Limin Wang, 'Community Programs and Women's Participation: The Chinese Experience', Working Paper No. 2622, The World Bank, Washington, DC, USA, June 2001, pp. 22 and 24–25.
93. IFAD, 2001, p. 110.

an agreed payment will be made, about whether deliveries will be made on time, etc. – is almost a defining characteristic of underdevelopment, and the cost of overcoming it is very high. Gordillo has said:

> What is important is to understand that rural development is not only a set of economic policies, but that it also has to take into account the political and social conditions that have been the result of the economic modernization and the democratization of our societies.
>
> ... the new approach for rural areas requires three central ingredients:
>
> • Provide incentives for opportunities to progress and improve well-being
> • Create certainty
> • Foster social cohesion [increasing social capital][94]

9.6 IMPLEMENTATION OF A STRATEGY

The work invested in developing an agricultural or rural strategy will be of little value if at least some of its major thrusts are not implemented. On the other hand, it usually is unrealistic to expect full implementation of all of the recommendations in the document. The process of formulating a strategy should itself include preparations for the implementation stage, and for that purpose it is helpful to be aware of the means of policy implementation while designing the policy proposals.

To reiterate what was said at the end of Chapter 2, a strategy is implemented through five different channels, as follows:

• New *legislation* (the legislative policy channel).
• *Administrative decisions and decrees* of the executive branch that alter the rules governing the economic environment for agriculture and change institutional structures (the administrative policy channel).
• Allocations of public investment, or *capital account funding*, some of which may come from external partners in development (the investment channel).
• Allocations of the *current account budget* of the government (the program channel).
• Voluntary participation in implementation *by the private sector and civil society* (the non-governmental channel).

It can be added to that earlier discussion that, after formal endorsement of a strategy, either by the Executive branch or by the Legislature – preferably by both – principal elements of the implementation process also include, in addition to those discussed in Chapter 2:

• Development of a medium-term implementation plan, under the aegis of which the annual implementation plans would be formulated.
• Establishment of an implementation secretariat or some other high-level committee designated as responsible for managing the execution and monitoring of the strategy.

The implementation plans should identify clearly the needed actions in all five implementation channels. The timing of the implementation plans needs to be co-ordinated with the timing of the budgetary cycle. Each implementation plan should contain a legislative action plan, a plan for Executive decrees and administrative actions that fall within the ambit of the Executive branch without budgetary implications, a statement of the requirements of the current budget, an investment plan, and an implementation plan worked out jointly with the representatives of the private sector or civil society regarding agreed actions in the non-governmental sector. The latter, of course, cannot be a binding document but it can help reinforce the coherence of the entire implementation effort.

---

94. Gustavo Gordillo de Anda, 'Un nuevo trato para el campo', paper delivered at the International Conference on Access to Land: Innovative Agrarian Reforms for Sustainability and Poverty Reduction, Bonn, Germany, May 19–23, 2001, p. 36 [author's translation].

If the investment plans are worked out on a medium-term basis, say for three years, then they can also feed into the programming exercises of international agencies that are supporting the country's development. Adapting their support to the priorities and reforms indicated in a national strategy would require a degree of flexibility on the part of the international development agencies that has not generally been demonstrated so far, but it would give concrete content to the rhetoric of participation. Taking that kind of co-operation one step further, agencies that prefer to make their disbursements conditional on specified reforms could tie them to specific and major steps in the implementation of the country's own national strategy. Even with the best of intentions, an implementation process always encounters many roadblocks along the way, and that kind of conditionality would increase the sense of urgency to overcome the roadblocks. It would be more palatable to the host government and society, and more effective in the longer run, to make it clear that the *conditionality is based on the country's own plans rather than on those developed elsewhere*.

Concern about the apparent ineffectiveness of traditional conditionality is becoming widespread. Jan Willem Gunning has expressed those doubts, and has proposed a different kind of solution, in the following words:

> There is now overwhelming evidence that aid is not effective in bringing about policy reform. I have argued that rather than redesigning the aid contract to make *ex ante* conditionality more effective, donors should switch to *ex post* conditionality (selectivity). Under selectivity the allocation of aid is tied to success.[95]

The solution is not to give up on *ex ante* conditionality, but rather *to have the loan and grant conditions designed, in effect, by the host country*. Asking a country to implement its own strategy, as a condition for disbursement, is a means of helping push that implementation process along.

At the same time, such conditionality is likely to find greater receptivity in the developing countries.

One of the keys to timely implementation is a carefully structured implementation monitoring system which requires frequent and rapid reporting to the implementation secretariat of progress and problems in each area. Equally, the secretariat needs to be endowed with powers to take special measures to accelerate the implementation process in areas in which it is found to be lagging – or to recommend such measures to the Cabinet. It can be useful *to involve in the implementation monitoring process persons who were involved in designing the policies contained in the strategy*, from inside and outside government. They often represent the best expertise in each area and, for that reason, are in a position to recommend modifications of the policies as needed, on the basis of feedback received from the implementation process.

For moving ahead in the legislative channel, usually teams or task forces of lawyers and other technical experts are assembled to draft legislation in the areas where the strategy's recommendations suggest that it is needed. Lawyers and other kinds of experts need to work together closely in this process. Lawyers are experts in legislative technique and therefore can indicate *how* to achieve specified aims, but it typically is the other technical experts who can indicate *what* those aims should be. Allowing lawyers to take sole command of the process of drafting legislation has sometimes resulted in outcomes that did not fulfill all of the postulated policy aims. The implementation secretariat is the appropriate body to oversee the labors of a legislative working group and, when the topic warrants it, submit the draft legislation for review by the entire participatory group that developed the strategy.

9.7 CONCLUDING OBSERVATIONS

The usefulness of a strategy will depend on many circumstances, including some or all of the fol-

---

95. Jan Willem Gunning, 'Rethinking Aid', in Boris Pleskovic and Nicholas Stern (Eds), *Annual World Bank Conference on Development Economics, 2000*, The World Bank, Washington, DC, USA, 2001, p. 141.

lowing factors: the quality of the work that goes into it, the extent of participation in its development, the timing of the document with respect to the electoral cycle, the political situation of the government at the time, and the support that the strategy generates in the international community. It is always a quixotic undertaking to believe that a document has the capacity to change people's lives, and there is never a shortage of people who will dismiss such an effort in advance as inherently irrelevant. However, if a strategic document faithfully reflects a consensus and the determination of those who participate in creating the consensus, it can have tangible consequences. History provides many such examples.

Working on a strategy is by no means a routine exercise in analysis or bureaucratic programming. It requires stepping out of the confines of traditional ways of developing planning documents in large institutions. It requires extensive communications among persons with different backgrounds. It requires a suspension of disbelief. It invariably is a learning experience for all of those involved, including the advisors. If a strategy is to be successful, those who participate in it have to come to believe in its worth. It is the result of a peculiar fusion of realism, analysis and hopes.

Above all, it requires persistence and a long-term commitment on the part of those involved. It is not only a technical economic process but also an effort in building a nation's social capital. The broader the process, then the greater its chances for producing successful outcomes.

DISCUSSION POINTS FOR CHAPTER 9

- An agricultural development strategy is a consistency framework for bringing together policy initiatives in an overarching logical structure that adopts a medium- or long-term view of the sector's prospects. A defining characteristic of a strategy is that it presents an integral approach to development issues and does not deal with them in isolation from each other.
- Although a strategy needs firm technical foundations, in order to be effective it also needs widespread support, especially from producers, government and international development agencies.
- When a strategy proposes several different kinds of policy reforms, benefitting different groups in society, it may be easier to overcome narrowly-based opposition to some of the reforms, than if the reforms were proposed individually. One of the roles of a strategy is to eliminate special privileges and create a level economic playing field.
- One of the roles of a strategy is to elevate the level of the national dialog on policies. It is an educational process for all participants, including technical advisors. Work on a strategy can provide a 'learning-by-doing' process for participants, in learning how to define the policy issues themselves in fruitful ways and how to pursue solutions in perhaps unaccustomed directions.
- In recent years, many efforts have been made to enhance the participatory character of development programs and policies, but citizen participation in formulating national or sectoral development strategies has not been so common.
- In addition to respecting basic principles of democracy, encouraging participation in the development of a strategy can be valuable for the following reasons:

(a) It improves the chances of attaining a national consensus on policy reforms.

(b) It strengthens the channels of national dialog, thus empowering citizens to participate more effectively in the resolution of future policy issues. It represents capacity building for civil society and the private sector in regard to national development policy issues.

(c) It helps develop more solid policies. Experience has shown that representatives of the private sector and NGOs not only can play a role of supporting consensus policies, but that they are also capable of making fundamental contributions for improving the quality of the reforms, even to the point of assisting in the drafting of proposed legislation.

(d) It helps improve the accountability and transparency of the policy making process.

(e) It helps empower the country in international dialogs, so that truly national priorities can serve as guides for programs of international assistance, instead of allowing national policies to be defined implicitly by the sum of conditionalities attached to international loans and grants.

- It is important to ensure that those who participate in such a process are sufficiently representative of a broad spectrum of that society. This includes women, the poor and other traditionally disadvantaged groups. Participants do not necessarily have to be formal representatives of associations or other recognized entities, but should be recognized in their communities and groups as thoughtful persons on policy issues.

- The risk of not reaching a consensus on strategic questions among members of civil society is real. However, if there are important disagreements in society on basic policy issues, the work on a strategy is a good forum for trying to heal them.

- A principal objective of a participatory process of strategy formulation is to create a sense of ownership of the product on the part of the participants. In good measure, achieving this objective depends on how the process is carried out. Another objective is to generate a sense of ownership by the country. This objective is particularly relevant in countries where many policies have effectively been imposed by international development agencies through the conditionality tied to financial assistance.

- The ownership question, for the country and for civil society, suggests limits on the extent to which international financial organizations and governments can directly sponsor participatory processes of strategy development. Ideally, both should contribute expertise to the process, and assist with the implementation of the strategy's recommendations, but civil society may be reluctant to commit itself to working under the aegis of one of those entities.

- How the participants are selected for the process also can be relevant to the issue of ownership. This is an issue that deserves full discussion with the participants themselves before their membership is fully decided.

- If a strategy is drafted by technical advisors and then presented to civil society for consultations, it is unlikely to create a sense of ownership among those who are consulted. The key to ownership is for persons from the country concerned, and civil society in particular, to take the lead in the process of drafting the strategy document.

- Technical quality control is an important issue in strategies, especially those in which civil society plays a leading role. It can be best promoted by capacity building, and that, in turn, requires a sustained working relationship between participants and technical advisors, in a 'learning-by-doing' process. On the other hand, often there is a considerable amount of latent technical talent in a developing country that has not previously been tapped into for work on policy issues.

- Civil society's capacity for analyzing policy issues should be institutionalized so that it may be sustained over the longer run. This is a topic that warrants the attention and support of international development institutions. In the final analysis, one of the most valuable roles for official development assistance is building national capacity for policy analysis and formulation.

- The organization of a strategy effort centers around task forces, or working groups, that carry out the analysis and write drafts of chapters. Managing the effort requires a co-ordinating committee, and sometimes there are two levels of co-ordination, one for day-to-day logistics and the other, in a larger group, for review of substantive issues and consistency requirements among the components of the strategy.

- Task forces can draw into the process talent from civil society that otherwise may not have been brought to bear on policy considerations.

- The principal challenges that are inherent in launching a participatory effort at developing an agricultural strategy are the following:

(a) To motivate a sufficient number of leading farmers and ranchers to commit themselves to the process and dedicate the time required to bring it to a successful conclusion. Since both agricultural entrepreneurs and peasant farmers have many claims on their time, it is not easy to satisfy this requirement.

(b) To reach beyond the group of persons who reside in the capital city and attain a degree of geographical representativeness in the participation. The logistics of this challenge can be difficult in large countries and countries with poor communications and transport networks.

(c) To overcome schisms that may be present in the participating group for reasons of political partisanship or socio-economic differences, in order to be able to put together a consensus position that has the implicit support of all or most major groupings in rural society.

(d) To create an environment in which civil society representatives jointly go forward to lobby for sound economic policies that will benefit the entire sector, instead of promoting narrow personal or sectarian interests.

(e) To avoid creating a parallel process of dialog that may weaken existing channels of communication and social processes rather than strengthen them.

(f) To avoid premature publicity about the participatory work in its drafting stage. Although some publicity about the process can be useful, to help keep the public informed, premature publicity about the content of the strategy runs the risk of becoming a 'lightning rod' for sharp, sometimes partisan, criticisms of the tentative policy formulations, even in the national press. The danger is that such criticism early in the effort could fracture the emerging consensus among participants and even derail the entire effort.

(g) To gain acceptance for the principal policy recommendations from the government of the day, or the next government if elections are due relatively soon.

- While a creative and feasible vision about the future is an essential element of a strategy, the means of making the vision come true eventually need to be stated in sufficient detail and rigor as to constitute an implementable program.

- A useful step in the transition from a vision statement to a program of policy proposals is to identify principal obstacles to the fulfillment of the vision. Another worthwhile step is for the task forces to develop descriptions of the existing policy framework, and that may not be a trivial task for those who have spent most of their professional careers working within the same policy framework. Working continuously within a given context makes it hard to step outside it and see alternatives.

- In a strategy, the logical links between diagnoses and solutions need to be made clear. One way to structure a strategy document is to ensure that for each sector, sub-sector or topic it includes the following elements:

(a) Basic characteristics of the sector, sub-sector or topic under review.

(b) Review of past and present policies for the area or topic.

(c) Major issues and constraints to be dealt with.

(d) Specific (subsidiary) objectives for the area or topic.

(e) Policy recommendations and supporting technical justifications. (The policies are developed logically as ways to overcome the constraints and satisfy the objectives.)

(f) Appendices: recommended legislative reforms and recommended investment program (not applicable for all areas).

- Human capital is the most strategic factor for agricultural development, especially as new technologies emerge, markets demand higher quality and safer products, and as consumer requirements for quality and delivery timing change. Learning how to continuously access information and assimilate is acquiring increasing importance in agriculture throughout the world. Equally, institutions and policies that facilitate such access are increasingly critical, but often Ministries of Agriculture are slow to recognize the importance of quality issues.

- Institutional development is one of the keys to agricultural growth. Solid institutions open the doors to new knowledge for farmers and allow them to enter into reliable arrangements for future delivery of inputs and products and deferred payments. When all transactions and payments have to be made on a spot basis, there is little scope for development. In most developing economies, institutions serving the rural sector need to be improved in terms of accountability and efficiency.

- The policy recommendations of a strategy should be consistent, both within the sector and across sectors. Implementation of one policy may require changes in another, complementary policy. Agriculture is especially dependent on the macroeconomic framework, which largely determines producer incentives. If there has not been enough exploration of macroeconomic policy options, an agricultural strategy may have to suggest some alternatives at the macro level.

- The most basic policy goal for an agricultural strategy is to improve incomes of rural households. Increasing production and productivity are important as means to achieve that goal, and policies that shift relative prices in favor of agriculture also contribute toward fulfillment of that goal.

- In all countries, the aim of an agricultural strategy is to generate sustainable growth that is widely shared, but producing a successful strategy requires creativity in finding concrete and detailed solutions to specific problems that are viable in their context.

- Historical experience and conceptual models of long-term agricultural development have emphasized the need to minimize government interventions in product markets and instead give priority to improving the functioning of factor markets, particularly in the areas of education training (human capital) and for the factor markets related to land, water, credit and technology – the latter being another dimension of human capital.

- The 'historical model' of Vernon Ruttan and Yujiro Hayami is one of the richest of that genre in policy implications. Their model emphasizes the fundamental role of technical innovation in agricultural development, and it also states that the nature of innovation is heavily influenced by relative factor prices and also by real prices of agricultural outputs.

- Agricultural production is only one part of a chain of related activities, stretching from input supply and technology development through production to post-harvest management, marketing and processing. Therefore, agriculture cannot prosper without forging solid linkages to markets, both domestic and international, and markets in turn are ever more demanding in terms of product quality and conditions of delivery. The primordial role of markets is as true for smallholder production as for larger farms.

- In the broadest sense, the most important factors for generating agricultural growth are adequate market and pricing conditions and sufficient productive capital. In the latter category human capital, as noted, is the most essential for improving development prospects.

- The key substantive orientations of agricultural development strategies may be summarized as follows:

(a) *Market Development and Pricing Policies*

— ***Market development policies***, including a country's international trade negotiations for agriculture (bilateral as well as multilateral), policies to ensure compliance with phytosanitary norms and food safety standards, product grading (especially for grains), export promotion measures, market information and testing efforts, finance for marketing and storage, training of extension agents in product quality issues and organic production, and related efforts.

— ***Pricing policy***, mostly at the macroeconomic level, to avoid declines in real agricultural prices, and to reverse at least partially large declines that may have occurred. Pricing policy also should aim for relatively uniform rates of effective protection among products within the sector, and between sectors.

— An *open trade regime* without, however, transmitting obvious distortions in the international arena to the domestic economy.

(b) *Human Capital Policies*

— Improving *rural education*, changing, if needed, the way in which such schools are managed.

— Extension and training programs in rural areas should incorporate elements of community and producer organization, especially as regards women. The ability to work together co-operatively is a key to success in many development efforts, especially those that involve penetrating new markets. This is the factor of *rural social capital*.

— The capacity for *agricultural research and extension* needs to be strengthened very considerably in most developing countries, with emphasis on the quality of staff and on participatory approaches.

— Human capital also is manifest in institutions and, as commented upon, *strengthening institutions is vital to agricultural progress*. In the most general form, it requires inculcating respect for the rule of law, and for property rights in all forms. *Sound governance* is a key to all aspects of agricultural development. In more specific form, institutional strengthening requires that they be made more *accountable and efficient*, and usually in developing agriculture that requires a greater degree of *decentralization of institutions* than now exists.

(c) *Physical Capital Policies for Agricultural Growth*

— Clarifying and buttressing *property rights*, including lease and rental rights,

and improving *access to land* without disrupting the security of property rights.

— *Investments in rural physical infrastructure*, especially those related to *irrigation, transportation, electrification and communications*.

• The *ways* in which agricultural policies are conceived, designed and implemented are major determinants of their results. The lessons of experience point to five key approaches that help ensure the effectiveness of policy reforms:

— *Reducing distortions in product and factor markets*.

— *Reinforcing the legal strength of contractual relationships*.

— According *special emphasis to women, small farmers and the rural poor* in the targeting of policies and programs.

— *Decentralization and participation*: an approach of devolution of public services, with privatization where appropriate, and encouraging farmer and community participation in the design and implementation of programs and policies.

— Emphasis on creating *viable institutions*.

• To advance the implementation process, a strategy should be accompanied by recommendations for legislative reform where appropriate.

• The integrated rural development approach followed in many countries has generally been a failure. The reasons include an adverse policy environment, lack of sufficient government commitment, lack of appropriate technology, neglect of institutional development, lack of beneficiary participation, and weak interagency co-ordination.

• Because of these concerns, among international agencies the emphasis in rural development has shifted to decentralized and participatory approaches, including 'demand-driven' investment programs in rural areas. Increasingly, communities are asked to take the lead in defining or selecting the programs they will participate in, and the role of local governments is being put in higher relief.

- Successful rural development also requires institutional decentralization, supportive policies in other sectors, and good governance.
- The effectiveness of rural investments can be increased by implementation of complementary programs of fiscal transfers to poor rural households. Such transfers are consistent with WTO rules on agricultural policy and represent a way to deliver incentives to smallholders and to producers of export products, whom it is otherwise difficult to target in programs of incentives.
- Direct transfers also are neutral with respect to cropping patterns, unlike most other forms of fiscal support for agriculture. If combined with an area-based land tax, the net effect would be progressive with respect to farm size.
- The administration of direct support measures requires considerable decentralization of the operations of ministries of agriculture.
- It can be argued that a substantial share of public investments in rural areas should be channeled into a few essential kinds of infrastructure, instead of being dispersed over many fields, and that those investments will enable rural families to make other kinds of choices that will raise their standard of living. International experience would appear to confirm that there are four priority areas: education, transport, electricity and communications. If agroeconomic conditions are favorable to irrigation, then it should be added to the list as a fifth priority, in view of the large potential that it holds for increased agricultural productivity.
- To the extent possible, approximate estimates of investment requirements in priority areas should be incorporated into an agricultural development strategy.
- Five principal classes of constraints hold back the efforts of rural families to improve their lot, as follows:

  — *Human capital*: low education levels and insufficient training in topics relevant to their work.
  — *Social capital*: low levels of community organization, producers' organization, etc., reflected in a inability to undertake productive efforts in co-operative or associative ways.

  — *Physical infrastructure*: inadequate roads, communications facilities, energy supplies and irrigation water.
  — *Institutional infrastructure*: weak agricultural research and extension, underdeveloped rural finance systems, inadequate rules and institutions for defining property rights and resolving conflicts over those rights, weak contract enforcement, and so forth.
  — *Private physical capital*: insufficient land and on-farm irrigation infrastructure; inadequate investment levels in livestock, tree crops and other productive capital. This is the constraint of insufficient farm-level physical capital for the poor.

From these constraints, five classes of rural development policies and programs may be designed, and priorities among them may be defined separately for each community. Each community's degree of progress along each of the five 'axes' can be assessed to determine the next priority for its programs of rural development.

- In the area of social capital, there is little doubt that the priority should be placed on removing gender disparities, on both equity and efficiency grounds. This issue can be addressed at the community level, through programs of education and awareness building and through economic empowerment of women, and through sector policies in the ways indicated in the foregoing chapters. It also can be addressed through national policies and legislation regarding inheritance rights, eligibility for land under agrarian reform, domestic violence, women's health and related topics.
- A strategy's value is limited if it is not at least partially implemented. Therefore, in the strategy development process it is worthwhile to take into account the ways in which policy reforms are implemented. There are five channels through which policies are put into effect:

  — New *legislation* (the legislative policy channel).
  — *Administrative decisions and decrees* of the executive branch that alter the rules govern-

ing the economic environment for agriculture and change institutional structures (the administrative policy channel).

— Allocations of public investment, or *capital account funding*, some of which may come from external partners in development (the investment channel).

— Allocations of the *current account budget* of the government (the program channel).

— Voluntary participation in implementation *by the private sector and civil society* (the non-governmental channel).

• A strategy should be followed by an implementation plan, and an implementation secretariat can be established to oversee the steps taken to

put it into effect. Those who participated in developing the strategy will have acquired the expertise needed to monitor its implementation and develop solutions to problems encountered along the way.

• The most effective kind of conditionality international assistance is that which requires a country to implement its own plans, particularly those developed in a participatory manner. Even with the best of intentions, hurdles always are encountered along the road to implementation, and therefore it can be useful for international agencies to specify the implementation of a nationally developed strategy, or key parts of it, as conditions for disbursement of external assistance.

Annex
National Economic Policies and Irrigation in Yemen[1]

1 WATER IN YEMEN'S DEVELOPMENT PROCESS

Owing to the rapid depletion of aquifers in the most populated areas of the country, and the present inadequacy of water supply to some major urban areas, water has become the most limiting constraint on Yemen's development process. The nature of the water constraint varies by region within the country, in part as a function of the location of aquifers. Pumping water from the very large Mukalla aquifer in the east to some of the major cities, including the capital Sana'a and Ta'iz, is probably too costly to be a realistic option, and so those cities will have to satisfy their water needs from closer sources, which are much more limited. On the other hand, in the long run Aden may be able to satisfy its water needs through desalinization, an option which is not available to other cities because their altitude would lead to excessively high pumping costs for the desalinized water.

For the highlands area of Yemen, including Sana'a, *all* of the aquifers of any scale are experiencing drawdowns, so interbasin transfers within that area cannot be considered a solution in the longer run.[2] In almost all regions, *a solution will require a drastic reduction in agricultural use of water, and probably in irrigation production itself*, for the following reasons:

- Agriculture accounts for approximately 90% of total water use.
- When present trends in water use are extrapolated, the foreseeable additional sources of water supply will not be sufficient to meet expected water demands by about 2010, without exhausting some principal aquifers, and on a per capita basis those demands are quite modest.
- The value added associated with using water in the production process is approximately 147 times greater in manufacturing than in agriculture.

1. This annex is adapted from a study developed by the author for the Yemeni Government and donor agencies at the end of 1995. Accordingly, some of its observations, particularly on policies, may be out of date. It is included here to illustrate some of the issues in managing scarce water resources and the scope of policies which may affect water use. The original study was issued as R. D. Norton, 'Economic Policies for Water Management in Yemen', Discussion Paper No. 1.1, Yemen Water Strategy, Multi-Donor Group on Yemen Water, The World Bank, UNDP and the Government of the Netherlands, Washington, DC, USA, December 1995. The author is grateful to Christopher Ward for helpful comments and guidance on the original study.
2. A possible exception to this statement would be the transfer of water from Marib to Sana'a, but that option would entail significant sacrifices on the part of the residents of the Wadi as Sudd and its environs, and it would not constitute a complete solution to the water problems of Sana'a.

Agricultural Development Policy Concepts and Experiences. R. D. Norton
© 2004 Food and Agriculture Organization of the United Nations
ISBNs: 0-470-85778-1 (HB)   0-470-85779-X (PB)   FAO Edition: 92-5-104875-4

In some well fields, water tables are reported to be declining by as much as 3 m per year (Sana'a North Basin), to 4 m per year (Sa'dah Basin), and even 6 m per year (Amran, Rada' Basin, and Wadi Adanah). These declines in water tables will force groundwater-based irrigated agriculture to close down in less than a generation in principal basins, or sooner if pumping costs rise to unmanageable levels because of ever-deeper wells. Indications are that an increase in the diesel price to world market levels would make irrigated production of almost all crops unprofitable, even at present well depths. The production of the crop *qat* is perhaps the exception to this statement.

In such circumstances, the main policy goals of an integrated water strategy would be likely to include the following:

- Making this transition out of agriculture occur before water supplies to urban areas also vanish.
- Minimizing the social and economic disruption attendant upon such a transition. This can include measures to 'stretch it out' in time (but consistent with the first goal), such as gradually increasing pumping costs, so that the numbers of farmers driven to give up agriculture each year can be handled through special transition programs. It can also include efforts to reduce the disruption through measures such as improving irrigation efficiency and making wastewater available for irrigation. Nevertheless, it clearly means implementation of strong measures to stimulate the creation of employment in the industrial sector and very innovative social and economic measures to support an economic transformation that would be virtually unprecedented in nature and scale.

2 DEMAND MANAGEMENT VERSUS WATER SUPPLY OPTIONS

Yemen is fragmented into numerous hydrological zones, and the predominance of groundwater as a source makes it exceptionally difficult to develop a national 'master plan' for managing supply and demand. Different approaches may be appropriate for different zones. In general, for each of the country's basins and aquifers, one of three scenarios exists: (i) periurban areas where there is competition between urban and rural users of water; (ii) rural areas where there is excess use of water and a decline in the water table; (iii) other rural areas where supply is adequate to meet uses without mining aquifers.

Periurban areas. The Sana'a Basin provides the most compelling example of the emerging water crisis. Water use in the basin was estimated at 178 Mm^3 in 1990, and present consumption trends indicate that it will reach 442 Mm^3 in 2010 without a significant change in policies. The annual natural recharge in the basin is estimated at 41.9 Mm^3, and so the gap to be filled, in order to avoid depletion of the aquifers, will reach about 400 Mm^3 in 15 years, more than twice the current *level* of use. This is equivalent to 12 698 l/s from pumps that run 24 h a day.

On the supply side, present water sources are dwindling. It is estimated that the output of government wells for Sana'a city will drop from 600 l/s to 100 l/s by 2010. Identified possible new sources of water for the city would supply 2350 to 2400 l/s, or less than a fifth of the expected demand. Some of them involve taking water away from agricultural uses outside the Sana'a Basin, and some will last as little as 10 years before being exhausted also.

These considerations give rise to two important implications. First, *a supply-side strategy alone will not solve the problem – nor will it even come close – so drastic reductions in the extraction of water will be necessary.* Secondly, *transfers of water between sectors will be necessary, and mechanisms are required which allow and encourage water to flow to the uses on which society places the highest value.* In this situation, in order to avoid conflicts, *it will be essential to manage agricultural demand in a fair and equitable manner and to ensure that agreement is reached on methods for reducing extraction for agriculture and to transfer water to urban uses.* Similar conclusions can be reached for the Ta'iz area, although there the chances are better of satisfying both rural and urban demands over the medium term, given proper water management.

Rural areas with excess water demand. The sharp rates of aquifer decline in some rural areas were noted above. In areas such as the Sa'dah Plain and Amran, the problem is localized, as there is no demand on the resource from outside of those areas. There are two possible resolutions: *either the community, with or without outside help, decides on a sustainable management plan that would entail a massive reduction in extractions, or the costs of extraction mount and the resource dwindles to the point where aquifers dry up, agricultural activity ceases, and wholesale out-migration occurs.*

Rural areas with balanced water supply and demand. An example is provided by the lower western wadis in Dhamar Governorate,[3] where the isolation of the communities and distance from markets make the use of diesel pumps uneconomic. Current rates of water use are sustainable but as incentives change a forward-looking vision will be needed to equip these communities to be able to manage their resources.

3 THE ORIGINS OF THE WATER CRISIS

The water crisis in Yemen has technological, social, institutional and economic origins. The *technological* aspect is the introduction of pumps, which began on a large scale in about 1970. Previously, most of the wells for irrigation and municipal water were hand-dug and thus were self-limiting in terms of the amount of water extracted. Now, the total number of tubewells is estimated to be more than 400 000 and growing.

The *social* aspects of the crisis include the country's rapid population growth (in excess of 3% per year) and the rapid increase in the consumption of qat, which is one of the most water-intensive crops. Nationally, its cultivation consumes annually about 480 Mm^3 of water, a figure which may be compared with an estimate of 986 Mm^3 for the total annual recharge of groundwater in the Northern Governorates.

The *institutional* dimension of the water crisis refers to uncertainty about water rights and the weakness of institutions for controlling the rate of groundwater exploitation. In addition, urban water systems have not proven to be as effective as would be desired in reducing water losses throughout the systems.

The *economic* origins of the crisis are rooted in policies that effectively provide incentives for excessive use of water. Principal examples include the following:

- Diesel fuel, used in most water pumps, has been priced at about 10% of its equivalent international level. Electricity, used in some pumps, also is subsidized compared to the cost of producing it, tariffs having been estimated at 17% of the break-even point for the Public Electricity Corporation.
- The Co-operative and Agricultural Credit Bank (CACB) has been lending for the purchase of water pumps at nominal interest rates of 9 to 11%, while the inflation rate for 1995 was in excess of 70%.
- International donors have provided concessional funding for water pumps.
- Urban water supplied by the public system has been priced at about YR 11/m^3 on average in Sana'a, and at comparable levels in Ta'iz. Privately supplied water has been priced at about YR 100/m^3 in Sana'a (and higher in periods of scarcity) and at about YR 250/m^3 in Ta'iz.
- Fruit, vegetables and qat, all highly water-intensive crops, are favored by import bans that raise their profit margins and hence their attractiveness for farmers.

From these few examples, it is evident that *current economic policies constitute a powerful engine pushing in the direction of exhaustion of Yemen's aquifers.* It would be unrealistic to expect that economic policy alone could solve the crisis, given that it has been created in part by strong technological, social and institutional forces. Nevertheless, economic policy will be a key to an eventual solution, as will a program of institutional strengthening. Equally vital will be an

3. These areas are described in DHV Consultants BV, *Environmental Profile of Dhamar Governorate*, Sana'a, Yemen, May 1990.

effort to involve the public in measures to alleviate the crisis; the required changes in way of life will be large, and so an understanding of the situation on the part of the public will be essential to its acceptance of reforms and its assistance in implementing them.

4 PRINCIPAL POLICY INSTRUMENTS AVAILABLE FOR WATER DEMAND MANAGEMENT

The policy instruments for water demand management can be grouped according to the specific aims of the policies. There are instruments that are designed to:

(1) Promote greater efficiency of water use in agriculture.
(2) Reduce water use in agriculture (whether through efficiency gains or otherwise).
(3) Transfer water out of agriculture to urban areas.
(4) Set aside water rights for aquifer recharge.
(5) Promote greater efficiency of urban water delivery systems.
(6) Promote water recycling.
(7) Promote industrial development and job creation, in order to attract people away from irrigated farming.
(8) Relocate industry and services to areas better endowed with water resources.
(9) Provide transition assistance for populations affected by the water crisis.

The instruments that can lead to a net reduction in agricultural water use without specifically reallocating it to another use (those in categories (1) and (2) above) include instruments that *subsidize the adoption of more efficient irrigation techniques*, those that *make pump-irrigated agriculture less profitable*, and those that *impose direct controls on pumping*.

More efficient irrigation techniques in Yemen include sprinkler systems, drip systems and microtube bubbler systems. Anecdotal evidence suggests that there has been little interest in adopting such systems to date. It should be pointed out that this *reluctance to adopt improved irrigation systems is to be expected in the present economic environment*. The financial returns to those systems are a function of the amount of water saved multiplied by the cost of the water to the farmer. The large subsidy on the price of diesel fuel effectively reduces the cost of groundwater to farmers to very low levels, and so the returns to improved systems are correspondingly low. In addition, the cost of the new systems is rather high, with their purchase and installation running from $3100/ha to $4400/ha.[4]

Therefore, it is likely that adoption of the improved techniques will occur on a larger scale only if a substantial subsidy is extended for the purchase of the equipment or the cost of pumping is raised. Given the fiscal situation, with government expenditures running at more than twice revenues, there would appear to be a strong presumption in favor of the latter option.

Another consideration also indicates that the potential for net water savings from improved irrigation systems is likely to be limited – the fact that in many areas it is water and not land that is the constraining factor on groundwater-based agricultural production at present. To the extent that water is saved by investments in new irrigation techniques, a rational decision by farmers would be to apply the water savings to an expansion of the cultivated area. Thus, in the end there may be little or no net savings of water from use of the new techniques. It is suggested that promotion of these techniques will be most successful under the following conditions:

(1) After there have been significant increases in the real price of diesel fuel.
(2) In areas where land also is a limiting factor for agricultural production.

On a national scale, these considerations suggest that it is not likely that improved irriga-

4. DHV Consultants BV, *Northern Region Agricultural Development Project, Technical Assistance for Engineering Services, YEM/87/015: Regional Water Management Plan*, United Nations Development Program and Yemen Republic, Ministry of Agriculture and Water Resources, Sana'a, Yemen, January 1994, p. 47.

tion systems will contribute in a major way to solving the water crisis, and so most of the emphasis should be placed on other approaches, including policy reform. (So far, most of the discussion of possible solutions to the water crisis has centered around proposals for promoting the improved irrigation techniques and around schemes for interbasin water transfers.)

Policy reforms that would make pump irrigation less profitable include higher prices of diesel fuel and electricity, higher tariffs and taxes on pumping equipment, reduced credit subsidies for pumps, and removal of the import restrictions on agricultural goods whose production is water-intensive. Reducing the credit subsidy can mean either raising interest rates for loans for pumps or simply eliminating that line of lending. Halting the government imports of pumps would go hand-in-hand with elimination of those credit lines. It is estimated that in 1995 the government imported 75% of the tubewell pumps, often with assistance from international donors.

Direct controls or taxes (resource use charges) on pumping could constitute an alternative approach. However, given the administrative requirements and costs of direct controls, such an approach would not be feasible without a very substantial strengthening of the institutions of water management.

5 BEARING THE COSTS OF CHANGE

Under policies which aim to reduce groundwater use by lowering the profitability of agriculture, farmers would bear the full cost of the transition. In order to spread the costs of change more equitably, such policies could be accompanied by others that provide compensation to farmers for giving up their water rights and also by programs that help create employment for them in the industrial sector and provide the requisite training.

When higher-value users purchase water rights from lower-value uses, water transfers occur without fiscal cost and those giving up the rights are compensated. There are examples in Yemen of such transactions. Near Sana'a, private water suppliers have purchased wells from farming

families, in order to supply water to expanding urban areas. Obstacles to the development of such transactions on a wider scale include the following:

- Lack of a clear definition of water rights and their marketability.
- Lack of a clear understanding of the true value of water on the part of farmers. Such an understanding could make them more willing participants in schemes for water saving and water reallocation.
- Lack of clear policies on the long-run role of the private sector in supplying water to urban areas.

It is likely that the government's policy objectives include not only achieving sustainability in the use of the natural resource base, but also achieving equity in apportioning the costs of adjustment. The third objective appears likely to be minimization of the role of the government and of the costs to it. A policy approach consistent with these objectives would consist of recognition of the role of private markets in water transfer and supply, and to set up an enabling framework that would allow the private sector to invest in water and manage it efficiently.

The necessity of compensating farmers for loss of their water rights has been recognized in the recent negotiations for transferring water out of the Rada' Basin and for transferring water to Ta'iz from well fields in villages near Ibb and Ta'iz. Therefore, the concept is not new; what are lacking so far are effective mechanisms for facilitating this kind of transfer and for making timely compensation to the rural population. An appropriate solution would consist of developing market mechanisms for effecting these transfers, given their efficiency as an allocative mechanism.

However, *even if such an approach were successful, transfers of water rights out of agriculture to meet increments in urban water demands will not alone solve the problem of overdraft of the highlands aquifers. It also will be necessary to purchase larger quantities of water rights and retire them from use, in order to bring total abstractions more in line with the natural rate of recharge. In this* case, the government clearly would have to finance the purchases of water, either directly or through private companies, since the purchased rights could not subsequently be sold.

There is another fundamental point, which bears on another aspect of policy. In order to avoid exhausting aquifers, *a relatively large share of the agricultural labor force will have to be relocated in non-agricultural occupations. This is not an option but rather a necessity.* If measures are not taken soon to encourage a more rapid intersectoral movement of labor, circumstances will force it upon the economy, as wells begin to dry up and the cost of deepening them becomes prohibitive. In this area, one of the most effective steps the government could take would be to promote the growth of the manufacturing industry, through measures such as reforming the investment law, privatizing State-owned industries, and improving the legal and regulatory framework.

6 SCHEMATIC PRESENTATION OF POLICY OPTIONS

A part of the effort at putting together a national water strategy will necessarily consist of choosing the most appropriate package out of a menu of optional policies like those described in the foregoing sections and others that may be developed through further analysis. To facilitate an overview of the options, Table A.1 lists them and indicates their qualitative impacts by sign (+ or −) on water savings, on the government budget, on the urban population, on the rural population based in rainfed areas, and on the rural population which uses pump agriculture. The urban population would benefit from *all* water saving measures through greater and more regular availability of water. For the sake of clarifying the mechanisms at work, the effects shown in the table are *only* the ones in addition to that general benefit.

As defined in the table, the rural rainfed population includes farm families dependent on both rainfed and spate agriculture. The effects noted with regard to population groups refer to changes in economic welfare. Some of the changes may be accompanied by an intersectoral shift of population, but the effect refers to the population in its sector of origin.

Two additional measures, besides those discussed above, have been added at the bottom of Table A.1: a reduction of the generalized subsidy on wheat, and retaining the subsidy on liquified petroleum, gas (LPG). As the notes to the table explain, *increasing the farmgate price of wheat would be important in order to raise the profitability of rainfed agriculture. It would be desirable to retain in rural areas as much as possible of the population which gives up tubewell agriculture, in order to minimize or stretch out the transition costs, and rainfed cultivation is one of the rural options.* For the same reason, it is important *not* to encourage additional out-migration from rainfed agriculture, in addition to that which is foreseen from tubewell agriculture. The present generalized subsidy on wheat, effected through a low and controlled price and a tariff exemption, could be replaced by one that is targeted on the poor, at a much reduced cost.

The LPG subsidy has been added to the list of options because the present trend of rural households, in shifting from fuelwood to gas for cooking, should be accelerated in order to conserve trees. Although the forestry sector is small in Yemen, it is important for retaining water and conserving soil, and also for inducing more rainfall. A subsidy of this nature could be justified through its externalities. Programs of rural reforestation are included in the list as well, but it should be noted that their success will depend in part on more widespread adoption of LPG for cooking.

7 FISCAL AND STRATEGIC ISSUES

The magnitude of the impending changes that will be brought about by the water crisis will place unusual demands on the system of policy planning and implementation, and on the process of building a consensus with the public. Government funds will be needed to ease the transition out of agriculture for thousands of rural families, through programs like purchasing water rights, training workers for new occupations, and expanding the coverage of the social safety net, and

Table A.1 Indicative matrix of the incidence of policies for saving water

Policy instrument	Net water saving	Government budget	Urban population	Rural rainfed population	Rural pump irrigation population
Subsidy: improved irrigation techniques	+	—			+
Eliminate credit subsidy on pumps (1)	+	+			—
Eliminate government import of pumps (2)	+				—
Higher tariffs on pumps and parts (3)	+	+			—
Raise diesel and electricity prices	+	+	—	—	—
Remove vegetable import controls	+		+		—
Remove qat import controls (4)	+		+		—
Restrictive licenses on drilling (5)	+				—
Raise fertilizer prices (6)	+	+		—	—
Promote market in water rights (7)					+
Purchase water rights for 'set-aside' (8)	+	—			+
Raise water price in urban systems (9)	+	+	—		
Improve water distribution systems (10)	+	—			
Wastewater treatment for irrigation	+	—			+
Privatize urban water supply	+	+	—		
Improved investment rules (11)	+		+	+	+
Vocational training (11)	+	—	+	+	+
Industrial investment subsidies (11)	+	—	+	+	+
Industrial location controls	+				
Subsidy on wheat (12)		+	—	+	
Subsidy on LPG and gas stoves (13)	a	—	+	+	+
Reforestation programs	a	—		+	+

Note: a, indicates would help augment water supply.

Notes on the mechanisms that lead to the results shown in the table

(1) In 1995, the CACB lent many millions of dollars per year for pumps, currently at nominal interest rates of 9 to 11 %, while inflation for 1995 was more than 70 %. A benefit to the government budget is shown as a consequence of abolishing this credit subsidy. A new policy for the CACB was being adopted, under which there would no longer be transfers from the Central Government.

(2) It was estimated that the government accounted for 75 % of the pumps imported each year, in large part through internationally financed projects.

(3) There are doubts as to whether higher tariffs on drilling equipment could be enforced under the existing institutional arrangements.

(4) In the past, qat was imported from Ethiopia. Knowledgeable Yemenis state that it would be a much cheaper source, even allowing for air freight costs, and it would be of better quality.

(5) There are doubts that the administrative capacity exists to enforce controls on drilling and pumping.

(6) Raising the fertilizer price would reduce agricultural water use via the indirect channel of reducing agricultural output. In 1995, the CACB was not charging farmers the full cost of handling and distributing fertilizer, and the implicit subsidy in the price to farmers was about 20 %. The benefit shown for the government budget would go to the CACB.

(7) Under market mechanisms for transferring water, farmers would receive monetary benefits for giving up their water use rights. Transferable water rights are not yet formally recognized in Yemen, but there have been instances of agreement between farmers and the government to provide the former with specified benefits in exchange for giving up part of their water to urban areas. It is anticipated that this model will be followed increasingly in the future.

(8) In addition to reallocating water to sectors where it generates higher value added per unit used, it will be necessary to retire from use considerable amounts of water in order to avoid depletion of aquifers.

(9) In Sana'a, urban households that receive water from private distribution systems pay 9 times the rate paid by households linked to public mains.

(10) For the Sana'a, it was estimated in 1995 that line losses could be as high as 50 % of the water put into the system.

(11) These measures promote increased industrial employment and thus a rural–urban population shift, resulting in a net water savings on average. The economic benefits to the urban population would take the form of more jobs and higher-paying jobs.

(12) Reducing the wheat subsidy for consumers would figure in the water picture in two important ways: (i) it would increase revenues available for financing water-saving programs and investments, and (ii) it would make rainfed agriculture more profitable. In the longer run, agriculture based on pumping groundwater will have to diminish sharply, and so rainfed agriculture will acquire increased importance for the rural population. Additionally, if rainfed agriculture is not made more profitable relative to irrigated agriculture, then the rural migrants attracted to new industries will come overwhelmingly from rainfed areas, and thus there will be little net water savings from an intersectoral reallocation of the labor force.

(13) Promoting household use of gas, a process already underway, will reduce the demand for fuelwood and thus help arrest the progressive deforestation, and that in turn would increase the capacity of watersheds. In addition to this effect, rural households would benefit directly through switching to a cheaper source of energy.

possibly for special transitional subsidies such as those for start-up investments in industry, improved irrigation systems, recycling of waste-water, repair of urban water mains, reforestation, and LPG and LPG stoves.

A substantial part of the funds needed for these kinds of transitional programs could be raised through reductions in existing subsidies in other areas. The main current subsidies include the following: petroleum products, wheat consumption, electricity, urban water, financial services, irrigation water and, to a lesser extent, fertilizer. *The total value of these subsidies in 1994 was about $882 million, or YR 46 746 million. They all were implicit subsidies in the sense that they represent amounts of potential revenue foregone.* In comparison, the official estimate for Central Government revenue in 1994 was YR 41 400 million.

The implicit unit subsidy on irrigation water is the difference between the pumping costs and the willingness of society to pay for the water in other uses. According to informal information gleaned from field observations, it may be assumed that pumping costs are approximately $0.10/m^3 in the highlands. The currently observed willingness to pay by urban households translates into a dollar value of $1.32/m^3 at the estimated average exchange rate for 1995 (of YR 75/$). Interestingly, the cost of supplying Sana'a from its marginal (i.e. most expensive) potential sources of water gives similar numbers: Upper Wadi Surdud well field, $1.46/m^3; Upper Wadi Surdud surface water, $1.42/m^3; Wadi Kharid dam, $1.05/m^3; Marib dam, $1.12/m^3.

Estimated highlands abstractions of groundwater in 1994 were running at about 500 Mm^3. To this figure should be added the abstractions in the Wadi Surdud of the Tihama, of 125 Mm^3, since that resource likely will be transferred to the Sana'a Basin. Using this quantity also for 1995, the total implicit subsidy for water in the highlands areas would be about $762.5 million. A small amount should be subtracted from that total to represent private urban water, but a larger amount should be added to represent the subsidy (at a lower unit rate) for irrigation water in the rest of the Tihama. Therefore, this global figure for the highlands is conservative but yet amounts

to about 18% of the GDP in 1995. (The pump irrigation water subsidy is inclusive of the parts of the subsidies on diesel and electricity that are associated with pumping.)

The significance of the irrigation subsidy is that *the present structure of incentives is very heavily biased in favor of overexploitation of groundwater resources.* It also serves to provide an approximate estimate of the fiscal costs involved if all groundwater rights were to be purchased from farmers in the highlands. A large part of that cost would be allocated to transport of the water to Sana'a, and the remainder would take the form of payments to farmers for their water rights.

Given that these potential fiscal costs would be large in comparison with the present government budget, it appears likely that *the government would have to opt for a combination of measures, partly inducing farmers to give up irrigation by raising the cost of pumping and partly offering them compensation for their water rights.* The higher the pumping costs, then the less money per unit of water would have to be offered to farmers, for their future income streams from irrigated farming decrease as pumping costs rise.

8 SUMMARY

National water resource planning in Yemen is inherently an intersectoral matter. To conclude, some of the principal intersectoral issues can be summarized briefly from the foregoing discussion, as follows:

(1) A solution to the water crisis will require use of both supply options and tools of demand management.
(2) Demand management would be most simply and effectively accomplished by bringing prices of diesel fuel and credit into line with true economic values.
(3) Transfers of groundwater rights from farming to urban uses are part of any feasible future scenario. Transfers can take the form of inducing farmers to give up groundwater irrigation by making it unprofitable, or by effectively purchasing their 'water rights', or by combining the two approaches.

(4) In addition to transfers, considerable amounts of groundwater rights will have to be retired from use, to bring abstractions to a sustainable level. Currently total abstractions, estimated at $2135\,Mm^3$ in the highlands, are running at more than twice the level of natural recharge of $986\,Mm^3$.

(5) Whatever set of policies is employed, it will be important to avoid inducing faster migration out of areas of rainfed agriculture. This may require a rethinking of the priorities of agricultural research and extension services as well as measures such as reducing the generalized wheat subsidy to consumers.

(6) Agricultural self-sufficiency clearly is not a viable long-run objective. Food security can be attained, however, by ensuring that foreign exchange is not rationed and that well-functioning food distribution channels are in place, including programs of targeted subsidies.

(7) Yemen's industrial development will have to be accelerated, along with a more advisable geographical location policy, as part of a solution to the water crisis.

(8) It will be necessary to share the costs of managing the water crisis, designing a package of policies so that neither farmers nor townspeople, nor the general taxpayers, shoulder all of the burden of the transition.

(9) Existing rural development programs should be re-evaluated if they are located in areas with aquifers that are going to dry up soon or whose waters will be tapped for urban needs. The result may be to shift the orientation of these programs toward rainfed agriculture or off-farm rural employment.

(10) Programs that promote greater irrigation efficiency are unlikely to reduce total water abstractions unless the price of diesel is raised from its present highly subsidized level.

(11) The adequacy of rural and urban safety net programs should be evaluated in light of the likely intersectoral movements of population as a result of the water crisis.

The challenge is great, possibly unprecedented in the annals of water management, but the policy tools do exist to handle it. Wisdom will be required in selecting the best mix of these tools, along with decisiveness in their implementation, and full participation of the public in the decision-making process will be necessary in order to effectively implement the program.

The indicative matrix shown in Table A.1 is presented for discussion purposes only. It shows the qualitative impact of policies, positive or negative, on the selected target variables. In the next phase of the work, there needs to be a *quantification* of the impact of these alternative policies, an *evaluation* of their feasibility, and a *reconciliation* of the tradeoffs between conflicting priorities.

Index

Abiad-Shields, Ghada *see* Meinzen-Dick, Ruth, Mendoza, Meyra, Sadoulet, Loic, Abiad-Shields, Ghada, and Subramanian, Ashok
ACLEDA (Cambodia), 289
Access to land, 21, 112, 113–14, 117, 119–20, 176–91, 185–91
Accountability
  agricultural policies, 29, 30
  financial governance, 328
ACP Protocol, 75
Acreman, Michael *see* Dorcey, Tony, Steiner, Achim, Acreman, Michael, and Orlando, Brett
Adam, Christopher, 37, 37f, 48f, 50f
Adams, Dale, W, 284, 297, 297f, 345
  see also Vogel, Robert C, and Adams, Dale, W
Adams, Geoffrey, 409f, 412f
Adams, M, 146, 148
Adams, Martin, and Howell, John, 141f, 143f, 146f, 148f
Adams, Richard, H, 461f
Adams, W M, 201, 201f, 219, 219f
Administered prices, 86–7
Africa
  agricultural strategy, 451
  cereal yield increase, 359, 390
  co-operatives, 174, 320
  collective farms, 169
  communal lands, 132–3
  credit, 320, 351
  extension services, 384–6, 408
  gender issues, 289, 329, 403
  HIV/AIDS, 385–6
  irrigation projects, 214, 216–19, 222, 223, 231, 236, 270
  land reform, 143, 144, 145–6, 148
  land tenure, 119, 123, 124, 129, 130, 151, 152, 153
  land titling, 115–16, 155, 169
  poverty alleviation, 449
  price and trade controls, 19
  research, 364, 365, 390, 403, 461
  State land ownership, 139–40
  technology development, 361, 461
  trade policy, 61
  water resources management, 207, 210
  see also North Africa; South Africa; Sub-Saharan Africa
Agreement on Agriculture (AoA), 59, 67
Agricultural development
  economic policy, 211
  economic transfers, 5–6, 7, 465–8
  educational role, 427–8, 444–6, 452, 456, 457, 458
  exchange rate policy, 77–80
  extension, 405, 457
  gender issues, 46–8, 185–91, 459, 472–3
  governance, 457, 464
  governmental role, 12, 25–32, 426, 429, 461, 465, 468
  holistic approach, 205–6, 207, 210, 404, 458, 464
  infrastructure, 18, 19, 20, 38, 456, 457–8, 468, 470, 471–2
  macroeconomic framework, 35–7, 38, 449, 458
  models, 448, 449, 451–5
  participation, 31, 428–42, 459, 462
  poverty alleviation priorities, 8, 18–19
  rural finance, 277–355
  strategies, 425–60
  sustainability, 418, 451
  trade-offs, 35–7, 38, 426–7, 489, 491
  see also Rural development
Agricultural knowledge and information system for rural development (AKIS/RD), 376–7

Agricultural Development Policy Concepts and Experiences. R. D. Norton
© 2004 Food and Agriculture Organization of the United Nations
ISBNs: 0-470-85778-1 (HB) 0-470-85779-X (PB) FAO Edition: 92-5-104875-4

Agricultural policy framework
  co-ordination, 20, 31, 32
  consistency, 31, 98, 443, 446–50
  controls, 19, 37, 86–9
  dissemination, 31–2
  extension, 405
  governmental role, 20, 21, 25–32, 465
  implementation, 30–2, 91–2, 443, 474–5
  incentives, 21, 443–4, 448, 454
  institutions, 20, 23, 25, 27, 29–30, 444–6, 454, 456,
    457, 459, 462, 470, 471–2
  legislative framework, 20, 30, 51–3, 459, 464, 474,
    475
  macroeconomic framework, 35–7, 38, 449, 458
  objectives, 22–5, 448–9
  participation, 31, 428–42, 459, 462
  policy instruments, 19–22, 36, 486–7
  principles, 23, 25
  productivity improvement, 360–1, 449
  reform sequencing, 91–2
  sectoral strategies, 15–17, 30, 447–8
  taxonomy, 21
Agricultural prices
  determinants, 56–8
  farm prices, 87–9, 94–9, 211
  fiscal policy, 80–6
  food security, 98–103, 202, 491
  indexes, 57
  sectoral policies, 86–98, 456
  see also Food prices; Pricing policies; Real prices;
    Relative prices
Agricultural productivity
  farm size, 125–7
  international competition, 7
  irrigation, 228–9, 260
  labor release, 4, 5, 7, 9, 18
  land ownership, 151, 183
  overall economic well-being, 4–6, 8, 9, 11
  policy context, 360–1
  pricing policies, 36–7
  research, 363, 396
  sectoral strategies, 16, 449
  sharecropping, 161
  smallholders, 6, 11
  subsidizing industrial development, 5, 6–7
  yield increase, 358–9, 390, 391, 453
Agricultural technology *see* Technology
Agriculture
  generalized support, 42–4
  strategic orientations, 455–8
  water management policies, 197–276
Agrochemicals, 96–7, 373–4
AGS (Colombia), 324–5

Ahmed, Raisuddin, 63f
Ahmed, Zia, U, 306–7, 306f
AIDS, extension, 378, 385–7, 414–16, 419
AKIS/RD (Agricultural knowledge and information
  system for rural development), 376–7
Aknazarov, F *see* Tashmatov, A, Aknazarov, F,
  Jureav, A, Khusanov, R, Kadyrkulov, K D,
  Kalchayev, K, and Amirov, B
Akshinbay, A *see* Baydildina, A, Akshinbay, A,
  Bayetova, M, Mkrytichyam, L, Haliepesova, A,
  and Ataev, D
Albania, group lending, 332
Alden Wily, Liz, 116, 116f, 130, 130f, 139, 139f, 141,
  144f, 145–6, 146f, 153, 153f, 154
Ali, M *see* Murgai, R, Ali, M, and Byerlee, D
Allocation
  land, 111–16
  water, 223–5, 227, 249, 486
Amanor, K, and Farrington, J, 382f
Amirov, B *see* Tashmatov, A, Aknazarov, F, Jureav,
  A, Khusanov, R, Kadyrkulov, K D, Kalchayev,
  K, and Amirov, B
Andalucía, Spain, 467
Anderson, Jock, R, 360–1, 363, 364–5, 366, 375–6,
  383, 388
Anderson, Jock, R *see* Picciotto, R, and Anderson,
  J R; Purcell–Dennis, L, and Anderson,
  Jock, R
Andhra Pradesh, 29f, 255
Angel, Amy, L *see* Norton, Roger, D, and Angel,
  Amy, L
Angell, Kenneth, J *see* Boomgard, James, J, and
  Angell, Kenneth, J
Angola, calorie consumption, 99f
Annual implementation plans, 31, 32, 474
Antholt, Charles, H, 365, 365f, 375–6, 382, 383, 383f,
  392, 409, 409f, 410f, 416–17, 417f
Anti-agricultural bias, 8, 82–3, 91, 448
Antichresis, 184, 303, 323
Anti-hoarding laws, 52
Anti-trust legislation, 97
Apex organizations, 324–6, 343
Appreciation, 75–6
Appropriation rights, 239
Area of land, 179–83, 232, 234, 358, 466
Argentina
  bulk handling of rice, 93
  collateral, 304
  commodity taxes, 80
  economic stagnation, 7
  exchange rate policy, 79
  extension services, 410–11
  fruit exports, 94

tax and negative economic growth, 8
water resources management, 262
Argwings-Kodhek, G *see* Jayne, TS, and Argwings-
    Kodhek, G
Arias, Ricardo *see* Norton, Roger, D, Arias, Ricardo,
    and Calderón, Vilma; Strasma, John, Arias,
    Ricardo, García, Magdalena, Meza, Daniel,
    Soler, René, and Umaña, R
Aristizábal, Gladys, Echenique, Jorge, Villalabos, Ruy
    de, and Fischer, Wolfram, 465f, 467f
Aryeetey, Ernest, Hettige, Hemamala, Nissanke,
    Machiko, and Steel, William, 280f
Ashby, J A, 402
Ashby, Jacqueline, A, Braun, Ann, R, Gracia, Teresa,
    Guerrero, María del Pilar, Hernández, Luis
    Alfredo, Quirós, Carlos Arturo, and Roa, José
    Ignacio, 367f, 394f, 402f, 403f
Asia
    farm support prices, 88
    pesticide use, 373–4
    selective credit, 340
    subsidizing industrial development, 5, 6–7
    water pricing, 226
    see also Central Asia; East Asia; South Asia; South
        East Asia
Askari, H, and Cummings, J T, 81f
Assets, 111, 310–11, 314–15, 337
Associative tenure, 121
Ataev, D *see* Baydildina, A, Akshinbay, A,
    Bayetova, M, Mkrytichyam, L, Haliepesova, A,
    and Ataev, D
Atwood, D A, 115, 115f, 119f, 124f, 125, 129, 133,
    133f, 151, 151f, 152, 152f, 158, 158f
Auction of contracts, 261
Auditing, 338, 339
Autarkic approach, 100–1
Aycrigg, Maria, 429f, 441f
Azerbaijan, land ownership, 173
    privatization of extension, 409
Azumi, Hatusya, 260

Babu, S *see* Pinstrup-Andersen, P, Pandya-Lorch, R
    and Babu, S
Badan Kredit Kecamatan (BKK) (Indonesia), 289,
    293, 331
Baffes, John, and Meerman, Jacob, 85f
Bagadion, B U, 257
Bagadion, B U, and Korten, F F, 257f
Bahal, R, 385
Bahal, R *see* Swanson, B E, Farmer, B J, and Bahal,
    R
Balisca, Al *see* Reardon, T, Stamoulis, K, Cruz, M E,
    Balisca, Al, and Berdegué, J

BANADESA (Honduras), 297
Bancafé (Honduras), 289, 322, 323, 338
Banco Caja Social, 338
Banco del Occidente (Honduras), 322
Banco Ganadero (Colombia), 322, 323, 338, 350
BancoSol (Bolivia), 289, 334
Banda, Diana, J *see* Chinene, Vernon, R N, Maimbo,
    Fabian, Banda, Diana, J, and Msune, Stemon, C
Banerjee, Abhijit, V, 110, 110f, 113, 113f, 143, 143f,
    149, 149f, 156f
    Besley, Timothy, and Guinnane, Timothy, 319f
    and Ghatak, Maitreesh, 164f
Bangladesh
    antichresis, 303
    extension, 416–17
    financial training, 297–8
    gender issues, 329, 348–9
    group lending, 332, 333–4
    land mortgage, 184
    loans, 279, 288–9, 289–90, 329
    microfinance institutions, 283, 292, 294, 295, 318,
        345, 351
    returns to irrigation, 199
    subsidized credit, 293
    water resources management, 226, 258
Bangladesh Rural Advancement Committee (BRAC)
    (Bangladesh), 349, 351
Bank for Agriculture and Agricultural Co-operatives
    (BAAC) (Thailand), 331, 340
Bank Dagang Bali, 336
Bank Rakyat, Indonesia, 282f, 289, 298, 307, 331,
    334, 336–7, 338, 339, 340
Bankruptcy law, 53
Banks
    agricultural policy co-ordination, 20
    branching, 312, 343
    cartels, 341
    commercial, 347–8
    explicit subsidies, 38
    farmers, 349–51
    indigenous, 285, 306–7
    land funds, 176–8
    lender of last resort, 313, 326
    loans, 52–3, 124, 125–6, 184, 287, 289
    mobile, 334
    regulations, 308–17, 313–15, 347
    rural finance, 308–17, 322–4
    supervision, 313–17, 315–17, 324, 326, 328, 335, 348
    see also Microfinance institutions; World Bank
Bardhan, Pranab, 28, 29f, 50, 50f
Barry, N, 293f, 297, 298f, 324f, 343, 343f
Basel Convention, 337
Basle Committee, 339

Basle Committee on Banking Supervision, 339f
Basque, Laurent, 78
Bassoco, Luz María, 305
Bassoco, Luz María, Cartas, Celso, and Norton, Roger, D, 305f
Bastidas, Elena, P, 268, 268f
Baydas, Mayada, Graham, Douglas, and Valenzuela, Liza, 347f
Baydildina, A, Akshinbay, A, Bayetova, M, Mkrytichyam, L, Haliepesova, A, and Ataev, D, 172f
Bayetova, M *see* Baydildina, A, Akshinbay, A, Bayetova, M, Mkrytichyam, L, Haliepesova, A, and Ataev, D
Bayoumi, T *see* Masson, P R, Bayoumi, T, and Samiei, H
Bedingar, Touba, 359
Bedingar, T *see* Masters, W A, Bedingar, T, and Oehmke, J F
Befus, David R *see* Reed, Larry R, and Befus, David, R
Beghin, John, C, and Fafchamps, Marcel, 446f
Beilock, Richard, 25
Beilock, Richard *see* Lele, Uma, Emerson, Robert and Beilock, Richard
Beintema, Nienke, 365
Beintema, N M *see* Pardey, P G, Roseboom, J, and Beintema, N M
Belarus, land ownership, 168, 174
Bender, Karen, 97
Bender, Karen L *see* Hill, Lowell, D, and Bender, Karen, L
Benjamin, M P, 286, 298–9, 304, 321, 333
Benjamin, M P Jr *see* Yaron, J, Benjamin, M P Jr, and Piprek, G L
Benor, Daniel, 383
Bentvelsen, Kitty, 269, 269f
Berdegué, J *see* Reardon, T, Stamoulis, K, Cruz, M E, Balisca, Al, and Berdegué, J
Berenbach, Shari, 300
Berenbach, Shari, and Churchill, Craig, 302f, 316f
Berlinsky, Julio, 64, 64f
Berry, Albert, 127
Berry, S, 129, 129f
Besley, Timothy *see* Banerjee, Abhijit, V, Besley, Timothy, and Guinnane, Timothy
Beynon, J, Jones, S and Yao, S, 91f
Binswanger, Hans, 58, 115, 120, 125, 127, 130, 133, 138, 144, 145, 146, 148, 154, 163, 164–5, 176, 181, 185, 241, 244, 451–2, 461–2
Binswanger, H P, 183f, 427f, 451f, 456f, 459f, 462f, 463f

Deininger, K, and Feder, G, 120f, 130f, 133f, 143f, 144f, 146f, 150f, 156f, 163f, 165f, 167f, 181f
and Elgin, M, 185f
McIntire, J, and Udry, C, 164c
Mundlak, Y, Yang, M-C, and Bowers A, 58f
see also Deininger, Klaus, and Binswanger, H; Rosegrant, Mark W, and Binswanger, H P
Binswanger, H P *et al* (citing Berry and Cline), 127f
Biotechnology, 28, 371, 395
Bird, Richard, 180f
Birkhaeuser, D, Evenson, R E, and Feder, G, 361f
Bishop, R C, 131
Bishop, R C *see* Ciriacy-Wantrup, S V, and Bishop, R C
Bledsoe, David, 173
Bledsoe, David *see* Giovarelli, Renee, and Bledsoe, David
Block, Steven, 9
Block, Steven, and Timmer, C Peter, 9f
Bolivia
    collateral, 304
    credit unions, 318
    microfinance institutions, 283, 307, 312, 334
    overgrazing, 133f
Bond financing, 327, 341, 351
Bonded warehouses, 303
Bonnen, James, T, 445, 445f
Boomgard, James, J, and Angell, Kenneth, J, 307f, 334f
Border protection, 59–62
Bos, M G, 232, 233
Bos, M G, and Wolters, W, 232f, 233f
Boserup, Ester, 113f, 118, 118f, 120
Boughton D *see* Reardon, T, Kelly, V, Crawford, E, Diagana, B, Dioné, J, Savadogo, K, and Boughton, D
Bowers, A *see* Binswanger, H, Mundlak, Y, Yang, M-C, and Bowers, A
Brandâo, Antonio Salazar P, and Carvalho, José, L, 64f
Braun, Ann R *see* Ashby, Jacqueline, A, Braun, Ann, R, Gracia, Teresa, Guerrero, María del Pilar, Hernández, Luis Alfredo, Quirós, Carlos Arturo, and Roa, José Ignacio
Braun, Joachim von *see* Zeller, Manfred, Schreider, Gertrud, Braun, Joachim von, and Heidhues, Franz
Brazil
    agricultural growth, 8, 462–3
    bulk handling of rice, 93
    extension, 388
    farm size and productivity, 127

industrial development, 5, 7
land ownership, 125
land reform, 145, 178
loans, 278, 279
research, 398
subsidized credit, 331
tariffs, 64
taxes on land, 183
Britain, land tenure, 118
Bromley, Daniel, 228, 235
Bromley, D W, 228f
see also Larson, B A, and Bromley, D W
Brooks, K
and Lerman, Z, 168f, 169f, 171f, 173f
see also Lerman, Zvi, Brooks, K, Csaki, C
Brown, David, 441
Brown, F Lee *see* Kneese, Allen, V, and Brown,
F Lee
Brown, L David, and Tandon, Rajesh, 441f
Bruce, J W, 162, 162f
Buckwell, A, and Davidova, S, 168f, 171f
Bulgaria, co-operatives, 168, 170, 171
Bulk products, 92–3
Burfisher, Mary, E, 43f, 60f
Burke, Jacob, 204
Burke, Jacob, J, and Moench, Marcus, H, 204f
Burkina Faso
gender issues, 46–7, 369–70, 403
waste water recycling, 213
Burnett, Jill *see* Klein, Brigitte, Meyer, Richard,
Hannig, Alfred, Burnett, Jill, and Fiebig,
Michael
Burt, Charles, 220
Burt, Charles *see* Plusquellec, Hervé, Burt, Charles,
and Wolter, Hans, W
Byerlee, Derek, 361, 372f, 394, 396f, 400
see also Echevarria, Rubén, G, Trigo, Eduardo, J,
and Byerlee, Derek; Maredia, M K, Byerlee,
D and Pee, P; Murgai, R, Ali, M, and
Byerlee, D
Byiringiro, Fidele, 126–7
Byiringiro, F, and Reardon, T, 127f

Cadastral surveys, 181
Cain, M, 167f
Calderón, Vilma *see* Norton, Roger, D, Arias,
Ricardo and Calderón, Vilma
Cambodia, microfinance institution, 289
Cameroon
collective farms, 137–8
gender issues, 46–7
private sector prejudice, 90

Canada
co-operatives, 52, 95
megatariffs, 59
Capacity building, 436–7
Capital
agricultural investment, 279–81, 455
minimum requirements, 309, 310–11, 314–15, 337
seed funding, 350
transfer, 5–6, 7
Capital account funding, 21–2, 31, 37, 474
Caribbean
agricultural strategy, 458
international trade, 59
Carlson, Gerald, A, Zilberman, David and
Miranowski, John, A, 122f
Cartas, Celso, 305
Cartas, Celso *see* Bassoco, Luz María, Cartas, Celso,
and Norton, Roger, D
Carvalho, José, L *see* Brandâo, Antonio Salazar, P,
and Carvalho, José, L
Castaldi, Juan Carlos *see* Gálvez, Gilberto, Colindres,
Miguel, González, Tulio Mariano, and Castaldi,
Juan Carlos
Castillo, L *see* Herdt, R W, Castillo, L, and
Jayasuriya, S
Cavallo, Domingo, and Mundlak, Yair, 8f
Celis, Rafael, 180, 182
Celis, Rafael *see* Strasma, John, and Celis, Rafael
Central America
collective farms, 169
directed credit, 288
income from cash crops, 17
price bands, 66–7
trade restrictions, 69
see also Latin America
Central Asia
market liberalization, 61
privatization, 50
State land ownership, 139
see also Asia; East Asia; South Asia; South East
Asia
Central Europe
co-operatives, 172–3
extension services, 412
subsidies, 71
see also Eastern Europe; Europe; Western Europe
Certificates of deposit, 45, 92, 93, 97, 303
CFA devaluation, 77
CGIAR research, 374
Chabot, Philippe, 50
Chabot, Philippe *see* Goletti, Francesco, and Chabot,
Philippe

Chain of causality, 75
Chambers, Robert, 382
Chamorro, Juan Sebastian, 126
Chamorro, Juan Sebastian *see* Deininger, Klaus, and
    Chamorro, Juan Sebastian
Chancellor, F, 236
Chancellor, F M
    Hasnip, N, and O'Neill, D, 222f, 236f
    and Hide, J M, 219f
Change
    sectoral strategies, 16
    water resources management, 487–8
Chaudhry, M J, 237, 237f
Chayanov, A V, 120F
Chen, Greg, 289f
Chen, Jiyuan, 135–6
Chen, Jiyuan *see* Niu, Ruofeng, and Chen, Jiyuan
Chenery, Hollis, 5
Chenery, Hollis, and Syrquin, Moises, 5f
Chile
    agricultural growth, 8, 77
    environmental protection, 245
    exchange rate policy, 77
    extension services, 410
    investment priorities, 469
    irrigation projects, 257, 269, 271, 462
    loans to women, 329
    price bands, 65
    small farmers, 44
    trade policy, 61, 70
    water resources management, 227, 229, 231, 257,
        258, 263
    water rights markets, 239, 241, 243, 244, 245
China
    co-operatives, 175
    collective farms, 135–7
    extension services, 410
    farm support prices, 88
    gender issues, 186, 473
    growth rate change, 12
    historical background, 3, 55–6
    incentives, 444
    intersectoral labor movement, 18
    irrigation projects, 215, 231, 260–1
    research investment, 28
Chinene, V R N, 125, 132, 142, 187
Chinene, Vernon, R N, Maimbo, Fabian, Banda,
    Diana, J, and Msune, Stemon, C, 118f, 125f,
    132f, 156f
Choksi, Armeane, M *see* Papageorgiou, Demetrios,
    Choksi, Armeane, M and Michaely, Michael
Chowdhury, Mrinal, K, and Gilbert, Elon, H, 416f

Christen, Robert Peck, 289f, 290, 308–9, 314–15,
    315f, 316–17, 330f, 332, 332f, 338
    Rhyne, E, Vogel, R C, and McKean, C 290f
    and Rosenberg, Richard, 309f, 316f, 317f, 319f, 324f
Chuma, Edward, 368
Chuma, Edward *see* Hagmann, Jürgen, Chuma,
    Edward, and Gundani, Oliver; Hagmann,
    Jürgen, Chuma, Edward, and Murwira, K
Churchill, Craig, 300
Churchill, Craig *see* Berenbach, Shari, and Churchill,
    Craig
CIALs (Local Agricultural Research Committees),
    367, 394, 402, 407
Ciriacy-Wantrup, S V, 131
Ciriacy-Wantrup, S V, and Bishop, R C, 131f
Civil society organizations (CSOs), 430–3, 434, 436,
    437–8, 439, 440–1, 465, 474
Clarkson, Max, 328–9
Clarkson, M, and Deck, M, 328f
Clay, Daniel C, and Reardon, Thomas, 162f
Cleaver, Kevin, 449f
Cline, William, 127
Co-operatives
    agriculture, 52, 165, 167–91, 171, 172–3, 174–6
    finance, 53, 310, 318–22
    processing facilities, 95
    see also Collective farms
Coady, David, Dai, Xinyi, and Wang, Limin, 473f
Coercive land reform, 142–3, 145, 148, 149
Coffee
    bond financing, 327
    export incentives, 68
    quotas, 81–2
Coffey, Elizabeth, 278f, 344f
Coleman, J R *see* Faruqee, R, Coleman, J R, and
    Scott, T
Colindres, Miguel *see* Gálvez, Gilberto, Colindres,
    Miguel, González, Tulio Mariano, and Castaldi,
    Juan Carlos
Collateral
    crops in storage, 303–4
    intangible, 347
    land, 124, 125–6, 302–3
    loans, 52–3, 302–5, 311, 314, 323, 330
    social, 302, 332
Collective farms, 45, 95, 116, 121, 134–9
    transfer of ownership, 167–74
    see also Co-operatives
Collier, P, 151
Colombia, 40f
    agricultural product chains, 96
    apex organizations, 324–5

collateral, 303
commodity exchange, 98, 351
gender issues, 187, 369
irrigation projects, 260
land reform, 149, 178–9
loans, 307–8
microfinance institutions, 303, 307–8, 322, 323, 338, 350
research, 367, 394, 395, 397, 463
subsidies, 40
tariffs, 62, 64
Commodity markets, 80–2, 98, 351
Common pool goods, 379
Communal lands, 121, 130–4, 154, 156–7
see also Customary land regimes; Traditional land rights
Communal water rights, 239–40
Comparative advantage, 16, 17–19, 39, 63, 360
Compensation, 85, 466
Competitiveness
governmental role, 26, 27
international trade, 7
pricing policies, 37
sectoral strategies, 16
subsidies, 39
tariffs, 62
Compulsory savings, 296, 336
Conflicts, distributive, 29
Consistency, 31, 98, 443, 446–50
Constantino, Luis, F see Kirmse, Robert, D, Constantino, Luis, F, and Guess, George, M
Consultative Group to Assist the Poorest (CGAP), 297
Consumer price index, 56–7
Consumer protection legislation, 52
Consumptive use rights, 240–1
Contracts
auction, 261
enforcement, 305–6
institutional reforms, 46
lease–purchase, 179
legal strength, 305–6, 459
share tenancy, 163–4
water resources management, 261, 266
women, 190
Coolier, P, 151f
Cornish, G, 219f
Corporate farms, 167–9, 170, 173, 175
Corruption, 29, 311
Costa Rica
apex organizations, 324
fruit exports, 94
land ownership, 125, 126

Costs
cost-effectiveness, 387, 390
irrigation project recovery, 228, 230, 236, 257, 270
operation and maintenance, 216, 228, 229–30, 234, 235, 236
research, 397
rural finance, 315, 338
rural–urban migration, 18
water rights markets, 243–5
Côte d'Ivoire
communal lands, 132
exchange rate policy, 77, 78
gender issues, 403
land tenure, 129, 151–2
Cotlear, Daniel, 118, 118f
Coulter, J, and Golob, P, 91f
Country Assistance Strategies (CASs), 429
Cox, Maximiliano, 462f, 463f
Crawford, E see Reardon, T, Kelly, V, Crawford, E, Diagana, B, Dioné, J, Savadogo, K, and Boughton, D
Credit
co-operatives, 53, 310, 318–22
directed, 288, 340, 346
informal, 285–6
information bureaus, 312, 347
land collateral, 124, 125–6
long-term, 344, 350–1
microcredit programs, 292, 294, 295, 335
peer monitoring, 332, 345
private sector, 304–5
risk management, 296, 303, 310–12, 324, 333, 337, 339
screening, 286–7, 332
subsidized, 6, 36, 38, 40, 287–8, 293, 316–17, 318, 331, 340, 346
supply and demand, 286
Credit markets, 278–84
Credit unions, 310, 316, 318–22
Crisis management, 309, 483, 484, 485–91
Crops
agricultural policies, 21, 22, 446–7
certificates of deposit, 45, 92, 93, 97, 303
cultivation techniques, 393–4, 419
grading standards, 303
higher-value products, 358
income generation, 17
insurance, 305
irrigation policies, 211, 259
liens, 45, 92, 302, 303
protection, 373
research, 393–4, 395, 401

taxation, 6, 80–2
yield increase, 358, 390, 391, 453
Crowley, Eve, 187, 187f, 188f
Cruz, M E *see* Reardon, T, Stamoulis, K, Cruz, M E, Balisca, Al, and Berdegué, J
Csaki, C *see* Lerman, Zvi, Brooks, K, Csaki, C
Cummings, J T *see* Askari, H, and Cummings, J T
Cummings, Ronald, 229
Cummings, R G, and Nercissiantz, V, 229f
Currency boards, 79
Current account budget, 22, 31, 37, 474
Currle, Jochen, and Schütz, Paul, 411f
Customary land regimes, 121, 124–5
    transfer to formal systems, 127–30, 150–7
    see also Communal lands; Traditional land rights
Czech Republic, privatization, 49

Dai, Xinyi *see* Coady, David, Dai, Xinyi, and Wang, Limin
Damhaug, Torbjorn *see* Sharma, Narendra P, Damhaug, Torbjorn, Grey, David and Okaru, Valentina
Damiani, Octavio, 94f, 455f
Dams, water assessments, 212–13
Datt, Gaurav *see* Ravallion, M, and Datt, Gaurav
Datt, Gaurav, 9–10, 12
David, Wilfred *see* McNally, William, David, Wilfred and Flood, David
Davidova, S *see* Buckwell, A, and Davidova, S
de Janvry, Alain, 144, 179
de Klerk, M J, 185
de Zeeuw, Henk, 367
Decentralization
    autonomy, 466–7
    extension services, 362, 406, 419, 454
    fiscal, 463–4
    government, 28, 29, 250–1, 252, 459, 463–4
    institutions, 457, 459
    land registration, 159, 181
    microfinance institutions, 300–1, 323, 340
    research, 45, 392–3, 396–7, 454
    rural development, 462–5, 466–8
    water resources management, 250–1, 252, 255–64
Deck, M *see* Clarkson, M, and Deck, M
Deck, Michael, 328–9
Deininger, Klaus, 10, 115, 120, 126, 127, 130, 137, 137f, 144, 145, 147, 148, 149, 149f, 154, 177, 178, 178f
    and Binswanger, H, 115f, 144f, 145f, 154f, 155f, 165f, 181f
    and Chamorro, Juan Sebastian, 126f
    and May, Julian, 177f
    Olinto, Pedro and Maartens, Miet, 147f, 148f

and Squire, Lyn, 10f
see also Binswanger, H P, Deininger, K and Feder, G
Deneke, K *see* Pelletier, D L, Deneke, K, Kidane, Y, Haile, B, and Negussie, F
Denizer, Cevdet *see* Melo, Martha de, Denizer, Cevdet and Gelb, Alan
DHV Consultants BV, 485f, 486f
Denmark, co-operatives, 52
Deposit insurance, 313
Depreciation, 74–5, 79
Deregulation, 50
Devaluation, 77–9
Development Assistance Committee (DAC), 73
Diagana, B *see* Reardon, T, Kelly, V, Crawford, E, Diagana, B, Dioné, J, Savadogo, K, and Boughton, D
Diaz-Bonilla Eugenio, 58–9
Diaz-Bonilla, E, and Reca, L, 59f
Diesel fuel subsidies, 485, 486–7, 491
Dinar, Ariel, 232
    and Subramanian, Ashok, 226f
    see also Tsur, Y, and Dinar, A
Dioné, J *see* Reardon, T, Kelly, V, Crawford, E, Diagana, B, Dioné, J, Savadogo, K, and Boughton, D
Diop, Jean-Marie, Jong, Marga de, Laban, Peter, and Zeeuw, Henk de, 368f
Directed credit, 288, 340, 346
Discrimination against agriculture, 8, 82–3, 91, 448
Distributive conflicts, 29
Diversion rights, 240–2
Dixon, John, 206
Dixon, John A, and Easter, K William, 206f
Dobrilovic, Stevan, 128f
Doggert, Clinton L Jr, 123, 123f
Dominican Republic
    collective units, 45
    commodity taxes, 80–1
    gender issues, 187, 369
    participatory processes, 430
    rural savings, 282
    water resources management, 258
Donnelly-Roark, P *see* Tikare, S, Youssef, D, Donnelly-Roark, P, and Shah, P
Donor funding, 310, 342–6
Donovan, C *see* Tschirley, D, Donovan, C, and Weber, M T
Dorcey, Tony, Steiner, Achim, Acreman, Michael, and Orlando, Brett, 212f
Doss, Cheryl, 370–1, 403–4
Doss, C R, and Morris, M L, 370f, 404f

Douglas, Roger, 31
Drainage systems, 215, 265
Dual-economy model, 5, 451
Due, J M, Mollel, N, and Malone, V, 384f
Dumping, 65
Duron, Guadalupe, 369
Duron, Guadalupe *see* Paris, Thelma, Feldstein, Hilary Sims, and Duron, Guadalupe
Dürr, Georg *see* Lamers, John, Dürr, Georg, and Feil, Petra

East Asia
  coercive land reform, 143
  export growth, 7, 61
  governmental role, 27, 344
  industrialization, 6–7
  price and trade controls, 19
  rural economy, 11–12
  urban–rural income differentials, 7
  women's education, 47
  see also Asia; Central Asia; South Asia; South East Asia
Easter, K William, 206, 229, 244, 245, 248
Easter, K William *see* Dixon, John, A, and Easter, K William; Hearne, Robert, R, Easter, K William
Eastern Europe
  co-operatives, 171, 172–3
  corporate farms, 168, 170
  extension services, 412
  holding companies, 51
  land tenure, 128, 173–4
  State land ownership, 49, 167
  subsidies, 41, 71
  see also Central Europe; Europe; Western Europe
Eastern Germany (former), land ownership, 169
Echenique, Jorge, 269f, 271, 271f
  see also Aristizábal, Gladys, Echenique, Jorge, Villalabos, Ruy de, and Fischer, Wolfram
Echevarria, Rubén G, 360, 360f, 368, 368f, 394, 398
  Trigo, Eduardo J, and Byerlee, Derek, 360f, 394f, 395f, 396f, 397f
Eckert, J, 123, 123f
Econometric models, 9
Economic association, 52
Economic efficiency
  agricultural policies, 85–6, 111, 114, 122, 151
  irrigation policies, 206, 207, 211, 228, 260
  land tenure policies, 111, 114, 122, 151
Economic growth, 4–9, 485
Economic policy *see* Rural finance
Economic stability, 35–6, 103–4, 342, 449

Economic sustainability, 23, 30
Economic transfers, 5–6, 7, 465–8
Ecuador
  bulk handling of rice, 90
  exchange rate policy, 78f
  gender issues, 268, 348
  land ownership, 125
  trade restrictions, 70
Education and training
  agricultural strategies, 427–8, 444–6, 452, 456, 457, 458, 468
  extension, 376–7, 388, 414, 419, 420, 457
  farm restructuring, 173
  finance management, 297–8, 329, 336, 344, 347, 351, 463
  gender issues, 189, 403, 473
  investment, 26
  poor rural families, 46
  subsidies, 41
  women, 47, 185, 473
Edwards, Sebastian, 80
Egypt, 40f
  credit unions, 318
  irrigation design, 222
  loans, 278
  subsidies, 40
El-Said, Moataz *see* Löfgren, Hans, and El-Said, Moataz
El Salvador
  agricultural strategy, 447, 454
  bond financing, 327
  coffee quotas, 81–2
  collective farms, 45, 95, 134–5, 170, 171
  corporations, 168
  exchange rate policy, 75, 76
  extension services, 409
  land funds/banks, 176, 177–8
  land titling, 158, 159
  price bands, 66–7
  rural–urban migration, 18
  tariffs, 61
Elgin, M, 185
Ellis, Frank, 89
Ellis, F, Senanayake, P, and Smith, M, 89f
Emergencies, 41, 74
Emerson, Robert, 25
Emerson, Robert *see* Lele, Uma, Emerson, Robert, and Beilock, Richard
Endowed research foundations, 398–9
Endowment of assets, 111
Engel's Law, 451
Environmental sustainability
  agricultural policies, 17–18, 23, 25

irrigation policies, 203–4, 206, 207, 223, 227, 487
land area expansion, 358
land tenure policies, 111–12, 114, 117
research, 365, 366, 396
subsidies, 41
water extraction, 204, 212, 245–6
Equity
investment capital, 281, 310, 331
irrigation policies, 199–200, 203, 204, 228, 487
land tenure policies, 111, 112, 114, 144, 163–4, 165, 166
water rights markets, 246–7
Eritrea, land tenure, 153
Estonia
economic transfers, 465
exchange rate policy, 76, 79–80
extension services, 409
farm titling, 45, 183
participatory processes, 435, 438, 442–3
productivity, 183
rediscount lines, 326–7
rural development strategy, 460–1
State land ownership, 140, 141, 169, 170
Estonia, Ministry of Agriculture, 80f, 170f
Ethiopia
collective farms, 137
food security, 101
irrigation projects, 203, 259–60
land rental, 162
share tenancy contracts, 163
State land ownership, 141
Europe
farm support prices, 88
price and trade controls, 19
see also Central Europe; Eastern Europe; Western Europe
European Union
agricultural support, 43–4
environmental sustainability, 115
exchange rate policy, 75
food prices, 56
food safety standards, 93
McSharry Plan, 85, 465
Evenson, R E, 363, 364
Pray, Carl, E, and Rosegrant, Mark, W, 364f
and Rosegrant, M W, 363f
Waggoner, P E, and Ruttan, V W, 26f
and Westphal, Larry E, 363f
see also Birkhaeuser, D, Evenson, R E, and Feder, G
Exchange rates
appreciation, 75–6

currency boards, 79
depreciation, 74–5, 79
devaluation, 77–9
export incentives, 7
irrigation development, 211
macroeconomic framework, 36, 58, 83, 84, 447
negative economic growth, 8–9
overvalued, 299
policy, 74–80, 209
role, 74–7
Excludability, 378–80
Explicit subsidies, 37, 38
Exports
agricultural prices, 57–8
crop taxation, 6, 80–2
East Asia, 7, 61
incentives, 67–8
macroeconomic framework, 36
subsidies, 6–7, 38, 60, 67, 457
substitutes, 84
tariffs, 58, 62, 67, 457
trade restrictions, 68–72
Extension, 357–60
administrative constraints, 388
agents, 376, 377–84, 388, 395, 409
agricultural strategies, 457
alternatives, 404–8
client orientation, 409–13
decentralization, 362, 406, 419, 454
farmer-led approaches, 388, 393, 407, 409–13, 418–19
funding, 377–84, 387–8, 405–6, 420
gender issues, 384–5, 413–14, 419
historical background, 375–7
HIV/AIDS effects, 385–6, 419
incentives, 382, 419
new approaches, 404–20, 417–20
new paradigm, 387–8
para-professionals, 407
participatory, 406–7, 412–13, 418, 419, 420
private sector, 361
services, 46, 362, 375–90, 404
subsidized, 380, 381
trends, 416–17
see also Research; Technology
Ewing, A see Lieberman, I W, Ewing, A, Mejstrik, J, Mukherjee, J, and Fidler, P (Eds)

Fafchamps, Marcel, 163f
see also Beghin, John, C, and Fafchamps, Marcel
Falcon, Walter, 449, 460

Falcon, Walter, P *see* Timmer, C Peter, Falcon, Walter, P, and Pearson, Scott, R
Family law, 190
Famines, 90
Fan, S, 400
Fan, S, Hazell, P, and Haque, T, 372f, 400f
FAO, 46f, 47f, 48f, 67f, 69f, 77f, 83f, 99f, 100f, 101f, 121f, 198f, 199f, 202f, 207f, 209f, 213f, 214f, 219f, 221f, 233f, 237f, 247f, 249f, 251f, 255f, 265f, 270f, 279f, 329f, 330f, 385f, 402f, 413f, 414f, 448f
  CFA devaluation, 77
  extension services, 384–5, 402–3, 413–14
  Farmer Field Schools, 374–5, 388, 394, 395, 407
  gender issues, 190, 329–30, 384–5, 402–3, 413–14
  irrigation, 199, 202, 206, 207, 208–9, 215, 219, 221
  Land Tenure Service, 176f, 177f, 183, 183f, 188f, 189, 189f, 191f
  undernourishment reduction, 99–100
  water policies, 213, 233, 234, 247, 250, 251, 255–6, 270
  and the World Bank, 377f
Farm prices, 87–9, 94–6, 211
Farmer, B J, 385
Farmer, B J *see* Swanson, B E, Farmer, B J, and Bahal, R
Farmer Field Schools (FFSs), 374–5, 388, 394, 395, 407
Farmers' associations, 392, 393, 405, 407–11, 458, 470, 471
Farms
  banking, 349–51
  collectives, 45, 95, 116, 121, 134–9
  management, 378, 419
  size, 125–7, 174, 399
  small, 45, 127, 136, 158, 160, 183–4, 200, 204
  see also Land, titling
Farrington, J, 377f, 404f, 407f
  see also Amanor, K, and Farrington, J
Farrington, John, 377, 404, 406–7
Faruqee, R, Coleman, J R, and Scott, T, 97f
Feder, Gershon, 120, 123, 125–6, 127, 128, 130, 151, 165–6
  and Feeny, D, 128f, 133f
  and Noronha, R, 123f, 125f, 126f, 128f, 130f, 154f, 166f
  and Onchan, Tongroj, 125f
  Onchan, Tongroj and Raparla, Tejaswi, 125f
  see also Binswanger, H P, Deininger, K, and Feder, G; Birkhaeuser, D, Evenson, R E, and Feder, G
Feeny, David, 128

Feeny, D *see* Feder, Gershon, and Feeny, D
Fei, John, 5
Fei, John, C H, and Ranis, Gustav, 5f, 451f
Feil, Petra *see* Lamers, John, Dürr, Georg, and Feil, Petra
Fekete, Ferenc, 137
Fekete, Ferenc, Fènyes, Tamas, I, and Groenewald, Jan, A, 137f
Feldstein, Hilary, 369
Feldstein, Hilary Sims *see* Paris, Thelma, Feldstein, Hilary Sims, and Duron, Guadalupe
Fènyes, Tamas, 137
Fènyes, Tamas, I *see* Fekete, Ferenc, Fènyes, Tamas, I, and Groenewald, Jan, A
Fertilizers, 96–7, 98, 403
Fidler, P *see* Lieberman, I W, Ewing, A, Mejstrik, J, Mukherjee, J, and Fidler, P (Eds)
Fiduciary responsibilities, 328
Fiebig, Michael, 309f, 309–10, 313, 313f, 317, 317f, 339, 346, 347, 350
  see also Klein, Brigitte, Meyer, Richard, Hannig, Alfred, Burnett, Jill, and Fiebig, Michael
Finance *see* Rural finance
Fiscal expenditure
  compensation, 85, 466
  decentralization, 463–4
  extension services, 410, 417
  irrigation projects, 203, 207, 258, 270–1, 488, 490
  macroeconomic framework, 35–6, 58, 84–5
  policy, 80–6
  subsidies, 37–45
  sustainability, 23, 203, 365, 387
Fischer, Stanley, 49–50, 50f
Fischer, Wolfram *see* Aristizábal, Gladys, Echenique, Jorge, Villalabos, Ruy de, and Fischer, Wolfram
Fisher, William *see* Trackman, Brian, Fisher, William, and Salas, Luis
Fishlow, Albert, 113f
Fisman, R, and Gatti, R, 29f
Fleisig, Heywood, 303f, 304, 304f, 306, 306f
Flood control, 198, 265, 271
Food
  aid, 72–4, 100
  assistance, 64, 73, 91, 102
  reserves, 89–92
  safety standards, 93–4
  security, 98–103, 202, 491
  subsidies, 9, 40f
  supply and demand, 57–8, 358
Flood, David *see* McNally, William, David, Wilfred, and Flood, David
Fonseca, Luz Amparo, 94f

Food prices
  financial food security reserves, 92
  poverty reduction, 11–12, 63–4, 449–50
  producer incentives, 55–8
  see also Agricultural prices; Pricing policies; Real
    prices; Relative prices
Forced savings, 296, 336
Forests, 182, 488
Former Eastern Germany, land ownership, 169
Former Soviet Union
  co-operatives, 171
  collective farms, 121, 167
  corporate farms, 168
  land tenure, 128, 173
  price controls, 80
  State land ownership, 49, 140
Forni, Nadia, 130–1, 131f, 133f, 141, 157, 157f
Foundation for International Community Assistance
    (FINCA), 329
Fox, James *see* Malhotra, Mohini, and Fox, James
France, water resources management, 225
Fraud, 311, 317
Free trade agreements, 12, 62, 98
Fruit exports, 94
Fry, Maxwell J, 285, 285f, 306–7, 307f, 340f, 340–1,
    341f, 342, 342f, 345
Fulginiti, Lilyan, 358–9, 360
Fulginiti, L E, and Perrin, R K, 358f
Fundación de la Mujer Campesina (FUNDELAM),
    348f
Furtado, Celso, 5, 5f, 7

Galán, Beatriz, B, 187f, 188f
Gale Johnson, D, 4
Gallagher, Kevin, D, 375f
Galvéz, Gilberto, 381–2
Gálvez, Gilberto, Colindres, Miguel, González, Tulio
    Mariano, and Castaldi, Juan Carlos, 381f
The Gambia
  agricultural strategy, 433
  village finance, 318
García, J, 62, 62f
García, Luis, E, 200f, 207f
García, Magdalena, 63, 102–3, 400
    Norton, Roger, Ponce Cámbar, Mario, and van
      Haeften, Roberta, 102f, 127f
    see also Norton, Roger, D, and Garcia, U,
      Magdalena; Schreiner, Dean, F, and García,
      U, Magdalena; Strasma, John, Arias, Ricardo,
      García, Magdalena, Meza, Daniel, Soler,
      René, and Umaña, R
Gardner, Bruce, L, 83f
Gariyo, Zie, 432f

Gavian, Sarah, 90–1
Gavian, S *see* Mellor, J, and Gavian, S
Gazmuri, S R *see* Rosegrant, Mark, W, Gazmuri,
    S R, and Yadav, S N
GDP, agricultural contribution, 8, 10, 11, 451
Gelb, Alan *see* Melo, Martha de, Denizer, Cevdet,
    and Gelb, Alan
Gender issues
  access to land, 185–91
  agricultural development, 46–8, 185–91, 459,
    472–3
  education, 189, 403, 473
  extension services, 384–5, 413–14, 419
  irrigation systems, 204, 207, 222, 235–6, 268–9, 403
  loans to women, 288–9, 289, 294, 295, 329–30, 348
  mainstreaming, 47
  poverty alleviation, 289–92
  research, 366, 367, 368–71, 391, 402–4
  rural development, 46–8, 185–91, 472–3
  rural finance, 329–30, 347, 348–9
  water pricing, 235–6
Geographic distribution, 301, 322, 392, 440, 491
Germany
  extension services, 411
  land ownership, 169
Gerpacio, R V *see* Pingali, P L, and Gerpacio, R V
Ghana
  communal lands, 132
  credit unions, 318
  gender issues, 370–1, 403–4
  traditional land transfer, 127
Ghatak, Maitreesh *see* Banerjee, Abhijit, V, and
    Ghatak, Maitreesh
Giehler, Thorsten, 344f
Gilbert, Elon H *see* Chowdhury, Mrinal K, and
    Gilbert, Elon H
Giovarelli, Renee, 173
Giovarelli, Renee, and Bledsoe, David, 174f
Gisselquist, David, 389
Gisselquist, D, and Grether, J-M, 389f
Globalization, 12, 30, 93–4, 395
Glosser, Amy J, 307f, 314f
Goldin, I, Knudsen, O and Mensbrugghe, D van der,
    43f
Goletti, Francesco, 50
  and Chabot, Philippe, 50f, 61f, 83f
  see also Minot, Nicholas, and Goletti, Francesco
Golob, P *see* Coulter, J, and Golob, P
Gómez Alfonso, Arelis (with Nan Barton and Carlos
    Castello), 325f
González, Tulio Mariano *see* Gálvez, Gilberto,
    Colindres, Miguel, González, Tulio Mariano,
    and Castaldi, Juan Carlos

González Orellana, Mauricio, 82f
Gonzalez-Vega, Claudio, 279f, 308f
Gordillo de Anda, Gustavo, 177, 178f, 473–4, 474f
Gorriz, Cecilia M, 256
  Subramanian, Ashok, and Simas, José, 256f
  see also Groenfeldt, David, and Gorriz, C
Government
  accountability, 29, 30
  agricultural development, 12, 25–32, 426, 429, 452, 461, 465, 468
  annual implementation plans, 31, 32
  co-ordination, 20, 31, 32
  consensus, 31
  credit co-operatives, 320
  decentralization, 28, 29, 250–1, 252, 459
  expenditure, 19, 37–44
  extension services, 41, 380–4, 406, 407, 418, 4209
  farm support prices, 87–9
  financial institutions, 317, 323, 330–1, 344–6
  food prices, 56
  gender issues, 269
  incentives, 61
  industrial development subsidies, 4–7, 486, 491
  institutions, 20, 23, 25, 27, 29–30, 115–16, 401–2
  irrigation projects, 261–2, 263, 271, 487–8, 490
  loan guarantees, 305
  managed deposit rates, 340–1
  participatory processes, 433, 441
  pest management, 373
  policies, 20, 21, 25–32
  post-reform investment, 28
  pricing agreements, 95–6
  private sector role, 27–8
  production and marketing management, 19
  programs, 21–2, 73, 74
  projects, 21–2, 73
  research, 390
  sectoral strategies, 15, 16, 452–3
  strategic reserves, 89, 91–2
  subsidies, 39–40, 296, 340
  technology transfer, 389
  urban-biased policies, 299
  water allocation, 224–5, 249
  water resources management, 249–52, 261, 265, 266, 269, 487
  see also Public sector; State intervention
Gow, Hamish R *see* Swinnen, Johan FM and Gow, Hamish R
Gracia, Teresa *see* Ashby, Jacqueline A, Braun, Ann R, Gracia, Teresa, Guerrero, María del Pilar, Hernández, Luis Alfredo, Quirós, Carlos Arturo, and Roa, José Ignacio

Graham, Douglas H, 318
  Nagarajan, Geetha, and Quattara, Korotoumou, 298f
  see also Baydas, Mayada, Graham, Douglas, and Valenzuela, Liza; Nagarajan, Geetha, Meyer, R L, and Graham, D H
Grain
  certificates of deposit, 45, 92, 93, 97
  market liberalization, 391–2
  silo privatization, 45, 49–50, 92, 168, 427
  storage policy instruments, 92–3
  strategic reserves, 89–92
Grameen Bank (Bangladesh), 283, 288–9, 289, 293, 295, 297–8, 318, 332, 333–4, 345, 349
Gravity irrigation, 221
Green Revolution, 362, 376, 396
Greenwald, B, 345
Greenwald, B, and Stiglitz, J, 345f
Grether, Jean–Marie, 389
Grether, J-M *see* Gisselquist, D, and Grether, J-M
Grey, David *see* Sharma, Narendra P, Damhaug, Torbjorn, Grey, David and Okaru, Valentina
Groenewald, Jan, 137
Groenewald, Jan A *see* Fekete, Ferenc, Fènyes, Tamas I, and Groenewald, Jan A
Groenfeldt, David, 260f, 262f
  and Gorriz, C, 266f
Groppo, Paolo *see* Munro-Faure, Paul, Groppo, Paolo, Herrera, Adriana, and Palmer, David
Groppo, Paulo, 110–11
Grosh, Margaret, 64f
Groundwater supplies, 203–4, 205, 212, 222–3, 484, 486, 487, 490–1
Group lending, 184, 302, 332–3, 345
Group titles, 133, 156
Guatemala
  credit co-operatives, 321
  land funds, 177
  tariffs, 62
Guerrero, María del Pilar *see* Ashby, Jacqueline A, Braun, Ann R, Gracia, Teresa, Guerrero, María del Pilar, Hernández, Luis Alfredo, Quirós, Carlos Arturo, and Roa, José Ignacio
Guess, George M *see* Kirmse, Robert D, Constantino, Luis F and Guess, George M
Guillaumont, Patrick, 86, 86f
Guinea, exchange rate policy, 78
Guinnane, Timothy *see* Banerjee, Abhijit V, Besley, Timothy, and Guinnane, Timothy
Gundani, Oliver, 368
Gundani, Oliver *see* Hagmann, Jürgen, Chuma, Edward, and Gundani, Oliver
Gunning, Jan Willem, 475, 475f

Guyana
  exchange rate policy, 75
  participatory processes, 428, 433–4, 435, 436, 437, 438, 439, 442–3
  privatization, 50, 428
  State land ownership, 141, 142
  trade restrictions, 69, 70
Guyana, Ministry of Finance, 141f

Hacienda system, 120, 121, 140
Haddad, Lawrence, 401
Haddad, Lawrence see Hazell, Peter, and Haddad, Lawrence
Hadley, D, 363–4
Hadley, D see Thirtle, C, Hadley, D, and Townsend R
Hag Elamin, N, 59, 60f
Hagmann, Jürgen, 368, 412
  Chuma, Edward, and Murwira, K, 412f
  Chuma, Edward, and Gundani, Oliver, 368f
  see also Moyo, Evison, and Hagmann, Jürgen
Haile, B see Pelletier, D L, Deneke, K, Kidane, Y, Haile, B, and Negussie, F
Haiti
  commodity taxes, 81
  environmental sustainability, 114
  tree rental, 122
Haliepesova, A see Baydildina, A, Akshinbay, A, Bayetova, M, Mkrytichyam, L, Haliepesova, A, and Ataev, D
Halter, A N see Officer, R R, and Halter, A N
Hanke, Steve H, and Schuler, Kurt, 79f
Hannig, Alfred see Klein, Brigitte, Meyer, Richard, Hannig, Alfred, Burnett, Jill, and Fiebig, Michael
Haque, T see Fan, S, Hazell, P, and Haque, T
Hardin, L S, 359f
Hashemi, Syed (with Maya Tudor and Zakir Hossein), 292f
Hasnip, N see Chancellor, F M, Hasnip, N, and O'Neill, D
Hayami, Yujiro, 148–9, 149f, 360, 453–5
  see also Ruttan, Vernon W, and Hayami, Yujiro
Hazell, Peter B R, 401
  and Haddad, Lawrence, 401f
  and Norton, Roger D, 365f
  see also Fan, S, Hazell, P, and Haque, T
Hearne, Robert R, 229, 244, 245, 248
Hearne, Robert R, and Easter, K William, 229f, 243f, 244f, 245f, 248f
Heber Percy, Robin, 103

Heidhues, Franz see Zeller, Manfred, Schreider, Gertrud, Braun, Joachim von, and Heidhues, Franz
Heiman, Amir see Zilberman, David, Yarkin, Cherisa, and Heiman, Amir
Heltberg, R, 126, 126f, 127
Heney, Jennifer, 347f
Henneberry, Shida Rastegari, 81f
Herdt, R W, 373–4
Herdt, R W, Castillo, L, and Jayasuriya, S, 373f
Hernández, Luis Alfredo see Ashby, Jacqueline A, Braun, Ann R, Gracia, Teresa, Guerrero, María del Pilar, Hernández, Luis Alfredo, Quirós, Carlos Arturo, and Roa, José Ignacio
Herrera, Adriana, 110–11, 144, 145
  Riddell, Jim and Toselli, Paolo, 143f, 144f, 145f, 154f
  see also Munro-Faure, Paul, Groppo, Paolo, Herrera, Adriana, and Palmer, David
Hettige, Hemamala see Aryeetey, Ernest, Hettige, Hemamala, Nissanke, Machiko, and Steel, William
Hide, J M see Chancellor, F M, and Hide, J M
Hill, Lowell, 97
Hill, Lowell D, and Bender, Karen L, 97f
Historical background
  agricultural policies, 3–4, 18
  extension, 375–7
  food prices, 55–6
  interest rates, 306
  irrigation, 197–8
  land reform, 142–3, 147
  land tenure, 110, 117–20
  lending practices, 277–8
HIV/AIDS, extension, 378, 385–7, 414–16, 419
Hoff, Karla, 164f, 286, 287
  and Stiglitz, Joseph, 41f, 287f, 345f
Holding companies, Eastern Europe, 51
Holistic approach, 205–6, 207, 210, 404, 458, 464
Holland
  appreciating exchange rate, 75
  research funding, 395
Hommes, Rudolf, 463f
Honduras
  agrarian reform, 134f
  coffee quotas, 81–2
  collective farms, 135, 138, 170, 171
  communal lands, 134
  corporations, 168
  crop insurance, 305
  extension services, 381–2
  farm size and productivity, 127

farm support prices, 88
farm titling, 45, 147, 183
food security, 102–3
governmental role, 465
information provision, 380–1
land rental, 162
land tenure, 156, 158
loans, 289, 297, 305
microfinance institutions, 289, 297, 322, 323, 338
NGO credit programs, 317
participatory processes, 427, 430–1, 434–5, 436, 438, 439, 440
price bands, 66–7
research, 398–9, 400
rural development, 462
State land ownership, 142
subsidies, 39, 50
tariffs, 63, 64
taxes on land, 179
women's access to land, 187
Howell, J, 146, 148
Howell, John *see* Adams, Martin, and Howell, John
Hristova, M, Maddock, N, 168f, 180f
Hubbard, Michael, 426
Hulme, David, 290, 291
Hulme, David, and Mosley, Paul, 291f
Human capital
agricultural strategies, 444, 455, 456, 457, 470, 471–2
development, 22, 362
extension, 418
farm restructuring, 173
financial institutions, 336
investment, 26, 281
policies, 21, 22
Hungary, collective farms, 137, 171
Hunger, 99–100, 111
Hushak, L J *see* Nagarajan, G, Meyer, R L, Hushak, L J
Hutchens, Adrian O, 200f
Hybrid seeds, 371–2, 394

IFAD, 46f, 394f, 470f, 471f, 473f
IFPRI, women's education, 47
Immink, Bodo, 440–1
Immink, Bodo, and Olagoke, Macauley (Eds), 441f
Implementation
research, 391–6
strategies and policies, 30–2, 91–2, 443, 474–5
Implicit subsidies, 37–8
Imports
agricultural prices, 57–8, 60

exchange rate policy, 77
phytosanitary controls, 70
substitutes, 5, 7, 84
tariffs and quotas, 37, 58, 62, 67, 70–1
trade restrictions, 68–71, 388–9
Incentives
agricultural policies, 21, 443–4, 448, 454
export, 7, 67–8
extension agents, 382, 419
financial staff, 340
loan repayment, 286–7, 332, 333–4
pricing policies, 11, 36
producers, 55–107
rural savings, 283–4
Income
access to land, 113–14
agricultural growth and poverty reduction, 10–11, 17, 449–50
production, 287–9
purchasing power, 22
rural improvement, 44–6
urban–rural differentials, 7, 43, 449
Income Generation for Vulnerable Groups Development (IGVGD), 292
India
anti-agricultural bias, 82–3
canal irrigation, 28–9
communal lands, 131
contributions of irrigation, 202
environmental sustainability, 396
exchange rate policy, 79
gender issues, 46, 186, 329
groundwater extraction, 204
interest rates, 307
irrigation projects, 201, 210, 233, 259
loans, 279, 338
microfinance institutions, 291, 298, 331
pest management, 394
poverty reduction, 9–10
research, 363, 364, 372, 400
share tenancy contracts, 164
taxes on land, 181
water resources management, 224, 233, 259, 266–7
water rights markets, 237
women's access to land, 186
Indonesia
extension, 388
farm support prices, 89
Farmer Field Schools, 375
irrigation projects, 220, 258
loans, 278, 289, 331

microfinance institutions, 289, 293, 298, 307, 309, 312, 331, 336–7, 340
pest management, 373
savings, 282, 334–5, 338
subsidized credit, 293
Industrial development
agricultural growth, 5, 6–7, 451–2
capital investment, 279–80
credit subsidies, 6
government subsidies, 4–7, 486, 491
India, 9–10
State ownership, 7
Industrialization, 6–7, 280, 299, 486, 491
Inflation, 74–5, 76, 79–80, 103, 283, 331, 341–2
Information provision
access to information, 380–1
extension, 376–7, 378, 379–81, 405
HIV/AIDS, 416
internet, 397, 408
management information systems, 311, 323, 334, 339
marketing, 26–7, 97, 378, 419, 455
price data, 95–6
research, 399
rural finance, 340
subsidies, 41
Infrastructure
agricultural policies, 18, 19, 20, 38, 456, 457–8, 468, 470, 471–2
fiscal policy, 80, 85
land reform, 145
pricing policies, 97
rural development, 26, 345, 400–1, 468, 471–2
rural–urban migration, 18
water allocation, 224
Inheritance law, 190
Insecticides, 373–4
Institutions
agricultural strategies, 444–6, 454, 456, 457, 459, 470, 471–2
apex organizations, 324–6, 343
co-operatives, 175
decentralization, 457, 459
efficiency, 85–6
financial, 283–6
gender issues, 191
governance, 297, 300–1, 310, 325, 328–9, 457
governmental role, 25, 27, 29–30
growth management, 301–2
irrigation projects, 248–54
land tenure transitions, 128–30
research, 361–3, 368, 397–8
rural development, 20, 401–2, 462, 470

rural finance, 281–2, 287, 289–99, 290–2, 300–2
subsidies, 40
sustainability, 23, 111, 115–16, 203, 207, 250
water resources management, 203, 207, 208, 209, 248–69
see also Microfinance institutions
Insurance, 305, 313
Intangible collateral, 347
Integrated pest management (IPM), 373–5, 394
Intellectual property rights (IPRs), 361, 371–2
Inter-American Development Bank, 126f, 199f, 210–11, 211f, 345–6, 462, 465
Interest rates
ceilings, 307–8, 346
historical background, 306
inflation, 341–2
loans, 315–16, 323, 331–3, 335
macroeconomic framework, 36
managed deposit rates, 340–1
market levels, 296–7
moneylenders, 286
regulations, 306–8
savings, 283
stabilization, 342
subsidized, 288, 316–17, 340
International agencies
agricultural development, 426, 462, 470–1
co-ordination, 31
gender issues, 269
participatory processes, 431
pest management, 373
International financial institutions (IFIs), 433
International Fund for Agricultural Development (IFAD), 455, 473
International Irrigation Management Institute, 199–200, 200f, 203, 215, 215f, 260
International Network on Participatory Irrigation, 231f
International Programme for Technology and Research in Irrigation and Drainage (IPTRID), 204–5, 205f, 232
International trade
agricultural development, 455
bond issues, 351
competition, 7, 12, 36, 37, 84, 360
foreign investment, 98
market development, 456
restrictions, 68–72
subsidies, 64–5
tariffs, 59, 60
Internet communication, 397, 408
Inverse marketing margin, 87

Investment
agricultural development, 279–81, 462, 464, 465–8, 468–9
annual implementation plans, 31, 32
conditionality, 475
demand-driven, 462, 464, 468
education, 26
foreign, 98
human capital, 26, 281
irrigation projects, 28–9, 213–14, 228, 230
macroeconomic framework, 36
planning, 31, 32, 469, 474–5
post-reform, 28
priorities, 468–9
private sector, 280–1, 361, 363, 364, 371–2, 397–8
public sector, 16, 21–2, 31, 280, 363, 364, 371–2, 397–8, 453
research, 26, 28
Iran, land tenure, 157
water supplies, 199
Irrigation management transfer (IMT), 250, 260, 266, 270
Irrigation
centralized control, 201
cost recovery, 228, 230, 236, 257, 270
demand management, 205, 210, 223–5, 228, 484–7, 490–1
design, 220–2, 265, 268
diesel fuel subsidies, 485, 486–7, 491
efficiency, 199–201, 203, 205, 213, 215–16, 486–7, 491
farmer ownership, 264
gender issues, 204, 207, 222, 235–6, 268–9, 403
governance, 249–52, 261–2
historical background, 197–8
institutional issues, 248–54
investment, 28–9, 213–14, 228, 230
large-scale vs small-scale, 45, 218–20, 225
leases, 261
local water management, 255–64
operation and maintenance costs, 216, 228, 229–30, 234, 235, 236, 260, 261–4, 266, 270
policies, 203–7, 211, 223–48
pressurized, 217
pricing systems, 232–6, 252
privatization, 51
rehabilitation vs new systems, 213–16, 222, 266
return flows, 235, 240–2
strategic issues, 207–23
subsidies, 38, 226, 228, 270–1
supplemental, 217–18
sustainability, 203–4, 206, 207, 223, 227, 250, 487
system types, 216–23

water assessments, 212–13
water crises, 485–91
water logging, 215
yield increase, 358
Islamic law, land rights, 119, 157
Israel, kibbutzim, 134, 137
Ivory Coast *see* Côte d'Ivoire

Jagannathan, N Vijay *see* Subramanian, Ashok, Jagannathan, N Vijay and Meinzen-Dick, Ruth
Jamaica, land ownership, 125, 126
Janvry, Alain de
Key, Nigel and Sadoulet, Elisabeth, 71f, 144f
Sadoulet, Elisabeth and Macours, Karen, 179f
Japan
food prices, 56
land reform, 148
megatariffs, 59
mixed economy, 30
Jardine, Veronica *see* Scobie, Grant M, and Jardine, Veronica
Java, land leasing, 161
irrigation efficiency, 199
Jayasuriya, S *see* Herdt, R W, Castillo, L, and Jayasuriya, S
Jayne, T S, and Argwings-Kodhek, G, 63f
Jiménez, E, and Sawada, Y, 429f
Jodha, N S, 131, 131f
Johnson, Gale, 12, 26, 29
Johnson, D G, 4f, 12f, 448f
Johnson, D Gale, 4f, 27f, 453f
Johnson, O E G, 151–2, 151f
Johnson, S H I, 258f
Johnson, Sam, 201f
Johnston, Bruce, 6, 11, 452
Johnston, Bruce F, and Mellor, John W, 6f, 452f
Joint-stock companies, 52
Jolly, Robert, 363
Jolly, Robert W *see* Mudahar, Mohinder S, Jolly, Robert W, and Srivastava, Jitendra P
Jones, S, 449, 449f
see also Beynon, J, Jones, S and Yao, S
Jong, Marga de *see* Diop, Jean-Marie, Jong, Marga de, Laban, Peter, and Zeeuw, Henk de
Jordan, irrigation projects, 209–10
Josling, Timothy, 59, 59f, 60
Judaeo–Christian tradition, 3
Junguito, R, 147
Junker estates, 120–1
Jureav, A *see* Tashmatov, A, Aknazarov, F, Jureav, A, Khusanov, R, Kadyrkulov, K D, Kalchayev, K, and Amirov, B

Kadyrkulov, K D *see* Tashmatov, A, Aknazarov, F,
    Jureav, A, Khusanov, R, Kadyrkulov, K D,
    Kalchayev, K, and Amirov, B
Kalchayev, K *see* Tashmatov, A, Aknazarov, F,
    Jureav, A, Khusanov, R, Kadyrkulov, K D,
    Kalchayev, K, and Amirov, B
Katz, Elizabeth, 187f, 369f
Kazakhstan
    exchange rate policy, 75
    irrigation efficiency, 200–1
    land reform, 172
    sequencing of reforms, 83, 91
Keith, S H *see* Valletta, W, Keith, S H, Norton, R D
Kelly, V *see* Reardon, T, Kelly, V, Crawford, E,
    Diagana, B, Dioné, J, Savadogo, K, and
    Boughton, D
Kemper, Karin E, 239, 254
Kenya
    credit schemes, 290, 351
    economic growth model, 9
    education, 445
    fertilizer market liberalization, 96–7
    gender issues, 46–7, 329, 385
    group lending, 333
    irrigation projects, 201, 218, 219
    land ownership, 125, 129, 151, 158
    policy reforms, 426
    subsidies, 63
    trade restrictions, 69
Key, Nigel, 144
Key, Nigel *see* Janvry, Alain de, Key, Nigel and
    Sadoulet, Elisabeth
Khan, Mahmood Hasan, 182
Khandker, Shahidur R, 292, 292f, 294, 294f, 295,
    299f, 302, 302f, 329, 348
Khusanov, R *see* Tashmatov, A, Aknazarov, F,
    Jureav, A, Khusanov, R, Kadyrkulov, K D,
    Kalchayev, K, and Amirov, B
Kibbutzim, 134, 137
Kidane, Y *see* Pelletier, D L, Deneke, K, Kidane, Y,
    Haile, B, and Negussie, F
Kikeri, S, Nellis, J, and Shirley, M, 49f
Kingsmill, William, 152
Kingsmill, William, and Rogg, Christian, 153f
Kinni, Amoul, 127f
Kirk, Michael, 115f, 119
    see also Ngaido, Tidiane, and Kirk, Michael
Kirmse, Robert D, Constantino, Luis F, and Guess,
    George M, 69f
Klein, Brigitte, Meyer, Richard, Hannig, Alfred,
    Burnett, Jill, and Fiebig, Michael, 279f, 323f,
    330f

Klerk, M J de, 185f
Klug, Heinz, 147, 147f
Kneese, Allen V, and Brown, F Lee, 239f
Knudsen, O
    and Scandizzo, P, 101f
    see also Goldin, I, Knudsen, O, and Mensbrugghe,
        D, van der
Kone, D, 213f
Korea *see* Republic of Korea; South Korea
Korten, F F, 257
Korten, F F *see* Bagadion, B U, and Korten, F F
Krueger, Anne, 5, 8
Krueger, Anne O, 5f, 7f, 8f
    Schiff, Maurice and Valdés, Alberto, 8f, 76f
Kumar, Shubh, K, 72f

Laarman, Jan G, 69, 69f, 70f
Laban, Peter *see* Diop, Jean-Marie, Jong, Marga de,
    Laban, Peter, and Zeeuw, Henk de
Labor
    agricultural productivity, 4, 5, 7, 9, 18, 279–80
    exchange, 174
    intersectoral movement, 18
    substitution, 453
Lamers, John, Dürr, Georg, and Feil, Petra, 409f, 411f
Land
    access to, 21, 112, 113–14, 117, 119–20, 176–91
    area expansion, 358
    ceilings on ownership, 166–7
    distribution, 111–16, 142–9
    leasing, 112, 125, 141–2, 158, 159–63
    markets, 116–17, 149–76, 164–7, 174, 176–85
    mortgages, 45, 166, 176, 178, 179, 184
    ownership, 117, 121–2, 125, 134–9, 142–9, 166–71,
        190–1, 309–10
    prices, 176
    registration, 115, 159, 181, 189, 190, 302
    rental, 45, 112, 122, 140–1, 158, 159–63, 179
    sales, 164–7, 174
    tax, 156, 179–83, 466
    titling, 45, 115–16, 125–6, 128–30, 133, 147, 150,
        153–9, 183–4, 190, 302, 446
    women's access, 185–91
Land funds/banks, 45, 176–8, 182, 191
Land reform, 142–9, 159, 173, 177, 178–9, 182
Land rights, 116, 121–30
    formalization, 128–30, 150–6
    historical trends, 117–20
    tenure security, 114, 119–20, 122–5
    transferability, 127–8, 151, 159, 161
Land tenure policies, 21, 109–95
    associative tenure, 121

gender issues, 330
importance, 110–11
objectives, 111–16
overview, 116–21
Landed estates, 120–1
Landlordism, 40, 78, 141, 142
Lao Tze, 3f
Larivière, Sylvain *see* Martin, Frédéric, Larivière, Sylvain, and Staatz, John M
Larson, B A, and Bromley, D W, 122f, 131f
Lastarria-Cornhiel, Susana, and Melmed-Sanjak, Jolyne (assistance from Beverly R Phillips), 164f
Latin America
agricultural strategy, 451, 458
extension services, 408
hacienda system, 120, 121, 140
international trade, 59
land reform, 143–4, 147, 446
price and trade controls, 19
research committees, 367, 394, 407
subsidizing industrial development, 5
taxes on land, 180
water resources management, 210, 263
women's access to land, 187
see also Central America
Lawrence, Pareena, 369, 403
Lawrence, P G, Sanders, J H, and Ramaswamy, S, 369f, 370f
Lease–purchase contracts, 179
Leasehold tenure, 112, 125, 141–2, 158, 159–63
Ledgerwood, Joanna, 285f, 294f, 319, 319f, 325–6, 326f, 330f
Legislative framework
agricultural policies, 20, 30, 51–3, 459, 464, 474
anti-monopoly, 95
anti-trust, 97
consumer protection, 52
contractual relationships, 305–6, 459
enforcement, 53
finance, 52–3, 299, 347
gender issues, 189, 190
land allocation, 116
land tenure, 157, 184
savings mobilization, 335
water rights markets, 237
water users associations, 266
see also Regulatory framework
Lele, Uma, 25
Lele, Uma, Emerson, Robert, and Beilock, Richard, 26f
Lemel, Harold, 141f
Lenders of last resort, 313, 326

Lerman, Zvi, 139, 139f, 172, 173f
Brooks, K, Csaki, C, 171f
see also Brooks, K, and Lerman, Z
Lesotho, land tenure, 123, 124
Less developed countries (LDCs)
agricultural productivity, 359
extension, 375–7
Lewa, Peter, M, 426
Lieberman, I W, 49
Lieberman, I W, Ewing, A, Mejstrik, J, Mukherjee, J, and Fidler, P (Eds), 49f
Liens, 45, 92, 302, 303
Lin, J Y, 136, 136f
Lithuania, privatization, 49
Litsinger, J A, 374, 374f
Llort, M, 431f *see also* Norton, Roger D, and Llort, M
Loans
adverse selection, 341–2
antichresis, 184
co-operatives, 53, 310, 318–22
collateral, 52–3, 124, 125–6, 302–5, 311, 314, 323
commercial banks, 347–8
costs, 338
credit markets, 278–84
documentation, 312, 314
financial institutions, 283–6, 290–2, 295, 310–13, 317–18
government guarantees, 305
groups, 184, 302, 332–3, 345
historical background, 277–8
interest rates, 315–16, 323, 331–3
land ownership, 124, 125–6
policies, 298
production vs income, 287–8
provisioning, 311–12
regulatory framework, 300–2
repayment, 286–7, 323, 332, 333–4
risk management, 296, 303, 310–12, 324, 333, 337, 339
targeted, 318
unsecured, 332
Local Agricultural Research Committees (CIALs), 367, 394, 402, 407
Local issues
irrigation, 217, 255–64
research committees, 367, 394, 402, 407
rural finance, 287, 317–18, 336
Löfgren, Hans, and El-Said, Moataz, 40f
Log exports, trade restrictions, 70
Lomé Convention, 75
López, R, 132, 132f
Liquified, petroleum gas (LPG), 488, 489, 490

Lyne, Michael, 161
Lyne, Michael, Roth, Michael, and Troutt, Betsy, 161f, 162f

Maartens, Miet *see* Deininger, Klaus, Olinto, Pedro, and Maartens, Miet
Macedonia, gender issues, 269
Macours, Karen *see* Janvry, Alain de, Sadoulet, Elisabeth and Macours, Karen
Macours, Karen, 179
Macroeconomic framework
  agricultural development, 35–7, 38, 449, 458
  anti-agriculture bias, 82–3, 448
  irrigation prices, 236
  pricing policies, 456
  producer incentives, 83–6, 103–4, 448
  real prices, 58
  rural finance, 289–90, 340–6, 346
  savings mobilization, 335
  technology development, 360–1
  water resources management, 209, 211
Madagascar
  anti-agricultural bias, 83
  sharecropping, 161
Maddock, N *see* Hristova, M, Maddock, N
Maertens, Miet, 147, 148
Magill, John H, 319–20, 320f, 321, 322f
Mahmood, M, 167f
Maimbo, Fabian *see* Chinene, Vernon R N, Maimbo, Fabian, Banda, Diana J, and Msune, Stemon C
Maize research, 394, 402
Malawi
  calorie consumption, 99f
  extension, 402, 413, 414
  gender issues, 329, 402, 413
  group lending, 332
  microfinance institutions, 291, 331
  participatory processes, 433
  private sector involvement, 50, 90–1
  technology development, 361, 402
Malaysia
  farm size and productivity, 127
  irrigation projects, 251–2, 261
Malhotra, Mohini, and Fox, James, 334f
Mali
  exchange rate policy, 78
  irrigation projects, 203, 223
  research, 393
  State land ownership, 141
  trade restrictions, 69
Malnutrition, 72, 90f, 100, 201
Malone, V *see* Due, J M, Mollel, N, and Malone, V

Maloney, William F, 61, 61f
Managed deposit rates, 340–1
Management information systems, 311, 323, 334, 339
Manufacturing tariffs, 59
Marcoux, A, 196
Maredia, Mywish, 361
Maredia, M K, Byerlee, D, and Pee, P, 361f, 364f, 390f, 395f
Market economy
  agricultural policies, 12, 20, 25–6, 30, 455, 458–9
  extension, 405
  failures, 444
  financial services, 298
  globalization, 12, 30, 93–4, 395
  interest rates, 296–7
  land reform, 149, 159, 173, 177, 178–9, 182
  legal framework, 52
  liberalization, 60–2, 67, 70–2, 89–92, 96
  pricing policies, 456–7
Marketing
  access policies, 21
  efficiency, 85–6
  information provision, 26–7, 97, 378, 419, 455
  research, 397, 398
  rural finance, 336–7
  State management, 19
Martin, Frédéric, Larivière, Sylvain, and Staatz, John M, 451f
Martin, Michael, 135
Martin, Michael, and Taylor, Timothy, G, 135f
Martínez Damián, Miguel A *see* Valdivia Alcalá, Ramón, Matus Gardea, Jaime A, Martínez Damián, Miguel A, and Santiago Cruz, Maria de J
Martínez, H P, 322f
Masson, P R, Bayoumi, T, and Samiei, H, 296f
Masters, William, 359
Masters, W A, Bedingar, T, and Oehmke, J F, 359f
Matus Gardea, Jaime A *see* Valdivia Alcalá, Ramón, Matus Gardea, Jaime A, Martínez Damián, Miguel A, and Santiago Cruz, Maria de J
Maxwell, Simon, 99f, 100, 102, 102f, 103
  and Percy, Robert Heber, 104f
May, Julian, 177
May, Julian *see* Deininger, Klaus, and May, Julian
McIntire, J, 164–5
McIntire, J *see* Binswanger, H P, McIntire, J, and Udry, C
McKean, Cressida, 290
McKean, C *see* Christen, Robert Peck, Rhyne, E, Vogel, R C, and McKean, C
McKee, Kate, 314

McNally, William, David, Wilfred and Flood, David, 95f
McSharry Plan, 85, 465
Mechanization, 453, 454
Megatariffs, 59
Mehra, Rekha, 196
Meinzen–Dick, Ruth, 133–4, 134f, 189, 189f, 223–5, 237, 237f, 251–2, 257, 258, 262, 264, 266, 270
   Mendoza, Meyra, Sadoulet, Loic, Abiad-Shields, Ghada, and Subramanian, Ashok, 234f, 252f, 257f, 258f, 259f, 263f, 264f, 265f
   and Rosegrant, Mark, W, 199f, 224f, 225f
   see also Subramanian, Ashok, Jagannathan, N Vijay, and Meinzen-Dick, Ruth
Mejstrik, J *see* Lieberman, I W, Ewing, A, Mejstrik, J, Mukherjee, J, and Fidler, P (Eds)
Mellor, John W, 6, 6f, 9, 9f, 11, 12, 12f, 90–1, 452
   and Gavian, S, 90f
   see also Johnston, Bruce F, and Mellor, John W
Melmed-Sanjak, Jolyne *see* Lastarria-Cornhiel, Susana, and Melmed-Sanjak, Jolyne (assistance from Beverly R Phillips)
Melo, Martha de, Denizer, Cevdet, and Gelb, Alan, 87f
Mendieta y Núñez, Lucio, 112f, 117f
Mendoza, Meyra *see* Meinzen-Dick, Ruth, Mendoza, Meyra, Sadoulet, Loic, Abiad-Shields, Ghada, and Subramanian, Ashok
Mensbrugghe, D van der *see* Goldin, I, Knudsen, O, and van der Mensbrugghe, D
Mesopotamia, 3
Mexico
   agricultural policy, 5
   collective farms, 138, 170
   contributions of irrigation, 202
   credit unions, 318
   crop insurance, 305
   economic transfers, 465
   education, 444
   environmental protection, 245
   exchange rate policy, 75, 78
   extension services, 381
   fruit exports, 94
   group lending, 332
   intersectoral labor movement, 18
   irrigation projects, 215, 262
   land allocation, 112
   land reform, 147
   loans, 279
   price stabilization, 104
   PROCAMPO, 85, 465
   research, 394–5, 395
   rural finance, 346

   subsidized credit, 287–8, 331
   tariffs, 60, 66
   water prices, 229
   water resources management, 254, 256, 262, 263, 266
   water rights markets, 237, 239, 241, 243, 245, 246–7
Meyer, Richard L, 318
   and Nagarajan, Geetha, 285f
   see also Klein, Brigitte, Meyer, Richard, Hannig, Alfred, Burnett, Jill, and Fiebig, Michael; Nagarajan, Geetha, Meyer, R L, and Graham, D H; Nagarajan, Geetha, Meyer, R L, Hushak, L J
Meza, Daniel *see* Strasma, John, Arias, Ricardo, García, Magdalena, Meza, Daniel, Soler, René, and Umaña, R
Michaely, Michael *see* Papageorgiou, Demetrios, Choksi, Armeane M and Michaely, Michael
Microcredit programs, 292, 294, 295, 335
Microenterprise programs, 331–2
Microfinance institutions (MFIs)
   accountability, 328
   aid, 279
   apex organizations, 324–6, 343
   auditing, 338, 339
   autonomy, 297, 330–1
   capital to asset ratios, 310–11, 314–15, 337
   commercial bank role, 347–8
   decentralization, 300–1, 323, 340
   delinquency management, 301
   development contributions, 293–5
   forced savings programs, 296
   funding, 290, 310, 316–17, 318, 323–4, 342–6
   governance, 297, 300–1, 310, 325, 328–9
   group lending, 184, 302, 332–3
   growth management, 301–2
   licensing, 323–4
   liquidity management, 311, 313, 314–15, 324, 336, 337–8, 350
   loans, 283–6, 290–2, 295, 310–13, 317–18
   local structures, 287, 317–18, 336
   management, 328, 330–40, 339
   market orientation, 298
   monitoring, 308–9, 317, 319, 326, 334, 340, 475
   objectives, 293, 294–5
   portfolio diversification, 338
   profitability, 299
   regulations, 315–17, 348
   reporting requirements, 312–13
   responsibilities, 328
   risk management, 296, 300–2, 303, 310–12, 324, 333–4, 337, 339
   self-supervision, 316
   sequencing, 337

staff incentives, 340
strategies, 328, 346–51
supervision, 315–17, 324, 326, 328, 335, 348
transparency, 328
voluntary savings, 313, 314, 335–6, 337
see also Rural finance
Micro-irrigation, 217
Middle East
groundwater supplies, 204
price and trade controls, 19
water resources management, 208, 210, 213, 262
women's education, 47
Migration, rural–urban, 18, 111–12, 113, 449
Milgrom, P, and Roberts, J, 446f
Ministries of Agriculture
gender issues, 47
HIV/AIDS information, 416
mobile banks, 334
policy co-ordination, 20
regional offices, 466–7
roles, 19
staff turnover, 25
Ministries of Economy/Trade, agricultural policy
co-ordination, 20
Ministries of Environment/Natural Resources,
agricultural policy co-ordination, 20
Ministries of Finance, agricultural policy co-
ordination, 20
Minot, Nicholas, and Goletti, Francesco, 70f
Miranowski, John A *see* Carlson, Gerald A,
Zilberman, David, and Miranowski, John A
Mkrytichyam, L *see* Baydildina, A, Akshinbay, A,
Bayetova, M, Mkrytichyam, L, Haliepesova, A,
and Ataev, D
Mobile banking, 334
Models, 5, 7, 9, 448, 449, 451–5
Moench, Marcus, 204
Moench, Marcus H *see* Burke, Jacob J, and Moench,
Marcus, H
Moldova, land ownership, 168, 172–3
Molecular biology, 371
Mollel, N *see* Due, J M, Mollel, N, and Malone, V
Moneylenders, 286, 304, 307
Monopolies
anti-monopoly legislation, 95
limitation, 27
privatization, 48, 49, 50
State agencies, 71, 80–1
Montiel, Peter J, and Ostry, Jonathan D, 74f
Moris, J R, 211, 214, 222, 223, 270
Moris, J R, and Thom, D J, 199f, 211f, 214f, 216f,
217f, 218f, 219f, 222f, 223f, 231f, 270f

Morris, M L *see* Doss, C R, Morris, M L
Morocco
land privatization, 133f
mobile banking, 334
volumetric water pricing, 233
water supplies, 199
Morris, Michael, 370–1, 403–4
Mortgages on land, 45, 166, 176, 178, 179, 184
Mose, L O *see* Omamo, S W, and Mose, L O
Mosley, Paul, 290, 291, 292f
see also Hulme, David, and Mosley, Paul
Moulin, A, 110f
Moyo, Evison, 412
Moyo, Evison, and Hagmann, Jürgen, 413f
Mozambique
calorie consumption, 99f
communal lands, 132–3, 153
credit unions, 318
food aid, 72–3
participatory processes, 429, 431, 437, 440
privatization, 50
rediscount lines, 326
Msune, Stemon C *see* Chinene, Vernon R N, Maimbo,
Fabian, Banda, Diana, J, and Msune, Stemon, C
Mtawali, K M, 91f
Mud, historical background, 3–4
Mudahar, Mohinder, 363, 390–1
Mudahar, Mohinder, S, Jolly, Robert, W, and
Srivastava, Jitendra P, 363f, 390f
Mukherjee, Joyita, 331f, 335, 335f, 338f
and Wisniwski, Sylvia, 337f, 338f, 340f
see also Lieberman, I W, Ewing, A, Mejstrik, J,
Mukherjee, J and Fidler, P (Eds)
Mundlak, Yair, 76f
see also Binswanger, H, Mundlak, Y, Yang, M-C,
and Bowers, A; Cavallo, Domingo, and
Mundlak, Yair
Munro–Faure, Paul, 110–11, 123–4
Munro-Faure, Paul, Groppo, Paolo, Herrera,
Adriana, and Palmer, David, 110f, 124f
Murgai, R, Ali, M, and Byerlee, D, 396f
Murray–Rust, D H, 220
Murray-Rust, D H, and Snellen, W B, 220f
Murshid, K A S, 184f, 303f
Murwira, K *see also* Hagmann, Jürgen, Chuma,
Edward, and Murwira, K
Mutua, Albert Kimanthi, 290, 290f, 333f
Myanmar, returns to irrigation, 199

NAFTA, tariffs, 60, 62
Nagarajan, Geetha, 318
Meyer, R L, Hushak, L J, 280f

Meyer, R L, and Graham, D H, 318f
see also Graham, Douglas H, Nagarajan, Geetha, and Quattara, Korotoumou; Meyer, Richard L, and Nagarajan, Geetha
Namibia
land reform, 146
State land ownership, 141
National agricultural research systems (NARS), 365, 372
National development, agricultural policies, 22–3, 25–32
Natural resources
agricultural use, 17
degradation, 396
gender analysis, 190–1
implicit subsidies, 37–8
management, 21, 190–1, 401
see also Water management policies
Naylor, Rosamund, 449, 460
Negussie, F *see* Pelletier, D L, Deneke, K, Kidane, Y, Haile, B, and Negussie, F
Nellis, J *see* Kikeri, S, Nellis, J and Shirley, M
Nepal
group lending, 332
irrigation system ownership, 230
water pricing, 226
water resources management, 258, 259, 262
Nercissiantz, Vahram, 229
Nercissiantz, V *see* Cummings, R G, and Nercissiantz, V
Neuchatel Group, 387, 388f, 404, 404f, 411–12, 419f, 420
New Zealand
exchange rate policy, 77
structural reform, 31
Ngaido, Tidiane, 119, 119f
and Kirk, Michael, 120f
NGOs
extension, 364, 406, 407, 408, 419–20
funding, 290, 316–17, 318, 323–4, 343, 463, 471–2
policy reforms, 429
see also Private sector
Nicaragua
governmental role, 465
irrigation projects, 271
irrigation pumps, 223
land ownership, 126
land reform, 147
NGO credit programs, 317
participatory processes, 435, 438
private sector involvement, 90
rediscount lines, 326–7

research, 392
rural development, 462
subsidies, 41
tariffs, 61, 64, 448
Niger
irrigation rehabilitation, 203, 214
traditional land transfer, 127, 154
Nigeria
exchange rate policy, 75, 84
intercropping, 266f
intersectoral labor movement, 18
irrigation projects, 202–3, 219
land rental, 161
loans to women, 329
trade restrictions, 69
Nikonov, Alexander, 150–1
Nikonov, A A, 151f
Nissanke, Machiko *see* Aryeetey, Ernest, Hettige, Hemamala, Nissanke, Machiko, and Steel, William
Niu, Ruofeng, 135–6
Niu, Ruofeng, and Chen, Jiyuan, 136f
Nkunya, G K, 128, 128f
Non-agricultural employment, 23, 460, 486, 488, 491
Non-consumptive use rights, 240
Non-governmental organizations *see* NGOs
Noronha, R, 123, 125–6, 127f, 165–6
see also Feder, G, and Noronha, R
Norse, David, and Saigal, Reshma, 132f
North Africa
water resources management, 208, 210, 213, 262
women's education, 47
see also Africa; South Africa; Sub-Saharan Africa
North America
corporate farms, 167
see also Canada; United States
North, Douglas C, 445–6, 445f
Norton, Roger D, 8f, 23f, 29f, 57f, 58f, 77f, 80f, 81f, 86f, 98f, 342f, 361–2, 362f, 381, 431f, 433f, 436f, 446f, 454f, 458f, 483, 483f
and Angel, Amy L, 18f, 447f
Arias, Ricardo and Calderón, Vilma, 114f
and Benito, C A, 64f
and Garcia U, Magdalena, 62f
and Llort, Mercedes, 45f, 135f
see also Bassoco, Luz María, Cartas, Celso, and Norton, Roger D; García, Magdalena, Norton, Roger, Ponce Cámbar, Mario, and van Haeften, Roberta; Hazell, Peter B R, and Norton, Roger D; Valletta, W, Keith, S H, Norton, R D
Norway, extension services, 411

Ntangsi, Joseph, 90, 138f
Núñez, Miguel Angel, 379
Nutrition levels, 22–3, 99–103, 295
Nygaard, David, 372–3
Nygaard, David see Yudelman, Montague, Ratta, Annu, and Nygaard, David

Objectives
  agricultural policies, 22–5, 448–9
  land tenure policies, 111–16
  rural finance, 287–95
  sectoral strategies, 16, 17, 22–3
  water management policies, 203–7
Oblitas, Keith, and Peter, J Raymond, 201f, 210f, 255f, 267f
OECD, 39f, 71f, 88f
  border protection, 60
  subsidy study, 43
Oehmke, James, 359
Oehmke, J F see Masters, W A, Bedingar, T, and Oehmke, J F
Officer, R R, and Halter, A N, 450f
Okaru, Valentina see Sharma, Narendra P, Damhaug, Torbjorn, Grey, David and Okaru, Valentina
Olagoke, Macaulay, 440–1
Oligopolistic financial systems, 341
Olinto, Pedro, 147, 148
Olinto, Pedro see Deininger, Klaus, Olinto, Pedro and Maartens, Miet
Olson, Douglas, 239, 254
Omamo, S W, and Mose, L O, 96f
Onchan, Tongroj, 125
Onchan, Tongroj see Feder, Gershon, and Onchan, Tongroj; Feder, Gershon, Onchan, Tongroj and Raparla, Tejaswi
O'Neill, D see Chancellor, F M, Hasnip, N, and O'Neill, D
Open access lands, 121
Organic production, 94
Orlando, Brett see Dorcey, Tony, Steiner, Achim, Acreman, Michael, and Orlando, Brett
Orozco y Berra, Manuel, 112f
Ostrom, Elinor, 265f
  Schroeder, Larry, and Wynne, Susan, 256f, 265f, 268f
Ostry, Jonathan D see Montiel, Peter J, and Ostry, Jonathan D
Otero, María, 284, 293
  and Rhyne, Elisabeth, 284f
  see also Rhyne, Elisabeth, and Otero, María; Rock, Rachel, and Otero, M; Rock, Rachel, Otero, M, and Rosenberg, R
Otsuka, K, 400–1, 400f

Overseas Developoment Institute, 73f, 74f
Overseas Development Institute (ODI), food aid, 73, 100
Oxfam, farmer-to-farmer visits, 407

Pakistan
  environmental sustainability, 396
  farm size and productivity, 126, 127
  irrigation efficiency, 199
  research, 396, 400
  water rights markets, 237
Palmer, David, 110–11
Palmer, David see Munro-Faure, Paul, Groppo, Paolo, Herrera, Adriana, and Palmer, David
Pandya-Lorch, R see Pinstrup-Andersen, P, Pandya-Lorch, R and Babu, S
Pan-territorial pricing, 87
Panama
  corporate farms, 167
  land titling, 446
  tariffs, 61
Papageorgiou, Demetrios, Choksi, Armeane M and Michaely, Michael, 77f
Parastatal reform, 27, 28
Pardey, P G, Roseboom, J, and Beintema, N M, 365f
Pardey, Philip, 365
Paris, Thelma, 369
Paris, Thelma, Feldstein, Hilary Sims, and Duron, Guadalupe, 369f
Participation
  agricultural policy, 31, 428–42, 459, 462
  challenges and risks, 439–42
  extension, 406–7, 412–13, 418, 419, 420
  investment needs, 469
  organization, 437–9
  process ownership, 432–4
  quality control, 436–7, 440–1
  reimbursement, 438
  rural development, 462
  task forces, 438, 442
  technology development, 366–8, 391–2, 393–5, 398, 402, 403
  water management policies, 249, 260, 265
Partnerships, 208
Patents, 371–2
Paternalism, 138, 139
Pathogens, 372–3
Paz Cafferata, Julio, 66, 66f, 457
Pearson, Scott R see Timmer, C Peter, Falcon, Walter P, and Pearson, Scott R
Pee, Peter, 361
Pee, P see Maredia, M K, Byerlee, D and Pee, P
Peer monitoring, 332, 345

Pelletier, D L, Deneke, K, Kidane, Y, Haile, B, and Negussie, F, 101f
Percy, Robert Heber *see* Maxwell, Simon, and Percy, Robert Heber
Perrin, Richard, 358–9, 360
Perrin, R K *see* Fulginiti, L E, and Perrin, R K
Peru
    collateral, 304
    collective farms, 135, 165
    fruit exports, 94
    intersectoral water transfer, 243
    investment priorities, 469
    land tenure, 118
    overgrazing, 133f
    pan-territorial pricing, 87
    research, 403
    rural savings, 281–2
    water resources management, 263
    water supplies, 199
Pest management, 372–5, 394, 396
Pesticides, 373–4
Peter, J Raymond *see* Oblitas, Keith, and Peter, J Raymond
Peters, G H, 168f, 172f, 169f, 171f
Philippines
    credit, 280–1
    irrigation efficiency, 199
    irrigation projects, 202, 257
    land reform, 147, 148
    loans, 278
    microfinance institutions, 312
    research, 400–1
    water resources management, 264, 267
Phytosanitary conditions
    import control, 70, 93
    product quality, 93–4, 456
Picciotto, R, 375–6, 383, 388
Picciotto, R, and Anderson, J R, 362f, 384f, 388f
Pinckney, Thomas C, 445, 445f
Pingali, P L, 374
    and Gerpacio, R V, 374f
    see also Rola, A C, and Pingali, P L
Pinstrup–Andersen, P, 359, 371–2, 372f
    Pandya-Lorch, R and Babu, S, 99f
Piprek, G L, 286, 298–9, 304, 321, 333
Piprek, G L *see* Yaron, J, Benjamin, M P Jr, and Piprek, G L
Pischke, J D Von, 326f
PL, 480, 64f
Plant diseases *see* Phytosanitary conditions
Platteau, J-P, 119, 119f, 123, 123f, 124, 124f, 129, 129f, 131, 132, 132f, 138f, 140f, 142, 150, 150f, 154, 154f, 161, 164–5, 165f, 174, 174f, 184f

Pluralistic extension system, 362, 406, 417
Plusquellec, Hervé, 220
Plusquellec, Hervé, Burt, Charles, and Wolter, Hans W, 220f, 221f, 226f, 233f
Poland, privatization, 49
Policy reform *see* Agricultural policy framework; Pricing policies; Trade policy
Pollution *see* Environmental sustainability
Ponce Cámbar, Mario *see* García, Magdalena, Norton, Roger, Ponce Cámbar, Mario, and van Haeften, Roberta
Pontius, John, Dilts, Russell, Bartlett, Andrew (Eds), 375f
Poor people
    education and training, 46
    financial services, 284
Posner, Richard A, 51, 52f
Poverty alleviation
    agricultural development, 9–12, 17, 18, 449–50, 460–74
    economic transfers, 465–8
    extension services, 381
    gender outreach, 289–92
    irrigation policies, 204–5, 207
    land tenure policies, 111–13, 145
    rental and sharecropping, 160
    research, 372, 399–402
    rural development, 460–74
    rural finance, 346
    sectoral strategies, 16, 449
    subsidies, 38–49, 228
    sustainability, 294
    tariffs, 63–4
    trade policy, 62–7
    urban, 9–12
    see also Rural poverty; Urban poverty
Pray, Carl E, 364
    and Umali-Deininger, D, 368f, 396f
    see also Evenson, R E, Pray, Carl E, and Rosegrant, Mark W
Pressurized irrigation, 217, 222
Price indexes, 56–7
Pricing policies
    administered prices, 86–7
    agricultural policies, 21, 97–8, 431, 454–5, 456–7
    controls, 19, 37, 86–9
    explicit subsidies, 38
    farm prices, 87–9, 94–6, 97–8, 211
    government endorsement, 95–6
    guaranteed prices, 87–9
    incentives, 11, 36
    international, 64–5
    irrigation water, 226–36, 252

macroeconomic framework, 35–7, 38, 449–50
market development, 456–7
pan-territorial pricing, 87
participation, 431
price bands, 65–7, 84
price reduction, 36–7
real prices, 23, 38
repression, 8
stabilization, 103–4, 449–50
subsidizing industrial development, 6
see also Agricultural prices; Food prices; Real
  prices; Relative prices
Priorities
  irrigation, 204, 269–71
  poverty alleviation, 8, 18–19
  research, 390–1, 392, 393
Private land rights, 121
Private sector
  credit, 304–5
  crop insurance, 305
  extension services, 380, 405–6, 419
  governmental role, 27–8
  investment, 280–1
  irrigation projects, 51, 230–2
  microfinance institutions, 322, 323, 325, 350
  policy reforms, 31, 429, 474
  research funding, 361, 363, 364, 371–2, 397–8
  silo privatization, 45, 49–50, 92–3, 168, 427
  strategic reserves, 90–1, 92
  technology, 453
  water supplies, 487
  see also NGO funding
Privatization, 45, 48–51, 92–3, 104, 173, 190, 200,
  350, 427, 428, 459
PROCAMPO (Mexico), 85, 465
Processing, 95, 378, 395, 431
Producers
  incentives, 55–107, 448
  ownership of processing facilities, 95
Production
  efficiency, 85–6, 111, 114, 122, 151
  income, 287–9
  restrictions, 312
  State management, 19
Productivity *see* Agricultural productivity
Profitability, rural finance, 293
Property rights, 27, 111, 119, 121–5, 182
Proportional rights, 239
Protectionism, 67, 84, 88–9
Prudential finance, 308–9, 337–40, 350
Prussia, junker estates, 120–1
Public goods, 26, 29

Public health, 203
Public sector
  corruption, 29
  extension services, 377–84, 387–8, 405–6, 416–17,
    418, 419, 453
  investment, 16, 21–2, 31, 280, 463–4
  research funding, 363, 364, 371–2, 397–8, 453
  subsidies, 39–40
  urban-biased policies, 299
  water resources management, 208, 225
  see also Government
Pump irrigation, 223, 258, 485, 486, 487, 488, 489,
  490
Purcell, Dennis, 360–1, 363, 364–5, 366
Purcell, Dennis L, and Anderson, Jock R, 361f, 363f,
  365f, 366f, 383f
Purchasing power, rural households, 22–3

Qamar, M Kalim, 362, 362f, 384, 384f, 385–6,
  385f, 387, 391–2, 392f, 409f, 410f, 414–15, 414f,
  415f
Quality control
  food safety standards, 93–4, 456
  participatory processes, 436–7, 440–1
  products, 455–6, 458, 468
  seeds, 391
Quattara, Korotoumou *see* Graham, Douglas H,
  Nagarajan, Geetha, and Quattara,
  Korotoumou
Quirós, Carlos Arturo *see* Ashby, Jacqueline A,
  Braun, Ann R, Gracia, Teresa, Guerrero, María
  del Pilar, Hernández, Luis Alfredo, Quirós,
  Carlos Arturo, and Roa, José Ignacio
Quotas, 58, 81–2

Raby, Namika, 267f
Ramaswamy, S, 369, 403
Ramaswamy, S *see* Lawrence, P G, Sanders, J H, and
  Ramaswamy, S
Ranis, Gustav, 5
Ranis, Gustav *see* Fei, John C H, and Ranis,
  Gustav
Rao, J M, 181f
Rao, Mohan, 180–1
Raparla, Tejaswi *see* Feder, Gershon, Onchan,
  Tongroj and Raparla, Tejaswi
Ratta, Annu, 372–3
Ratta, Annu *see* Yudelman, Montague, Ratta, Annu,
  and Nygaard, David
Ravallion, M, 9–10, 12, 289
  and Datt, Gaurav, 10f
  and Sen, B, 289f

Ravenscroft, Neil, 160, 160f, 161f, 162f
Real prices, 23f, 56–7, 58, 92
   see also Agricultural prices; Food prices; Pricing
     policies; Relative prices
Reardon, Thomas, 126–7
   Kelly V, Crawford E, Diagana, B, Dioné, J,
     Savadogo, K, and Boughton D, 78f
   Stamoulis, K, Cruz, M E, Balisca, Al, and
     Berdegué, J, 460f
   see also Byiringiro, F, and Reardon, T; Clay, Daniel
     C, and Reardon, Thomas
Reca, Lucio, 58–9
Reca, L *see* Diaz-Bonilla, E, and Reca, L
Recessional irrigation, 216
Recycled water, 213, 486, 490
Rediscount lines, 326–7, 345–6, 351
Reed, Larry R, and Befus, David R, 322f
Registration
   land, 115, 159, 181, 189, 190, 302
   water, 240
Regulatory framework
   banks, 308–17, 313–15, 347, 350
   costs, 315
   implicit subsidies, 38
   interest rates, 306–8
   land allocation, 116, 117
   land tenure, 157
   loans, 286
   macroeconomic policies, 35–6
   preventive, 309–13
   privatization, 50
   protective, 309, 313–17
   rural finance, 298–9, 300–17, 323, 347, 350
   savings mobilization, 335
   technology transfer, 389
   warehouses, 97
   water management policies, 249–50, 253
   see also Legislative framework
Relative prices, 56, 63, 64–5, 77, 81
   see also Agricultural prices; Food prices; Pricing
     policies; Real prices
Relief food aid, 73
Renewable water supplies, 198–9
Renkow, M, 399, 399f, 400, 401–2, 402f
Rent-seeking behavior, 40, 78, 141, 142
Rental
   land, 38, 45, 112, 122, 140–1, 158, 159–63, 179
   price effects, 82
   trees, 122
Republic of Korea *see also* South Korea
   education investment, 26
   tariffs, 62

Research, 359–60, 363–75
   adaptive, 363
   agricultural strategy, 457
   applied, 363
   autonomy, 390, 397
   biotechnology, 28, 371, 395
   budget reduction, 397
   cost recovery, 397
   decentralization, 45, 392–3, 396–7, 454
   efficiency, 365
   explicit subsidies, 38
   funding, 361, 363, 364, 371–2, 395, 397–9, 453–4,
     463
   gender issues, 366, 367, 368–71, 391, 402–4
   implementation, 391–6
   information provision, 399
   institutional structures, 397–8
   internet liaison, 397
   investment, 26, 28, 359
   management, 396–8
   marketing, 397, 398
   modalities, 390–1
   molecular biology, 371
   new directions, 390–404
   participatory, 41, 366–8, 391–2, 393–5, 398, 402,
     403
   post-harvest management, 395
   poverty alleviation, 372, 399–402
   priorities, 390–1, 392, 393
   risk avoidance, 365–6, 391
   rural development, 461
   science-driven, 391
   strategic, 363
   see also Extension; Technology
Resources
   management policies, 21, 190
   see also Natural resources
Restrictions
   land sales, 165–7
   production, 312
   trade, 68–72, 388–9
   zoning, 122
Retail lenders, 325, 326
Reutlinger, S, 72f
Rhodes, G F Jr, and Sampath, R K, 229f, 233f
Rhyne, Elisabeth, 284, 290, 293
   and Otero, Maria 293f, 298f, 324f, 332f
   see also Christen, Robert Peck, Rhyne, E, Vogel, R
     C, and McKean, C; Otero, Maria, and Rhyne,
     Elisabeth
Rice, E B, 199f, 222f, 234f, 236, 236f, 262–3, 263f,
     265f, 270

Rice
  pest management, 373–4
  varietal research, 403
Riddell, James C, 160f
  see also Herrera, Adriana, Riddell, Jim, and
      Toselli, Paolo
Riddle, J, 144, 145
Risk management
  microfinance institutions, 33–4, 296, 300–2, 303,
      310–12, 324, 337, 339
  participatory processes, 439–42
  research, 365–6, 391
Rivalry, 378–80
Rivera, William, 381, 388, 406, 408
Rivera, William M, 377f, 381f, 388f, 406f,
      411f
Roa, José Ignacio see Ashby, Jacqueline A, Braun,
      Ann R, Gracia, Teresa, Guerrero, María del
      Pilar, Hernández, Luis Alfredo, Quirós, Carlos
      Arturo, and Roa, José Ignacio
Roberts, J see Milgrom, P, and Roberts, J
Robertson, A F, 124, 124f
Robinson, Marguerite, 282, 334–5
Robinson, Marguerite S, 282f, 298f, 335f
Rock, Rachel, 293f, 296, 314, 315f
  Otero, M, and Rosenberg, R, 313f, 337f, 342f
  and Otero, M, 316f
Rodrik, Dani, 29–30, 30f, 61, 61f, 62, 69, 79f
Rogg, Christian, 152
Rogg, Christian see Kingsmill, William, and Rogg,
      Christian
Rola, A C, 374
Rola, A C, and Pingali, P L, 374f
Romania, land ownership, 173
Roseboom, Johannes, 365
Roseboom, J see Pardey, P G, Roseboom, J, and
      Beintema, N M
Rosegrant, Mark W, 223–5, 227, 230, 241, 244, 363,
      364
  and Binswanger, H P, 229f, 237f, 240f, 241f, 245f,
      251f, 255f, 261f
  Gazmuri S, R, and Yadav, S N, 224f, 227f, 239f,
      241f, 242f, 244f, 246f
  see also Evenson, R E, Pray, Carl E, and
      Rosegrant, Mark W; Evenson, R E, and
      Rosegrant, M W; Meinzen-Dick, Ruth, and
      Rosegrant, Mark W
Rosenberg, Richard, 308, 315, 316–17, 326f
  see also Christen, Robert Peck, and Rosenberg,
      Richard; Rock, Rachel, Otero, M, and
      Rosenberg, R
Rotating savings and credit associations (ROSCAs),
      318

Roth, Michael, 162, 162f
  see also Lyne, Michael, Roth, Michael, and Troutt,
      Betsy
Rural development
  agricultural policies, 19–20, 469–74
  decentralization, 462–5, 466–8
  economic transfers, 5–6, 7, 465–8
  gender issues, 46–8, 185–91, 472–3
  holistic approach, 205–6, 207, 210, 404, 458, 464
  human capital formation, 281
  institutions, 20, 401–2, 462, 470, 471–2
  irrigation projects, 269–71
  land tax, 182
  macroeconomic framework, 36
  markets, 26–7
  non-farm employment, 23
  poverty alleviation, 460–74
  projects, 461–2
  sectoral strategies, 16
  State land distribution, 185
  sustainability, 418, 451
  see also Agricultural development
Rural finance
  apex organizations, 324–6, 343
  banks, 308–17, 322–4
  capital requirements, 309, 310–11, 314–15, 337
  co-operatives, 175
  external funding, 342–6
  feasibility studies, 310
  financial services, 284–6, 298
  financing cap, 346
  food aid, 72–3, 74
  food security reserves, 92
  fraud, 311, 317
  gender issues, 329–30, 347, 348–9
  group lending, 184, 302, 332–3, 345
  intermediation, 295–9
  irrigation management, 234–5
  land ownership, 176, 178
  land reform, 147, 158
  legislative framework, 52–3
  long-term, 344, 350–1
  macroeconomic framework, 289–90, 340–6, 346
  management, 297–8, 330–40
  market segmentation, 280, 287
  new approaches, 346–8
  NGO funding, 290, 316–17, 318, 323–4, 343
  objectives, 293
  ownership requirements, 309–10
  policy objectives, 287–95
  prudential aspects, 308–9, 337–40, 350
  refinancing, 344
  regulatory framework, 298–9, 300–17, 323, 347

repayment enforcement, 286–7, 332
self-sufficiency, 293, 322, 325, 346
start-up funding, 343, 350
strategies, 328, 346–51
structural considerations, 317–30
sustainability, 279, 281–2, 290, 293, 295–9, 300,
   346
tribunals, 305
water resources management, 485
write-off policies, 311
see also Loans; Microfinance institutions; Savings
Rural populations
  communal lands, 132
  financial services, 278, 286–7
  spending patterns, 9
Rural poverty
  access to land, 176–85
  agricultural development priorities, 8, 18–19, 449,
    459, 460–74
  food security, 98–103
  land tenure policies, 111–13, 145
  nutrition levels, 22–3, 99–103
  pricing and taxation effects, 6, 11
  subsidy elimination, 185
  tariff exemptions, 63–4, 66–7
  women, 185, 186, 189
Russia
  collateral, 304
  collective farms, 172
  corporate farms, 168, 171
  farm support prices, 88
  land ownership, 150–1, 169, 171–4
  privatization, 49, 173
Rutherford, Stuart, 284, 284f
Ruttan, Vernon W, 7, 7f, 360, 453f, 453–5, 459–60,
  460f
  and Hayami, Yujiro, 453f
  see also Evenson, R, Waggoner, P E, and Ruttan,
    V W
Rwanda
  farm size and productivity, 126–7
  research, 367

Sabluk, Peter, 171
Sadoulet, Elisabeth, 144, 179
Sadoulet, Elisabeth *see* Janvry, Alain de, Key, Nigel
  and Sadoulet, Elisabeth; Janvry, Alain de,
  Sadoulet, Elisabeth and Macours, Karen
Sadoulet, Loic *see* Meinzen-Dick, Ruth, Mendoza,
  Meyra, Sadoulet, Loic, Abiad-Shields, Ghada,
  and Subramanian, Ashok
Sagardoy, Juan A *see* Vermillion, D L, and Sagardoy,
  Juan A

Sahel governments, exchange rate policy, 78
Saigal, Reshma *see* Norse, David, and Saigal,
  Reshma
Salas, Luis *see* Trackman, Brian, Fisher, William, and
  Salas, Luis
Sale of agricultural land, 164–7, 174
Salinity, 215, 396
Samiei, H *see* Masson, P R, Bayoumi, T, and
  Samiei, H
Sampath, R K, 226f, 229, 232, 233, 247, 247f
  see also Rhodes, G F Jr, and Sampath, R K
Sanders, John, 369, 403
Sanders, J H *see* Lawrence, P G, Sanders, J H, and
  Ramaswamy, S
Santiago Cruz, Maria de J *see* Valdivia Alcalá,
  Ramón, Matus Gardea, Jaime A, Martínez
  Damián, Miguel A, and Santiago Cruz, Maria
  de J
Savadogo, K *see* Reardon, T, Kelly V, Crawford E,
  Diagana, B, Dioné, J, Savadogo, K, and
  Boughton D
Savings
  agricultural credit, 279–84
  co-operatives, 53, 310, 318–22
  forced, 296, 336
  incentives, 283–4
  interest rates, 283, 331
  mobilization, 283–4, 295–6, 298, 307, 317–18,
    334–7, 347
  voluntary, 313, 314, 335–6, 337
Sawada, Y *see* Jiménez, E, and Sawada, Y
Sayad, J, 342, 342f
Scandizzo, P *see* Knudsen, O, and Scandizzo, P
Schama, Simon, 56f
Schejtman, Alexander, 468f
Scherr, S, 393, 393f
Schiff, Maurice, 8, 299
  and Valdés, Alberto, 76f
  see also Krueger, Anne O, Schiff, Maurice and
  Valdés, Alberto
Schleifer, Andrei, 123f, 128f
Schreider, Gertrud *see* Zeller, Manfred, Schreider,
  Gertrud, Braun, Joachim von, and Heidhues,
  Franz
Schreiner, Dean F, and García U, Magdalena, 103f
Schreiner, Dean, 63, 102–3, 400
Schroeder, Larry *see* Ostrom, Elinor, Schroeder,
  Larry, and Wynne, Susan
Schuh, G E, 147
Schuler, Kurt *see* Hanke, Steve H, and Schuler, Kurt
Schultz, Theodore W, 26, 26f, 376, 444, 444f
Second-tier finance, 45, 176–7, 324–5, 326, 343, 345
Schütz, Paul *see* Currle, Jochen, and Schütz, Paul

Scobie, Grant M, and Jardine, Veronica, 78f
Scott, T *see* Faruqee, R, Coleman, J R and Scott, T
Sectoral strategies
  agricultural prices, 86–98
  comparative advantage, 16, 17–19
  consistency, 447–8
  constraints, 16
  HIV/AIDS, 415–16
  objectives, 15–17, 22–3
  research, 372
  rural finance, 301, 346
  water resources management, 209, 212–13, 248–54
Security
  food prices, 92, 98–103, 202, 491
  land rights, 114, 119–20, 122–5
Seed capital, 350
Seeds, improvement, 28, 371–2, 391
Selective credit, 340
Self-employment, 294
Self-sufficiency, 44, 86, 293, 322, 325, 346, 471
Seligson, MA, 126, 126f
Selvaggio, Kathleen, 433, 433f
Sen, B, 289
Sen, B *see* Ravaillon, M, and Sen, B
Senanayake, P, 89
Senanayake, P *see* Ellis, F, Senanayake, P, and Smith, M
Senauer, B, 103, 103f
Senegal
  exchange rate policy, 77
  land rental, 161, 162
  land tenure, 129, 140
  water resources management, 253, 257
Sequeira, Gustavo, 147f
Seshamani, V, 93, 93f, 98f
Settlers rights, 141
Shaffer, James D, 444
Shah, P *see* Tikare, S, Youssef, D, Donnelly-Roark, P, and Shah, P
Shah, T, 237, 237f
Share ownership, 156, 167, 170, 171, 308, 310, 324, 326, 328, 350
Sharecropping, 158, 160–4
Sharma, Manohar, 329f, 349f
Sharma, Narendra P, 202–3, 208, 210, 218, 219, 247, 253–4, 270, 348–9
Sharma, Narendra P, Damhaug, Torbjorn, Grey, David and Okaru, Valentina, 203f, 207f, 211f, 219f, 247f, 252f, 253f, 254f, 260f, 270f
Shepherd, Andrew, 166f, 210f
Shirley, M *see* Kikeri, S, Nellis, J and Shirley, M

Simas, José, 256
Simas, José *see* Gorriz, Cecilia M, Subramanian, Ashok, and Simas, José
Slovakia, privatization, 49
Small farms, 45, 127, 136, 158, 160, 183–4, 200, 204
Smallholders
  development strategies, 6, 11, 182, 183
  economic transfers, 466–7
  forced land sales, 164–5
  irrigation priorities, 204, 269–71
  land titles, 302–3
  research, 372, 399–400
Smith, Adam, 4, 4f
Smith, Lawrence D, 27, 28f, 393, 463–5, 464f, 465f
Smith, Marisol, 89
Smith, M *see* Ellis, F, Senanayake, P, and Smith, M
Snellen, W B, 220
Snellen, W B *see* Murray-Rust, D H, and Snellen, W B
Social capital, 30–1, 470, 471
Social collateral, 302, 332
Social factors
  agricultural policies, 22
  irrigation policies, 206, 228
  rural–urban migration, 18
  sustainability, 23
  water pricing, 227–8
Soil
  agricultural strategy, 446–7
  historical background, 3–4
  research, 374, 395–6
  sustainability, 114–15
Soler, René *see* Strasma, John, Arias, Ricardo, García, Magdalena, Meza, Daniel, Soler, René, and Umaña, R
Somalia
  land rental, 162
  private sector involvement, 90–1
South Africa
  land tenure, 146, 153, 178
  loans, 351
  see also Africa; North Africa; Sub-Saharan Africa
  water supplies, 199
South Asia
  land reform, 149
  price and trade controls, 19
  water rights markets, 237
  women's education, 47
  see also Asia; Central Asia; East Asia; South East Asia

South East Asia
  industrialization, 6–7
  irrigation pricing, 236
  rural economy, 11–12
  water resources management, 262–3
  see also Asia; Central Asia; East Asia; South Asia
South Korea *see also* Republic of Korea
  canal irrigation, 28–9
  co-operatives, 174
  credit unions, 318
  extension services, 381
  land reform, 138, 148
  managed deposit rates, 341
Soviet Union *see* Former Soviet Union
Spain
  rural development, 467
  water resources management, 243, 253
Spate irrigation, 216
Spoor, M, 90f
Sprinkler irrigation, 217, 486
Squatters, 142, 185
Squire, Lyn, 10
Squire, Lyn *see* Deininger, Klaus, and Squire, Lyn
Sri Lanka
  farm support prices, 89
  irrigation design, 220
  land tenure, 156
  water resources management, 258
Srivastava, Jitendra, 363
Srivastava, Jitendra P *see* Mudahar, Mohinder S,
  Jolly, Robert W, and Srivastava, Jitendra P
Staatz, John M *see* Martin, Frédéric, Larivière,
  Sylvain, and Staatz, John M
Stamoulis, K *see* Reardon, T, Stamoulis, K, Cruz,
  M E, Balisca, Al, and Berdegué, J
Stakeholders, 208, 265, 328, 329, 405, 406, 427, 433
Stanfield, J David, 158f
Staple foods
  food security, 99, 100
  prices, 55–6, 58
  production, 22
  research, 403
  trade policy, 60–1
Start-up funding, 343, 350, 490
State intervention
  banks, 287, 316–17, 322, 346, 350
  farm support prices, 88–9
  land rights, 116, 121–2, 130, 157
  leases, 159–60, 179
  monopolies, 72
  ownership, 19, 36, 38, 134, 137, 138, 139–42,
    167–71

privatization, 48–51, 167–8, 350, 427, 428
  rural distribution, 185
  rural finance, 278–9, 294
  transfer of ownership, 167–71
  see also Government
Static welfare loss, 39
Steel, William *see* Aryeetey, Ernest, Hettige,
  Hemamala, Nissanke, Machiko, and Steel,
  William
Steiner, Achim *see* Dorcey, Tony, Steiner, Achim,
  Acreman, Michael, and Orlando, Brett
Stiglitz, Joseph E, 25, 25f, 27, 27f, 103, 103f, 281,
  281f, 286, 287, 317f, 344, 344f, 345
  see also Greenwald, B, and Stiglitz, J; Hoff, Karla,
    and Stiglitz, Joseph
Strasma, John, 176, 176f, 179, 179f, 180, 182
  Arias, Ricardo, García, Magdalena, Meza, Daniel,
    Soler, René, and Umaña, R, 177f
  and Celis, Rafael, 180f, 182f
Strategies
  agricultural development, 425–60
  consistency, 31, 443, 446–50
  drafting process, 434–5, 439, 441
  foreign elements, 30
  grain reserves, 89–92
  implementation, 30–2, 91–2, 443
  irrigation development, 211–23, 488, 490
  objectives, 448–9
  orientations, 450–60
  quantification, 16
  research, 363
  'road maps', 15
  rural development, 460–74
  rural finance, 328, 346–51
  sectoral, 15–17, 447–8
  structure, 442–6
  technical justification, 16
  water resources management, 207–11, 484, 488,
    490
Structural reforms, 31, 85–6, 317–30
Sturzenegger, Adolfo (with Wylian Otrera), 81f
Sub-Saharan Africa
  exchange rate policy, 78
  extension, 387–8
  HIV/AIDS, 386
  land tenure, 119, 124, 150, 153
  research, 363–4
  sharecropping, 161
  structural reform, 86
  water rights markets, 247
  women's education, 47
  see also Africa; North Africa; South Africa

Subramanian, Ashok, 254, 256
  Jagannathan, N Vijay, and Meinzen-Dick, Ruth,
    232f, 252f, 254f
  see also Dinar, Ariel, and Subramanian, Ashok;
    Gorriz, Cecilia M, Subramanian, Ashok, and
    Simas, José; Meinzen-Dick, Ruth, Mendoza,
    Meyra, Sadoulet, Loic, Abiad-Shields, Ghada,
    and Subramanian, Ashok
Subsidies
  anti-economic mentality, 40, 448
  credit, 6, 36, 38, 40, 287–8, 293, 316–17, 318, 331,
    340, 346
  crop insurance, 305
  diesel fuel, 485, 486–7, 491
  elimination, 293, 296–7, 488, 489
  explicit, 37, 38
  exports, 6–7, 38, 60, 457
  extension services, 380, 381
  fiscal expenditure, 37–44
  foreign exchange, 5
  implicit, 37–8
  industrial development, 4–7, 486, 491
  interest rates, 288, 316–17, 340
  international, 64–5
  irrigation projects, 38, 226, 228, 270–1
  land reform, 177, 178
  OECD study, 43
  pros and cons, 38–42
  regressive, 40, 382, 410
  rural poverty, 185
  self-sustaining growth, 44, 86, 471
  transitional, 39f, 70, 347, 486, 490, 491
Sudan
  exchange rate policy, 78
  irrigation projects, 219
Sugar pricing, 75f
Supplemental irrigation, 217–18
Supply and demand
  credit, 286
  food prices, 57–8, 358
  water resources management, 205, 210, 223–5, 227,
    228, 484–7, 490–1
Surface flow irrigation, 217
Sustainability
  fiscal, 23, 203, 365, 387
  irrigation policies, 203, 250
  policy principles, 23, 25
  poverty alleviation, 294
  research, 365, 366, 396
  rural development, 418, 451
  rural finance, 279, 281–2, 290, 293, 295–9, 300,
    346

subsidies, 44
  see also Environmental sustainability
Swanson, B E, 385
Swanson, B E, Farmer, B J, and Bahal, R, 385f
Swinnen, Jo, 56
Swinnen, Johan F M
  and Gow, Hamish R, 39f
  and Zee, Frans A van der, 56f
Syrquin, Moises, 5
Syrquin, Moises *see* Chenery, Hollis, and Syrquin,
  Moises

T & V (Training and Visit system), 376, 383–4,
  416–17
Taiwan
  education investment, 26
  land reform, 148
  State land ownership, 142
  water resources management, 258
Takavarasha, T, 98, 98f
Tandon, Rajesh, 441
Tandon, Rajesh *see* Brown, L David, and Tandon,
  Rajesh
Tanzania
  education, 445
  gender issues, 384
  grain market liberalization, 91
  irrigation rehabilitation, 214
  land tenure, 153
Tanzi, Vito, 463f
Tariff-rate quotas (TRQs), 58
Tariffs
  agricultural strategies, 431, 448
  base period, 60
  binding, 65
  developing countries, 62–7
  export, 58, 62, 67, 457
  import, 37, 58, 62, 67, 70–1
  irrigation water, 226
  macroeconomic framework, 36, 58
  manufacture, 59
  megatariffs, 59
  substitutes, 84
  trade policy, 58–67
  uniformity of rates, 62–7
Tashmatov, A, Aknazarov, F, Jureav, A, Khusanov,
  R, Kadyrkulov, K D, Kalchayev, K, and
  Amirov, B, 160f
Taxation
  area-based land tax, 179–83, 466
  commodity-specific, 6, 448
  export crops, 6, 80–2

industrial development, 5
land ownership, 156, 466
macroeconomic framework, 36
negative economic growth, 8–9, 359
price effects, 82
subsidies, 37
tax credits, 7
Taylor, Timothy, 135
Taylor, Timothy G *see* Martin, Michael, and Taylor, Timothy G
Technology, 357–424
appropriateness, 365–8
demand-driven, 391–2, 395, 406, 413
environmental sustainability, 396
farmer-led development, 393, 394–5
mechanization, 453, 454
participatory, 366–8, 391–2, 393–5, 398, 402, 403
pricing policies, 37
rural development, 461
transfer, 376, 388–90, 395, 405
water crisis, 485–6
see also Extension; Research
Téllez, L, 156f
Tendler, Judith, 388, 388f
Thailand
extension, 416
irrigation efficiency, 199
land ownership, 125–6
loans, 278
microfinance institutions, 331, 340
traditional land transfer, 128
water resources management, 263
Third-party rights, 240–2, 244
Thirtle, C, 363–4
Thirtle, C, Hadley, D, and Townsend R, 363f, 396f
Thobani, M, 234–5, 235f, 237, 237f, 238–9, 239f, 240, 240f, 242, 242f, 246, 247, 247f, 248, 248f
Thom, D J, 211, 214, 222, 223, 270
Thom, D J *see* Moris, J R, and Thom, D J
Thompson, R L, 459, 459f, 460f
Tiered pricing, 232
Tikare, S, 441
Tikare, S, Youssef, D, Donnelly-Roark, P, and Shah, P, 429f
Time-Life Books editors, *The Rise of Cities*, 4f
Timmer, C Peter, 9, 10, 11, 11f, 12, 11-12f, 42, 42f, 448, 449f, 452, 452f, 456–7
Falcon, Walter P, and Pearson, Scott R, 450f
see also Block, Steven, and Timmer, C Peter
Tinbergen, Jan, 89, 89f
Togo, credit co-operatives, 320–1

Toll goods, 379
Topouzis, Daphne, 415–16, 415f, 416f
Toselli, P, 144, 145
Toselli, Paolo *see* Herrera, Adriana, Riddell, Jim, and Toselli, Paolo
Touray, Abdou, 433f
Townsend, R, 363–4
Townsend R *see* Thirtle, C, Hadley, D, and Townsend R
Trackman, Brian, Fisher, William, and Salas, Luis, 159f
Trade liberalization, 60–2, 67, 70–2
economic growth model, 9
grain, 89–92
input markets, 96–7
South East Asia, 6–7
Trade policy, 58–74, 62–7, 209, 360–1, 456–7
protection, 59–62
restrictions, 68–72, 388–9
Tradeable water rights, 224, 230, 231, 236–48, 266, 487–8, 490–1
Traditional issues
finance, 318
irrigation, 216, 217, 222
Judaeo–Christian, 3
land rights, 127–8, 157
see also Communal lands; Customary land regimes
Training *see* Education and training
Training and Visit (T & V) system, 376, 383–4, 416–17
Transitional issues
land titling, 128–30
subsidies, 70, 347, 486, 490, 491
Treasury bills, 341
Tree rental, 122
Tribal land transfer, 127–8
Trigo, Eduardo J, 394
Trigo, Eduardo J *see* Echevarria, Rubén G, Trigo, Eduardo J, and Byerlee, Derek
Trinidad and Tobago, State land ownership, 141
Troutt, Betsy *see* Lyne, Michael, Roth, Michael, and Troutt, Betsy
Tsakok, Isabelle, 81f
Tschirley, D, Donovan, C, and Weber, M T, 73f
Tsur, Yacov, 232
Tsur, Y, and Dinar, A, 232f, 233f
Tun Wai, U, 285f
Tunisia
share tenancy contracts, 163
volumetric water pricing, 233
water resources management, 258–9

Turkey
   interest rates, 342
   T & V system, 383
Turkmenistan, irrigation efficiency, 200, 215

Udry, C, 164–5
Udry, C *see* Binswanger, H P, McIntire, J, and
      Udry, C
Uganda
   gender issues, 46
   land ownership, 151–2, 153
   participatory processes, 432
   women's access to land, 187
Ukraine, land ownership, 171, 172–4
Umali–Deininger, D, 358, 358f, 378–80, 378f, 381,
      387, 387f, 408f, 410–11, 411f, 412
   see also Pray, Carl E, and Umali-Deininger, D
Umaña, R *see* Strasma, John, Arias, Ricardo, García,
      Magdalena, Meza, Daniel, Soler, René, and
      Umaña, R
United Nations, Land Tenure Service, 141f
Undernourishment, 99–100
United States
   co-operatives, 174
   environmental protection, 245–6
   food prices, 56
   institutions, 30
   land tenure, 123
   megatariffs, 59
   water rights, 239, 241, 242–3, 244, 245
Urban development
   agricultural contributions, 9–12, 23, 24, 451–2
   bias, 299
   poverty alleviation, 9–12
   rural–urban migration costs, 18
Urban–industrial impact model, 7
Uruguay
   bulk handling of rice, 93
   collateral, 304
   research, 395, 398
Uruguay Round's Agreement on Agriculture, 59, 67
Usufruct rights, 117, 141, 152, 153, 184, 224, 238

Valdés, Alberto, 8, 44f, 63, 63f, 76, 76f, 77f, 299
   see also Krueger, Anne O, Schiff, Maurice and
      Valdés, Alberto; Schiff, Maurice, and Valdés,
      Alberto
Valdivia Alcalá, Ramón, Matus Gardea, Jaime A,
      Martínez Damián, Miguel A, and Santiago
      Cruz, Maria de J, 85f
Valenzuela, Liza *see* Baydas, Mayada, Graham,
      Douglas, and Valenzuela, Liza

Valletta, W, Keith, S H, Norton, R D, 162f
van de Walle, Dominique, 289–90, 468
van der Zee, Frans, 56
van Haeften, Roberta *see* García, Magdalena,
      Norton, Roger, Ponce Cámbar, Mario, and
      van Haeften, Roberta
van Koppen, B, 246, 268, 268f, 269f, 270
Van Tuijl, Willem, 214, 214f, 215–16, 216f, 222, 222f
van Veldhuizen, Laurens, 367, 393
Varietal research, 393–4, 395, 403, 453
Vegetable exports, 94
Velayet, Dashowuz, 200
Veldhuizen, L van, Waters-Bayer, A, and Zeeuw, H
      de, 366f, 367f
Venezuela, bulk handling of rice, 93
Vermillion, D L, 250, 255, 260, 260f, 266, 270
   and Sagardoy, Juan A, 225f, 250f, 265f
Vietnam
   land reform, 130f
   market liberalization, 70
   returns to irrigation, 199
   water resources management, 265
Villalabos, Ruy de *see* Aristizábal, Gladys, Echenique,
      Jorge, Villalabos, Ruy de, and Fischer, Wolfram
Villanueva, D, and Mirakhor, A, 342f
Vogel, Robert C, 281, 282, 282f, 283f, 290, 297f, 345
   and Adams, Dale W, 345f
   see also Christen, Robert Peck, Rhyne, E, Vogel,
      R C, and McKean, C
Vollrath, Thomas, 10, 78, 82
Vollrath, T L, 79f, 82f, 458f, 459f
Volumetric pricing, 232–4, 235
Voluntary savings, 313, 314, 335–6, 337
Von Pischke, J D, 326
Vyas, Vijay S, 112, 112f, 160, 160f

Wade, R, 28–9, 28f
Waggoner, P E *see* Evenson, R, Waggoner, P E and
      Ruttan, V W
Waible, H, 374f
Wai, U Tun, 285
Waibel, H, 374
Walle, Dominique van de, 290f, 468f
Wang, Limin *see* Coady, David, Dai, Xinyi, and
      Wang, Limin
Wartime land reform, 146, 148
Waste water recycling, 213, 490
Waters-Bayer, A *see* Veldhuizen, L van, Waters-Bayer,
      A, and Zeeuw, H de
Water management policies, 197–276
   allocation, 223–5, 227, 249
   conflict resolution, 266

contracts, 261, 266
demand, 205, 210, 223–5, 227, 228, 484–7, 490–1
groundwater supplies, 203–4, 205, 212, 222–3, 484, 486, 487, 490–1
holistic approach, 205–6, 207, 210
institutional issues, 203, 207, 208, 209, 248–69
intersectoral, 212–13, 242–3
investment, 471
objectives, 203–7
participation, 249, 260, 265
pricing, 226–36, 243–5, 247–8
recycling, 213, 486, 490
registration, 240
review, 213
strategies, 207–11, 484
supply, 484–5
water use, 485–6, 489
Yemen, 208, 211, 216, 224, 227, 483–91
Water rights markets, 224, 230, 231, 236–48, 266, 487–8, 490–1
Water supply companies, 261
Water users associations (WUAs), 220, 230, 233–4, 244–5, 246, 252, 254, 255–68
Waterlogging, 215, 222, 396
Waters–Bayer, Ann, 367
Watersheds, 206
Watts, R, 139f, 141f
Weber, M T *see* Tschirley, D, Donovan, C, and Weber, M T
Weeds, 373
Wen, Simei, 444
Wenner, Mark, 41f, 345–6, 345f, 346f
Western Europe
    environmental sustainability, 115
    mixed economy, 30
    technology, 376
    see also Central Europe; Eastern Europe; Europe
Westphal, Larry E *see* Evenson, R E, and Westphal, Larry E
Wightman, John A, 374, 374f
Wisniwski, Sylvia *see* Mukherjee, Joyita, and Wisniwski, Sylvia
Wolter, Hans, 220
Wolter, Hans W *see* Plusquellec, Hervé, Burt, Charles, and Wolter, Hans W
Wolters, W, 232, 233
Wolters, W *see* Bos, M G, and Wolters, W
Women
    access to land, 185–91
    agricultural development, 46–8, 185–91, 459, 472–3
    education, 47, 185, 374

extension services, 384–5, 413–14, 419
HIV/AIDS effects, 386
irrigation policies, 204, 207, 222, 235–6, 268–9
land rights, 116, 152
legislation, 189, 190
loans, 288–9, 289, 294, 295, 329–30, 348
research, 366, 367, 368–71, 391, 402–4
World Bank
    agricultural development, 10–11, 429, 447, 458, 460, 461, 462
    exchange rate policy, 76
    extension services, 410
    financial services, 285
    gender analysis, 47
    industrial credit subsidies, 6
    irrigation projects, 209–10, 215, 260
    land reform, 144, 167, 169
    land rights, 137, 150, 154
    land titling, 115, 169
    participatory projects, 433, 441
    poverty alleviation, 10–11, 429, 460
    price fluctuations, 450
    private sector development, 27–8
    rediscount lines, 326
    research, 364–5, 397
    rural development, 460, 461
    rural finance, 282, 320–1, 346
    T & V system, 376, 383
    taxes on land, 182
    water resources management, 212–13, 225, 226, 250, 251, 252, 255, 259
    water rights markets, 236–7, 243
World Bank, The, 7f, 46f, 47f, 49f, 137f, 150f, 155f, 167f, 169f, 188f, 198f, 208f, 210f, 212f, 213f, 226f, 227f, 229f, 236f, 250f, 251f, 252f, 255f, 258f, 259f, 260f, 262f, 285f, 318f, 320f, 321f, 332f, 361f, 367f, 447f, 450f, 458f, 460f, 461f
    Action Strategy, 464f
    Latin America and Caribbean Office, 158f, 308f
    Latin America and the Caribbean Region, 319f
    Latin America and Caribbean Regional Projects Department, 215f
    Operations Evaluation Department, 210f, 433f, 462f
World Council of Credit Unions (WOCCU), 310, 321
World Development Report, 1990 (World Bank), 7–8
World Food Program (WFP), 74
World Trade Organization (WTO), 42, 65, 66, 67, 84
Wynne, Susan *see* Ostrom, Elinor, Schroeder, Larry, and Wynne, Susan

Yadav, S N *see* Rosegrant, Mark W, Gazmuri, S R, and Yadav, S N

Yang, M-C *see* Binswanger, H, Mundlak, Y, Yang, M-C, and Bowers A

Yao, S *see* Beynon, J, Jones, S, and Yao, S

Yarkin, Cherisa *see* Zilberman, David, Yarkin, Cherisa, and Heiman, Amir

Yaron, Jacob, 278–9, 279f, 286, 288f, 289f, 297f, 298–9, 321, 331, 333

    Benjamin, M P Jr, and Piprek, G L, 278f, 287f, 288f, 299f, 302f, 303f, 328f, 331f, 332f, 334f, 343f, 344f, 346f

Yemen

    forestry, 488

    groundwater supplies, 484, 486, 487, 490–1

    industrial development, 486, 491

    integrated water strategy, 484, 488

    irrigation efficiency, 486–7, 491

    rainfed population, 488, 489, 491

    subsidies, 485, 486–7, 488, 490

    water allocation, 224, 486

    water crisis, 483, 484, 485–91

    water pricing, 227

    water resources management, 208, 211, 216, 483–91

    water rights transfer, 487–8, 490–1

    water supply and demand, 484–7, 490–1

Youssef, D *see* Tikare, S, Youssef, D, Donnelly-Roark, P, and Shah, P

Yudelman, Montague, 372–3

Yudelman, Montague, Ratta, Annu, and Nygaard, David, 373f

Zambia

    calorie consumption, 99f

collective farms, 139

communal lands, 132

grain storage policy, 93

land tenure, 118, 125, 129, 155

pan-territorial pricing, 87

policy volatility, 98

State land ownership, 141, 142

technology development, 361

trade restrictions, 69

women's access to land, 187

Zee, Frans A van der *see* Swinnen, Johan F M, and Zee, Frans A van der

Zeeuw, Henk de *see* Diop, Jean-Marie, Jong, Marga de, Laban, Peter, and Zeeuw, Henk de; Veldhuizen, L van, Waters-Bayer, A, and Zeeuw, H de

Zeller, Manfred, Schreider, Gertrud, Braun, Joachim von, and Heidhues, Franz, 284f, 325f

Zhang, Xiaoshan, 175, 175f

Zimbabwe

    calorie consumption, 99f

    coercive land reform, 143, 146, 147, 148

    communal lands, 131–2

    economic policies, 98

    extension services, 412

    group lending, 332

    HIV/AIDS effects, 415

    irrigation projects, 223

    research, 368

    water supplies, 199

Zilberman, David

    Yarkin, Cherisa, and Heiman, Amir, 371f

    see also Carlson, Gerald A, Zilberman, David, and Miranowski, John A

Zoning restrictions, 122

Printed in the United States
by Book-mart

Printed in the United States
By Bookmasters